THOMAS D. BROCK *Professor of Microbiology, Indiana University*

Biology of microorganisms

Prentice-Hall, Inc. ENGLEWOOD CLIFFS, NEW JERSEY

Biological Science series
William D. McElroy
and Carl P. Swanson, editors

THOMAS D. BROCK   *Biology of microorganisms*
© *1970 by Prentice-Hall, Inc., Englewood Cliffs, New Jersey*

Printed in the United States of America
13-076851-0
Library of Congress Catalog Card Number 76-79113
Current printing   10 9 8 7 6 5 4 3 2

Prentice-Hall International, Inc., London
Prentice-Hall of Australia, Pty. Ltd., Sydney
Prentice-Hall of Canada, Ltd., Toronto
Prentice-Hall of India Private Ltd., New Delhi
Prentice-Hall of Japan, Inc., Tokyo

This book has been set in Galaxy,
a film version of a contemporary
sans-serif typeface, with headings
and captions in italic
Century Expanded.

Biology of microorganisms

Preface

This book was designed as a modern textbook for courses in introductory microbiology at the undergraduate level. It should also appeal to anyone interested in obtaining a general knowledge of microbiology and microorganisms. The emphasis throughout is on microorganisms—their functional, ecological, and evolutionary relationships—and on the activities of interest to man that they carry out. Although a considerable amount of material on applied microbiology is presented, it is integrated into chapters on fundamental principles rather than confined to separate chapters. In this way the applications of microbiology illuminate and make the fundamentals more interesting. Eucaryotic microorganisms have been more emphasized than they usually are in microbiology books; instead of being relegated to small isolated chapters, these organisms are discussed wherever relevant, and a special effort has also been made to contrast procaryotic and eucaryotic microorganisms. In addition, separate chapters treat those aspects of eucaryotes which are quite distinct from procaryotes: cell structure, genetics, taxonomy, and life cycles.

Although the book presents the experimental bases for microbiological concepts and facts, material that would be more useful in the laboratory portion of a microbiology course is not given: The aim is not to use space for material that can better be discussed elsewhere.

A special word should be said about the last two chapters, describing the structure, function, and ecology of groups of microorganisms. Although these chapters may be read straight through, another approach is perhaps preferable: to read specific sections in conjunction with earlier chapters. For instance, sections dealing with medically important groups of microorganisms could be read together with Chapter 15 on host-parasite relationships, and sections on autotrophic bacteria could be read with Chapter 6 on the autotrophic way of life.

It is assumed that the reader of this book will have had some acquaintance with general chemistry and general biology as well as at least a superficial acquaintance with some aspects of molecular biology. Although some of this background material is presented in sufficient depth to make the microbial material understandable, the available space has been used as much as possible to discuss microorganisms themselves. There are many good books that clearly and concisely describe the concepts of molecular biology; one who lacks the necessary background should turn to one of these for guidance.

This book could not have been written without the aid of a number of persons. Since no one can be expert on all phases of a subject as large as microbiology, I have welcomed the help of the publishers, who asked a number of specialists to review preliminary versions of various chapters. In order to ensure an objective evaluation, the anonymity of these outside reviewers has been maintained so that I cannot thank them individually; but their assistance has been invaluable, and I hereby thank them collectively.

Illustrative material has been generously provided by many individuals and firms, whose names are cited near the appropriate illustrations; and I am also indebted to those who assisted in the preparation of original drawings or reviewed them. In addition, the courtesy of authors and publishers in allowing me to use previously published figures and tables is gratefully acknowledged. Virtually all the line drawings were prepared especially for this book, and I thank the artist, Russell Peterson, for converting the sketches devised by Louise Brock into final artwork.

The staff at Prentice-Hall, especially John R. Riina and David R. Esner, have provided invaluable aid throughout production, as has the designer, Betty Binns. Competent secretarial and bibliographic assistance has been provided by Linda Detwiler, Mary Swarthout, Nancy Doemel, Sue Bott, Pat Holleman, Bonnie Hodler, and Jane Griffith-Jones. I should also like to express my appreciation to the staff of the Biology Library of Indiana University, without whose cooperation this book could not have been written. Errors and omissions are, of course, my own responsibility, and I should greatly appreciate receiving comments, suggestions, and corrections.

Finally, although it is customary for an author to thank his wife for help and support, these thanks here must go much further: Louise has been involved in a detailed way in the making of this book from its inception and could almost have been included as a coauthor. She has outstandingly handled all the myriad details that must be attended to if such a book as this is to be published, and without her the book might never have been completed.

Thomas D. Brock
Bloomington, Indiana,
and West Yellowstone, Montana

Contents

Contents

ix

Contents

Contents

xi

Biology of microorganisms

Microbiology is the study of microorganisms, a large and diverse group of free-living forms that exist as single cells or cell clusters. Being free-living, microbial cells are distinct from the cells of animals and plants since the latter are able to live not alone in nature but only in characteristic groups. A single microbial cell is generally able to carry out its life processes of growth, respiration, and reproduction independently of other cells, either of the same kind or of different kinds. Although there are exceptions to this statement, which we will consider later, this definition provides a basis for our introduction to microorganisms.

1.1 Microorganisms as cells

The theory that the cell is the fundamental unit of all living matter is one of the great unifying theories of biology. A single cell is an entity, isolated from other cells by a cell wall or membrane and containing within it a variety of subcellular structures, some of which are found in all cells and some of which are variable in their occurrence. All cells have certain chemical characteristics in common in that they all contain proteins, nucleic acids, lipids, and polysaccharides. Because these chemical components are common throughout the living world, it is thought that all cells have descended from some common ancestor, a primordial cell. Through millions of years of evolution, the tremendous diversity of cell types that exist today have arisen. Cells vary enormously in size, from submicroscopic bacteria too small to be seen even with the light microscope, to the 170- by 135-mm ostrich egg, the largest single cell known. Microbial cells show a narrower but also extensive size range. Some microbial cells are much larger than human cells. The single-celled protozoan *Paramecium* is 4,800 times the weight of the human red blood cell.

Although each kind of cell has a definite structure and size, cells should not be viewed as unchanging bodies: A cell is a dynamic unit constantly undergoing change and replacing its parts. Even if it is not growing, a cell is continually taking materials from its environment and working them into its own fabric. At the same time, it perpetually discards into its environment cellular materials and waste products. A cell is thus an open system, forever changing yet generally remaining the same.

1.2 Microbial diversity

Although microbial cells have much in common, there are in reality two basic plans of cellular architecture, which differ from each other in many

one

Introduction

fundamental ways. These two kinds of microbial cells are called *pro-caryotes* and *eucaryotes*. Bacteria and blue-green algae are procaryotes, whereas all other algae, fungi, and protozoa are eucaryotes. The most important difference between procaryotes and eucaryotes is in the structure of the nucleus. The eucaryote has a true nucleus (*eu-* means "true"; *karyo-* is the combining form for "nucleus"), a membrane-bound structure within which are the chromosomes that contain the hereditary material. The procaryote, on the other hand, does not have a true nucleus or chromosomes, and its hereditary material is contained in a single naked DNA molecule. There are also many other structural differences between procaryotes and eucaryotes that will be covered in more detail in Chapters 2 and 3. At present it is enough to know that these differences exist and that they are so fundamental as to make us believe they reflect an evolutionary divergence in the early history of life. Since the cells of higher animals and plants are all eucaryotic, it is likely that eucaryotic microorganisms were the forerunners of higher organisms, whereas procaryotes represent a branch that never evolved past the microbial stage.

Because of the diversity of microorganisms, it is useful to give different organisms names, and to do this we must have ways of telling one organism from another. After close study of the structure, composition, and behavior of a microorganism, we can usually recognize a group of characteristics by which it can be distinguished from most other forms. This set of characteristics then becomes the defining attributes of that organism, and a name can be given to it. Microbiologists use the binomial system of nomenclature first developed for plants and animals. The *genus* is like a surname (for example, Murphy) and includes a number of related organisms, whereas each different type of organism within the genus has a *species* name, which is like a given name (for example, John). Genus and species names are always used together to describe a specific type of organism, whether it be a single cell or a group of such cells. Thus three specific types of organism go under the common name of "red bread mold" (see the marginal table). Usually the names used come from Latin and Greek and indicate some characteristic of the organism. For instance, *Saccharomyces cerevisiae* is the species of beer yeast. Yeasts convert sugar into alcohol, and *saccharo-* means "sugar." A yeast is a fungus, and the combining form *-myces* derives from the Greek word for fungus. The word *cerevisiae* derives from the Latin word for "beer." Unfortunately, it is rarely possible to break down the name of a microorganism as easily as in this example, so that even one who knows Latin and Greek would have little success in translating many species names into English. Although we shall try to keep the number of species names to a minimum in this book, the student should be prepared to familiarize himself with at least the most important names (and their spellings!).

Genus	Species	Common name
Neurospora	crassa	Red bread mold
Neurospora	sitophila	Red bread mold
Neurospora	tetrasperma	Red bread mold

In addition to genus and species, higher orders of classification are used. Thus related genera are grouped by their similarities into a family, families into orders, orders into classes, and classes into phyla:

Kingdom Plant
   Phylum Eumycophyta
     Class Ascomycetaceae
       Order Sphaeriales
         Family Sordariaceae
           Genus *Neurospora*
             Species *crassa*

Many biologists divide the living world into two groups, plants and animals. Are microorganisms plants or animals? There seems to be no doubt that some microorganisms, such as the chlorophyll-containing algae, are plants, whereas many protozoa seem quite animallike. Yet we run into difficulties in some cases. For instance, the organism *Euglena gracilis* is a chlorophyll-bearing organism and thus is clearly a plant, yet after certain drug treatments it loses its chloroplasts and never regains them; forever thereafter its offspring live as animals. Although fungi lack chlorophyll, in many ways they are more closely related to the algae than they are to the protozoa. The procaryotic organisms, as a group, are so different from either animals or plants that it would seem foolish to try to place them in one group or the other. One solution proposed by some microbiologists is to collect all microorganisms together in a separate group, the Protista, of equal status with plants and animals. Yet to do this ignores the fact that many algae clearly are more related to plants than they are to animals, and many protozoans have more in common with animals than with plants. Perhaps it would be best if a final judgment were not made at this time; after the reader has become more acquainted with microorganisms, he may make his own decision.

Viruses are not cells. They lack many of the attributes of cells, of which the most important is that they are not dynamic open systems. A single virus particle is a static structure, quite stable and unable to change or replace its parts. Only when it is associated with a cell does a virus acquire some attributes of a living system. Whether or not a virus is to be considered alive will depend on how life itself is defined. We will reserve further discussion of this interesting question until Chapter 10.

1.3   The discovery of microorganisms

Although the existence of creatures too small to be seen with the eye had long been suspected, their discovery was linked to the invention of the

Figure 1.1 Early drawings by Robert Hooke (1664) of a blue mold growing on the surface of leather: the round structures contain spores of the mold; the lower drawings are of mold growing on the surface of an aging and deteriorating rose leaf. One of the first microscopic descriptions of microorganisms. (From R. Hooke: Micrographia or Some Physiological Descriptions of Minute Bodies Made by Magnifying Glasses with Observations and Inquiries thereupon. London: Royal Society, 1665.)

microscope. Robert Hooke described the fruiting structures of molds in 1664 (Figure 1.1), but the first person to see microorganisms in any detail was the Dutch amateur microscope builder Antonie van Leeuwenhoek, who used simple microscopes of his own construction (Figure 1.2). Leeuwenhoek's microscopes were extremely crude by today's standards, but by careful manipulation and focus he was able to see organisms as small as bacteria. He reported his observations in a series of lively letters to the Royal Society of London, which published them in English translation. Drawings of some of Leeuwenhoek's "wee animalcules" are shown in Figure 1.3. His observations were confirmed by other workers, but progress in understanding the nature of these tiny organisms came slowly. Only in the nineteenth century did improved microscopes become

Figure 1.2 A replica of the kind of microscope used by Leeuwenhoek. The base was a piece of leather, with the lens inserted into the small hole at the left side. The object to be viewed was placed on the small pointed wire attached to the screw, and the object was then moved back and forth by turning the screw. (Replica courtesy of Archives of the American Society for Microbiology.)

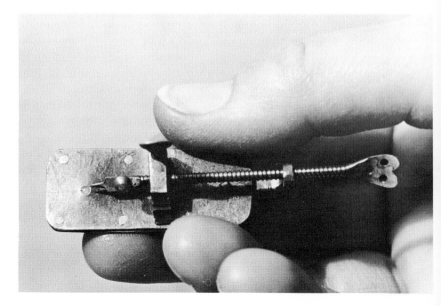

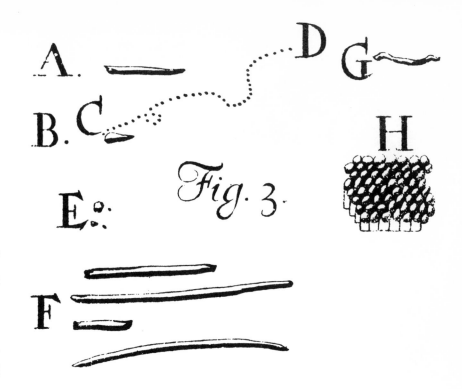

Figure 1.3 Leeuwenhoek's drawings of bacteria, published in 1684. Even from these crude drawings we can recognize several kinds of common bacteria. Bacteria lettered A, C, F, and G are rod-shaped; E, spherical or coccus-shaped bacterium; H, coccus-shaped bacteria in packets. [From "An abstract of a Letter from Mr. Anthony Leevvenhoeck at Delft, dated Sep. 17, 1683," Phil. Trans. Roy. Soc. London 14:568 (1684).]

available, and as a result of the industrial revolution they became more widely distributed. At all stages of its history, the science of microbiology has taken the greatest steps forward when better microscopes have been developed, for these enable scientists to penetrate ever deeper into the mysteries of the cell. Indeed, most of our knowledge of the detailed structure of microbial cells has come only with improvements in electron microscopy in the past ten years! Figure 1.4, which shows the yeast cell as seen with different microscopes, well illustrates this point.

Microbiology as a science did not develop until the latter part of the nineteenth century. This long delay occurred because certain basic techniques for the study of microorganisms needed to be devised. In the nineteenth century investigation of two perplexing questions laid the foundation of microbiological science: (1) Does spontaneous generation occur? (2) What is the nature of contagious disease? Study of these two questions went hand in hand, and sometimes the same people worked on both. By the end of the century both questions were answered, and the science of microbiology was firmly established as a distinct and growing field.

1.3 *The discovery of microorganisms*

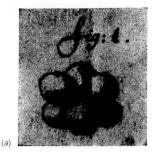

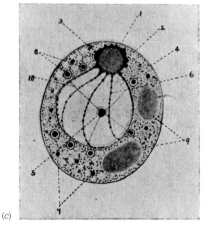

(a)

(b)

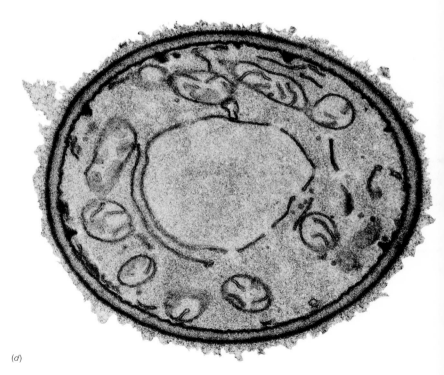

Figure 1.4 *The yeast cell as it seemed to different observers. The great increase in our understanding of cell structure came with improvement in our microscopes. (a) Leeuwenhoek's model of yeast, dating from 1680. The six globules represent six separate cells in a cluster. Note the complete absence of any cellular detail. [From Committee of Dutch Scientists (ed.): "A letter of June 14, 1680," in The Collected Letters of Antoni van Leeuwenhoek. Amsterdam: Swets & Zeitlinger, Ltd., 1948.] (b) Pasteur's drawings of yeast, made in 1860, showing the budding process by which yeasts grow. The contrast of the outer cell wall and inner cytoplasm is distinct. The large objects in the cytoplasm are vacuoles. [From L. Pasteur: Ann. Chim. Phys. 58:323 (1860).] (c) drawing of the idea of a yeast cell in 1910. The greater detail inside the cell derives partly from improved microscopy and partly from the use of dyes that increase contrast and stain particular structures. However, some of the labeled structures are probably artifacts. [From H. Wager and A. Peniston: Ann. Bot. 24:45 (1910).] (d) Photograph of a yeast cell as seen with the electron microscope. Only a thin section of the cell can be seen with this microscope. The cell is first treated with chemicals that preserve the structure and stain particular components. [From S. F. Conti and T. D. Brock: J. Bacteriol. 90:524 (1965).]*

(c)

(d)

The basic idea of spontaneous generation can easily be understood. If food or nutrient infusions are allowed to stand for some time, they putrefy, and when the putrefied material is examined microscopically, it is found to be teeming with bacteria. Where do these bacteria come from since they are not seen in fresh food? Some people said that the bacteria developed from seeds or germs that had entered the food from the air, whereas others said that they arose spontaneously. If spontaneous generation did occur, this would mean that life could arise from something nonliving, but many people could not imagine something so wonderful and complex as a living cell arising spontaneously from dead materials.

The most powerful opponent of spontaneous generation was the French chemist Louis Pasteur, whose work on this problem was the most exacting and convincing. Pasteur first showed that there were structures present in air that resembled closely the microorganisms seen in putrefying materials. He did this by passing air through guncotton filters, the fibers of which stop solid particles. After the guncotton was dissolved in a mixture of alcohol and ether, the particles that it had trapped fell to the bottom of the liquid and were examined on a microscope slide. Pasteur found that in ordinary air there exists continually a variety of solid structures ranging in size from 0.01 mm to more than 1.0 mm. Many of these structures resemble the spores of common molds, the cysts of protozoa, or various other microbial cells. As many as 20 to 30 of these organized bodies were found in 15 liters of ordinary air. Since these organized bodies present in the air could not be distinguished from the organisms found in much larger numbers in putrefying materials, Pasteur concluded that the organisms found in putrefying materials originated from the organized bodies present in the air, which are constantly being deposited on all objects. If this conclusion were correct, it would mean that, if a food or nutrient infusion were so treated as to destroy all the living organisms contaminating it, then it should not putrefy.

In order to eliminate contaminants, Pasteur used heat since it had already been established that heat effectively killed living organisms. In fact, many workers had shown that, if a nutrient infusion was sealed in a glass flask and heated to boiling, it never putrefied. The proponents of spontaneous generation criticized such experiments by declaring that fresh air was necessary for spontaneous generation and that the air itself inside the sealed flask was affected in some way by heating so that it would no longer support spontaneous generation. Pasteur skirted this objection simply and brilliantly by constructing a swan-necked flask, now called the

"Pasteur flask" (Figure 1.5). In such a flask putrefying materials can be heated to boiling; after the flask is cooled, air can reenter although the bends in the neck prevent particulate matter, bacteria, or other microorganisms from getting in. Material sterilized in such a flask did not putrefy, and no microorganisms ever appeared as long as the neck of the flask remained intact. If the neck was broken, however, putrefaction occurred and the liquid teemed with living organisms. This simple experiment effectively settled the controversy of spontaneous generation.

Killing all bacteria or germs is a process we now call "sterilization," and the procedures that Pasteur and others used were eventually carried over into microbiological research. The study of spontaneous generation thus led to the development of effective sterilization procedures, without which microbiology as a science could not have developed.

It was later shown that flasks and other vessels could be protected from contamination by cotton stoppers, which still permit the exchange of air. The principles of aseptic technique, developed so effectively by Pasteur, are the first procedures learned by the novice microbiologist. Food science also owes a debt to Pasteur since his principles are applied in the canning and preservation of many foods.

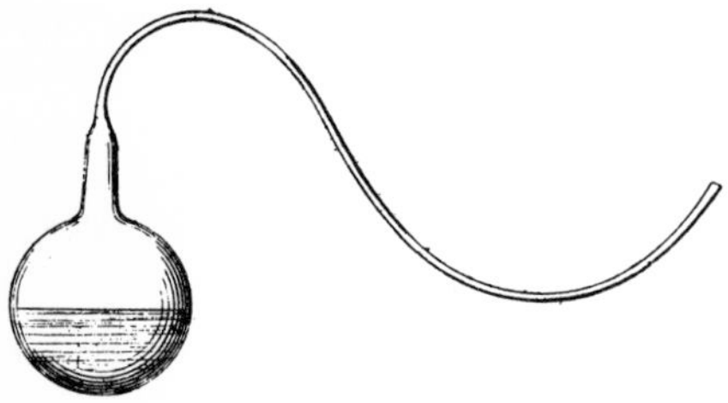

Figure 1.5 Pasteur's drawing of a swan-necked flask. In his own words: "In a glass flask I placed one of the following liquids which are extremely alterable through contact with ordinary air: yeast water, sugared yeast water, urine, sugar beet juice, pepper water. Then I drew out the neck of the flask under a flame, so that a number of curves were produced in it. . . . I then boiled the liquid for several minutes until steam issued freely through the extremity of the neck. This end remained open without any other precautions. The flasks were then allowed to cool. Anyone who is familiar with the delicacy of experiments concerning the so-called "spontaneous" generation will be astounded to observe that the liquid treated in this casual manner remains indefinitely without alteration. The flasks can be handled in any manner, can be transported from one place to another, can be allowed to undergo all the variations in temperature of the different seasons, the liquid does not undergo the slightest alteration. . . . after one or more months in the incubator, if the neck of the flask is removed by a stroke of a file, without otherwise touching the flask, molds and infusoria begin to appear after 24, 36, or 48 hours, just as usual, or as if dust from the air had been inoculated into the flask. . . . At this moment I have in my laboratory many highly alterable liquids which have remained unchanged for 18 months in open vessels with curved or inclined necks." [From L. Pasteur: Ann. sci. nat., ser. 4, 16:5 (1861), and T. D. Brock (ed.): Milestones in Microbiology. Englewood Cliffs, N.J.: Prentice-Hall, Inc., 1961.]

The German scientists Ferdinand Cohn and Robert Koch later described a special structure, the bacterial spore (Figure 1.6), which is highly resistant to heat and cannot be killed easily by mere boiling. Bacterial spores were not present in the materials Pasteur used to disprove the theory of spontaneous generation, but they are common on plant materials and in the soil. To kill spores temperatures above boiling usually are needed, and to achieve these temperatures in aqueous materials it is necessary to heat them under pressure, as in a pressure cooker or autoclave.

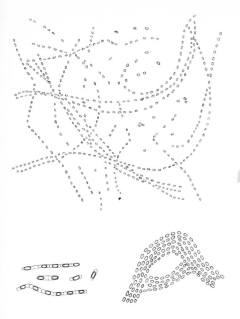

Figure 1.6 *Drawings by Koch and Cohn of bacterial spores, published in 1876. The spores first form in long rows within the bacterial chains but are eventually liberated. The organism is Bacillus anthracis, the causal organism of anthrax in cattle and man. Spores of harmless bacteria, such as Bacillus subtilis, are much the same. [From R. Koch: Beitr. Biol. Pflanzen 2:277 (1876), and F. Cohn: Beitr. Biol. Pflanzen 2:247 (1876).]*

1.5   The germ theory of disease

Proof that microorganisms could cause disease (which had long been surmised) provided one of the greatest impetuses for the development of the science of microbiology. Indeed, even in the sixteenth century it was thought that something could be transmitted from a diseased person to a well person to induce in the latter the disease of the former. Many diseases seemed to spread through populations and were called "contagious"; the thing which did the spreading was called the "contagion." After the discovery of microorganisms, it was more or less widely held that these organisms might be responsible for contagious diseases though proof was lacking. In 1845 M. J. Berkeley provided the first clear demonstration that microorganisms caused diseases by showing that a mold was responsible for Irish potato blight (Figure 1.7). Discoveries by Ignaz Semmelweis and Joseph Lister provided some evidence for the importance of microorganisms in causing human diseases, but it was not until the work of Koch, a physician, that the germ theory of disease was placed on a firm footing.

In his early work, published in 1876, Koch studied *anthrax*, a disease of cattle, which sometimes also infects man. Anthrax is caused by a spore-forming bacterium now called *Bacillus anthracis*, and the blood of an animal infected with anthrax teems with cells of this large bacterium. Koch established by careful microscopy that the bacteria were always present in the blood of an animal that had the disease. However, he knew that the mere association of the bacterium with the disease did not prove that it caused the disease; it might instead be a result of the disease. Therefore Koch demonstrated that it was possible to take a small amount of blood from a diseased animal and inject it into another animal, which in turn became diseased and died. He could then take blood from this second animal, inject it into another, and again obtain the characteristic disease symptoms. By repeating this process as often as 20 times, successively transferring small amounts of blood containing bacteria from one animal to another, he proved that the bacteria did indeed cause anthrax:

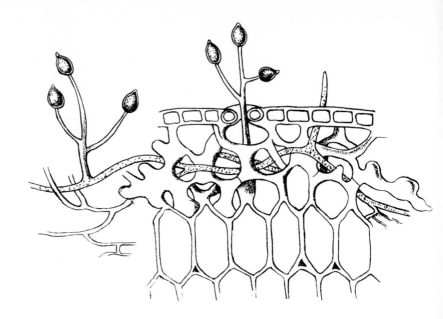

Figure 1.7 Berkeley's drawing of a disease-causing microorganism, done in 1846. The disease was Irish potato blight, which was responsible for the great famine in Ireland. The drawing shows the manner in which the fungus grew around and through the cells of the potato leaf. The round structures at the tips of filaments are spore bodies, which liberate the spores that transmit the fungus infection to other plants. Compare with Hooke's drawing in Figure 1.1. [From M. J. Berkeley: J. Roy. Hort. Soc. London 1:9 (1846).]

the twentieth animal died just as rapidly as the first; and in each case Koch could demonstrate by microscopy that the blood of the dying animal contained large numbers of the bacterium.

Koch carried this experiment further. He found that the bacteria could also be cultivated in nutrient fluids outside the animal body, and even after many transfers in culture the bacteria could still cause the disease when reinoculated into an animal. Bacteria from a diseased animal and bacteria in culture both induced the same disease symptoms upon injection. On the basis of these and other experiments Koch formulated the following criteria, now called "Koch's postulates," for proving that a specific type of bacterium causes a specific disease:

1 The organism should always be found in animals suffering from the disease and should not be present in healthy individuals.

2 The organism must be cultivated in pure culture away from the animal body.

3 Such a culture, when inoculated into susceptible animals, should initiate the characteristic disease symptoms.

4 The organism should be reisolated from these experimental animals and cultured again in the laboratory, after which it should still be the same as the original organism.

Koch's postulates not only supplied a means of demonstrating that specific organisms cause specific diseases but also provided a tremendous impetus for the development of the science of microbiology by stressing laboratory culture.

In order successfully to identify a microorganism as the cause of a disease, one must be sure that it alone is present in culture (that is, it must be a *pure culture*). With material as small as microorganisms, it is not easy to be sure that a culture is pure. Even a very tiny sample of blood or animal fluid may contain several kinds of organisms that may all grow together in culture. Koch realized the importance of pure cultures and developed several ingenious methods of obtaining them, of which the most useful is that involving the isolation of single ''colonies.'' Koch observed that when a solid nutrient surface such as a potato slice is exposed to air and then incubated, bacterial colonies developed, each having a characteristic shape and color. He inferred that each colony had arisen from a single bacterial cell that fell on the surface, found suitable nutrients, and began to multiply. Because the solid surface prevented the bacteria from moving around, all of the offspring of the initial cell had remained together, and when a large enough number was present, the mass of cells became visible to the naked eye. He assumed that colonies with different shapes and colors were derived from different kinds of organisms. When the cells of a single colony were spread out on a fresh surface, many colonies developed and each had the same shape and color as the original colony.

Koch knew that this discovery provided a simple way of obtaining pure cultures since streaking mixed cultures on solid nutrient surfaces spread the various organisms so far apart that the colonies they produced did not mingle. Many organisms could not grow on potato slices; so he devised semisolid media, in which gelatin was added to a nutrient fluid such as blood serum, in order to solidify it. When the gelatin-containing fluid was warmed it liquefied and could be poured out on glass plates; upon cooling the solidified medium could be inoculated. Later *agar* (a material derived from seaweed) was found to be a better solidifying agent than gelatin, and this substance is widely used today.

In the twenty years following the formulation of Koch's postulates, the causal agents of a wide variety of contagious diseases were discovered and isolated. This led to the development of successful treatments for the prevention and cure of contagious diseases and contributed to the development of modern medical practice. The impact of Koch's work has been felt throughout the world and especially in Western civilization, where the present-day standards of health are the highest in all the history of mankind.

It is important to remember, however, that not all diseases are caused by microorganisms; many are inherited or are due to deficiencies in diet and to other deleterious influences of the environment. Further, the microorganism is only one factor in the disease. It is a necessary cause, but it is not sufficient in itself. To produce disease, the microbe must infect a sensitive host, and not all hosts are equally susceptible. The state of health of the host, its general vigor, the presence or absence of specific immunity—all influence the outcome of the infection. Infectious disease is not some kind of thing, or entity; it is a process in which host and microbe interact.

It is easy to get the idea that, because some microbes cause diseases, all microbes are harmful. This is far from the truth. Most microbes are probably beneficial to man (see Chapters 14 and 16) or are at least harmless, and it is only the rare organism that causes disease.

1.6 The microbial environment

Whether life as we know it is a unique phenomenon existing only on planet Earth or whether it is a universal phenomenon existing on many distant planets is an unsolved question, which we shall consider in some detail near the end of this book (Chapter 17). That the earth is an excellent environment on which life could evolve and develop, however, is clear from the rich diversity of living organisms now present on our planet. Of all living organisms, there are none more versatile than the microorganisms. In no environment where higher organisms are present are microorganisms absent, and in many environments devoid of or inimical to higher organisms microorganisms exist and even flourish. Because microorganisms are usually invisible to the naked eye, their existence in an environment is often unsuspected; yet without them higher organisms would quickly disappear from the earth.

The geologist divides the earth into three zones, the lithosphere, the hydrosphere, and the atmosphere; to these zones we can add the biosphere. The *lithosphere* is the solid portion of earth, composed of solid and molten rocks and soil. Microbes are not common on rocks but make up an important part of the soil. The *hydrosphere* represents the aqueous environments of the earth—the oceans, lakes, rivers, and so on—and in these microorganisms are quite common, being found throughout the oceans as well as in fresh-water habitats from the tropics to the poles. The *atmosphere* comprises the gaseous phase of the earth, which is relatively dense near the surface but thins to nothing in the upper atmosphere. Although microbes are carried around the world on winds and other air currents, they do not actually reproduce in the atmosphere. The *biosphere* represents the mass of living organisms found in a thin belt at the earth's surface. Living organisms have had a profound influence on the earth itself, being responsible for almost all of the oxygen found in the atmosphere as well as for the enormous deposits of oil, coal, and sulfur found underground. Living organisms and their corpses provide excellent microbial environments, and thus we find large populations of microbes associated with higher plants and animals.

Although the earth provides suitable environments for microbial growth, we do not find the same organisms everywhere. In fact, virtually every environment, no matter how slightly it differs from others, probably has its own particular complement of microorganisms that differ in major or minor ways from organisms of other environments. Because microorganisms are small, their environments are also small. Within a single handful of soil or cupful of water many microenvironments exist, each providing conditions suitable for the growth of a restricted range of microorganisms. When we think of microorganisms living in nature, we must learn to ''think small.''

New environments for microorganisms continually are being created. Some of these result from natural processes, such as the formation of a new volcanic island or the creation of a new lake after an earthquake. Most new environments created today are man-made, however. The pollution of streams and lakes, the clearing of forests, the planting of exotic crops, and the introduction of new pesticides and fertilizers all create new microbial environments. Whether these environments are natural or man-made, microorganisms are able to establish themselves. Often the creation of a new environment makes possible a further step in evolution, and a new organism results.

Thus we should not be surprised by the great diversity of microbial life we find on planet earth, which only reflects the great diversity of habitats within which microbes can grow and evolve. Microorganisms are not passive inhabitants, however; through their activities they affect in various ways the environments in which they live. We have already mentioned that some organisms cause diseases, effects far more harmful than we would have predicted from their small sizes. Other deleterious changes, such as food spoilage, souring of milk, deterioration of clothing and dwellings, and corrosion of metal pipes, also occur primarily through microbial action. But microorganisms play many beneficial roles in nature, as, for example, in soil formation. They are responsible for most of the decomposition of dead animal and plant bodies and thus help in returning important plant nutrients to the soil. Many microorganisms that live in the intestinal tracts of animals synthesize certain vitamins, thus making their hosts independent of the need for these vitamins in their diets. Without microorganisms animals such as cows, sheep, and goats would be unable to digest the cellulose of grass and hay and hence would be unable to survive on earth. On the whole, the beneficial effects of microorganisms far outweigh their harmful ones. We will have much to say about this in succeeding chapters.

1.7 The contemporary study of microorganisms

Microbiology has come a long way from the days of Koch and Pasteur. Today it is one of the most sophisticated of the biological sciences, and has greatly influenced biology as a whole. Because of the special laboratory requirements for the study of microorganisms, microbiology is an independent discipline, but it is first of all a biological science.

Microorganisms have played a great role in recent years as model systems for the study of basic biological processes. Much of our understanding of molecular biology has come from studies with microbes, and thus one finds that microbiology and molecular biology are often grouped in one field. Yet there is a distinct difference between how molecular biologists and microbiologists study microbes. The former is interested in microbes as models and selects for his work those organisms which are simplest and easiest to study for his purposes. The microbiologist, on the other hand, is interested in microbes as organisms and studies them because of the things they do, in both natural and man-made environments. Thus the microbiologist does not restrict himself to the study of simple microbes, but studies what he considers to be the important microbes. In this text we will concentrate as much as possible on the second approach. The reader is probably already familiar with many of the ideas of molecular biology, either from previous courses or through reading magazines and newspapers. Even though he does not necessarily have a detailed understanding of them, he will at least have heard of RNA, DNA, proteins, and genes. There are many good modern books available at an elementary level, both paperbacks and short texts, that explain these biological entities with clarity. One may find it of advantage to read from time to time from one of these inexpensive books to fill in gaps in his knowledge.

1.8 The subdisciplines of microbiology

As knowledge increases it becomes ever more necessary for scientists to specialize and study only a small portion of a field. No one can know everything. Specialization has led to the spontaneous development of subdisciplines of microbiology, each of which has its own object for study. These various ways of dividing the field are shown in Table 1.1.

It should be pointed out, however, that all these categories are some-

Table 1.1   Subdisciplines of microbiology

Taxonomically oriented	Habitat oriented	Problem oriented
Virology	Aquatic microbiology	Microbial ecology
Bacteriology	Soil microbiology	Pathogenic microbiology
Phycology (or Algology)	Marine microbiology	Agricultural microbiology
Mycology		Industrial microbiology
Protozoology		Geomicrobiology

what arbitrary and may overlap each other. Sometimes a scientist will work in two subdisciplines at the same time, as for instance an aquatic microbiologist who studies only algae. The important thing to remember is that the scientists working on all of these fields are microbiologists and ultimately relate their work to the broader problems of general microbiology.

Supplementary readings

Brock, T. D. (ed. and trans.)   *Milestones in Microbiology.* Englewood Cliffs, N.J.: Prentice-Hall, Inc., 1961.

Bulloch, W.   *The History of Bacteriology.* London: Oxford University Press, 1938.

Davis, B. D., R. Dulbecco, H. N. Eisen, H. S. Ginsberg, and W. B. Wood, Jr.   *Microbiology.* New York: Harper & Row, Publishers, 1967.

Dobell, C. (ed. and trans.)   *Antony van Leeuwenhoek and His "Little Animals."* London: Constable & Co., Ltd., 1932. (Reprinted in paper covers. New York: Dover Publications, Inc., 1960.)

Dubos, R. J.   *Louis Pasteur, Free Lance of Science.* Boston: Little, Brown and Company, 1950.

Frobisher, M.   *Fundamentals of Microbiology,* 8th ed. Philadelphia: W. B. Saunders Co., 1968.

Hooke, R.   *Micrographia.* London: Royal Society, 1665. (Reprinted in paper covers. New York: Dover Publications, Inc., 1961.)

Kluyver, A. J., and C. B. van Niel   *The Microbe's Contribution to Biology.* Cambridge, Mass.: Harvard University Press, 1956.

Schierbeek, A.   *Measuring the Invisible World: The Life and Works of Antoni van Leeuwenhoek.* New York: Abelard-Schuman Limited, 1959.

Stanier, R. Y.   "Toward a definition of the bacteria," in I. C. Gunsalus and R. Y. Stanier (eds.), *The Bacteria: A Treatise on Structure and Function,* vol. V, p. 445. New York: Academic Press, Inc., 1964.

Stanier, R. Y., M. Doudoroff, and E. A. Adelberg   *The Microbial World,* 2nd ed. Englewood Cliffs, N.J.: Prentice-Hall, Inc., 1963.

Thimann, K. V.   *The Life of Bacteria,* 2nd ed. New York: The Macmillan Company, 1963.

part one

Microbial

structure

and function

The most important discoveries of the laws, methods, and progress in Nature have nearly always sprung from the examination of the smallest objects that she contains.

    J. B. Lamarck

This part introduces the student first to fundamental aspects of how microbial cells are constructed, with Chapter 2 discussing in more detail procaryotic cells and Chapter 3 eucaryotic cells. We begin the study of cell structure with an examination of cells and cell parts, using both the light and the electron microscopes. However, the structures that we see microscopically are constructed of chemical building blocks, and an important part of Chapters 2 and 3 explains cell structure in chemical terms. The rest of Part 1 deals with fundamental aspects of cellular function. Chapter 4 discusses the way in which cells obtain energy and considers enzymes and enzymatic reactions. Chapter 5 concerns the biosynthetic reactions catalyzed by enzymes that are involved in synthesizing important chemical building blocks. Also discussed is nutrition since some knowledge of this is necessary for understanding how the cell obtains from its environment the building blocks that it is unable to synthesize. Chapter 6 focuses on energy transformations of organisms that obtain their energy from nonorganic energy sources, such as light or inorganic chemicals. One should be prepared for a relatively stiff dose of chemistry in Part 1 because it is with the language of chemistry that we must explain cell function. Do not be overwhelmed by the chemistry; the basic concepts are relatively easy. However, the importance of understanding the material in Part 1 cannot be overemphasized since it presents the notions fundamental to most of the succeeding parts.

In this chapter we present current ideas on the structure of representative procaryotic cells. We discuss the techniques used for studying cell structure, the kinds of structures that can be seen, and the chemical nature of these structures. Our ability to see structural details of cells depends greatly upon the tools available to us. Early microscopes were crude, and knowledge of cell structure was correspondingly limited. Today we have available highly sophisticated microscopes, which have opened up new vistas. Although we see structural details in terms of images created by our instruments, we must remember that underlying these structures are the chemical components of which they are a part. Cells are built up as are houses, by connecting simple building blocks in various ways to create more complex structures. We could learn how a house is constructed by tearing it down bit by bit and examining the building blocks as they fall apart. In an analogous way we can use this method to discover how cells are put together. Our goal is to describe the chemical building blocks of various cell structures and how they are connected. Since a knowledge of structure is basic to understanding cell function, the information in this chapter lays a cornerstone on which we will build material in succeeding chapters.

2.1 Microscopical methods

The compound microscope   The compound microscope, which became widely available in the midnineteenth century and which was of crucial importance for the evolution of microbiology as a science, is with minor modifications still the mainstay of routine microbiological research. A compound microscope (Figure 2.1) has two lenses, the *objective lens*, which is placed close to the object to be viewed, and the *ocular lens*, which is placed close to the eye. The primary enlargement of the object is produced by the objective lens, and the image so produced is transmitted to the ocular, where the final enlargement occurs (Figure 2.2). The compound microscope is able to achieve considerably greater powers of magnification than is a microscope constructed of only a single lens. The latter, called a ''simple microscope,'' is used mainly in hand lenses and magnifying glasses.

In addition to magnification, an important property of a microscope is its *resolving power*. This is the ability to show as distinct and separate two points that are close together; and hence the greater the resolving power, the greater the definition of an object. Microscopes with high resolving power are especially good for viewing small structures. Resolving power in a compound microscope depends on the wavelength of light

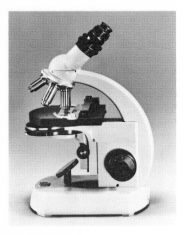

Figure 2.1 Photograph and diagram of a modern light microscope. The light source can be either built in (as in the diagram) or external. (Courtesy of Carl Zeiss, Inc., New York.)

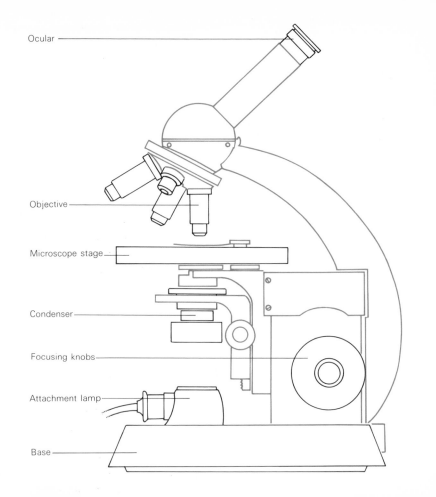

Ocular

Objective

Microscope stage

Condenser

Focusing knobs

Attachment lamp

Base

used and an optical property of the lens known as *numerical aperture* (Figure 2.3). Since the wavelength of light is usually fixed, the resolution of an object is a function of the numerical aperture; the larger the aperture, the smaller the object resolved. There is a rough correspondence between magnification of an objective and its numerical aperture so that lenses with higher magnification usually have higher numerical apertures. (The value of the aperture is printed on the side of the lens.) One factor in addition to the lens construction that affects numerical aperture is the medium through which the light passes. As long as the objective is sep-

$40\times$	$\times$	$10\times$	$=$	$400\times$
Magnification of objective		Magnification of ocular		Total magnification

Figure 2.2 The total magnification of a compound microscope is calculated by multiplying the magnification of the objective by the magnification of the ocular.

Diameter of smallest resolvable object
(resolving power)

$$= \frac{(0.6) \text{ wavelength of light used}}{\text{numerical aperture of objective}}$$

Figure 2.3

arated from the object by air, its numerical aperture can never be greater than 1.0; to achieve numerical apertures greater than this the objective must be immersed in a liquid of higher refractive index than that of air. Oils of various kinds are used, and lenses designed for use with oil are called *oil-immersion lenses*. The numerical aperture of a high-quality oil-immersion lens is between 1.2 and 1.4. Although oil-immersion lenses are usually of higher magnification than are lenses designed for dry use, they need not be so. An oil-immersion lens used in air presents a very unsatisfactory image, and it should never be employed without the use of oil.

According to optical theory, the highest resolution possible in a compound light microscope will permit the visualization of an object whose diameter is about 0.2 μm (micrometer), or 0.0002 mm (1 μm is thus 0.001 mm).* If an object 0.2 μm in diameter is magnified 1,000 times, it would appear to the eye as an object 0.2 mm in diameter, which can easily be seen. With the compound microscope, magnifications greater than 1,000 are attainable by choice of appropriate oculars but provide no further resolution of the object (this is called "empty magnification," since it adds nothing to the resolution).

Most microscopes used in microbiology have oculars that magnify about 10× and objectives of various sizes, usually 10× (100× total magnification), 40× (400× total), and 90× or 100× (oil immersion, 900× or 1,000× total). The lower power lens is used for scanning the specimen to locate objects of interest, the 40× lens permits the detailed visualization of large microorganisms such as algae, protozoa, and fungi, and the 90× or 100× lens is used for viewing bacteria and small eucaryotic microorganisms.

The illumination system of a microscope is also of considerable importance, especially when high magnifications are used. The light entering the system must be focused on the specimen, and a *condenser lens system* is used for this purpose. By raising or lowering the condenser the plane of focus of the light can be altered and a position can be chosen that leads to precise focus. The condenser lens system also has an *iris diaphragm*, which controls the diameter of the circle of light as it leaves the condenser system. The purpose of the iris diaphragm is not to control the intensity of light falling on the object but to ensure that the light leaving the condenser system just fills the objective lens. If the iris diaphragm is too large, some of the light will pass not into the objective but around it and will be of no use. If the light is too bright, it should not be

---

* A *micrometer* is the same as a *micron*, symbolized by μ. By recent international agreement, the word micron is being abandoned and replaced by micrometer. The student should remember that, whenever he sees the symbol μ in front of a unit, it means that the unit should be divided by one million. Thus 1 μm is one-millionth of a meter, or 10^{-6} m.

reduced by altering the position of the condenser or the iris diaphragm but by using neutral density filters or by decreasing the voltage to the lamp itself. It cannot be too strongly emphasized that the proper adjustment of the light is crucial to good microscopy, especially at higher magnifications.

Two other factors related to the lens system used are depth of field and area of field. *Depth of field* means the thickness of the specimen in focus at any one time, and is always greater with low power than with high power. Under oil immersion the depth of field is very shallow, usually less than 1 μm. The *area of field* is represented by the diameter of the specimen that is within view. Area of field is always greater at low power than at high power, and it is for this reason that low-power lenses are useful for scanning a slide.

Staining procedures   The size of most bacterial cells is such that they are difficult to see well with the light microscope. The main difficulty is the lack of contrast between the cell and the surrounding medium, and the simplest way of increasing contrast is through the use of dyes. Dyes may also be used to localize specific structures in the cell or to distinguish among different kinds of cells.

Cells are usually treated to coagulate protoplasm before staining, a process called ''fixing.'' For bacteria, heat fixing is most commonly used, although fixing can also be brought about by chemicals such as formaldehyde, acids, and alcohols. After fixation, if dye is added, further structural changes in protoplasm do not take place. Fixation is usually performed on cells that have been dried on a slide, the slide is then treated with the fixation agent, and the staining process follows immediately. Fixation usually causes cell shrinkage, and staining often makes cells appear larger than they really are so that size measurements on cells that have been fixed or stained cannot be carried out with much accuracy.

Dyes are organic compounds having some specific affinity for cellular materials (Figure 2.4). Many commonly used dyes are positively charged molecules (cationic) and combine strongly with such negatively charged cellular constituents as nucleic acids and acidic polysaccharides. Examples of cationic dyes are methylene blue, crystal violet, and basic safranin. Other dyes are negatively charged molecules (anions) and combine with such positively charged cellular constituents as many proteins. These include eosin, acid fuchsin, and Congo red. Another group of dyes is called *fat-soluble;* dyes in this group combine with fatty materials in the cell and are often used to reveal the location of fat droplets or deposits. An example of a fat-soluble dye is Sudan black.

Some dyes will stain well only after the cell has been treated with

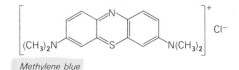

Methylene blue

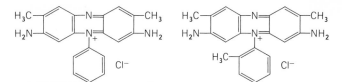

Safranin O, a mixture of dimethyl and trimethyl phenosafranin

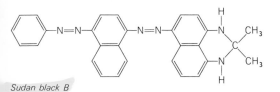

Sudan black B

Figure 2.4

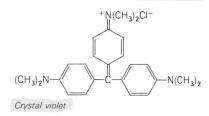

Crystal violet

another chemical, which itself is not a dye. Such treatment is called *mordanting;* a common mordant is tannic acid. The mordant combines with a cellular constituent and alters it in such a way that the dye will now attach.

If one desires merely to increase the contrast of cells for microscopy, simple staining procedures suffice. Methylene blue is a good simple stain that works on all bacterial cells rapidly and does not produce such an intense color that all cellular details are obscured. It is especially helpful in examining natural samples for the presence of bacteria since most of the noncellular material is not dyed.

Differential staining; the Gram stain One may often use a procedure that does not stain all cells equally, a process called *differential staining.* The most widely used differential stain is the *Gram stain*, named for the Danish bacteriologist Christian Gram, who developed it. On the basis of their reaction to the Gram stain, bacteria can be divided into two groups, Gram-positive and Gram-negative. Because of its importance in bacterial taxonomy and because it indicates fundamental differences in cell-wall structure of different bacteria (see Section 2.3), we shall describe the Gram stain here in some detail. Heat-fixed cells on a slide are stained first with a crystal violet solution (other basic dyes are not so effective), and are then washed to remove excess dye. At this stage all cells, both Gram-positive and Gram-negative, are stained blue. The slide is then treated with an iodine (I_2)–potassium iodide (KI) solution. The active ingredient here is I_2; the KI merely renders the I_2 water-soluble. The I_2 enters the cells and forms a water-insoluble complex with the crystal violet. Again both Gram-positive and Gram-negative cells are equally affected. Decolorization is then performed using either alcohol or acetone, substances in which the I_2–crystal violet complex is soluble. Some organisms (Gram-

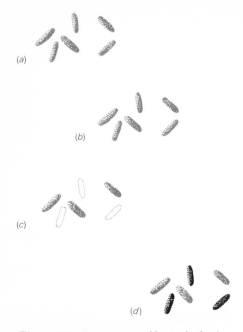

(a)

(b)

(c)

(d)

Figure 2.5 Appearance of bacteria during various steps in the Gram stain: (a) after staining with crystal violet—all cells dark violet; (b) after addition of iodine—no change in color; (c) after decolorization with alcohol—Gram-positive organisms remain dark violet, Gram-negative organisms are decolorized; (d) after counterstaining with safranin—Gram-negative organisms are red, Gram-positive organisms are dark violet.

positive) are not decolorized, whereas others (Gram-negative) are. The essential difference between these two cell types is thus in their resistance to decolorization; and as will be discussed later (Section 2.3), this resistance is probably due to the fact that Gram-positive cells, because of their thicker cell walls, are not permeable to the solvent. After decolorization Gram-positive cells are still blue, but Gram-negative cells are colorless. In order to visualize the Gram-negative cells, a counterstain is used. Usually this is one of the red acidic dyes, such as safranin or acid fuchsin. After treatment with counterstain, the Gram-negative cells are red, whereas the Gram-positive cells remain blue (Figure 2.5).

Two crucial aspects of the Gram stain should be noted. (1) The crystal violet treatment *must* precede the iodine treatment. Iodine itself has little affinity for the cells. (2) The decolorization must be carried out without much water present to avoid decolorization of Gram-positive cells. The decolorization process must be short, and precise timing is essential for satisfactory results. Finally, Gram-positivity is not really an all-or-nothing phenomenon. Some organisms are more Gram-positive than others and some organisms are Gram-variable, being Gram-positive at some times and Gram-negative at others. The Gram stain is one of the most important staining procedures in the bacteriological laboratory and is one with which the student should be well acquainted.

Another commonly used differential staining procedure, the *acid-fast method,* is of importance mainly with the genus *Mycobacterium,* which contains the causal organism of tuberculosis.

Other staining procedures   *Negative staining* is the reverse of the usual staining procedure: The cells themselves are left unstained, but the background is stained. The cells are thus seen in outline. The substance used for the negative staining is an opaque material that has no affinity for cellular constituents and merely surrounds the cells, such as india ink (which is a suspension of colloidal carbon particles) or nigrosin (a black, water-insoluble dye). Negative staining is a satisfactory way of increasing the contrast of cells for light microscopy, but it is most useful in revealing the presence of capsules around bacterial cells (see Figure 2.37).

Other staining methods are available that can be used to reveal the presence of specific cell constituents, such as flagella, spores, capsules, cell walls, centers of respiratory activity, and so on. These will be discussed later in the sections which deal with these topics. Staining methods are of great utility but must always be used with caution, as they may be quite misleading. Dye molecules occasionally form precipitates or aggregates which resemble actual cellular structures, but which are completely artificial formations induced by the dye itself. Such structures are called "artifacts," and great care must be taken to be sure that one is not misled into believing that an artifact is a truly existent structure.

Phase microscopy   The phase microscope makes it possible to see small cells easily even without staining. Cells differ in refractive index from their surrounding medium, and this difference can be used to create an image with a much higher degree of contrast than can be obtained with the normal light microscope (Figure 2.6). In Figure 2.7 the images obtained with unstained preparations by normal and phase microscopy are compared; the difference in contrast is clearly seen. Phase microscopy makes it possible to visualize cells more easily in the living state, and thus help to avoid the creation of artificial conditions, such as those introduced by staining. In bacteriological laboratories, the phase microscope has virtually replaced the light microscope as a research tool.

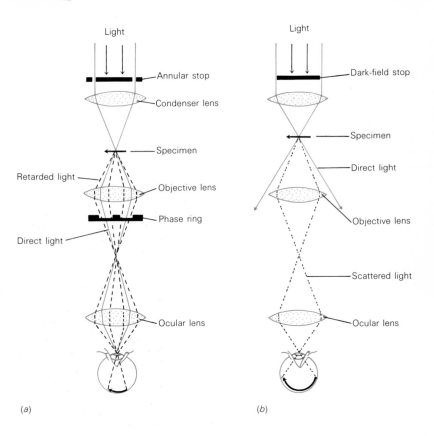

Figure 2.6 Light paths of phase-contrast (a) and dark-field (b) microscopes. Images seen by these two optical systems are shown in Figure 2.7.

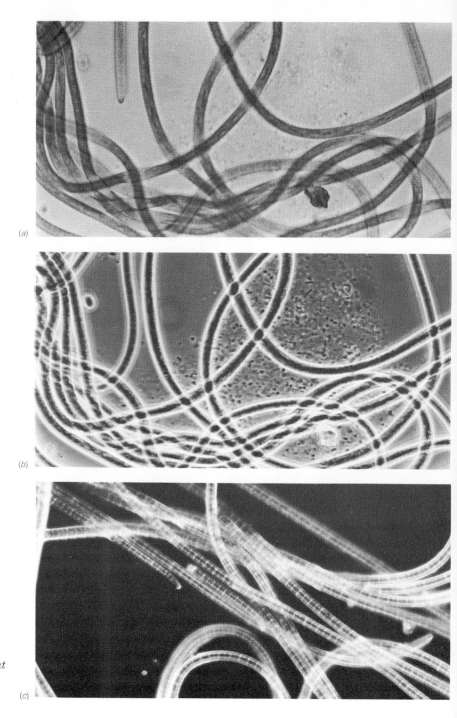

(a)

(b)

(c)

Figure 2.7 Photomicrographs (magnification, 580×) of a blue-green alga (Oscillatoria) with associated bacteria: (a) bright field microscopy—bacteria are poorly visible; (b) same field by phase-contrast microscopy—bacteria are clearly visible; (c) photomicrograph by dark-field microscopy of Oscillatoria without bacteria—note that internal structure is invisible but that the cylindrical shape of the filaments is clearly shown.

Other light microscopes A *dark-field microscope* is an ordinary light microscope whose condenser system has been modified to direct light at the specimen from the sides so that only light scattered by the specimen is passed to the ocular and visualized (Figure 2.6). By this arrangement, the specimen appears light on a dark background (Figure 2.7). Dark-field microscopy makes possible the observation in the living state of particles and cells that are otherwise below the limits of resolution of the light microscope, although few structural details are visible. Dark-field microscopy has been most widely used in the study of small motile cells such as *Treponema pallidum*, the spirochete that causes syphilis, which is invisible by ordinary light microscopy. The *fluorescence microscope* is used to visualize specimens that fluoresce, either because of the presence within them of natural fluorescent substances, or because they have been treated with fluorescent dyes. Fluorescence is the property many chemicals have of emitting light of one color upon excitation by light of another color. In the fluorescence microscope the exciting light is eliminated by a filter placed between the objective and the ocular, so that only the emitted light is seen. The use of the fluorescence microscope in immunology is discussed in Chapter 15.

Electron microscope In order to study the internal structure of procaryotes, use of an electron microscope is essential. With this microscope, electrons are used instead of light rays, and electromagnets function as lenses (Figure 2.8). When electrons pass through the specimen some of them are scattered, thus creating an image which is visualized on an electron-sensitive screen. The wavelength of the radiation of electrons is much shorter than that of visible light, and since the resolving power of a microscope is proportional to the wavelength of radiation used, the resolution obtained with the electron microscope is much greater than that obtained with the light microscope. Whereas with the light or phase microscope the smallest structure that can be seen is about 0.2 μm, with the electron microscope objects of 0.001 μm can readily be seen.* With the electron microscope it is possible to see many substances of even molecular size. However, because of the nature of this instrument, only very thin objects can be examined: If one is interested in seeing internal structure even a single bacterium is too thick to be viewed directly. Consequently, special techniques of thin sectioning are needed in order to prepare specimens for the electron microscope. For sectioning, cells must first be fixed and dehydrated, the latter usually performed by transferring the cells to an organic solvent. After

* For such small dimensions, the unit *nanometer* (nm) is often used. The prefix *nano-* means that the unit following should be multiplied by 10^{-9}. The older unit millimicron (mμ) is identical with the nanometer.

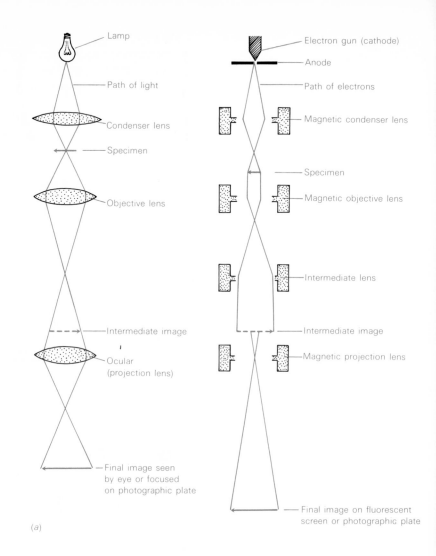

Lamp

Path of light

Condenser lens

Specimen

Objective lens

Intermediate image

Ocular
(projection lens)

Final image seen
by eye or focused
on photographic plate

(a)

Electron gun (cathode)

Anode

Path of electrons

Magnetic condenser lens

Specimen

Magnetic objective lens

Intermediate lens

Intermediate image

Magnetic projection lens

Final image on fluorescent
screen or photographic plate

dehydration the specimen is embedded in plastic and from this plastic thin sections are cut using a special ultramicrotome, usually with a diamond knife. A single bacterial cell, for instance, may be cut into five or six very thin slices, which are then examined individually with the electron microscope. To obtain sufficient contrast, the preparations are treated with special electron-microscope stains, such as osmic acid, permanganate, uranium, lanthanum, or lead. These materials are composed of atoms of high atomic weight, and because of this they scatter

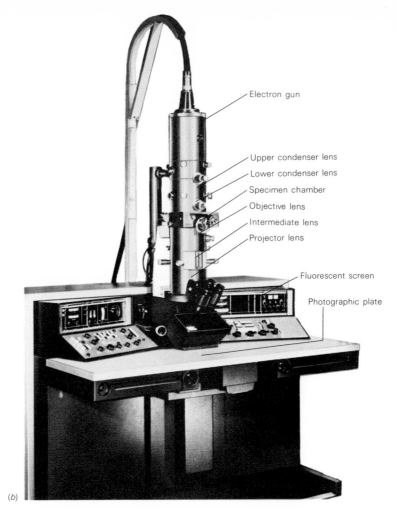

Figure 2.8   (a) Comparison of the path of light rays in a light microscope with the path of electrons in an electron microscope. In the electron microscope the beam is focused with magnets that are analogous to the lenses of the light microscope. In the electron microscope the final image is displayed on a fluorescent screen or on a photographic plate. (The right-hand diagram is adapted from Operating Manual, EMU-3 Electron Microscope. Camden, N.J.: RCA Scientific Instruments.) (b) A modern electron microscope. The positions of the various components should be compared with the diagram of (a). (Courtesy of RCA Scientific Instruments.)

Electron gun

Upper condenser lens
Lower condenser lens
Specimen chamber
Objective lens
Intermediate lens
Projector lens

Fluorescent screen
Photographic plate

(b)

electrons well. Cellular structures stained with one of these materials show greatly increased contrast, and hence are better seen.

Another way to achieve contrast with the electron microscope is negative staining. The same principle applies as in negative staining with the light microscope. A substance is used which does not penetrate the structure but which scatters electrons. One of the most commonly used negative stains for electron microscopy is phosphotungstic acid. Examples of negatively stained preparations are shown in Figures 2.21 and 2.32 and

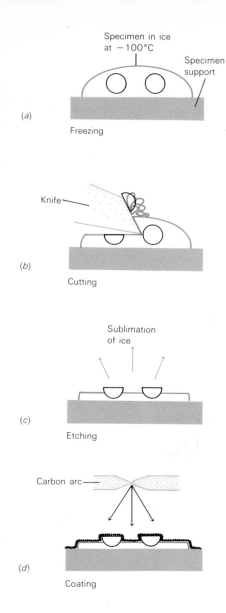

(a) Freezing

Specimen in ice at −100°C

Specimen support

(b) Cutting

Knife

(c) Etching

Sublimation of ice

(d) Coating

Carbon arc

Figure 2.9 Steps in specimen preparation by freeze-etching. An example of an electron micrograph of a freeze-etched sample is shown in Figure 3.24. [From H. Moor: "Freeze etching," Balzers High Vacuum Report 2:1 (1965). Santa Ana, Calif.: Balzers High Vacuum Corp.]

in several of the figures in Chapter 10. To examine the surfaces of cells, *carbon replicas* may be prepared by evaporating on the surface of the cells a thin coating of carbon, which conforms to the cell's surface contours and when stripped off is thin enough to be viewed directly with the electron microscope [Figure 18.11(*a*)]. Another recently developed procedure for electron microscopy is *freeze-etching*, which was designed to prevent the formation of artifacts by eliminating chemical fixation and embedding. As shown in Figure 2.9, the specimen to be examined is frozen without chemical treatment, and the frozen block is cut with a diamond knife in such a way that portions of the surfaces of cells are chipped away. Carbon replicas of these surfaces are then made and examined, with the net result that one sees surface or internal structures of cells (Figure 3.24). Most cellular structures seen in thin sections of chemically fixed specimens are also seen in freeze-etched material, suggesting that these structures are not artifacts.

Another recent instrument for examining surface structures is the *scanning electron microscope*. The material to be studied is coated with a thin film of heavy metal such as platinum. The electron beam is directed down on the specimen and scans back and forth across it. Electrons scattered by the heavy metal activate a viewing screen to produce an image. With the scanning electron microscope even fairly large specimens can be observed and the depth of field is very good. An example of the image obtained in this manner is given in Figure 3.5.

Electron microscopy is a highly developed art, and with this technique we are able to see cellular structures that cannot be seen in any other way. Many electron micrographs will be used in this book.

2.2 Size and form of procaryotes

Cell size Although most procaryotic cells are small there is a wide variation in size among different organisms. The smallest procaryotic cells are members of the Mycoplasma group, and are almost too small to be seen with the light microscope. The largest procaryotic cells, blue-green algae, are over 500 times as large in diameter as Mycoplasma, and can almost be seen with the naked eye.

Measuring the length of a microscopic object is essentially the same as measuring the length of any object; the object is placed next to a ruler of known length, and the dimension is read off directly. The ruler used in microscopy is the *ocular micrometer*, a small glass disc containing a ruler engraved on it (Figure 2.10). The ocular micrometer must be calibrated for each microscope and for each objective of the micro-

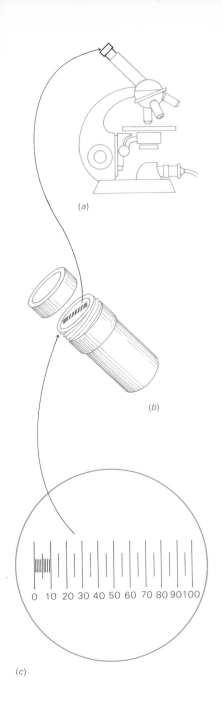

(a)

(b)

(c)

Figure 2.10   Ocular micrometer disc (c) used for measuring cell size. The disc fits into the ocular (b), which in turn is positioned in the light microscope (a). The cell is placed under the microscope so that the dimension to be measured extends along the length of the ruled micrometer.

scope; once calibrated, the lines on the ocular micrometer will be equivalent to micrometers or parts of micrometers. Size is now measured by moving the cell so that it is optically adjacent to the rule of the ocular, and its size is read off directly. It is quite easy to measure the sizes of a large number of cells rather quickly in this way. For spherical cells only the diameter need be measured, but for nonspherical cells both length and width must be measured. The range of cell size in procaryotes is illustrated in Figure 2.11.

Cell shape   Most bacteria have distinctive cell shapes, which remain more or less constant, although shape is influenced to some extent by the environment (see Figure 2.12 for representative shapes). Sphere-shaped bacteria are called *cocci* (singular, *coccus*), whereas cylindrical-shaped cells are called *rods*. Some bacteria are spirally or helically shaped, such as *Vibrio, Spirillum,* or the spirochetes. The shape of a cell definitely affects its behavior and stability. Thus cocci, being round, become less distorted and hence more resistant to drying than do rods or spirals. Rods, on the other hand, have more surface exposed per unit volume than cocci do and thus can more readily take up nutrients from dilute solutions. The spiral forms are usually motile and move by a corkscrew motion, which means that they meet with less resistance from the surrounding water than do motile rods, in the same way that it is easier to drive a screw than a large nail into hardwood.

Cell arrangement   When cells divide they often remain attached to each other, and the manner of attachment is usually characteristic of the organism, being related to the type of cell division. Thus many coccus-shaped organisms form chains (*Streptococcus* is the best-known example) as a result of their having divided always in the same plane. The length of such chains may be short (2 to 4 cells) or long (over 20 cells). Some cocci divide along two planes at right angles to each other, leading to the formation of sheets of cells; the largest and most regular sheets (tablets) are formed by the organism *Lampropedia. Sarcina* is an example of a third group of cocci, which divide in three planes and form cube-shaped packets. If there is no pattern to the orientation of successive division planes, an irregular clump will be formed; such a random cluster is characteristic of *Staphylococcus.*

Oscillatoria princeps, 37 × 5.25 μm

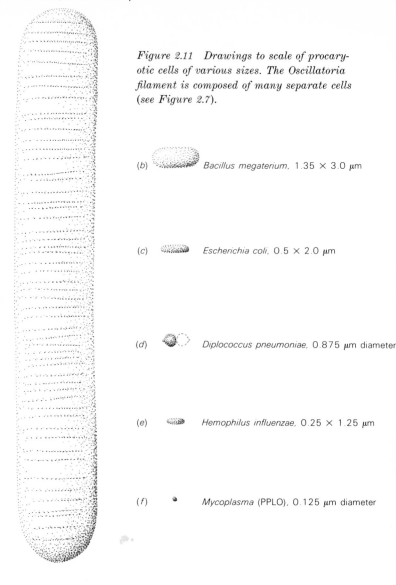

Figure 2.11 Drawings to scale of procary-otic cells of various sizes. The Oscillatoria filament is composed of many separate cells (see Figure 2.7).

(b) Bacillus megaterium, 1.35 × 3.0 μm

(c) Escherichia coli, 0.5 × 2.0 μm

(d) Diplococcus pneumoniae, 0.875 μm diameter

(e) Hemophilus influenzae, 0.25 × 1.25 μm

(f) Mycoplasma (PPLO), 0.125 μm diameter

(a)

Rods always divide in only one plane and hence form chains (such as in many *Bacillus* species) but not more complicated arrangements. Spirally shaped organisms also divide in only one plane, but they usually separate immediately and do not form chains. Many blue-green algae and their colorless relatives form long filaments composed of individual cells, such as are shown in the diagram of *Oscillatoria* (Figure 2.11). Because the cells remain attached throughout their width, the filaments are different from chains of rods, and the entire filament should be thought of as a single multicellular structure. Some filamentous bacteria lack cross walls (for example, the Actinomycetes; see Chapter 18); these can also be considered as long multinucleate cells. In most cases the cell arrangement is characteristic of the individual organism, but it may be influenced by environmental factors. Thus, some streptococci form very long chains under conditions of magnesium deficiency, but short chains when magnesium is plentiful. At one time cell arrangement was thought to be a key characteristic useful in classifying bacteria, but we now know that it is so variable a trait that it should be considered only a secondary characteristic.

Chemical composition The most important single chemical component of bacteria, as of all organisms, is water. At least 75 percent of the weight of a cell is water; often the percentage is considerably higher. Water is the solvent in which the organic and inorganic constituents of the cell are dissolved, and it provides the medium through which the various cellular substances move and interact. Not all the cell water is free; significant amounts may be bound to various cell constituents (that is, the cell materials are said to be *hydrated*), although it is difficult to determine exactly what proportion of water is bound.

Cells contain a variety of inorganic minerals, of which the most important are potassium, sodium, magnesium, calcium, iron, zinc, phosphorus, and sulfur. When a suspension of cells is dried on a microscope slide and then placed in a very hot oven, all of the organic materials are burned, leaving only the minerals behind. When examined under the microscope, the outline of each cell is still evident, showing that minerals are distributed throughout the cell, although often the concentration is greater in the region of the cell wall.

Although minerals are important for the life of each cell, the main building blocks that fit together to make the fabric of the cell are organic materials connected in various ways to make *macromolecules* or *polymers*. Each polymer is composed of a definite group of small organic molecules, called monomers, which are combined in a distinct way. Although a few substances unique to procaryotes are known, procaryotic cells on the whole are composed of the same building blocks as other cells. The precise characteristics and structure of any cell are determined not so much by the chemical group as by the exact manner in which these building blocks are connected. Just as an architect can design a wide variety of buildings from the same bricks, so can the same monomers form a variety of cells.

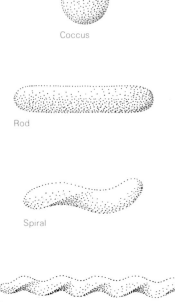

Figure 2.12 Representative cell shapes in procaryotes.

Coccus

Rod

Spiral

Spiral-helix

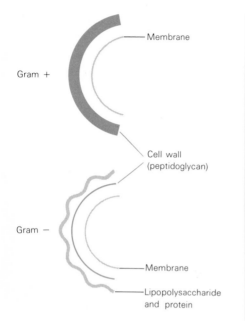

Gram +

Membrane

Cell wall
(peptidoglycan)

Gram −

Membrane

Lipopolysaccharide
and protein

Figure 2.13   Comparison of the cell-enve-lope structure of Gram-positive and Gram-negative bacteria. The whole outer layer of the cell is frequently called the "cell enve-lope," and the rigid peptidoglycan layer is called the "cell wall."

Figure 2.14   Electron micrograph of a thin section of a Gram-positive bacterium, Bacillus subtilis. Magnification, 29,900×. (Courtesy of Stanley C. Holt.)

One of the most important structural features of the procaryotic cell is the cell wall, which confers rigidity and shape. The procaryotic cell wall is chemically quite different from that of any eucaryotic cell and this is one of the features distinguishing procaryotic from eucaryotic organisms. The cell wall is difficult to visualize well with the light microscope but can readily be seen in thin sections of cells with the electron microscope. Gram-positive and Gram-negative cells differ considerably in the thickness of their cell walls as is shown in a diagram (Figure 2.13) and in the three electron micrographs of Figures 2.14 and 2.15. The Gram-negative cell wall is a thin structure sandwiched between the cell membrane and an outer layer composed of lipopolysaccharide and protein, while the Gram-positive cell wall is much thicker and lacks the outer lipopolysaccharide layer.

The chemical composition of the cell wall can be determined only from purified walls, from which all other cellular constituents have been removed. The first step in this purification is usually to break the cells mechanically by shaking them violently with glass beads. As breaks are made in the cell wall, the internal constituents (proteins, nucleic acids, soluble constituents, and so on) come out, leaving behind the empty hulls. By various chemical treatments, any materials attached to the cell walls are removed. Preparing clean cell walls from Gram-negative bacteria is difficult because the outer lipopolysaccharide complex adheres tightly. Purified cell-wall preparations retain the structure of the whole bacterial cell, showing that the cell wall itself confers shape to the whole cell.

Purified cell walls of both Gram-negative and Gram-positive bacteria are very similar in chemical composition. They contain two sugar deriva-

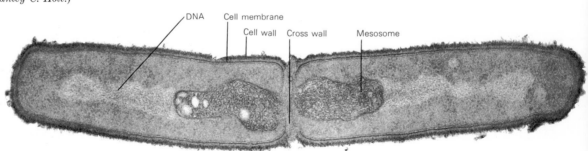

DNA   Cell membrane
Cell wall   Cross wall   Mesosome

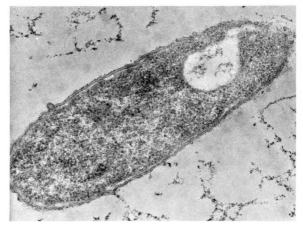

(a)

Figure 2.15   Electron micrographs of thin sections of <u>Gram-negative bacteria</u>, showing the cell-wall structure: (a) whole cell of Escherichia coli (magnification, 40,000×) (courtesy of R. G. E. Murray); (b) detail of the cell wall of Leucothrix mucor. Compare the structure of the cell wall to that of the Gram-positive bacterium in Figure 2.14. (Magnification, 165,000×.) (T. D. Brock and S. F. Conti, unpublished.)

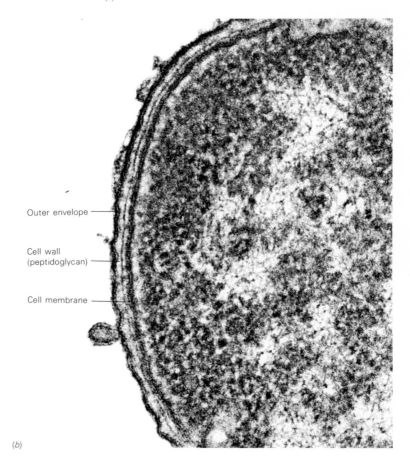

Outer envelope

Cell wall (peptidoglycan)

Cell membrane

(b)

2.3   Cell wall

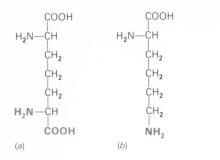

Figure 2.16 (a) *Diaminopimelic acid;* (b) *lysine.*

tives, N-acetylglucosamine and N-acetylmuramic acid, and a small group of amino acids, consisting of L-alanine, D-alanine, D-glutamic acid, and either lysine or diaminopimelic acid (DAP) (Figure 2.16). These constituents are connected to form a repeating structural unit, which is called the peptidoglycan* (Figure 2.17), and the peptidoglycan chains are cross-linked to form the three-dimensional cell-wall structure, as shown in Figure 2.18. Such a repeating structure is in reality a large macromolecule, the size of the cell, which the German biochemist Wolfgang Weidel called a "bag-shaped macromolecule."

In the intact cell, the cell wall is often associated with other constituents that do not confer rigidity but which affect in other ways the properties of the cell envelope. As noted, walls of Gram-negative bacteria have lipopolysaccharide attached, and in these organisms the complex outer layer of peptidoglycan and lipoprotein is collectively called the *cell envelope*. Gram-positive bacteria often have attached to or associated with the cell wall acidic macromolecules called *teichoic* acids (from the Greek word *teichos*, meaning "wall"). Teichoic acids contain repeating units, either glycerol or ribitol, connected by phosphate esters, and usually have other sugars and D-alanine attached (Figure 2.19). Because

*The repeating unit of the cell wall is sometimes called "mucopeptide" instead of peptidoglycan.

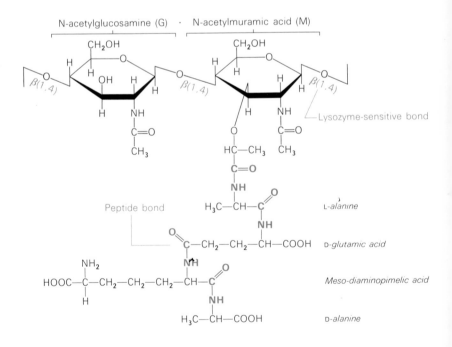

Figure 2.17 *Structure of one of the repeating units of the peptidoglycan cell-wall structure. The structure given is that formed in Escherichia coli and most other Gram-negative bacteria. In some bacteria, other amino acids are found.*

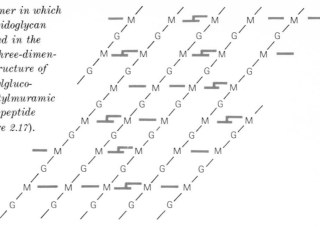

Figure 2.18 Manner in which the repeating peptidoglycan units are connected in the formation of the three-dimensional cell-wall structure of E. coli. G, N-acetylglucosamine; M, N-acetylmuramic acid; heavy lines, peptide cross links (Figure 2.17).

Figure 2.19 Glycerol teichoic acid from Lactobacillus casei.

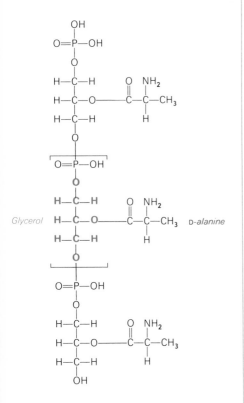

they are negatively charged, teichoic acids are partially responsible for the negative charge of the cell surface as a whole. Not all teichoic acids are associated with the cell wall; some are on the cell membrane.

Are the structural differences between the cell walls of Gram-positive and Gram-negative bacteria responsible in any way for the Gram-stain reaction? Recall that in the Gram reaction an insoluble crystal violet–iodine complex is formed inside the cell and that this complex is extracted by alcohol from Gram-negative but not from Gram-positive bacteria. Gram-positive bacteria have very thick cell walls, which become dehydrated by the alcohol. This causes the pores in the walls to close, thus preventing the insoluble crystal violet–iodine complex from escaping. The Gram reaction is not related directly to the bacterial cell-wall chemistry, however, since yeasts, which have a thick cell wall but of an entirely different chemical composition (see Chapter 3), are also Gram-positive organisms. Thus it is not the chemical constituents but the physical structure of the wall that confers Gram-positivity.

In addition to conferring shape, the cell wall is essential in maintaining the cell's integrity. Most bacterial environments have solute concentrations considerably lower than the solute concentration within the cell. Since water passes from regions of low solute concentration to regions of high solute concentration, a process called *osmosis*, there is a constant tendency throughout the life of the cell for water to enter, and the cell would swell and burst were it not for the rigidity of the cell wall. This can be dramatically illustrated by treating a suspension of bacteria with the enzyme *lysozyme*. Lysozyme is found in tears, saliva, other body fluids, and egg white. It hydrolyzes the cell-wall polysaccharide (Figure 2.17), thereby weakening the wall. Water then enters, the cell swells,

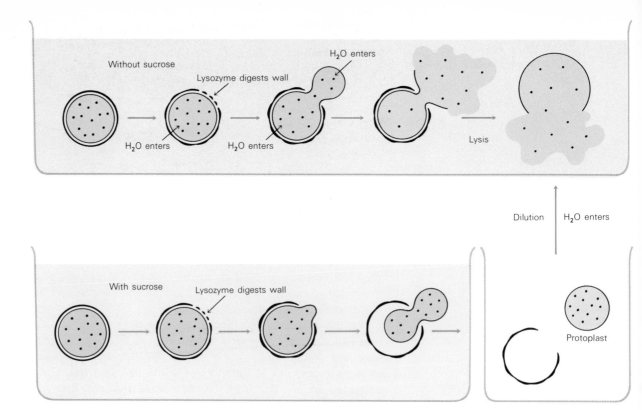

Figure 2.20 *Stabilization of protoplast by sucrose and lysis upon dilution.*

and the protoplast pushes out and bursts, a process called *lysis* (Figure 2.20). The release of a protoplast as seen with the electron microscope is shown in Figure 2.21.

If the proper concentration of a nonpermeable solute, such as sucrose, is added to the medium, the solute concentration outside the cell balances that inside. Under these conditions, lysozyme still digests the cell wall, but lysis does not occur, and an intact protoplast* is formed (Figure 2.20). If such sucrose-stabilized protoplasts are diluted in water, lysis occurs immediately. Osmotically stabilized protoplasts are always spherical in liquid medium, even if derived from rod-shaped organisms, further emphasizing that the shape of the intact cell is conferred by the cell wall.

---

*Strictly defined, a *protoplast* is a structure completely devoid of cell wall and consists of the cell membrane and all intracellular components. Spherical, osmotically sensitive bodies can also be formed in which the cell wall is only partially removed. Such spherical structures with some cell wall fragments still attached are usually called *spheroplasts*, to distinguish them from true protoplasts.

Figure 2.21 *Release of protoplast from a Bacillus stearothermophilus cell as a result of partial lysis of the cell wall with lysozyme.
Electron micrograph of a whole cell made by negative staining
with phosphotungstic acid. Because of the unusual stability of the
cell membrane of this thermophilic bacterium, osmotic lysis of the
protoplast does not occur, thus making the micrograph possible.
Magnification, 18,000×. [From D. Abram: J. Bacteriol. 89:855 (1965).]*

If the medium has a higher solute concentration than that in the cells,
water flows out (the cells become dehydrated) and the protoplast col-
lapses, a process called *plasmolysis*. This is one basis for protecting
foods from bacterial spoilage by curing them with strong salt or sugar
solutions (see Chapter 8).

When a cell enlarges during the division process, new cell-wall synthesis must take place, and this new wall material must be added in
some way to the preexisting wall. This process can occur in different
ways, as shown in Figure 2.22. In order for the new material to be
inserted into the cell-wall fabric, small openings in the macromolecular
structure of the wall must develop. These openings are created by enzymes produced within the cell that are similar to lysozyme. The new
wall material is then added across the openings. In nongrowing cells
spontaneous lysis can occur, a result of the cell's own enzymes' hydro-
lyzing the cell wall without concomitant new cell-wall synthesis—a
process called *autolysis*. Certain antibiotics, such as *penicillin* and
cycloserine, inhibit the synthesis of new cell-wall material in growing cells
and induce cell lysis. The action of these drugs is discussed in Chapter 9.

Figure 2.22 *Localization of new cell-wall
synthesis during cell division. In the cocci
new cell-wall synthesis is localized at only
a few points, whereas in rod-shaped bacteria it occurs at several locations along
the cell wall. [Adapted from R. M. Cole:
Bacteriol. Rev. 29:326 (1965).]*

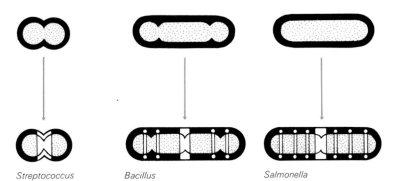

Streptococcus   Bacillus   Salmonella

Although most bacteria cannot survive without their cell walls, a few organisms are able to do so; these are the Mycoplasma, a group of organisms that cause certain infectious diseases. Mycoplasma are essentially free-living protoplasts, and are probably able to survive without cell walls partly because they live in osmotically protected habitats, such as the animal body. These organisms will be discussed in more detail in Chapter 18.

Although the cell wall may seem continuous enough when viewed in the electron microscope, it is actually full of pores, through which water and various chemical materials pass. The cell wall is a barrier only to such very large molecules as proteins and nucleic acids, and particles such as viruses.

Some organisms are able to undergo changes in shape and can exist in more than one form; these are called *pleomorphic*. Often the changes are random, but cells of the genus *Arthrobacter* show regular changes from rod-shaped to coccus-shaped forms (see Chapter 18). Similarly, *Azotobacter* and *Myxococcus* rods frequently transform into spherically shaped cells called "microcysts" (see Chapter 18). These changes in shape reflect aspects of cell-wall structure that are not yet understood.

2.4 Cell membrane

The cell membrane can be seen in the electron micrographs of Figures 2.14 and 2.15 as the inner thin layer at the cell periphery. Membrane systems of some bacteria are simple, whereas in photosynthetic procaryotes they are often much more complex (Figure 2.23), with a large

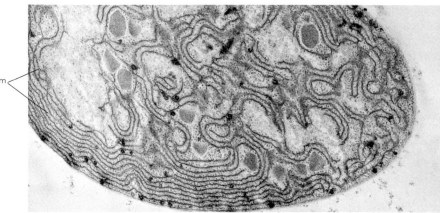

Photosynthetic membrane system

Figure 2.23   Electron micrograph of a blue-green algal cell (Anabaena azollae), showing the extensive array of photosynthetic membranes (thylakoids). Magnification, 10,000×. [From N. J. Lang: J. Phycol. 1:127 (1965).]

number of distinct layers. Many bacteria have a special membranous structure called a *mesosome* (meaning "middle body"), which is well shown in Figures 2.14, and 2.29. This seems to be associated in some way with the formation of the bacterial cross wall.

Bacterial membranes can be isolated when sucrose-stabilized protoplasts are lysed by dilution. The contents of the cell go into suspension and the cell membrane remains as a thin structure called a "ghost," which can be treated to remove adhering contaminants. The chemical structure of the cell membrane is better known in eucaryotes than in procaryotes and will be discussed in more detail in Chapter 3. The membrane plays various roles in the cell. It is the main barrier to the passage of large molecules in and out of the cell, and it also helps impede the movement of water and solutes. It probably has pores, but they are too small to be seen readily even with the electron microscope. Their existence can be inferred from the fact that small solute molecules do pass in either direction through the membrane, although large molecules do not. In addition to sustaining cell permeability, the procaryotic cell membrane plays a key role in cell respiration, as the enzymes associated with this process are part of the membrane (see Chapter 4). The chlorophyll and the photosynthetic machinery of photosynthetic organisms are also associated with membranes (see Chapter 6).

The interior of the cell consists of an aqueous solution of salts, sugars, amino acids, vitamins, coenzymes, and a wide variety of other soluble materials, which is called the *cell pool*. When the permeability barrier of the cell is destroyed, most of these materials are able to leak out, and only substances too large to pass through the cell-wall pores are retained. Components enter the pool either as nutrients taken up from the environment or as materials synthesized from other constituents in the cell.

2.5 Ribosomes

In many of the electron micrographs in this chapter, small dark particles can be seen within the cytoplasm. These particles, which are part of the protein-synthesizing machinery of the cell, are called ribosomes, because they are ribonucleic acid (RNA)–containing bodies (-*some* means "body"). They are comprised of about 60 percent RNA and 40 percent protein and are about 20 nm in diameter. Ribosomes can easily be isolated from ruptured bacteria by high-speed centrifugation and have been extensively studied. The ribosomes from procaryotic microorganisms are slightly smaller and lighter in weight than those of eucaryotes. Ribosomes have characteristic sedimentation constants: procaryotic ribosomes have a sedimentation constant of 70 S, while those of eucaryotes have a constant of 80 S (see Chapter 3).* When examined with the electron microscope, each ribosome particle is seen to consist of two parts of unequal size (Figure 2.24). These are distinct entities which are separable by physical means and can be reassociated. It is thought that proteins of each particle are attached to the RNA and that the whole unit is wrapped up into a globular structure. In the intact cell ribosomes often occur in aggregates of various sizes called *polyribosomes* (see Figure 2.25). When cells are broken open, the polyribosomes fragment easily, and the individual ribosomes float free. Thus, unless special precautions are taken, only single ribosomes are seen in cell extracts. The role of ribosomes in protein synthesis is discussed in Chapter 7.

*The sedimentation constant is a physical parameter of a particle measured with an analytical ultracentrifuge, and is a function of both molecular size and shape. Sedimentation constants are usually given in units called "Svedbergs," abbreviated S, which are named after the Swedish scientist who pioneered in analytical ultracentrifugation. Unless otherwise specified, the sedimentation constant is measured in water at 20°C.

Figure 2.24   Subunit structure of a ribosome particle. The 50-S and 30-S particles probably exist free in the cell and come together to form a 70-S particle on the messenger RNA.

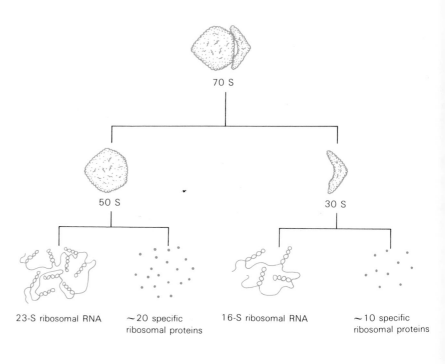

70 S

50 S   30 S

23-S ribosomal RNA   ~20 specific   16-S ribosomal RNA   ~10 specific
                        ribosomal proteins   ribosomal proteins

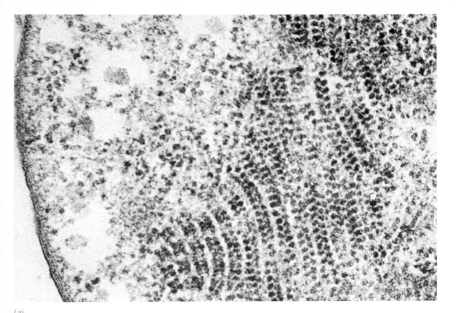
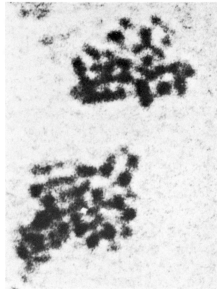

(a)

(b)

Figure 2.25 *(a) Extensive array of 70-S ribosome particles in an E. coli cell. Such pictures can be taken only with techniques that preserve the structure during preparation for the electron microscope. Magnification, 98,000×. (Courtesy of Elizabeth W. Kingsbury and Herbert Voelz.) (b) Enlargement of two arrays of ribosome particles (polyribosomes) in Bacillus cereus. Magnification, 470,000×. [From R. M. Pfister and D. G. Lundgren: J. Bacteriol. 88:1119 (1964).]*

2.6 Nuclear region—DNA

Procaryotic organisms do not possess a true nucleus as do eucaryotic organisms. In the electron micrographs of thin sections, the nuclear region is seen as a weakly contrasting area that contains thin fibrillar material, but which lacks a definite membrane (Figure 2.14). The thin fibrillar material is deoxyribonucleic acid (DNA), the genetic material of the cell (Figure 2.26). Each DNA molecule is composed of two strands complementary to each other because of the specific pairing of adenine with thymine and guanine with cytosine. When a bacterial protoplast is lysed gently, this DNA pours out of the cell in a long, single fiber (Figure 2.27), probably a single molecule, whose length when stretched out would be very many times the length of the cell itself. For instance, the length of an *Escherichia coli* cell is about 2 μm, whereas the length of the *E. coli* DNA is 1,100 to 1,400 μm! This means that within the cell

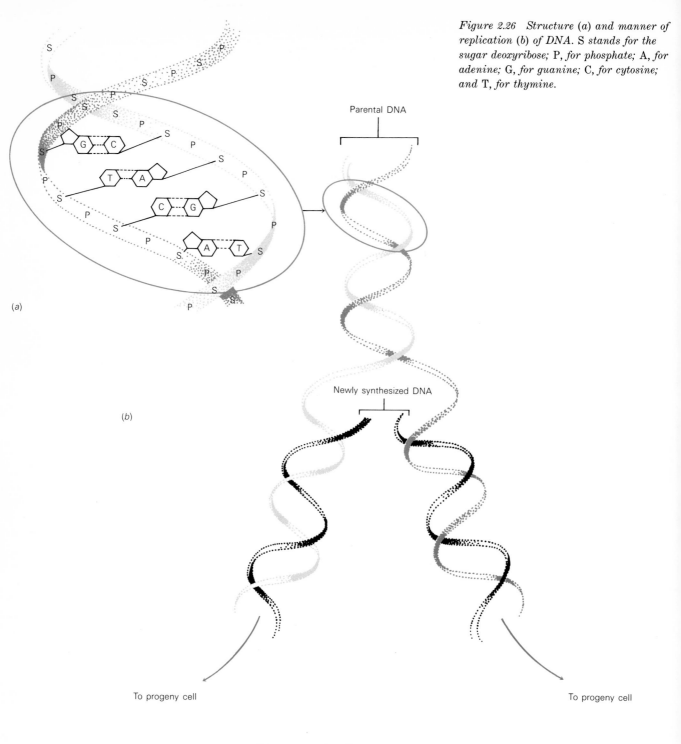

Figure 2.26 Structure (a) and manner of replication (b) of DNA. S stands for the sugar deoxyribose; P, for phosphate; A, for adenine; G, for guanine; C, for cytosine; and T, for thymine.

Parental DNA

(a)

(b)

Newly synthesized DNA

To progeny cell

To progeny cell

The procaryotic cell

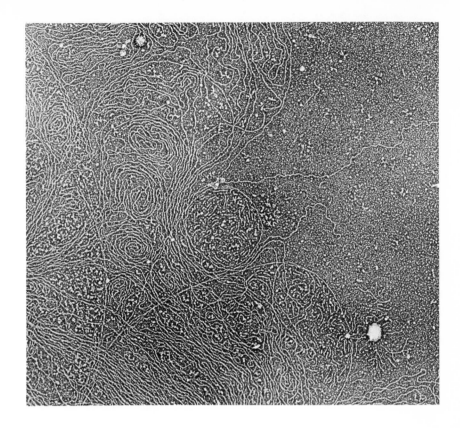

Figure 2.27 Electron micrograph of DNA fibrils released upon gentle lysis of a Micrococcus lysodeikticus protoplast. Note the small number of free ends. A fragment of the lysed cell is seen in the lower right. The DNA is rendered visible by complexing with a basic protein and by shadowing with a heavy metal. (Courtesy of A. K. Kleinschmidt.)

the DNA must be highly folded. In *E. coli* this single DNA molecule is probably circular, with no free ends. Because of its long, thin structure, it is very difficult to isolate a complete, unbroken molecule of DNA—most purified preparations are composed of many short fragments of the original molecule. Sometimes more than one nuclear region is seen in a single cell, but each of these probably contains only a single DNA molecule. The genetic functions of DNA are described in Chapters 11 and 12.

Cell division   Cell division requires the doubling of all cell constituents and their orderly partitioning into the two daughter cells. The first step in cell division is the doubling of the DNA complement of the nuclear body (Figure 2.26), which occurs by the separation of the two strands and the replication along each strand of a new complementary strand. Because of the enormous length of the DNA, the cell faces a problem in partitioning the doubled DNA between the two daughters. It is thought that the DNA molecule is attached at some point to the cell

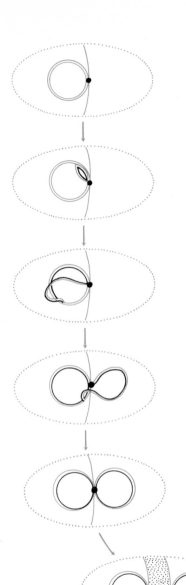

membrane, and after replication, each one of the double strands, while still attached, swings toward one of the two cell halves (Figure 2.28). Only after this step has been accomplished can the new cross wall form. At the time the cross wall forms, a mesosome of the membrane system can be seen attached to the growing cross wall (Figure 2.29), and it is likely that this mesosome plays some role in the synthesis of the cross wall. The DNA is probably attached to the other end of the same mesosome, as can also be seen in Figure 2.29.

This highly coordinated process, _DNA doubling/DNA partitioning/cross wall formation_, is the equivalent of the mitotic process (see Chapter 3) in eucaryotic organisms. Each step must occur in proper sequence, else chaos would result. For instance, if cross-wall formation took place before DNA partitioning, one cell of the pair would completely lack DNA and would be unable to divide. Such DNA-less cells are occasionally formed, apparently when something goes wrong with the machinery that coordinates the division process.

In procaryotes, as opposed to eucaryotes (Chapter 3), DNA synthesis occurs almost continuously from one cell division to the next and ceases only during the time of DNA partitioning.

2.7 Other cell structures

Flagella and motility Many bacteria are motile, and this ability to move by their own power is usually due to the presence of a special

Figure 2.28 Replication and partitioning of DNA during cell division, showing how precise control is effected by attachment of DNA to the growing cell wall. [Adapted from J. Cairns: Cold Spring Harbor Symp. Quant. Biol. 28:43 (1963) and from F. Jacob, S. Brenner, and F. Cuzin: Cold Spring Harbor Symp. Quant. Biol. 28:329 (1963).]

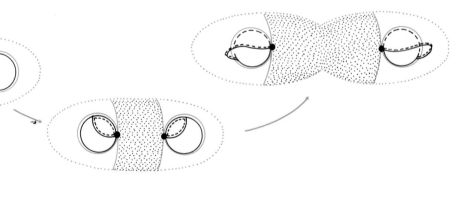

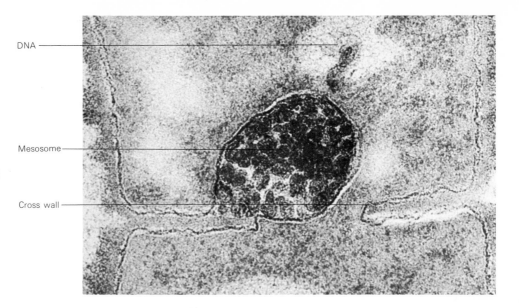

DNA

Mesosome

Cross wall

Figure 2.29 Cross-wall formation in Bacillus megaterium and
relationship of the mesosome to
the developing wall. It is thought
that the DNA is also attached to
the mesosome. Magnification,
103,000×. [From D. J. Ellar,
D. G. Lundgren, and R. A. Slepecky: J. Bacteriol. 94:1189
(1967).]

organelle of motility, the *flagellum* (plural, *flagella*). Bacterial flagella are
long, thin appendages free at one end and attached to the cell at the
other end. They are so thin (about 20 nm) that a single flagellum can
never be seen directly with the light microscope, but only after staining
with special flagella stains. One of the most common flagella stains employs the dye basic fuchsin with tannic acid as a mordant. The mordant
promotes the attachment of dye molecules to the flagellum, and a crust
or precipitate forms along the length of the flagellum, making it visible
with the light microscope (Figure 2.30). Flagella are also readily seen
with the electron microscope if they are coated with a thin film of heavy
metal, a process called shadowing (Figure 2.31).

Flagella are arranged differently on different bacteria. Polarly flagellated
organisms have a single flagellum attached at one end of the cell (Figure
2.30). Occasionally a tuft of flagella may arise at one end of the cell, an
arrangement called "lophotrichous" (*lopho-* means "tuft"; *trichous*
means "hair"). In *peritrichous* (*peri-* means "around") *flagellation* the
flagella are not localized, but grow from many places on the cell surface.
The type of flagellation is often used as a characteristic in the classification of bacteria (Chapter 18). Flagella are not straight but helically
shaped; when flattened, they show a constant length between two adjacent bends, called the wavelength. The flagellum's wavelength is a
constant characteristic for each species.

2.7   Other cell structures

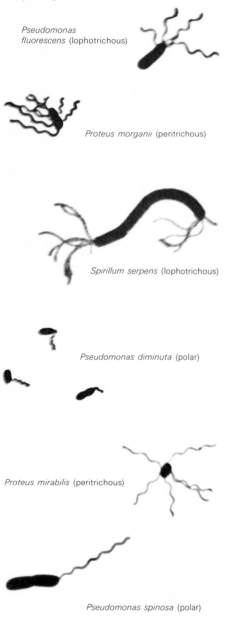

Pseudomonas fluorescens (lophotrichous)

Proteus morganii (peritrichous)

Spirillum serpens (lophotrichous)

Pseudomonas diminuta (polar)

Proteus mirabilis (peritrichous)

Pseudomonas spinosa (polar)

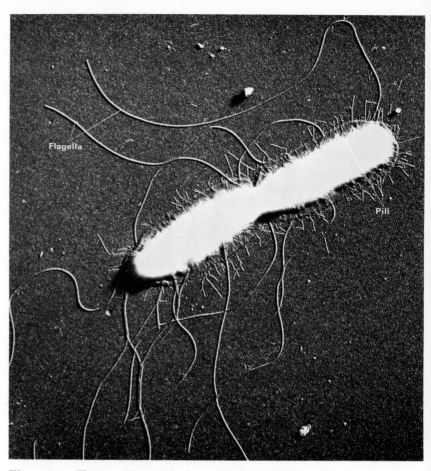

Flagella

Pili

Figure 2.31 Electron micrograph of a metal-shadowed whole cell of Salmonella typhosa, showing flagella and pili. Magnification, 15,000×. [From J. P. Duguid and J. F. Wilkinson: "Environmentally induced changes in bacterial morphology," in G. G. Meynell and H. Gooder (eds.), Microbial Reaction to Environment (Eleventh Symposium, Society for General Microbiology). London: Cambridge University Press, 1961.]

The site of attachment of a flagellum is closely associated with the cell membrane. This is shown by the fact that the flagella still remain attached when a flagellated cell is converted into a protoplast, and thus in a normal cell, the flagellum must pass through the cell wall. By careful study with the electron microscope, it is possible to see that the base of the flagellum is attached to a special membrane unit just inside the cell membrane (Figures 2.32 and 2.33).

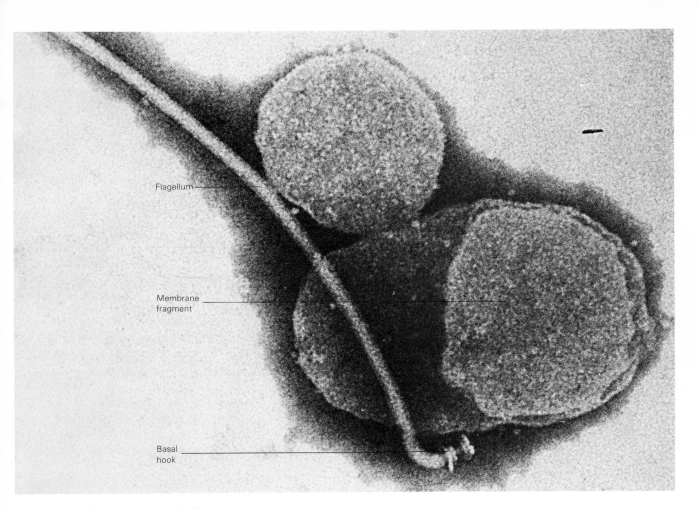

Flagellum

Membrane
fragment

Basal
hook

Figure 2.32   Electron micrograph of a preparation of Rhodospirillum molischianum negatively stained with phosphotungstic acid. The bacteria were partially lysed by treatment with lysozyme, uncovering the basal hook of the flagellum, here in association with a membrane fragment. Magnification, 240,000×. [From G. Cohen-Bazire and J. London: J. Bacteriol. 94:458 (1967).]

Figure 2.33   An interpretive drawing of the probable manner of attachment of a flagellum. Compare with Figure 2.32. (Courtesy of G. Cohen-Bazire.)

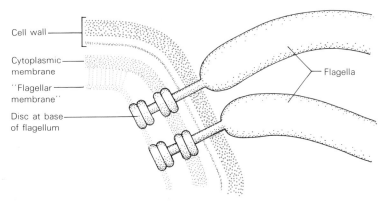

Cell wall

Cytoplasmic
membrane

"Flagellar
membrane"

Disc at base
of flagellum

Flagella

2.7   *Other cell structures*

49

Flagella can easily be removed from an organism by stirring a cell suspension rapidly, such as in a food blender. The shearing forces strip off the flagella, but the viability of the cells is unaffected. Once removed from the cell, the flagella can be purified by centrifugation.

Bacterial flagella are composed of protein subunits; the protein is called *flagellin*. When a flagella preparation is acidified, the protein subunits come apart and one obtains a clear homogeneous protein solution. On neutralization of the acidity, the protein subunits can reassociate and reform the intact flagella. The amino-acid composition of flagellin is somewhat atypical; there are lower amounts of sulfur-containing and aromatic amino acids than in most cellular proteins, whereas aspartic and glutamic acids occur more frequently.

The role of flagella in motility is now well established, as may be seen by the following lines of evidence: (1) Removing flagella by rapid stirring leads to a loss of cell motility; when such cells are allowed to resynthesize flagella, motility is regained. (2) Under some growth conditions flagella are not synthesized, and such unflagellated cells are nonmotile. When growth conditions allow flagellar synthesis, motility resumes. (3) One can isolate from motile organisms mutant strains (see Chapter 11) that lack flagella and are nonmotile. By genetic exchange (see Chapter 12) the mutant gene can be replaced by the wild-type gene, and, at the same time flagella appear on the cell, motility is restored.

These results show that in flagellated organisms flagella and motility are always associated. Do flagella move? It is impossible to see the movement of a single flagellum since it is invisible with the light microscope and flagellar staining methods kill the cell. (Electron microscope preparations must be completely dry, hence movement is impossible.) However, a flagellar tuft, as in a lophotrichous organism, is visible with the dark-field microscope, and the movement of this tuft can be watched and correlated with cell movement. The exact manner in which the flagellum moves remains unknown. Perhaps flagellar movement results from the contraction and expansion of the flagellum, through movement of the flagellin molecules, but there is no evidence for this.

The motions that polar and lophotrichous organisms make are different from those made by peritrichous organisms. Peritrichously flagellated organisms generally move and rotate in a straight line in a slow, stately fashion. Polar organisms, on the other hand, spin around rapidly and dash from place to place.

The fastest known rate of movement of a motile bacterial cell is about 50 μm/sec. This would be equivalent to 0.0001 mi/hr for macroorganisms, which is not a very impressive figure when compared with the rates at which even rowboats can be moved through the water. Probably the main reason the rate is not faster is that water is a viscous medium, and the effects of viscosity are much greater on small objects than on large ones. Thus the work of moving a bacterial cell through the water is roughly analogous to the work it would take to move a rowboat through thick molasses syrup. Considered in these terms, the amazing thing about bacterial movement is that it occurs at all! Eucaryotic cells (see Chapter 3) move considerably faster, a fact consonant with their larger sizes.

Motility is more common in rod-shaped than in spherical bacteria, and this is also probably due to hydrodynamic factors. An elongated object moves through the water much more easily than does a spherical object since it is subject to much less viscous drag and is much easier to steer. Anyone who has tried to paddle a wash tub across a pond can appreciate the problem that a motile coccus would have.

When watching motile bacteria under the microscope, it is important to distinguish between vital movement and another process, *Brownian motion*, which is exhibited even by nonliving matter. Brownian motion is a random agitation exhibited by all small particles, and is due to the molecular movement of the water molecules themselves, which is transferred to the particles. A particle exhibiting Brownian motion moves around in a haphazard manner but does not make any headway in the water. A motile bacterium, on the other hand, moves in a directed manner through the water. Only small particles exhibit Brownian motion; thus large bacteria, yeasts, and larger cells show this phenomenon slightly or not at all.

Most motile bacteria are able to change their movements in response to environmental stimuli, exhibiting what are called tactic movements. Thus, for example, photosynthetic bacteria (see Chapter 6) move in response to light (*phototaxis*). Imagine a photosynthetic bacterium moving in a spot of light. As it courses around through the liquid, it may by accident wander out of the light field. Immediately, the cell stops and reverses its direction, and thus reenters the light. If by chance it should leave the light field at a later time, it would exhibit the same response. Suppose now we examine a cell that is completely removed from the light spot. It moves back and forth through the dark liquid, showing no especial attraction to the light. (After all, it has no way of knowing the light spot is even there.) But if, in the course of its movement, it should wander into the light spot, it now remains in the light, and whenever it leaves the light spot, it reverses direction. Thus, even though bacteria do not swim towards the light, most of the cells end up in the light spot, and only a few remain in the dark. In addition to being able to distinguish light from dark, these bacteria can distinguish bright light from dimmer light. It is not the absolute amount of light to which they respond, but to differences in light intensity; phototactic bacteria can dis-

Figure 2.34 *Distribution and synthesis of flagella during cell division: (a) polarly flagellated organism; (b) peritrichously flagellated organism.*

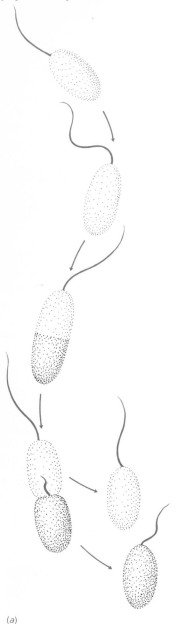

tinguish between two light sources that differ in intensity by only 5 percent. Tactic movements are also exhibited by many bacteria in response to differences in concentrations of chemicals, a phenomenon called *chemotaxis*. Sometimes the response is positive (towards the chemical) and other times negative (away from the chemical). Tactic responses are obviously advantageous to a motile organism in its natural environment, as they enable the organism to leave unfavorable environments and to remain in favorable ones. The manner in which the organism is able to control its flagellar movement in responding to a tactic stimulus is not known. Obviously, some coordinating system within the cell must exist or tactic movement would not be possible.

When a cell divides, the two daughter cells must acquire in some way a full complement of flagella. In polarly flagellated organisms, the process of cell division probably occurs as shown in Figure 2.34(a), the

(a)

(b)

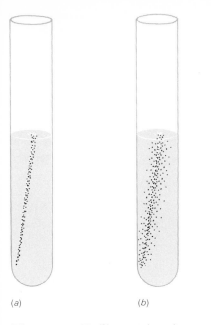

Figure 2.35 Motility test in soft agar: (a) nonmotile organisms; (b) motile organisms.

new flagellum forming at the location where cell division has just occurred. In a monopolarly flagellated cell, the two poles of the cell probably differ in some way so that the flagellum is formed at one pole and not at the other.

In peritrichously flagellated organisms, the relation of cell division to flagella synthesis probably occurs as shown in Figure 2.34(b), the preexisting flagella being distributed equally between the two daughter cells, and new flagella being synthesized and filling in the gaps.

Since motility is often used in classifying bacteria, it is important to be aware of the proper way of studying this process in the laboratory. Microscopic examination is done using a wet mount, preferably with a phase microscope since stained preparations cannot be used. Only freshly grown cultures should be examined since in old bacterial cultures many of the cells may have lost their flagella or may be only weakly motile. The medium and conditions of culture are important, as often organisms do not make flagella under conditions in which they show otherwise normal growth. Microscopic examination is a tedious way of checking the motility of large numbers of cultures. A much simpler nonmicroscopic method is with the use of soft agar of about 0.3 percent concentration. This agar concentration is sufficient to form a soft gel without hindering motility. The medium is inoculated by a single stab straight down the center of the tube, and during incubation motile cells swarm out from the line of inoculation. Through successive cell divisions they form a diffuse region of growth (Figure 2.35). If the culture is nonmotile, swarming will not occur and growth will be confined to a compact band at the center of the tube. Another way of doing this is to prepare soft agar in a petri dish, and inoculate the culture at a single spot in the center. If the organism is motile, it will smarm out from this center, and through continual cell divisions a diffuse region of growth will cover the whole plate.

Spirochetes move in a manner quite different from that described above. The spirochete cell has wrapped around its periphery a bundle of fibers, the *axial filament* (Figure 2.36), which is attached at the two ends of the cell. When these fibers contract, the flexible spirochete cell is thrown into a helical configuration (Figure 2.36). Movement is thought to occur by alternate expansion and contraction of the filament, leading to increase and decrease in the helix of the cell; the spirochete moves through water somewhat in the manner of a snake.

Many procaryotic organisms are motile but have no flagella. These organisms exhibit *gliding motility* and move only when in contact with a surface. Representative gliding bacteria are the myxobacteria, many blue-green algae, and certain filamentous bacteria that seem related to the blue-green algae but are colorless. (Some eucaryotic organisms also

2.7 Other cell structures

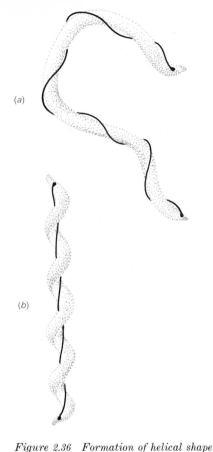

(a)

(b)

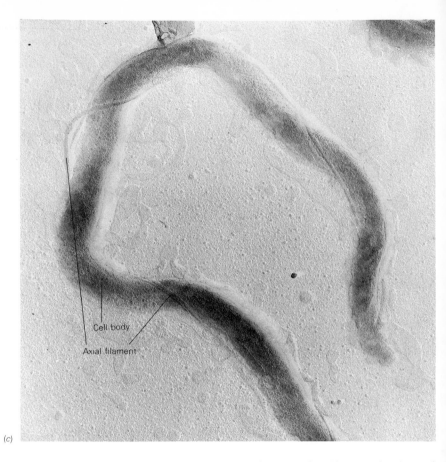

Cell body

Axial filament

(c)

*Figure 2.36 Formation of helical shape
(b) in a spirochete as a result of contraction
of the axial filament (black). The filament
is relaxed in (a). (c) Electron micrograph
of a metal-shadowed preparation of Spirochaeta stenostrepta, showing the position
of the axial filament. Magnification,
38,500×. [From S. C. Holt and E. Canale-
Parola: J. Bacteriol. 96:822 (1968).]*

show gliding motility, as is discussed in Chapter 3.) The mechanism of
gliding motility is completely unknown. It must involve in some way an
interaction between the cell and the surface, since movement ceases as
soon as the cell is removed from the surface and becomes suspended in
water. Gliding organisms also show tactic movements similar to those
flagellated organisms manifest.

Pili A structure that is somewhat analogous to the flagellum but
which is not involved in motility is the *pilus* (plural, *pili*).* Pili (Figure
2.31) are considerably shorter than flagella and are much more numerous on the cell. They may be chemically similar to flagella. Not all organisms have pili, and the ability to produce them is an inherited trait. On
some organisms there is more than one kind of pilus. One interesting

*There is another word sometimes used to describe these structures, *fimbria* (plural,
fimbriae).

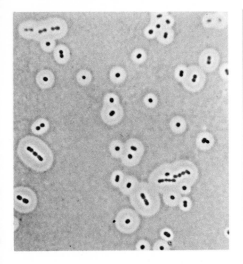

Figure 2.37 Demonstration of the presence of a capsule by negative staining with india ink observed under phase-contrast microscopy (Bacterium anitratum). (Courtesy of Elliot Juni.)

type involved in bacterial conjugation, the sex pilus, is described in Chapter 12. The functions of other types of pili are not known for certain, but in some cases they may enable the organism to stick to inert surfaces.

Capsules, slime layers, and holdfasts Some bacteria and blue-green algae secrete on their surfaces slimy or gummy materials, which can sometimes be seen by the use of negative stains (Figure 2.37). When the material is arranged in a compact manner around the cell surface it is called a *capsule,* whereas when it is loosely attached so that if forms only a diffuse layer it is called a *slime layer.* Capsules and slime layers are usually composed of polysaccharides, polypeptides, or polysaccharide-protein complexes (Figure 2.38).

The ability to produce a capsule is an inherited property of the organism, but capsules are not absolutely essential cellular components, as mutant strains without capsules are able to grow normally in pure culture. Also, enzymes are known which specifically hydrolyze capsular materials, and such treatment does not kill the cell. Capsules often are formed only when an organism is grown on certain culture media. For

Figure 2.38 Chemical structures of representative capsular materials in (a) Streptococcus (hyaluronic acid), (b) Pneumoccus type 3 (glucose-glucuronic acid), (c) Bacillus anthracis (poly-D-glutamic with peptide bond γ), and (d) Leuconostoc [dextran (α-1-6-poly-D-glucose)].

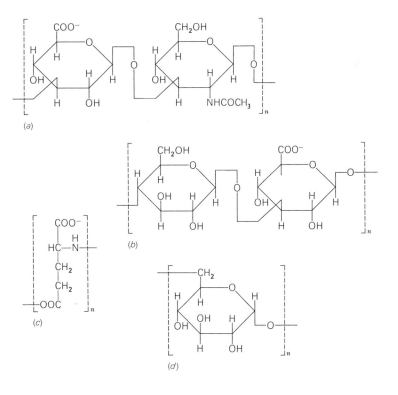

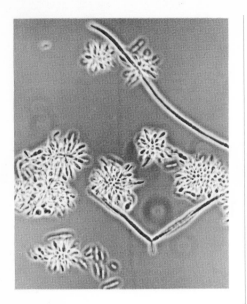

Figure 2.39 *Rosettes of Leucothrix mucor formed as a result of synthesis of holdfast at the poles of the cells. Magnification, 455×.*

instance, *Leuconostoc mesenteroides* forms its dextran capsule only when grown on sucrose. This occurs because the enzyme that produces dextran (dextran sucrase) uses sucrose as the substrate.

Although capsules are not essential cellular structures, they probably have survival value for the organism in nature. The best example of this is shown by the bacteria that cause pneumonia in man, the pneumococci. Strains of pneumococci pathogenic to man always have capsules, which make it difficult for phagocytes of the host to ingest the bacteria and destroy them (see Chapter 15). Capsulated pneumococci are therefore able to grow well in the human body and cause disease, whereas noncapsulated pneumococci are nonpathogenic. In soil and water bacteria, capsules may protect the organisms against ingestion by protozoa.

In some organisms, an adhesive slime is produced not over the whole surface of the cell, but only at one localized region, usually at one end. Such a localized slime layer is called a *holdfast* because it enables the cell to attach to a surface specifically at one end. Occasionally a group of cells with holdfasts will attach to each other, leading to the formation of a star-shaped cluster or rosette (Figure 2.39).

Inclusions and storage products When examined under the microscope, many bacteria have granules, aggregates, or other inclusions within the cells, which are often mistaken for nuclei. The nature of these inclusions differs in different organisms, but they almost always function in the storage by the cell of energy or structural building blocks. Inclusions can often be seen directly with the light microscope without special staining, but their contrast can usually be increased by using dyes. Inclusions often show up very well with the electron microscope.

In procaryotic organisms, one of the most common inclusion bodies consists of *poly-β-hydroxybutyric acid* (PHB) (Figure 2.40), a compound with a name more complex than its structure. Beta-hydroxybutyric acid units are connected by ester linkages, forming the long PHB polymer,

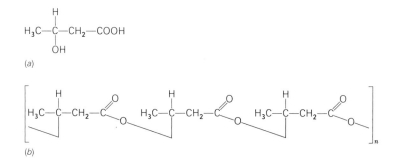

Figure 2.40 *Structure of (a) β-hydroxybutyric acid and (b) poly-β-hydroxybutyric acid.*

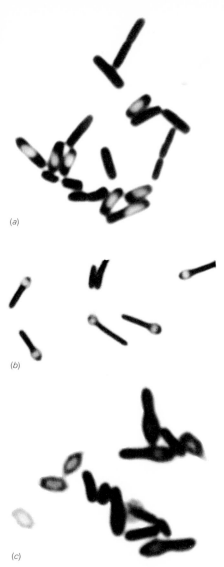

(a)

(b)

(c)

Figure 2.41 Light photomicrographs illustrating several types of endospore morphologies: (a) central spores, sporangium wall not enlarged; (b) terminal spores, sporangium wall enlarged; (c) subterminal spores, sporangium wall enlarged. (Courtesy of The Wellcome Research Laboratories Anaerobic Bacteriology Department, Beckenham, Kent, England.)

and these polymers aggregate into granules. With the electron microscope the positions of these granules can often be seen as light areas that do not scatter electrons. The granules have an affinity for fat-soluble dyes such as Sudan black and can be identified tentatively with the light microscope by staining with this compound. Poly-β-hydroxybutyric acid can be positively identified by extraction from the cell and chemical analysis. The PHB granules are a storage depot for carbon and energy (see Section 4.4).

Another storage product is *glycogen,* which is an α-1,4 polymer of glucose subunits. Glycogen granules are usually smaller than PHB granules, and can only be seen with the electron microscope, but the presence of glycogen in a cell can be detected in the light microscope because the cell appears a red-brown color when treated with dilute iodine, due to a glycogen-iodine reaction. Some cells deposit a starchlike polymer that is also an α-1,4 polymer of glucose, but which differs from glycogen in the lengths of the polymer chains and degrees of branching. When it is treated with iodine, the starch appears deep blue in color.

Many microorganisms accumulate large reserves of inorganic phosphate in the form of granules of *polyphosphate.* These granules are stained by many basic dyes; one of these dyes, toluidin blue, becomes reddish violet in color when combined with polyphosphate. This phenomenon is called *metachromasy* (color change), and granules that stain in this manner are often called *metachromatic granules.* *

Many organisms store energy and carbon in the form of *fats.* Fats are water-insoluble and usually collect in the cell in droplets, which can often be seen with the phase microscope, or can be made visible for normal light microscopy by the use of a fat-soluble dye, such as Sudan black. Sulfur bacteria and a few others produce granules of *sulfur;* the discussion of these will be reserved for Chapter 6.

2.8   *Bacterial endospores*

As noted in Chapter 1, one of the most important microbiological discoveries was that bacterial spores exist since knowledge of such remarkably heat-resistant forms was essential for the development of adequate methods for the sterilization not only of culture media but also of foods and other perishable products. Although many organisms other than bacteria form spores, the bacterial endospore is unique in its degree of heat resistance. Such spores are characteristically produced by two

---

* Another name for them is *volutin* granules.

Figure 2.42 Comparison of the fine structure of vegetative cell and endospore of Bacillus species: (a) vegetative cell, magnification 41,000× [from D. J. Ellar, D. G. Lundgren, and R. A. Slepecky: J. Bacteriol. 94:1189 (1967)]; (b) endospore after liberation from sporangium, magnification 158,000× [from B. J. Moberly, F. Shafa, and P. Gerhardt: J. Bacteriol. 92:220 (1966)].

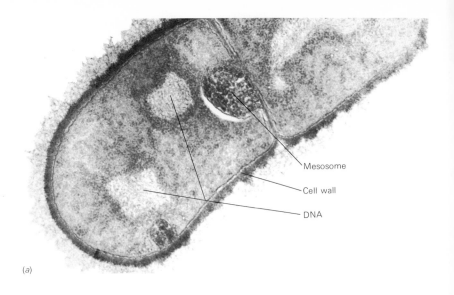

Mesosome

Cell wall

DNA

(a)

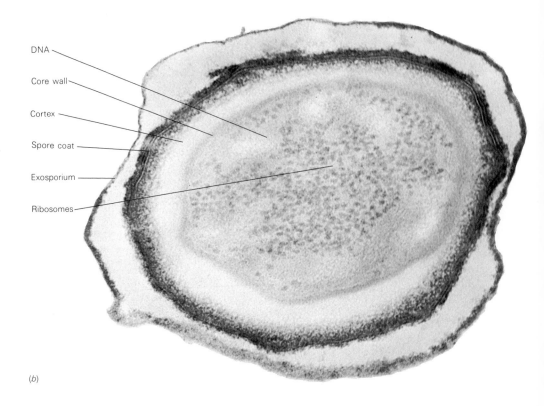

DNA

Core wall

Cortex

Spore coat

Exosporium

Ribosomes

(b)

genera of bacteria, *Bacillus* and *Clostridium*, both of which are widely distributed, especially in soil.

Endospores (so called because the spore is formed within the cell) are readily seen under the phase microscope as strongly refracting bodies (Figure 2.41). Spores are very impermeable to dyes, so that occasionally they are seen as unstained regions within cells that have been stained with basic dyes such as methylene blue. To stain spores specifically, special spore-staining procedures must be used. The structure of the spore as seen with the electron microscope is vastly different from the structure of the vegetative cell; these two types are compared in Figure 2.42. The spore's structure is much more complex than that of the vegetative cell in that it has many layers. The outermost layer is the *exosporium*, a thin, delicate covering. Within this is the *spore coat*, which is composed of a layer or layers of wall-like material; below the spore coat is the *cortex*, a region consisting of many concentric rings. Like the cell wall, the cortex is composed of peptidoglycan. Inside the cortex are the usual cell wall, cell membrane, nuclear region, and so on. Thus the spore differs structurally from the vegetative cell primarily in the kinds of structures found outside the cell wall.

One chemical substance that is characteristic of spores but not of vegetative cells is *dipicolinic acid* (DPA)* (Figure 2.43). This substance has been found in all spores examined, and it is probably located primarily in the cortex. Spores are also high in calcium ions, most of which are associated with the cortex, probably in combination with dipicolinic acid. There is good reason to believe (see Chapter 8) that the association of calcium and dipicolinic acid is important in the unusual heat resistance of bacterial spores.

The structural changes that occur during the conversion of a vegetative cell to a spore can be studied readily with the electron microscope. Under certain conditions (to be described in Chapter 7), instead of dividing, the

*One should not confuse dipicolinic acid, DPA, with diaminopimelic acid, DAP.

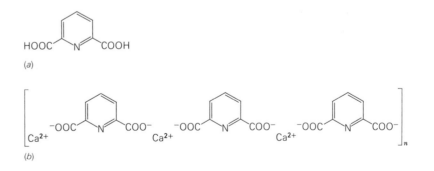

Figure 2.43 Structure of (a) dipicolinic acid and (b) calcium–dipicolinic acid complex.

(a)

(b)

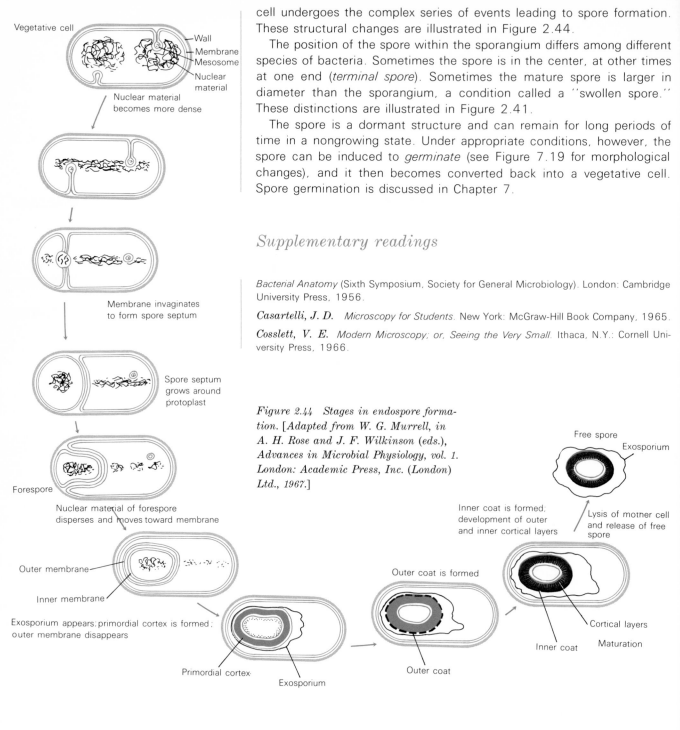

cell undergoes the complex series of events leading to spore formation. These structural changes are illustrated in Figure 2.44.

The position of the spore within the sporangium differs among different species of bacteria. Sometimes the spore is in the center, at other times at one end (*terminal spore*). Sometimes the mature spore is larger in diameter than the sporangium, a condition called a "swollen spore." These distinctions are illustrated in Figure 2.41.

The spore is a dormant structure and can remain for long periods of time in a nongrowing state. Under appropriate conditions, however, the spore can be induced to *germinate* (see Figure 7.19 for morphological changes), and it then becomes converted back into a vegetative cell. Spore germination is discussed in Chapter 7.

Supplementary readings

Bacterial Anatomy (Sixth Symposium, Society for General Microbiology). London: Cambridge University Press, 1956.

Casartelli, J. D.   Microscopy for Students. New York: McGraw-Hill Book Company, 1965.

Cosslett, V. E.   Modern Microscopy; or, Seeing the Very Small. Ithaca, N.Y.: Cornell University Press, 1966.

Vegetative cell

Wall
Membrane
Mesosome
Nuclear material

Nuclear material becomes more dense

Membrane invaginates to form spore septum

Spore septum grows around protoplast

Forespore

Nuclear material of forespore disperses and moves toward membrane

Outer membrane
Inner membrane

Exosporium appears; primordial cortex is formed; outer membrane disappears

Primordial cortex
Exosporium

Outer coat is formed
Outer coat

Inner coat is formed; development of outer and inner cortical layers

Lysis of mother cell and release of free spore

Free spore
Exosporium

Cortical layers
Inner coat
Maturation

Figure 2.44   Stages in endospore formation. [Adapted from W. G. Murrell, in A. H. Rose and J. F. Wilkinson (eds.), Advances in Microbial Physiology, vol. 1. London: Academic Press, Inc. (London) Ltd., 1967.]

Gunsalus, I. C., and R. Y. Stanier (eds.) *The Bacteria: A Treatise on Structure and Function,* vol. I, *Structure.* New York: Academic Press, Inc., 1960.

Hartley, W. G. *How to Use a Microscope.* Garden City, N.Y.: Natural History Press, 1964.

Kay, D. H. (ed.) *Techniques for Electron Microscopy.* Philadelphia: F. A. Davis Co., 1965.

Martin, H. H. "Biochemistry of bacterial cell walls," *Ann. Rev. Biochem.* 35, pt. 2: 457 (1966).

Rogers, H. J., and H. R. Perkins *Cell Walls and Membranes.* London: E. & F. N. Spon Ltd., 1968.

Salton, M. R. J. "Structure and function of bacterial cell membranes," *Ann. Rev. Microbiol.* 21:417. (1967).

Salton, M. R. J. *The Bacterial Cell Wall.* New York: Elsevier Publishing Co., 1964.

Sussman, A. S., and H. O. Halvorson *Spores: Their Dormancy and Germination.* New York: Harper & Row, Publishers, 1966.

Supplementary readings

three

The eucaryotic cell

Eucaryotic cells (cells with true nuclei) are structurally much more complex than procaryotic cells, and show to varying degrees localization of cellular functions in distinct membrane-bound intracellular structures (*cell organelles*). Examples of such intracellular organelles, to be discussed in detail below, are nuclei, mitochondria, and chloroplasts. In addition to complexity of structure, processes of cell division and sexual reproduction are considerably more complex in eucaryotes than in procaryotes. The restriction of certain functions to certain types of organelles makes it easier for the experimenter to localize specifically these functions within the cell. The increased structural complexity of eucaryotes is not accompanied by an increased chemical complexity and, in fact, in a few cases the chemical constitutions of structures in eucaryotes are even simpler than the equivalent structures in procaryotes. For instance, the cell walls of many eucaryotes are composed of simple glucose polymers, whereas, as we learned in Chapter 2, the cell walls of procaryotes are complex peptidoglycan polymers.

The cells of all higher organisms, both plant and animal, are eucaryotic. Among the microorganisms, fungi, protozoa, and algae (except the blue-green algae) are eucaryotic. There is a tremendous diversity of cell types and cell functions in this vast collection of eucaryotic microorganisms, but all have in common the fact that many of their cellular functions are located in or on intracellular organelles. An important point to which we will return at the end of this chapter is that a cell is either procaryotic or eucaryotic—there does not seem to be any middle ground. This suggests that the evolution of the first eucaryotic cell from a procaryotic one probably involved a single large evolutionary step.

Size and shape The range of sizes found in eucaryotic cells is much greater than that in procaryotes. Some eucaryotic cells are as small as the larger procaryotes, whereas others are many times larger. Most eucaryotic cells have fairly regular shapes, and these are often quite complex. Cell shapes in the algae and protozoa are especially interesting (Figure 3.1), with a geometrical beauty often lacking in the cells of higher plants and animals.

The formation of filaments, which occurs occasionally in procaryotes, is much more common in eucaryotes, especially in algae and fungi.

3.1 Cell wall

Most algae and fungi, like higher plants, have cell walls that confer shape and rigidity to their cells. These walls are usually much thicker than those of procaryotic cells and can often be seen even with the light microscope, although electron microscopy is necessary to obtain any detail. In the electron micrograph of Figure 3.2 the thick cell wall of a yeast cell is well illustrated. The isolation and purification of eucaryotic cell walls involves much the same procedures already described for procaryotes: mechanical breakage, differential centrifugation, extensive washing, extraction with solvents, acids, or bases, and treatment with various enzymes. Such treatments remove associated proteins, lipids, and polysaccharides that are not necessarily involved in rigidity.

The chemical composition of the rigid layer of eucaryotic cell walls is usually much simpler than that of procaryotes. In the algae, the cell wall is composed of cellulose, a polymer consisting only of glucose units (Figure 3.3). A single cellulose polymer can have up to 8,000 glucose units, corresponding to a length of 4 μm, and a number of these macromolecules intertwine to form *microfibrils*, which then associate and overlap to frame the structure of the cell wall (Figure 3.4). Although the cell walls of most algae and all higher plants are composed of cellulose, two groups of algae, the diatoms and the chrysophytes, have walls composed of silica. In the diatoms the walls are beautifully sculptured and are highly complex in geometry, as is shown in Figure 3.5. One group of algae, the coccolithophores, has only thin cellulose walls to which are attached intricate scales composed of calcium carbonate (Figure 3.6). Associated with the rigid cell-wall structure in algae are other polysaccharides, such as mannans (mannose polymers), xylans (xylose polymers), and pectic substances (polymers of uronic acids).

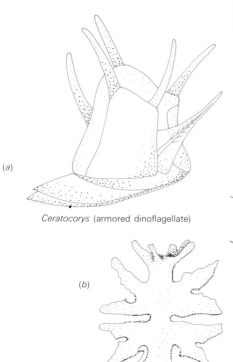

(a)

Ceratocorys (armored dinoflagellate)

(b)

Micrasterias (desmid)

(c)

Dinobryon (colonial flagellate)

(d)

Stephanodiscus (centric diatom)

(e)

Gonyaulax
(dinoflagellate)

(f)

Coccolithophore

Figure 3.1 *Diagrams of cell shapes of various eucaryotic microorganisms: (a–f) algae; (g–i) protozoa. Compare these with the much simpler cell shapes of procaryotes in Figure 2.11.*

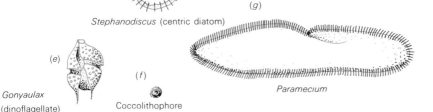

(g)

Paramecium

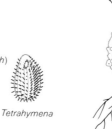

(h)

Tetrahymena

(i)

Foraminiferan

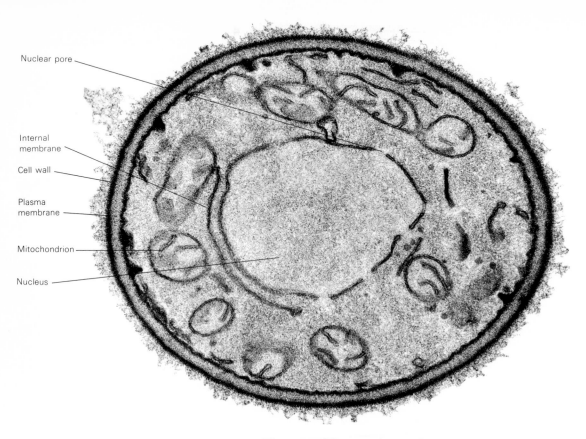

Nuclear pore

Internal
membrane

Cell wall

Plasma
membrane

Mitochondrion

Nucleus

*Figure 3.2 Electron micrograph of a thin section of a yeast cell.
Magnification, 39,000×. [From S. F. Conti and T. D. Brock:
J. Bacteriol. 90:524 (1965).]*

*Figure 3.3 Structures of (a) β-1, 3- and
(b) β-1, 4-glucan (cellulose) components of
the rigid cell-wall layers of many eucaryotic microorganisms.*

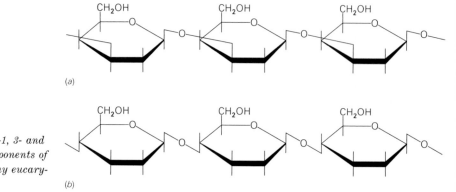

Figure 3.4 (a) Cell-wall organization of the mature wall of the filamentous green alga *Chaetomorpha melagonium*. Note the regular arrangement of the cellulose microfibrils in alternating layers. Magnification, 15,000×. [From E. Frei and R. D. Preston: *Proc. Roy. Soc.*, Ser. B, 155:55 (1961).]
(b) Cell-wall organization of the first-formed primary cell wall of the filamentous green alga *Valonia*. Note the random arrangement of the cellulose microfibrils. Magnification, 9,200×. [From F. C. Steward and K. Mühlethaler: *Ann. Bot.*, New Ser., 17:295 (1953).]

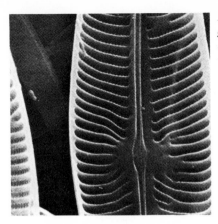

Figure 3.5 Scanning electron micrograph of the rigid outer layer (sometimes called a frustule) of a diatom. Magnification, 3,000×. (© Patrick Echlin and Cambridge Scientific Instruments, Cambridge, England.)

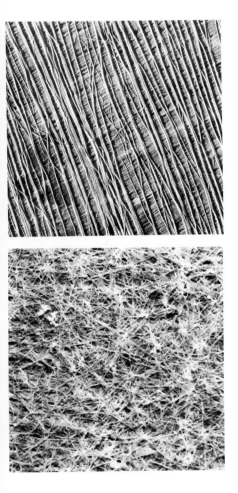

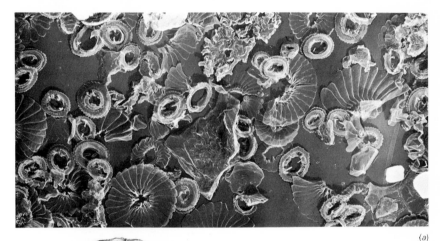

(a)

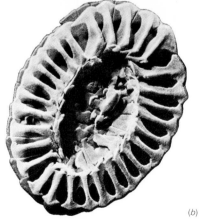

(b)

Figure 3.6 (a) Low-power electron micrograph showing the complex frustules of coccolithophores. These coccoliths are composed primarily of calcium carbonate. The sample photographed was from material collected from an ocean-bottom sediment in deep tropical waters. (Courtesy of Annika Sanfilippo.) (b) High-power electron micrograph of the rigid outer layer of a coccolithophore (*Coccolithus huxleyi*). Magnification, 17,000×. [From N. Watabe and K. M. Wilbur: *Limnol. Oceanogr.* 11:567 (1966).]

Cellulose is found in the cell walls of certain lower fungi. In other lower fungi and in the higher fungi (yeasts, mushrooms, and most molds) the basal cell-wall structure is composed of a glucose polymer in which the glucose units are connected in β-1,3 instead of β-1,4 linkage (Figure 3.3). Although the difference between β-1,3 and β-1,4 linkage might seem trivial, it actually has profound effect on the properties of the polymer. Unlike cellulose, the β-1,3-glucan cannot form crystallike microfibrils but only more amorphous structures. Thus in the wall of the yeast cell (Figure 3.2), there is no indication of cell-wall microfibrils. In most fungus cell walls there are also β-1,6-glucose linkages, which may be involved in the cross-linking function necessary for the formation of the three-dimensional wall structure, in a manner analogous to the peptide cross links found in procaryotes. In addition to β-1,3-glucan, the cell walls of many fungi also contain *chitin*, a polymer of N-acetylglucosamine molecules in β-1,4 linkage. It is of interest that chitin is also the key constituent of the rigid exoskeleton of insects, crustaceans, and certain other groups of animals. The relative importance of chitin and β-1,3-glucan in conferring rigidity on fungal cell walls is not completely understood. Yeasts, for instance, have very little or no chitin, and chitin is usually absent in fungi that have cellulose walls.

Enzymes (β-1,3- and β-1,6-glucanases) that attack the cell-wall glucan of yeasts and fungi are widely distributed in nature. One of the best sources of such enzymes is the stomach juice of the large land snail *Helix pomatia*, and snail enzymes have been used to produce protoplasts of yeasts and filamentous fungi.

Although most protozoa do not have cell walls, the fact that most of them maintain distinct nonspherical shapes indicates that they have some type of surface layer that provides an element of rigidity. The chemical composition of this skeletal layer is poorly understood. In some cases a protein-polysaccharide matrix (pseudochitin) is present, but true chitin is rare in protozoa. The dinoflagellates, a group of organisms of uncertain classification (claimed, in fact, by both protozoologists and algologists!) have cellulose walls, as have a few ciliates. Calcium carbonate shells or tests are widespread among the foraminifera, and the shells of radiolaria and of a few other groups are composed of silica; one strange group of radiolaria even has a shell composed of strontium sulfate ($SrSO_4$). However, these shells can hardly be called cell walls in the usual sense of the word; the foraminifera and radiolaria can leave their cell walls, move around, and then later form new walls. In addition to its protective and supportive role, the skeletal layer of protozoa provides a point for attachment of contractile fibrils, analogous to muscles, which function in locomotion.

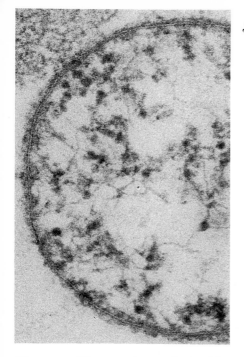

Figure 3.7 Electron micrograph of a thin section of a plasma membrane. Note the distinct double tract, which arises because the osmic acid used to stain the membrane combines preferentially with the polar groups that face away from the hydrophobic central area. Magnification, 130,000×. Compare with Figure 3.9. (Courtesy of Walther Stoeckenius.)

Plasma membrane   The plasma membrane is a thin structure that completely surrounds the cell as a bag or sphere; it is located just inside the cell wall. The plasma membrane can generally be visualized in thin sections studied with the electron microscope, and a representative example is seen in Figure 3.2. To prepare the plasma membrane for study with the electron microscope, the cells must first be treated with osmic acid or some other electron-dense material that combines with components of the membrane. By careful high-resolution electron microscopy, the plasma membrane appears as two thin lines separated by a lighter area (Figure 3.7). The basic building blocks of the plasma membrane are phospholipids and proteins; the former are mixed glycerol esters of long, water-insoluble fatty acid residues and ionic, water-soluble phosphate esters (Figure 3.8). Phospholipid molecules in water disperse themselves in such a way that on the one hand the water-insoluble (hydrophobic) groups associate together, and on the other hand the ionic (hydrophilic) groups associate, leading almost automatically to the formation of a double-layered membrane, or bimolecular leaflet, as shown in Figure 3.9. Protein molecules then adsorb to the ionic groups of the phospholipid, and water molecules congregate in a more or less ordered structure around the outside of the bimolecular leaflet. Double-layered membranes of this type can be formed experimentally in the laboratory, using pure solutions of phospholipid alone, suggesting that the protein need not play a basic structural role, although it probably modifies the structure of the membrane in subtle ways.

So far as can be told from the studies at hand, the membranes of all organisms, both procaryotes and eucaryotes, have the same basic groundwork. Minor differences (which may be functionally important) arise from differences in the kinds of phospholipids and proteins used in the construc-

Figure 3.8 Structure of lecithin, a phospholipid. The fatty acids (in color boldface type) can be of various chain lengths and structures and are nonionic and water-insoluble. In phospholipids other than lecithin, choline (in black boldface type) is replaced by other groups such as amino acids, sugars, or amines. These are ionic and soluble in water.

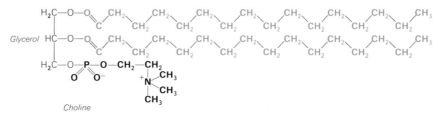

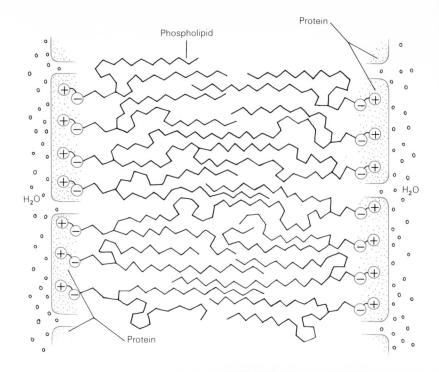

Phospholipid

Protein

H₂O

H₂O

Protein

Figure 3.9   Diagram of the manner in which phospholipids and proteins associate in the formation of a membrane. The hydrophobic groups are directed toward the center and the ionic portions of the phospholipids are directed outwards, where they associate with protein. The presence of protein is not essential for the formation of the membrane, but probably adds considerably to its stability. Compare with Figure 3.7.

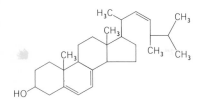

Figure 3.10   Structure of a typical sterol, yeast ergosterol.

tion of the membrane. One rather notable difference in the chemical composition, however, is that the eucaryotes have sterols (Figure 3.10) in their membranes, whereas sterols are rare or absent in procaryotes. Although the function of sterols in the cell membrane is not completely understood, one idea is that they play a role in stability of the membrane structure. As is noted in Chapter 9, certain antibiotics that react with sterols are active against eucaryotes but not against procaryotes.

The plasma membrane as a permeability barrier   The plasma membrane as we described it above might appear to be static. In actuality, the phospholipids of the membrane are probably in constant motion and possibly cause pores in the membrane to open and close. When a pore is open, water and many uncharged molecules (nonelectrolytes) dissolved in water might go through; whereas when the pore is closed, penetration of water-soluble materials is impossible. The size of these pores is small, about 0.4 to 0.5 nm* and hence only small molecules can penetrate rapidly. However, if the permeability barrier is broken by damaging the membrane in some way, larger molecules penetrate about

* Dimensions this small are often expressed in units called *angstroms*, named after a Swedish physicist. One angstrom (Å) is equal to 0.1 nm; thus 0.4 nm equals 4 Å.

as rapidly as small ones. With ionized molecules, such as organic acids, amino acids, inorganic salts, and so on, there is a further barrier to penetration because ionized molecules are repelled by the electrical charge on the surface of the membrane. On the other hand, fat-soluble substances, such as hydrocarbons and uncharged fatty acids, penetrate cells readily by becoming dissolved in the lipid phase of the membrane. We see then that the selective permeability properties of cell membranes are determined in great part by their chemical composition and structure. However, membranes also have special mechanisms for the penetration of specific substances that are needed by the cell but which would otherwise penetrate with difficulty (see Chapter 5). Thus the plasma membrane is both a barrier to penetration of materials in general and an agent for selective transport or uptake of materials.

Internal membrane systems   In addition to the plasma membrane, eucaryotic cells have an extensive array of internal membranes (Figure 3.2). Such a membrane system, called the *endoplasmic reticulum,* was first discovered in the cells of higher animals but is now known to be present in most organisms.

Although the functions of the endoplasmic reticulum have been little studied in microorganisms, it is known that these membranes often function in cells of higher animals as a surface upon which the ribosomes and other components of the protein-synthesizing machinery are attached. The endoplasmic reticulum probably also provides a series of communication channels between the cell surface and interior cell structures, making possible the more efficient movement of materials. In eucaryotic cells, with their larger size and increased complexity, such a communication system would be of more significance for efficiently maintaining cell function than it would be in the small, structurally simple procaryotes.

Vacuoles   Vacuoles are membrane-bound bodies of low density often seen in the cytoplasm of eucaryotic microorganisms (Figure 3.2). Vacuoles seem to be most frequent in aging cells and are generally absent from cells in division; they contain within them a concentrated solution of salts, amino acids, sugars, and other constituents.

Vacuoles have been studied most extensively in protozoa, in which they play important roles in cell function. Two kinds of vacuole are recognized: *food vacuoles,* which are involved in food digestion, and *contractile vacuoles,* which aid in osmotic regulation and in the excretion of waste products. Many protozoa obtain their food in the form of particles, which they ingest and then digest intracellularly. The ingested particles become surrounded by a membrane, forming the food vacuole (Figure 3.27), into which digestive enzymes are secreted. In some protozoa the

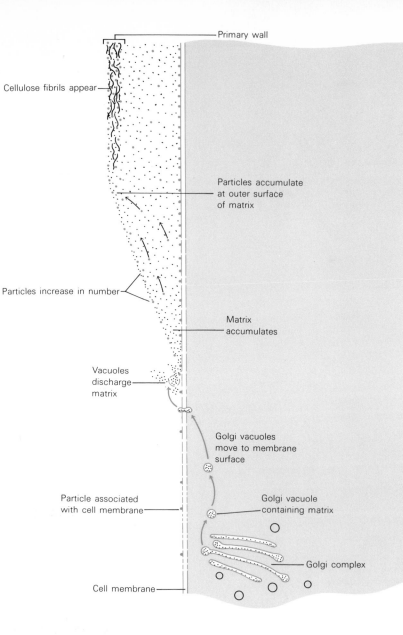

Primary wall

Cellulose fibrils appear

Particles accumulate
at outer surface
of matrix

Particles increase in number

Matrix
accumulates

Vacuoles
discharge
matrix

Golgi vacuoles
move to membrane
surface

Particle associated
with cell membrane

Golgi vacuole
containing matrix

Golgi complex

Cell membrane

*Figure 3.11 The function of the Golgi
membrane complex in the formation of the
cellulose cell wall of an alga.*

food vacuoles move around in the cytoplasm of the cell, reaching eventually an "anal" region, where undigested material is expelled. The vacuole then disappears.

Contractile vacuoles, which are found in most freshwater protozoa and in some marine forms, function in the excretion of waste products and water. They are called contractile because they can be seen to enlarge by taking up waste products from surrounding cytoplasm and decrease in size (actually they collapse completely) by releasing their contents to the outside of the cell. Their role in excretion of water in freshwater protozoa seems related to the fact that the cytoplasm of these organisms is always at a higher osmotic pressure than is their surroundings. Therefore, water is continually flowing into the cell and the excess water is taken into the contractile vacuole to be expelled to the outside. The membrane of the protozoal vacuole resembles closely the plasma membrane and is probably similar in chemical composition.

Lysosomes   Lysosomes are membrane-bound bodies in which are localized a variety of digestive enzymes that destroy foreign particles. It is thought that these enzymes are membrane-bound so that they will not come into contact with and digest vital cell structures. Lysosomes may surround invading particles and envelop them, and digestion then takes place inside the lysosome. Hence lysosomes could be considered analogous to the food vacuoles of protozoa. Lysosomes of phagocytic cells play an important role in the resistance of the animal body to infection (see Chapter 15).

Golgi bodies   Golgi bodies are large aggregates of membranes that are often seen in certain regions of a cell. They probably have a number of roles, but one recently well-established role is the part they play in cell-wall synthesis in plants. The precursors of the cellulose microfibrils and the other materials of the cell wall may be synthesized in the Golgi bodies and transferred to the growing cell wall (Figure 3.11). In animal cells the function of the Golgi apparatus is to package materials such as enzymes for export out of the cell.

3.3   Mitochondria

In eucaryotic cells the processes of respiration and oxidative phosphorylation (see Chapter 4) are localized in special membrane-bound structures, the *mitochondria* (singular, *mitochondrion*). Mitochondria may have many shapes, but most often they are rod-shaped structures about 1 μm. in diameter by 2 to 3 μm long. Hence they should be visible with the light microscope, although often it is difficult to distinguish mitochondria from other cell organelles of similar size. Certain dyes, such as Janus green B, stain mitochondria more or less specifically; Janus green B is a specific stain because it is maintained in the oxidized (colored) form by the respiratory activity of the mitochondrion, whereas in the rest of the cell it becomes reduced to the colorless form. It is also possible to see mitochondria in unstained preparations with the phase microscope by using special mounting and preparing techniques. However, the best way to see mitochondria well, and the only way to see structural details within mitochondria, is with the electron microscope. Figure 3.2 shows mitochondria in an electron micrograph of a thin section of a yeast cell. In addition to the outer membrane mitochondria usually possess a complex system of inner membranes; these inner membranes, called *cristae*, are unique to mitochondria and do not occur in other structures of similar size (Figure 3.12). With high-resolution electron microscopy of the cristae it is possible to see that these membranes are of complex structure, containing small round particles attached to the membrane by short stalks. The various biochemical components involved in respiration, which are localized both in the membrane itself and in the small stalked particles, will be described in Chapter 4.

Isolation and purification of mitochondria were essential in order to demonstrate that they are indeed the sites of cellular respiration. A broken

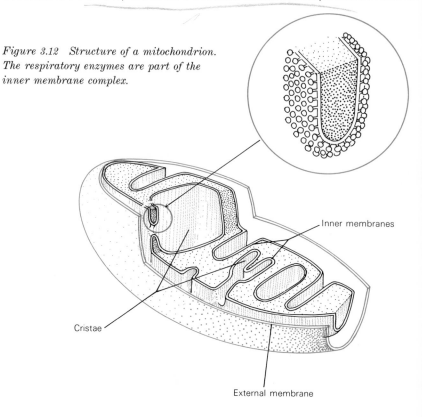

Figure 3.12 Structure of a mitochondrion. The respiratory enzymes are part of the inner membrane complex.

Inner membranes

Cristae

External membrane

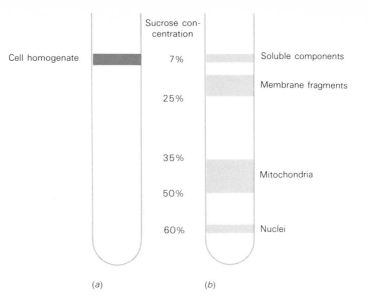

Figure 3.13 Use of sucrose-gradient centrifugation to separate various cell organelles: (a) before centrifugation; (b) after centrifugation. The organelles become banded in the gradient at positions determined by their respective densities.

cell preparation contains a mixture of all cell organelles; the separation of mitochondria from nuclei and other cell constituents is best done by sucrose-gradient centrifugation (illustrated in Figure 3.13). The mitochondrial fraction so obtained can then be studied biochemically or structurally.

Mitochondria are found in largest numbers in actively respiring cells. For instance, yeasts grown in the absence of air possess only poorly developed mitochondria, whereas those grown in the presence of air have many well-developed mitochondria. In animals, organs requiring large amounts of energy, such as the flight muscles of insects, have large numbers of mitochondria in their cells, whereas organs with little energy demand have cells with few mitochondria. Mitochondria produce adenosine triphosphate (ATP), the main mediator of chemical energy in the cell (this is discussed in detail in Chapter 4).

In recent years it has been shown that mitochondria contain a DNA fraction that probably determines the inheritance of at least some of the properties of mitochondria (see Chapter 13). Mitochondria also contain their own protein-synthesizing machinery in the form of ribosomes and other components; they reproduce by a division process. Hence, to a considerable extent the mitochondria retain an autonomy of function within the cell, although their growth and reproduction is precisely integrated with all other phases of cell growth and reproduction. An idea discussed for about as long as mitochondria have been known is that they are derived from bacteria. Proponents hold that the mitochondria descended from intracellular bacteria that became completely dependent on the cell for survival and lost their ability to live independently. Further discussion of this fascinating idea will be reserved for Chapter 13.

3.4 Chloroplasts

Chloroplasts are green, chlorophyll-containing organelles that are found in all eucaryotic organisms able to carry out photosynthesis (see Chapter 6). Chloroplasts of many algae are quite large and hence are readily visible with the light microscope (Figure 3.14). The size, shape, and number of chloroplasts varies extensively throughout the different groups of algae; some representative examples are shown in Figure 3.15.

Electron microscopy reveals that the fine structure of the chloroplast

Figure 3.14 Photomicrograph of the spiral-shaped chloroplast of the alga Spirogyra. This photograph was taken with red light, which is absorbed strongly by chlorophyll, in order to emphasize the chloroplast and deemphasize cytoplasmic constituents. Magnification, 570×.

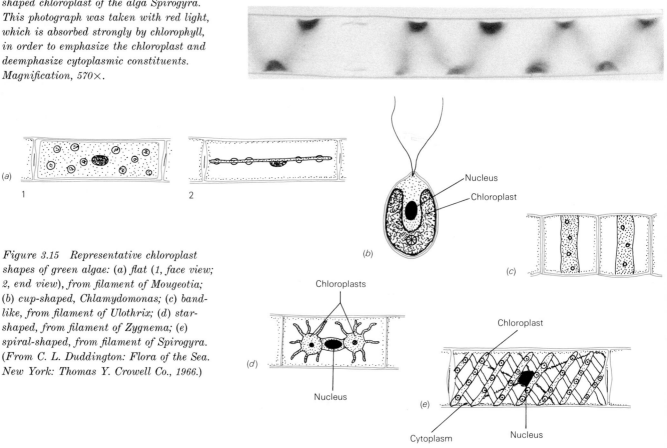

Figure 3.15 Representative chloroplast shapes of green algae: (a) flat (1, face view; 2, end view), from filament of Mougeotia; (b) cup-shaped, Chlamydomonas; (c) band-like, from filament of Ulothrix; (d) star-shaped, from filament of Zygnema; (e) spiral-shaped, from filament of Spirogyra. (From C. L. Duddington: Flora of the Sea. New York: Thomas Y. Crowell Co., 1966.)

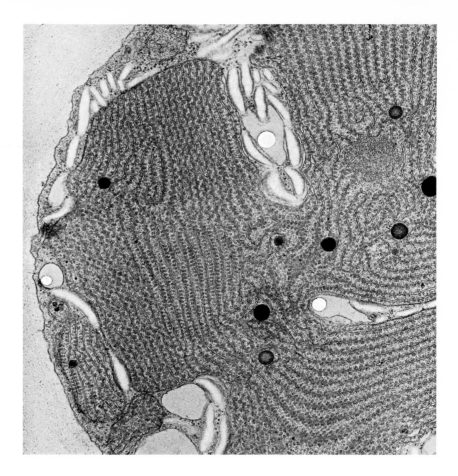

Figure 3.16 Electron micrograph of a portion of the chloroplast of a red alga, Porphyridium cruentum. Note the extensive array of chloroplast membranes arranged in parallel layers. Since this is a cross section the membranes are seen in profile; each is in reality a flat sheet. Magnification, 58,000×. (Courtesy of Mercedes R. Edwards.)

has some resemblance to that of the mitochondrion (Figure 3.16). Each chloroplast has an outer unit membrane, and within this is a large number of short, closely packed internal membranes (called *photosynthetic lamellae*, or *thylakoids*) with which the chlorophyll is associated. A number of these internal membranes aggregate to form a series of closely packed discs or lamellae; the manner in which the chlorophyll molecules are probably associated with these membranes is illustrated in Figure 3.17.

Chloroplasts are released when cells are ruptured and can be isolated and purified by differential centrifugation. All of the chlorophyll of the cell is present in the purified cholorplasts, and these isolated structures are able to carry out the complete process of photosynthesis for a limited period of time. In some algae the chlorophyll content of the chloroplasts is related to the light intensity at which the alga has been grown. At high

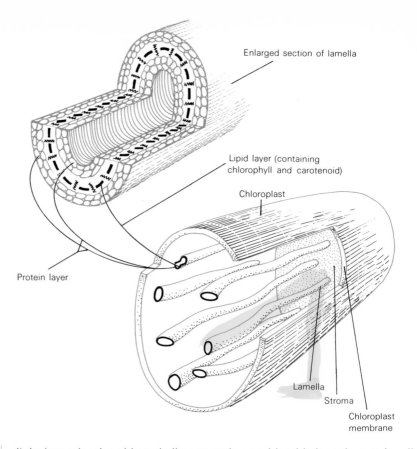

Enlarged section of lamella

Lipid layer (containing
chlorophyll and carotenoid)

Chloroplast

Protein layer

Lamella

Stroma

Chloroplast
membrane

*Figure 3.17 Internal structures of a
chloroplast. In the type of chloroplast diagrammed here, the photosynthetic membranes are tubes rather than flat sheets as
in Figure 3.16.*

light intensity the chlorophyll content is considerably less than at low light
intensity, and the extent of lamellar structure is also reduced when the
chlorophyll content is lower. Since chlorophyll forms an integral part of
the lamellar membrane, the close correlation between degree of lamellar
development and amount of chlorophyll seems understandable.

Chloroplasts, like mitochondria, have a certain degree of autonomy in
the cell. Ribosomes, DNA, and other components of the protein-
synthesizing machinery have been found in chloroplasts. Aspects of
chloroplast heredity will be discussed in Chapter 13.

3.5 Movement

Almost all protozoa, most algae, and many fungi exhibit mechanical activity or movement of some kind. Two basic kinds are distinguished: *cyto-*

plasmic streaming, which is the movement of cytoplasm within stationary cells, and *motility,* whereby the cell moves itself through space. The analogy between mechanical activity of microorganisms and that of muscle cells in higher animals has often been made, and it is likely that the underlying mechanisms may be similar. Yet muscle activity and mechanical activity in microorganisms differ in many details.

Cytoplasmic streaming and amoeboid movement Cytoplasmic streaming is readily detected with the light microscope in the larger eucaryotic cells by observing the behavior of various particles such as chloroplasts or mitochondria; they are seen to move en masse in a definite pattern, suggesting that they are being carried passively by the movement of the cytoplasm. In many algae the mass of cytoplasm *rotates* in a definite pattern just inside the cell surface, the rate of movement remaining relatively constant with time. In other cases a ''fountain'' type of movement is seen, in which the cytoplasm moves in one direction through the center of the cell and moves in the opposite direction down the periphery of the cell. In some fungi cytoplasmic flow is always toward the tips of the growing filaments, a pattern called *tidal movement.* These contrasting types of movement are illustrated in Figure 3.18. Rates of cytoplasmic streaming vary greatly, depending on the organism and environmental conditions; values from 2 μm/sec to greater than 1,000 μm/sec have been recorded.

In cells without walls cytoplasmic streaming can result in *amoeboid movement,* so called because it is the movement characteristic of amoebae. Cytoplasm flows forward because the tip of the pseudopodium is less contracted and hence less viscous, while the rear is contracted and viscous; thus cytoplasm takes the path of least resistance. Amoeboid motion requires a solid surface along which the protoplasm can move. A more massive but comparable type of motion is exhibited by a group of organisms called acellular slime molds (also called myxomycetes). At one stage of their life cycle, these are naked masses of multinucleated protoplasm resembling a large number of amoebae fused together. This whole protoplasmic mass moves slowly across a surface (see Chapter 19). Slime molds are favorite organisms for the demonstration and study of cytoplasmic streaming.

Flagella Most eucaryotic microorganisms move through the activity of special organelles of motion, the flagella and cilia. Some organisms have only a single flagellum; others have more than one. The number and arrangement of flagella is an important characteristic in classifying various groups of algae, fungi, and protozoa (see Chapter 19.) *Flagella* (singular, *flagellum*) are long filamentous structures attached to one end of the cell,

Figure 3.18 Different types of cytoplasmic streaming: (a) rotational, (b) fountain, and (c) tidal movement.

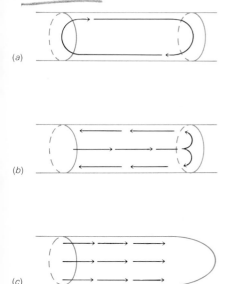

(a)

(b)

(c)

which move in a whiplike manner to impart motion to the cell. It should be emphasized that flagella of eucaryotes are quite different in structure from those of procaryotes (see Section 2.7), even though the same word is used for both. Without exception, flagella of eucaryotes contain two central fibers surrounded by nine peripheral fibers, with each of the latter being composed of two subfibrils (Figure 3.19). The whole unit is surrounded by a membrane and the fibrils are embedded in an organic matrix. Flagellar movement is probably due to contraction of the fibrils in a coordinated way, which is probably brought about by an energy-generating system similar to that which occurs in muscle. Eucaryotic flagella are large enough to be seen with the light microscope, and it is relatively easy to observe their movements (Figure 3.20). The manner in which the flagellum moves determines the direction of cellular movement, as is illustrated in Figure 3.21. The rate of movement of flagellated microorganisms ranges from 30 to 250 μm/sec. Recently it was possible to show that even flagella separated from cells can move if ATP (see Chapter 4) is applied. This important advance should now make it possible to study in greater detail the biochemistry of movement in microorganisms.

Figure 3.19 (a) Electron micrograph of a cross section of the flagellum of the zoospore of the fungus Blastocladiella emersonii showing the outer sheath, the outer nine fibrils and the central pair of single fibrils. Magnification, 132,000×. (Courtesy of Melvin S. Fuller.) (b) Interpretive diagram of the arrangement of the fibrils in a flagellum. The outer sheath is not shown. [From D. J. Paolillo, Jr.: Trans. Amer. Micros. Soc. 86:428 (1967).]

Sheath

Outer fibrils

Central fibrils

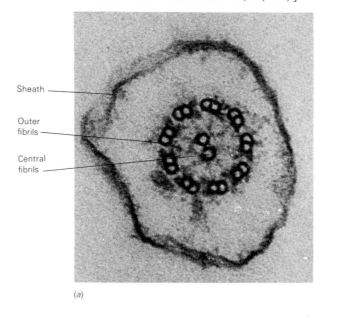

(a)

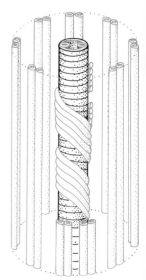

(b)

The eucaryotic cell

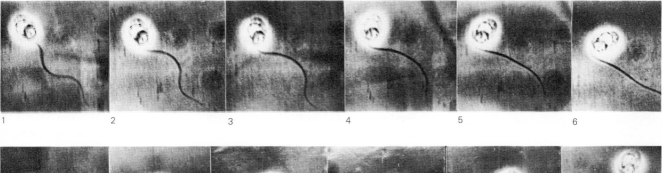

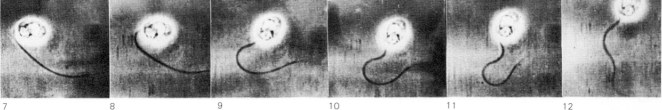

Figure 3.20   High-speed photomicrographs illustrating the turning movement of a zoospore of Blastocladiella emersonii. The whole sequence takes place in 0.25 sec. Magnification, 810×. (Courtesy of C. A. Miles.)

Figure 3.21   Manner in which flagella of different types impart motility to cells: (a) Wave directed to the rear pushes cell in opposite direction; type of movement found in dinoflagellates and animal spermatozoa. (b) Wave directed forward pulls the cell in the direction opposite to the wave; type of movement found in trypanosomes and other flagellated protozoa. (c) Flagellum contains stiff lateral projections called "mastigonemes." Wave directed to the rear pulls the cell in the same direction as the wave; type of movement found in chrysophytes. [From T. L. Jahn and E. C. Bovee: in J. Chen (ed.), Research in Protozoology, vol. 1, New York: Pergamon Press, 1967.]

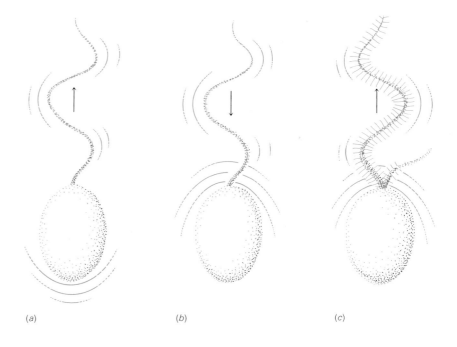

(a)   (b)   (c)

3.5   Movement

79

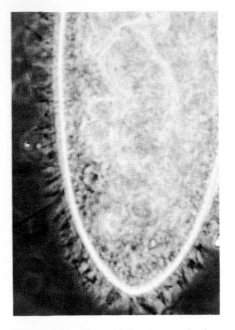

Cilia *Cilia* (singular, *cilium*) are similar to eucaryotic flagella in fine structure but differ in being shorter and more numerous; also, they are distributed around the surface of the cell (Figure 3.22). In microorganisms, cilia are found primarily in one group of protozoa, appropriately called the ciliates. (Cilia are widespread among the cells of higher animals, however.) A single protozoan cell of a *Paramecium* species has between 10,000 and 14,000 cilia. These organelles operate like oars, beating about 10 to 30 strokes per minute. They do not act in unison but usually beat in a coordinated fashion, with a wave of motion passing over the cell. During the power stroke the cilium is held rigid and is thrust backward; on the recovery stroke the tip is bent backward and the cilium is brought forward (Figure 3.23). The rate of movement of ciliates varies from 300 to 2,500 μm/sec for different species. Thus ciliated organisms move much more rapidly than flagellated ones.

Figure 3.22 Phase photomicrograph of a Paramecium cell, showing the large number of cilia arranged over the surface of the cell. Magnification, 1,235×.

3.6 The nucleus and cell division

Nuclear structure One of the distinguishing characteristics of a eucaryote is that its genetic material (DNA) is organized in chromosomes located in a membrane-bound structure, the nucleus. In many eucaryotic

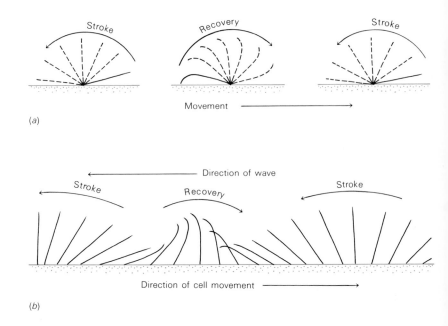

Figure 3.23 The beat of a single cilium (a) and the coordinated movement of a group of cilia on the surface of the cell (b).

cells the nucleus is a large organelle many microns in diameter, which is easily visible with the light microscope even without staining. In smaller eucaryotes, however, special staining procedures often are required to see the nucleus. The most specific nuclear-staining method is the *Feulgen* procedure, named after one of its discoverers. Other nuclear stains (well described in books on microtechnique) also reveal the presence of DNA although they are less specific for DNA than is the Feulgen procedure. In electron micrographs much nuclear detail can be seen; note particularly the nuclear membrane in Figure 3.2. Such micrographs should be contrasted with those showing the nuclear region of procaryotes (see Chapter 2).

Nuclear size differs widely among eucaryotes, varying roughly in proportion to cell size. Thus small cells have, in general, small nuclei, and large cells have large nuclei. Cells without nuclei usually do not divide, although they may continue to respire, synthesize proteins, and carry out other functions.

Close examination of the fine structure of the nucleus reveals a nuclear membrane quite similar in appearance to that of the plasma membrane. However, the holes or pores of the nuclear membrane are clearly visible along its surface. Pores can be seen in Figure 3.2 but are better illustrated in Figure 3.24 (a freeze-etched preparation). The pores may permit the

Figure 3.24   Electron micrograph of the nucleus of a yeast cell by the freeze-etch process. This is a surface view of the nuclear membrane and shows the large number of nuclear pores. The small granules are ribosomes. Magnification, 48,000×. This figure should be compared with Figure 3.2, which shows a yeast nucleus in cross section. [From H. Moor: "Freeze etching," Balzers High Vacuum Report 2:1 (1965). Balzers High Vacuum Corp., Santa Ana, Calif.]

passage in and out of the membrane of macromolecules and large particles. For instance, in eucaryotes the components of the ribosomes are synthesized in the nucleus but function in the cytoplasm, hence pores are necessary so that these particles can leave the nucleus.

A structure often seen within the nucleus is the *nucleolus.* This is an area that stains differently from the rest of the nucleus because of its high content of RNA; it is now known that the nucleolus is the site of ribosomal RNA synthesis. The ribosomes are at least partially constructed in the nucleolus.

All of the DNA of procaryotes is contained in a single molecule of free DNA (see Chapter 2). In eucaryotes, on the other hand, the DNA is present in more complex structures, the *chromosomes,* of which there are always more than one per nucleus. "Chromosome" means "colored body," for the chromosomes were first seen as structures colored by the use of certain stains. Many chromosome stains involve dyes that react strongly with basic (that is, cationic) proteins called *histones*, which in eucaryotes often are attached to the DNA. The DNA of procaryotes is not complexed with histones; thus this is one of the distinguishing characteristics between procaryotes and eucaryotes.* Chromosomes also usually contain small but probably significant amounts of RNA.

Chromosomes of cells not actively dividing are extended and quite thin and hence are difficult to visualize. At other times (mainly during nuclear division or mitosis, which is discussed below) they are contracted, making the histone density quite high and the chromosomes easy to see. In some microorganisms, such as yeast, the chromosomes are very small and difficult to stain. Thus for many years it was thought that yeasts might not have chromosomes, but it now seems likely that yeasts do resemble other eucaryotes in this respect.

The arrangement of DNA in a chromosome is still under investigation. Each chromosome contains a long double helix of DNA, to which the histones are attached, but it is not certain that this long molecule is continuous. The DNA content per nucleus varies from species to species, in much the same way as does nuclear size. In addition, the chromosome number also varies greatly, from no fewer than two to many hundreds.

Although the nuclear membrane confers some degree of selective permeability on the nucleus, this selectivity is not so great as that for the plasma membrane since large molecules of proteins and nucleic acids pass through the pores in the nuclear membrane. The chemical composition of the nuclear membrane is different from that of the plasma membrane, having different kinds of phospholipids.

* By convention, the DNA molecule of procaryotes is also often called a chromosome, despite its simpler structure.

One group of protozoa, the ciliates, has two kinds of nuclei, the *micronucleus* and the *macronucleus,* whose functions are discussed in Chapter 19.

Cell division cycle Cell division in eucaryotes is a highly ordered process in which the key events are (1) DNA synthesis, which leads to doubling of the genetic material of the cell, (2) nuclear division, by a process called mitosis, (3) cell division, which involves the formation of a cell membrane separating the two parts of the divided cell, and (4) cell separation, by which the two cells become detached from each other and acquire independent existences. These stages in cell division are illustrated in Figure 3.25.

In our discussion of procaryotes we stated that DNA synthesis occurs

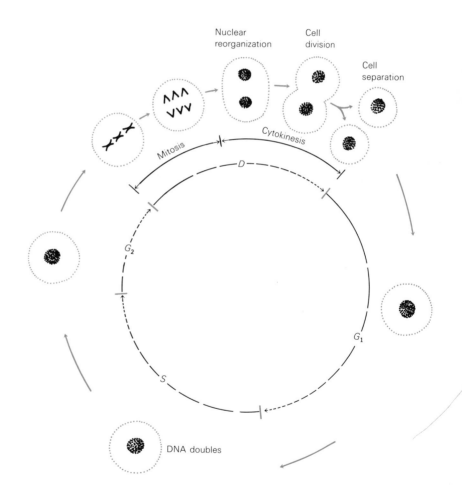

Figure 3.25   The cell division cycle in a eucaryote. S, period during which DNA synthesis takes place; D, period of mitosis and cell division; G₁, growth period before S; G₂, growth period after S. The lengths of the various phases vary from organism to organism. In some organisms, G₂ is absent.

Radioactive substance

Organisms

Incubate (allow radioactivity to be incorporated)

Stop the incorporation, fix the organisms, separate them from fluid, wash, and place on microscope slide coated with adhesive

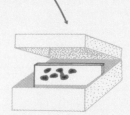

In the darkroom

Cover specimen with thin coating of special photographic emulsion

Place slide in light-tight container for several days; during this time the radioactivity "exposes" the emulsion directly above it

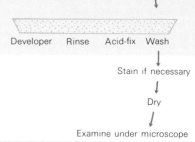

Developer Rinse Acid-fix Wash

Stain if necessary

Dry

Examine under microscope

throughout most of the cell-division cycle (see Chapter 2). This is not true for eucaryotic cells, which synthesize DNA during only about one-third to one-half of the cell-division cycle. For many studies it is convenient to relate the phases of cell division to the time at which DNA is synthesized; consequently the following phases are recognized (Figure 3.25): G_1, the interval between the end of the previous cell division and the beginning of DNA systhesis; S, the interval of DNA synthesis; G_2, the period after DNA synthesis up until the first visible sign of cell division; D, the division period itself, most of which is occupied by the events of mitosis. The only part of the cell cycle during which there are obvious changes in appearance of the nucleus is at mitosis. However, mitosis occupies only a small portion of one cell-division cycle, usually about 10 percent of the total time.

One of the key phases of the cell-division cycle, DNA synthesis, can be studied by the special technique of *autoradiography,* using tritium-labeled thymidine. Thymidine is incorporated only into DNA, and it can be made radioactive by replacing some of the hydrogen atoms with tritium atoms. Tritiated thymidine will be incorporated into DNA if it is present during the time of DNA synthesis, but it will not be incorporated into preformed DNA. Radioactive DNA can be detected in single cells by obtaining a photographic image of the radiations (autoradiography) (Figure 3.26). A typical autoradiograph of an amoeba that has incorporated tritiated thymidine is shown in Figure 3.27. The nucleus is heavily labeled, as is shown by the presence over it of dark silver grains, whereas the cytoplasm is unlabeled. That DNA synthesis in the nucleus is confined to the chromosomes can be shown by obtaining similar autoradiographs of cells that have been exposed to tritiated thymidine and then allowed to proceed to mitosis, at which time the chromosomes are most readily visible. Auto-

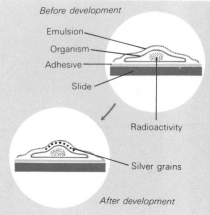

Before development

Emulsion
Organism
Adhesive
Slide

Radioactivity

Silver grains

After development

Unexposed silver halide has been removed, while exposed and developed silver grains remain in the layer of transparent gelatin; these grains are superimposed upon the source of radioactivity

Figure 3.26   Steps in the autoradiographic procedure.

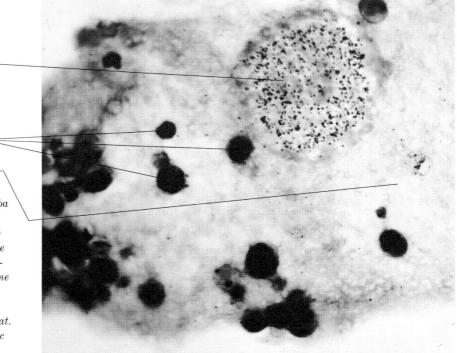

Nucleus with
silver grains

Food vacuoles

Cytoplasm

Figure 3.27 Autoradiograph of an amoeba incubated with tritiated thymidine. Note the large number of silver grains over the nucleus and the absence of grains over the cytoplasm. The black bodies are food vacuoles which have been stained with toluidine blue. Magnification, 1,000×. [From D. M. Prescott: in R. J. C. Harris (ed.), "Cell Growth and Cell Division," Symp. Internat. Soc. Cell Biol., vol. 2. New York: Academic Press, 1963.]

radiographs show that the radioactive label is confined to the chromosome structure. Autoradiographic experiments permit the timing of DNA synthesis in relation to the division cycle and the measuring of the length of the *S* period.

Mitosis   The main function of mitosis is to ensure the orderly partitioning of the DNA-bearing chromosomes. The problem is simply stated: Each cell has a specific number of chromosomes, and after DNA duplication a copy of each chromosome must go to each of the daughter cells. Some mechanism must exist to avoid irregularities in chromosome partitioning, and this is the main function of mitosis. Several distinct stages can be recognized during mitosis (Figure 3.28). Mitosis requires that the nuclear membrane break down and reform; thus at metaphase the nucleus itself no longer exists. However, order is maintained through the agency of the spindle and spindle fibers; this grouping of chromosomes and spindle is called the *mitotic apparatus*.

Cell separation   In microorganisms with cell walls, the final stage of cell division is the formation of the cross wall that separates the two

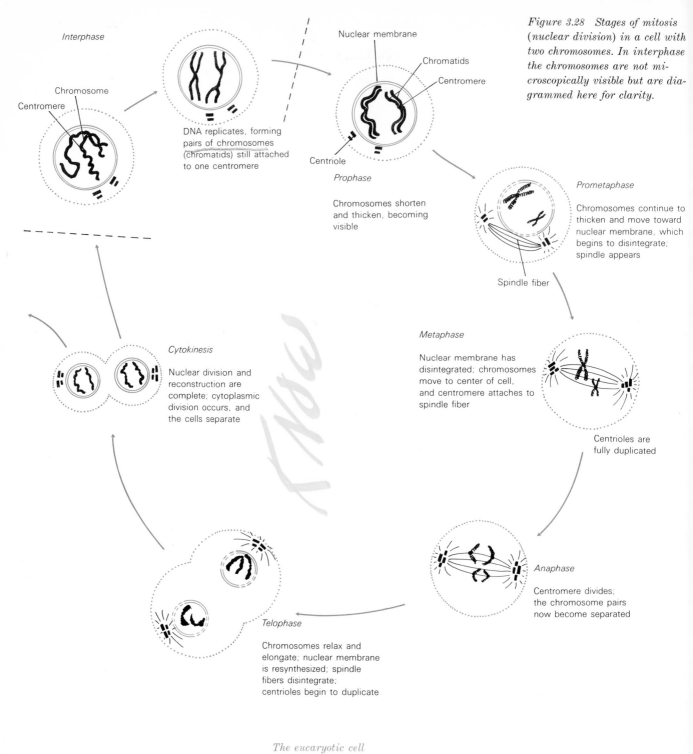

Interphase

Centromere

Chromosome

DNA replicates, forming
pairs of chromosomes
(chromatids) still attached
to one centromere

Nuclear membrane

Chromatids

Centromere

Centriole

Prophase

Chromosomes shorten
and thicken, becoming
visible

Prometaphase

Chromosomes continue to
thicken and move toward
nuclear membrane, which
begins to disintegrate;
spindle appears

Spindle fiber

Metaphase

Nuclear membrane has
disintegrated; chromosomes
move to center of cell,
and centromere attaches to
spindle fiber

Centrioles are
fully duplicated

Anaphase

Centromere divides;
the chromosome pairs
now become separated

Cytokinesis

Nuclear division and
reconstruction are
complete; cytoplasmic
division occurs, and
the cells separate

Telophase

Chromosomes relax and
elongate; nuclear membrane
is resynthesized; spindle
fibers disintegrate;
centrioles begin to duplicate

*Figure 3.28 Stages of mitosis
(nuclear division) in a cell with
two chromosomes. In interphase
the chromosomes are not microscopically visible but are diagrammed here for clarity.*

daughter cells. Wall formation can begin either at the center of the cell, proceeding toward the outside (centrifugal), or at the outside, proceeding toward the center (centripetal) (Figure 3.29).

In addition to the doubling of the number of nuclei during cell division, there must also be some mechanism for doubling the complement of cell organelles, such as mitochondria and chloroplasts. It is most likely that these organelles increase in number by division processes more or less under their own control, and hence these are, at least to a great extent, independent of the nuclear division cycle.

The nuclear processes involved in sexual reproduction (meiosis) are considered in Chapter 13.

3.7 Comparisons of the procaryotic and eucaryotic cell

We may summarize this chapter and Chapter 2 by drawing comparisons between the procaryotic and eucaryotic cell. It should be clear by now that there are profound differences in the structures of these two cell types. These differences we summarize in Table 3.1.

One difference between procaryotes and eucaryotes that deserves emphasis is the size of the ribosomes. Procaryotes have ribosomes with a sedimentation constant of 70 S, which are composed of two subunits having constants of 50 S and 30 S. (That the sum of 50 and 30 in this instance is 70 rather than 80 is due to the sedimentation constant's being a function not only of the density of a particle but also of its volume.) Cytoplasmic ribosomes of eucaryotes have a sedimentation constant of 80 S, and these are composed of two subunits with sedimentation constants

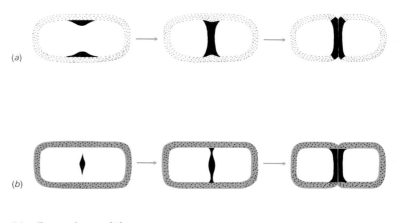

Figure 3.29   Cell-wall growth after mitosis: (a) centripetal wall growth, characteristic of microorganisms; (b) centrifugal wall growth, characteristic of higher plants.

Table 3.1   Comparisons of the procaryotic and eucaryotic cell

	Procaryotes	Eucaryotes
Nuclear body	No nuclear membrane; no mitosis	True nucleus, nuclear membrane; mitosis
DNA arrangement	Single molecule, not complexed with histones	In several or many chromosomes, usually complexed with histones
Composition of membranes	Lack sterols	Contain sterols
Respiratory system	Respiratory system part of plasma membrane or of mesosome; mitochondria absent	Present in membrane-bound organelles, the mitochondria
Photosynthetic apparatus	Photosynthetic apparatus in organized internal membranes; chloroplasts absent	Present in membrane-bound organelles, the chloroplasts
Ribosome size	70 S	80 S, except for ribosomes of mitochondria and chloroplasts, which are 70 S
Cytoplasmic movement	Cytoplasmic streaming rare or absent	Cytoplasmic streaming often occurs
Cell wall	Chemically complex cell wall structure (peptidoglycan)	Where present, composed of simple organic or inorganic materials
Flagella	Submicroscopic size; each flagellum composed of one fibril of molecular dimensions	Microscopic size; each flagellum composed of 20 fibrils in a distinct pattern, $2 \times 9 + 2$
Sexual reproduction	Fragmentary process; no meiosis; usually only portions of genetic complement reassorted (see Chapter 12)	Regular process; meiosis; reassortment of whole chromosome complement (see Chapter 13)
Vacuoles	Rarely present	Often present

of 60 S and 40 S. The ribosomes of the mitochondria and chloroplasts of eucaryotes are 70 S, however.

Although Table 3.1 emphasizes the great *structural* differences between procaryotes and eucaryotes, we should remember that these groups of organisms are chemically quite similar. Both contain proteins, nucleic acids, polysaccharides, and lipids of similar composition, and both use the

The eucaryotic cell

same kinds of metabolic machinery. The only consistent *chemical* differences found so far have been (1) the presence of sterols in the membranes of eucaryotes but not of procaryotes and (2) cell-wall chemistry. Thus, it is not in their building blocks that procaryotes differ most strikingly from eucaryotes but in how these building blocks have been put together.

At present it is believed that procaryotic organisms evolved first and that eucaryotes arose from procaryotes. Although this idea has not been proved, most of the existing evidence favors it. One theory holds that the intracellular organelles of eucaryotes are derived from infection by procaryotic cells, which then became obligately restricted to their new environment. Mitochondria, then, would have derived from bacteria, and chloroplasts from blue-green algae. The fact that both mitochondria and chloroplasts seem to have independent heredities (see Chapter 13) and have 70 S ribosomes as do procaryotes rather than 80 S ribosomes as does eucaryotic cytoplasm makes this hypothesis attractive. However, one should be cautioned that all ideas on the evolutionary relationships between procaryotes and eucaryotes are speculative.

Supplementary readings

Bracker, C. E., Jr. "Ultrastructure of fungi." *Ann. Rev. Phytopathol.* 5:343 (1967).

Chance, B. (ed.) *Energy-Linked Functions of Mitochondria* (First Colloquium, Eldridge Reeves Johnson Foundation for Medical Physics). New York: Academic Press, Inc., 1963.

DeRobertis, E. D. P., W. W. Nowinski, and F. A. Saez *Cell Biology* (*General Cytology*, 4th ed.). Philadelphia: W. B. Saunders Co., 1965.

Gibbons, I. R. "The biochemistry of motility." *Ann. Rev. Biochem.* 37:521 (1968).

Giese, A. C. *Cell Physiology,* 2nd ed. Philadelphia: W. B. Saunders Co., 1962.

Jahn, T. L., and E. C. Bovee "Movement and locomotion of microorganisms." *Ann. Rev. Microbiol.* 19:21 (1965).

Lehninger, A. L. *The Mitochondrion: Molecular Basis of Structure and Function.* New York: W. A. Benjamin, Inc., 1964.

Rogers, H. J., and H. R. Perkins *Cell Walls and Membranes.* London: E. & F. N. Spon Ltd., 1968.

Salton, M. R. J. *Microbial Cell Walls* (Ciba Lectures in Microbial Biochemistry). New York: John Wiley & Sons, Inc., 1961.

Zeuthen, E. (ed.) *Synchrony in Cell Division and Growth.* New York: Interscience Publishers, 1964.

The ability to utilize and transform energy is one of the most fundamental properties of living systems. Energy occurs in a number of forms. *Mechanical* energy is developed, for example, during cellular movement, beating of flagella and cilia, cytoplasmic streaming, movement and reorganization of intracellular structures such as mitochondria and chloroplasts, and alteration of cell shape. *Electrical* energy is produced when electrons move from one place to another; it is usually expressed as a flow of current between two points due to a difference in voltage. All living organisms produce electrical energy as an essential part of their cellular activity. *Electromagnetic* energy occurs in the form of *radiations,* and in biology the most significant is that from visible or near-visible light, such as radiations from the sun. Light is the primary energy source for photosynthetic organisms, and can also be used for certain functions by many nonphotosynthetic organisms. Some organisms, called *bioluminescent,* produce light energy. *Chemical* energy is the energy which can be released from organic or inorganic compounds by chemical reactions, and is the primary source of energy for nonphotosynthetic organisms. Cells also store chemical energy in the form of certain storage products; this energy can later be released. *Thermal* energy or heat is energy due to molecular agitation; the faster the molecules in a system are moving, the hotter the system is. All organisms produce heat as a part of their normal energy-transformation processes. Since heat cannot be used as a primary energy source by living organisms, that energy released by an organism as heat represents wasted energy. *Atomic* energy is contained within the structure of atoms themselves and is released in the form of atomic radiations. It is not utilizable by living organisms, but on the contrary it may cause damage to them.

Energy can be converted from one form to another, but energy can be neither created nor destroyed. Energy can be expressed in various kinds of units, but since all forms of energy are interconvertible, a single unit could be used for all forms of energy. In biology the most commonly used energy unit is the calorie (cal), a unit of heat. For instance, 1 cal of heat is equivalent to 860 watt-hours of electrical energy or 4.18 joule of mechanical energy. Energy conversion is a major property of living organisms and as long as an organism is functioning, such conversions are occurring. But since the conversion of energy from one form to another is never completely efficient, some energy is always lost as heat.

Oxidation-reduction reactions   The primary energy source for many organisms is chemical energy supplied by either organic or inorganic compounds. In this chapter we will be concerned with some aspects of how organic compounds are utilized as sources of energy by living organisms. Using a chemical substance as an energy source always involves

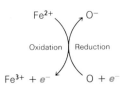

lose $\bar{e}$ —

Fe²⁺ and O⁻ diagram:

Fe^{2+} O^-

Oxidation | Reduction

$Fe^{3+} + e^-$ $O + e^-$

Figure 4.1 An example of a coupled oxidation-reduction reaction. The release of an electron from one substance (reduction) depends on the presence of another substance to accept this same electron (oxidation).

lower pot. donate to any sub. with less $\bar{e}$

what is called an *oxidation-reduction reaction:* the energy source becomes oxidized, while another substance becomes reduced. Although some oxidation-reduction reactions involve oxygen, many do not; instead of oxygen transfer, the real basis of an oxidation-reduction reaction is electron transfer. For example, when *ferrous* iron loses an electron, $Fe^{2+} \longrightarrow Fe^{3+} + e^-$, it becomes oxidized and is converted into *ferric* iron. By itself, however, this reaction would not occur because an *electron acceptor* is needed to take up the electron. Oxygen could function as the electron acceptor, $O + e^- \longrightarrow O^-$, as a result of which it becomes negatively charged and is said to be *reduced*. Thus we see that in oxidation-reduction reactions we must consider two reactants, one of which serves as the *electron donor* and becomes oxidized, and the other of which serves as the *electron acceptor* and becomes reduced (Figure 4.1). In terms of energy, the electron donor is also an *energy source,* whereas the electron acceptor is not an energy source, although it is essential in the coupled reaction. Once the electron donor has been oxidized it is usually no longer an energy source but may serve as an electron acceptor.

Chemical substances vary in their tendencies to give up electrons and become oxidized. It is convenient to express this tendency of a compound to give up electrons as its "oxidation-reduction potential," which is measured electrically in reference to a standard substance, usually hydrogen (H_2). The more reduced a material is, the more energy it contains, and the greater the tendency it has to donate electrons. The oxidation-reduction potential of $\frac{1}{2}H_2 \longrightarrow H^+ + e^-$ is -0.42 volt (V), that of $Fe^{2+} \longrightarrow Fe^{3+} + e^-$ is $+0.77$ V, and that of $O^- \longrightarrow O + e^-$ is $+0.82$ V. In principle, a substance of lower oxidation-reduction potential can donate electrons to any substance more oxidized than it is, and receive electrons from any substance more reduced than it is. Thus, most materials can be either electron donors (reductants) or electron acceptors (oxidants), depending upon what other materials they react with. Because oxidation-reduction reactions involve the transfer of electrons and are measured experimentally by electrical means, the oxidation-reduction potentials are usually expressed by an electrical unit, the volt (V). The term "oxidation-reduction" is often shortened to "redox" for use in such phrases as "redox potential" or "redox reaction."

Many dyes undergo color changes when oxidized or reduced. For example, methylene blue is colored when oxidized and colorless when reduced. If this dye is added to a cell suspension that is actively converting energy, electrons may be transferred to the dye, and its color disappears. This is one of the more dramatic examples of an oxidation-reduction reaction. Some dyes are colorless when oxidized and colored when reduced, such as the tetrazoliums. Different dyes have different redox potentials, and in series these dyes can be used to measure the oxidation-reduction potential of a system. This is done by determining which dyes of the series become oxidized and which reduced when added to the system.

Many electron-transfer reactions in organisms are accompanied by the transfer of hydrogen. For instance, in the oxidation of ethanol to acetaldehyde,

$$H_3C-\overset{\overset{\displaystyle H_2}{|}}{C}-OH \longrightarrow H_3C-\overset{\overset{\displaystyle H}{|}}{C}=O + 2H^+ + 2e^-$$

Ethanol   *Acetaldehyde*

two hydrogen atoms are removed from the ethanol, and for this reason the reaction is called *dehydrogenation* (the ethanol is said to be dehydrogenated). Since two electrons are also removed, the ethanol is oxidized. Every dehydrogenation is also an oxidation, and ethanol (the hydrogen donor) is hence also an electron donor. The hydrogen atoms that are released are transferred to an acceptor (called the *hydrogen acceptor*), which will accept hydrogen atoms only if it has also received electrons, and the hydrogen acceptor is thus also an electron acceptor. For instance, when oxygen accepts two electrons and a hydrogen atom, it becomes converted to the hydroxyl radical:

$$O + 2e^- \longrightarrow O^{2-}$$
$$H^+ + O^{2-} \longrightarrow OH^-$$

4.1 Enzymes

Chemical reactions such as oxidation and reduction can occur spontaneously, but often the rates are slow. This is because molecules must be *activated* before they can react with each other, and at low temperatures only a small percentage of the molecules of a given kind are in the activated state. If the temperature of the system is increased, more of the molecules become activated, and the chemical reaction will proceed more rapidly. Living organisms must carry out these same reactions at the lower temperatures compatible with life. Molecules can be activated at low temperatures by using *catalysts,* substances that promote chemical reactions without themselves being changed in the end. The catalysts of living organisms are proteins called *enzymes,* which promote specific reactions or groups of related reactions. Thus enzymes have two fundamental properties, *catalysis* and *specificity.*

Our knowledge of the structure and function of enzyme proteins has increased greatly in recent years. As might be anticipated from the wide variety of reactions catalyzed by enzymes, there is a tremendous diversity of protein structure in enzymes. The building blocks of a protein are the amino acids, which are connected in a specific polypeptide sequence to form the *primary structure* of the protein [Figure 4.2(a)]. Most of the properties of the enzyme relate ultimately to this primary structure. The polypeptide chain folds in a specific way, which is controlled by the primary structure, and after folding usually assumes a spherical or globular shape [Figure 4.2(c)]. Sometimes, but not always, several of these globular units combine to form larger aggregates that represent the completed and functional enzyme [Figure 4.2(d)].

—gly—ala—his—val—lys—gly—ala—phe—

(a)

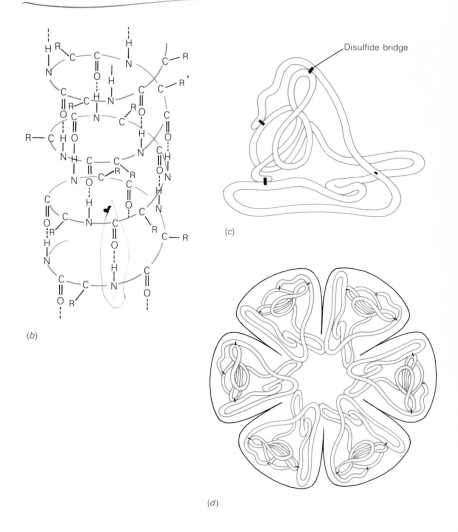

(b)

(c)

(d)

Figure 4.2 Protein structure: (a) peptide chain—short segment of the primary structure; (b) secondary structure—the peptide chain twists into a helix, in which the C=O from one turn forms a hydrogen bond with the N—H of the turn below; (c) tertiary structure—the helix folds into a globular configuration, the various parts held together by ionic, hydrophobic, and hydrogen bonds and by disulfide bridges; (d) quaternary structure—several globular subunits combine to form the final protein. If this were an enzyme, the active site might overlap portions of several of the subunits.

Disulfide bridge

Enzyme reactions   Specific folding leads to the creation of a region on the surface of the molecule that is involved in its reactive and catalytic properties; this region is called the *active site*. The compound undergoing reaction, the substrate (*S*), combines specifically with a site on the enzyme surface (*E*).

$$E \quad + \ S \quad \rightleftharpoons \quad ES$$

Enzyme   Substrate   Enzyme-substrate complex

This binding of substrate to enzyme is temporary, and if the reaction does not occur, the substrate dissociates from the enzyme. When the enzymatic reaction does occur, the substrate is converted into a product (*P*) and the product is released from the surface of the enzyme:

$$ES \longrightarrow E + P.$$

Often an enzymatic reaction involves two substrates, S_1 and S_2. The second reactant also has a specific combining site on the enzyme that is close to but not identical to that of the first. When both substances are on the enzyme surface at the same time, they may react and two products may be formed, one derived from each substrate:

$$E + S_1 + S_2 \rightleftharpoons ES_1S_2 \longrightarrow E + P_1 + P_2$$

In the above example, the enzyme was pictured as catalyzing the reaction between two substrates. In many cases only a single substrate is involved, which undergoes reaction with a nonprotein portion of the enzyme (the *prosthetic group*). Usually these prosthetic groups are permanently attached to the enzyme, having been built into the enzyme at the time it was synthesized. If an enzyme has a prosthetic group (not all do), the protein part of the enzyme is called the *apoenzyme*.

Enzymes are usually named for their substrates and for the reactions which they catalyze. Thus, ''glucose oxidase'' is an enzyme that catalyzes the oxidation of glucose, which is a good example of a simple oxidation reaction:

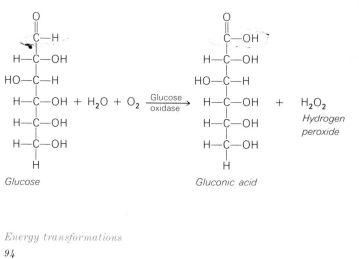

Glucose   Gluconic acid

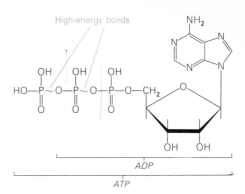

High-energy bonds

Figure 4.3 Structure of the oxidation-reduction coenzyme nicotinamide adenine dinucleotide (NAD). In NADP, another phosphate group is present, as indicated. Both NAD and NADP undergo oxidation-reduction as shown.

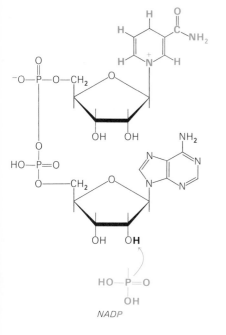

NAD

NADP

NAD**H**

Oxidized / Reduced

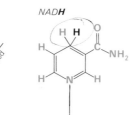

ADP

ATP

Figure 4.4 Structure of adenosine triphosphate (ATP). The two terminal phosphates are held to the molecule by ester linkages called **high-energy phosphate bonds** *because more energy is liberated upon their hydrolysis than it is with ordinary phosphate ester linkages.*

Many enzyme reactions require the concomitant action of *coenzymes*. These are carriers of electrons, hydrogen atoms, or chemical groups that are distinguished from prosthetic groups in that coenzymes are not permanently attached to the enzyme protein but combine with it only transiently during the enzymatic reaction. Many oxidation-reduction reactions involve the coenzymes NAD (nicotinamide adenine dinucleotide) or NADP (NAD-phosphate).* These coenzymes (Figure 4.3) become alternately oxidized or reduced by accepting or donating an electron and a hydrogen atom. The reduced forms are abbreviated NADH and NADPH† and will be so written subsequently.

Another coenzyme that plays an important role in energy metabolism is adenosine triphosphate, abbreviated ATP (Figure 4.4). Adenosine triphosphate functions as a carrier of phosphate and also of energy due to

* In the older literature, these coenzymes were given different names:
NAD = coenzyme I = DPN (diphosphopyridine nucleotide);
NADP = coenzyme II = TPN (triphosphopyridine nucleotide).
† Oxidation and reduction of NAD and NADP involve two electrons and two hydrogen atoms, the latter coming from H_2O. When the coenzymes are reduced, one of the hydrogen atoms is combined with the coenzyme molecule, so that they can appropriately be written "NADH" and "NADPH." The other hydrogen remains in solution as a hydrogen ion, H^+, but it will play a part in any subsequent oxidation of NADH. It is more accurate to write the reduced forms of these coenzymes as "NADH + H⁺" and "NADPH + H⁺" but since this is a rather awkward notation it generally is not used. One should remember that, in order to balance any reaction involving NADH or NADPH, a second hydrogen atom should be used.

4.1 Enzymes

95

the fact that two terminal phosphate groups are attached to the rest of the molecule by what are called *high-energy phosphate bonds*. During many oxidation-reduction reactions in the cell, ATP is synthesized from adenosine diphosphate (ADP) (as will be discussed below) and energy that is derived from oxidation-reduction is conserved in the high-energy bond. Energy in ATP can be utilized for energy-requiring functions of the cell, such as motility, biosynthesis, and growth.

Measurement of enzyme activity   An important procedure in microbiology and biochemistry is the measurement of enzyme activity. Rarely can an enzyme protein be detected directly; therefore enzyme activity is usually assayed by measuring the effect it has on a substrate or coenzyme or by estimating the amount of a product released. For instance, in the glucose oxidation reaction (page 94) glucose oxidase could be detected by measuring the rate of disappearance of glucose or oxygen or the rate of appearance of gluconic acid or hydrogen peroxide.

Often an enzyme that employs NAD as a coenzyme is assayed by measuring the oxidation or reduction of NAD. Since NADH absorbs light at 340 nm, whereas NAD does not, the rate of oxidation of NADH to NAD can be measured by plotting the rate at which light absorption at 340 nm disappears; the reduction of NAD to NADH can be assayed by measur-

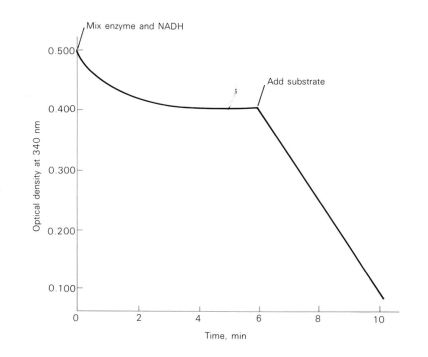

Figure 4.5   Measurement of NADH oxidation by change in light absorption at 340 nm (near ultraviolet). The initial change in absorbance is due to the reaction of NADH with other materials in the cell extract. The rate of decrease of absorbance after adding the substrate (6 to 10 min) is a measure of the amount of enzyme present. If less enzyme were present, the absorbance would decrease less rapidly. (From H.-U. Bergmeyer, Methods of Enzymatic Analysis, Weinheim/Bergstr.: Verlag Chemie, GmbH, and New York: Academic Press, 1963.)

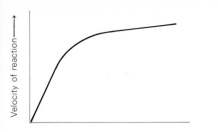

Figure 4.6 Effect of concentration of enzyme on the rate of an enzymatic reaction. The substrate concentration is held constant. When the enzyme concentration is high, the reaction rate is limited by how fast the substrate can diffuse to the enzyme.

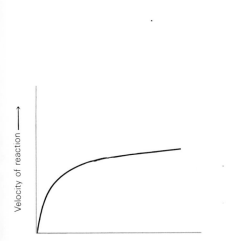

Figure 4.7 Effect of substrate concentration on the rate of an enzymatic reaction. The enzyme concentration is held constant. When the substrate concentration is high, the reaction is limited because all of the active sites on the enzyme molecules are saturated with substrate.

ing the rate of increase of light absorption at 340 nm (Figure 4.5). Sometimes an oxidation-reduction dye such as methylene blue can be used in an enzyme assay; the change in intensity of the color provides a measure of enzyme activity. The goal is always to find an assay method that is simple, reproducible, specific, and sensitive. Methods involving the appearance or disappearance of colored materials are always desirable, since these color changes can be measured so easily and accurately with laboratory photometers.

The rate of a reaction has been found to be proportional to the enzyme concentration: the more enzyme present, the faster the reaction, and the quicker the substrate disappears and product appears. This relationship is illustrated graphically in Figure 4.6. It is usual to define a unit of enzyme activity as representing the amount of enzyme that converts a given amount of substrate in a given time period. For example, one unit of enzyme might be that amount which causes the change of 1 micromole (μmole) of substrate per minute under standard conditions. The rate of the enzyme reaction will be influenced by the substrate concentration, as is shown in Figure 4.7. When the substrate concentration is low, the rate of the reaction is proportional to the substrate concentration, but eventually a point is reached at which further increase in substrate concentration does not lead to further increase in reaction; the enzyme is said to be "saturated" with substrate. This saturation level is an expression of the degree of affinity that the enzyme has for the substrate.

Extraction and purification of enzymes   Rarely can an enzyme be assayed with certainty in living or whole cells, since penetration of the substrate through the cell membrane is often restricted. Thus, in order to assay and study enzymes, *cell-free extracts* must be prepared. Such extracts are made by breaking the cells in a way that disrupts the cell wall and membrane but does not inactivate the enzyme. This can be accomplished by such methods as grinding cells with abrasives (such as powdered glass or alumina), by the use of ultrasonics, or by digestion of the cells by a lytic enzyme such as lysozyme. Many enzymes are unstable in cell-free extracts, and much experimentation may be required to obtain a suitable enzyme preparation.

Once an active cell-free extract is prepared, the enzyme usually can be purified, using various techniques of protein chemistry. Occasionally the enzyme can be crystallized; this usually indicates that the enzyme has been highly purified. Although much work can be done with crude extracts or on only partly purified enzymes, studies of the chemical structure and mechanism of action of the enzyme require material that has been highly purified.

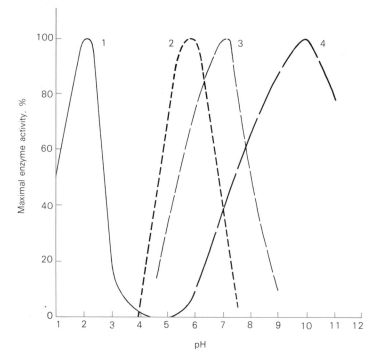

Figure 4.8 Effect of pH on activity of several different enzymes. Each enzyme has a characteristic pH optimum: (1) pepsin; (2) glutamic decarboxylase; (3) salivary amylase; (4) arginase (From J. S. Fruton and S. Simmonds, General Biochemistry, 2nd ed. New York: John Wiley & Sons, Inc., 1958.)

Effect of environment on enzyme reactions Various environmental factors influence the rate of enzyme action. Figure 4.8 shows how *pH* may affect several different enzymatic reactions; note that there is a pH range in which each enzyme acts best (the pH optimum), and that the rate of reaction decreases at higher or lower pH values than optimum. Some enzymes have broad pH optima, others have narrow ones.

The effect of *temperature* on an enzyme reaction is shown in Figure 4.9; here we see that the rate of reaction increases progressively as the temperature increases, until the optimum is reached. Above this temperature, rate of reaction drops sharply, usually because the enzyme is inactivated at high temperature.

Many enzymes require *metal ions* for activity, and the appropriate ion must be present if the reaction is to proceed. Such essential ions are called "activators" or "cofactors"; Na^+, K^+, Mg^{2+}, Ca^{2+}, and Zn^{2+} are common metal-ion cofactors. In some cases the metal ion is an integral part of the protein structure of the enzyme, whereas in others the ion is only transiently associated with the enzyme.

Occasionally a cell contains two enzymes that act on the same substrate. In such situations the two enzymes often differ in optimum temperature or pH, substrate affinity, or nature of the product formed. One

Figure 4.9 *Effect of temperature on the activity of several different enzymes:* (1) *formic hydrogenlyase of a bacterium that grows optimally at 25°C;* (2) *formic hydrogenlyase of Escherichia coli, a bacterium that grows optimally at 35°C;* (3) *amylase of Thermoactinomyces vulgaris, a bacterium that grows optimally at 55°C;* (4) *aldolase of Thermus aquaticus, a bacterium that grows optimally at 70°C.* [*Curves 1 and 2 from J. Upadhyay and J. L. Stokes: J. Bacteriol. 85:177 (1963). Curve 3 from M. J. Kuo and P. A. Hartman: Can. J. Microbiol. 13:1157 (1967). Reproduced by permission of the National Research Council of Canada. Curve 4 from T. D. Brock and H. Freeze, unpublished.*]

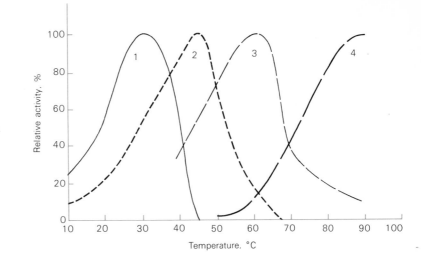

useful method for distinguishing different enzymes that act on the same substrate is *electrophoresis* on starch or acrylamide gel (Figure 4.10). Electrophoresis involves subjecting the preparation to an electrical field; enzyme molecules will move toward the positive pole of this electrical field if negatively charged or toward the negative pole if positively charged. The rate of movement will be proportional to the charge density of the molecule, which in turn will be affected by the pH at which the electrophoresis is carried out.

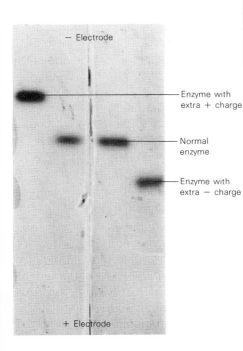

Figure 4.10 *Starch-gel electrophoresis of tryptophan synthetase. The enzyme protein was placed at the top of the gel and migrates towards the positive electrode. After electrophoresis the location of the enzyme was revealed by using a dye that combines specifically with proteins. Also shown is the migration of enzyme that had either an extra negative or an extra positive charge due to substitution of a single amino acid in the protein. The starch-gel electrophoresis technique is an extremely sensitive way of detecting small differences in amino acid composition of related proteins. [From C. Yanofsky, J. Ito, and V. Horn: Cold Spring Harbor Symp. Quant. Biol. 31:151 (1966).]*

4.1 *Enzymes*

Coupled enzyme systems   The presence of a catalyst is not always sufficient to insure that a reaction will occur. Some reactions proceed only if energy is put into the system; such reactions are called *endergonic.* Other reactions occur spontaneously and liberate heat; these are called *exergonic.* An analogy to the flow of water is appropriate: water flowing downhill spontaneously is like an exergonic reaction. Water can flow uphill, however, only if it is pulled or pushed in some way—this would be comparable to an endergonic reaction. In organisms an endergonic enzymatic reaction that would normally not proceed may be coupled with a subsequent exergonic reaction so that the energy released from the exergonic reaction could be used to "pull" the endergonic reaction.

Sometimes two or more enzymes that participate in a sequence of reactions occur together as parts of a single unit. A simple example is the pyruvate decarboxylase system of *Escherichia coli* (Figure 4.11), in which the important organic acid pyruvate is converted to the acetic acid derivative acetyl-coenzyme A (acetyl-CoA) through a series of three enzymatic reactions, each involving a separate coenzyme. In this aggregate not only are there three enzymes associated but also several molecules of each

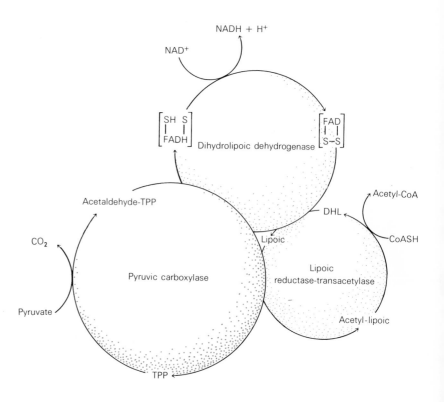

Figure 4.11 Pyruvate decarboxylase, an integrated enzyme complex. The reaction sequence from pyruvate to acetyl-CoA proceeds in three steps, each catalyzed by a separate enzyme. The product of one enzyme is passed immediately to the next without being released from the complex, thus greatly increasing the efficiency of the sequence. DHL stands for dihydrolipoic acid; TPP, for thiamine pyrophosphate.

enzyme are linked in the complex. One advantage of such an enzyme aggregate is that the product of the first reaction is immediately in contact with the next enzyme that will act on it, greatly increasing the efficiency of the reaction sequence.

4.2 Energy release in biological systems

The oxidation of chemical substances by living organisms involves a step-wise release of energy, which results from an ordered series of enzymatic reactions. In this way much of the energy of the compound is conserved instead of being converted into heat. A wide variety of chemical substances can be utilized by different organisms as energy sources, although many forms make use of only a limited number of such sources. The use of inorganic compounds as energy sources is discussed in Chapter 6; in the rest of this chapter we consider the enzymatic reactions involved in using organic compounds as sources of energy.

The complete oxidation of any organic compound yields CO_2 as an end product. As the most completely oxidized form of carbon, CO_2 itself can never be an energy source. The energy available from the complete oxidation of an organic compound depends on the closeness of its oxidation level to that of CO_2. For instance, when one mole of ethanol (CH_3CH_2OH) is oxidized to CO_2, 326 kcal of energy are released, whereas when one mole of acetic acid (CH_3COOH) is oxidized, 269 kcal are released. The difference arises because acetic acid is more oxidized than ethanol and hence has less energy available. In order for the substance to be completely oxidized, the available electrons of the compound must be transferred to an electron acceptor. One of the most common electron acceptors is O_2, which upon accepting electrons becomes reduced to water. In the absence of O_2, certain other inorganic compounds can be used (see Section 4.5) by some organisms as electron acceptors.

4.3 Anaerobic fermentation

Even in the absence of an external electron acceptor, many organisms can still oxidize some organic compounds with the release of energy, a process called *anaerobic fermentation* because it occurs in the absence of O_2 or air. Under these conditions only partial oxidation of the organic compound occurs and only a small amount of the energy is released, the rest remaining in the products. These partial oxidations involve the use of the same substance as both electron donor and electron acceptor, which means that some molecules of the compound are oxidized and others are reduced. As an example, yeast oxidizes glucose in the absence of oxygen in the following way:

$$C_6H_{12}O_6 \longrightarrow 2CH_3CH_2OH + 2CO_2 + 57 \text{ kcal}$$

Glucose	Ethanol	Carbon	Energy
(intermediate	(reduced	dioxide	
oxidation	product)	(oxidized	
level)		product)	

Note that some of the carbon atoms end up in CO_2, a more oxidized form than glucose, while other carbon atoms end up in alcohol, which is more reduced (that is, it has more hydrogens and electrons per carbon atom) than glucose.* This chemical balance between the glucose fermented and the alcohol and CO_2 produced was first determined for yeast by Louis Pasteur. It is possible to make an extract of yeast cells that will carry out this complete alcoholic fermentation of glucose since all of the enzymes necessary are present in the extract in soluble form.† When the purified enzymes are mixed, the same products are obtained in the same proportions as those that occur in cells. A number of separate enzymatic reactions occur during the alcoholic fermentation; these are shown in Figure 4.12.

In the stepwise breakdown of glucose in alcoholic fermentation two coenzymes, ATP and NAD, are involved. Initially, the glucose is *phosphorylated* by ATP, yielding glucose-6-phosphate. Phosphorylation reactions of this sort often occur preliminary to oxidation. When ATP is converted to ADP, energy is utilized because the organic phosphate bond in glucose-6-phosphate is at a lower energy level than was the phosphate bond of ATP. (The energy used at this step will be regained later in the reaction sequence.) The initial phosphorylation of glucose *activates* the

---

* The degree of oxidation or reduction of a carbon compound is conveniently expressed by its oxidation state, which is determined by counting the number of oxygen and hydrogen atoms. Each hydrogen atom is given a value of $-\frac{1}{2}$, and each oxygen is given a value of $+1$. Glucose, $C_6H_{12}O_6$, has 12H (value -6) and 6O (value $+6$) so that its oxidation state is $(-6) + (+6) = 0$. Ethanol, CH_3CH_2OH, has 6H (value -3) and 1O (Value $+1$), with an oxidation state of -2, and CO_2 has an oxidation state of $+2$.

To calculate an oxidation-reduction balance, the number of molecules of each product is multiplied by its oxidation state. For instance, in calculating the oxidation-reduction balance for the alcoholic fermentation, we may put 2 molecules of ethanol at $-2 = -4$ and 2 molecules of CO_2 at $+2 = +4$. Thus in alcoholic fermentation by yeast there is a balance between oxidized and reduced products.

† It is of historical interest that the word *enzyme* derives from a Greek word pertaining to the action of yeast, so derived because the first cell-free biochemical reaction (alcoholic fermentation of glucose) made use of a preparation made from an extract of yeast. It was later discovered that a number of separate entities, called enzymes, were required for the reaction, rather than a single yeast component as was first thought.

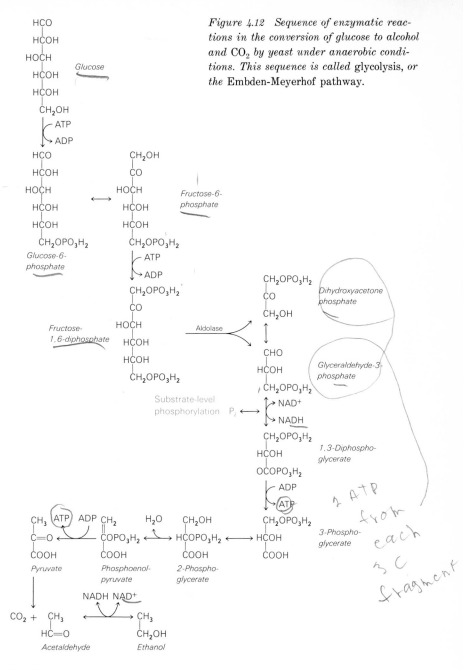

Figure 4.12   Sequence of enzymatic reactions in the conversion of glucose to alcohol and CO_2 by yeast under anaerobic conditions. This sequence is called glycolysis, or the Embden-Meyerhof pathway.

HCO
HCOH
HOCH
HCOH
HCOH
CH₂OH

Glucose

ATP
ADP

HCO
HCOH
HOCH
HCOH
HCOH
CH₂OPO₃H₂

Glucose-6-phosphate

⟷

CH₂OH
CO
HOCH
HCOH
HCOH
CH₂OPO₃H₂

Fructose-6-phosphate

ATP
ADP

CH₂OPO₃H₂
CO
HOCH
HCOH
HCOH
CH₂OPO₃H₂

Fructose-1,6-diphosphate

Aldolase

CH₂OPO₃H₂
CO
CH₂OH

Dihydroxyacetone phosphate

CHO
HCOH
CH₂OPO₃H₂

Glyceraldehyde-3-phosphate

Substrate-level phosphorylation   Pᵢ ⟷   NAD⁺   NADH

CH₂OPO₃H₂
HCOH
OCOPO₃H₂

1,3-Diphosphoglycerate

ADP
ATP

CH₃
C=O
COOH

Pyruvate

ATP   ADP

CH₂
COPO₃H₂
COOH

Phosphoenolpyruvate

H₂O

CH₂OH
HCOPO₃H₂
COOH

2-Phosphoglycerate

⟷

CH₂OPO₃H₂
HCOH
COOH

3-Phosphoglycerate

CO₂ +   CH₃
HC=O

Acetaldehyde

NADH   NAD⁺

CH₃
CH₂OH

Ethanol

Overall reaction:
Glucose + 2 phosphate + 2ADP ⟶ 2 ethanol + 2CO₂ + 2ATP + 2H₂O
Energy released: 57 kcal/mole glucose fermented

4.3   Anaerobic fermentation

molecule for the subsequent reactions. An isomerization and another phosphorylation lead to the production of *fructose-1,6-diphosphate,* which is a key intermediate product in the breakdown process. The enzyme *aldolase* now catalyzes the splitting of fructose-1,6-diphosphate into two three-carbon molecules, glyceraldehyde-3-phosphate and dihydroxyacetone phosphate.* Note that as yet there has been no actual oxidation since all of the reactions proceeded without any electron transfer, although two high-energy phosphate bonds from ATP have been used.

The first oxidation reaction now occurs in the conversion of glyceraldehyde-3-phosphate to 1,3-diphosphoglyceric acid. In this reaction, the coenzyme NAD accepts two electrons and becomes converted into NADH, while inorganic phosphate is converted into an organic form. As opposed to the organic phosphate bond of hexose phosphates, one phosphate bond of diphosphoglyceric acid represents the synthesis of a new high-energy phosphate bond. The energy that would otherwise have been released as heat in this oxidation is thus conserved. This manner of synthesizing new high-energy phosphate bonds is called *substrate-level phosphorylation* to distinguish it from two other ways of synthesizing high-energy phosphate bonds, which will be discussed later. Further reactions shown in Figure 4.12 lead ultimately to the synthesis of *pyruvic acid* and to the transfer of the energy of the high-energy phosphate bonds to ADP, forming ATP. Two ATP molecules are used initially to phosphorylate the sugar, and four ATP molecules are synthesized (two from each three-carbon fragment), so that the net gain is two ATP molecules per molecule of glucose oxidized. The energy content of a high-energy bond of ATP is about 10 kcal per mole, and 57 kcal per mole of energy is released during the alcoholic fermentation of glucose. Thus, about 35 percent of the energy released from glucose is retained in the high-energy bonds of ATP, and the rest is lost as heat.

In the above reactions, NAD has been reduced to NADH. The cell has only a limited supply of NAD, and if all of it were converted to NADH, the oxidation of glucose would stop. This "roadblock" is overcome by the oxidation of NADH back to NAD through reactions involving the conversion of pyruvic acid to ethanol and CO_2. The first step is the decarboxylation of pyruvic acid to acetaldehyde and CO_2; electrons are then transferred to acetaldehyde from NADH, leading to the formation of ethanol and NAD. The NADH that had been produced earlier is thus oxidized back to NAD.

In any energy-yielding process oxidation must balance reduction, and there must be an electron acceptor for each electron removed. In this

---

* The aldolase reaction is completely reversible and can catalyze the synthesis of fructose-1, 6-diphosphate from the two three-carbon molecules. Another enzyme catalyzes the interconversion of dihydroxyacetone phosphate and glyceraldehyde-3-phosphate.

example the reduction of NAD at one enzymatic step is coupled with its oxidation at another. The final products, CO_2 and ethanol, also are in oxidation-reduction balance (see footnote, page 102).

The ultimate result of this series of reactions is the net synthesis of two high-energy phosphate bonds, two molecules of ethanol, and two molecules of CO_2. For the yeast cell the crucial product is ATP, which is used in a wide variety of energy-requiring reactions, ethanol and CO_2 being merely waste products. These latter substances would hardly be considered waste products by man, however. For the distiller and brewer the anaerobic fermentation of glucose by yeast is a means of producing ethanol, the crucial product in alcoholic beverages; and for the baker the desired product is CO_2, which is essential in the rising of bread dough.

The biochemical pathway described above was unraveled after many years of effort by a large number of investigators. Initially, all that was known was that a cell-free extract of yeast would convert glucose to ethanol and CO_2 in the absence of air. Then it was found that in the absence of phosphate, the fermentation would stop, suggesting that some phosphorylated compound might be an intermediate in the sequence. When glucose-6-phosphate was discovered, it provided a partial solution; but how was glucose-6-phosphate formed? A major advance came when ATP and ADP were detected. These compounds are present in the cell in only very small amounts, and therefore their discovery required considerable biochemical work. Later it was shown that an enzyme, hexokinase, catalyzed the reaction of glucose and ATP, leading to the formation of glucose-6-phosphate and ADP. Further work disclosed fructose-6-phosphate, fructose-1,6-diphosphate, and the enzymes that formed them. Another important advance came when NAD (then known as Coenzyme I) was discovered. Soon after, iodoacetic acid was found to inhibit glucose fermentation, and in the presence of this substance glyceraldehyde-3-phosphate accumulated. Glyceraldehyde-phosphate was concluded to be an intermediate in glucose fermentation, derived from the breakdown of fructose-1,6-disphosphate. Other inhibitors came to light that blocked specifically other steps. About this time enzyme purification became a well-established biochemical practice, and the various enzymes in the pathway were eventually separated. Today, anaerobic fermentation in yeast is considered to be one of the simplest biochemical sequences. Its discovery, however, went hand in hand with the development of the techniques and procedures that now find wide usage in the study of any new biochemical pathway. Currently we have an almost complete understanding of many of the biochemical pathways occurring in some microorganisms—and to a great extent we owe this understanding to principles and concepts learned by the study of anaerobic fermentation in yeast. (The sequence of reactions from glucose to pyruvic acid is sometimes

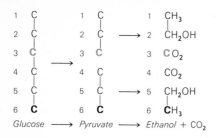

$$\text{Glucose} \longrightarrow \text{Pyruvate} \longrightarrow \text{Ethanol} + CO_2$$

Figure 4.13 The fate of the carbon atoms in different positions of the glucose molecule after passing through the Embden-Meyerhof pathway. The outcome shown can be verified by using glucose labeled in specific locations with radioactive carbon 14 and determining the site of the label in the product. This figure should be studied in conjunction with the reaction sequence shown in Figure 4.12.

called *glycolysis;* it is also called the Embden-Meyerhof pathway, after two of the principal discoverers).

By the use of radioactive tracers, one can demonstrate whether or not the Embden-Meyerhof pathway occurs in an organism. As shown in Figure 4.13, if the carbon in position 1 (or 6) of glucose is labeled with radioactive carbon 14 the radioactivity will end up in ethanol, whereas if the carbon in position 3 is labeled the radioactivity will end up in CO_2. Other mechanisms for the breakdown of glucose do not give this same labeling pattern.

The reactions proceeding from glucose to pyruvic acid, described above, occur in a wide variety of microorganisms, but the resulting pyruvic acid may be processed further in a number of different ways. Many bacteria, as well as higher animals, carry out the reaction *Pyruvic acid + NADH* $\longrightarrow$ *lactic acid + NAD,* with the end product thus being lactic acid instead of alcohol and CO_2. Other bacteria form acetic, succinic, or other organic acids, alcohols such as butanol, and ketones such as acetone. We will discuss some of these variant fermentations in Chapter 18. Under aerobic conditions, pyruvic acid is usually oxidized to CO_2 and H_2O by way of the citric acid cycle described later in this chapter.

Many compounds other than glucose can be fermented. These include most sugars, many amino acids, certain organic acids, purines, pyrimidines, and a variety of miscellaneous products. The biochemical reactions involved in the fermentation of some of these compounds are discussed in Chapters 5 and 18. For a compound to be fermentable it must be sufficiently oxidized so that it can be reduced and hence serve as an electron acceptor. Another requirement is that the compound be convertible to an intermediate that can partake in substrate-level phosphorylation. Nonfermentable compounds include hydrocarbons, fatty acids, and other highly reduced compounds. In some cases, an organic compound cannot be fermented unless another organic compound is present as the electron acceptor. Mixed fermentations of this type are common in the genus *Clostridium* and are discussed in Chapter 18.

4.4 Respiration

We have discussed above the oxidation of glucose molecules by a fermentative pathway that functions in the absence of an external electron acceptor. Only a small amount of the energy present in the substrate is liberated, with a net yield of two ATP molecules per glucose molecule fermented. However, if an external electron acceptor is present, all the substrate molecules can be oxidized to CO_2, and a yield of *thirty-eight* ATP molecules per glucose unit is possible. Oxidation of an energy source with an external electron acceptor is called *respiration*. The most important electron acceptor in respiration is gaseous oxygen (O_2), which becomes reduced and is converted to water. As in fermentation, the electrons released during the oxidation of the substrate are usually transferred initially to the coenzyme NAD. Respiration differs from fermentation specifically in the manner in which the reduced NADH is oxidized. The electrons from the reduced coenzyme, NADH, instead of being transferred to an intermediate such as pyruvic acid, are transferred to oxygen through the mediation of an electron transport system, forming oxidized NAD and H_2O.

Electron transport systems Electron transport systems have two basic functions: (1) to accept electrons from the electron donor and transfer them to the electron acceptor (O_2), and (2) to conserve some of the energy that is released during electron transfer by effecting the synthesis of ATP.

Key components of electron transport systems are the flavoproteins and cytochromes, which act as electron carriers. *Flavoproteins* are proteins containing a derivative of riboflavin (Figure 4.14); the flavin portion, which is bound to a protein, is the prosthetic group, which is alternatively reduced as it accepts electrons and becomes oxidized when the electrons are passed on.

The *cytochromes* are iron-containing porphyrin rings attached to proteins (Figure 4.15). They undergo oxidation and reduction through loss or gain of an electron by the iron atom at the center of the cytochrome:

Cytochrome-Fe^{2+} $\rightleftharpoons$ cytochrome-Fe^{3+}
(reduced)   (oxidized)

The cytochromes were discovered as colored components of animal cells; their nature and function were first worked out by the British scientist David Keilin through studies on yeast. Cytochromes can be detected

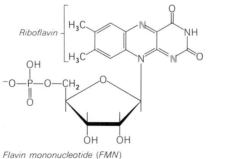

Figure 4.14   Structure of the oxidation-reduction coenzyme FMN (flavin mononucleotide).

Flavin mononucleotide (FMN)

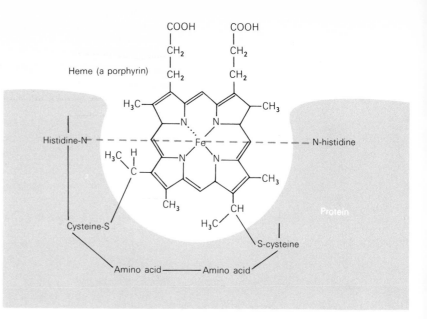

Heme (a porphyrin)

Figure 4.15   Structure of the porphyrin (heme) prosthetic group of cytochrome c, indicating the manner in which the porphyrin is connected to the apoprotein.

directly in suspensions of cells with a spectroscope, which measures the absorption by the cells of light of specific colors (Figure 4.16). By chemical fractionation it can be shown that the absorption bands are a composite, due to the presence in the cell of several distinct cytochromes, each with its own absorption characteristics. The involvement of the cytochromes in respiration was demonstrated by the changes in the absorption lines that occurred under aerobic and anaerobic conditions (Figure 4.16). Cytochromes are present in virtually all plants and animals, but it was the discovery of their oxidation and reduction in yeast cells that first provided evidence of their key role in cellular respiration.

Figure 4.16   Absorption bands of oxidized and reduced cytochromes of yeast, as seen in living whole cells in a microspectroscope. Note that each of the cytochromes a, b, and c absorbs at a different wavelength and that the absorption bands disappear when the cells are aerated. It was this observation that first suggested to Keilin that cytochromes might be involved in cellular respiration.

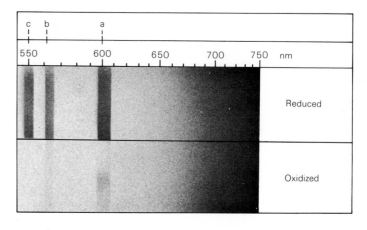

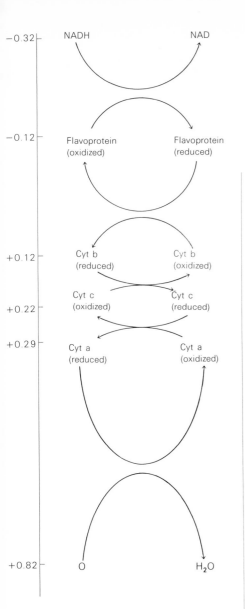

Figure 4.17 *The integrated cytochrome system leading to the oxidation of NADH by O_2. Electrons are passed from the more electronegative to the less electronegative components, each component becoming successively reduced as it accepts electrons, and oxidized as it passes them to the next component. The oxidation-reduction potential, shown in volts, of each component is a measure of its degree of electronegativity.*

Several cytochromes are known, differing in their oxidation-reduction potentials (Figure 4.17). One cytochrome can transfer electrons to another that has a higher oxidation-reduction potential and can accept electrons from a more reduced cytochrome. The different cytochromes are designated by letters, such as cytochrome a, cytochrome b, cytochrome c. Recent work has shown that the cytochromes of one organism may differ slightly from those of another, so that we now have designations such as cytochrome a_1, cytochrome a_2. A given organism usually has a limited group of cytochromes of different oxidation-reduction potentials.

The flavoproteins and cytochromes are associated in an organized particulate structure, the *electron transport particle*. Within this particle, electrons pass efficiently from one component to the other and ultimately to O_2, as shown in Figure 4.17.

Oxidative phosphorylation Much of the energy released by the transfer of electrons from NADH to O_2 is conserved through the synthesis within the electron transport particle of ATP by a process called *oxidative phosphorylation*. This synthesis of ATP should be contrasted with substrate-level phosphorylation described earlier in this chapter. In oxidative phosphorylation, ATP synthesis is coupled with electron transport and oxygen uptake. The biochemical mechanism of oxidative phosphorylation is not completely understood but probably involves the formation from inorganic phosphate of a high-energy phosphate bond on the electron transport particle, and the binding of this high-energy phosphate to ADP, yielding ATP. The rate of oxidative phosphorylation is studied experimentally by measuring the rate of oxygen uptake and the rate of phosphate conversion to ATP. A ratio is then calculated—phosphate uptake / oxygen uptake—the so-called P/O ratio. With NADH as the electron donor, this ratio is about 3; that is, three molecules of ATP are synthesized for each atom of oxygen taken up and each molecule of NADH oxidized. With electron donors other than NADH, the P/O ratio may differ. Thus, succinate oxidation, which involves the direct donation of electrons from succinate to flavoprotein without the mediation of NAD, leads to a P/O ratio of 2.

Recall that, in fermentation, NADH oxidation was not coupled with ATP synthesis, and the energy contained within the NADH did not become available to the cell. The most important aspect of oxidative phosphorylation is that it provides the organism with a means of deriving energy from the oxidation of NADH. Of course, oxidative phosphorylation is contingent on the presence in the environment of oxygen gas or another suitable electron acceptor. Aerobic organisms can thus make much more ATP than fermentative organisms from the same amount of energy source, and hence can synthesize much more cell material. Actually, many organic compounds cannot be utilized fermentatively because they cannot enter into reactions in which substrate-level phosphorylation can occur, and hence cannot participate in this type of ATP synthesis. On the other hand, many of these compounds can be utilized as energy sources aerobically since they can reduce NAD to NADH and make possible ATP synthesis through oxidative phosphorylation.

Certain chemicals, such as dinitrophenol, act as uncoupling agents, inhibiting oxidative phosphorylation without inhibiting electron transport. In the presence of dinitrophenol, NADH oxidation and oxygen uptake occur normally, but no ATP synthesis occurs, thus resulting in energy wastage.

Inhibitors of electron transport   Various chemicals inhibit electron transport. *Carbon monoxide* (CO) combines directly with the terminal cytochrome, cytochrome oxidase, and prevents the attachment of oxygen; *cyanide* (CN^-) and *azide* (N_3^-) bind tightly to the iron of the porphyrin ring of the cytochromes and prevent its oxidation-reduction function; the antibiotics *antimycin A* and *oligomycin* inhibit electron transport between specific cytochromes. These agents occasionally are useful in studying the mechanism of oxidative phosphorylation. In microbiology they are most useful as selective inhibitors of aerobic organisms since anaerobes not using the cytochrome system are unaffected. Thus, sodium azide is often added to culture media to selectively isolate lactic-acid bacteria (see Chapter 18) since these bacteria lack a cytochrome system and are able to grow in the presence of sodium azide, whereas most other bacteria are not.

Localization of the electron transport system   By fractionating cells and determining in which fraction the cytochromes are, one can determine in what cellular structures the electron transport system is localized. In procaryotes the electron transport system is associated with the plasma membrane system. The electron transport particles are either attached to or are an integral part of the membrane, for when whole membranes are isolated, all of the cytochromes are found in this fraction. In eucaryotic organisms, the cytochromes and electron transport system are found in the mitochondria. Indeed, isolated mitochondria take up oxygen and carry out oxidative phosphorylation at rates similar to the rate in the intact cell. Oxidative phosphorylation occurs in the cristae of the mitochondrion, and the energy present in the newly synthesized ATP becomes available for biosynthetic reactions in the cytoplasm.

Citric acid cycle   Now that we have described the characteristics of the electron transport system, we can consider its function in the oxidation of pyruvic acid, the key intermediate in the oxidation of glucose (see Section 4.3). Pyruvic acid retains most of the energy present in the glucose, and most aerobic organisms are able to oxidize pyruvic acid completely to CO_2 through a series of steps called the *citric acid cycle* (one of the intermediates is citric acid).* The NADH formed in the citric acid cycle is reoxidized by way of an electron transport system.

A diagram of the citric acid cycle is shown in Figure 4.18. Pyruvic acid is first decarboxylated, leading to the production of one molecule of NADH and an acetyl radical coupled to coenzyme A (CoA). Acetyl-coenzyme A (abbreviated acetyl-CoA) (Figure 4.19) is an activated form of acetate, the acetyl-CoA bond being a high-energy bond. In addition to being a key intermediate in the citric acid cycle, acetyl-CoA also plays many important biosynthetic roles (see Chapter 5). The acetyl group of acetyl-CoA combines with the four-carbon compound oxalacetic acid, leading to the formation of citric acid, a six-carbon organic acid, the energy of the high-energy acetyl-CoA bond being used to drive this synthesis. Dehydration, decarboxylation, and oxidation reactions proceed, and two CO_2 molecules are released. Ultimately oxalacetic acid is regenerated, and can serve again as an acetyl acceptor; for this reason the process is called a cycle. The net result of the citric acid cycle is the complete oxidation of pyruvic acid to CO_2 with the production of four molecules of NADH. Each of the NADH molecules can be oxidized back to NAD through the electron transport system, producing three ATP molecules and four H_2O molecules per molecule of NADH oxidized. In addition, the oxidation of succinic acid to fumaric acid involves substrate-level phosphorylation and donation of electrons to the flavin of an electron transport particle without the mediation of NAD, producing three more molecules of ATP. Thus, a total of 15 ATP molecules are synthesized for each turn of the cycle. In the oxidation of glucose, two molecules of pyruvic acid are formed from each glucose molecule, so that 30 molecules of ATP can be synthesized in the citric acid cycle. Also, the two NADH molecules produced during glycolysis can be reoxidized by the electron transport

* The citric acid cycle is sometimes also called the Krebs cycle, after one of its discoverers, Sir Hans Krebs.

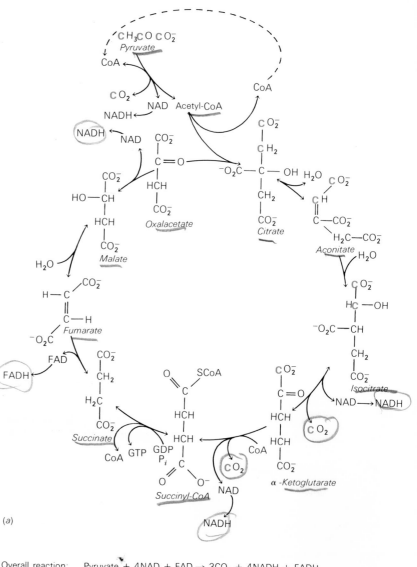

(a)

Figure 4.19 Structure of CoA and acetyl-CoA.

Adenine
|
Ribose
|
Phosphate
|
Phosphate
|
Pantetheine S—C
Coenzyme A   CH₃
Acetyl-CoA

Overall reaction:	Pyruvate + 4NAD + FAD → 3CO$_2$ + 4NADH + FADH
	GDP + phosphate → GTP
	GTP + ADP → GDP + ATP
Oxidative phosphorylation:	4 NADH ≡ 12 ATP
	FADH = 2 ATP
	15 ATP

(b)

Table 4.1 *ATP yields from the anaerobic and aerobic oxidation of glucose*

	Glycolysis	*Citric acid cycle*	*Total*
Anaerobic organisms	2 ATP	Not operative	2 ATP
Aerobic organisms	2 ATP (substrate-level phosphorylation)		
	6 ATP (oxidative phosphorylation)	30 ATP	38 ATP

system, yielding six more molecules of ATP. Finally, two molecules of ATP are produced by substrate-level phosphorylation during the conversion of glucose to pyruvic acid, so that aerobes can form 38 ATP molecules from glucose breakdown, in opposition to only two molecules of ATP produced anaerobically (Table 4.1). If we again assume that the high-energy phosphate bond of ATP is about 10 kcal/mole, this means that 380 kcal of energy could be converted to high-energy phosphate bonds in ATP by the complete oxidation of glucose to CO_2 and H_2O. Since the total amount of energy available from the complete oxidation of glucose is about 688 kcal/mole, this means that respiration is about 55 percent efficient. This is somewhat higher than the 35 percent efficiency of fermentation, although fermentative organisms obtain only about 5 percent as much ATP per glucose oxidized as do respiratory organisms.

In addition to its function as an energy-yielding mechanism, the citric acid cycle provides two key intermediates for amino acid synthesis (see Chapter 5), α-ketoglutarate and oxalacetate.

Respirable compounds   Many organic compounds that cannot be fermented can be broken down through respiration because the reoxidation of NADH and the production of ATP in respiring organisms can occur in electron transport systems. All that is required is the existence of an enzyme that will transfer electrons from the compound to NAD. Respirable compounds that cannot be fermented include hydrocarbons, fatty acids, and many alcohols.

It is not necessary that a respirable compound be oxidized completely to CO_2 and H_2O to serve as an energy source, although in a high proportion of cases complete oxidation does occur, usually through the citric acid cycle after conversion of the compound to pyruvate or acetyl-CoA.

Energy storage compounds   Adenosine triphosphate is only a short-term energy currency, and if it is not utilized, it is quickly hydrolyzed to

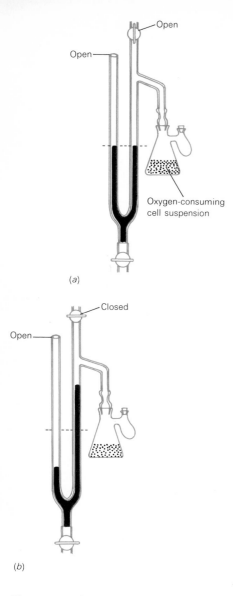

(a)

(b)

Figure 4.20   A manometer and its use in measuring oxygen uptake: (a) Initial position of fluid in manometer. (b) After stopcock is closed, the fluid level changes because of oxygen consumption in the flask.

ADP by reactions not yielding energy. For long-term storage of energy, most organisms produce organic polymers that can later be oxidized for the production of ATP. The glucose polymers starch and glycogen are produced by many microorganisms, both procaryotic and eucaryotic, whereas poly-β-hydroxybutyrate (PHB) is produced by many procaryotes. These polymers often are deposited within the cells in large granules that can be seen with the light or electron microscope (see Chapter 2). In the absence of an external energy source, the cell may then oxidize its energy-storage material and thus be able to maintain itself even under starvation conditions.

Measurement of respiration in cells   During aerobic respiration O_2 is consumed, and the rate of respiration can be determined by measuring the rate of oxygen uptake. This can be done by chemical methods that are specific for oxygen, but the instrument most commonly used is the manometer, a device for detecting changes in gas pressure (Figure 4.20). As O_2 is consumed, the pressure drops, and the fluid in the manometer rises; the CO_2 released during respiration is absorbed by an alkali trap, so that the pressure changes are due only to O_2 uptake. Respiration can also be studied by measuring the rate of CO_2 release; and with certain modifications of the manometer technique, both O_2 uptake and CO_2 production can be measured at the same time. The ratio of CO_2 produced to O_2 consumed (CO_2/O_2) is called the *respiratory quotient* (RQ) and is calculated as follows:

$$C_6H_{12}O_6 + 6O_2 \longrightarrow 6CO_2 + 6H_2O \qquad 6CO_2/6O_2; \ RQ = 1.0$$
Glucose

For carbohydrates and other compounds at the oxidation level of carbohydrate the RQ is 1.0, for alcohols, amino acids, and other compounds more reduced than carbohydrate the ratio is less than 1.0, and for some organic acids and other compounds more oxidized, it is greater than 1.0.

The amount of oxygen taken up will obviously be higher when the cell density is high than when it is low. Oxygen-uptake measurements are standardized by expressing the rate of O_2 uptake as microliters (μliters) of O_2 per hour per milligram of dry weight of cell material, a value called the Q_{O_2}. Instead of dry weight, cell nitrogen or some other measure of cell mass is sometimes used. Different organisms show different Q_{O_2} values, even with the same substrate. Smaller cells usually show higher Q_{O_2} values than larger cells, presumably because small size allows faster diffusion of substrate and oxygen.

Respiration can occur in the absence of an added energy source if the cells utilize an internal reserve of energy such as polysaccharide or fat.

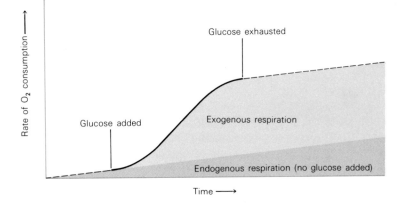

Figure 4.21 *Comparison of endogenous respiration using internal energy reserves and exogenous respiration using a substrate (glucose) added to the medium.*

This is called *endogenous respiration,* to distinguish it from *exogenous respiration* in which an external energy source is used (Figure 4.21). The rate of endogenous respiration is dependent on the prehistory of the cells; well-fed cells will have considerable energy reserves and high endogenous respiration, whereas starved cells may have a very low endogenous respiration. Both kinds of cells may respond equally well to the addition of an exogenous energy source.

Relationship between respiration and fermentation   Many organisms are able to carry out both respiratory and fermentative metabolism of the same compound. One factor that determines which type of metabolism will occur is of course the presence or absence of the electron acceptor O_2. Another factor is the concentration of respirable compound. For instance, if a high concentration of a compound such as glucose is present, the organism often ferments the compound even if O_2 is present, with the resulting accumulation of fermentation products such as acids and alcohols. When the glucose is exhausted, the organism then switches to a respiratory metabolism and utilizes the fermentation products previously released. The biochemical factors controlling these changes in metabolism are complex and are still but poorly understood.

4.5   Anaerobic respiration

Although O_2 is the most common electron acceptor in the oxidation of NADH in electron transport systems, some organisms can use organic or

Table 4.2   Electron acceptors used in respiration

	Electron acceptor	Reduced products

Anaerobic respiration	Inorganic:
	    Nitrate (NO_3^-)
		Nitrite (NO_2^-), nitrous oxide
		    (N_2O), nitrogen (N_2)

	Sulfate (SO_4^{2-})	Hydrogen sulfide (H_2S)
	Carbon dioxide (CO_2)	Methane (CH_4)

	Organic:
	    Fumarate
	        (HOOC—CH=CH—COOH)
		Succinate
		    (HOOC—CH_2—CH_2—COOH)

Aerobic respiration	O_2	H_2O

inorganic compounds instead (Table 4.2). Oxidation with these alternative electron acceptors is called *anaerobic respiration*. One of the most common of these alternative acceptors is nitrate, NO_3^-, which becomes converted into more reduced forms of nitrogen, NO_2^-, N_2O and N_2, a process called *denitrification*. The nitrate molecule combines with a terminal cytochrome of the electron transport system and accepts electrons, becoming converted to nitrite, which latter compound usually accumulates until all the nitrate is used up. After this, nitrite is reduced further to nitrogen gases. Since nitrite can be measured by a sensitive chemical assay, the ability of organisms to reduce nitrate is usually determined by observing whether nitrite accumulates when the organism is grown anaerobically in the presence of nitrate. In most organisms nitrate reduction is strongly inhibited by O_2, and it is thought that NO_3^- and O_2 compete for the same electrons from the electron transport particle. Most bacteria that reduce nitrate are called facultative since they will transfer electrons to O_2 if it is present or to NO_3^- when O_2 is absent. The gaseous products of denitrification, if produced in soil, escape to the atmosphere. Hence denitrification is considered a detrimental process agriculturally since it results in a loss of nitrogen from the soil (see Chapter 16).

Another electron acceptor used by some bacteria is sulfate (SO_4^{2-}), which is reduced to H_2S. Sulfate-reducing bacteria are usually strict anaerobes, unable to grow on or use O_2, and often they are even killed by O_2. The biochemistry of sulfate reduction is discussed in Chapter 5, and the ecological importance of sulfate-reducing bacteria is discussed in Chapter 16. The methanogenic bacteria are a group of strict anaerobes that use CO_2 as an electron acceptor, reducing it to methane (CH_4). Certain members of this group are important inhabitants of the rumen (see Chapter 14) and of anaerobic sewage-treatment plants (see Chapter

16). Some bacteria (for example, *Streptococcus faecalis*) can use such organic compounds as fumarate as alternative electron acceptors.

4.6   Summary

We have learned in this chapter that utilizing organic compounds as sources of energy involves oxidation-reduction reactions mediated by enzymes. A key component of oxidation-reduction is nicotinamide adenine dinucleotide (NAD), which through its ability to be alternately oxidized and reduced carries electrons from the organic energy source to an electron acceptor. Coupled with oxidation-reduction reactions is the synthesis of high-energy phosphate bonds in adenosine triphosphate (ATP). As we will see in Chapters 5 and 7, energy of ATP is used in various biosynthetic reactions of the cell that in turn lead to the synthesis of new cell material and to growth. Thus, ATP can be viewed as a form of energy currency by the cell. However, ATP is only a short-lived energy reserve, and for long-term energy storage, organic polymeric compounds such as starch, glycogen, or PHB are used. To be utilized as an energy source, the organic compound must be able to give up electrons and become oxidized, and since each oxidation reaction must be accompanied by a reduction, there must be an electron acceptor to take up the electrons from the energy source. The most widely occurring electron acceptor is oxygen, but O_2 can serve as an electron acceptor only if it can be activated by such means as a cytochrome system present in an electron transport particle. When O_2 accepts electrons it becomes reduced to H_2O. The utilization of O_2 as an electron acceptor is called *respiration;* it can be measured by the uptake of oxygen gas by the respiring organism or system. Other electron acceptors that can replace O_2 are the inorganic compounds nitrate, sulfate, and CO_2. The utilization of these electron acceptors in place of O_2 is called anaerobic respiration and is restricted to relatively few groups of bacteria. Organic compounds can also be utilized as energy sources in the absence of O_2 or inorganic electron acceptor by a process called fermentation. In this process, an organic compound serves as both electron donor and electron acceptor. In the fermentation of glucose by yeast, for instance, some of the carbons of glucose are oxidized to CO_2, whereas other carbons are reduced to alcohol. Although NAD is involved in fermentation, an electron transport system is not. Whereas in respiration all of the potential chemical energy of an organic compound can be released, in fermentation it cannot. Thus energy and ATP yields in a fermentation such as the alcoholic fermentation by yeast discussed earlier in this chapter:

$$\text{Glucose} \longrightarrow 2 \text{ ethanol} + 2CO_2 + 2 \text{ ATP}$$

are much lower than those in respiration,

$$\text{Glucose} + 6O_2 \longrightarrow 6CO_2 + 6H_2O + 38 \text{ ATP}$$

Supplementary readings

Chance, B., W. D. Bonner, Jr., and B. T. Storey "Electron transport in respiration," *Ann. Rev. Plant Physiol.* 19:295 (1968).

Chance, B. (ed.) *Energy-linked Functions of Mitochondria* (First Colloquium, Eldridge Reeves Johnson Foundation for Medical Physics). New York: Academic Press, Inc., 1963.

Dixon, M., and E. C. Webb *Enzymes,* 2nd ed. New York: Academic Press, Inc., 1964.

Fruton, S., and S. Simmonds *General Biochemistry,* 2nd ed. New York: John Wiley & Sons, Inc., 1958.

Gunsalus, I. C., and R. Y. Stanier (eds.) *The Bacteria: A Treatise on Structure and Function,* vol. II, *Metabolism.* New York: Academic Press, Inc., 1961.

Hastings, J. W. "Bioluminescence," *Ann. Rev. Biochem.* 37:597 (1968).

Keilin, D. *The History of Cell Respiration and Cytochrome.* London: Cambridge University Press, 1966.

Lehninger, A. L. *Bioenergetics: The Molecular Basis of Biological Energy Transformations.* New York: W. A. Benjamin, Inc., 1965.

Lehninger, A. L. *The Mitochondrion: Molecular Basis of Structure and Function.* New York: W. A. Benjamin, Inc., 1964.

Mahler, H. R., and E. H. Cordes *Biological Chemistry.* New York: Harper & Row, Publishers, 1966.

Mandelstam, J., and K. McQuillen (eds.) *Biochemistry of Bacterial Growth.* Oxford: Blackwell Scientific Publishers, 1968.

Rose, A. H. *Chemical Microbiology,* 2nd ed. London: Butterworth & Co. (Publishers), Ltd., 1968.

Umbreit, W. W., R. H. Burris, and J. F. Stauffer *Manometric Techniques: A Manual Describing Methods Applicable to the Study of Tissue Metabolism,* 4th ed. Minneapolis: Burgess Publishing Co., 1964.

Wood, W. A. "Carbohydrate metabolism," *Ann. Rev. Biochem.* 35, pt. 2:521 (1966).

Nutrients are substances required by living organisms for growth and function. Some microorganisms require a large number of substances for growth, whereas other microorganisms need only a few. The nutritional requirements of a microorganism are determined in part by its genetic makeup and in part by factors in its environment. Nutrients can be divided into two classes: (1) necessary nutrients, without which a cell cannot grow, and (2) useful but dispensable nutrients, which are used if present but which are not essential. Some nutrients are the building blocks from which the cell makes macromolecules and other structures, while other nutrients serve only as energy sources without being incorporated directly into the cellular material; sometimes a nutrient can play both roles. The required substances are sometimes divided into two groups, macronutrients and micronutrients, depending upon whether they are required in large or small amounts. It is easy to detect when a macronutrient is required, merely because so much of it is needed. Often micronutrients are required in such small amounts that it is impossible to measure exactly how much is required; indeed, one may not even suspect that the particular micronutrient is present in the medium in which the organism is growing.

The major elements of the cell, C, H, O, N, S, and P, occur in nature rarely or never as simple elements but as compounds whose constituents are covalently linked to other elements. To obtain the needed elements, the cell must be able to carry out chemical reactions on the compounds in which the elements are present. Not all chemical compounds are equally susceptible to chemical attack, and not all organisms can attack every compound. Much of this chapter will be concerned with the various mechanisms by which organisms transform compounds of carbon, nitrogen, sulfur, and phosphorus. *Metabolism* refers to the array of chemical reactions occurring in cells, which lead to transformations of various compounds. The processes involved in the degradation of compounds, usually encompassing energy transformations, are collectively called *catabolism;* in contrast, reactions involved in the building up or biosynthesis of compounds are called *anabolism*.

5.1 Carbon metabolism and nutrition

Hexoses and disaccharides The most common carbon sources of microorganisms are the simple hexose carbohydrates such as glucose, galactose, or fructose. These carbohydrates are often energy sources as well; oxidizing glucose to provide energy was discussed in Chapter 4.

Glucose, the most common hexose both in the cell and in nature, is often the biosynthetic precursor of other hexoses. In the conversion of one sugar to another, the coenzyme form of the sugar is usually an intermediate. For example, the coenzyme form of glucose, uridine diphosphoglucose, is formed in the following series of reactions:

Glucose + ATP $\longrightarrow$ glucose-6-phosphate $\longrightarrow$
   glucose-1-phosphate + UTP (uridine triphosphate) $\longrightarrow$ UDP-glucose + PP$_i$

The synthesis of galactose involves the conversion of UDP-glucose to UDP-galactose by an enzyme, *epimerase:*

UDP-glucose $\xrightleftharpoons{\text{NAD}}$ UDP-galactose

UDP-galactose is then converted to galactose-1-phosphate:

UDP-galactose $\rightleftharpoons$ galactose-1-phosphate + UMP

Utilizing *galactose* as a carbon and energy source involves first the formation of galactose-1-phosphate from galactose and ATP by a specific galactokinase. The above enzymes operate in the reverse direction to further transform galactose-1-phosphate to glucose-6-phosphate. Glucose-6-phosphate is then oxidized through the normal glycolytic pathway (see Chapter 4). The transformations of many other sugars and sugar derivatives involve nucleotide sugar intermediates. Thus, transformations of mannose involve guanine-diphosphomannose (GDP-mannose), and those with rhamnose involve thymidine-diphosphorhamnose.

Many microorganisms can use *disaccharides* for growth. Disaccharides are composed of two monosaccharides connected by a glycosidic linkage. Before a disaccharide is utilized, it is usually converted into its component monosaccharides through the function of specific enzymes (Figure 5.1). *Lactose* is the main sugar in milk, and its utilization by microorganisms is of considerable economic importance because milk-souring organisms produce lactic acid from lactose. The utilization of lactose involves the enzyme β-galactosidase (Figure 5.1), which is of considerable interest in molecular biology because its genetics and mechanism of synthesis have been widely studied in *Escherichia coli* (see Chapters 7 and 12). Many bacteria living in the mammalian intestine form β-galactosidase, which enables them to metabolize some of the lactose that reaches the intestinal tract. *Sucrose,* the common sugar of higher plants, is usually first hydrolyzed to its component monosaccharides by the enzyme *invertase* (sometimes called *sucrase*).

Carbohydrates such as glucose, galactose, and mannose are important structural components in cell walls and other cell constituents. If sugars are not present in an organism's environment they must be synthesized from noncarbohydrate precursors, a process called *gluconeogenesis.* In

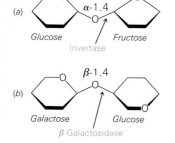

Figure 5.1 Two representative disaccharides and the enzymes that cleave them: (a) sucrose and (b) lactose.

this process, two molecules of pyruvate are converted by a reversal of the glycolytic pathway (Section 4.3) to fructose-1,6-diphosphate, using energy from ATP. Fructose-1,6-diphosphate is then transformed to fructose-6-phosphate by a specific hydrolytic enzyme (fructose-1,6-diphosphatase), the fructose-6-phosphate is isomerized to glucose-6-phosphate, and this last compound is then changed into glucose by another hydrolytic enzyme (glucose-6-phosphatase). It might seem strange that the cell has two alternate systems—one utilizing and one forming glucose—and one might wonder how the relative rates of these opposing processes are controlled. The key enzyme is fructose-1,6-diphosphatase, whose activity is inhibited by high concentrations of its substrate. Thus when there is much carbohydrate present, the phosphatase will be inhibited, and the pathway will proceed in the direction of sugar utilization. When the sugar concentration is low, the inhibition of the enzyme will be abolished and sugar synthesis will occur.

Pentoses Five-carbon sugars (pentoses) are carbon and energy sources for many microorganisms. One pathway for ribose utilization, shown in Figure 5.2, involves xylulose-5-phosphate as the key intermediate in a reaction sequence during which two molecules of ATP are formed for each molecule of pentose fermented, the end products being acetate and lactate. Pentoses other than ribose are fermented in a similar fashion, first being converted to xylulose-5-phosphate. The key enzyme is *phosphoketolase*, the enzyme that splits xylulose-5-phosphate into acetyl-phosphate and glyceraldehyde-3-phosphate; this reaction is essentially irreversible, and a high-energy phosphate bond is formed.

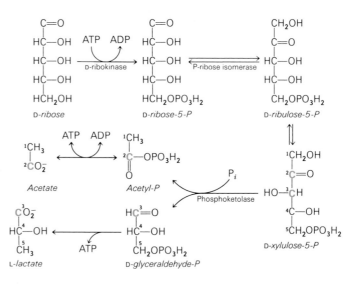

Figure 5.2 Pathway for the utilization of ribose, a typical pentose.

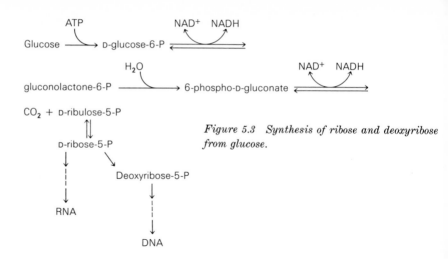

Figure 5.3   Synthesis of ribose and deoxyribose from glucose.

The pentoses ribose and deoxyribose are essential components of the nucleic acids RNA and DNA respectively. In their absence from the environment, these sugars must be formed in the cell, either from hexoses or from noncarbohydrate precursors. Synthesizing ribose from glucose involves the reactions shown in Figure 5.3; deoxyribose is then synthesized from ribose through a reductive reaction. There are many other transformations of pentoses, some of considerable complexity. The involvement of a pentose cycle in CO_2 fixation in photosynthesis is discussed in Chapter 6.

Polysaccharides   Many polysaccharides are insoluble, and the cell cannot incorporate them directly into its cytoplasm. Instead, enzymes are excreted by the cell to hydrolyze the polysaccharides, and breakdown occurs outside the cell. The monosaccharide or disaccharide units then pass into the cell. The breakdown of two polysaccharides, starch and cellulose, is shown in Figure 5.4. Although both of these polysaccharides

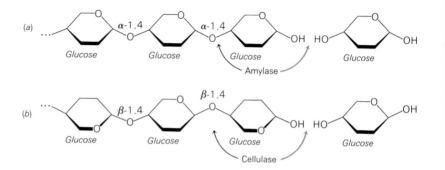

Figure 5.4   The breakdown of the polysaccharides starch (a) and cellulose (b).

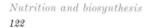

Figure 5.6 Synthesis
of glucose polysac-
charide from UDPG
(uridine-diphospho-
glucose).

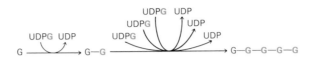

Figure 5.5 Demonstration of hydrolysis of starch by colonies of Bacillus subtilis. After incubation, the plate was flooded with Lugol's iodine solution. Where starch hydrolysis has occurred, the characteristic purple color of the starch-iodine complex is absent. Hydrolysis of starch occurs at some distance from the bacterial colonies because of the production of extracellular amylase.

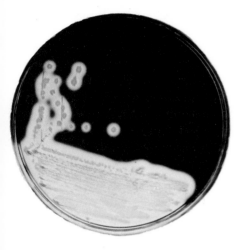

are composed of glucose units they are connected differently, and this profoundly affects their properties. Cellulose is much more insoluble than starch and is usually less rapidly digested. Cellulose, one of the most abundantly occurring polysaccharides in nature, forms long fibrils (Section 3.1), and organisms that digest cellulose are often found closely associated with them. Many fungi are able to digest cellulose; these are mainly responsible for the decomposition of plant materials on the forest floor. Cellulose digestion is restricted to only a few groups of bacteria, however, of which the myxobacteria, clostridia, and actinomycetes (see Chapter 18) are the most common. Anaerobic digestion of cellulose is carried out by a few *Clostridium* species, which are common in lake sediments, animal intestinal tracts, and systems for anaerobic sewage digestion. Starch is digestible by many fungi and bacteria; this is illustrated for a laboratory culture in Figure 5.5.

Agar is another polysaccharide that is of considerable interest because of its frequent use in solid media for culturing microorganisms. Agar occurs naturally in many seaweeds but is not found in nonmarine plants. This polysaccharide, composed of galactose and galacturonic acid units, can be digested by some marine microorganisms; when a colony of agar-digesting organisms grows on an agar plate, a small depression is produced around the colony, which then slowly sinks into the agar.

The *synthesis* of polysaccharides is not merely a reversal of the reactions involved in breakdown. A nucleotide sugar (for example, UDP-glucose) is the starting material, and sugar units are added stepwise to the end of a polysaccharide chain, as is shown in Figure 5.6. Energy to drive the synthesis is obtained by hydrolysis of high-energy sugar-phosphate bonds in the nucleotide sugar. Some of the most interesting polysaccharides synthesized by microorganisms are those which form the capsules of pneumococcus, the bacterium that is frequently the cause of pneumonia in man. These polysaccharides are usually composed of two or more sugars or sugar derivatives in alternating sequence; a separate enzyme catalyzes addition of each kind of sugar, and the sugar sequence forms by the successive action of the separate enzymes (Figure 5.7).

Figure 5.7 Formation of polysaccharide heteropolymer by stepwise addition of two different sugars.

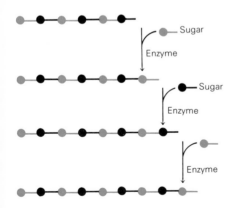

Organic acid metabolism   Many microorganisms can utilize the organic acids produced through the citric acid cycle (discussed in Section 4.4) as carbon and energy sources. However, many microorganisms that have a normal citric acid cycle cannot take up these same organic acids from the environment and use them for growth, probably because they cannot penetrate into the cell.

The utilization of two- or three-carbon acids as carbon sources cannot occur by means of the citric acid cycle alone since this cycle can continue to operate only if the four-carbon acid oxalacetate is regenerated at each turn of the cycle (see Chapter 4) and any removal of carbon compounds for biosynthetic reactions would prevent the completion of the cycle. In the utilization of acetate, this difficulty is circumvented through the operation of the *glyoxylate cycle* (Figure 5.8), so called because glyoxylate is a key intermediate. After two turns of the cycle, four acetates are consumed and one molecule of hexose and two molecules of CO_2 are produced. Two key enzymes of this cycle that are not involved in the citric acid cycle are *isocitratase,* which splits isocitrate to succinate and glyoxylate, and *malate synthetase,* which converts glyoxylate and acetyl-CoA to malate.

To be used, acetate must be first converted to acetyl-CoA. One mechanism for doing this that is common in microorganisms involves the following sequence of reactions:

$$\text{Acetate} + \text{ATP} \xrightleftharpoons{\text{Acetate kinase}} \text{acetyl-phosphate} + \text{ADP}$$

$$\text{Acetyl-phosphate} + \text{CoA} \xrightleftharpoons{\text{Phosphotransacetylase}} \text{acetyl-CoA} + P_i$$

Note that this conversion requires the energy from one high-energy bond of ATP Another mechanism for the synthesis of acetyl-CoA from acetate involves the enzyme acetate thiokinase, the overall reaction for which can be summarized as below:

$$\text{Acetate} + \text{ATP} \rightleftharpoons \text{acetyl-AMP (adenosine monophosphate)} + \text{PP}$$
$$\text{Acetyl-AMP} + \text{CoA} \rightleftharpoons \text{acetyl-CoA} + \text{AMP}$$

In this case, both high-energy bonds of ATP are used since AMP is formed instead of ADP.

The utilization as a carbon source of three-carbon compounds, such as pyruvate or compounds converted to pyruvate (for example, lactate or carbohydrates), also cannot occur through the tricarboxylic acid cycle alone. The oxalacetate needed to keep the cycle going is synthesized from pyruvate by the addition of a carbon atom from CO_2. In some organisms this step is catalyzed by the enzyme pyruvate carboxylase:

$$\text{Pyruvate} + \text{ATP} + CO_2 \longrightarrow \text{oxalacetate} + \text{ADP} + P_i$$

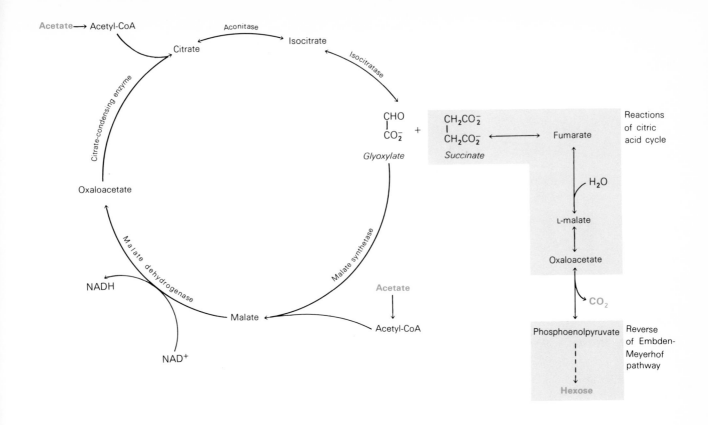

Sum after 2 turns of the cycle: 4 Acetate → 1 hexose + 2CO$_2$
(8C → 6C + 2C)

Figure 5.8   Reactions of the glyoxylate cycle leading to the synthesis of hexose from acetate.

whereas in others it is catalyzed by phosphoenolpyruvate carboxylase:

Phosphoenolpyruvate + CO$_2$ ⟶ oxalacetate + P$_i$

These reactions replace oxalacetate that is lost when compounds of the citric acid cycle are used in biosynthesis, and the cycle can continue to function.

Such reaction sequences as the glyoxylate cycle and those involving the carboxylating enzymes, which serve to supplement the catabolic reactions of the tricarboxylic acid cycle, are called *anaplerotic* (from the Greek for "filling up").

5.1   *Carbon metabolism and nutrition*

Fatty acid and hydrocarbon metabolism Fats are esters of glycerol and *fatty acids*. Some microorganisms produce enzymes called *lipases*, which hydrolyze the ester linkages between glycerol and the fatty acids (Figure 5.9), or *phospholipases*, which hydrolyze phospholipids (Figure 5.10). Certain bacterial phospholipases are of considerable practical importance as toxins in specific infectious diseases (see Chapter 15). The fatty acids released by lipase action are oxidized by a process called beta oxidation, in which two carbon atoms at a time are split off (Figure 5.11). The acetyl-CoA formed by β-oxidation is then oxidized by way of the citric acid cycle. The attack on lipids by microorganisms in foods is of considerable significance in the food industry, especially since many of the fatty acids liberated by microbial lipases have undesirable odors.

Fatty-acid synthesis occurs by the stepwise buildup of long chains of two-carbon fragments from acetyl-CoA, but involves an entirely different set of enzymes from those of fatty-acid breakdown (Figure 5.12). Initially, the acetyl group of acetyl-CoA is transferred to a small protein which

Figure 5.9   Mechanism of action of a lipase.

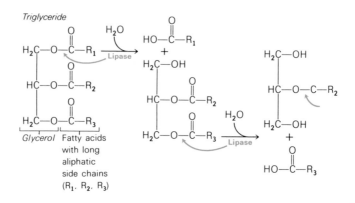

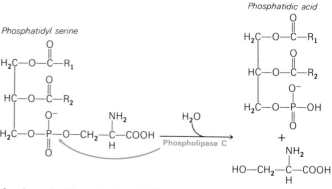

Figure 5.10   Mechanism of action of a phospholipase.

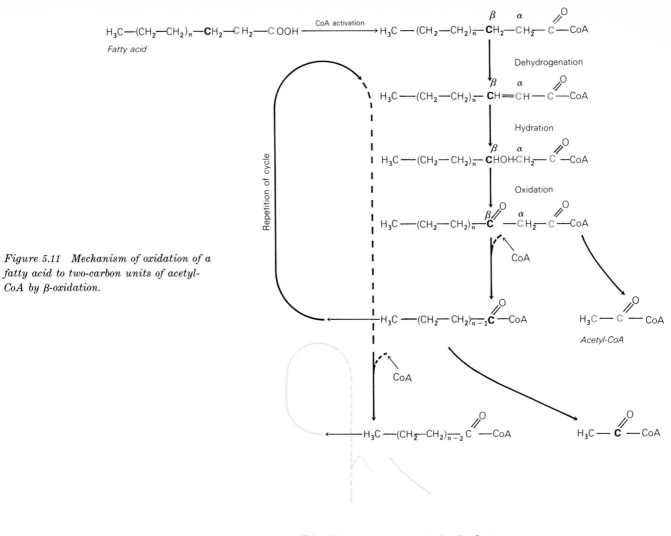

$$\text{H}_3\text{C}-(\text{CH}_2-\text{CH}_2)_n-\text{CH}_2-\text{CH}_2-\text{COOH} \xrightarrow{\text{CoA activation}} \text{H}_3\text{C}-(\text{CH}_2-\text{CH}_2)_{\overline{n}}-\overset{\beta}{\text{CH}_2}-\overset{\alpha}{\text{CH}_2}-\text{C}-\text{CoA}$$

Fatty acid

Dehydrogenation

Hydration

Oxidation

CoA

CoA

Acetyl-CoA

Repetition of cycle

Figure 5.11   Mechanism of oxidation of a fatty acid to two-carbon units of acetyl-CoA by β-oxidation.

Sum, where n = 6:

$$\text{H}_3\text{C}-(\text{CH}_2)_{12}-\text{CH}_2-\text{CH}_2-\text{COOH} \longrightarrow 8\ \text{H}_3\text{C}-\underset{\text{O}}{\text{C}}-\text{CoA}$$

$$[16\text{C} \longrightarrow 8(2\text{C})]$$

serves as a carrier, called *acyl carrier protein* (ACP). The acetyl-ACP is attached to the enzyme *fatty-acid synthetase* throughout the series of reactions in which successive two-carbon fragments are added and reduced. Interestingly, although the fatty-acid chain is increased two carbons at a time, the two carbons actually come from the three-carbon malonyl portion of the compound *malonyl-ACP*. Malonyl-ACP itself is synthesized from malonyl-CoA; the latter is derived from acetyl-CoA and

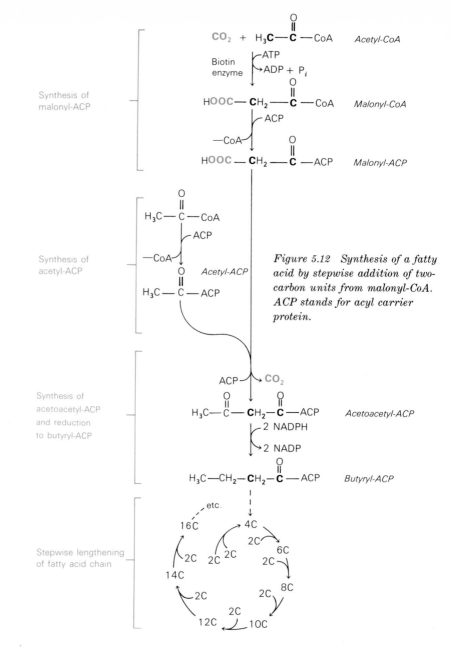

Figure 5.12 Synthesis of a fatty acid by stepwise addition of two-carbon units from malonyl-CoA. ACP stands for acyl carrier protein.

carbon dioxide in an enzymatic reaction dependent on the vitamin *biotin* (Figure 5.13). Malonyl-ACP can be viewed as an activated form of acetyl-ACP since the presence of the additional carboxyl makes the methyl group of acetyl-ACP more reactive. The buildup of a fatty acid proceeds

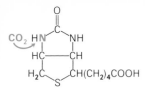

Figure 5.13 Biotin, the vitamin involved in activation of CO_2 in fatty-acid synthesis.

by the stepwise addition of two of the carbons of malonyl-ACP, with the release of the third carbon as CO_2 (Figure 5.12). The acetyl fragment added to the end of the chain is then reduced in a series of reactions involving NADPH. In the first turn of the cycle, butyryl-ACP is formed, and this then accepts a second acetyl unit from malonyl-ACP. The fatty-acid chain is thus lengthened two carbons at a time. It may be for this reason that the most common fatty acids found in organisms are those with even numbers of carbon atoms, as for instance palmitic acid ($C_{16}H_{32}O_2$), stearic acid ($C_{18}H_{36}O_2$), and oleic acid ($C_{18}H_{34}O_2$).

It is of interest that CO_2 through its involvement in malonic acid synthesis is an essential component in fatty-acid biosynthesis even though carbon from CO_2 does not appear in the final product. This explains a peculiar observation that was made years ago in studying the nutrition of microorganisms: CO_2 was found to be required for growth of certain nonphotosynthetic organisms, but this requirement could be replaced by a fatty acid, such as oleic acid. Also, biotin, which is a required growth factor for some microorganisms, and which is involved in the CO_2 reaction, could also be replaced by a fatty acid. Since there is a small amount of CO_2 in normal air, it is usually unnecessary to provide CO_2 for laboratory cultures. However, some pathogenic bacteria, such as *Neisseria meningitidis,* require higher concentrations of CO_2 than are present in normal air, and this must be taken into account when attempting to isolate such organisms.

Unsaturated fatty acids contain one or more double bonds. In most aerobic organisms the formation of the double bond involves a reaction requiring molecular oxygen. The exact mechanism of this reaction is not understood, but the requirement for O_2 explains why, when microorganisms such as yeast are cultured under anaerobic conditions, an unsaturated fatty acid, such as oleic acid, must be added to the medium. In anaerobic and some aerobic bacteria unsaturated fatty acids are synthesized through dehydration of a hydroxy acid, a process not involving O_2:

$$\underset{\underset{\displaystyle H \quad H \quad H}{|\quad\;\;|\quad\;\;|}}{\overset{\overset{\displaystyle H \quad OH \ H}{|\quad\;\;|\quad\;|}}{R-C-C-C-COOH}} \longrightarrow R-CH=CH-CH_2-COOH + H_2O$$

This difference in the manner in which unsaturated fatty acids are synthesized is thought to be of considerable evolutionary significance and will be discussed in this context in Chapter 17.

Other fatty acids found in microorganisms include branched chain fatty acids and fatty acids with a cyclopropane ring:

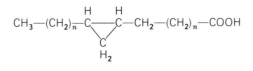

The significance of these unusual fatty acids is not understood.

Sterols are a class of complex ring compounds that are usually classified with lipids. As was noted in Section 3.2, sterols are found in eucaryotic but generally not in procaryotic organisms. The biosynthesis of the sterol skeleton is shown in Figure 5.14(a). These complex rings are built up by joining acetyl groups to form isopentenyl units through a complex series of steps. Molecular oxygen is required; hence sterols cannot be

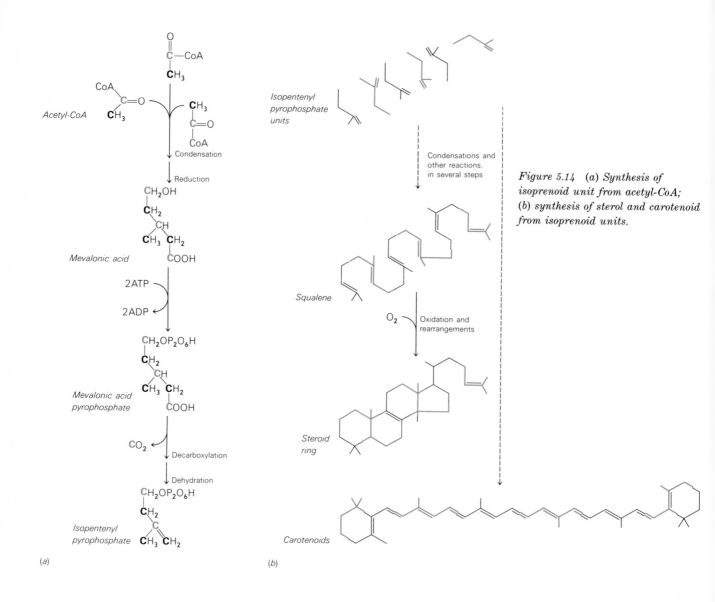

Figure 5.14 (a) Synthesis of isoprenoid unit from acetyl-CoA; (b) synthesis of sterol and carotenoid from isoprenoid units.

(a)

(b)

synthesized under anaerobic conditions; when yeast, for example, is grown without air, ergosterol or a related compound must be added to the medium. Isopentenyl units are also the precursors of carotenoids, yellow lipidlike pigments found in many organisms [Figure 5.14(b)].

Hydrocarbons are organic carbon compounds containing only carbon and hydrogen, which are highly insoluble in water. Low-molecular-weight hydrocarbons are gases, whereas those of higher molecular weight are liquids or solids at room temperature. Among hydrocarbons there is a tremendous variation in chain length, degree of branching, and number of double bonds. Only relatively few kinds of microorganisms (for example, *Nocardia, Pseudomonas, Mycobacterium,* and certain yeasts and molds) can utilize hydrocarbons for growth. This is strictly an aerobic process: in the absence of O_2, hydrocarbons are completely unaffected by microbes. This accounts for the fact that hydrocarbons have remained unchanged for millions of years as petroleum deposits, which are the richest natural sources of these compounds. The steps in the utilization of hydrocarbons are not completely known, but the ultimate products probably are acetate units which are then oxidized through the citric acid cycle. As soon as petroleum resources are brought to the surface and exposed to air, oxidation of the hydrocarbons begins and the materials are eventually decomposed (see Chapter 16). In recent years there has been considerable interest in growing microorganisms on hydrocarbons to produce microbial cells as ''food'' for animals or man.

Some microorganisms can synthesize long-chain hydrocarbons as well as break them down, but the mechanism of synthesis is not known. Perhaps they are derived from fatty acids by decarboxylation and reduction. The simplest hydrocarbon, methane, is synthesized in large amounts by one group of anaerobic bacteria, the methane bacteria, which use CO_2 as a terminal electron acceptor (see Section 4.5).

5.2   Amino acid metabolism

Amino acid utilization   When amino acids are used as carbon sources, usually the first transformation to take place is the removal of the amino group, converting the amino acid into an organic acid. Reactions that add or remove amino groups involve the coenzyme pyridoxal phosphate (Figure 5.15). The aldehyde group of this coenzyme reacts with the amino group, forming a compound the organic chemist calls a ''Schiff base.'' The amino group so activated can then be transferred to a keto acid or can be released as ammonia.

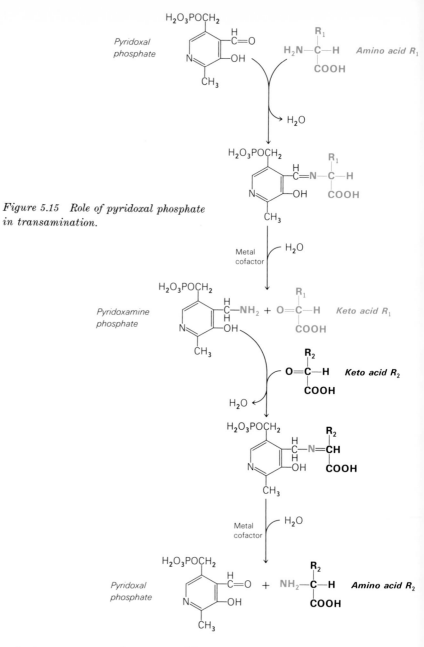

Figure 5.15 Role of pyridoxal phosphate in transamination.

Amino groups can be removed in a variety of ways. The simplest way, and probably the most common in microorganisms, is by *transamination,* in which the amino group is exchanged for the keto group of a keto acid

(Figure 5.16), yielding a different amino acid. Another way is through *dehydrogenation*, as is illustrated for glutamic acid. Dehydrogenation is an oxidation-reduction reaction involving the use of a coenzyme, usually NAD. In dehydrogenation the amino acid is likewise converted into a keto acid but, in contrast to transamination, the amino group is not conserved in another amino acid but is converted to ammonia. Since NADH is produced during dehydrogenation, this process is an energy-yielding reaction. Amino acids can also be utilized after *decarboxylation*, in which the carboxyl group is converted to CO_2 and the amino acid to an amine, or by *deamination*, in which the amino group is converted to ammonia and the amino acid to an organic acid. Since each of these ways of acting upon an amino acid requires the presence in the organism of specific

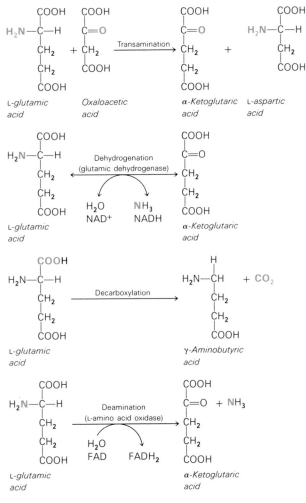

Figure 5.16 Amino acid transformations.

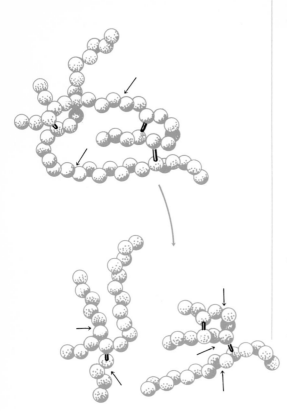

enzymes, different organisms may break down a single amino acid in different ways. However, a single organism may have more than one mechanism.

Protein utilization Proteins are usable only after they are broken down to small peptides or individual amino acids through the activity of enzymes called *proteases* (Figure 5.17). Some proteases specifically hydrolyze the peptide linkage between particular amino acids, whereas other proteases are nonspecific and hydrolyze any peptide bond. The amino acids released by protease action are utilized as carbon sources by one of the mechanisms described above.

Not all microorganisms can utilize proteins for growth. Among those that can are many pathogenic microorganisms such as streptococci and staphylococci, soil microorganisms such as bacilli, and food-spoilage organisms such as certain species of *Pseudomonas*. Some fungi can digest and grow on keratin, a specialized, highly insoluble protein that is the main component of animal hair; this property is often found among fungi that infect the skin (see Chapter 19).

Amino acid synthesis There are 20 amino acids common to proteins, and meeting the requirements for these constitutes a major problem for an organism. Those forms that cannot obtain some or all amino acids

Figure 5.17   Stepwise hydrolysis of a protein by protease.

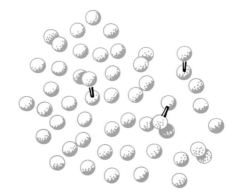

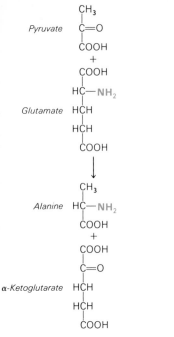

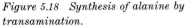

Figure 5.18 Synthesis of alanine by transamination.

preformed from the environment must synthesize them from nonamino sources. The amino acids in proteins are α amino acids in L configuration. An amino acid could be represented schematically as:

$$R—C—COOH$$

with the R group varying from one amino acid to another. Two problems are involved in the synthesis of the amino acids: the synthesis of the "carbon skeleton," and the synthesis and attachment of the amino group. In some cases, the carbon skeleton is synthesized first and the amino group added afterward, usually by transamination, as is shown in the synthesis of alanine from pyruvate (Figure 5.18). Amino acids whose carbon skeletons are derived directly from intermediates of the citric acid cycle are aspartic acid (from oxalacetate) and glutamic acid (from α-ketoglutarate). These two amino acids, in turn, are precursors of some of the other amino acids. In the glutamate family are proline and arginine, while in the aspartate family are threonine, methionine, lysine, and isoleucine (Figure 5.19). The benzene ring of the aromatic amino acids

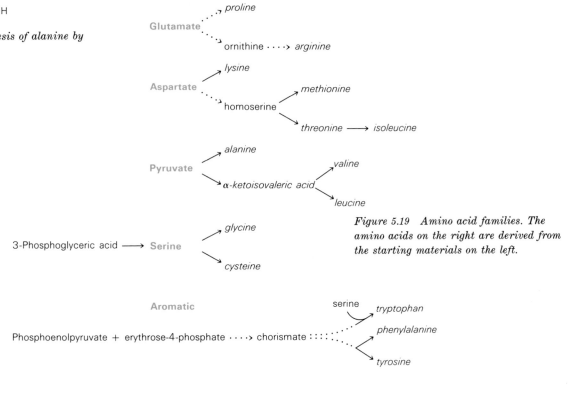

Figure 5.19 Amino acid families. The amino acids on the right are derived from the starting materials on the left.

tyrosine and phenylalanine is derived from the sugar erythrose-4-phosphate and phosphoenolpyruvic acid. Tryptophan, which has an indole nucleus, is derived from the aromatic compound anthranilic acid.

Although most amino acids are synthesized by the same pathways in all organisms, *lysine* is unique in that some organisms use one pathway and others use an entirely different one. The evolutionary significance of this is discussed in Chapter 17. These two alternative routes to lysine, called the α-aminoadipic pathway and the diaminopimelic pathway, are illustrated in Figure 5.20. Interestingly, the starting material for both pathways is the same, α-ketoglutaric acid.

Figure 5.20   Comparison of the α-aminoadipic and the diaminopimelic acid pathways for the synthesis of lysine. For an outline of the organisms using each pathway, see Table 17-7.

α-Aminoadipic pathway	Diaminopimelic pathway
α-Ketoglutaric	α-Ketoglutaric
Homocitric	Aspartic
Homoaconitic	Aspartic-β-semialdehyde
Homoisocitric	Piperidine-2,6-dicarboxylic acid
Oxaloglutaric	2,3-Dihydrodipicolinic acid
α-Ketoadipic	N-succinyl-ε-keto-L-α-aminopimelic acid
α-Aminoadipic	N-succinyl-L-α,ε-diaminopimelic acid
α-Aminoadipic-ε-semialdehyde	

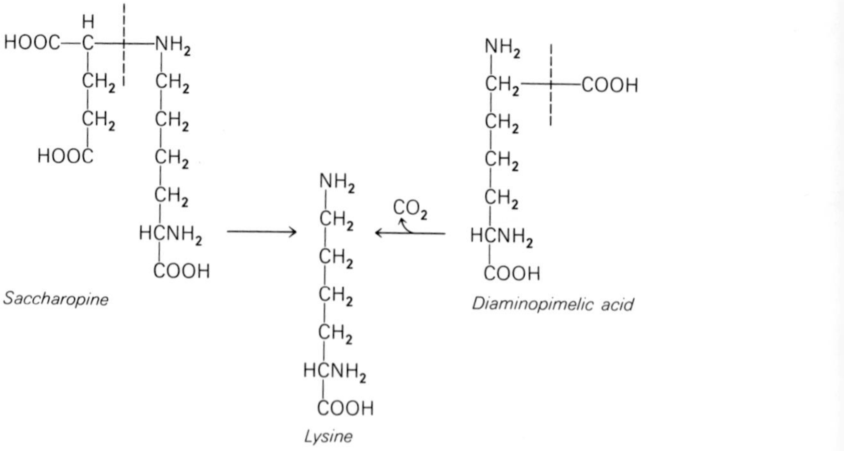

Saccharopine

Diaminopimelic acid

Lysine

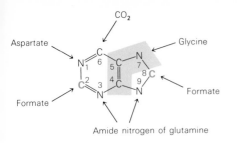

Figure 5.21 The basic precursors of the purine skeleton.

D amino acids   Although the amino acids of proteins are in L configuration, procaryotes produce D amino acids as components of their cell walls. The synthesis of D-alanine has been studied in some detail and we know that it is synthesized from L-alanine by the enzyme alanine racemase; pyridoxal phosphate is a cofactor and a probable transient intermediate is pyruvate.

The utilization of a D amino acid involves the activity of D-amino acid oxidase, a flavin-containing enzyme that causes the formation of an aldehyde, ammonia, and CO_2. The aldehyde is then oxidized to an organic acid and processed through a pathway handling organic acids.

5.3   Purine and pyrimidine metabolism

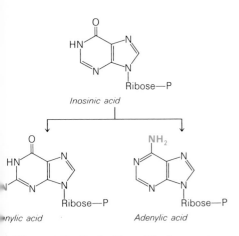

Figure 5.22 Derivation of the key purines guanine and adenine.

Synthesis   Purines and pyrimidines are components of nucleic acids, as well as of many vitamins and coenzymes (for example, ATP, NAD, thiamine). The *purine* ring is built up almost atom by atom, using carbons and nitrogens derived from amino acids, CO_2, and formate, which are added stepwise to a ribose-phosphate starting material (Figure 5.21). The first purine ring formed is that of inosinic acid, which serves as an intermediate for the formation of the two key purine derivatives, adenylic acid and guanylic acid, as is shown in Figure 5.22.

The origin of the atoms of the *pyrimidine* ring is shown in Figure 5.23. In contrast to the synthesis of the purine ring, a complete pyrimidine ring (orotic acid) is built up before the sugar is added. After the addition of ribose and phosphate to orotic acid, the other pyrimidines are successively synthesized (Figure 5.24).

Although most animals and man can synthesize both purine and pyrimidine rings, many microorganisms, such as the lactic acid bacteria, lack this ability. These purine- and pyrimidine-requiring organisms often derive their nutrients from animals or animal products such as milk. So many environments provide purines and pyrimidines that it is not surprising to find an organism that has dispensed with its own synthesizing mechanism.

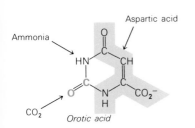

Figure 5.23 Basic precursors of the pyrimidine skeleton.

Utilization   There are a number of microorganisms that can utilize the nucleic acids as sources of carbon, nitrogen, and energy. Initially, the nucleic acids are hydrolyzed to nucleotides by nucleases. There are many nucleases; one group is specific for RNA and another for DNA, but even within these groups there are nucleases which show different specificities. For instance, one ribonuclease will hydrolyze only the ribose-phosphate connected to a purine, another preferentially catalyzes the hydrolysis of sugar-phosphate linkages attached to pyrimidines, and a third is even more specific and acts only on internucleotide linkages adjacent

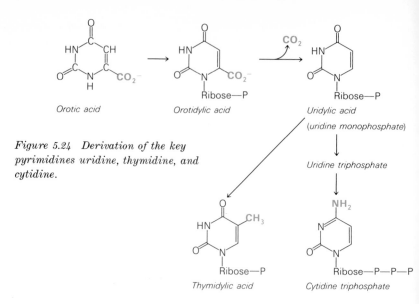

Orotic acid   Orotidylic acid   Uridylic acid
(uridine monophosphate)

Uridine triphosphate

Thymidylic acid   Cytidine triphosphate

Figure 5.24   Derivation of the key pyrimidines uridine, thymidine, and cytidine.

to the purine guanine. These nucleases are extracellular enzymes and can act on nucleic acids at a distance from the organism. This makes it possible for the organism to utilize high-molecular-weight nucleic acids that cannot pass through the cell membrane; only the low-molecular-weight nucleotide hydrolysis products need penetrate the cell. The pathogenic pus-forming bacteria of the genera *Streptococcus* and *Staphylococcus* produce nucleases in large amounts. In the pimple or boil where these organisms grow, there is a large amount of tissue destruction, and through the production of nucleases these bacteria are able to use as nutrients the nucleic acids liberated from the dead cells (see Chapter 15). Some soil bacteria such as certain *Bacillus* species also produce large amounts of nucleases.

The purine and pyrimidine bases liberated by the hydrolysis of nucleic acids can be used directly as carbon, nitrogen, and energy sources by certain bacteria. The degradation of the bases occurs not by a reversal

Figure 5.25   Degradation of a purine ring.

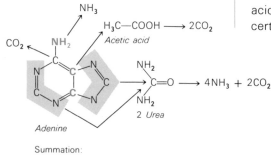

Adenine

Summation:
Adenine $\longrightarrow$ 5NH$_3$ + 5CO$_2$

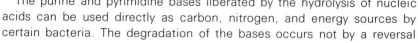

β-Alanine

Uracil

Summation:
Uracil $\longrightarrow$ 2NH$_3$ + 4CO$_2$

Figure 5.26   Degradation of a pyrimidine ring.

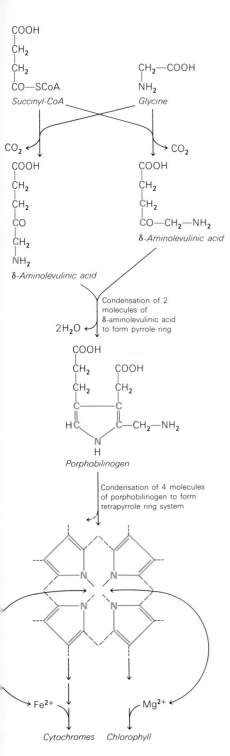

COOH
|
CH$_2$
|
CH$_2$
|
CO—SCoA
Succinyl-CoA

CH$_2$—COOH
|
NH$_2$
Glycine

CO$_2$

CO$_2$

COOH
|
CH$_2$
|
CH$_2$
|
CO
|
CH$_2$
|
NH$_2$
δ-Aminolevulinic acid

COOH
|
CH$_2$
|
CH$_2$
|
CO—CH$_2$—NH$_2$
δ-Aminolevulinic acid

2H$_2$O

Condensation of 2 molecules of δ-aminolevulinic acid to form pyrrole ring

COOH
|
CH$_2$ COOH
| |
CH$_2$ CH$_2$
| |
C   C
|| ||
HC   C—CH$_2$—NH$_2$
\   /
N
|
H
Porphobilinogen

Condensation of 4 molecules of porphobilinogen to form tetrapyrrole ring system

Fe^{2+} Mg^{2+}

Cytochromes   *Chlorophyll*

of the biosynthetic pathways but by entirely different routes. The end products of purine degradation are NH$_3$ and CO$_2$; urea and acetic acid are intermediates (Figure 5.25). The end products of pyrimidine degradation are also NH$_3$ and CO$_2$, but the intermediate is β-alanine (Figure 5.26).

5.4 Porphyrin ring

The porphyrin ring is found in pigments involved in energy-generating systems: the cytochromes (see Chapter 4) and the chlorophylls (see Chapter 6). Porphyrin is built up of four pyrrole units connected in a ring (the *tetrapyrrole ring*). Different porphyrins are formed by modifications of the side chains of the ring.

The biosynthesis of pyrrole and the tetrapyrrole ring is shown in Figure 5.27. The starting materials, succinyl-CoA and glycine, condense to form δ-aminolevulinic acid, two molecules of which condense, making porphobilinogen, a pyrrole. Four molecules of porphobilinogen condense into the tetrapyrrole ring system, and subsequent steps add side chains. The addition of a metal ion to the center of the ring (Mg^{2+} for chlorophyll and Fe^{2+} for heme) occurs near the end of the sequence. So far as is known, the early steps in porphyrin biosynthesis are the same in all organisms. As we will see in Chapter 17, the synthesis of the porphyrin ring was a significant event in early evolution and, by making photosynthesis possible, has had profound effects on subsequent evolutionary events.

5.5 Control of biosynthetic versus degradative processes

In the preceding pages we have discussed both the biosynthesis and the utilization of carbohydrates, organic acids, fatty acids, amino acids, purines, and pyrimidines. It may have been noted that in all cases the pathway for the degradation of each substance was different from a mere reversal of the pathway for biosynthesis. Different intermediates and different enzymes were involved. This is perhaps one of the most surprising and interesting discoveries to arise from studies on microbial metabolism. By utilizing different pathways for synthesis and for breakdown, an organism is able to keep its signals straight. Imagine the dilemma of a

Figure 5.27 Steps in the synthesis of the pyrrole ring and the porphyrin ring.

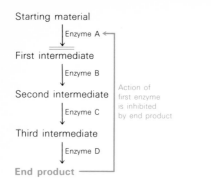

Starting material
  | Enzyme A
First intermediate
  | Enzyme B
Second intermediate
  | Enzyme C   Action of
                      first enzyme
Third intermediate   is inhibited
                      by end product
  | Enzyme D
End product

Figure 5.28 Feedback inhibition. The activity of the first enzyme of the pathway is inhibited by the end product, thus controlling production of end product.

cell using glutamic acid as a sole carbon source if it followed the same pathway for the synthesis and breakdown of this amino acid: Since glutamic acid is a major constituent of all proteins, it must be available in considerable amounts for protein synthesis; yet at the same time glutamic acid must continually be broken down if it is to produce the energy needed for growth. The difficulty is avoided if separate pathways are used for breakdown and synthesis. A further means of control is by using the two different coenzymes NAD and NADP, the former being involved in degradative reactions and the latter in biosynthetic reactions.

Another problem which an organism has is that of controlling the rate of synthesis of the various amino acids, purines, pyrimidines, and other constituents that serve as the building blocks of cellular structures. Some of these building blocks are required in larger amounts than others, and ideally the rate of synthesis should balance the rate of utilization, so that excess amounts of these materials are not built up. One mechanism of control is *feedback inhibition:* The amino acid or other end product of a biosynthetic pathway inhibits the activity of the first enzyme in this pathway. Thus, as the end product builds up in the cell, its further synthesis is inhibited. If the end product is used up, however, synthesis can resume (Figure 5.28). In cases in which two amino acids are derived from a common precursor each exerts feedback inhibition on the first enzyme

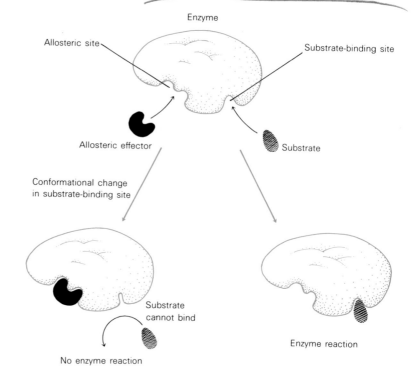

Figure 5.29 Mechanism of enzyme inhibition by allosteric effector. When the effector combines with the allosteric site, the conformation of the enzyme is altered so that the substrate can no longer bind.

Enzyme

Allosteric site   Substrate-binding site

Allosteric effector   Substrate

Conformational change
in substrate-binding site

Substrate
cannot bind

Enzyme reaction

No enzyme reaction

that is unique to its own pathway. In this way, neither amino acid interferes with the synthesis of the other.

How is it possible for the end product to inhibit the activity of an enzyme that acts on a compound quite unrelated to it? This occurs because of a property of the inhibited enzyme known as *allostery*. An allosteric enzyme has two important sites, the active site, where the substrate binds, and the allosteric site, where the inhibitor (sometimes called an ''effector'') binds. An inhibitor binding at the allosteric site leads to a change in structure or conformation of the enzyme molecule so that the substrate no longer binds efficiently at the active site (Figure 5.29). Allosteric enzymes are very common in both biosynthetic and degradative pathways, and usually occur at branch points in the pathways. It is assumed that such enzymes have developed because efficiently controlling the rates of enzyme activity is of advantage to the organism in enabling it to adapt to changing environments.

5.6   *Inorganic nitrogen metabolism*

Nitrate and ammonia   Nitrogen can be obtained by microorganisms from either inorganic or organic forms. The most common inorganic nitrogen sources are nitrate and ammonia. One way in which *ammonia* is utilized is by means of the reversible action of glutamic dehydrogenase (illustrated in Figure 5.16). Glutamic acid, once formed, can then serve as an amino donor for the synthesis of other organic nitrogen compounds, and thus is a key intermediate in the nitrogen transformations of the cell. It can, through transamination reactions (see Section 5.2), donate the amino group for the synthesis of all other amino acids that are required by the cell.

When *nitrate* is utilized as a nitrogen source, it is first reduced to ammonia. The first step in this process is the reduction of nitrate to nitrite by the enzyme nitrate reductase, a molybdenum-containing flavoprotein. Although subsequent steps in the conversion of nitrite to ammonia are not completely understood, we know that the final product, ammonia, is assimilated with the mediation of glutamic dehydrogenase. The reduction of nitrate to ammonia discussed here is called *assimilatory* nitrate reduction, in contrast to *dissimilatory* nitrate reduction (Section 4.5), the process occurring in denitrifying bacteria, in which the function of nitrate is that of an alternative electron acceptor to O_2 in oxidative phosphorylation. In assimilatory nitrate reduction phosphorylation does not occur. Different enzymes are involved in assimilatory and dissimilatory reductions, yet another example of the distinction between degradative and biosynthetic reactions.

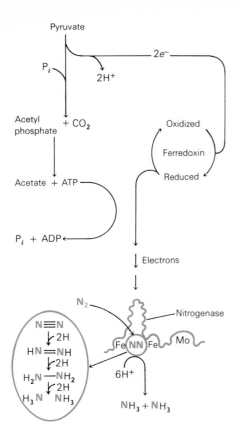

Overall: $6(H) + N_2 \rightarrow 2NH_3$

$ATP \rightarrow ADP + P_i$

Figure 5.30 Nitrogen fixation: steps in reduction of N_2 to NH_3.

Nitrogen fixation   Nitrogen gas (N_2) can be used as a nitrogen source by only certain genera of bacteria and blue-green algae—and possibly by some fungi. Gaseous nitrogen is utilized by a process called *nitrogen fixation*, which requires a considerable amount of energy since N_2 is very stable. Nitrogen is activated by an enzyme called nitrogenase, which contains molybdenum. That molybdenum is essential for nitrogen fixation is shown by the fact that nitrogen-fixing bacteria require molybdenum when they are using N_2 as a nitrogen source but not when they are using NH_3. The electrons for the reduction of N_2 come from NADPH, and are first transferred to a nonheme iron protein called *ferredoxin*, which has an oxidation-reduction potential of -0.317 (almost the same as that of NADPH). Electrons from ferredoxin are then transferred to nitrogenase and ultimately to N_2 (Figure 5.30). The first stable product of N_2 reduction is NH_3, which can then be converted to organic nitrogen through reactions such as those involving glutamic dehydrogenase, discussed above. Nitrogen oxides, especially N_2O, are powerful inhibitors of nitrogen fixation; N_2O competes with N_2 for its specific site on nitrogenase.

It has been recently found that nitrogenase is not specific for N_2 but will also reduce cyanide [$(CN)^-$], acetylene ($HC{\equiv}CH$), and several other compounds. The reduction of these compounds probably serves no useful purpose to the cell but provides the experimenter with a simple way of measuring the activity of nitrogen-fixing systems. It is fairly easy to measure the reduction of acetylene to ethylene ($H_2C{=}CH_2$), and this technique is now used to detect nitrogen fixation in unknown systems. Previously, it was not easy to prove that an organism fixed N_2; indeed, many claims for nitrogen fixation in microorganisms have been erroneous. The growth of an organism in a medium to which no nitrogen compounds have been added does not mean that the organism is fixing nitrogen from the air since traces of nitrogen compounds often occur as contaminants in the ingredients of culture media or drift into the media in gaseous form or as dust particles. Even distilled water may be contaminated with ammonia. One method of proving nitrogen fixation is to show a net increase in the total nitrogenous content of the medium plus its organisms after incubation; the assumption would be that the increased nitrogen could come only from N_2 from the air. A more sensitive procedure is to use the isotope of nitrogen, ^{15}N, as a tracer.* The gas phase of a culture is enriched with ^{15}N, and after incubation the cells and medium are digested, the ammonia produced being distilled off and assayed for its ^{15}N content. If there has been a significant production of ^{15}N-labeled NH_3, it is proof of nitrogen fixation. However, the acetylene-reduction method

---

* [15]N is not a radioactive isotope but a stable isotope, and its presence must be detected with the mass spectrometer, an expensive and rather cumbersome instrument.

mentioned above is an even more sensitive way of measuring nitrogen fixation and is rapidly replacing the more difficult ^{15}N method.

Table 5.1 lists a variety of organisms that have been shown definitely to fix nitrogen. In soil the most important nitrogen-fixing organisms are anaerobes, the clostridia. In aquatic environments nitrogen fixation by blue-green algae is probably of greatest significance. Symbiotic nitrogen fixation, which involves the joint participation of bacteria and plants, is discussed in Chapter 14.

Many microorganisms can utilize more than one kind of nitrogen source. When several nitrogen sources are presented to a microorganism at the same time, one of them is usually preferred to the others. For example, amino acids are broken down before ammonia, and the latter before nitrate. Nitrogen-fixing organisms use N_2 only if there is no other utilizable nitrogen source. Experimentation has shown that a microorganism generally uses first the nitrogen sources on which it must do the least amount of work, and only after the easily utilizable nitrogen sources are consumed does the microorganism turn to substances more difficult to assimilate.

The nitrogen content of different environments varies considerably, both in quantity and in kinds of compounds. In general, the nitrogen requirements of specific organisms reflect the environment in which they are

Table 5.1   Nitrogen-fixing organisms

Symbiotic N$_2$-fixing agents		Free-living N$_2$-fixing agents			
		Aerobes		*Anaerobes*	
Leguminous plants	*Nonleguminous plants*	*Heterotrophs*	*Photoautotrophs*	*Heterotrophs*	*Photoautotrophs*
Many genera of the subfamilies Mimosoideae, Papilionatae, Caesalpinioideae in association with a bacterium of the genus *Rhizobium*	*Alnus, Elaeagnus, Myrica, Ceanothus, Comptonia, Shepherdia, Hippophaë, Coriaria, Casuarina, Discaria* in association with a filamentous microorganism, probably an actinomycete *Psychotria* in association with the bacterium *Klebsiella* Many lichens in association with blue-green algae	Bacteria: *Azotobacter* spp., *Azotomonas, Pseudomonas* spp., *Spirillum, Klebsiella* spp., *Pseudomonas methanitrificans, Beijerinckia, Nocardia* spp. Yeasts: *Rhodotorula, Pullularia*	Blue-green algae (heterocyst-formers only): *Nostoc, Calothrix, Anabaena, Fischerella, Mastigocladus, Chlorogloea, Tolypothrix, Stigonema, Scytonema, Cylindrospermum*	Bacteria: *Clostridium* spp. *Aerobacter aerogenes,* *Bacillus polymyxa, Desulfovibrio desulfuricans, Achromobacter* spp.	Bacteria: *Chromatium, Chlorobium, Rhodospirillum rubrum, Rhodomicrobium vanniellii, Rhodopseudomonas* spp., *Methanobacterium*

living. In aquatic environments nitrogen is low and is mainly in the form of nitrate and ammonia. It is not surprising therefore that aquatic organisms generally grow well on media with low nitrogen content and do not require amino acids. The animal body is high in nitrogen and in amino acids; bacteria isolated from the animal often need amino nitrogen and cannot use nitrate or ammonia. This is merely a general rule; there are exceptions. The ecological significance of nitrogen fixation and other nitrogen transformations carried out by microorganisms is discussed in Chapter 16.

5.7 Sulfur metabolism

Although the two sulfur-containing amino acids cysteine and methionine are utilized as sulfur sources by many microorganisms, most microorganisms are also able to use inorganic sulfate (SO_4^{2-}) as the sole source of sulfur to synthesize not only these amino acids but also the sulfur-containing vitamins (thiamine, biotin, and lipoic acid). The assimilation of sulfate first involves its activation, using energy from ATP (Figure 5.31). Subsequently, the sulfate radical attached to phosphoadenosine phosphosulfate (PAPS) is reduced to sulfite (SO_3^{2-}) by an enzyme that uses NADPH as the electron donor. Sulfite is then reduced to hydrogen sulfide (H_2S) by another enzyme using NADPH. The conversion of H_2S to organic sulfur occurs by reaction with the amino acid serine (Figure 5.32). Other organic sulfur compounds can later be synthesized from the reduced sulfur of cysteine. Mammals cannot use sulfate as a sole sulfur source because they lack the enzymes for sulfate reduction; thus they must obtain their organic sulfur preformed in their foodstuffs.

Sulfate-reducing bacteria (for example, *Desulfovibrio*), which use sulfate as a terminal electron acceptor (Section 4.5), carry out dissimilatory sulfate reduction, producing H_2S as the end product. In these bacteria the

Figure 5.31 Steps in activation and reduction of sulfate to H_2S.

1 Synthesis of active sulfate:

$$ATP + SO_4^{2-} \rightleftharpoons \text{adenosine phosphosulfate} + PP_i$$
$$ATP + \text{adenosine phosphosulfate} \longrightarrow \text{phosphoadenosine phosphosulfate} + ADP$$

2 Reduction of active sulfate to sulfite:

NADPH NADP

$$\text{Phosphoadenosine phosphosulfate} \longrightarrow \text{phosphoadenosine phosphate} + SO_3^-$$

3 Reduction of sulfite to hydrogen sulfide:

NADPH NADP

$$SO_3^- \longrightarrow H_2S + H_2O$$

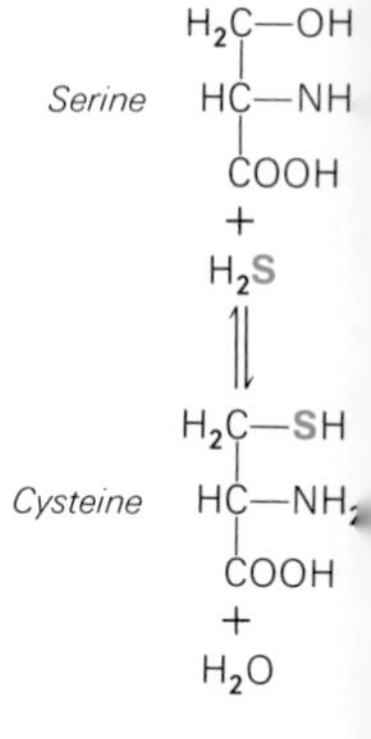

Figure 5.32 Production of organic sulfur from H_2S.

reduction of sulfate occurs with adenosine phosphosulfate (APS) rather than PAPS. This is another example of the concept presented above (see Section 5.5) that synthetic and degradative processes in microorganisms proceed by alternative pathways.

Most terrestrial and aquatic environments are high in sulfate; sulfur-deficient environments are quite rare in nature.

5.8 Phosphorus metabolism

Phosphorus occurs in nature in the form of organic and inorganic phosphates and is utilized by microorganisms primarily for synthesizing nucleotides and nucleic acids. Probably most or all microorganisms utilize inorganic phosphate for growth. One of the most important reactions in the conversion of inorganic to organic phosphate is the glyceraldehyde phosphate dehydrogenase reaction of the Embden-Meyerhof pathway, which was discussed in Section 4.3.

Since inorganic phosphate readily forms precipitates with calcium, magnesium, iron, and other minerals, much of the phosphate in natural environments is in insoluble forms and hence is unavailable to organisms. Lack of phosphate is more likely to limit microbial growth in aqueous environments than is any other factor.

Organic phosphate compounds occur very often in nature, and they are utilized as phosphate sources through the action of *phosphatases,* which hydrolyze the organic phosphate ester. Two types of enzymes are known, acid and alkaline phosphatases, distinguished by the optimum pH for activity. Phosphatases, which are present in nearly all organisms, are often localized between the cell wall and membrane in a region called the *periplasm* (Figure 5.33). Thus situated, they are in an excellent position to act on external phosphates.

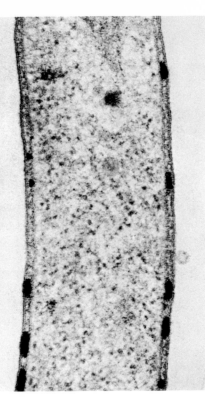

Figure 5.33   Electron micrograph showing localization of a phosphatase, the enzyme ATPase, at the cell periphery of Myxococcus xanthus. Before preparation for electron microscopy the cells were incubated with ATP and a lead (Pb+) compound. The phosphate liberated by the ATPase combined with the Pb+ to form insoluble and electron-dense lead phosphate. The dark bodies, which are sites of lead phosphate accumulation, are inferred to be the sites of ATPase localization. The same technique can be used to locate other phosphatases. Magnification, 78,000×. [From H. Voelz and R. O. Ortigoza: J. Bacteriol. 96:1357 (1968).]

5.9 Mineral nutrition

A variety of minerals are required by organisms for growth. These can be separated into two groups, macronutrient minerals and micronutrient minerals or trace elements. Potassium, magnesium, calcium, and iron are the most common *macronutrient minerals.*

Potassium is universally required. A variety of enzymes, including some of those involved in protein synthesis, are specifically activated by potassium. In laboratory cultures, potassium can be replaced by rubidium, its heavier relative in the periodic table, but not by sodium, its lighter relative.

Magnesium functions to stabilize ribosomes, cell membranes, and nucleic acids, and this element is also required for the activity of many enzymes. Thus relatively large amounts of magnesium are required for growth. Interestingly, Gram-positive bacteria require about 10 times more magnesium than do Gram-negative bacteria.

Calcium ions play a key role in the heat stability of bacterial spores (see Chapter 8) and may also be involved in the stability of the cell wall. Even though calcium and magnesium are closely related in the periodic table, calcium cannot replace magnesium in many of its roles in the cell.

Iron is required by virtually all organisms. It is essential in the cytochromes and in ferredoxin as an electron carrier in oxidation-reduction.

Sodium is required by some but not all organisms, and its need may reflect the environment; for example, seawater has a high sodium content, and marine microorganisms generally require sodium for growth, whereas closely related fresh-water forms may be able to grow in the complete absence of sodium.

The element *silicon* is required in the form of silicate by one specific group of algae, the diatoms, whose cell walls are composed of silica, a form of glass.

The *trace-element* requirements of microorganisms are difficult to determine experimentally. Even though the concentration of an element in the culture medium may have been reduced to such an extent that it can no longer be assayed chemically it may still be present in sufficient quantity to meet an organism's requirement for growth. Contaminating amounts of trace elements may come from the glassware, distilled water, culture ingredients, and even from a cotton plug. Often, only after the glassware has been scrupulously cleaned and the culture ingredients highly purified is it possible to demonstrate a trace-element requirement. The trace elements commonly required by most microorganisms are zinc, copper, cobalt, manganese, and molybdenum. These metals function in enzymes or coenzymes. The trace element *cobalt* is needed only for the formation of vitamin B_{12}, and if this vitamin is added to the medium, cobalt may no longer be needed. *Zinc* plays a structural role in certain enzymes in that it helps hold together protein subunits in the proper configuration for enzyme activity. *Molybdenum* and *copper* play an oxidation-reduction role in certain enzymes. *Manganese* is an activator of many enzymes that act on phosphate-containing compounds, substituting in these enzymes for magnesium; however, manganese probably has other functions in enzyme catalysis.

''Growth factors'' are specific organic compounds that are required in very small amounts and which cannot be synthesized by the cell. Substances frequently considered growth factors are vitamins, amino acids, purines, pyrimidines, and a few other organic compounds.

Vitamins   The first growth factors to be discovered and studied in any detail are those that we now call vitamins. Table 5.2 lists commonly needed vitamins and their functions. Most vitamins function in living organisms in forming coenzymes; for instance, the vitamin nicotinic acid is a part of the coenzyme NAD (Figure 5.34). Some microorganisms have very complex vitamin requirements, whereas others require only a few vitamins. The lactic acid bacteria, which include the genera *Streptococcus* and *Lactobacillus,* are renowned for their complex vitamin requirements, which are even greater than those of man. Cobalamin (Vitamin B$_{12}$) is often required by aquatic organisms, bacteria as well as algae. Over half of the algae tested, both fresh-water and marine, require some form of

Table 5.2   Commonly required vitamins

	Function
p-Aminobenzoic acid	Precursor of folic acid
Folic acid	One-carbon metabolism; methyl group transfer
Biotin	Fatty acid biosynthesis; β-decarboxylations; CO$_2$ fixation; CO$_2$ release
Cobalamin (B$_{12}$)	Reduction of and transfer of single carbon fragments
Lipoic acid	Transfer of acyl groups in decarboxylation of pyruvate and α-ketoglutarate
Nicotinic acid (niacin)	Precursor of NAD; electron transfer in oxidation-reduction reactions
Pantothenic acid	Precursor of CoA; activation of acetyl and other acyl derivatives
Riboflavin	Electron transport
Thiamine (B$_1$)	α-Decarboxylations; transketolase
Vitamin B$_6$ (pyridoxal-pyridoxamine group)	Amino acid and keto acid transformations
Vitamin K group; quinones	Probably involved in electron transport between NAD and cytochrome c

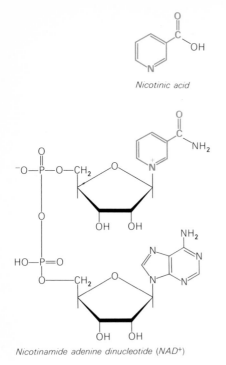

Nicotinic acid

Nicotinamide adenine dinucleotide (NAD⁺)

Figure 5.34 Relationship of the vitamin nicotinic acid to the coenzyme NAD.

vitamin B_{12}, and often thiamine and biotin as well. Fungi, on the other hand, never require vitamin B_{12}, and do not synthesize it; the reactions in which B_{12} are involved in bacteria are carried out by different pathways in fungi.

Other growth factors Many microorganisms require specific amino acids. Inability to synthesize an amino acid is related to the lack of the enzymes that are needed for its synthesis. The required amino acids can usually be supplied either as the free amino acids or in small peptides. When the peptide enters the cell, it is hydrolyzed by a peptidase to the component amino acids.

Purines and pyrimidines, which are the building blocks of nucleic acids and coenzymes, are growth factors that are essential for a number of microorganisms.

Some organisms require the porphyrin ring or one of its derivatives. Recall that chlorophyll, the cytochromes, and the hemoglobin of animals all contain porphyrins. A porphyrin requirement explains why certain pathogenic organisms (for example, *Hemophilus influenzae*) can be cultured only on media containing red blood cells, since one of the major constituents of red cells is hemoglobin, which contains the porphyrin heme. Heme-requiring bacteria lack the ability to synthesize porphyrins and hence are restricted to environments where blood is present. Since a porphyrin is needed for the synthesis of the cytochrome system, the respiratory apparatus of these bacteria becomes deranged when they are cultured in the absence of heme.

Another interesting example of a growth-factor requirement concerns the bacteria that live in the rumen of cows, sheep, and other ruminants (see Chapter 14). Many of these organisms were difficult to cultivate in the laboratory until it was found that they required specific growth factors present in rumen fluid. These growth factors were subsequently identified as 4-carbon to 6-carbon branched and straight-chain fatty acids. As far as is known, these growth factors are required only by rumen organisms.

Even organisms that do not require growth factors may use them if they are present in the environment. Thus if *E. coli* is provided with amino acids in the culture medium, it ceases to make these amino acids until the external material is gone (see Chapter 7).

Microbiological assay of growth factors Microorganisms are very useful tools for the assay of small amounts of vitamins, amino acids, and other growth factors and hence have been widely employed for the examination of foods, pharmaceuticals, and other preparations. A microbiological assay has the virtues of specificity, sensitivity, and simplicity. It also makes possible the assay of compounds whose chemical properties are unknown. To perform the assay, a culture medium is used in which all substances needed by the microorganism for growth are supplied, except for the substance to be assayed. Under these conditions the amount of growth obtained after incubation is proportional to the concentration of the limiting growth factor (Figure 5.35).

Microbiological assays were formerly used for amino acids, but modern chemical methods are as sensitive and are more rapid. Unlike microbiological assays, however, chemical assays do not distinguish between the D and L optical isomers of an amino acid. Some vitamins (for example, thiamine, riboflavin) can also be assayed more conveniently by chemical means, but vitamins such as B_{12} and biotin, which are active at extremely low concentrations, must still be assayed microbiologically. Microbiological assays using marine bacteria have recently been introduced for the assay of the vitamin content of sea water. Sea water contains only minute traces of vitamins, the existence of which was unsuspected until microbiological assays were used. Indeed, until vitamins were discovered in sea water it was difficult to explain why many vitamin-requiring microorganisms could survive in the sea.

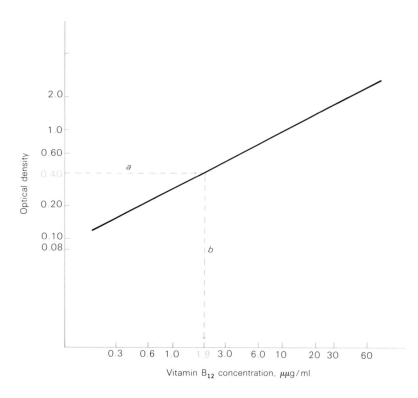

Figure 5.35 *Relationship between vitamin B_{12} concentration in the culture medium and growth of a vitamin B_{12}-requiring organism. This figure illustrates the construction of a standard curve from which the B_{12} concentration of an experimental sample can be calculated, as shown by the lines a and b. The optical density of the culture is proportional to the amount of growth.*

5.11 Nutrient media for different organisms

In general, the nutrient requirements of microorganisms reflect the natural environments in which they live. Thus, in developing a culture medium for a newly isolated microorganism, the experimenter is often aided by a knowledge of the habitat of an organism. However, a culture medium does not duplicate the natural environment since the goal of laboratory culture is to obtain for experimental work populations of cells usually much larger than those found in nature. During the past 50 to 75 years there has been an enormous amount of research on the development of culture media for microorganisms, and literally thousands of formulas have been devised. Some culture media are general-purpose formulations, suitable for a variety of organisms, whereas others are especially tailored to individual species. In addition to the necessary nutrients, a culture medium also contains a buffer to control the pH. The development of new culture media in many cases led to the discovery of hitherto unsuspected growth factors. If a synthetic medium that contains all known growth factors does not support growth of an organism, some unknown factor is probably missing. A natural material such as a plant or animal extract may then be added in searching for this unknown factor. If growth is restored, the extract can be fractionated chemically and the new factor isolated and its structure determined. Among the factors discovered in this way are biotin, lipoic acid, and pantothenic acid; first discovered as essential for certain microorganisms, these were later found to be growth factors for certain higher organisms also.

5.12 Permeability and nutrition

As was discussed in Chapters 2 and 3, the cell membrane is a barrier through which solutes pass in either direction. Since nutrients must enter the cell through the membrane, the permeability properties of the membrane are important in cell nutrition. The penetration of the membrane by a compound can occur in either of two ways, by passive uptake or by active transport (Figure 5.36). In passive uptake, the compound distributes itself in a nonspecific manner so that its concentration is the same inside the cell as outside.

However, microorganisms (and cells in general) are not limited to passive uptake of nutrients but can absorb materials, even against concentration gradients, by specific mechanisms called *active transport* proc-

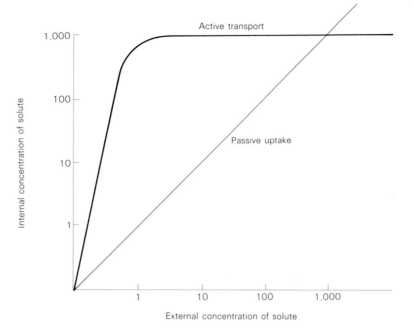

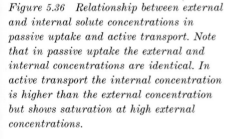

Figure 5.36 Relationship between external and internal solute concentrations in passive uptake and active transport. Note that in passive uptake the external and internal concentrations are identical. In active transport the internal concentration is higher than the external concentration but shows saturation at high external concentrations.

esses. The necessity for active transport mechanisms in microorganisms can readily be seen. Most microbes live in environments whose concentrations of salts and other nutrients are many times lower than concentrations within the cells. If passive uptake were the only type to occur, these cells would not be able to acquire the proper concentrations of solutes. Active transport mechanisms overcome this problem by enabling the cell to accumulate solutes against a concentration gradient, as seen in Figure 5.36. With low external concentrations, a solute accumulation against a concentration gradient can occur only through the expenditure of energy, probably from ATP. Figure 5.36 shows another distinction between active and passive uptake. The former shows a saturation effect: Once a certain level within the cell is reached, no more substance is taken up, even though the external concentration is greatly increased. In passive uptake, on the contrary, the internal concentration continues to increase with the external concentration.

Active transport mechanisms show marked specificity. In a group of very closely related substances some may be transported rapidly, some slowly, and others not at all. Further, there may be competition for uptake; for instance, L-alanine, L-serine, and glycine, all closely related amino acids, compete with each other for uptake, as do L-valine, L-leucine

5.12 Permeability and nutrition

and L-isoleucine. This is one reason why amino acid imbalance in a culture medium leads to poor growth of organisms requiring them—an excess of one amino acid prevents the uptake of the other. A similar kind of competition for uptake is also shown among related cations, such as Na⁺, K⁺, and Rb⁺.

It is thought that in many cases of active transport specific membrane proteins called *permeases* are involved. There is good evidence that in the bacterium *E. coli* a β-galactoside permease functions in the active transport of lactose and other galactosides. The uptake of galactosides in this organism shows all the characteristics of an active-transport system: concentrative uptake, saturation, competition between related compounds, and high specificity. In addition, certain genetic mutants of *E. coli* lack the active-transport system, although they still take up galactosides by passive processes. These mutants will grow on media with high but not low concentrations of lactose. Since mutation leads to alteration in structure and function of specific proteins, it has been reasonably hypothesized that these mutations affect a specific active-transport enzyme, the galactoside permease. A few other cases of such permeability mutants are known. One important point is that as far as can be told the permease enzyme functions only in the active-transport process and not in other metabolic processes. Good evidence exists to suspect that membrane transport proteins carry sugars, amino acids, and inorganic ions (for example, sulfate, Na⁺, and K⁺) into the cell. Membrane transport proteins are located at the periphery of the cell, either on the plasma membrane or sandwiched between the membrane and the cell wall. Each membrane transport protein binds specifically the substance whose transport it catalyzes; a likely model for how a membrane transport protein functions is illustrated in Figure 5.37.

The above discussion has referred only to the uptake of lower molecular-weight substances. Some microorganisms can take up large molecules such as proteins and nucleic acids. This ability is most highly developed in those protozoa which lack cell walls and thus have no rigid barrier against the entrance of these large molecules. Protozoa can also take up macromolecules in solution by a process called *pinocytosis* (derived from a Greek word meaning "to drink"). Fluid droplets are sucked into a channel formed by the invagination of the cell membrane, and when portions of this channel are pinched off, the fluid is enclosed within a membranous vacuole (Figure 5.38). Any solutes or macromolecules dissolved in the fluid are thus taken directly into the cell. Pinocytosis differs from active transport in that it is a relatively nonspecific process. Further, protozoa are able to ingest particulate material by a process called *phagocytosis* (from a Greek word meaning "to eat"). Whereas diffusion, osmosis, and active transport are continuous and more or less specific processes,

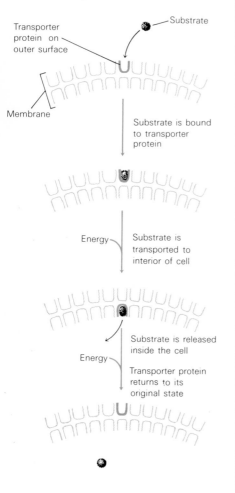

Figure 5.37 Model for how a transporter protein might function in transfer of a solute molecule from outside to inside.

Transporter protein on outer surface

Substrate

Membrane

Substrate is bound to transporter protein

Energy

Substrate is transported to interior of cell

Substrate is released inside the cell

Energy

Transporter protein returns to its original state

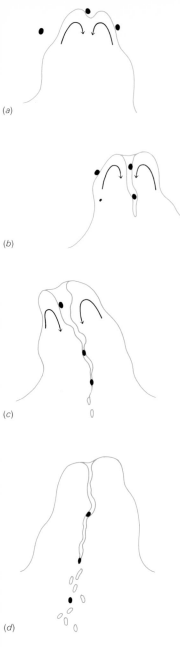

(a)

(b)

(c)

(d)

Figure 5.38 Pinocytosis: the uptake of proteins or other macromolecules through the plasma membrane.

phagocytosis is a process that has been found to be both discontinuous and nonspecific.

In eucaryotes, the chloroplasts, mitochondria, nuclei, vacuoles, and other internal organelles have their own membrane systems. Thus once a solute has penetrated the outer cell membrane it may still have to traverse additional membranes before it reaches its site of action. Little is known about the permeability properties of the membranes of these interior organelles, although the membranes appear to have larger pores and to be less restrictive.

Supplementary readings

Burris, R. H. "Biological nitrogen fixation," *Ann. Rev. Plant Physiol.* 17:155 (1966).

Cohen, G. N., and J. Monod "Bacterial permeases," *Bacteriol. Rev.* 21:169 (1957).

Davis, B. D., R. Dulbecco, H. N. Eisen, H. S. Ginsberg, and W. B. Wood, Jr. *Microbiology.* New York: Harper & Row, Publishers, 1967.

Guirard, B. M., and E. E. Snell "Nutritional requirements of microorganisms," in I. C. Gunsalus and R. Y. Stanier (eds.), *The Bacteria: A Treatise on Structure and Function,* vol. IV, p. 33. New York: Academic Press, Inc., 1962.

Gunsalus, I. C., and R. Y. Stanier (eds.) *The Bacteria: A Treatise on Structure and Function,* vol. II, *Metabolism.* New York: Academic Press, Inc., 1961.

Gunsalus, I. C., and R. Y. Stanier (eds.) *The Bacteria: A Treatise on Structure and Function,* vol. III, *Biosynthesis.* New York: Academic Press, Inc., 1962.

Hardy, R. W. F., and R. C. Burns "Biological nitrogen fixation," *Ann. Rev. Biochem.* 37:331 (1968).

Horecker, B. L. "The biosynthesis of bacterial polysaccharides," *Ann. Rev. Microbiol.* 20:253 (1966).

Kates, M. "Biosynthesis of lipids in microorganisms," *Ann. Rev. Microbiol.* 20:13 (1966).

Mahler, H. R., and E. H. Cordes *Biological Chemistry.* New York: Harper & Row, Publishers, 1966.

Mandelstam, J., and K. McQuillen (eds.) *Biochemistry of Bacterial Growth.* Oxford: Blackwell Scientific Publishers, 1968.

McKenna, E. J., and R. E. Kallio "The biology of hydrocarbons," *Ann. Rev. Microbiol.* 19:183 (1965).

Reeves, H. C., R. Rabin, W. S. Wegener, and S. J. Ajl "Fatty acid synthesis and metabolism in microorganisms," *Ann. Rev. Microbiol.* 21:225 (1967).

Rose, A. H. *Chemical Microbiology,* 2nd ed. London: Butterworth & Co. (Publishers), Ltd., 1968.

Truffa-Bachi, P., and G. N. Cohen "Some aspects of amino acid biosynthesis in microorganisms," *Ann. Rev. Biochem.* 37:79 (1968).

Wood, W. A. "Carbohydrate metabolism," *Ann. Rev. Biochem.* 35, pt. 2:521 (1966).

The organisms we have discussed up until now obtain their energy from organic compounds. Such organisms are called *heterotrophs* or *organo-trophs* (*-troph* is a combining form meaning "nourishing").

Autotrophs, on the other hand, are organisms that can obtain all of their energy from sources other than organic compounds and in addition are able to obtain the carbon they need for cellular biosynthesis from CO_2. In this chapter we will discuss first the energy metabolism of autotrophs and then go on to study the manner in which they are able to use CO_2 as sole carbon source. One large group of autotrophs obtains energy for cellular metabolism from light; these organisms are called *photosynthetic*. Other autotrophs, called *lithotrophic* or *chemosynthetic*, obtain their energy from the oxidation of inorganic compounds, such as H_2S, NH_3 and Fe^{2+}.

6.1   *Photosynthesis*

One of the most important biological processes on earth is *photosynthesis*, which in essence is the conversion of light energy by the cell into chemical energy that can then be used for the forming of cellular constituents from CO_2. The ability to photosynthesize is dependent on the presence of special green pigments, the *chlorophylls*, which are found in plants, algae, and some bacteria. Photosynthesis has been studied for many years in higher plants, but in recent years our knowledge of this process has been greatly advanced through biochemical studies on algae and photosynthetic bacteria. It has been found that photosynthetic reduction of CO_2 consists of two more or less distinct sets of reactions, the *light reactions*, in which light energy is converted into chemical energy, and the *dark reactions*, in which this chemical energy is used to reduce CO_2 to organic compounds.

Both ATP and reducing power (NADPH) are needed for the dark reactions; the light reactions are concerned with how energy from light is converted into chemical energy in these two substances. Photosynthetic bacteria use light only in forming ATP; they obtain their reducing power from constituents of their environment, such as H_2S and organic compounds. In the anaerobic environments in which these bacteria live, such reduced compounds are usually available to them. Green plants and algae, on the other hand, are not able to use H_2S and other compounds to acquire reducing power, which they obtain instead by splitting H_2O, at the same time producing O_2 (Figure 6.1). As we shall see, this important difference between photosynthetic bacteria and other photosynthetic organisms has its basis in significant differences in the functions of the photosynthetic apparatus.

six

The autotrophic way of life

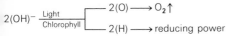

$$2(OH)^- \xrightarrow[\text{Chlorophyll}]{\text{Light}} \left[\begin{array}{l} 2(O) \longrightarrow O_2\uparrow \\ 2(H) \longrightarrow \text{reducing power} \end{array} \right.$$

Figure 6.1 Production of O_2 and reducing power from H_2O.

Figure 6.2 Chlorophyll a.

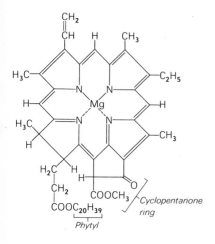

Pigments   Visible light of different wavelengths is seen by the eye as different colors; but it is not the color of light that is significant biologically, but its energy content. Shorter wavelength light has a higher energy content per photon than longer wavelength light, but for *any* light energy to be effective photochemically, it must be absorbed. A black body absorbs light of all wavelengths equally, but living organisms contain specific pigments that absorb light selectively at some wavelengths and not at others. Not all light that is absorbed may actually initiate chemical reaction; some of the absorbed energy may be dissipated as heat or as fluorescence.

As we have noted, photosynthesis occurs only in organisms that possess chlorophyll, and it seems reasonable therefore to conclude that chlorophyll is related to the assimilation of light energy. Chlorophyll is a porphyrin, as are the cytochromes, but chlorophyll contains a *magnesium* atom instead of an iron atom at the center of the porphyrin ring. Another difference is that chlorophyll is not bound to a protein as are the porphyrins of the cytochromes. Instead, chlorophyll has a lipid-soluble residue and is found associated with the lipid layer of photosynthetic membranes. The structure of chlorophyll a, the principal chlorophyll of higher plants and of most algae, is shown in Figure 6.2. Chlorophyll a is green in color because it absorbs red and blue light preferentially and transmits green light. The spectral properties of any pigment can be best expressed by its *absorption spectrum,* which indicates the degree to which the pigment absorbs light of different wavelengths. The absorption spectrum of an ether extract of chlorophyll a is given in Figure 6.3. This figure shows the strong absorption by chlorophyll a of red light (maximum absorption at 665 nm) and blue light (maximum at 430 nm).

There are a number of chemically different chlorophylls that are distinguished by their absorption spectra. Chlorophyll b, for instance, absorbs maximally in the red region at 645 nm rather than at 665 nm. Many plants have more than one chlorophyll, but the most common are chlorophylls a and b. Among the procaryotes, the blue-green algae have chlorophyll a, but photosynthetic bacteria have chlorophylls of different structure, which are called *bacteriochlorophylls.* Bacteriochlorophyll from purple photosynthetic bacteria absorbs maximally at 770 nm and its color is not purple, but green. Other bacteriochlorophylls absorb maximally at 660 nm and 870 nm.

Although it is relatively easy to discover which wavelengths of light are absorbed by a living organism it is more difficult to determine which wavelengths are actually initiating significant photochemical reactions in the organism. Roles of different pigments are determined through an *action spectrum,* in which the quantitative effect of selected wavelengths of light on the biological process is measured. This action spectrum is

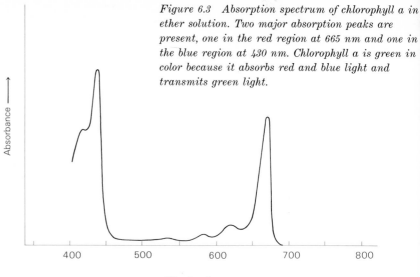

Figure 6.3   Absorption spectrum of chlorophyll a in ether solution. Two major absorption peaks are present, one in the red region at 665 nm and one in the blue region at 430 nm. Chlorophyll a is green in color because it absorbs red and blue light and transmits green light.

then compared with the absorption spectrum of pigments in the organism; in this way it is possible to infer which of several pigments in an organism might be responsible for the reaction.

The involvement of chlorophyll in photosynthesis can be shown from the following observations: (1) Only organisms that possess chlorophyll photosynthesize; (2) if a chlorophyll-containing organism is cultured under conditions that make it lose its chlorophyll, it cannot photosynthesize, and when it regains its chlorophyll, it regains at the same time its ability to photosynthesize; (3) genetically altered organisms (mutants) that have lost the ability to make chlorophyll do not photosynthesize, but if such mutant strains are converted by genetic means back to ones that produce chlorophyll, they regain the ability to photosynthesize; (4) the wavelengths of light that are absorbed best by chlorophyll are most effective for photosynthesis, whereas those not absorbed are not effective. Thus it is well established that chlorophyll is the primary material in the plant responsible for the conversion of light energy to chemical energy.

The absorption spectrum given in Figure 6.3 was measured on chlorophyll dissolved in ether. It is also possible to measure the absorption spectrum of chlorophyll while still in its natural state inside cells. The results indicate that the absorption maximum of chlorophyll within cells is at a higher wavelength than it is in an extract. A physical chemist would tell us that this observation means that the chlorophyll molecules within the cell are not in simple solution but are associated in complex aggregates or arrays.

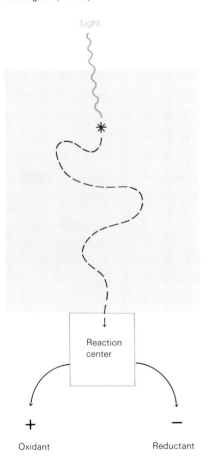

Figure 6.4 The photosynthetic unit and its associated reaction center. A packet of energy, absorbed by chlorophyll in the light-harvesting system, migrates to the reaction center, where it promotes the separation of oxidant and reductant. (From R. K. Clayton: Molecular Physics in Photosynthesis. Boston: Blaisdell Publishing Co., 1965.)

Just where is the chlorophyll located inside of the cell? We noted in Section 3.4 that algae have special chlorophyll-containing intracellular organelles, the chloroplasts. Even here, the chlorophyll is not found uniformly distributed throughout the chloroplast but is found only in association with the lamellar membrane structures of the chloroplast. These photosynthetic membrane systems are sometimes called *thylakoids*. Within these membranes, the chlorophyll molecules are associated in groups consisting of about 200 to 300 molecules. The chlorophyll molecules are stacked in such a way that they interact with each other. Light energy absorbed in one chlorophyll molecule can then be passed to another, and another, and thus travel through the whole unit until it reaches the *reaction center*, where it is converted into chemical energy by processes discussed below. In photosynthetic procaryotes, both bacteria and blue-green algae, chloroplasts are not present, but chlorophyll is found in extensive internal membrane systems (see Chapters 2 and 18). In bacteria, about 40 chlorophyll molecules probably comprise a single reaction center.

Light reactions and photophosphorylation Although light energy comes in waves, its energy is transferred to chemical substances that absorb it in discrete units called *quanta* (singular, *quantum*). When a chlorophyll molecule absorbs a quantum of light, the chlorophyll undergoes a change in properties (excitation); the energy from the light is now present in the excited chlorophyll molecule. Excitation results in an electron of the chlorophyll molecule being driven off, and the chlorophyll molecule itself becomes positively charged: $Chl + light \longrightarrow Chl^+ + e^-$ At any one time within the photosynthetic unit, many chlorophyll molecules undergo this type of transition. The electrons so released migrate through the photosynthetic unit to the reaction center and transfer energy to a special *reaction-center chlorophyll* (Figure 6.4). At the reaction center, a charge separation (equivalent to an oxidation-reduction reaction) occurs. The electron is the energy source and the positively charged chlorophyll unit is the electron acceptor. In photosynthetic bacteria, electron flow after charge separation is relatively simple (Figure 6.5): The electron moves through an electron transport system, being passed successively to ferredoxin, ubiquinone, cytochrome b and cytochrome c, and thence back to the positively charged unit. In the step between cytochrome b and cytochrome c, ATP synthesis occurs in a manner similar to that described for oxidative phosphorylation (Section 4.4); since the energy used is light however, this process is called *photophosphorylation*. In essence, the electron has gone around a cycle, starting from chlorophyll and ending with chlorophyll, and one molecule of ATP has been synthesized. This cyclic process by which energy from light is converted into chemical energy of ATP is therefore called "cyclic phosphorylation." Note that no NADP has been reduced in these reactions.

As mentioned, algae differ from photosynthetic bacteria in that they split water to provide the reducing power for cellular syntheses. This requires the cooperative action of light reactions in two electron transport systems, one called light reaction I (which is similar to that occurring in photosynthetic bacteria), and the other called light reaction II (which operates best with light at shorter wavelengths—that is, with light quanta of higher energy) (Figure 6.6). In light reaction II light absorbed by chlorophylls in the photosynthetic unit again leads to a charge separation. Now, however, an electron is transferred from the hydroxyl radical of water to the positive charge in the chlorophyll unit, leading to the formation of O_2. The electron derived from water now migrates through an electron transport system, until it reaches a special cytochrome, cytochrome f. (Cytochrome f is involved only in photosynthesis, whereas the other cytochromes also

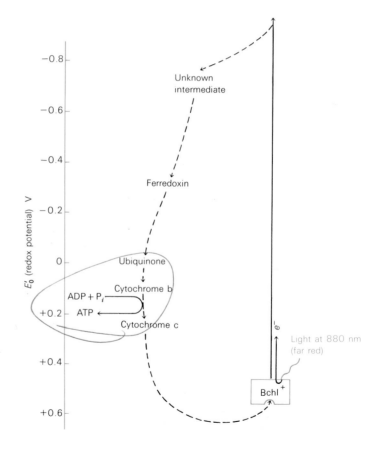

Figure 6.5 *Electron flow in bacterial photosynthesis. Upon excitation by far-red light, an electron from bacteriochlorophyll (Bchl) is passed to an electron acceptor at a high negative potential. The electron travels through an electron transport chain and returns to chlorophyll. In the step between cytochromes b and c, ATP is synthesized. The whole process is called cyclic photophosphorylation. [Adapted from A. W. Frenkel and K. Cost: "Photosynthetic phosphorylation," in M. Florkin and E. H. Stotz (eds.), Comprehensive Biochemistry, vol. 14, Biological Oxidations. Amsterdam: Elsevier Pub. Co., 1966.]*

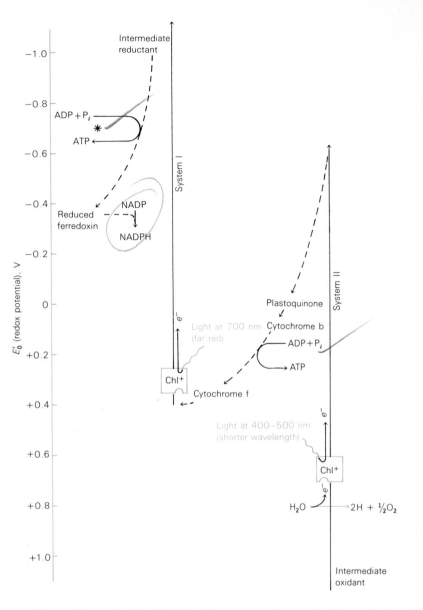

Figure 6.6 Electron flow in green-plant photosynthesis. Two light reactions are shown, with light system I, using far-red light, being similar to that of bacterial photosynthesis. In light system II, blue light is used preferentially. Through the combined action of systems I and II, NADP is reduced to NADPH by the action of electrons derived from water, and O_2 is formed. Two sites of ATP synthesis are shown, although the proposed site (marked by an asterisk) of phosphorylation in light system I is uncertain. [Adapted from A. W. Frenkel and K. Cost: "Photosynthetic phosphorylation," in M. Florkin and E. H. Stotz (eds.), Comprehensive Biochemistry, vol. 14, Biological Oxidations. Amsterdam: Elsevier Pub. Co., 1966.]

participate in respiration.) During the passage from cytochrome b to cytochrome f, one high-energy phosphate bond in ATP is synthesized. The reduced cytochrome f is now reoxidized in light system I, the electron from cytochrome f being transferred to NADP by way of the positive charge created by light system I. A second photophosphorylation probably occurs in light system I. We thus see that in algae and green plants, *two* cooperative light reactions are required, and the net result is the synthesis of two ATP molecules, one NADPH, and the splitting of water with the formation of molecular oxygen. Because the electron flow is one-way rather than cyclic, ATP synthesis with combined light reactions I and II is called *noncyclic photophosphorylation*. Plants can also carry out light reaction I alone to synthesize ATP, but in this case no reducing power is made.

That two light reactions occur in green plant photosynthesis was one of the more surprising results of work on photosynthesis in the last ten years. The discovery arose from studies on the effect of monochromatic (single-wavelength) light on rate of photosynthesis. It was found that far-red light (greater than 680 nm) was quite inefficient in bringing about oxygen evolution in algae, but if a second beam of shorter wavelength light were superimposed on the first, the efficiency of the far-red light was enhanced. It seemed as if light of shorter wavelength was actually making it possible for the algae to use the longer wavelength light more efficiently. As might be expected, this enhancement phenomenon has not been found in photosynthetic bacteria. These possess only one light system, which is similar to light system I, and which does not yield O_2 during photosynthesis.

Note that in plant photosynthesis both ATP and reduced NADP are synthesized during the light reactions, whereas in bacterial photosynthesis only ATP is synthesized. How then do photosynthetic bacteria obtain the NADPH they need for the reduction of CO_2 to organic carbon? As stated earlier, photosynthetic bacteria live in anaerobic environments, where H_2S or other reduced compounds are present. It is likely that the photosynthetic bacteria use these directly in the synthesis of NADPH. For instance:

$$H_2S + NADP \longrightarrow S + NADPH$$

The photosynthetic bacteria that use H_2S as a source of reducing power produce sulfur granules and are usually called photosynthetic sulfur bacteria. The production of S from H_2S by these bacteria is analogous to the production of O_2 from H_2O by green plants. (In some photosynthetic sulfur bacteria the accumulated S is further oxidized to SO_4^{2-} so that S accumulation is only transient.)

Some photosynthetic bacteria are not able to use H_2S but use organic compounds as sources of reducing power. In these nonsulfur photosynthetic bacteria the synthesis of NADPH is not fully understood but probably involves a process called reversed electron transport (discussed in Section 6.2).

The differences between photosynthesis in bacteria and in algae and plants are summarized in Table 6.1. In Chapter 2 we emphasized the

Table 6.1　Differences between plant and bacterial photosynthesis

	Plant photosynthesis	Bacterial photosynthesis
Organisms	Blue-green algae Eucaryotic algae Higher plants	Purple sulfur bacteria Green sulfur bacteria Nonsulfur purple bacteria
Chlorophyll type	Chlorophyll a 　(absorbs in red) Some have chlorophyll 　b, c, d, or e	Bacteriochlorophylls 　(some absorb in far red)
Light system I *(cyclic photophosphorylation)*	Present	Present
Light system II *(noncyclic photo-* 　*phosphorylation)*	Present	Absent
Produce O_2	Yes	No
Source of reducing power	H_2O	H_2S, other sulfur compounds, 　organic compounds

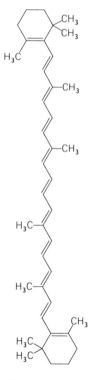

Figure 6.7　A typical carotenoid,
β-carotene.

structural similarities between blue-green algae and other procaryotes, but now we see that, as far as the biochemistry of photosynthesis is concerned, blue-green algae resemble the eucaryotes more than they do the bacteria.

Accessory pigments　Although chlorophyll is obligatory for photosynthesis, most photosynthetic organisms have other pigments that are involved, at least indirectly, in the capture of light energy. The most widespread accessory pigments are the *carotenoids,* which are almost always found in photosynthetic organisms. Carotenoids are insoluble in water but soluble in organic solvents; the structure of a typical carotenoid is shown in Figure 6.7. Carotenoids have long hydrocarbon chains with alternating C—C and C=C bonds, an arrangement called a conjugated double bond system. As a rule, carotenoids are yellow in color and absorb light in the blue region of the spectrum. The carotenoids are usually closely associated with chlorophyll in the photosynthetic apparatus, and there are approximately the same number of carotenoid as of chlorophyll molecules. Carotenoids do not act directly in photosynthetic reactions, but they may transfer the light energy they capture to chlorophyll, and this energy may thus be used in photophosphorylation in the same way as is the light energy captured directly by chlorophyll.

Blue-green algae, red algae, and a few others contain *biliproteins,* which are accessory pigments that are red or blue in color. The red pigment, called *phycoerythrin,* absorbs light most strongly in the wavelengths

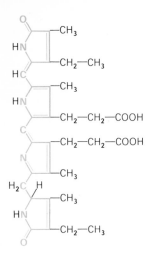

Figure 6.8 A typical phycobilin. This compound is an open-chain tetrapyrrole. The structure shown is the prosthetic group, or chromatophore, of phycocyanin, a proteinaceous pigment found in most blue-green and red algae.

around 550 nm, whereas the blue pigment, *phycocyanin,* absorbs most strongly at about 600 nm. These biliproteins contain open-chain tetrapyrrole rings called *phycobilins* (Figure 6.8), which are coupled to proteins. A very efficient energy transfer, approaching 100 percent, occurs from biliprotein to chlorophyll. The biliproteins occur in clusters attached to the lamellar membranes of the photosynthetic apparatus, where they are closely linked to the chlorophyll-containing system, so that energy transfer is efficient.

The light-gathering function of the accessory pigments would seem to be of obvious advantage to the organism. Light from the sun is distributed over the whole visible range; yet chlorophylls absorb well in only a part of this spectrum. By having accessory pigments, the organism is able to capture more of the available light. Another function of accessory pigments, especially of the carotenoids, is as photoprotective agents. Bright light can often be harmful to cells, in that it causes various photooxidation reactions that can actually lead to destruction of chlorophyll and of the photosynthetic apparatus itself. The accessory pigments absorb much of this harmful light and thus provide a shield for the light-sensitive chlorophyll. Since photosynthetic organisms must by their nature live in the light, the photoprotective role of the accessory pigments is an obvious advantage.

6.2   *Energy from the oxidation of inorganic energy sources*

Some bacteria can obtain their energy from the oxidation of inorganic compounds. These organisms are called *chemosynthetic* or *lithotrophic* (literally, ''rock-eating''). They can obtain all of their carbon from CO_2 by processes very similar to those used by photosynthetic organisms (see Section 6.3), but the mechanism by which they produce ATP is quite different from that of the photosynthesizers. In lithotrophs ATP is generated by oxidative phosphorylation during the oxidation of the inorganic energy source.

Hydrogen   Some bacteria can use hydrogen gas as an energy source: $2H_2 + O_2 \longrightarrow 2H_2O$. The hydrogen is first activated by an enzyme called *hydrogenase,* and the hydrogens are transferred to NAD. The NADH thus formed donates electrons to an electron transport particle, and ATP is synthesized by oxidative phosphorylation. All such hydrogen bacteria are also able to use a certain number of organic compounds as energy sources.

Figure 6.9 A Thiothrix rosette. The filaments of the rosette are packed with sulfur granules that can be seen as shiny, highly refractive material. Magnification, 480×.

Sulfur Many reduced sulfur compounds can be used as energy sources by a variety of *colorless sulfur bacteria* (called "colorless" to distinguish them from the chlorophyll-containing sulfur bacteria discussed above). The most common sulfur compounds used as energy sources are H_2S, free sulfur (S), and thiosulfate ($S_2O_3^{2-}$), which are electron donors for oxidative phosphorylation. The first oxidation product of H_2S is sulfur, a highly insoluble material. Some bacteria deposit sulfur inside the cell (Figure 6.9), whereas others deposit it extracellularly. The sulfur deposited as a result of the initial oxidation is an energy reserve, and when the supply of H_2S has been depleted, additional energy can be obtained from the oxidation of sulfur to SO_4^{2-}. The production of SO_4^{2-} leads to highly acidic conditions, with pH values sometimes below 2.0. Sulfur-oxidizing bacteria are unusually resistant to these acidic conditions and are commonly found in nature in acidic environments. The ability of sulfur-oxidizing bacteria to produce sulfuric acid is sometimes employed in agricultural practice in alkaline soils; powdered sulfur is plowed into the soil, and sulfur bacteria naturally present in the soil oxidize it and reduce the soil pH to values more suitable for agricultural crops. The biochemistry of sulfur oxidation is discussed in Section 18.11.

Iron The oxidation of iron from the ferrous to the ferric state is also an energy-yielding reaction for a few bacteria. Only a small amount of energy is available from this oxidation, and for this reason the bacteria must oxidize large amounts of iron in order to grow. Ferric iron forms a very insoluble hydroxide [$Fe(OH)_3$] in water, and this insoluble iron precipitates around the bacterial cell as oxidation proceeds. Iron bacteria thus become ensheathed in an iron corset.

One complication for iron bacteria is that ferrous iron also oxidizes spontaneously in air at neutral pH values. Thus if iron bacteria lived at neutral pH, they would have to compete with atmospheric oxygen to oxidize iron. Autotrophic iron bacteria living at neutral pH have not been demonstrated. Under acidic conditions, ferrous iron is not oxidized spontaneously, but there is a special group of acid-tolerant iron bacteria (genus *Ferrobacillus*) that can carry out iron oxidation and grow. The water that drains coal-mining dumps is often acidic and contains ferrous iron, and it is in these acid mine waters that *Ferrobacillus* proliferates.

Nitrogen The most common inorganic nitrogen compounds used as energy sources are *ammonia* (NH_3) and *nitrite* (NO_2^-), which are oxidized to nitrate by certain bacteria (*nitrifying bacteria*). One group of organisms oxidizes ammonia to nitrite, and another group oxidizes nitrite to nitrate; the complete oxidation of ammonia to nitrate is carried out by members of these two groups acting in sequence.

Nitrifying bacteria are widespread in the soil, and their significance in soil fertility and in the nitrogen cycle is discussed in Chapter 16; their taxonomy is discussed in Section 18.11.

Carbon monoxide Another interesting group of bacteria obtains energy from the oxidation of carbon monoxide (CO) to CO_2. Since CO is very toxic for most aerobes because it inhibits the cytochrome system, CO-oxidizing bacteria obviously must have some means of circumventing CO toxicity. Unfortunately, these bacteria have been little studied.

Energy yields in lithotrophs Table 6.2 summarizes the energy yields from the oxidations of various inorganic energy sources. The amount of ATP formed is directly proportional to the amount of energy released in a given oxidation. Since reactions that provide little energy yield little ATP, the organisms using these reactions are, in turn, able to synthesize only small amounts of cell substance. However, since these organisms utilize energy sources that are not available to other organisms, they are able to survive in nature. Lithotrophs use NADH rather than NADPH for the reduction of CO_2. Lithotrophs must thus produce reduced NAD in addition to ATP if they are to grow on CO_2 as a sole carbon source. Although NADH can be readily produced from H_2 and H_2S, certain of the inorganic energy sources have redox potentials higher than that of NAD:

	Redox potential
$Fe^{2+} \longrightarrow Fe^{3+}$	+ 0.77
$NO_2^- \longrightarrow NO_3^-$	+ 0.42
$NADH \longrightarrow NAD$	− 0.32
$H_2 \longrightarrow H^+$	− 0.42

If the redox potential of an energy source is higher than that of NADH, there is no way in which its oxidation can be *directly* coupled to the reduction of NAD to NADH. To circumvent this difficulty, these organisms reduce NAD by a process called *reversed electron transport*. Ordinarily, electrons flow from NADH to O_2 through the cytochrome system, generating ATP. In reversed electron transport, electrons flow to NAD, with the *consumption* of ATP and the formation of NADH. Thus ATP formed by oxidative phosphorylation is used to synthesize the NADH needed for CO_2 fixation. A similar system occurs in certain photosynthetic bacteria.

Note that the transformations of inorganic elements discussed here are essentially the reverse of those discussed for anaerobic respiration in Section 4.5. In anaerobic respiration oxidized forms of the elements are reduced using energy usually derived from organic compounds, whereas in lithotrophic oxidation the reduced forms of the elements are oxidized

Table 6.2 Energy yields from the oxidation of various inorganic energy sources

$CO + \frac{1}{2}O_2 \longrightarrow CO_2 + 74$ kcal
$H_2S + \frac{1}{2}O_2 \longrightarrow H_2O + S + 41.5$ kcal
$S + 1\frac{1}{2}O_2 + H_2O \longrightarrow H_2SO_4 + 118$ kcal
$HNO_2 + \frac{1}{2}O_2 \longrightarrow HNO_3 + 17$ kcal
$NH_4^+ + 1\frac{1}{2}O_2 \longrightarrow NO_2^- + 2H^+ + H_2O +$ 66 kcal
$H_2 + \frac{1}{2}O_2 \longrightarrow H_2O + 56$ kcal
$Fe^{2+} + H^+ + \frac{1}{4}O_2 \longrightarrow Fe^{3+} + \frac{1}{2}H_2O +$ 40 kcal

using O_2 as an electron acceptor. By the combined action of lithotrophs and organisms carrying out anaerobic respiration, *cycles* of transformation can occur. Thus:

$$H_2S \longrightarrow S \longrightarrow SO_4^{2-} \quad \text{(lithotrophic)}$$
$$SO_4^{2-} \longrightarrow H_2S \quad \text{(anaerobic respiration)}$$

These cycles of key inorganic elements will be discussed in detail in Chapter 16.

6.3 Autotrophic CO_2 fixation

Although most organisms can obtain some of their carbon for cellular biosynthesis from CO_2, autotrophs can obtain *all* of their carbon from CO_2. The overall equation for CO_2 fixation in algae and green plants is:

$$6CO_2 + 6H_2O \longrightarrow C_6H_{12}O_6 + 6O_2$$

where $C_6H_{12}O_6$ should be viewed as any compound at the oxidation level of carbohydrate. As discussed earlier, the O_2 formed during photosynthesis is derived from H_2O. We are concerned here with the biochemical reactions of CO_2 reduction to the oxidation level of carbohydrate, a process called CO_2 fixation. All of the reactions of CO_2 fixation will occur in complete darkness, using ATP and reducing power (NADPH) generated either during the light reactions of photosynthesis or during lithotrophic oxidation of inorganic compounds. All autotrophs so far examined have a special pathway for CO_2 reduction, the Calvin cycle (named after its discoverer, Melvin Calvin). This cycle is described in abbreviated form later, but first let us consider some of the key individual reactions.

Enzyme reactions of the Calvin cycle   The first step in CO_2 reduction is the reaction catalyzed by the enzyme *carboxydismutase,* which involves a reaction between CO_2 and ribulose-diphosphate (Figure 6.10) leading

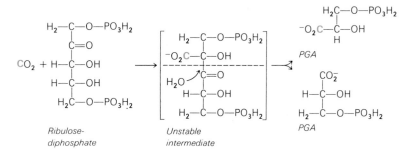

Figure 6.10 Carboxydismutase reaction (the enzyme is also called ribulose-diphosphate carboxylase).

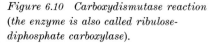

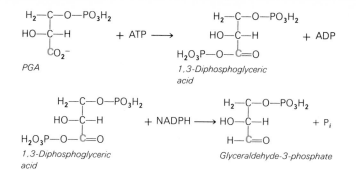

Figure 6.11 Steps in the conversion of PGA to glyceraldehyde-3-phosphate. Note that both ATP and NADPH are required. These reactants are synthesized during the light reactions of photosynthesis, as shown in Figure 6.6.

to the formation of two molecules of 3-phosphoglyceric acid (PGA), one of which contains the carbon atom from CO_2 (PGA constitutes the first *stable* intermediate in the CO_2 reductive process). The carbon atom in PGA is still at the same oxidation level as it was in CO_2; the next two steps involve the reduction of PGA to the oxidation level of carbohydrate (Figure 6.11). In these steps, *both* ATP and NADPH are required: the first is involved in the reaction that activates the carboxyl group, the second in the actual reduction itself. The carbon atom from CO_2 is now at the reduction level of carbohydrate $(CH_2O)_n$, but only one of the carbon atoms of glyceraldehyde-phosphate has been derived from CO_2, the other two having arisen from the ribulose-diphosphate. Since autotrophs can grow on CO_2 as a sole carbon source, there must be a way by which carbon from CO_2 can become incorporated into the other positions in glyceraldehyde-phosphate. Further, the ribulose-diphosphate which was used up in the carboxydismutase step must be regenerated. The remainder of the reactions of the Calvin cycle are concerned with these matters. The series of enzyme reactions leading to the synthesis of ribulose-diphosphate is shown in summary form in Figure 6.12. In this series of reactions five molecules of glyceraldehyde-phosphate (or its equivalent, dihydroxyacetone-phosphate) are used up, and three molecules of ribulose-diphosphate are formed; three molecules of ATP are also required, but no NADPH, as there were no reductive steps.

In a single carboxylation reaction, $1 CO_2$ (1C) and 1 ribulose-diphosphate (5C) are capable of yielding two molecules of PGA (2 × 3C), which become converted to two molecules of glyceraldehyde-phosphate. Since we need five molecules of glyceraldehyde-phosphate (or their equivalent, five molecules of dihydroxyacetone-phosphate) to complete one turn of the cycle, this means that three carboxylation reactions must take place for each turn of the cycle. Since each carboxylation yields two glyceraldehyde-phosphates, three carboxylations would yield six, but we need only five for regeneration of ribulose-diphosphate. The sixth glyceraldehyde-

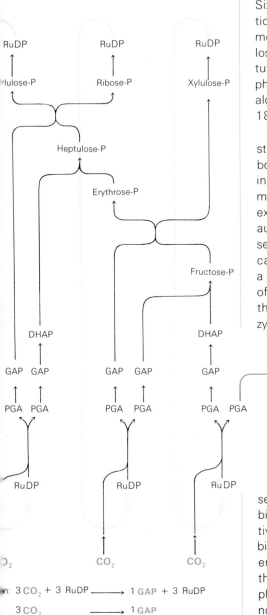

RuDP   RuDP   RuDP

⬥lulose-P   Ribose-P   Xylulose-P

Heptulose-P

Erythrose-P

Fructose-P

DHAP   DHAP

GAP GAP   GAP GAP   GAP → GAP

PGA PGA   PGA PGA   PGA PGA

RuDP   RuDP   RuDP

O₂   CO₂   CO₂

n: $3\,CO_2 + 3\,RuDP \longrightarrow 1\,GAP + 3\,RuDP$

$3\,CO_2 \longrightarrow 1\,GAP$

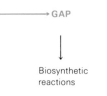

Biosynthetic reactions

phosphate thus represents a net gain; three CO_2 molecules have been in effect converted into one glyceraldehyde-phosphate.

Let us consider now the complete balance sheet for the conversion of three molecules of CO_2 into one molecule of glyceraldehyde-phosphate. Six ATP molecules and six NADPH molecules are required for the reduction of six molecules of PGA to glyceraldehyde-phosphate, and three ATP molecules are required for the conversion of ribulose-phosphate to ribulose-diphosphate. Thus six NADPH and nine ATP are required to manufacture one glyceraldehyde-phosphate from CO_2. Since two glyceraldehyde-phosphates can give rise to one hexose sugar molecule (through the aldolase reaction and gluconeogenesis—see Chapter 5), 12 NADPH and 18 ATP are required to make one hexose molecule from CO_2.

Some of the enzymes of the Calvin cycle were already known from studies on pentose metabolism (Section 5.1), but the key enzyme, carboxydismutase (or ribulose-diphosphate carboxylase), which is involved in the initial CO_2 fixation reaction, does not function in normal pentose metabolism. This enzyme has been found in all photosynthetic organisms examined, plants, algae, and bacteria. It is also found in chemosynthetic autotrophic bacteria, such as the sulfur, iron, and nitrifying bacteria. The second key enzyme is glyceraldehyde-phosphate dehydrogenase, which catalyzes the reductive step of the Calvin cycle. It should be noted that a glyceraldehyde-phosphate dehydrogenase is crucial to the fermentation of glucose in the glycolytic pathway (Chapter 4); however, the photosynthetic enzyme differs in that it uses NADPH instead of NADH as a coenzyme. This distinction is of considerable significance, as it allows for the

Figure 6.12 Abbreviated Calvin cycle. Three turns of the cycle are necessary to synthesize 1 glyceraldehyde-3-phosphate from $3CO_2$. RuDP stands for ribulose-1,5-diphosphate; DHAP, for dihydroxyacetone phosphate; GAP, for glyceraldehyde-3-phosphate; PGA, for 3-phosphoglyceric acid.

separate functioning of the photosynthetic process, which is essentially biosynthetic and reductive, and the glycolytic process, which is degradative and oxidative. This is a further example of how cells separate their biosynthetic and degradative pathways (Section 5.5). Two additional enzymes that may be specific for the Calvin cycle are the phosphatases that convert the diphosphates of sedoheptulose and fructose to the monophosphates. All of the other enzymes of the Calvin cycle are found in nonautotrophic organisms as well.

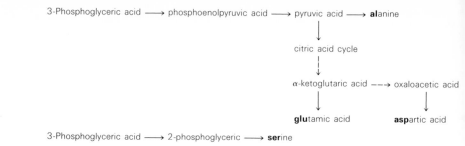

Figure 6.13 Synthesis of amino acids from 3-phosphoglyceric acid.

Biosynthetic reactions Most photosynthetic organisms live in nature in alternating light and dark regimes, but the primary products of the light reactions, ATP and NADPH, are short-lived and are quickly used up. Photosynthetic organisms circumvent this difficulty by converting CO_2 into energy-rich storage products during the light cycle, then using these during the dark cycle. In algae, storage products are usually carbohydrates such as sucrose or starch, whereas in photosynthetic bacteria the main storage product is poly-β-hydroxybutyric acid. In the light, much of the CO_2 fixed goes into storage products first, and these substances are then broken down for energy and biosynthesis. Some amino acids, however, may be synthesized more directly from PGA (Figure 6.13).

6.4 Ecology of photosynthetic organisms

Photosynthetic organisms are found in nature almost exclusively in areas where light is available. Thus caves, deep ocean waters, turbid rivers, the interiors of animals, and other permanently dark habitats are usually devoid of photosynthetic organisms, although these places are usually well colonized by nonphotosynthesizers. Photosynthetic microorganisms are most abundant in aquatic environments where light penetration is good, but they are also found on the surfaces of soils and rocks. In a lake or ocean the depth to which light will penetrate in sufficient amounts to allow for the growth of photosynthetic organisms will vary with the turbidity of the water; in very clear waters light penetration for several hundred meters is possible. However, the spectral quality of the light changes with depth, water absorbing red light more effectively than blue light. This selective absorption of red light puts a special limitation on photosynthetic bacteria, whose chlorophyll absorption maxima are in the far-red region. In fact, at wavelengths beyond 900 nm, the absorption of light by water is so strong that none of it ever reaches the organisms.

All species of photosynthetic organisms absorb light selectively at certain wavelengths, and the light absorbed by one organism is of course not available to another. Species have adjusted their light-absorbing properties so that they can capture light rays not used by others. Figure 6.14 shows the absorption spectra of several different microorganisms measured directly in vivo, without extraction of the pigments, and we can see that some organisms absorb light well in regions where other organisms absorb light poorly and that these differences are of ecological significance. For instance, since algae are aerobes, they grow in surface waters and absorb much of the light in the blue and red regions. Photosynthetic bacteria are anaerobes, and these organisms must grow in deep-lying waters and on

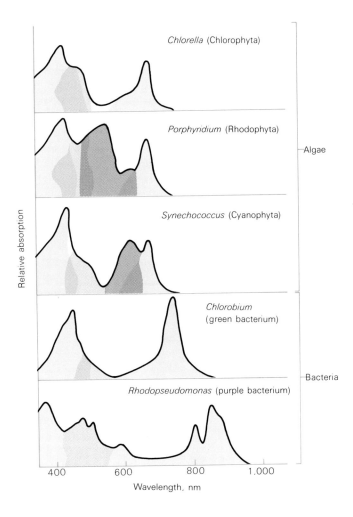

Figure 6.14 Absorption spectra of whole cells, resulting from the presence of chlorophylls and accessory pigments. (The chlorophylls are in color; accessory pigments, light and dark gray.) Note that the photosynthetic bacteria absorb at longer wavelengths than the algae, and are thus able to utilize light that the algae have not absorbed. [From R. Y. Stanier and G. Cohen-Bazire: in Microbial Ecology (Seventh Symposium, Society for General Microbiology). Cambridge, England: Cambridge University Press, 1957.]

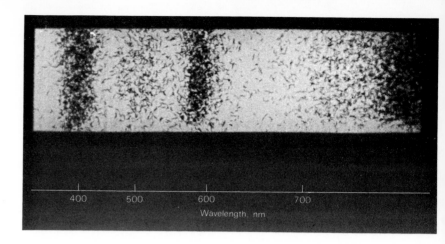

Figure 6.15 Phototactic accumulation of the photosynthetic bacterium Thiospirillum jenense at light wavelengths at which its pigments absorb. A light spectrum was displayed on a microscope slide containing a dense suspension of the bacteria; after a period of time, the bacteria had accumulated selectively and the photomicrograph was taken. The wavelengths at which accumulations occur are those at which bacteriochlorophyll a absorbs. (Courtesy of Norbert Pfennig.)

the surface of muds, where anaerobic conditions prevail. If they are to survive, they must use the light the algae allow to pass—light mainly in the far-red and infrared regions of the spectrum. Thus the different absorption spectra of various groups of photosynthetic microorganisms have an ecological basis.

The global significance of autotrophs is emphasized by the fact that they are the agencies by which CO_2 is converted into organic compounds that then become available as food for heterotrophic organisms, such as animals, fungi, and most bacteria (see Chapter 16).

Many photosynthetic microorganisms swim towards the light, a process called *phototaxis* (Section 2.7). The advantage of phototaxis to a photosynthetic organism is that it allows the organism to orient itself for most efficient photosynthesis; indeed, if a light spectrum is spread across a microscope slide on which are motile photosynthetic bacteria, the bacteria accumulate at those wavelengths at which their pigments absorb (Figure 6.15).

6.5 Comparison of autotrophs and heterotrophs

The energy metabolism and the terminology used in describing the energy relationships of autotrophs and heterotrophs are indicated in Table 6.3, which emphasizes the similarities and differences among various groups. The two factors to be considered in comparing these organisms are the nature of the energy source and the nature of the electron donor.

Table 6.3   Comparison of autotrophs and heterotrophs

	Electron donor	
	Inorganic (autotroph)	*Organic (heterotroph)*

Oxidizable substance
(chemotroph[a])
	Chemolithotrophs:
	    Hydrogen bacteria
	    Colorless sulfur bacteria
	    Nitrifying bacteria
	    Iron bacteria
		Chemoorganotrophs:
		    Animals
		    Most bacteria
		    Fungi
		    Protozoa

Energy source: light
(phototroph)
	Photolithotrophs:
	    Green plants
	    Algae
	    Purple sulfur bacteria
	    Green sulfur bacteria
		Photoorganotrophs:
		    Nonsulfur purple
		        bacteria

[a] For chemotrophs, the electron donor and the energy source are the same material.

Among the autotrophs some organisms (called *facultative autotrophs*) can also grow heterotrophically with organic compounds as energy sources. Other autotrophs are completely unable to grow heterotrophically, and these organisms are called obligate autotrophs. Obligate autotrophy is something of a biochemical puzzle, since it is hard to imagine why a phototroph or lithotroph could not utilize glucose, for instance, if it were added to its environment. This puzzle is compounded by the fact that some autotrophs can assimilate organic compounds as carbon sources, but not as energy sources. Recent work suggests that obligate autotrophs lack certain of the key enzymes in carbohydrate metabolism, and thus cannot couple the breakdown of an organic compound with the generation of ATP, even though they can convert the organic compound into many of the carbon compounds needed for cellular biosynthesis.

One group of photosynthetic bacteria, the Athiorhodaceae (nonsulfur purple bacteria), show either phototrophic or heterotrophic metabolism, depending on environmental conditions. These organisms are photosynthetic under anaerobic conditions and heterotrophic under aerobic.

Supplementary readings

Bassham, J. A., and M. Calvin   The Path of Carbon in Photosynthesis. Englewood Cliffs, N.J.: Prentice-Hall, Inc., 1957.

Calvin, M., and J. A. Bassham   The Photosynthesis of Carbon Compounds. New York: W. A. Benjamin, Inc., 1962.

Clayton, R. K.   *Molecular Physics in Photosynthesis*. New York: Blaisdell Publishing Co., 1965.

Frenkel, A. W., and K. Cost   "Photosynthetic phosphorylation," p. 397 in M. Florkin and E. Stotz (eds.), *Comprehensive Biochemistry,* sec. 3, vol. 14. New York: Elsevier Publishing Co., 1966.

Gest, H., A. San Pietro, and L. P. Vernon (eds.)   *Bacterial Photosynthesis* (Symposium on Bacterial Photosynthesis). Yellow Springs, Ohio: Antioch Press, 1963.

Gibbs, M.   "Photosynthesis," *Ann. Rev. Biochem,* 36, pt. 2:757 (1967).

Hind, G., and J. M. Olson   "Electron transport pathways in photosynthesis," *Ann. Rev. Plant Physiol.* 19:249 (1968).

Jensen, S. L.   "Biosynthesis and function of carotenoid pigments in microorganisms," *Ann. Rev. Microbiol.* 19:163 (1965).

Kamen, M. D.   *Primary Processes in Photosynthesis*. New York: Academic Press, Inc., 1963.

Lees, H.   *Biochemistry of Autotrophic Bacteria*. London: Butterworth & Co. (Publishers), Ltd., 1955.

Peck, H. D., Jr.   "Energy-coupling mechanisms in chemolithotrophic bacteria," *Ann. Rev. Microbiol.* 22:489 (1968).

Pfennig, N.   "Photosynthetic bacteria," *Ann. Rev. Microbiol.* 21:285 (1967).

Thomas, J. B.   *Primary Photoprocesses in Biology*. New York: John Wiley & Sons, Inc., 1965.

part two
Microbial
growth and
its control

Now that we have an understanding of energy and biosynthetic reactions of microorganisms, we can proceed to a discussion of microbial growth processes and their control. In Chapter 7 we first consider quantitative aspects of the growth of microbial populations and various aspects of the microbial growth curve. We then turn to the synthesis of the macromolecules DNA, RNA, and protein in relation to growth and consider how macromolecular synthesis is controlled so that orderly synthesis of all components is assured. Chapter 7 closes with a discussion of how cell differentiation is related to macromolecular synthesis and growth, using the bacterial endospore as an example. In Chapter 8 we study the effects on microbial viability and growth of various chemical and physical factors of the environment. We show how these can influence the distribution and survival of microorganisms in nature, and we see how, by a judicious selection of environmental conditions, growth of microorganisms in industrial processes can be controlled. In Chapter 9 we are concerned with the effects of various antimicrobial agents on growth. After a general discussion of selective toxicity of certain chemicals, we study antibiotics, antiseptics, disinfectants, and other chemical agents, their chemical natures, and their modes of action. Part 2 completes the portions of the book that are generally considered to encompass the subdiscipline of microbial physiology.

In this chapter we examine microbial growth and its relation to macro-molecular synthesis. ''Growth'' is defined as an increase in the quantity of cellular constituents and structures. When growth occurs in the absence of cell division, it usually results in an increase in the size and weight of the cell. In most microorganisms growth is followed by cell division; this results in an increase in cell number, the new cells formed eventually attaining the same size as the original cell.

It is useful to distinguish between the growth of individual cells and the growth of populations of cells. The growth of a cell results in an increase in its size and weight and is usually a prelude to cell division. Population growth, on the other hand, results in an increase in the *number* of cells as a consequence of cell growth and division. Because of their small size, the study of cell growth in microbes is difficult; most growth studies in microbiology deal with populations. However, since population growth is dependent on cell growth, let us first discuss the latter briefly.

7.1 Cell growth

In organisms that divide by binary fission, such as most bacteria, unicellular algae, fission yeasts, and most protozoa, the entire constituents of the parent cell are equally partitioned among the two offspring (Figure 7.1). In such organisms it is impossible to distinguish one of the two cell progeny from the other, and both cells are usually called daughters. In budding organisms such as most yeasts, however, the situation is quite different. The original components of the parental cell remain within it, and the newly synthesized constituents appear in the daughter cell (Figure 7.1). The parental cell can continue to generate new cells by continued budding, but a scar forms where the daughter cell has originated from the parent cell. New buds never appear in the region of the scar, and after the parent has generated 20 to 25 new cells, its surface is so covered with scars that it can no longer bud. Thus a cell that divides by budding ages, whereas a cell that divides by fission does not. Filamentous fungi also age since they grow only from the tips of the filaments; the regions extending backward from the tip are therefore progressively older (see Figure 7.1). The phenomenon of cell age is of considerable interest in that it may aid in understanding aging processes in higher organisms.

In Figure 7.2 the rates of increase of volume and weight of a cell are shown and correlated with the microscopic events through two cell divisions. The rate of increase of cell weight is seen to be practically identical throughout both division cycles, whereas volume increases only until the time of cell division and then ceases until the two cells have cleaved, at

seven

Growth, macro-molecular synthesis, and differentiation

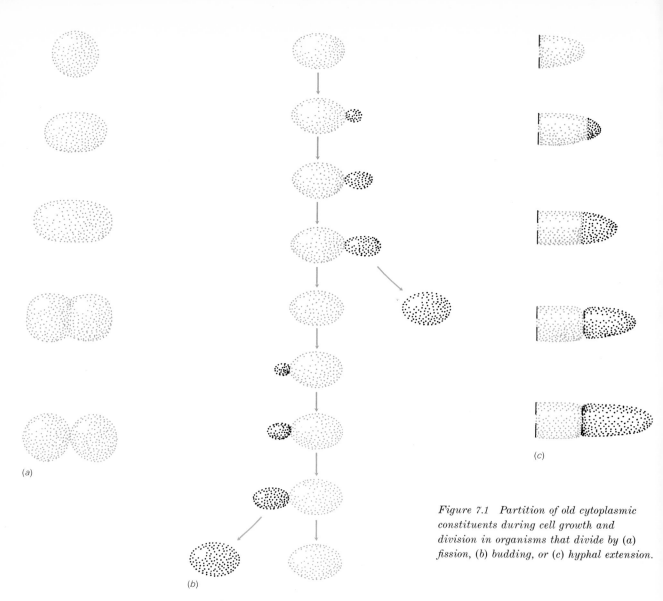

Figure 7.1 Partition of old cytoplasmic constituents during cell growth and division in organisms that divide by (a) fission, (b) budding, or (c) hyphal extension.

which time each of the daughter cells begins again to increase in volume. It is the increase in cell weight rather than cell volume that most closely represents growth, as the weight of a cell reflects its complement of cytoplasmic contents. Note that the rate of increase of cell weight is approximately linear; this should be contrasted with the quite different shape of the growth curve of a population of cells, as will be discussed below.

Although the growth of a population of cells results from growth of the individuals within that population, the quantitative relationships of population growth may be quite different from those of cell growth. Whereas individual cells grow approximately linearly with time, populations of cells grow exponentially. *Exponential growth* is a consequence of the fact that when a cell divides each daughter cell in turn divides and produces two new cells, so that at each division period the population doubles (Table 7.1). If the number of cells is plotted versus time on an arithmetic graph, the line obtained curves upward at a progressively increasing rate, whereas when the logarithm of the number of cells is plotted, a straight line results (Figure 7.3). Population growth of this type is called exponential growth, and is characteristic of unicellular organisms.

The rate of exponential growth is usually expressed as the *generation time*, or doubling time, which is the time it takes for the population to double. In the example of Table 7.1, the generation time is 30 min. Alternatively, the growth rate can be expressed as the number of generations per hr; for the data of Table 7.1, this would be two generations. Growth rates vary widely among organisms. The fastest generation times known are about 10 min, which are found in a few bacteria, whereas many bacteria show generation times of 30 to 60 min. At the other extreme, some slow-growing protozoa and algae show generation times of 24 hr or more. The doubling times of individual cells may vary quite widely since some cells will take much longer to divide than the average, whereas others will divide more quickly than the average. A population's generation time is the average value of the generation times of the individual cells in the population.

The simplest way to determine generation time is from a graph in which the logarithm of the number of cells is plotted against time; the time for the population to double is read directly from the graph (Figure 7.4). [*]

[*] In microbial populations, cell numbers are often of such magnitude that it is difficult to express them easily. To simplify the handling of these large numbers, the microbiologist uses exponents of 10. Thus, 10^6 is equivalent to 1,000,000, 10^8 to 100,000,000, and so on. Cell numbers can then be expressed as 2×10^8 (200,000,000), 5×10^7 (50,000,000), 6.5×10^3 (6,500), and so on, where the exponent of 10 represents the number of zeros to the right of the decimal point. When making graphs of growth curves, it is convenient to use five-cycle semilogarithmic graph paper since this makes it possible to represent on one sheet of paper an increase in cell number of 10^5 times, a 100,000-fold increase. When adding two figures expressed in exponents of 10, we must first write out the numbers or reduce them to the same power of 10:

$$2.5 \times 10^7 = 25,000,000 = 25 \times 10^6$$
$$\underline{3.0 \times 10^6 = 3,000,000 = 3.0 \times 10^6}$$
$$2.8 \times 10^7 = 28,000,000 = 28 \times 10^6$$

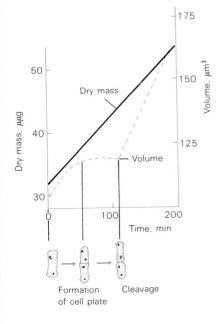

Figure 7.2 *Rate of growth of a single cell of a fission yeast. The volume is calculated from the microscopic dimensions of the cell, and the dry mass is measured with an interference microscope. Note that the increase in dry mass is linear with time, even throughout the period of cell division.* [*Adapted from J. M. Mitchison: Exptl. Cell Res. 13:244 (1957).*]

Table 7.1 Exponential growth of a population of cells from a single cell with generation time of 30 min

Time, hr	Cell no.	Cell no., $\log_{10}$
0	1	0
0.5	2	0.301
1	4	0.602
1.5	8	0.903
2	16	1.204
2.5	32	1.505
3	64	1.806
3.5	128	2.107
4	256	2.408
4.5	512	2.709
5	1,024	3.0103
.	.	.
.	.	.
.	.	.
10	1,048,576	6.021

Figure 7.3 Growth rate of a population of unicellular organisms as plotted on arithmetic and logarithmic scales.

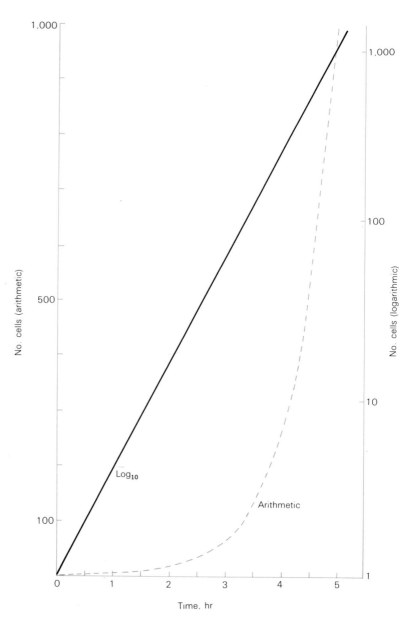

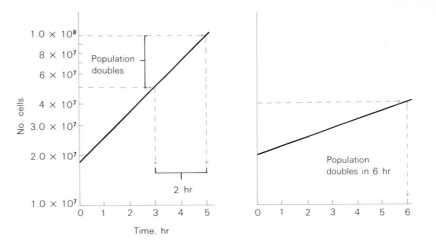

Time, hr

Figure 7.4 Method of estimating the generation time of an exponentially growing population.

One of the characteristics of exponential growth is that the rate of increase in cell number is slow initially, but it increases at an ever faster rate. This results, in the later stages, in an explosive increase in cell numbers. A practical implication of exponential growth is that, if a non-sterile product such as milk is allowed to stand at room temperature, a few hours during the early stages would not be detrimental, whereas standing for the same length of time during the later stages of exponential growth would be disastrous.

7.3 Measuring microbial number and weight

Cell number The number of cells in a population is measured most directly by counting under the microscope, a method called the *direct microscopic count*. The small size of most microbes and their large population densities make it necessary to use special chambers such as the Petroff-Hausser or hemocytometer chambers to count the number of cells in a sample. Direct microscopic counting is tedious, but it is an important procedure in many studies on microbial growth. However, it has certain limitations: (1) Dead cells cannot usually be distinguished from living cells. This is especially true with bacteria, which are so small that any cellular changes that might accompany death are hard to distinguish. (2) Small cells are difficult to see under the microscope and some cells may be missed. (3) Only dense cell suspensions can be counted, because there must be a reasonable number of cells in the small portion of the population that can be seen under the microscope. The smallest population size that can be counted is determined in part by cell size. With cells the size of most bacteria, populations must number 10^6 (1,000,000 cells) per milliliter before any will be seen. (4) High accuracy is hard to achieve.

To avoid the tedium and some of the limitations of direct microscopic counting, *electronic cell counters* have been invented. These counters were first devised for the routine counting of red blood cells in hospital laboratories, and although they are expensive, they have been found to be very useful for counting cells in studies on microbial growth since they are accurate and provide results almost immediately. In electronic counting, the electrical resistance of the fluid within a small hole is measured. As a cell passes through the hole, the resistance increases sharply, and this increase is recorded as a pulse on an electronic scaling device. A known small volume of liquid containing the cells to be counted is allowed to pass through the hole, and each cell is counted as it passes. The electronic counter cannot distinguish between microbial cells and inert particles, hence the suspending fluid must be very clean. This instrument can also be modified to measure cell volume as well as cell number.

The methods described above count both living and dead cells. In many cases we are interested in enumerating only live cells. A living cell is defined as one that is able to divide, and *viable cell counting* is usually done by determining the number of cells in a population capable of dividing and forming colonies.

Viable counting is done so frequently in both research and applied microbiology that an understanding of this method is important for all microbiologists. Although the organisms are usually placed on agar in petri plates, there are special techniques permitting the use of fluid media. The plate count is very sensitive because in principle any viable cell, when placed on an appropriate medium, will give rise to a colony. Indeed, no other procedure, microscopic or otherwise, can detect with certainty the presence of a single viable cell in a large volume of sample. In addition to sensitivity, plate counting also allows for the positive identification of the organism counted, as the colony that is formed can serve as the inoculum for a pure culture, which can be identified taxonomically. Furthermore, since different organisms often produce colonies of different shape, size, texture, and color, several kinds of organisms can be counted in a mixture.

Disadvantages of the viable count method are these: (1) Since one does not actually see the cell that gave rise to the colony, one cannot be absolutely certain that a colony arose only from a single cell. Two or more adjacent cells may give rise to only a single colony. (2) A satisfactory culture medium must be available on which the cells to be counted can grow. Often higher viable counts will be found on one medium than on another one. Environmental conditions of growth, temperature, pH, and so on, must be adequate. (3) The organism must be able to grow on a solid surface such as agar in order that the progeny remain in discrete colonies rather than become dispersed. (Viable counting can be done using liquid culture media by a technique called the *most probable number* method, but this procedure is much more complicated and time consuming than the plate-count method.) (4) The number of colonies appearing will depend on the length of incubation time. Some cells will begin to divide sooner than others, and the colonies they initiate will be visible first, so that one must incubate long enough to be certain that all colonies that will appear have increased to a size large enough to be seen. (5) Viable counting is slow since it almost always requires at least 24 hr for colonies to reach full size, and it may occasionally require several days to a week. In contrast, direct microscopic counting and electronic cell counting can provide data in a few minutes at most. (6) The number of colonies on the agar plate must not be too high. With petri plates of standard size and with bacteria that produce colonies of about 1 to 2 mm (an average size), no more than 300 colonies should be present on each plate. If plates are crowded some colonies will overlap others, and there will not be sufficient nutrients to support the growth of all viable cells present. (7) Neither must the number of colonies on the plate be too small since this leads to large statistical error. In practice, less than 30 colonies per plate is usually considered undesirable. (8) To achieve the appropriate colony number, a sample to be counted may have to be diluted in some way. Several different dilutions must be made and plated separately, in order to obtain one dilution in the right range. Since the volume of sample plated is usually between 0.1 and 1.0 ml and this sample should contain between 30 and 300 colonies, this means that if viable counts are to be done on large populations (for example, 10^6 to 10^9 cells per ml), the samples must be highly diluted. Dilutions are not only time consuming, but in addition, each time a dilution is made it increases the chance of error. (9) Viable counts are usually subject to much error, and accurate counts require great attention to standardization of all aspects of the technique.

After considering the above list of limitations, the student may wonder whether viable counting procedures are worth it. However, as indicated, the viable count provides information that can be obtained in no other way. In food, dairy, medical, and sanitation microbiology, viable counts are done frequently, often very frequently. To eliminate most of the complications described above, microbiology laboratories use rigidly standardized methods. The media, incubation conditions, dilution procedures, counting procedures, analysis, and interpretation of results have all been carefully developed through many years of work, so that the results in one laboratory are comparable to those in another.

Measurement of microbial mass   For many studies, especially those dealing with the biochemistry of growth processes, we are interested in determining the mass of the population rather than the number of cells present. Mass can be measured directly by determining either the dry or the wet weight of an aliquot separated from the population by centrifugation. Dry weight is usually about 20 to 25 percent of the wet weight, and is more directly related to cell mass than is wet weight because cells may absorb or release water in amounts that may not be specifically related to protoplasmic mass. Instead of the measurment of protoplasmic weight, total cell nitrogen or protein content may be assayed. This is a more direct measure of the biologically active protoplasmic mass than is dry weight since the latter includes the weight of the cell wall, which in many organisms is a large and variable fraction of the dry weight. Another chemical measurement that may be of considerable importance in studies on growth is the assay of DNA content since this substance plays a central role in cellular growth. In addition to direct chemical methods, it is possible to estimate cell mass indirectly by determining the metabolic acitvity of cells. Oxygen uptake has been used for aerobic organisms because the rate of oxygen uptake is usually proportional to cell mass. For anaerobic organisms, production of CO_2 or of some fermentation product such as lactic acid can be measured. The rate of dye reduction (see Chapter 4) could also be used as a measure of metabolic activity.

The average weight of a single cell can be calculated by measuring the weight of a large number of cells and dividing by the total number of cells present, as measured by direct microscopic count. Procaryotic cells range in dry weight from less than 10^{-15} g to greater than 10^{-11} g, and the cells of eucaryotic microorganisms range from about 10^{-11} g to about 10^{-7} g. (It would take 10^{12} cells weighing 10^{-12} g each to equal 1 g total weight.)

A cell suspension in a test tube looks turbid because each cell scatters light; the more cells present, the more the suspension will scatter light. *Turbidity measurements* are widely used in studies on microbial growth because they are simple, direct, and nondestructive. The turbidity of a cell suspension is more closely related to cell weight than to cell numbers, and hence provides a quick way of estimating increase in cell mass in cultures. To relate turbidity to cell mass a calibration curve must first be constructed for the organism in question since different organisms scatter light to different degrees. A variety of instruments are available for measuring turbidity. In the absence of such instruments, turbidity can be estimated visually simply by holding a tube up to the light and matching its turbidity to one of a series of tubes with known light-scattering properties.

Chemical and turbidimetric methods are much less sensitive than those involving cell counting and require either relatively large population densities or large volumes of cell suspension. However, in much of the work on the physiology of microbial growth such methods are very useful. Indeed, with filamentous organisms, they are essential since cell counting is otherwise difficult or impossible. It should be remembered that any method of measuring cell mass is somewhat arbitrary.

7.4 The growth cycle of populations

A typical growth curve for a population of cells is illustrated in Figure 7.5. This growth curve can be divided into several distinct phases, called the *lag phase, exponential phase, stationary phase,* and *death phase.*

Lag phase   When a microbial population is inoculated into a fresh medium, growth usually does not begin immediately, but only after a period of time. This is called the *lag phase,* which may be brief or extended, depending on conditions. If an exponentially growing culture is inoculated into the same medium at the same temperature a lag is not seen, and exponential growth continues at the same rate. However, if the inoculum is taken from an old (stationary phase) culture a lag usually occurs even if all of the cells in the inoculum are alive. This is because

Figure 7.5 Typical growth curve for a bacterial population.

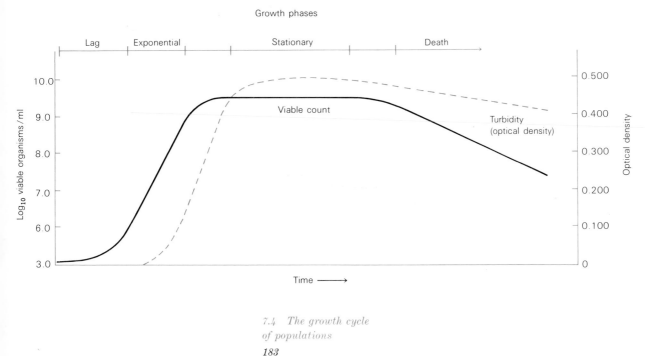

Growth phases

7.4   The growth cycle
of populations

183

the cells are usually depleted of various essential coenzymes or other cell constituents and time is required for resynthesis. A lag also ensues when the inoculum consists of cells that have been damaged (but not killed) by treatment with heat, radiation, or toxic chemicals, due to the time required for the cells to repair the damage.

A lag is also observed when a population is transferred from a rich culture medium to a poorer one. This happens because for growth to occur in a particular culture medium the cells must have a complete complement of enzymes for the synthesis of the essential metabolites not present in that medium. On transfer to a new medium, time is required for the synthesis of the new enzymes.

Exponential phase   The exponential phase has already been discussed and was shown to be a consequence of the fact that each cell in a population divides into two cells.

For a given organism, growth rate varies with environmental conditions and with the composition of the culture medium. In a rich medium with many preformed cell constituents, growth rate is usually faster than in a poor medium, in which the cell must synthesize these things for itself. If an essential nutrient (that is, one that the cell cannot synthesize from other constituents) is reduced to a low amount, growth rate will be low, and as the concentration of this essential nutrient is increased, growth rate will increase (see Figure 7.6). As the concentration of the essential nutrient keeps increasing, eventually a concentration will be reached at which no further increase in growth rate occurs. The point has now been reached where this particular nutrient no longer limits growth but, on the contrary, some other constituent of the medium has now become the limiting factor. (This forms the basis for the use of some microbiological assay procedures: see Chapter 5.)

In the yeast industry, baker's yeast is grown on molasses as a source of carbon and energy. It has been found that, if all the molasses is added at once, some of the excess sugar is converted by the cells to alcohol and hence is wasted. In the early phases of exponential growth, when the cell densities are low, only a small amount of sugar is needed to support the maximum growth rate. But as time proceeds, the densities increase, and more sugar is needed. What is done in practice is to add the molasses at an exponential rate that parallels the exponential growth rate of the yeast.

Stationary phase   In a closed culture vessel a population cannot grow indefinitely at an exponential rate. Growth limitation occurs either because the supply of an essential nutrient is exhausted or because some toxic metabolic product has accumulated. The period during which growth of

Figure 7.6   Relation of growth rate to concentration of limiting nutrient. (From R. Y. Stanier, M. Doudoroff, and E. A. Adelberg: The Microbial World, 2nd ed. Englewood Cliffs, N.J.: Prentice-Hall, Inc., 1963.)

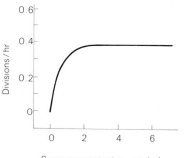

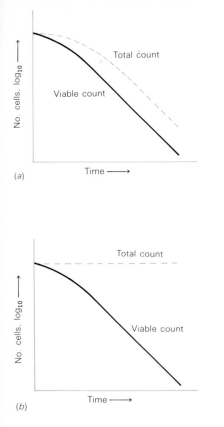

(a)

(b)

Figure 7.7 Change in total count and viable count during the stationary phase (a) if lysis occurs or (b) if lysis does not occur. (From R. Y. Stanier, M. Doudoroff, and E. A. Adelberg: The Microbial World, 2nd ed. Englewood Cliffs, N.J.: Prentice-Hall, Inc., 1963.)

a population ceases is called the *stationary phase* (Figure 7.5). Thus it is the activity of the organism itself that induces the changes in the medium leading to the stationary phase.

The population density reached at the stationary phase is called the *maximum crop*. The maximum crop is determined by a number of factors, including: (1) the species of organism and its efficiency in converting nutrients to cell mass; (2) the culture medium, especially the limiting-nutrient concentration; (3) environmental conditions (pH, temperature, and so on).

Death phase If incubation continues after a population reaches the stationary phase, the cells may remain alive and continue to metabolize, but often they die. If the latter occurs, the population is said to be in the *death phase*. As shown in Figure 7.7, during the stationary phase the direct microscopic count may remain constant but the viable count slowly decreases. In some cases death is accompanied by cell lysis, leading to a decrease in the direct microscopic count (Figure 7.7).

Continuous culture For many studies it is desirable to keep an organism growing at an exponential rate. A simple way of doing this is to keep diluting the culture into fresh medium as growth occurs. Automatic devices have been designed, in which fresh medium is added continuously at a slow, constant rate to a culture vessel; an overflow device removes surplus culture fluid plus organisms. One such device, called a *chemostat*, permits the control of both population density and growth rate at the same time. This continuous culture device (see Figure 7.8) acts by making one essential nutrient (for example, carbon source, nitrogen source, growth factor) a limiting factor, while all other nutrients are present in excess. Fresh culture medium with a limiting amount of the essential nutrient is pumped into the culture continuously, and at the same time surplus organisms and medium leave the vessel so that the volume remains constant. Since the supply of limiting nutrient is continuously replenished, the population in the chemostat continues to grow. Both growth rate and population density can be precisely controlled. The rate at which growth takes place is determined by the rate at which fresh medium is pumped in; the more rapid the rate of medium addition, the more rapid the growth rate. The population density, on the other hand, is controlled by the concentration of the limiting nutrient in the entering medium. In the chemostat we have a situation called a *steady state* since growth is occurring at a constant rate while population density neither increases nor decreases and at any one time the situation is exactly the same as it was at any other time. Such a time-independent state should be contrasted with time-dependent exponential growth, which occurs in a closed culture tube.

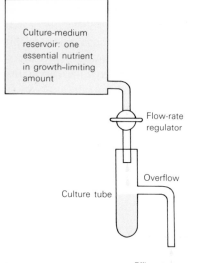

Figure 7.8 *Chemostat, a continuous-culture device. The population density is controlled by the concentration of the limiting nutrient, and the growth rate is controlled by the flow rate, which can be arbitrarily set.*

Figure 7.9 *Synchronous and asynchronous (random) growth. In synchronous growth, the bulk of the population divides at about the same time, whereas in random growth, divisions are occurring in different cells at widely differing times, even though the time required for each cell to divide is approximately the same. The synchronous population does not remain in synchrony indefinitely, but gradually becomes converted to a randomly growing population.*

Synchronous growth Even though a population may be growing at a constant rate, the cells do not all divide at the same time. If we begin with a single cell and follow the divisions of both this cell and of its offspring, we would find that initially the cell divisions of the offspring are reasonably well synchronized. As time proceeds, however, the synchrony disappears because of random variations in the time it takes for a cell to divide. Therefore in an exponentially growing culture under ordinary conditions only a small percentage of cells are actually dividing during any one period of time. For studies on the biochemical events related to cell

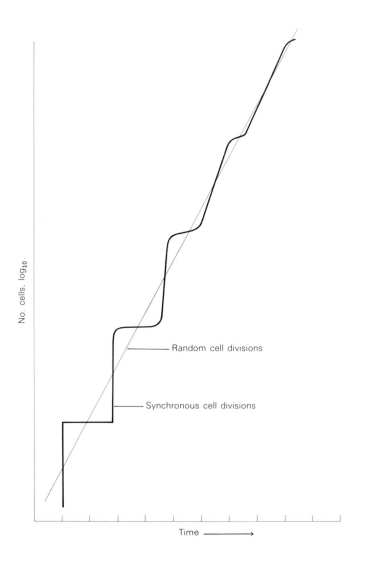

division it is desirable to produce populations in which growth is synchronous so that most or all of the cells divide within the same time period. The difference between nonsynchronous and synchronous growth of a population is shown in Figure 7.9. Synchrony in large populations can sometimes be induced by changing the chemical or physical properties of the environment, but the most common way is to select from the population that fraction of the cells that have just completed division, which are all at essentially the same stage of growth. This fraction consists of the smallest cells that can be selected by special filtration or centrifugation methods. When inoculated into fresh medium, these cells will undergo several synchronous divisions before asynchrony gradually returns.

7.5 Growth in nature

Growth kinetics in natural populations may resemble those in pure culture, but usually they are much more complex. This is because microbes in nature usually constantly compete with each other for limited food sup-

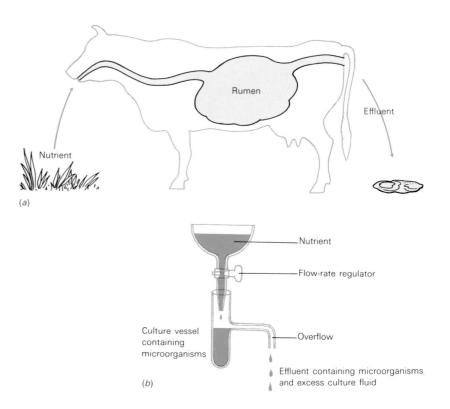

Figure 7.10 Microbial growth in the rumen of the cow (a) is analogous to growth in the chemostat (b).

Figure 7.11   Changes in RNA, DNA, and
protein content of a population during
growth: (a) balanced growth, (b) step-up
conditions. The arrow indicates the time
at which extra nutrient was added to
speed up growth. Note that, during the
step-up, the rate of RNA synthesis
increases first, followed by increases in
rates of protein and DNA syntheses.
[Adapted from N. O. Kjelgaard, O. Maaloe,
and M. Schaechter: "The transition
between different physiological states
during balanced growth of Salmonella
typhimurium," J. Gen. Microbiol. 19:607
(1958). Cambridge, England: Cambridge
University Press, 1958.]

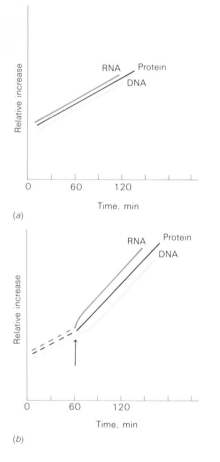

(a)

(b)

plies and because frequently they are consumed by parasites or predators.
Exponential growth in nature is rare; the growth curve more often resembles that in a chemostat. Let us imagine a bacterial population growing
in the intestinal tract of an animal (Figure 7.10). As food and digestive
products flow through the intestine, bacterial cells are carried down the
tract and are expelled with the fecal material, but the cells expelled are
replaced by new cells formed during growth. The population level in the
intestine itself remains relatively constant since wash-out is balanced by
growth. The generation time for *Escherichia coli* in the intestinal tract has
been calculated to be about 12 hr (two generations a day), whereas in pure
culture, *E. coli* grows much faster, often with generation times of 20 to
30 min. Conditions for microbial growth in nature are usually less favorable than those provided in pure cultures.

7.6   Macromolecular synthesis and growth

Balanced and unbalanced growth   During exponential growth, all biochemical constituents are being synthesized at the same relative rates
[Figure 7.11(a)], a condition called *balanced growth*. Figure 7.11(b) shows
what happens to the growth rates and the synthesis rates of RNA, DNA,
and protein if the culture is transferred to a different medium, in which
growth can occur at a faster rate than previously. (These are called
"step-up" conditions.) Immediately upon transfer to the richer medium
the rate of RNA synthesis increases, and somewhat later the rates of DNA
and protein synthesis also increase. The rate of cell division also steps up
after a longer lapse of time, and eventually the rates of synthesis of all
components are in balance again. In the initial period after the transfer
to the new medium we have conditions of *unbalanced growth* since not
all cell constituents are being synthesized at the same rate. If the reverse
experiment (a "step down") is performed, the rate of RNA synthesis
decreases immediately, while the rate of DNA and protein synthesis and
the rate of cell division continue at the previous (more rapid) rate, later
decreasing. These results suggest that, since the rate of RNA synthesis
is the first to change during step-up or step-down conditions, it may be
the key factor controlling growth rate.

Although there are several kinds of RNA in the cell (see below), the
largest fraction consists of RNA in ribosomes. Recall that ribosomes function in the synthesis of proteins (Section 2.5). During the step-up conditions, there is an increased rate of synthesis of ribosomes, leading to a
greater number of ribosomes per cell. Since protein synthesis occurs on
ribosomes, it is thus understandable that the rate of protein synthesis

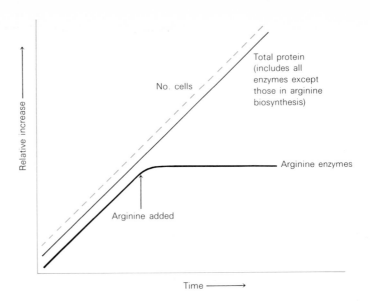

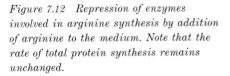

Figure 7.12 Repression of enzymes involved in arginine synthesis by addition of arginine to the medium. Note that the rate of total protein synthesis remains unchanged.

should start to change after the rate of ribosome synthesis increases.

Enzyme induction and repression In addition to factors such as those we have described that control the rate of synthesis of proteins in general, we can consider factors controlling the synthesis of specific enzyme proteins. Not all enzymes are synthesized by the cell at all times. The composition of the culture medium greatly influences the synthesis of specific enzymes. Often the enzymes catalyzing the synthesis of a specific product are not produced if this product is present in the medium. For instance, the enzymes involved in the formation of the amino acid arginine are synthesized only when arginine is not present in the culture medium; external arginine inhibits or *represses* the synthesis of these enzymes. As can be seen in Figure 7.12, if arginine is added to a culture growing exponentially in a medium devoid of arginine, growth continues at the previous rate, but the formation of the enzymes involved in arginine synthesis stops. Note that this is a specific effect, as the syntheses of all other enzymes in the cell are found to continue at the same rates as previously.

Enzyme repression is a very widespread phenomenon in microorganisms—it occurs during synthesis of a wide variety of enzymes involved in the biosynthesis of amino acids, purines, and pyrimidines. In almost all cases it is the final product of a particular biosynthetic pathway that represses the enzymes of this pathway. In these cases repression is quite specific, and the repressor usually has no effect on the synthesis of enzymes other than those involved in a single biosynthetic pathway. The value to the organism of enzyme repression is obvious since repression effectively insures that the organism does not waste energy synthesizing enzymes that it does not need.

A phenomenon complementary to repression is *enzyme induction*, the synthesis of an enzyme only when a substance is present. Figure 7.13 shows this process in the case of the enzyme β-galactosidase, which is involved in the utilization of the sugar lactose. If lactose is absent from the medium the enzyme is not synthesized, but synthesis begins almost immediately after lactose is added. Often enzymes involved in the breakdown of carbon and energy sources are inducible. Again, one can see the value to the organism of such a mechanism, as it provides a means whereby the organism does not synthesize an enzyme until it is needed.

The substance that initiates enzyme induction is called an *inducer*, and a substance that represses enzyme production is called a *repressor*; these substances, which are always small molecules, are often collectively called *effectors*. Not all inducers and repressors are substrates of the enzymes involved. The chemist can synthesize analogs of these substances that still induce or repress although they are not substrates of the enzyme. For instance, thiomethylgalactoside (TMG) is an inducer of β-galactosidase even though it cannot be hydrolyzed by the enzyme

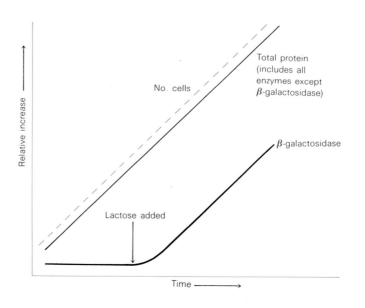

Fig. 7.13 Induction of the enzyme β-galactosidase upon the addition of lactose to the medium. Note that the rate of total protein synthesis remains unchanged.

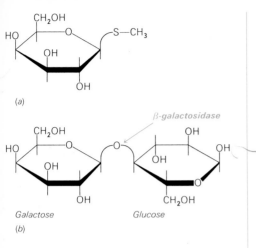

(a)

(b)

Galactose   Glucose

β-galactosidase

Figure 7.14 Structure of thiomethyl-β-galactoside (TMG) (a), a gratuitous inducer of β-galactosidase, an enzyme that acts on lactose (b). Although not a substrate for the enzyme, TMG is an excellent inducer of its synthesis.

(Figure 7.14). In nature, however, inducers and repressors are probably normal cell metabolites.

Mechanism of protein synthesis Even though enzyme inducers and repressors affect enzyme synthesis in opposite ways, it is now thought that they act very similarly. To understand this, we must first discuss the processes involved in the synthesis of specific proteins. The properties of each protein or enzyme of the cell are determined ultimately by its amino acid sequence. There are 20 different amino acids commonly found in proteins, and often 200 or more amino acid residues occur per protein molecule. How is the cell able to ensure that for each separate enzyme, the proper amino acids are connected in the correct order? The amino acid sequence of each protein is specified by a gene, which is a portion of the DNA molecule, and it is the sequence of purine and pyrimidine bases within this gene that *codes* for the amino acid sequence of the protein. The details of the genetic code will be discussed in Chapter 11, but the important point to understand for the present discussion is that the cell uses a triplet code, in which a sequence of three purine and pyrimidine bases codes for a single amino acid. We are concerned here with the mechanism by which the code in the base sequence of DNA determines the amino acid sequence of the protein.

The sequence of DNA purine and pyrimidine bases is first transcribed into a complementary sequence of purine and pyrimidine bases on an RNA molecule (Figure 7.15), and this RNA, containing the information for the amino acid sequence, is called informational or *messenger RNA* (mRNA). The mRNA now combines with a group of ribosomes to form a polyribosome (see Figure 2.25), and it is on this polyribosome that protein synthesis will occur.

How is the sequence of purine and pyrimidine bases in the RNA *translated* into the amino acid sequence of the protein? It has been well established that the amino acids themselves have no way of recognizing their triplet codes, and that this recognition step is accomplished through the agency of adaptor molecules consisting of RNA, called *transfer RNA* (tRNA). Each amino acid is first activated by a specific amino acid activating enzyme, using energy derived from ATP, and the activated amino acid is attached to a specific tRNA (Figure 7.15), there being at least one tRNA corresponding to each amino acid. The tRNA has on it a small sequence of purine and pyrimidine bases that *recognize* the complementary sequence on the mRNA coding for the amino acid. Each tRNA molecule and its associated activating enzyme thus can be viewed as the real translator of the genetic code since it is this system that recognizes both the genetic code and the amino acid and ensures that the proper

Figure 7.15 Mechanism of protein synthesis. By means of an amino acid-activating enzyme, the carboxyl end of an amino acid is attached to a specific transfer RNA (tRNA). The tRNA carrying this activated amino acid recognizes and attaches to its codon on the messenger RNA (mRNA), which is associated with a ribosome. At the peptide-forming sites the carboxyl end of the peptide chain is shifted from its tRNA to the amino end of the adjacent activated amino acid, thereby lengthening the peptide chain by one amino acid and releasing a tRNA. Although the entire process occurs on a single ribosome, many ribosomes can utilize the mRNA at the same time. The peptide chain remains attached to each successive tRNA to which it is shifted, and the newly liberated tRNA is free to accept another amino acid and reenter the cycle.

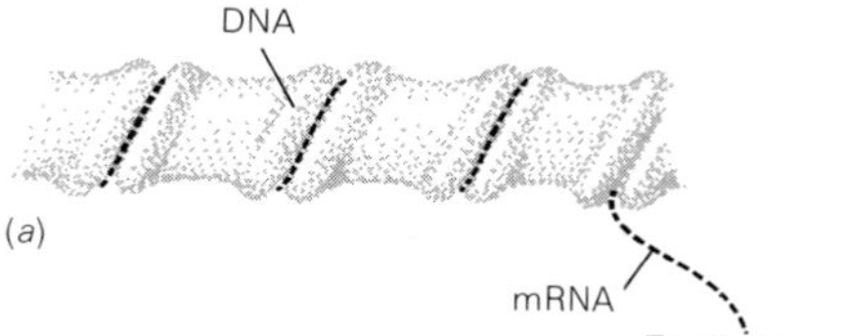

DNA

(a)

mRNA

To ribosome

mRNA

Beginning

Carboxyl end
Amino end

Amino acid

tRNA

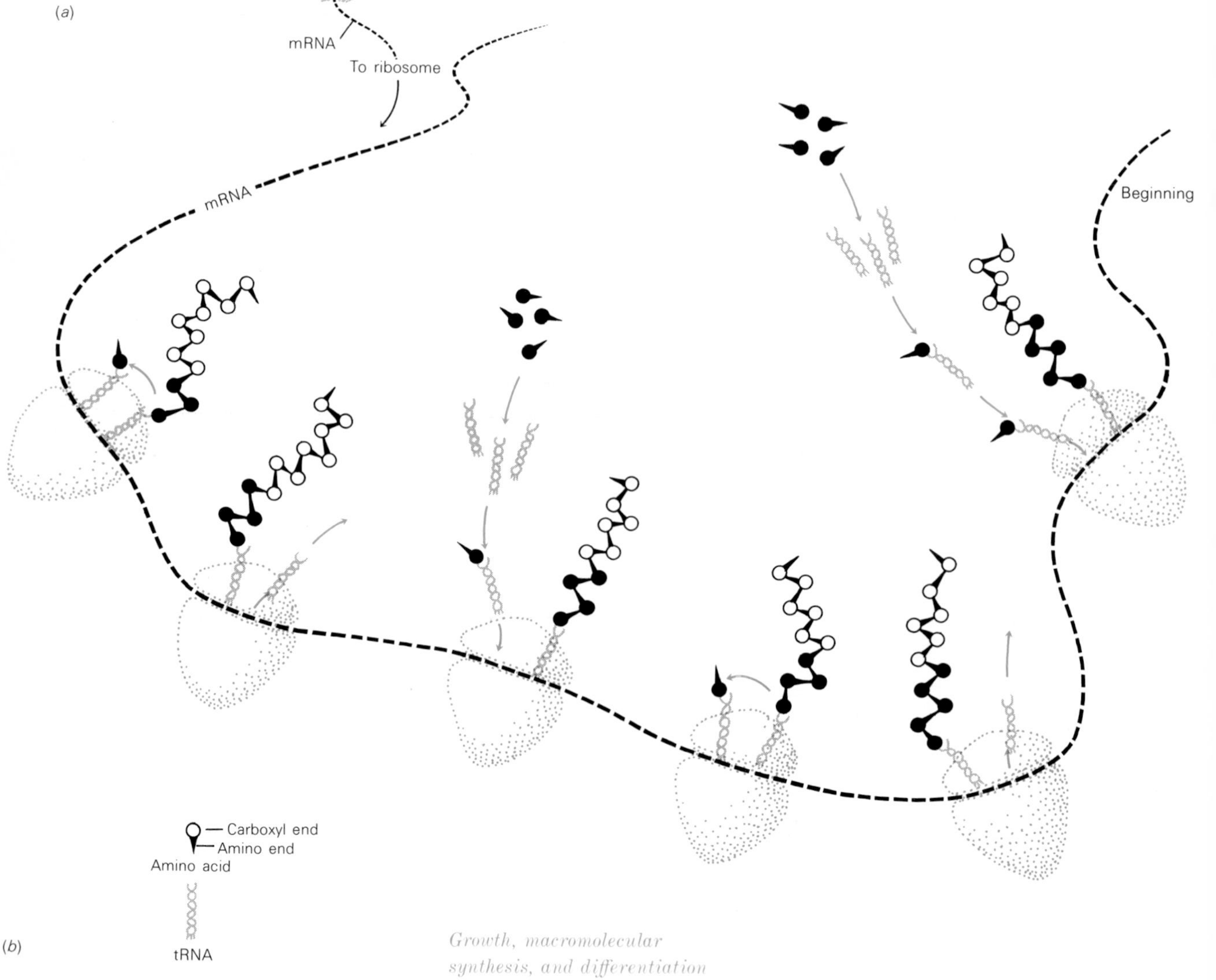

(b)

amino acid is used at the proper place in the growing polypeptide chain. The tRNA with attached amino acid now associates with the mRNA adjacent to the growing polypeptide chain on the polyribosome (Figure 7.15). As each amino acid is added, the tRNA to which it had been attached comes free, and can return to pick up another molecule of the same amino acid for addition to the polypeptide at a later place in the chain. Ultimately the polypeptide chain is completed, is released from the polyribosome, and folds into a specific configuration, which gives it its properties as an enzyme. The mRNA can then be used for the synthesis of another molecule of the same protein.

Mechanism of induction and repression   Let us now consider how the syntheses of inducible and repressible enzymes are controlled. In the case of a repressible enzyme the addition of the repressor inhibits further synthesis of the enzyme, and it is thought that the repressor (for example, arginine) combines with a specific *repressor protein* (sometimes called an *aporepressor*) that is present in the cell [Figure 7.16(*a*)]. The repressor protein is probably an allosteric protein (Figure 5.29), its configuration being altered when the repressor combines with it. This altered repressor protein can then combine with a specific region of the DNA at the initial end of the gene, the so-called *operator region*, where the synthesis of mRNA is initiated. The synthesis of the mRNA is blocked, and the protein specified by this mRNA thus cannot be synthesized.

For induction the situation is reversed. The specific repressor protein is thought to be active in the absence of the inducer, completely blocking the synthesis of the mRNA, but when the inducer is added it combines with the repressor protein and inactivates it. The inhibition of mRNA synthesis being overcome, the enzyme can be made [Figure 7.16(*b*)]. Induction and repression thus both have the same underlying mechanism, the inhibition of the synthesis of mRNA by the action of specific repressor proteins that are themselves under the control of specific small-molecule inducers and repressors. To emphasize the similarity between these two processes, induction is sometimes called *derepression*.

It should be emphasized, however, that not all enzymes of the cell are inducible or repressible. Those that are synthesized continuously are called *constitutive*. In fact, even inducible or repressible enzymes may become constitutive by genetic mutation. Two kinds of constitutive mutants are known: those that no longer produce the repressor protein, and those whose operator regions are altered so that they no longer respond to the action of the repressor protein. Although the net result of both of these mutations is the same, the two kinds of mutants can be distinguished by genetic tests (see Chapter 12). When the synthesis of a group of related enzymes is controlled by the same operator and repressor system, adding a single repressor or inducer to the medium alters the synthesis of all these related enzymes. A cluster of genes controlled by the same operator is called an *operon* (see Chapter 12).

A less specific type of enzyme repression is that called *catabolite repression*. In this phenomenon the syntheses of a variety of unrelated enzymes are inhibited when cells grow in a medium that contains a carbohydrate such as glucose as the main energy source. Catabolite repression is often called the *glucose effect* because glucose was the first substance shown to initiate it. Catabolite repression is thought to arise because some of the carbohydrate is converted into a common catabolite that is the real in-

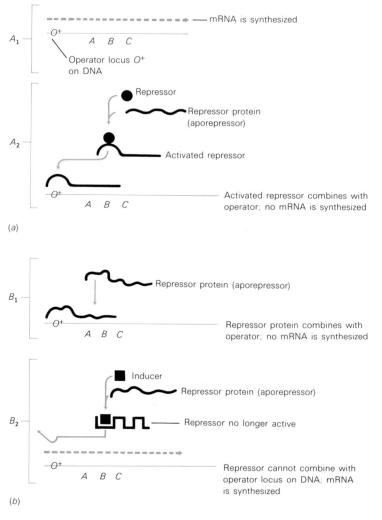

(a)

(b)

Figure 7.16   Roles of repressor in enzyme repression and induction: (a) repression, (b) induction. A_1, normal transcription; A_2, repression. B_1, inhibition of transcription; B_2, induction of transcription.

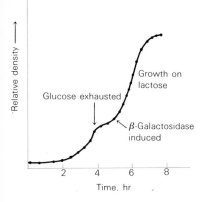

Figure 7.17 Diauxic growth on mixture of glucose and lactose. Glucose represses the synthesis of β-galactosidase. After glucose is exhausted, a lag occurs until β-galactosidase is synthesized, and then growth can resume on lactose.

ternal repressor. There is evidence that glucose-6-phosphate, which is formed from many carbohydrates, is an intracellular catabolite repressor; probably there are others.

One consequence of catabolite repression is that it can lead to so-called *diauxic* growth if two energy sources are present in the medium at the same time and if the enzyme needed for the utilization of one of the energy sources is subject to catabolite repression. Growth first occurs on the one energy source, and there is then a temporary cessation before growth is resumed on the other energy source. This phenomenon is illustrated in Figure 7.17 for growth on a mixture of glucose and lactose. The enzyme β-galactosidase, which is responsible for the utilization of lactose, is inducible, but its synthesis is also subject to catabolite repression. Thus, as long as glucose is present in the medium β-galactosidase is not synthesized; the organism grows only on the glucose and leaves the lactose untouched. When the glucose is exhausted, catabolite repression is abolished. After a lag β-galactosidase is synthesized, and growth on lactose can occur.

It might be asked whether enzyme induction and repression are worth all the trouble to an organism. Does it really matter much whether one or a few enzymes are synthesized when they are not needed? In the long view, yes. The synthesis of an unnecessary enzyme requires the expend-iture of energy that otherwise could be channeled into useful activities, and because of this the growth rate is usually slightly slower when an inducible or repressible enzyme is synthesized than when one is not. Thus, if two organisms in competition were otherwise identical except that one synthesized an enzyme all the time and the other synthesized it only when the inducer was present, the latter would eventually replace the former. When we remember that microorganisms have been competing with each other for millions of years, we can readily see the evolutionary advantages of repression and induction.

Induction and repression were discovered first in bacteria and have been studied most extensively in these organisms. Both phenomena are also found in eucaryotic microorganisms, as well as in higher animals and plants, although generally to a less dramatic extent.

As mechanisms for controlling metabolic rates, enzyme induction and repression should be contrasted with feedback inhibition, which was discussed in Section 5.5. Feedback inhibition is a mechanism essentially for immediate control of metabolic rate since it is a process that acts on preformed enzyme molecules. Enzyme induction and repression, on the other hand, are mechanisms that become significant only after consider-able periods of time. They can consequently be viewed as coarse control mechanisms whereas feedback inhibition is a fine control mechanism; the interaction of both types of mechanisms leads to precise control of cellu-lar function.

7.6 Macromolecular synthesis
and growth

Induction and repression are simple examples of *cellular differentiation;* the induced or repressed cell differs from its forerunner. Many microorganisms have evolved much more complicated differentiation mechanisms that include not only changes in enzyme content but changes in cellular structure as well. The study of cellular differentiation is often called *morphogenesis.* *

One of the most interesting examples of cellular differentiation is that of endospore formation in bacteria, during which the vegetative cell is converted into a nongrowing, heat-resistant structure, the endospore. The differences between the endospore and the vegetative cell are profound (Figure 2.42 and Table 7.2), and sporulation involves a very complex series of events, which are now just being elucidated. Bacterial sporulation does not occur when cells are dividing exponentially, but only when

* *Morpho-* means "form or structure"; *genesis* means "the origin of."

Table 7.2   Differences between bacterial endospore and vegetative cell

	Vegetative cell	*Spore*
Structure	Typical Gram-positive cell	Thick spore cortex Spore coat Exosporium (some species)
Microscopic appearance	Nonrefractile	Refractile
Chemical composition:		
Calcium	Low	High
Dipicolinic acid	Absent	Present
PHB	Present	Absent
Polysaccharide	High	Low
Protein	Low	High
Parasporal crystalline protein (some species)	Absent	Present
Sulfur amino acids	Low	High
Enzymatic activity	High	Low
Metabolism (O_2 uptake)	High	Low or absent
Macromolecular synthesis	Present	Absent
mRNA	Present	Low or absent
Heat resistance	Low	High
Radiation resistance	Low	High
Resistance to chemicals, acids, and so on	Low	High
Stainability by dyes	Stainable	Stainable only with special methods
Action of lysozyme	Sensitive	Resistant

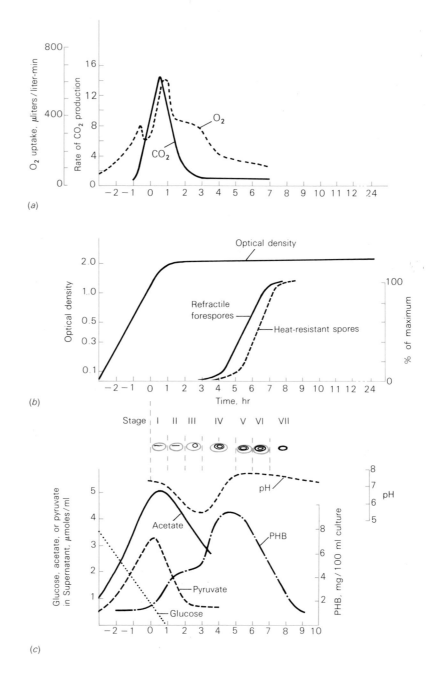

Figure 7.18 Biochemical and morphological changes during endospore formation in an aerobic bacillus. The events are initiated by the exhaustion of the energy source (glucose) from the medium (time zero). [(a) From H. O. Halvorson: J. Appl. Bacteriol. 20:305 (1957), and R. S. Hanson, V. R. Srinivason, and H. O. Halvorson: J. Bacteriol. 85:451 (1963). (b) Adapted from W. G. Murrell: in A. H. Rose and J. F. Wilkinson (eds.), Advances in Microbial Physiology, vol. 1. London: Academic Press, Inc. (London) Ltd., 1967. (c) From H. M. Nakata and H. O. Halvorson: J. Bacteriol. 80:801 (1960), and L. A. Kominek and H. O. Halvorson: J. Bacteriol. 90:1251 (1965).]

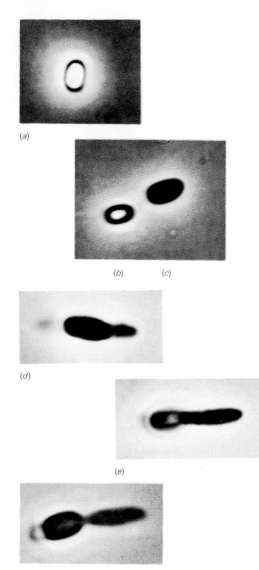

Figure 7.19 Photomicrographs showing the sequence of events during endospore germination in Clostridium pectinovorum: (a) ungerminated spore; (b, c) refractility is lost; (d–f) swelling and outgrowth. [From J. F. M. Hoeniger and C. L. Headley: J. Bacteriol. 96:1835 (1968).]

growth ceases due to the exhaustion of an essential nutrient. For instance, if a culture growing on glucose as an energy source exhausts the glucose in the medium, vegetative growth ceases and several hours later spores begin to appear. If more glucose is added to the culture just at the end of the growth period, sporulation is inhibited. Glucose probably prevents sporulation through catabolite repression (see above), inhibiting the synthesis of the specific enzymes involved in forming the spore structures. Glucose is not the only spore repressor; many other energy sources can also prevent spore formation. We thus see that growth and sporulation are opposing processes. It seems reasonable to assume that the elaborate control mechanisms in spore-forming bacteria ensure that in nature sporulation will occur when conditions are no longer favorable for growth.

The sequence of events in the sporulation process is shown in Figure 7.18. During exponential growth, glucose is fermented and fermentation acids accumulate, leading to a drop in pH. When glucose is exhausted, the population converts from fermentative to oxidative growth owing to the synthesis of enzymes of the citric acid cycle (see Section 4.4), and the organic acids are now used as energy sources. At this point there is a marked increase in oxygen uptake, and an increase in pH due to the uptake into the cells of the organic acids. At about this stage, the spore septum begins to appear. There is now a considerable breakdown of proteins within the cytoplasm in the region outside the developing spore, and the amino acids released are used in the synthesis of spore proteins and as additional energy sources. From now on, all the energy for sporulation must come from endogenous sources since those in the medium are exhausted. This energy comes from both protein and poly-β-hydroxybutyrate (PHB). New mRNA molecules, which are probably responsible for the synthesis of the new spore enzymes, are formed throughout the stages of the sporulation process until the mature spore is completed. The whole process of spore formation, from the time exponential growth ceases, takes many hours.

A spore is able to remain dormant for many years, but it can convert back into a vegetative cell (*spore germination*) in a matter of minutes (Figure 7.19). This process involves two steps: the cessation of dormancy, and outgrowth. The first is initiated by some environmental trigger such as heat. A few minutes of heat treatment at 60 to 70°C will often cause dormancy to cease. Once this happens outgrowth can occur if nutrients are present in the medium. The first indications of spore germination are loss in refractility of the spore, increased stainability by dyes, and marked decrease in heat resistance. The spore visibly swells and its coat is broken. The new vegetative cell now pushes out of the spore coat, a process called *outgrowth,* and begins to divide. The spore wall and spore coat eventually disintegrate through the action of lytic enzymes.

Supplementary readings

Attardi, G. "The mechanism of protein synthesis," *Ann. Rev. Microbiol.* 21:383 (1967).

Cellular Regulatory Mechanisms, Cold Spring Harbor Symposium on Quantitative Biology, vol. 26. 1961.

Gunsalus, I. C., and R. Y. Stanier (eds.) *The Bacteria: A Treatise on Structure and Function,* vol. IV, *The Physiology of Growth.* New York: Academic Press, Inc., 1962.

Kornberg, A., J. A. Spudich, D. L. Nelson, and M. P. Deutscher "Origin of proteins in sporulation," *Ann. Rev. Biochem,* 37:51 (1968).

Lamanna, C., and M. F. Mallette *Basic Bacteriology: Its Biological and Chemical Background,* 3rd ed. Baltimore: The Williams & Wilkins Co., 1965.

Maaløe, O., and N. O. Kjeldgaard *Control of Macromolecular Synthesis: A Study of DNA, RNA, and Protein Synthesis in Bacteria.* New York: W. A. Benjamin, Inc., 1966.

Mandelstam, J., and K. McQuillen, (eds.) *Biochemistry of Bacterial Growth.* Oxford: Blackwell Scientific Publishers, 1968.

Rose, A. H. *Chemical Microbiology,* 2nd ed. London: Butterworth & Co. (Publishers), Ltd., 1968.

Sussman, M. "Development phenomena in microorganisms and in higher forms of life," *Ann. Rev. Microbiol.* 19:59 (1965).

Synthesis and Structure of Macromolecules, Cold Spring Harbor Symposium on Quantitative Biology, vol. 28. 1963.

Watson, J. D. *Molecular Biology of the Gene.* New York: W. A. Benjamin, Inc., 1965.

Zeuthen, E. (ed.) *Synchrony in Cell Division and Growth.* New York: Interscience Publishers, 1964.

The activities of microorganisms are greatly affected by the chemical and physical conditions of their environments. An understanding of environmental influences helps us to explain the distribution of microorganisms in nature and makes it possible for us to devise methods for controlling microbial activities and for destroying undesirable organisms. Not all organisms respond equally to a given environmental factor. In fact, some environmental condition that may be extremely harmful to one organism may actually be beneficial to another. However, organisms can tolerate some adverse conditions under which they might not grow, and hence we must distinguish between the effects of these conditions on the viability of an organism and their effects on growth, differentiation, and reproduction.

A population will not be able to maintain itself indefinitely in a given environment unless it is able to grow. This explains in part the distribution of organisms in nature, and can serve as a basis for devising ways to prevent microbial growth in various practical situations. Effects of environment on viability are of most concern when we are interested in destroying unwanted organisms.

8.1 Temperature

Temperature is one of the most important environmental factors influencing the growth and survival of organisms. Temperature can affect living organisms in either of two opposing ways: (1) As temperature rises, chemical and enzymatic reactions in the cell proceed at more rapid rates and growth is faster. (2) Proteins, nucleic acids, and other cellular components are sensitive to high temperatures and may become irreversibly inactivated. It is usually observed, therefore, that, as the temperature is increased within a given range, growth and metabolic function increase up to a point where inactivation reactions set in. Above this point cell functions fall sharply to zero. Thus we find that for every organism there is a *minimum temperature* below which growth no longer occurs, an *optimum temperature* at which growth is most rapid, and a *maximum temperature* above which growth is not possible (Figure 8.1). The optimum temperature is always nearer the maximum than the minimum. These three temperatures, often called the *cardinal temperatures,* are generally characteristic for each type of organism, but are not completely fixed, as they can be modified by other factors of the environment such as pH and nutrition. Further, the cardinal temperatures of different microorganisms differ widely, and some have temperature optima as low as 5 to 10°C and some as high as 70 to 75°C. However, the temperature range

eight

The micro-organism in its environment

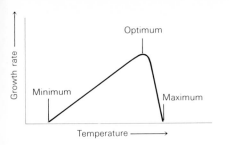

Figure 8.1 Effect of temperature on growth rate.

throughout which growth occurs is even wider than this, from below freezing ($-12°C$) up to boiling ($100°C$). No single organism will grow over this whole temperature range, however. The usual range for a given organism is about 30 to 40 degrees.

Cold environments Much of the world has fairly low temperatures. The oceans, which make up over half of the earth's surface, have an average temperature of $5°C$, and the depths of the open oceans have temperatures of around 1 to $2°C$. Vast land areas of the Arctic and Antarctic are permanently frozen, or are unfrozen only for a few weeks in summer. These cold environments are rarely sterile, and some microorganisms can be found alive and growing at any low temperature at which liquid water still exists. Even in many frozen materials there are usually microscopic pockets of liquid water present where microbes can grow—only at temperatures below -18 to $-20°C$ are such pockets absent.

Organisms that are able to grow at low temperatures are called *psychrophiles* or *cryophiles,* the accepted definition being that a psychrophile is an organism that is able to grow at $0°C$. However, even though they grow at $0°C$, many psychrophiles have optimum temperatures for growth at around 20 to $25°C$ (which optima many of them seldom or never experience in nature but meet only when they are cultured in the laboratory).

Psychrophilic algae are occasionally seen on the surfaces of snowfields and glaciers (Plate 1: Figure 8.2) in such large numbers that they impart a distinct red or green coloration to the surface. The alga is usually a species of *Chlamydomonas,* and the red color is due to a pigment present in the spores. Diatom algae are often found growing at temperatures of -1 to $0°C$ on the underside of floating sea ice in the Antarctic; only a small amount of light penetrates the ice, and the diatoms are adapted to grow in this very dim light.

Meat, milk and other dairy products, cider, vegetables, and fruits, when stored in refrigerated areas, provide excellent habitats for the growth of psychrophilic organisms. The growth of bacteria and fungi in foods at low temperatures can lead to changes in the quality of the food and eventually to spoilage. The lower the temperature, the less rapidly does spoilage occur; only when food is solidly frozen is microbial growth impossible. The great development of the frozen food industry in recent years was possible because of the much greater keeping qualities of frozen foods over merely refrigerated foods.

It is not completely understood how some organisms can grow well at low temperatures that prevent the growth of other organisms. At least one factor is that psychrophiles, as opposed to other organisms, have enzymes that are able to catalyze reactions more efficiently at low temperatures.

Perhaps as a consequence of this, the enzymes of psychrophiles are very sensitive to higher temperatures, being rapidly inactivated at temperatures of 30 to 40°C. Another feature of psychrophiles is that their active-transport processes function well at low temperatures, thus making it possible for the organisms to effectively concentrate essential nutrients. It is thought that this efficient low-temperature transport mechanism is due to peculiarities of the cell membrane, and psychrophiles seem to have a higher content of unsaturated fatty acids in the cell membrane than do other organisms.

Although freezing prevents microbial growth it does not always lead to microbial death. In fact, freezing is one of the best ways of keeping many microbial cultures for later study, and is used quite extensively in research laboratories. In general, large cells are more sensitive to freezing than are small cells, and those without cell walls are more sensitive than are those with walls. Even with organisms that can withstand freezing, some cells are usually killed when a culture is frozen, probably due to the formation of ice crystals in the protoplasm and consequent destruction of cellular integrity. If a cell survives the initial freezing it can then remain viable in the frozen state for long periods of time. This is really a state of suspended animation because cellular functions are completely cut off, and only when the culture is thawed can metabolism begin again. The long-term survival of the intestinal bacterium *Escherichia coli* in the frozen state was shown when scientists in the Antarctic recently discovered one of the huts used by the British explorer Sir Ernest Shackleton in 1913. Samples of frozen fecal material from this hut, when thawed and cultured, yielded viable *E. coli* cells.

In practice it is found that the lower the temperature at which a frozen culture is kept, the better it is preserved. Thus dry-ice temperature ($-70°C$) is better than ordinary deep-freeze temperature ($-20°C$); and liquid-nitrogen temperature ($-195°C$) is even better. In the case of animal cells, ordinary freezing, even if done rapidly, usually causes cell death, but if the cells are suspended in a medium containing about 15 percent glycerol and then cooled very slowly to dry-ice or liquid-nitrogen temperatures, many of the cells survive and can be maintained in the frozen state for long periods of time. Apparently glycerol alters the nature of the ice crystals so that they have less destructive effect on cellular integrity. With this procedure it is possible to preserve blood cells for transfusion or some tissues and organs for transplantation.

Warm environments   Microorganisms with optima in the range of 25 to 40°C are called *mesophiles*. Local heating from the sun can lead to temperatures in soils, plants, or cold-blooded animals of over 30°C, but the most common microbial environments with temperatures over 30°C

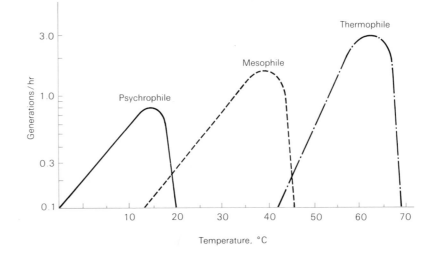

Figure 8.3 Relation of temperature to growth rates of a psychrophile, a mesophile, and a thermophile.

are the bodies of warm-blooded animals, the birds and mammals. These body temperatures range in different species from about 30 to 40°C (that of man is 37°C) but each species of warm-blooded animal maintains a remarkably constant temperature throughout its life, rarely varying more than 1°C, except during illness. The skin temperature is usually a few degrees lower than is the internal temperature, and it usually varies more than the internal temperature because it is influenced by the environmental temperature and the presence of hair, feathers, or other covering.

The temperature optimum of a microorganism living in a warm-blooded animal, whether it is a normal resident or a disease-causing organism, is usually very close to the normal temperature of the animal. For example, E. coli from man has a temperature optimum of about 39°C. However, many mesophiles will grow (but slowly) in culture at temperatures below 20°C. The minimum temperature of mesophiles is around 10°C, below which even slow growth is impossible. A comparison of the growth rates of a mesophile and a psychrophile at different temperatures is given in Figure 8.3. This figure shows that not only does the mesophile grow at higher temperatures than does the psychrophile, but also that the growth rate at its optimum temperature is faster, due to the fact that rates of enzymatic and chemical reactions speed up at higher temperatures.

High-temperature environments Organisms that grow at temperatures above 45 to 50°C are called *thermophiles*. Temperatures as high as these are found in nature only in certain restricted areas. For example, soils subject to full sunlight are often heated to temperatures above 50°C at midday, and darker soils may become warmed even to 70°C, although

Table 8.1 *Approximate upper temperature limits for different kinds of microorganisms[a]*

Organism	Temperature, °C
Most eucaryotic algae	40
Protozoa	45–50
Cyanidium caldarium (eucaryotic alga)	56
Fungi	60
Procaryotic (blue-green) algae	73–75
Bacteria	> 90

[a]Upper temperatures except for fungi determined from observations on hot-spring thermal gradients.

a few inches under the surface the temperature is much lower. Fermenting materials such as compost piles and silage usually reach temperatures of 60 to 65°C. However, the most extreme high-temperature environments are found in nature in association with volcanic phenomena.

Many hot springs have temperatures around boiling, and steam vents (fumaroles) may reach 130 to 140°C. Hot springs occur throughout the world but are especially concentrated in western United States, New Zealand, Iceland, Japan, the Mediterranean region, Indonesia, Central America, and central Africa. The area with the largest single concentration of hot springs in the world is Yellowstone National Park, Wyoming, where over 10,000 springs are located.

As the water overflows the edges of the spring and flows away from the source, it gradually cools, setting up a thermal gradient. Along this gradient, microorganisms develop (Plate 2: Figure 8.4), with different species within different temperature ranges. By studying the species distribution along such thermal gradients and by examining hot springs and other thermal habitats at different temperatures around the world it is possible to determine the upper temperature limits for each kind of microorganism (Table 8.1). From this information we conclude that: (1) procaryotic organisms in general are able to grow at temperatures higher than those at which eucaryotes can grow; (2) nonphotosynthetic organisms are able to grow at higher temperatures than can photosynthetic forms; (3) structurally less complex organisms can grow at temperatures higher than can more complex organisms.

The lower temperature limit for thermophily is arbitrary and depends somewhat on the kind of organism. Green algae that grow at 35°C are often called thermophiles, whereas thermophilic protozoa grow above 40°C (range, 40 to 45°C) and thermophilic fungi above 55°C (range, 55 to 60°C). Only blue-green algae that grow above 50°C would probably be called thermophiles, whereas for bacteria 55°C would be the lower limit.

Bacteria can grow over the complete range of temperatures in which life is possible, but no one organism can grow over this whole range. As mentioned above, each organism is limited to a restricted range of perhaps 30°C, and it can grow well only within a still narrower range. Even within the thermophilic group of bacteria there are differences. Some thermophilic bacteria have temperature optima for growth at 55°, others at 60°, and still others at 70°. Although thermophilic bacteria live in nature at temperatures as high as 85°C or over (Plate 2: Figure 8.5), no organisms have been cultured as yet that have optima greater than 70 to 75°C. The relationship of growth rate to temperature for a thermophilic bacterium is shown in Figure 8.3.

How can organisms survive and grow at these high temperatures? First of all, their enzymes and other proteins are much more resistant to heat than are those of mesophiles, just as the proteins of mesophiles are more stable to heat than are those of psychrophiles. Furthermore, the protein-synthesizing machinery (that is, ribosomes and other constituents) of thermophiles, as well as such structures as the cell membrane, are likewise more resistant.

Why is it that some groups of microorganisms have representatives that will grow at high temperatures, whereas others do not? We do not really know, and can only note that those groups that grow at the highest temperatures are those that are simplest in cellular structure, the procaryotes.

Thermophilic bacteria sometimes cause problems in industrial and food processes that use high temperatures. For example, in pasteurizing milk (see below), temperatures of 60 to 70°C are used, which although they are lethal to most bacteria are actually optimal for some thermophiles. Thermophilic bacteria can also cause difficulties in food-canning processes.

Destruction of microorganisms by heat Let us now consider the effects of temperature on viability. As temperature rises past the maximum temperature for growth, lethal effects become apparent. As shown in Figure 8.6, death from heating is an exponential function and occurs more rapidly as the temperature is raised. These facts have important practical consequences. If we wish to kill every cell, that is, to *sterilize* a population, it will take longer at lower temperatures than at higher temperatures.

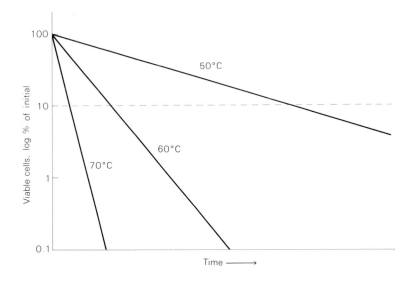

Figure 8.6 Effect of temperature on viability of a mesophilic bacterium.

It is thus necessary to select a time and temperature which under stated conditions will sterilize.

Water boils at 100°C at the atmospheric pressure of sea level. Sterilization times can be dramatically reduced by using temperatures higher than 100°C, which can be obtained by heating at increased pressure in devices such as autoclaves and pressure cookers. It may take hours to sterilize at 100°C, but only 15 to 20 min at 121°C. The inoculum size does not affect the rate of kill, but naturally does affect the time necessary for sterilization since the greater the number of organisms, the longer it takes for all of them to be killed. It is difficult to measure accurately exactly when sterilization has taken place since the presence of only a few viable cells subjects quantitation processes to great error. Most often sterilization is gauged by measuring the killing rate (which is given by the slopes of the lines in Figure 8.5) or by measuring the length of time in which a given fraction (for example, 90 percent) of the cells are killed. In the canning industry, however, it is very important to estimate the time it takes to achieve complete sterilization since even a few viable cells left in a container may be able to initiate growth and spoil the product. To be sure that a container has been completely sterilized and to provide a safety margin, the vessel and contents must be heated for a time longer than absolutely necessary to sterilize. However, since heating can alter the flavor of the food, it should not be carried out too much longer than necessary.

The precise mechanism by which heat kills organisms is not known—indeed, different mechanisms may exist for different organisms or even in the same organism at different temperatures. One theory for which some experimental support exists is that heat damages the plasma membrane and allows the essential cell constituents to escape.

Not all organisms are equally susceptible to heat. As might be expected, psychrophiles are the most heat sensitive, followed by mesophiles; thermophiles are the most heat resistant. Some thermophiles actually grow best under conditions that rapidly kill mesophiles and psychrophiles. Bacterial endospores are much more heat resistant than are vegetative cells of the same species. Since endospores are nearly always present in foods and other substances to be sterilized, heating times must be chosen that are long enough to kill all spores present. Spores of thermophiles are more heat resistant than are spores of mesophiles.

Why bacterial endospores are able to survive conditions of heating that rapidly kill vegetative cells is not completely understood, but probably involves several factors. Recall (Table 7.2) that endospores are high in calcium which is complexed with dipicolinic acid (DPA) (Figure 2.43). It is well established that both these materials are involved in some way in heat resistance since spores that lack DPA or those that have had their calcium replaced by strontium or barium are not so resistant to heat. The calcium-DPA complex is present mainly in the spore cortex and perhaps plays a role in stabilizing the cortex structure. The importance of an intact spore cortex in heat resistance is shown by the fact that, when germination occurs, loss of refractility (which is due to dissolution of the cortex) and decrease in heat resistance parallel each other. Another factor in heat resistance is the state of water in the protoplasmic constituents. Spore constituents are known to be less hydrated than are those of vegetative cells. Since the heat stability of proteins is much greater in the dry state than when hydrated, it is likely that this lack of hydration helps confer heat resistance.

The nature of the medium in which heating takes place also influences the killing process of vegetative cells or spores. Microbial death is more rapid at acidic pH values, and for this reason acid foods such as tomatoes, fruits, and pickles are much easier to sterilize than more neutral foods such as corn and beans. High concentrations of sugars, proteins, and fats usually increase the resistance of organisms to heat, while high salt concentrations may either increase or decrease heat resistance, depending on the organism. Dry cells (and spores) are more heat resistant than moist ones; for this reason, heat sterilization of dry objects always requires much higher temperatures and longer times than does sterilization of moist objects.

The transfer of heat to the object to be sterilized is not an instantaneous process. A bulky object requires a longer time to become thoroughly heated than does a small object, and the interior of an item may not reach sterilizing temperature so fast as does the periphery.

Pasteurization Pasteurization is a process using mild heat to reduce the microbial populations in milk and other foods that are exceptionally heat sensitive. It is named for Louis Pasteur, who first used heat for controlling the spoilage of wine. It is not synonymous with sterilization since not all organisms are killed. Originally, pasteurization of milk was used to kill pathogenic bacteria, especially the organisms causing tuberculosis, brucellosis, and typhoid, but it was discovered that the keeping qualities of the milk were also improved. Today, since milk rarely comes from cows infected with the pathogens mentioned above, pasteurization is used primarily because it improves the keeping qualities.

Pasteurization of milk is usually achieved by passing the milk continuously through a heat exchanger where its temperature is raised quickly to 71°C (161°F) and held there for 15 sec and then is quickly cooled, a process called *flash pasteurization*. Occasionally pasteurization is done by heating milk in bulk at 63 to 66°C (145 to 150°F) for 30 min and then quickly cooling. Flash pasteurization is more satisfactory in that it alters the flavor less and can be carried out on a continuous-flow basis, thus making it adaptable to large dairy operations.

8.2 Water

All organisms require water for life. Most microorganisms can grow only if liquid water is present, but some filamentous fungi are able to extract the water they need directly from the air. These fungi are common in very humid tropic air and often cause serious damage to cloth and leather goods, as well as to cameras and other optical instruments (Figure 8.7). The exact means by which they extract moisture from the air is not known; however, these fungi do grow faster in liquid media than in humid air.

Many microorganisms are remarkably resistant to drying. Small cells are usually more so than are large cells, round cells are more resistant than rod-shaped cells, and cells with thick cell walls (for example, Gram-positive microorganisms) are more resistant than those with thin walls. The spirochete *Treponema pallidum,* which causes syphilis, is a long cell with a thin wall and is so sensitive to drying that it dies almost instantly in air. On the other hand, *Mycobacterium tuberculosis,* which causes tuberculosis, is very resistant to drying because of its thick cell wall, which contains a heavy lipid coating. Bacterial spores, the sexual spores of algae and fungi, and the cysts of protozoa, resist drying much more than do the vegetative cells. Resistance to drying prolongs the survival of microorganisms that are dispersed through the air. A dry cell is metabolically dormant, and in the absence of heat or other external influences it may remain dormant but viable for long periods of time, and revive quickly upon the introduction of moisture.

Freeze-drying (lyophilization) is a laboratory procedure often used in the preservation of microbial cultures for future use. A suspension of cells, usually in milk or other protective agent, is frozen quickly in a dry ice–acetone bath and then dried from the frozen state by sublimation of the water under high vacuum. In this way, some of the destructive effects of drying can be avoided and even organisms that would be killed if dried in other ways can survive freeze-drying.

Drying is probably the oldest method of food preservation, and can be used for milk, meats, fish, vegetables, fruits, and eggs. Drying can be achieved simply through sunlight, but more often it involves the use of artificial heat in kilns, drums, and spray driers. Drying by heat always leads to changes in the flavor, texture, and color of foods, often of a drastic nature. Recently, freeze-drying of foods has been introduced. This is fairly expensive, but the food retains more of its natural characteristics.

Figure 8.7 Growth of a filamentous fungus on a camera lens during prolonged storage in a humid tropical environment. (Courtesy of Dr. W. Clark, Eastman Kodak Company, Rochester, N.Y.)

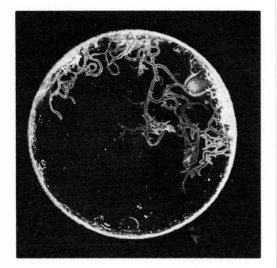

8.3 Salts, sugars, and other solutes

When a semipermeable membrane separates two aqueous solutions that have different concentrations, water will diffuse across the membrane towards the solution of higher concentration. The plasma membrane behaves as a semipermeable membrane, and many of the cell constituents dissolved in the cytoplasm do not readily penetrate this membrane. Microorganisms ordinarily live in rather dilute media, and because of the greater concentration of solutes inside than outside the cell, water tends to flow in. If unimpeded, the flow of water into the cell would make it swell and burst, but this tendency is counterbalanced by the tension pressure of the cell envelope (Section 2.3). It is possible to adjust the solute concentration outside the cell so that it exactly balances that inside, and at this point there will be no net movement of water in either direction. If the solute concentration of the medium is raised still further, water will now flow out of the cell. At high solute concentrations, the cell becomes dehydrated, and resembles a dried cell (Figure 8.8). This is the principle used in preserving foods by placing them in high concentrations of sugars and salts. Although solutes act mainly as dehydrating agents, such substances are not chemically inert and may have other effects on protoplasmic constituents. For instance, at high concentrations salts such as sodium chloride cause the precipitation of proteins, and if they should penetrate the cell, they would cause a drastic coagulation of the protoplasm.

The effect of increased salt concentration on the growth of microorganisms varies with the species. For instance, *Staphylococcus aureus* is able to grow in media containing 6.5 percent NaCl, whereas *E. coli* is inhibited by much lower concentrations. Because of these differences in salt sensitivity, a medium with 6.5 percent NaCl can be used to culture *S. aureus* selectively from intestinal contents where *E. coli* is the predominant species.

Sea water contains about 3.5 percent NaCl, plus small amounts of many other minerals and elements, and microorganisms found in the sea

Figure 8.8 Plasmolysis as a result of immersion of cells in a solution containing a high concentration of solute.

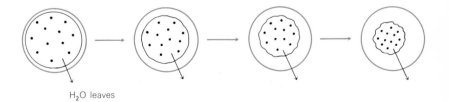

H$_2$O leaves

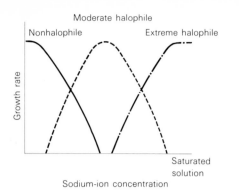

Figure 8.9 *Effect of sodium ion concentration on growth of microorganisms of different salt tolerances.*

usually require NaCl at concentrations similar to those of sea water. However, their growth is usually inhibited by higher or lower concentrations of salt (Figure 8.9). Marine microorganisms require Na^+ for the stability of the cell membrane and in addition many of their enzymes require Na^+ for activity; this requirement is specific for Na^+—it cannot be replaced by K^+ or another related ion.

Salt is often produced commercially by evaporating sea water in large basins, called "salt pans," where the NaCl concentration may reach 25 percent before crystallization occurs. Microorganisms that can grow in these brines are usually pink or red in color and are responsible for the brilliant coloration of salt pans (Plate 3: Figure 8.10). They are mostly bacteria, but the alga *Dunaliella salina* also lives under these conditions. Similar kinds of bacteria grow in food products preserved in salt; growth of red bacteria in salted fish leads to a pink discoloration.

Natural habitats containing high concentrations of salt are saline lakes such as the Great Salt Lake, Utah, and many smaller lakes of the Great Basin of western United States. Similar lakes are found in enclosed basins in other parts of the world; the best known are the Dead Sea, in what was formerly Palestine, and the Caspian Sea in Russia. Many saline lakes have an ionic composition similar to that of concentrated sea water, with NaCl predominating. Such lakes are called thalassohaline (*thalasso-* is a combining form meaning "sea"); the Great Salt Lake is in this category. Other saline lakes have an ionic composition quite different from that of concentrated sea water. The Dead Sea, for instance, has $MgCl_2$ as the predominating ionic constituent. These differences influence markedly the numbers and kinds of organisms that can grow in these waters. At the high salt concentrations found in saline lakes, the only organisms that grow well are certain bacteria and algae (Plate 3: Figure 8.11).

Microorganisms that require NaCl for growth are called *obligate halophiles,* whereas those which will grow in a NaCl solution but do not require it (for example, *S. aureus*) are called *facultative halophiles.* Among the obligate halophiles are organisms that show different degrees of NaCl requirement (Figure 8.9); those which grow only when the NaCl concentration approaches saturation are called *extreme halophiles.*

An extreme halophile, *Halobacterium,* has been much studied. It is a pink-colored Gram-negative rod, which is unable to form spores. Its specific requirement for Na^+ cannot be satisfied by replacement with the chemically related ion K^+. The Na^+ concentration inside the cells of this bacterium is similar to that outside, which means that the cells do not become dehydrated. When the NaCl concentration of the medium is reduced, the rod-shaped cells become spherical; if the concentration is lowered still further, the cells lyse. This is not a direct osmotic effect but is due to the fact that the cell wall of *Halobacterium* is stabilized by Na^+

ions: When there is insufficient Na^+ the wall breaks and the cell lyses. Many enzymes of *Halobacterium* require high Na^+ concentrations for activity, whereas the enzymes of nonhalophiles usually do not require Na^+. The extreme halophiles differ from other bacteria in other ways. For instance, the cell wall of *Halobacterium* apparently varies from that of other bacteria in that it lacks muramic acid. The ribosomes of *Halobacterium* also seem to be unusual, since they require high concentrations of K^+ for stability (ribosomes of other organisms have no K^+ requirement). The nutrition of *Halobacterium* is complex, and the organism requires many amino acids and vitamins for growth. Even under the best conditions, *Halobacterium* grows very slowly, showing a generation time of about 7 hr.

High concentrations of sugar are rarely found in natural environments, but occur in many preserved foods. There are a few yeasts and filamentous fungi that can grow in very sugary solutions. These organisms do not *require* high concentrations of sugar since they grow faster when the sugar concentration is lower. Food spoilage by these organisms is a slow process because of very slow growth rates, but once the sugar concentration has been sufficiently reduced through their activities, the rate of spoilage increases. Rarely are these sugar-tolerant fungi harmful to man, so that the discoloration and unsightly appearance of the moldy food are the main problems. Since they are usually aerobes, the fungi can be controlled in foods by excluding air; this is the function of the paraffin seal that is used on the home-preserved jam or jelly.

8.4 Hydrostatic pressure

A long column of water exerts pressure on the bottom of the column because of the weight of water; this is called *hydrostatic pressure*. In nature, high hydrostatic pressure is found only at the depths of the sea, and here it increases linearly with depth. Some parts of the ocean have depths of six miles, where the hydrostatic pressure is 1,000 times that at the surface. Such pressures completely inhibit the growth of most microorganisms, but the mechanism of this inhibition is unknown. In contrast, some bacteria that live on the ocean floor cannot grow at normal pressures; these are called *barophilic*. They must be cultured in the laboratory in special chambers under high pressure, which presents numerous difficulties as it is essential that all manipulations also be done under high pressure. Even more difficult is transferring a sample of water or mud from the ocean floor to the pressure chamber without decompression. Because of these technical problems, research work on barophiles has been limited.

8.5 Acidity and pH

The acidity or alkalinity of a solution is expressed by its pH value on a scale in which neutrality is pH 7 (Table 8.2). Those pH values which are less than 7 are acidic and those greater than 7 are alkaline. It should be

Table 8.2   The pH scale

	Concentration of hydrogen ions, g/liter	pH	Example
Increasing acidity	10^{-0}	0	
	10^{-1}	1	Gastric fluids
			Volcanic soils & waters
	10^{-2}	2	Lemon juice
			Acid mine drainage
	10^{-3}	3	Vinegar
			Rhubarb
			Peaches
	10^{-4}	4	Acid soil
			Tomatoes
	10^{-5}	5	American cheese
			Cabbage
	10^{-6}	6	Peas
			Corn, salmon
			Shrimp
Neutral	10^{-7}	7	Pure water
	10^{-8}	8	Seawater
	10^{-9}	9	Very alkaline natural soil
			Alkaline lakes
	10^{-10}	10	Soap solutions
Increasing alkalinity	10^{-11}	11	Household ammonia
	10^{-12}	12	Lime (saturated solution)
	10^{-13}	13	
	10^{-14}	14	

remembered that pH is a logarithmic function; thus a solution that has a pH of 5 is 10 times as acidic as one with a pH of 6.

The pH values of solutions are usually measured with commercial pH meters using glass electrodes. A wide variety of such instruments is available with different degrees of accuracy. Since many dyes show color changes at different pH values, pH can be measured simply although less accurately with these dyes. Indicator dyes can be added directly to the liquids to be measured or they can be incorporated into filter paper (which is then called "pH paper") and a small drop of the liquid to be measured is then added to the paper. Indicator dyes are often added to culture media to reveal acid production by bacteria. In natural habitats pH values range widely; representative examples are given in Table 8.2.

Microbial growth and pH range Each organism has a pH range within which growth is possible, and each usually has a well-defined optimum pH (Figure 8.12). Most natural environments have pH values between 5 and 9, and organisms with optima in this range are most common. Only a few species can grow at pH values of less than 2 or greater than 10.

Most yeasts and fungi are found to grow best in slightly acid media (pH 5 to 6). Many algae are tolerant of mild acidity, and a few are extremely acidophilic, whereas most blue-green algae grow optimally under slightly alkaline conditions. Although most bacteria grow best under neutral or slightly alkaline conditions, there are a few extremely acidophilic bacteria, of which the best example is *Thiobacillus thiooxidans*, which has been reported to grow at pH 0, or about the value of 1 N H_2SO_4!

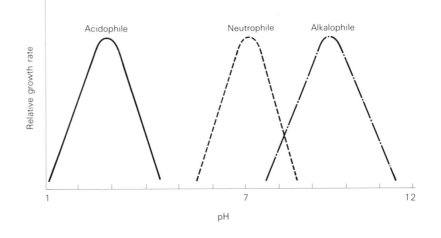

Fig. 8.12 Growth rates at different pH values of microorganisms with various pH tolerances.

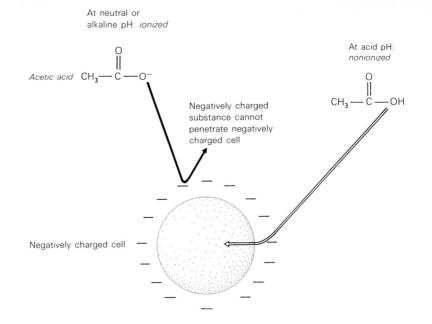

At neutral or
alkaline pH: *ionized*

At acid pH:
nonionized

Acetic acid

Negatively charged
substance cannot
penetrate negatively
charged cell

Negatively charged cell

Figure 8.13 Effect of pH on the penetration of an organic acid into a cell.

Although microorganisms are found in habitats over a wide pH range, the pH within their cells is probably close to neutrality. In an acid environment, the organism can maintain a pH close to neutrality either by keeping H$^+$ ions from entering, or by actively expelling H$^+$ ions as rapidly as they enter. Which mechanism is actually used is not known.

In addition to direct effects of pH on cells, there are indirect effects due to the ionization of organic compounds in the culture medium, since the nonionized form of most compounds can penetrate cells more readily than can the ionized form (Figure 8.13). With bases the situation is just the opposite: they will be nonionized at alkaline pH values.

Modification of pH by organisms   Organisms themselves through their activities alter the pH values of their environments. For instance, bacteria that ferment glucose to produce lactic acid (Section 4.3) will lower the pH of their environments, often as much as two pH units. Yeasts reduce pH by actively excreting H$^+$ ions into the medium. Because most ionized fermentation products are acidic, organisms are more likely to lower than to raise the pH of their environments. When the pH is raised through microbial action, it is usually because ammonia is released through the deamination of amino acids or other nitrogen-containing organic compounds. Environmental pH can also be altered by selective removal of substances from the environment. For instance, fungi growing on ammonium nitrogen (for example NH_4Cl) reduce the pH of the culture medium when removing NH_4^+, whereas those growing on a nitrate salt (for example, $NaNO_3$) raise the pH when removing the nitrate ion (NO_3^-); the physiological explanations for these changes are not known.

Because of such pH changes, it is difficult experimentally to keep the pH of a culture medium constant during growth. One method of doing this is to measure the pH at intervals during growth and then add alkali or acid as needed to return the pH to the desired level. It is often possible to keep the pH fairly constant by adding buffers to the culture medium; phosphate is a common buffer used to maintain the pH of media near neutrality.

Pickled or fermented foods   Acid is often used to prevent the growth of microorganisms in perishable foods, by a process usually called *pickling*. Foods commonly pickled are cucumbers (sweet, sour, and dill pickles), cabbage (sauerkraut), milk (buttermilk), and some meats and fruits. The food can be made acid either by adding vinegar or by allowing acidity to develop directly in the food through microbial action. In the manufacture of buttermilk, microorganisms play two roles. First, they produce lactic acid from the lactose of milk, causing a pH drop that induces coagulation of the casein and formation of a curd. In addition to acid, microorganisms also produce characteristic flavors that add to the quality of buttermilk. Originally milk contains a wide variety of microorganisms that can grow and cause undesirable changes. *Escherichia coli,* pseudomonads, bacilli, and many other bacteria find the nutrients in milk an excellent habitat. Once one of these organisms becomes established, its explosive increase leads to the development of detrimental conditions and the milk must be discarded. In the production of fermented milk products, conditions are used that promote the development of the lactic acid bacteria. By producing large amounts of lactic acid, these organisms lower the pH of the milk and make it no longer favorable for the development of other bacteria. Since the lactic acid bacteria are in general more tolerant to acidic conditions than are the other bacteria, once they gain a foothold they can continue to maintain themselves, even in the face of extensive competition from other organisms. A fermented milk has quite good keeping qualities compared to raw milk, but although this was the original reason for manufacturing these products, at present the keeping qualities are not so important as the flavor that develops through microbial action. Other fermented milks include sour cream, yogurt, acidophilus milk, and Bulgarian buttermilk. Fermented milks produced in some other regions of the world include kefir (Caucasus Mountains area), kumiss (Southern Russia), taette (Scandinavian countries), and skyr (Iceland).

Lactic acid bacteria also play important roles in the manufacture of cheeses. Their formation of acid causes the initial curd from which the cheese is manufactured, and some of their other metabolic products add to its final flavor.

A number of foods other than dairy products are prepared through the action of microorganisms. In some cases the action of the microbial agent results only in the production of desired flavors, while in other cases the fermentation acids produced serve as preservative agents. The microorganisms most commonly involved in food fermentations are the lactic acid bacteria, the acetic acid bacteria, and the propionic acid bacteria. In most cases the acids formed are derived from sugars occurring in the foods. The microorganisms involved in the fermentation may be present in the raw food when it is harvested from the field, or they may be added as starter cultures by the food plant operator.

When green vegetables such as cabbage, lettuce, spinach, cucumbers, and so on, are shredded or chopped and allowed to stand, a lactic acid fermentation usually develops from the action of naturally occurring lactic acid bacteria on the sugars in the liberated plant juices. Salt is usually added to the plant materials to prevent the growth of other bacteria, the lactic acid bacteria being more resistant to inhibition by salt than are most spoilage organisms. The amount of acid that develops depends on the amount of sugar in the food. If the sugar content is low, sugar can be added to increase the acid output.

Vinegar is also a microbial product, made by the action of acetic acid bacteria on alcoholic juices. The conversion of ethanol to acetic acid is an aerobic process:

$$CH_3CH_2OH + O_2 \longrightarrow CH_3COOH + H_2O$$

The most common vinegar is cider vinegar made from apple juice. The apple juice first undergoes an alcoholic fermentation brought about by yeast. The alcohol is then oxidized in the presence of O_2 to acetic acid by bacteria of the genus *Acetobacter*. In home methods of manufacture the alcoholic cider is placed in a barrel lying on its side, which has a hole open to the air. The *Acetobacter* develops as a thin film on the surface of the liquid and slowly oxidizes the alcohol to acetic acid. In commercial methods (Figure 8.14), the alcoholic juice is passed through a large vat of wood shavings aerated from the bottom. The bacteria develop on the wood shavings, whose large surface area provides a highly aerobic environment. As the juice passes down through the shavings the alcoholic juice is quickly oxidized to acetic acid. In starting a new acetic acid generator the *Acetobacter* must be established on the shavings by transfer from a previous tank. Once this has been done the process can be carried

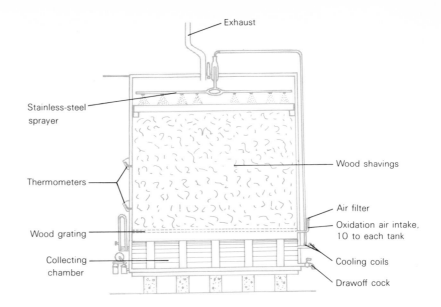

Exhaust

Stainless-steel sprayer

Thermometers

Wood grating

Collecting chamber

Wood shavings

Air filter

Oxidation air intake, 10 to each tank

Cooling coils

Drawoff cock

Figure 8.14 Diagram of a vinegar generator. The alcoholic juice is allowed to trickle through the wood shavings, and air is passed up through the shavings from the bottom. Acetic acid bacteria (Acetobacter species) develop on the wood shavings and convert alcohol to acetic acid. The acetic acid solution accumulates in the collecting chamber and is removed periodically. The process can be run in a semicontinuous fashion. (Courtesy of Food Engineering.)

out more or less continuously, with liquid being supplied from the top and vinegar drawn off from the bottom. The raw fermented vinegar is usually filtered and clarified, pasteurized to avoid later fungus growth, and diluted to a standard acidity, usually around 5 percent. Vinegar may also be made from wine (alcoholic grape juice) or from alcohol derived from a wide variety of sugary or starchy materials.

8.6 Oxidation-reduction potential

The concept of oxidation-reduction potential was presented in Chapter 4. The redox potential of the environment can fluctuate widely due to variations in availability of oxygen and to the activities of living organisms. Environments of low redox potential (that is, anaerobic environments) include: muds and other sediments of lakes, rivers, and oceans; bogs and marshes; water-logged soils; canned foods; intestinal tracts of animals; the oral cavity of animals, especially around the teeth (see Chapter 14); many sewage-treatment plants; deep underground areas such as in oil pockets; the sources of springs and other underground waters. In most of these habitats, the low redox potential is due to the activities of organisms, mainly bacteria, that consume oxygen during respiration. If no replacement oxygen is available, the habitat becomes anaerobic.

Anaerobic microorganisms Many microorganisms can grow in both aerobic and anaerobic environments; these are called *facultative*. Microorganisms that can grow only in anaerobic environments, on the other hand, are called *obligate anaerobes*. In aerobic habitats, obligate anaerobes cannot grow, and often they are killed. Why this should occur is not really understood. So far as is known, obligate anaerobes are found only in two groups of microorganisms, the bacteria and the protozoa. The best known obligately anaerobic bacteria belong to the genus *Clostridium*, a group of Gram-positive spore-forming rods. Clostridia are widespread in soil, lake sediments, and intestinal tracts, and are often responsible for the spoilage of canned foods. Other obligately anaerobic bacteria are found in the genera *Bacteroides*, *Fusobacterium*, *Ruminococcus*, and a few species of *Streptococcus*. Even among obligate anaerobes, the sensitivity to oxygen varies; some organisms are able to tolerate traces of oxygen whereas others are not.

The maintenance of anaerobic conditions in culture media requires great care, as even traces of oxygen may inhibit bacterial growth. Redox dyes such as methylene blue and resazurin are used to indicate the redox potential of the media since these dyes are colorless when reduced and colored when oxidized. Dissolved oxygen is driven out of the culture tube during heat sterilization and can be prevented from returning by use of an oxygen-impermeable seal, such as a rubber stopper. For the most oxygen-sensitive organisms, all operations such as inoculation and culture transfer must be carried out under a stream of nitrogen or helium gas, and culture tubes, bottles, and flasks must be incubated in a nitrogen atmosphere. The best way to obtain isolated colonies for pure culture is by the use of roll tubes, in which a thin film of agar is distributed around the walls of the tube by rolling it during the hardening process at the same time maintaining an oxygen-free atmosphere within the tube. Roll tubes are then sealed with an oxygen-impermeable rubber cap (not all kinds of stoppers are impermeable to oxygen). For incubation of small numbers of tubes or plates, anaerobic jars are used. Oxygen is removed from these by burning illuminating gas or H_2 inside the sealed container. It is much easier to grow obligate anaerobes in mixed culture with facultative aerobes because the latter consume traces of oxygen and maintain reducing conditions. The physiology of obligate anaerobes is poorly understood because of the difficulties in culturing them in the laboratory.

Facultative organisms include a wide variety of bacteria, such as members of the enteric bacteria, lactic acid bacteria, and many bacteria pathogenic to animals and man. Some protozoa, yeasts, and other fungi are also facultative. The physiology of facultative organisms is often profoundly different when they are grown anaerobically instead of aerobically. The cytochromes and other components of the respiratory system are

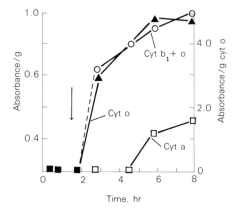

Figure 8.15 Effect of aeration on the synthesis of cytochromes a, b_1 and o in the facultative aerobe Staphylococcus aureus. At the time indicated by the arrow air was introduced into the culture. [From F. E. Frerman and D. C. White: J. Bacteriol. 94:1868 (1967).]

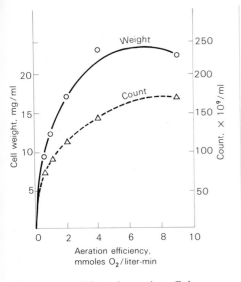

Figure 8.16   Effect of aeration efficiency on the cell yield of Serratia marcescens. Aeration efficiency was controlled by altering the rate at which air was supplied to the flasks. [From C. G. Smith and M. J. Johnson: J. Bacteriol. 68:346 (1954).]

often missing or greatly reduced in anaerobically grown facultative bacteria, and when oxygen is introduced, synthesis of these components rapidly resumes (Figure 8.15). Some organisms require a small amount of O_2 but are inhibited by larger amounts; these are called *microaerophilic*.

Aerobic microorganisms   Many bacteria and most fungi, algae, and protozoa are obligate aerobes, completely unable to grow in the absence of O_2. Oxygen is required for two purposes: It is a terminal electron acceptor in respiration (Section 4.4), and it is necessary for the biosynthesis of sterols and unsaturated fatty acids (Section 5.1). Because oxygen is not readily soluble in water, special means must often be provided to achieve adequate aeration in liquid cultures. Cotton plugs or other stoppers that keep out contaminants without impeding air movement are used, and rapid stirring of the liquid, either by shakers or by motor-driven stirring devices, is essential. The effect of aeration on the growth of an aerobic bacterium is shown in Figure 8.16. Poor aeration is a very frequent factor limiting laboratory growth of aerobic bacteria, especially when using tubes. However, when pure O_2 is used instead of air, many organisms are inhibited. The explanation of this surprising O_2 toxicity is still lacking. The problems of achieving appropriate aeration in industrial fermentors are discussed later in this chapter.

8.7   Radiation

Electromagnetic radiation   Two kinds of radiations are of biological interest: electromagnetic and ionizing. *Electromagnetic radiations* that have effects on living organisms include ultraviolet, visible, and infrared radiations (see Chapter 6). Ultraviolet (UV) radiation is of considerable microbiological interest because of its frequent lethal action on microorganisms. Although the sun emits intense UV radiation of all wavelengths, only that of longer wavelengths penetrates the atmosphere of the earth and reaches its surface. The UV radiations of short wavelengths, which are extremely lethal, do not penetrate through the earth's atmosphere (see Chapter 17); therefore microorganisms that may be wafted to high altitudes or dragged into space on the outside of rockets would be quickly killed by the UV there. Shorter-wavelength UV radiation is used in germicidal lamps and is usually lethal to microorganisms. The purine and pyrimidine bases of nucleic acids and the aromatic amino acids of proteins absorb UV radiation strongly, and this absorbed radiation may kill the cell. It is now well established that killing of cells by UV radiation of shorter wavelengths is due primarily to its action on DNA. The genetic effects of UV radiation on DNA are discussed in Chapter 11.

Dry cells are more resistant to the effects of UV than are wet cells, and spores are more resistant than are vegetative cells. Pigmented cells resist UV radiation better because most pigments absorb it, and thus less of the radiation reaches the sensitive nuclear material. Since death results mainly from effects on DNA, it is understandable that diploid or multi-nucleate cells are harder to kill than are haploid or uninucleate cells (Figure 8.17). Germicidal UV radiation passes very poorly through many kinds of glass, and not at all through opaque objects, so that its sanitary uses are limited, being restricted to the sterilization of organisms from air, where penetration is less of a problem, or from surfaces such as toilet seats and counter tops. Since UV light from germicidal lamps is very damaging to the eyes, it must be used with caution.

Many microorganisms have enzymatic mechanisms to repair damage induced in DNA by UV radiation. Damage in a given region of the DNA molecule is usually only on one of the two strands, and enzymes exist that will hydrolyze the phosphodiester bonds at each end of the damaged region and thus excise the altered material. Other enzymes then catalyze the insertion of new nucleotides that are complementary to the un-damaged strand. One kind of repair enzyme works only when the damaged

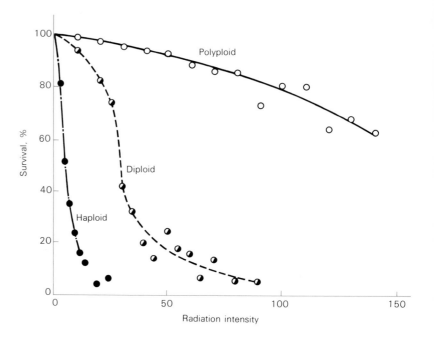

Figure 8.17. Effect of radiation intensity on the survival of haploid, diploid, and polyploid yeast. Note that the greater the number of chromosome complements per cell, the more resistant the cells are to radiation. [From C. S. Tobias: in J. J. Nicholson (ed.), Symposium on Radio-biology: The Basic Aspects of Radiation Effects on Living Systems. New York: John Wiley & Sons, Inc., 1962.]

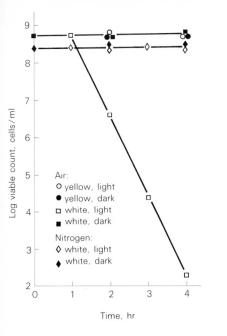

Figure 8.18 Effect of sunlight on the viability of white and yellow-colored Sarcina lutea. Note that death occurs in the white strain but not in the pigmented strain, from which it is inferred that the yellow pigment confers protection from sunlight. Killing does not occur anaerobically, but only in the presence of air. From this it is deduced that light-induced killing is a photooxidation reaction requiring oxygen. [From M. M. Mathews and W. R. Sistrom: Nature 184:1892 (1959).]

cell is activated by visible light in the blue region of the spectrum; recovery of UV radiation damage through action of these enzymes is called *photoreactivation* (Section 11.4). Other repair enzymes work in the absence of light (*dark reactivation*). Because of the existence of these repair enzymes, a given dose of ultraviolet radiation will cause death only in the event that the damage it induces is greater than that which can be repaired.

Visible light of sufficient intensity may actually lead to cell death. This is due to a process called *photooxidation,* in which the light absorbed by pigments in the cell causes inactivation of enzymes or other sensitive components when O_2 is present. In the absence of O_2, photooxidation cannot occur and the absorbed light is dissipated harmlessly (Figure 8.18). The photooxidative reaction brought about by visible light should be contrasted with that effected by UV radiation, in which O_2 is not required. Normal cell pigments such as the cytochromes and flavins can catalyze the photooxidation reactions induced by visible light. The destructive effects of sunlight on microorganisms are usually due to such photooxidative effects. Some microorganisms have special protective pigments, usually carotenoids, which prevent photooxidation because they are situated in the cell membrane. Here they absorb light and prevent it from reaching the sensitive regions of the cell. It is probably for this reason that many airborne bacteria are pigmented. Very few pathogenic bacteria survive long in bright sunlight, but *Staphylococcus aureus,* an orange-pigmented organism, does; it is significant to note that this pathogen is often airborne.

The role of visible light in photosynthesis and phototaxis was discussed in Chapter 6. Visible light is also involved in a number of other processes in microorganisms. Many fungi are induced to undergo sexual reproduction by light; this is especially true of mushroom-type fungi. Many filamentous algae grow toward the light; this process is called *phototropic* growth.

Ionizing radiation This is of much shorter wavelength than visible light, and has a much greater energy content. Examples of ionizing radiations are X rays, cosmic rays, and emanations from radioactive materials. Ionizing radiation does not kill by directly affecting the cell constituents but by inducing in the medium reactive chemical radicals (free radicals) (Figure 8.19). These free radicals can react with and inactivate sensitive macromolecules in the cell. Ionizing radiation can act on all cellular constituents, but death usually results from effects on DNA. Dark reactivation of damage due to ionizing radiation occurs, but photoreactivation is absent.

Ionizing radiation is being used to some extent to sterilize materials

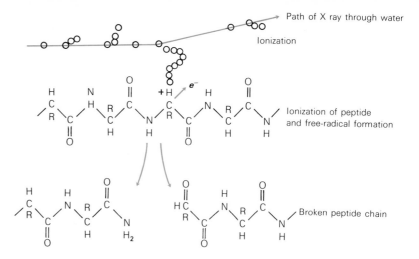

Figure 8.19 Production of a free radical by X ray passing through water and its action in inducing breakage of a peptide bond. (In part, from R. B. Setlow and E. C. Pollard: Molecular Biophysics. Reading, Mass.: Addison-Wesley Publishing Co., Inc., 1962.)

such as drugs, food, and other items that are heat sensitive. Since it penetrates solid objects well, it can sterilize prepackaged items. Ionizing radiation has also been used in sterilizing foods, especially meats. Although this process has been used to some extent by the military, it has not found routine use in civilian markets, partly because of cost factors and partly because ionizing radiation sometimes produces off-flavors and discoloration.

8.8 Interaction of environmental factors

When discussing the environment of an organism, we must consider all the various factors, physical and chemical. For simplicity, we have had to discuss each factor separately, but it should be emphasized that no environmental factor acts independently of the others and indeed one factor may greatly affect the action of another. For example, temperature may alter the response of an organism to low or high pH or affect its nutrient requirements. Although it is usual to study microorganisms which are growing under optimal conditions in the laboratory, we cannot assume that organisms also live under optimal conditions in nature. Natural environments are usually changing, and rarely is an environment optimal. Thus, our precisely controlled laboratory conditions may be optimal for the organism, but actually very unnatural.

A number of industrial processes make use of microorganisms; this necessitates understanding and applying the principles of microbial growth and physiology we have discussed in this chapter and the preceding ones. The manner in which the industrial process is carried out is determined by the nature of the desired product, the kind of organism involved, and various economic considerations. Industrial products are of three basic types. (1) The desired product is the microbial cells themselves, such as baker's or food yeast; (2) the desired material is a metabolic product of the microorganisms, such as an organic acid or alcohol, and antibiotic, an amino acid, a vitamin, an enzyme, or similar compound; (3) the organism forms a product of economic importance more or less directly from another substance (*bioconversion*).

Although we have used "fermentation" to describe the processes involved in the anaerobic oxidation of an organic energy source in the absence of an external electron acceptor (see Chapter 4), the industrial microbiologist uses the word in a different way. To him, any industrial process, whether involving a true fermentative metabolism or not, is a fermentation. Correspondingly, the large-scale vessel in which the process is carried out is called a *fermentor* (Figure 8.20). There are a number of problems in large-scale fermentation that the laboratory microbiologist does not meet. First, the volume of an industrial fermentor is usually huge. In most processes, vessels of 5,000- to 15,000-gal capacity are used. A 15,000-gal fermentor is usually as high as a three-story building. In vessels of such size, the cost of the ingredients of the culture medium is not incidental, and consequently cheap materials are used, such as corn steep liquor, distiller's solubles, molasses, and soybean meal. The media used are thus rather ill-defined, and much effort must be spent in finding a culture medium of industrial utility that permits the microbe to synthesize the same product it has made under laboratory conditions.

With such large volumes, even sterilization of the medium is difficult. Heat transfer into the center of a large vessel is slow, and consequently long heating times are necessary. Often such long heating times destroy heat-labile organic constituents of the medium or produce toxic materials. The medium can be sterilized directly in the fermentor by passing steam under pressure through a metal jacket that surrounds the outside of the vessel, or by continuous sterilization, in which the medium is passed through a heated chamber into the already sterilized fermentor. In some

Figure 8.20   An industrial fermentor installation. Only the upper portion of each fermentor is seen, the rest of the tank being below the floor. (Courtesy of Eli Lilly and Company, Indianapolis, Ind.)

Table 8.3 Outline of various microbial fermentations

Fermentation	Starting material	Organism	Conditions
Acids and alcohols:			
Acetic	Alcoholic juices	*Acetobacter*	Aerobic
Lactic	Malt extract; milk	*Lactobacillus*	Semianaerobic; pH controlled between 5.5 and 6.0
Citric	Molasses	*Aspergillus niger*	Aerobic; yields increased in media deficient in iron, zinc, and manganese
Acetone and butanol	Corn steep liquor; molasses	*Clostridium acetobutylicum*	Anaerobic
Glycerol	Molasses	*Saccharomyces cerevisiae*	Anaerobic; alkaline pH or addition of sulfite necessary to achieve good yields
Amino acids, vitamins, growth factors:			
Gibberellic acid	Sugar-containing substrates	*Gibberella fujikuroi*	Aerobic
Glutamic acid	Molasses	*Micrococcus glutamicus*	Aerobic; special mutants provide higher yields
	Corn steep liquor	*Brevibacterium flavum*	Aerobic
Lysine	Molasses	*Micrococcus glutamicus* (homo-serine-requiring mutant)	Aerobic
Riboflavin	Corn steep liquor; distiller's solubles	*Ashyba gossypii*	Aerobic
Vitamin B_{12}	Malt extract; corn steep liquor; distiller's solubles	*Streptomyces* species	Aerobic; addition of cobalt increases yields
Pharmacological agents:			
Dextran (α-1,6-glucose polymer)	Sucrose-containing substrates	*Leuconostoc mesenteroides*	Semianaerobic
Alkaloids	Infected rye kernels	*Claviceps purpurea*	Production occurs in field-grown rye plants
Steroid: 11-α-hydroxy progesterone	Progesterone	*Rhizopus nigrificans*	Aerobic; two-step process: progesterone added to fully grown culture

cases, the growth of the desired microbe is so rapid that contaminants cannot compete and sterile conditions are not necessary. This is true for the use of yeast in producing alcohol for distillation.

Many industrial fermentations are aerobic, and considerable attention must be given to providing adequate air to the fermentor. We have seen that the solubility of oxygen in water is limited. To achieve adequate aeration of a large fermentor, two procedures are used simultaneously: (1) The liquid is stirred violently by a large impeller; (2) air is introduced into the bottom of the fermentor through a metal tube with a series of fine holes, so that air is dispersed in fine bubbles. The smaller the bubble, the more rapidly does the oxygen pass into solution, and therefore the better the aeration. Naturally, the air entering the vessel must be sterile, but it is no easy matter to sterilize such large volumes of air. Usually a

Fermentation	Starting material	Organism	Conditions
Antibiotics:			
Penicillin	Corn steep liquor + sugar	*Penicillium chrysogenum*	Aerobic
Streptomycin	Soymeal extract	*Streptomyces griseus*	Aerobic
Tetracycline and derivatives	Peanut-meal extract; corn step liquor	*Streptomyces* species	Aerobic; low iron content
Erythromycin	Soymeal extract; corn steep liquor	*Streptomyces erythreus*	Aerobic; propionate and higher fatty acids increase yields
Polyene antibiotics	Soymeal extract	*Streptomyces* species	Aerobic; fatty acids increase yields
Bacitracin	Various complex media	*Bacillus licheniformis*	Aerobic; manganese increases yields
Beverages:			
Beer	Malt extract, usually with extracts of other grains (wheat, rice, corn) added	*Saccharomyces cerevisiae*	Anaerobic; temperature, yeast strain, and time of incubation vary with type of beer
Wine	Grape juice; occasionally other juices	*Saccharomyces cerevisiae*, var. *ellipsoideus*	Anaerobic; temperature and time of incubation vary with type of wine
Enzymes:			
Amylase	Corn meal; wheat bran	*Aspergillus* species	Aerobic; moist inoculated bran is tumbled in special drums
Invertase (sucrase)	Cane sugar (sucrose); salts	*Saccharomyces cerevisiae*	Aerobic; sugar added exponentially
Microbial cells:			
Baker's yeast	Molasses	*Saccharomyces cerevisiae*	Aerobic; sugar added exponentially
Mushrooms	Soil; horse manure	*Agaricus bisporus*	Special soil-containing beds; cool temperature

large air filter made out of steel wool packed in a long tube is used; as the air passes through the tube the microbes entrained in it impinge on the steel wool and are removed. To further ensure sterility, the air is also passed through a copper tube heated to a high temperature, where incineration of contaminants will occur.

Great care must be taken to keep the fermentor scrupulously clean. Any culture medium left within the fermentor or within its plumbing is a potential source of nutrients for the growth of contaminants, which may be difficult to eliminate.

A large fermentor is usually inoculated with a late-exponential-phase culture added in a volume that is 5 to 10 percent of the total volume of fluid in the fermentor. Thus, a 15,000-gal fermentor would require an inoculum of about 1,500 gal, a rather large volume itself. The inoculum

8.9 Control of microbial environment in industrial processes

is usually built up in a series of stages, starting with the initial culture growing on an agar slant that is used to inoculate an Erlenmeyer flask of 100- to 500-ml capacity. The culture obtained here is then diluted into a series of larger Erlenmeyer flasks, which are in turn inoculated into a small fermentor, perhaps of 100-gal capacity. The contents of this fermentor are then transferred to the 1,500-gal fermentor and from there to the 15,000-gal tank. At each stage it must be ascertained that the culture is behaving normally and that it is not contaminated. Quality control of the inoculum is one of the most important contributions of the industrial microbiologist to the fermentation process.

During fermentation itself, careful control of conditions within the fermentor is necessary. The heat produced during metabolism can lead to marked increases in temperature of the vessel, and to counteract this and to maintain a proper temperature, cold water is circulated through the fermentor jacket. Vigorous aeration usually leads to foaming, which is reduced by adding an antifoam agent such as lard oil or a silicone, usually through an automatic foam control system.

Control of pH, which is often desirable, is also done automatically. The pH inside the vessel is continuously monitored with a glass electrode. If the pH deviates from the desired range, concentrated acid or base is pumped in to bring the pH back to normal. The course of the fermentation is continuously recorded by monitoring dissolved oxygen and pH, and by removing samples from time to time for microscopic examination or assay for the desired product. Experience dictates when the fermentation has run its course, and harvest then begins. In industrial microbiology, the fermentor contents at this stage are called the ''beer,'' no matter what the product may be.

Harvesting consists of processing the fermentor contents in such a way as to remove the desired product. Microbial cells are removed by filtration or centrifugation. If the clarified liquid (the ''clear beer'' in industrial terminology) contains the desired product, it must then be processed to remove and purify this constituent. Since this moves us into the realm of chemical engineering, we will not discuss the further processing here except to say that the ease of this processing is often determined by the nature of the culture medium used; the industrial microbiologist may thus find himself with the task of discovering a more easily processed culture medium to substitute for one which he has already found best suited for the fermentation.

It should also be noted that a fermentation process that works admirably in the laboratory in flasks or in small-scale fermentors may work poorly or not at all under presumably similar conditions in large tanks. The reasons for this are often unknown but they require that the industrial microbiologist spend considerable time on the problem of *scale-up*. For this stage, the industrial company usually has a *pilot plant* in which the experimentation necessary for converting a laboratory process to an industrial one is carried out.

The goal of an industrial fermentation is to convert a cheap raw material into a useful end product. As such, the microbiological process must compete with synthetic chemical processes that may be able to perform the same task. Experience has shown that when the desired end product is a relatively simple organic molecule, chemical synthesis is usually preferable; however, if a complex molecule is desired, as for example one with many asymmetric carbon atoms or unstable intermediates, the microbial process is preferred. Table 8.3 lists some of the currently successful industrial fermentations. The successful management of industrial fermentation requires the combined efforts of chemical engineers and of microbiologists well trained in microbial physiology. Although at one time the microbiological aspects of industrial fermentations were carried out more or less by trial and error, currently there is great use of scientific experiment, based on modern concepts of microbial physiology.

Supplementary readings

Amerine, M. A., and R. E. Kunkee "Microbiology of winemaking," *Ann. Rev. Microbiol.* 22:323 (1968).

Brock, T. D. "Life at high temperatures," *Science* 158:1012 (1967).

Brock, T. D. *Principles of Microbial Ecology.* Englewood Cliffs, N.J.: Prentice-Hall, Inc., 1966.

Casida, L. E., Jr. *Industrial Microbiology.* New York: John Wiley & Sons, Inc., 1968.

Gunsalus, I. C., and R. Y. Stanier (eds.) *The Bacteria: A Treatise on Structure and Function,* vol. IV, *The Physiology of Growth.* New York: Academic Press, Inc., 1962.

Hollaender, A. (ed.) *Radiation Biology,* vol. I, pts. 1 and 2, *High Energy Radiation.* New York: McGraw-Hill Book Company, 1954.

Hollaender, A. (ed.) *Radiation Biology,* vol. II, *Ultraviolet and Related Radiations.* New York: McGraw-Hill Book Company, 1956.

Prescott, S. C., and C. G. Dunn *Industrial Microbiology,* 3rd ed. New York: McGraw-Hill Book Company, 1959.

Rainbow, C., and A. H. Rose (eds.) *Biochemistry of Industrial Micro-organisms.* New York: Academic Press, Inc., 1963.

Rose, A. H. (ed.) *Thermobiology.* New York: Academic Press, Inc., 1967.

ZoBell, C. E. *Marine Microbiology: A Monograph on Hydrobacteriology.* Waltham, Mass.: Chronica Botanica Co., 1946.

nine

Antimicrobial agents

An *antimicrobial agent* is a chemical that kills or inhibits the growth of microorganisms. Such a substance may be either a synthetic chemical or a natural product produced by a microorganism or other living being. Antimicrobial agents that are natural products are called *antibiotics*, although some antibiotics that were first discovered in nature are now synthesized in the laboratory.

Most antimicrobial agents are effective in extremely low concentrations, of the order of 1 to 10 parts per million (1 to 10 μg/ml). This is of special significance in practice, since it means that the agent can still be active even after it is highly diluted; hence it is possible to apply the agent in effective concentrations to animals or plants by injection or spray.

An important feature of many antimicrobial agents is *selective toxicity*, which means that the agent is more active on some kinds of organisms than on others. Some agents are highly specific and affect only a few organisms, others act on whole groups of organisms, whereas still others are even more unselective in their action. Agents that act selectively on disease-causing organisms without affecting human tissue are of course of most medical interest. The idea of selective toxicity was first advanced by the German scientist Paul Ehrlich and arose from his studies on the selective staining properties of certain dyes. Ehrlich observed that some dyes stain microorganisms but not animal or human tissue. He assumed that if a dye did not stain a tissue, the dye molecules were unable to combine with any of the cell constituents. He then reasoned that if such a dye had toxic properties it should not affect the animal cells because it could not combine with them, but it should attack the microbial cells. In an infected animal, chemicals of this sort should behave like ''magic bullets,'' striking the pathogen but missing the host. It was of course not necessary that the chemical be a dye; it only need be selective in its binding properties. Ehrlich proceeded to test large numbers of chemicals for selectivity and discovered the first chemotherapeutic agents, of which *Salvarsan,* a specific drug for the cure of syphilis, was the most famous.

Differences in selectivity frequently are seen in the action of chemicals on procaryotic and eucaryotic cells. When we consider the large number of differences between procaryotes and eucaryotes, selectivity of this sort should not be surprising. Within the procaryotes themselves, selectivity is often observed in the differential action of antimicrobial agents on Gram-positive and Gram-negative bacteria. Selectivity may also occur between plant and animal cells and between eucaryotic microorganisms and multicellular organisms. Complicated animals with nervous systems, musculatures, and so on, provide more possibilities for inhibition by chemicals than do less complicated organisms; thus a wide variety of substances affect higher animals without affecting microorganisms, lower animals, and plants. For control of infectious disease, a substance is needed that selectively attacks microorganisms without harming human cells; such a substance is called a *chemotherapeutic agent*. However, a substance need not be selective to be useful. Many chemicals are quite unselective but can be used effectively on nonliving objects; phenols, for example, are toxic to human beings but are the active ingredients of many commercial disinfectants.

Antimicrobial agents can affect cells in a variety of ways. Agents that kill cells are usually given names ending with *-cide* (for example, bactericide, fungicide, and so on), whereas those that inhibit growth without directly killing the cells are given names ending with *-stat* (for example, bacteriostat, fungistat, and so on). Agents that actually cause cell lysis are called *lytic* agents (for example, bacteriolytic).

9.1 Quantification of antimicrobial action

It is often desirable to compare the relative activities of a group of antimicrobial agents. Quantification of antimicrobial activity can be done by measuring the smallest amount of agent needed to inhibit growth or to kill a test organism. A suitable way to measure the inhibition of growth is to determine the *minimum inhibitory concentration* (MIC) of each agent. This is done by preparing a series of culture tubes, each containing medium with a different concentration of antimicrobial agent, and inoculating all the tubes with the same organism. The lowest concentration of agent that completely prevents appearance of turbidity is noted, and this concentration is known as the MIC. This simple and effective procedure is often called the *tube dilution technique*. The MIC is not an absolute constant, as it is affected by such factors as inoculum size, composition of medium, incubation time, conditions of incubation such as temperature, pH, and redox potential, and nature of the test organism. If all conditions are rigorously standardized, it is possible to compare a variety of antibiotics or other agents and determine which are most active. Alternatively, we can compare the activities of a single agent against a variety of microorganisms.

Another commonly used method of studying the quantitative aspects of antimicrobial action is the *agar diffusion method*. A molten agar culture medium is evenly inoculated with a suspension of the test organism and poured into petri plates. Known amounts of the antimicrobial agent are placed on filter-paper discs, which are then laid on the surface of the plates (Figure 9.1). During incubation the agent diffuses from the filter

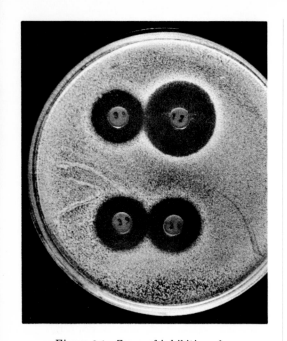

Figure 9.1   Zones of inhibition of Staphylococcus aureus formed around filter-paper discs containing the antibiotic penicillin. The agar plate had been inoculated with a culture of the test organism before the discs were applied. Three discs contain antibiotic of the same concentration and the fourth contains oxacillin, a more potent derivative of penicillin. (From V. Lorian: Antibiotics and Chemotherapeutic Agents in Clinical and Laboratory Practice. Springfield, Ill. Charles C Thomas, Publisher, 1966.)

paper into the surrounding zone, a concentration gradient is set up, and at some distance from the disc the MIC is reached. As the organism grows it forms a turbid layer, except in the region where the concentration of agent is above the MIC; here a zone of inhibition is seen. The size of the zone of inhibition is determined by the sensitivity of the organism, the nature of the culture medium and of incubation conditions, the rate of diffusion of the agent, and the concentration of the agent on the filter-paper disc. Some antibiotics diffuse much more readily in agar than do others, and hence it is not always possible to conclude that the antibiotic that gives the widest zone is the most active one against the test organism. For instance, substances of larger molecular weight diffuse less rapidly than do those of lower molecular weight, and some substances may diffuse poorly because they become bound to the agar; in both cases smaller zones of inhibition are produced.

In clinical medicine, in order to decide on the proper antibiotic for therapy, it is often necessary to determine to which one a pathogenic microorganism is most sensitive. Once the pathogen has been isolated in pure culture from the patient, its sensitivity to a variety of antibiotics can then be determined by either of the two techniques described above. Since the sensitivity of an organism to a group of antibiotics can easily be determined by agar diffusion, this method is most widely used.

It is often necessary to *assay* the concentration of an inhibitor that is present in unknown concentration. This is essential, for instance, in the industrial production of antibiotics and in clinical medicine, where it is often necessary to measure the inhibitor concentration in blood, urine, or other body fluids or tissues. Because the chemical composition of the agent may not be known at first or because the concentrations to be assayed are usually low, chemical methods cannot be used and only microbiological assays have sufficient sensitivity. Agar diffusion methods are most widely employed, as they require only small amounts of sample and equipment to carry out and are rapid. To assay an unknown concentration, one must first have a standard, that is, a pure sample of the inhibitor. A standard curve is prepared, in which zone sizes are determined with known concentrations of the standard. When the zone size is plotted against the logarithm of the inhibitor concentration, a straight line is usually obtained (Figure 9.2). Several dilutions of the unknown sample are then placed on filter-paper discs, and by measuring the zone sizes after incubation and comparing these with the standard, the concentration of unknown can be determined. There are many technical pitfalls associated with this technique, but when properly done the results can be reasonably precise.

To measure the lethal effect of an antimicrobial agent, viable counts

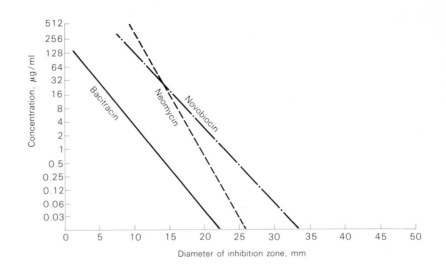

Figure 9.2 Relationship between antibiotic concentration and size of inhibition zone. Note that the antibiotic concentrations are plotted on a logarithmic axis. The slopes of the lines are different for the several antibiotics due at least partially to differences in rate of diffusion of the antibiotics through the agar. (From V. Lorian: Antibiotics and Chemotherapeutic Agents in Clinical and Laboratory Practice. Springfield, Ill.: Charles C Thomas, Publisher, 1966.)

Figure 9.3 The rate of killing of a microbial population by a chemical agent. Note that the rate is exponential.

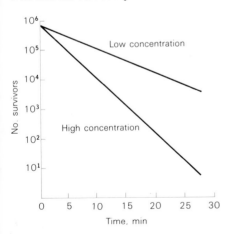

must be made. The agent is added to a suspension of cells and samples are removed at intervals to determine the fraction of surviving cells. One problem in studying the lethal effects of chemicals on cells is that the chemical must be inactivated or removed from the mixture before the viable count is performed, otherwise its presence in the medium used to determine viability may lead to growth inhibition and give a false picture of the number of viable cells. Sometimes the agent can be effectively removed by simple dilution of the mixture to the point where a growth-inhibitory concentration is no longer present, or the cells may be separated from the mixture by centrifugation or filtration techniques. Occasionally a substance is available that neutralizes the chemical, and this can then be added to the mixture or to the agar plate on which the viable count is to be performed. Very often the binding of the inhibitor to the cell is irreversible. This means that even after dilution or other procedure to separate cells from the agent, sufficient amounts of it may be bound to prevent further growth. In effect, a cell bound irreversibly with a drug is dead, although if we knew some means for displacing the drug from the cell we might enable the cell to recover and grow again. A cell population is not instantaneously killed by a chemical agent but death usually occurs at an exponential rate (Figure 9.3). This has important practical consequences, for it means that in order to achieve sterilization one must wait long enough for all cells to be killed.

In the rest of this chapter we discuss representative types of antimicrobial agents, their uses and modes of action.

9.1 Quantification of antimicrobial action

231

In Section 5.10 we discussed growth factors and defined them as specific chemical substances required in the diet because the organism cannot synthesize them. Substances exist that are related to growth factors, which act to block in some way the utilization of the growth factor. These "growth-factor analogs" usually are structurally similar to the growth factors in question but are sufficiently different so that they cannot duplicate the work of the growth factor in the cell. The first of these to be discovered were the *sulfa drugs*, the first modern chemotherapeutic agents to specifically inhibit the growth of bacteria; they have since proved highly successful in the treatment of certain diseases. These drugs were discovered by the German biologist Gerhard Domagk in the early 1930s in the course of testing a large variety of synthetic organic chemicals for ability to cure streptococcal infections. The first active compound was Prontosil (Figure 9.4), which was fairly effective in curing infected mice but which, surprisingly, had no effect against streptococci in culture. This difference was shown to be due to the fact that Prontosil is broken down in the animal body into sulfanilamide (Figure 9.4), which is the actual active antistreptococcal agent. With the discovery of this fact it was possible to embark on a program of synthesis based on the sulfanilamide structure, which yielded the large number of sulfonamide drugs we know today. D. D. Woods in England later showed that *p*-aminobenzoic acid (Figure 9.5) specifically counteracted the inhibitory action of sulfanilamide, and he also showed that the streptococci required *p*-aminobenzoic acid for growth. In fact, *p*-aminobenzoic acid is a structural part of the essential vitamin folic acid (Figure 9.5), and sulfanilamide acts as a competitive inhibitor of the enzyme that is involved in coupling *p*-aminobenzoic acid to make folic acid. The reason sulfonamides are active against bacteria and not against higher animals is that the latter utilize preformed folic acid present in the diet instead of synthesizing their own. The concept that a chemical substance could act as a competitive inhibitor of an essential growth factor has had far-reaching effects on chemotherapeutic research, and today analogs are known for various vitamins, amino acids, purines, pyrimidines, and other compounds. Despite this variety, the sulfonamide drugs still stand out as the most useful growth-factor analogs, being employed in the treatment of many infectious diseases.

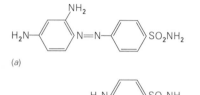

(a)

(b)

Figure 9.4 (a) *Prontosil and* (b) *sulfanilamide.*

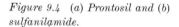

(a)

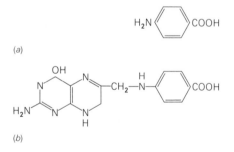

(b)

Figure 9.5 (a) *p-Aminobenzoic acid and* (b) *folic acid.*

9.3 Antibiotics

An antibiotic is usually defined as a chemical produced by one microorganism that is able to kill or inhibit the growth of other microorganisms.

Literally thousands of antibiotics have been discovered but only a few have practical use.

Microorganisms producing antibiotics All kinds of microorganisms produce antibiotics. Antibiotic-producing organisms occur very frequently in natural habitats such as soil, water, and mud, although producers probably are found most frequently in the soil. Many organisms produce more than one antibiotic, and some make as many as five or six. Among the bacteria, many antibiotic producers are found in the genera *Bacillus* and *Streptomyces*. Among the fungi, antibiotic producers are common in the genera *Penicillum* and *Aspergillus*. Table 9.1 lists some of the widely used antibiotics and the organisms that produce them. Antibiotics produced by algae and protozoa have as yet found little use.

Table 9.1   Common Antibiotics

	Taxonomic group of antibiotic-producing organisms
Penicillins and cephalosporins	Fungus
Griseofulvin	Fungus
Polypeptides:	
Bacitracin	Bacillus
Polymyxins and circulin	Bacillus
Actinomycins	Streptomyces
Polyenes:	
Nystatin	Streptomyces
Filipin	Streptomyces
Amphotericin B	Streptomyces
Glutarimides:	
Cycloheximide	Streptomyces
Chloramphenicol	Streptomyces
Tetracyclines:	
Tetracycline	Streptomyces
Oxytetracycline	Streptomyces
Chlortetracycline	Streptomyces
Macrolides:	
Spiramycin	Streptomyces
Carbomycin	Streptomyces
Erythromycin	Streptomyces
Oleandomycin	Streptomyces
Novobiocin	Streptomyces
Aminoglycosides:	
Streptomycin	Streptomyces
Dihydrostreptomycin	Streptomyces
Neomycin	Streptomyces
Kanamycin	Streptomyces

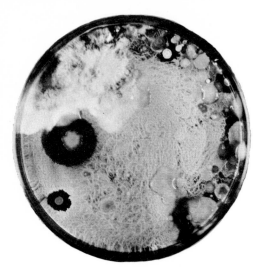

Figure 9.6 *Colonies of soil microorganisms in close proximity. The clear zones surrounding several colonies demonstrate antagonism. (Courtesy of Eli Lilly and Company, Indianapolis, Ind.)*

Antagonism among different microbes is often seen on agar plates on which a number of diverse microbial types are forming colonies close to each other (Figure 9.6). Proof that such antagonism results from the production of an antibiotic, rather than from simple competition for nutrients, requires the chemical isolation and purification of the antibiotic and the demonstration that the purified material has the same kind of antagonistic activity against the sensitive organism as did the original antagonist. This condition has been met numerous times, and there can no longer be any question that one of the most important factors in the antagonistic relationships seen on agar plates is the production and action of antibiotics.

Antibiotic structure Chemically, antibiotics are a diverse group, and although not just any organic chemical can have antibiotic activity, there seems to be no one type of chemical structure requisite. However, antibiotics can be grouped in families with similar chemical structures (Table 9.1).

The sensitivity of microorganisms to antibiotics varies; Gram-positive bacteria are usually more sensitive to antibiotics than are Gram-negative forms, although some antibiotics act only on Gram-negative bacteria. Broad-spectrum antibiotics act on a wide variety of organisms, whereas narrow-spectrum antibiotics have more restricted ranges of activity (see Figure 9.7). The great difference in antibiotic sensitivity of procaryotes and eucaryotes is noteworthy. Penicillin, for instance, acts only on procaryotes, and even in massive doses has no effect on the growth of eucaryotic cells. Cycloheximide, on the other hand, is very active against eucaryotes but has no effect on procaryotes. These differences are dramatic examples of selective toxicity, referable to the profound differences in cellular organization between procaryotes and eucaryotes.

Figure 9.7 *Range of activity of several antibiotics against various microbial groups. [Adapted from V. J. Cabasso: BioScience 17:796 (1967).]*

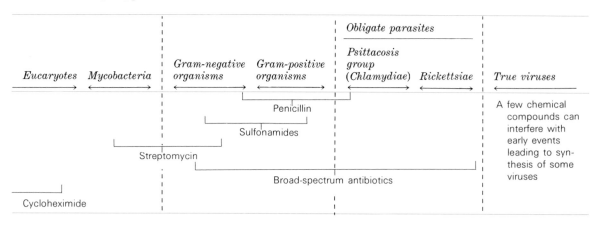

Occasionally it is possible to change the spectrum of an antibiotic by modifying its chemical structure. The most dramatic example of this has been the penicillins and cephalosporins, whose structure can be visualized in two parts, a basal structure (6-aminopenicillanic acid) and a side chain connected to the basal structure by a peptide bond. (Figure 9.8). The basal structure is constant, but the side chain can vary widely. The most common naturally occurring derivative is benzyl penicillin (penicillin G), but many other derivatives are produced naturally by various fungi, or have been synthesized using the basal structure. The differences in biological activity of several penicillins and cephalosporins are shown in Figure 9.8. Note that benzyl penicillin is highly active against Gram-positive bacteria

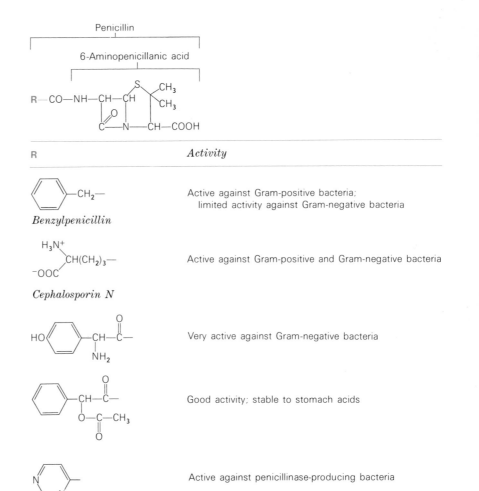

Figure 9.8 Structure of the penicillin "nucleus" and a few of the many possible side chains. Note that several properties of the penicillins are determined by the nature of the side chain.

Figure 9.9 Method of testing a microbial isolate for antibiotic production. Four of the six test organisms are sensitive to the antibiotic produced by the isolate. (Courtesy of Eli Lilly and Company, Indianapolis, Ind.)

but shows considerably less activity against Gram-negative organisms; cephalosporin N, on the other hand, shows less potency against Gram-positive organisms than against Gram-negative forms.

The search for new antibiotics   Seeking new antibiotics has occupied the time of a large number of microbiologists in pharmaceutical companies for the past generation. The most widely used procedure is what is called the *screening approach*. A large number of possible antibiotic-producing organisms are isolated from natural environments, and each is tested against a variety of organisms to see if antagonisms exist. The battery of test organisms usually is large and includes Gram-positive, Gram-negative, and acid-fast bacteria, yeasts and other fungi, and perhaps protozoa and algae. Antitumor agents are sought using tumor tissue cultures as test organisms, and virus-infected animal-cell cultures are used to screen for antiviral agents.

A convenient procedure in examining for antibacterial agents is to streak the possible antibiotic producer along one chord of an agar plate and incubate the plate for a few days to allow the organism to grow and produce antibiotic. Then a series of test bacteria are streaked at right angles to the chord, and the plate is reincubated and examined for zones of inhibition (Figure 9.9). Since different antibiotics attack different groups of organisms, it is desirable that any antibiotic screening program use a wide range of test organisms to make certain that few antibiotic-producing organisms are missed. It should be obvious that appropriate medium and culture conditions permitting the production of the antibiotic are necessary since environment can greatly affect antibiotic production.

Once antibiotic activity is detected in a new isolate, the microbiologist will determine if the agent is new or old. With so many antibiotics already known, the chances are good that the antibiotic is already known. Often simple chemical methods will permit the characterization and identification of the antibiotic. If the agent appears new, larger amounts are produced and the antibiotic purified and sometimes crystallized. Finally, the antibiotic would be tested for therapeutic activity, first in infected animals, later perhaps in man (see Chapter 15).

Most screening programs concentrate on antibiotics produced by the genus *Streptomyces* (Section 18.6), a group of organisms that has yielded most of the medically useful antibiotics now known (for example, chloramphenicol, streptomycin, neomycin, erythromycin, cycloheximide, tetracycline, and so on). Antibiotics researchers therefore are very familiar with this group. *Streptomyces* occur predominantly in the soil, and most screening programs collect soil samples from a wide variety of geographical areas, under the assumption that soils from new places may yield new antibiotic producers. However, it does not follow that soils from distant parts will yield more new antibiotics than would local soils. In fact, since the wind-borne spores of *Streptomyces* are widely dispersed, species producing the same antibiotics are found in similar kinds of soils throughout the world.

We have stated that organisms often produce more than one antibiotic, and sometimes as many as five or six. Multiple antibiotic production presents special problems, as activity against a test organism may be due to the concerted action of the several agents. The different antibiotics may all be closely related chemicals, or they may differ markedly. They may all be active against a single test organism, or they may show different spectra of activity. The presence of multiple antibiotics presents further problems and challenges for the microbiologist and chemist: Which antibiotic in the mixture is likely to be medically useful, how can its production be favored over the others, and how can it be purified? Antibiotics research presents endless puzzles and fascination for the microbiologist; the interesting work has only begun when a new antibiotic is discovered.

Mode of action of antibiotics   At present we have a considerable understanding of the molecular bases for antibiotic activity and selective toxicity. However, it was first necessary to understand the processes involved in normal cell function and growth; only then was it possible to see how these processes are altered by antibiotic action. In recent years, our knowledge of the molecular basis of microbial growth and function has advanced dramatically (see Chapter 7), and these advances have made possible an increased understanding of the mode of action of antibiotics.

Antibiotic molecules can affect cells only if they are able to bind to vital cell constituents since there are no long-range forces that enable antibiotics to act at distances. Antibiotic molecules are able to bind selectively to certain cell constituents, such as enzymes, ribosomes, nucleic acids, and so on, without binding to other constituents, just as substrate molecules are bound selectively by the enzymes that act on them (Chapter 4). It is this selective binding that is at the root of useful antibiotic activity.

Several antibiotics affect *cell wall synthesis*. The structure of the basal cell wall (peptidoglycan) of procaryotes was described in Section 2.3. Recall that this is a structure containing polysaccharide chains cross-linked by peptides that contain D-amino acids, the latter being found only in the cell wall of bacteria and not in cell proteins. The antibiotic *cycloserine* (Figure 9.10) is an analog of D-alanine and prevents its synthesis; in the absence of this essential cell-wall amino acid further cell-wall synthesis stops, and lysis occurs. *Penicillin* also specifically inhibits cell-wall synthesis; it acts

Figure 9.10   (a) Cycloserine and (b) D-alanine.

(a)

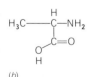

(b)

by inhibiting the enzyme that carries out the cross-linking function connecting two peptides on adjacent polysaccharide chains. Other antibiotics that act on the formation of cell walls are vancomycin and ristocetin. Inhibition of cell-wall synthesis is often followed by cell lysis because the autolytic enzymes that play a role in cell-wall growth (Section 2.3) can continue to function and eventually create holes in the cell wall, through which the protoplasm extrudes. In nongrowing cells, the autolytic enzymes usually are not functioning, so that penicillin and cycloserine usually do not cause lysis of nongrowing cells; active multiplication is necessary for their activity. This has important consequences in the clinical use of penicillin; nongrowing bacteria may persist in foci in the body during penicillin treatment, and may begin to multiply after therapy has been stopped. Antibiotics that inhibit cell-wall synthesis cause cell death by effecting cell lysis; when lysis is prevented by adding an osmotic stabilizer such as sucrose to the medium (see Figure 2.20) a spheroplast is produced. If the antibiotic is later removed, cell-wall synthesis can resume and the cell can again grow and divide. The cell-wall structure of eucaryotic microorganisms is quite different from that of procaryotes (Section 3.1), and antibiotics such as penicillin and cycloserine have no effects on eucaryotes.

Some antibiotics are able to combine with *cell membranes* and destroy the permeability barrier. *Polymyxin, circulin, gramicidin,* and *tyrocidin* act in this way; polymyxin B can serve as a representative example. Its structure, shown in Figure 9.11, is such that it has a positively charged region and a lipid-soluble region. The bacterial cell membrane is high in phospholipid (a negatively charged substance), and polymyxin molecules combine with the membrane in such a way that the positively charged part of the polymyxin molecule associates with the negatively charged part of the phospholipid, and the lipid-soluble region of polymyxin combines with the rest of the phospholipid. This leads to a disruption of the cell-membrane structure and causes loss of semipermeability, leakage of cytoplasmic constituents, and cell death. Metals such as Mg^{2+} can block polymyxin action by combining with the negative phospholipid groups, thus preventing the approach of the polymyxin molecules. Polymyxin is especially active against Gram-negative bacteria which, it will be recalled, have high lipid content in their cell envelopes, and the antibiotic is thus clinically useful for controlling such infections as those caused by *Pseudomonas* species. The antibiotic is also fairly toxic to human beings and hence must be used with caution.

One large group of antibiotics, the *polyenes,* act on the cell membranes of fungi and other eucaryotes but have no effect on the membranes of procaryotes. Representative examples are *nystatin, filipin,* and *amphotericin.* Recall that sterols are essential constituents of the cell membranes

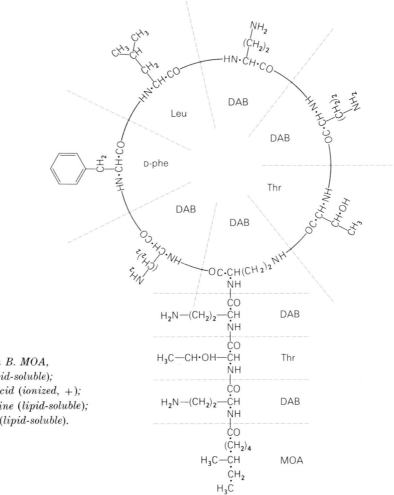

*Figure 9.11 Polymyxin B. MOA,
methyloctanoic acid (lipid-soluble);
DAB, diaminobutyric acid (ionized, +);
thr, threonine; leu, leucine (lipid-soluble);
D-phe, D-phenylalanine (lipid-soluble).*

of eucaryotes (Section 3.2). The polyene antibiotics combine specifically with sterols, and in so doing alter the permeability barrier. Unfortunately, these antibiotics attack the membranes of animal cells at only slightly higher concentrations than those needed to kill fungal cells and hence can be used internally against fungal infections in man only under carefully controlled conditions, although they can be used on the skin fairly safely.

A number of medically useful antibiotics act by inhibiting *protein synthesis*. Recall that in balanced growth (Section 7.6) the syntheses of RNA, DNA, and protein proceed in parallel. When an antibiotic that inhibits protein synthesis is added to an exponentially growing culture, RNA and DNA syntheses continue but protein synthesis stops completely

(Figure 9.12). Included in this category of specific inhibitors of protein synthesis are *chloramphenicol, streptomycin, neomycin, kanamycin,* the *tetracyclines, erythromycin,* and *puromycin* (see Figure 9.13). Protein synthesis occurs on ribosomes, and most of these antibiotics inhibit protein synthesis by combining in some way with ribosomes and altering their function. However, not all act in precisely the same way.

Recall that procaryotic ribosomes are composed of two subunits, which have sedimentation constants of 50 S and 30 S (see Section 2.5). Streptomycin, neomycin, and kanamycin combine specifically with the 30-S subunits and inhibit ribosome function by causing the wrong amino acids to be inserted into the growing polypeptide chain. Thus, in the presence of these antibiotics faulty (hence nonfunctional) proteins are synthesized (see Chapter 11). In organisms that have developed resistance to these antibiotics, a change has taken place in the 30-S component so that it is no longer affected by the antibiotics. Chloramphenicol and erythromycin combine with and inhibit the function of the 50-S component of the ribosome.

The antibiotics mentioned above show marked selective toxicity, inhibit-

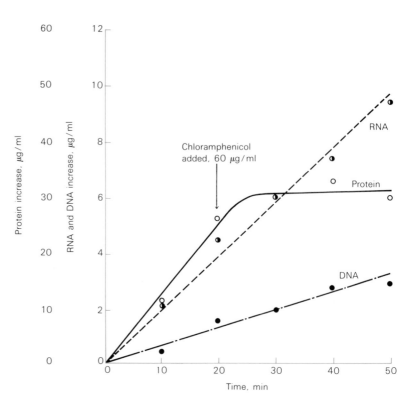

Figure 9.12 Specific inhibition of protein synthesis by chloramphenicol. At the time indicated by the arrow, chloramphenicol was added to the growing culture. RNA and DNA syntheses continued unchecked, but protein synthesis was immediately inhibited. [From C. L. Wisseman, Jr., J. E. Smadel, F. E. Hahn, and H. E. Hopps: J. Bacteriol. 67:662 (1953).]

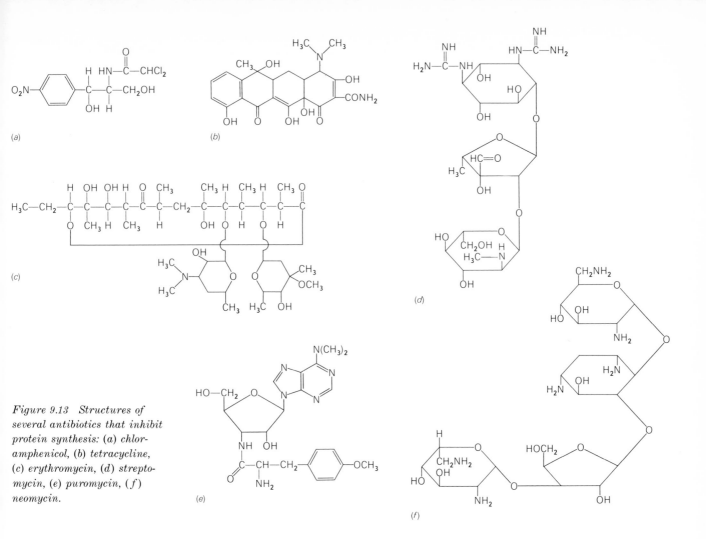

Figure 9.13 Structures of several antibiotics that inhibit protein synthesis: (a) chloramphenicol, (b) tetracycline, (c) erythromycin, (d) streptomycin, (e) puromycin, (f) neomycin.

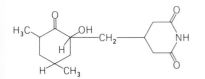

Figure 9.14 Cycloheximide, an antibiotic that inhibits protein synthesis in eucaryotes.

ing protein synthesis in procaryotes but having little or no effect on eucaryotes. Since ribosomes and a protein-synthesizing machinery are present in all organisms, how is this selectivity brought about? Recall that the ribosomes of procaryotes have sedimentation constants of 70 S whereas the cytoplasmic ribosomes of eucaryotes are 80 S. The antibiotics discussed above are able to combine with 70-S ribosomes but not with 80-S ribosomes. Thus, they do not affect the functioning of the 80-S ribosomes, which are responsible for the bulk of protein synthesis in eucaryotes. We learned, however, that in eucaryotes, 70-S ribosomes are found in the chloroplasts and mitochondria, and protein synthesis in these

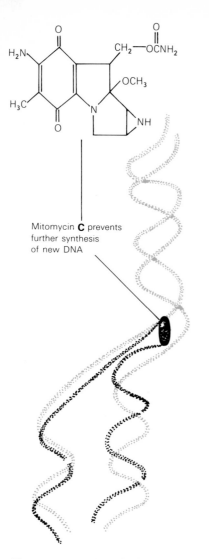

Mitomycin **C** prevents
further synthesis
of new DNA

Figure 9.15 Action of mitomycin C in inhibiting DNA synthesis. The drug reacts chemically with the two strands of the double helix and prevents their unwinding.

organelles *is* affected by the antibiotics. Since ribosomes in organelles make only a small proportion of the total protein, these antibiotics affect eucaryotes only when present at high concentrations.

Cycloheximide (Figure 9.14), an antibiotic that inhibits the functioning only of 80-S ribosomes, is highly inhibitory to eucaryotes but has no effect on procaryotes.

The antibiotic *puromycin* inhibits protein synthesis of both procaryotes and eucaryotes. It does not combine directly with the ribosomes but instead is an analog of the tRNA that brings the amino acid to the ribosome. Since tRNA molecules are the same in both procaryotes and eucaryotes, there is no selective toxicity, and puromycin is of little medical use.

Some antibiotics and synthetic chemicals combine with and alter the function of *nucleic acids,* although they rarely show selective activity. Effects on DNA are most significant since once DNA is destroyed, there is no way in which the cell can replace it. If an RNA molecule were inactivated, however, new RNA could be made on the DNA template.

The antibiotics *mitomycin* and *porfiromycin* react with DNA, usually with the guanine residues, and form links from one strand of the molecule to the other. These cross links act like safety pins to prevent the unzipping of the DNA double strand, which is essential if DNA is to be replicated (Figure 9.15). Since the cell cannot break the cross links, these drugs not only inhibit DNA synthesis but also effectively kill the cells because continued cell division cannot occur unless there is DNA synthesis. *Actinomycin,* on the other hand, combines with DNA in yet another way: It attaches to the side of the DNA molecule where mRNA synthesis takes place (Figure 9.16) and thus prevents this synthesis, but this drug has little effect on DNA synthesis. Since continued mRNA synthesis is necessary to ensure protein synthesis, actinomycin action eventually prevents protein synthesis.

Antibiotics and other drugs that affect DNA show little selective toxicity; only if the organism is impermeable to the drug is it resistant. In general, animal cells are very permeable to these agents, which therefore are quite toxic both to animals and to man. For this reason they are not used in controlling microbial infections; however, they have found some use in attempts to kill cancer cells in the body, although they must be very carefully monitored in order to avoid serious toxic effects on normal cells.

From the above discussion we can see that selective toxicity can manifest itself in various ways. At present we have a fair amount of knowledge concerning the molecular mechanisms of antimicrobial action, and we can see how these actions at the molecular level might lead to inhibition of growth and cell death. An understanding of these matters may help in

(a)

Figure 8.2 (a) Snowbank with red coloration due to presence of algae, from Grand
Teton National Park, Wyoming. (b) Vertical section through snowbank, showing
the subsurface distribution of a mixed
population of snow algae of the genera
Chlamydomonas, Scotiella, and Chodatella,
from Mt. Batchelor, Oregon. Light penetrates to considerable depths in the snow.
(Courtesy of Herbert Curl, Jr.) (c) Photomicrograph of red-pigmented resting spores
of the snow alga Chlamydomonas nivalis.
The red coloration of snowbanks occurs
when resting stages predominate; if vegetative cells predominate, the coloration
would be green. (Courtesy of Herbert
Curl, Jr.)

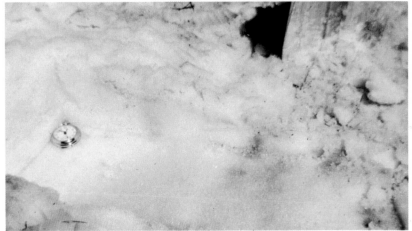
(b)

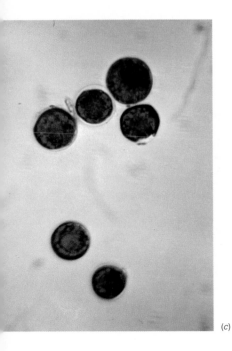
(c)

Plate one

Figure 8.4 Characteristic V-shaped pattern formed by blue-green algae at the upper temperature limit for growth, 73 to 75°C. The vee develops because the water cools more rapidly at the edges than in the center of the channel. (a) Effluent of Columbia Spring, Heart Lake Geyser Basin, Yellowstone National Park, Wyoming; (b) effluent of small unnamed spring in the Lower Geyser Basin, Yellowstone National Park.

Figure 8.5 Pinkish masses of nonphotosynthetic bacteria growing at temperatures of 85 to 88°C in the effluent of a hot spring (unnamed spring of Lower Geyser Basin, Yellowstone National Park). Nonphotosynthetic procaryotes are able to develop at temperatures considerably higher than photosynthetic procaryotes such as blue-green algae.

Figure 8.10 *Extensive development of pink halophilic bacteria in a saltern where solar salt is being prepared near San Diego, California. When the seawater is sufficiently concentrated, sodium chloride precipitates out. The halophilic bacteria develop only at salt concentrations approaching saturation.*

Figure 8.11 *Aerial view of Great Salt Lake, showing the heavy development of halophilic microorganisms. The reddish area, which is north of the railroad causeway, is more saline than the greenish area, and the salinity difference controls the nature of the organisms. The red area probably contains Dunaliella salina; and the green area, D. viridis. These algae are flagellated chlorophytes related to Chlamydomonas. (Photograph by Paul A. Zahl, © 1967 by National Geographic Society.)*

Plate three

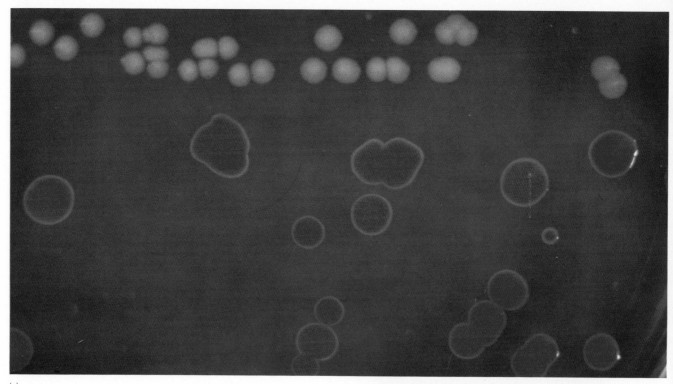

(a)

Figure 11.5   (a) Colonies of wild-type Serratia marcescens (red) and a white mutant; (b) back mutation of a white mutant of S. marcescens to the wild type. Note the occasional red colony among the predominantly white colonies.

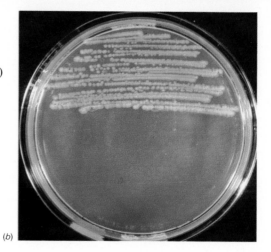

(b)

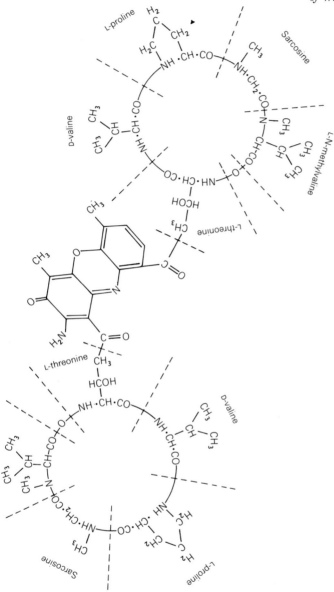

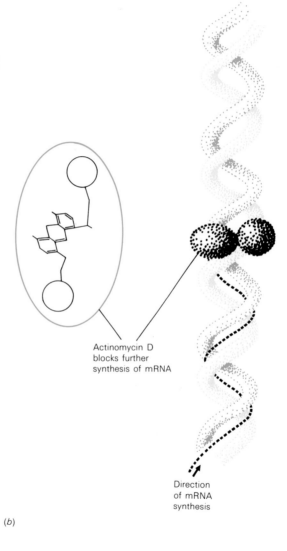

Figure 9.16   Action of actinomycin D in the inhibition of RNA synthesis. The antibiotic (a) attaches to the groove in DNA (b) where RNA synthesis is occurring. [Based on the interpretation of W. Müller and D. M. Crothers: J. Mol. Biol. 35:251 (1968).]

Actinomycin D
blocks further
synthesis of mRNA

Direction
of mRNA
synthesis

(b)

(a)

Table 9.2   Summary of modes of actions of antibiotics

	Specificity
Cell-wall synthesis:	
D-*cycloserine*	Procaryotes
Penicillin	Procaryotes
Vancomycin	Procaryotes
Membrane function:	
Candicidin	Eucaryotes
Circulin	Procaryotes, eucaryotes
Filipin	Eucaryotes
Gramicidin	Procaryotes
Nystatin	Eucaryotes
Polymyxin	Procaryotes, eucaryotes
Valinomycin	Procaryotes, eucaryotes
Deoxyribonucleic acid metabolism:	
Actinomycin	Procaryotes, eucaryotes
Mitomycin	Procaryotes, eucaryotes
Purine and pyrimidine synthesis:	
Azaserine	Procaryotes, eucaryotes
6-Diazo-5-oxo-L-norleucine (DON)	Procaryotes, eucaryotes
Psicofuranine	Procaryotes
Protein synthesis:	
Chloramphenicol	Procaryotes
Cycloheximide	Eucaryotes
Erythromycin	Procaryotes
Kanamycin	Procaryotes
Neomycin	Procaryotes
Puromycin	Procaryotes, eucaryotes
Streptomycin	Procaryotes
Tetracycline	Procaryotes
Respiration:	
Antimycin	Eucaryotes
Oligomycin	Eucaryotes

discovering or designing new antimicrobial agents. It may also help physicians to use these agents more effectively in treating infectious diseases.

Table 9.2 summarizes briefly the modes of actions of various antibiotics.

Antibiotic resistance   Many organisms are able to develop resistance to antibiotics through genetic changes (Figure 9.17; see also Chapter 11). Antibiotic resistance is usually a quantitative phenomenon; the resistant organism may still be affected if higher concentrations of the agent are used. Antibiotic resistance is thus expressed in terms of a change in the MIC. Development of resistance to one antibiotic generally does not result in concomitant resistance to other antibiotics unless they are

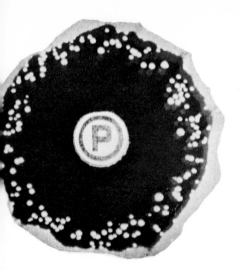

Figure 9.17 Penicillin-resistant mutants of S. aureus developing within the inhibition zone. Note that most of the resistant colonies occur in the region of the zone where the antibiotic concentration is lowest. As is discussed in Chapter 11, such mutants were already present in the inoculum and were not induced by the antibiotic itself. (From V. Lorian: Antibiotics and Chemotherapeutic Agents in Clinical and Laboratory Practice. Springfield, Ill.: Charles C Thomas, Publisher, 1966.)

closely related. If an organism is resistant to a second antibiotic as a result of being resistant to the first, this is called cross resistance.

Several mechanisms of antibiotic resistance are known. (1) The sensitive structure (for example, cell wall, enzyme, or similar component) may be lacking in the resistant form. This is seen in certain forms, called L-forms (Section 18.16), that have acquired resistance to penicillin by losing the need to synthesize a cell wall. (2) The cellular structure that is the antibiotic's target may undergo an alteration in structure so that it no longer binds the antibiotic but is still able to carry out its normal function. (3) The resistant organism may be impermeable to the antibiotic because of some surface barrier or lack of a transport system. If this is so, then the antibiotic may still act on the sensitive structure if the permeability barrier is altered or destroyed. (4) The organism may be able to modify the antibiotic to an inactive form. This is very frequently the case in penicillin resistance. A group of enzymes (penicillinases) hydrolyze the side chain of the penicillin molecule. Penicillinases are produced by a wide variety of bacteria and have definitely been shown to confer antibiotic resistance. However, some penicillins are less susceptible to penicillinase activity than are others (see Figure 9.8), and thus are able to act on otherwise penicillin-resistant bacteria. Penicillins that are resistant to penicillinase action have different side chains, and in recent years many of these have been prepared chemically and are now widely used in medicine. The development of antibiotic resistance has had important medical consequences, as will be discussed in Chapter 15.

A related phenomenon is the development, in some cases, of *antibiotic dependence*. In this case, the organism is not only resistant to the antibiotic but actually *requires* it as a growth factor. Streptomycin and neomycin dependence develop fairly frequently, and dependence on chloramphenicol has also been reported. The precise molecular mechanisms of antibiotic dependence are not known, but in the case of streptomycin, one theory is that as a result of mutation the 30-S ribosome is sufficiently altered that it can no longer function normally unless it is complexed with the antibiotic.

The role of antibiotics in nature The widespread occurrence of antibiotic-producing microorganisms in the soil and other natural environments has led to the hypothesis that antibiotic production is an important factor in the competition between microorganisms. Does this mean, however, that antibiotics play roles in antagonistic relationships in nature, for instance, in the soil? It is possible that antibiotic production is a laboratory artifact and does not occur in nature since antibiotics are usually produced best when organisms are growing in media rich in organic materials, such as are provided in the laboratory; when media reflect the nutrient-poor conditions often found in soil, however, antibiotic production is usually minimal or absent. Thus it is conceivable that laboratory cultural conditions induce organisms to form antibiotics, which they would not normally produce in the soil. However there is good evidence that under some conditions certain antibiotics can be formed in natural soils. Further research on the roles of antibiotics in nature is urgently needed.

Bacteriocines A *bacteriocine* is a substance that acts only upon a restricted range of organisms closely related to the producing organism, in contrast to antibiotics, which often act on unrelated or only distantly related organisms. For instance, strains of *E. coli* produce a variety of bacteriocines that act on other strains of the same species, but which have no activity against other genera of bacteria. Almost all genera of bacteria are known to produce bacteriocines active against other members of the same genus. They are usually named in reference to the genus or species that produces them. Thus the bacteriocines of *E. coli* are called *colicines*, those of *Bacillus megaterium* are called *megacines*, those of *Pasteurella pestis* are called *pesticines*, and so on. The organism producing the bacteriocine is always resistant to its action even though closely related organisms, even of the same species, may be highly sensitive. Bacteriocines are almost always complex protein molecules that adsorb specifically to sensitive cells and do not adsorb to resistant cells. After adsorbing to sensitive cells, the bacteriocine initiates events leading to death of the cell; in at least two instances this killing is known to be due to effects on the cell membrane. Although bacteriocines have not as yet found any medical use, their highly specific action makes their study of considerable interest. The genetics of bacteriocine production has been widely investigated and is discussed briefly in Chapter 12.

No data are available on the role of bacteriocines in nature, but it seems possible that they may be involved in competitive relationships among closely related organisms. Since close relatives are likely to have similar nutritional and environmental requirements, they usually compete for a common pool of resources. If one of the two is a bacteriocine producer, it seems reasonable to assume that it would be more successful in competition.

9.4 Germicides, disinfectants, and antiseptics

A wide variety of chemical substances are used to kill or prevent the growth of microorganisms in primarily nonmedical situations or are used medically only on the external parts of the body. A *germicide* is any chemical agent that kills microorganisms and could be either an antiseptic or a disinfectant. An *antiseptic* is defined as a chemical substance that inhibits growth or kills microorganisms, which is sufficiently nontoxic so that it can be applied to the skin or mucous membranes, although it is not necessarily safe enough to be taken internally. A *disinfectant* is an agent that kills most microorganisms (but not necessarily their spores) and is distinguishable from an antiseptic by the fact that it is not safe for application to living tissue, but can be used to reduce the microbial load of inanimate objects.

The above terms are in wide use, but most of them are somewhat arbitrary. Phenol, for instance, was originally used on the skin by Joseph Lister and thus was an antiseptic; today phenol is considered much too toxic to apply to the body, but it still is commonly employed as a disinfectant. Table 9.3 lists various classes of antiseptics and disinfectants and indicates some of their important uses.

9.5 Chemical sterilization

Many materials that would be damaged by heat sterilization can be sterilized chemically. Examples include some plastics, surgical instruments, and books and paper products. Also, hospital rooms and other large chambers may need to be sterilized, and here also heat would be impractical. Although at one time formaldehyde found wide usage, today

ethylene oxide $\mathrm{H_2C\overset{\textstyle{\diagdown\!\diagup}}{\underset{O}{}}CH_2}$ is most commonly used. At room temperature

ethylene oxide is a gas, but it liquefies at $10.8°C$. It is highly reactive and reacts with amino, sulfhydryl, carboxyl, and hydroxyl groups of organic compounds. Since such groups are present in proteins, nucleic acids, and other cellular constituents, cell death is probably brought about by general inactivation of a wide variety of essential macromolecules. Pure ethylene oxide is vigorously flammable in air; it is usually used commercially as a 10 percent mixture with 90 percent CO_2, in which proportions it is not flammable. For sterilizing small objects, pressure chambers analogous to autoclaves are used. The gas is introduced into the sealed chamber and penetrates rapidly to all parts of the objects to be sterilized. However, the object must not have regions impenetrable to gas, as the agent sterilizes only those surfaces it comes into contact with. Sterilization with ethylene oxide is a slow process; depending on the concentration and the load of microorganisms, treatment time may range from 4 hr to overnight. Bac-

Table 9.3 Classes of antiseptics and disinfectants and their uses

	Chemical Action	Sensitive sites in cell	Uses	Comments
Metals: Bichloride of mercury ($HgCl_2$)	SH groups of proteins	SH enzymes	Disinfectant	Activity neutralized by SH compounds. organic matter
Organic mercurials: merthiolate, mercurochrome, and so on	$R-Hg + HS-R' \rightarrow$ $R-Hg-S-R'$	SH enzymes	Antiseptic on skin	Activity neutralized by SH compounds. organic matter
Silver nitrate ($AgNO_3$)	Protein precipitant	Enzymes	Eyes of newborn to prevent gonorrhea	Activity neutralized by organic matter
Copper sulfate ($CuSO_4$)	Unknown	Unknown	Algicide (swimming pools and water supply)	Bordeaux mixture ($CuSO_4$ + lime). one of the first antimicrobial agents (midnineteenth century)
Organic copper compounds			Fungal diseases of plants	
Halogens: I_2 + KI (Lugol's solution)	Iodination of tyrosine in proteins $I_2 + HO-\!\!\bigcirc\!\!-R \longrightarrow HO-\!\!\bigcirc\!\!-R$ (with I substituents)	Enzymes requiring tyrosine residues for activity	Antiseptic on skin; disinfectant of medical instruments: purification of drinking water (halazone tablets)	Relatively nontoxic on skin: toxic internally
Cl_2 *gas*	Oxidizing agents	Enzymes. other proteins	Water purification	Activity neutralized by organic matter
$Ca(OCl)_2$, NaOCl			General disinfectant in food and dairy industries	
Chloramines (derivatives of NH_2Cl*)*			General disinfectant in food and dairy industries	

Chemical	Mode of action	Site of action	Use	Comments
Alcohols: *Ethanol* (CH_3CH_2OH)	Lipid solvent and protein denaturant	Cell membrane	Antiseptic on skin	Most active in presence of some water (70% ethanol in water is preferred concentration)
Isopropanol ($CH_3CHOHCH_3$)			Disinfectant of medical instruments	Not active against endospores and *Mycobacterium tuberculosis*
Phenols: HO—⬡ *(halogen, alkyl, and HO groups in ortho and para positions enhance activity)*	Surface active agents	Cell membrane	Disinfectants	Neutralized by organic materials
Bis-phenols: hexachlorophene	Unknown	Unknown	Antiseptic soaps and lotions; body deodorants	Bacteriostatic rather than bactericidal. Retains activity in presence of organic matter
Cationic detergents: *Quaternary ammonium compounds* *(R groups are hydrophobic)*	Surface active agents	Cell membrane	Antiseptic soaps and lotions for skin; disinfectants for medical instruments, food and dairy equipment	Neutralized by phospholipids, metal ions, low pH

terial spores as well as vegetative cells are killed although spores are, of course, considerably more resistant. Since ethylene oxide is quite toxic to man, it must always be used in a closed container, taking precaution to prevent it from coming into contact with human beings. Another gas that has found some use as a sterilizing agent is ozone (O_3), a substance that is also formed naturally through the action ultraviolet radiation on O_2.

9.6   Chemical food preservatives

Chemicals to be used as *food preservatives* must meet very stringent requirements concerning nontoxicity. Because foods are consumed in large amounts and over extended periods of time, a preservative must cause no chronic toxicity or accumulative effect. If a substance does show signs of toxicity, it must then be shown to be completely destroyed by cooking, and the substance can be used only in foods that will never be eaten raw. Because of a long history of successful use, certain preservatives are considered safe, while new agents must be given extensive toxicity tests before use. In addition, the preservatives should be water soluble, odorless, colorless, and tasteless. Food preservatives need only be bacteriostatic since they will be continuously present in the food. Since foods are often stored for periods of time, the agent should also be chem-

Table 9.4   Chemical food preservatives

	Food	Organisms inhibited
Sodium or calcium propionate	Bread	Bacteria and fungi
Sodium benzoate	Carbonated beverages	Bacteria and fungi
	Fruit juices	Bacteria and fungi
	Pickles	Bacteria and fungi
	Margarine	Bacteria and fungi
	Preserves	Bacteria and fungi
Sorbic acid	Citrus products	Fungi
	Cheese	Fungi
	Pickles	Fungi
	Salads	Fungi
Sulfur dioxide	Dried fruits and vegetables	Fungi
Chlortetracycline	Chicken	Bacteria
	Fish	Bacteria
Formaldehyde *(from food-smoking process)*	Meat	Bacteria and fungi
	Fish	Bacteria and fungi

ically stable. Actually, many food ingredients will neutralize the action of chemical preservatives. Table 9.4 lists commonly used chemical food preservatives and their uses.

Supplementary readings

Albert, A. *Selective Toxicity,* 3rd ed. New York: John Wiley & Sons, Inc., 1965.

Biochemical Studies of Antimicrobial Drugs (Sixteenth Symposium, Society for General Microbiology). London: Cambridge University Press, 1966.

Davis, B. D., and D. S. Feingold Antimicrobial agents: mechanism of action and use in metabolic studies," in I. C. Gunsalus and R. Y. Stanier (eds.), *The Bacteria: A Treatise on Structure and Function,* vol. IV, p. 343. New York: Academic Press, Inc., 1962.

Gottlieb, D., and P. D. Shaw (eds.) *Antibiotics,* vol. I, *Mechanism of Action.* New York: Springer-Verlag, 1967.

Gottlieb, D., and P. D. Shaw (eds.) *Antibiotics,* vol. II, *Biosynthesis.* New York: Springer-Verlag, 1967.

Kavanagh, F. (ed.) *Analytical Microbiology.* New York: Academic Press, Inc., 1963.

Lawrence, C. A., and S. S. Block (eds.) *Disinfection, Sterilization, and Preservation.* Philadelphia: Lea & Febiger, 1968.

Reddish, G. F. (ed.) *Antiseptics, Disinfectants, Fungicides and Chemical and Physical Sterilization,* 2nd ed. Philadelphia: Lea & Febiger, 1957.

Sykes, G. *Disinfection and Sterilization.* London: Chapman & Hall, Ltd. 1965.

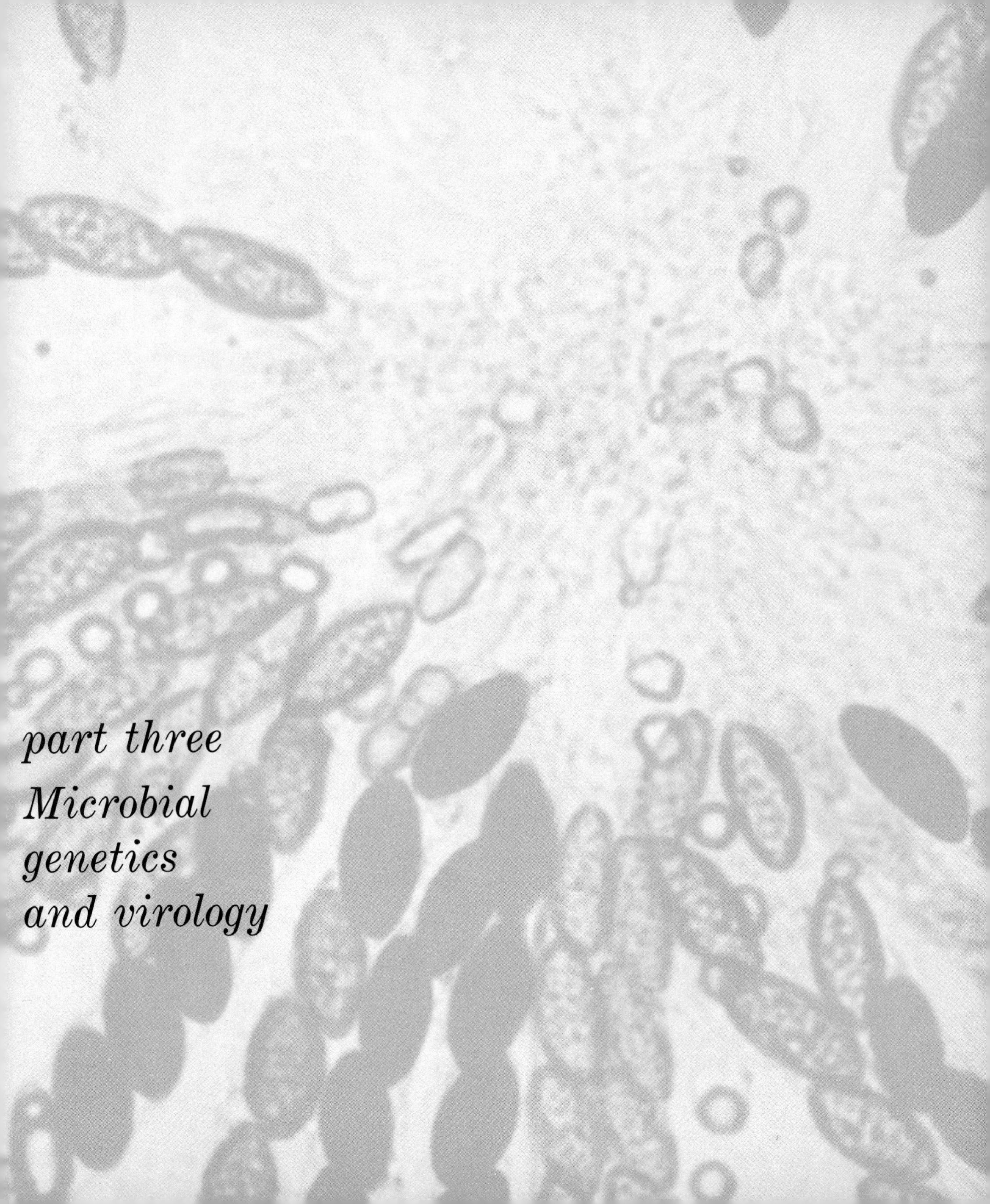

part three
Microbial genetics and virology

In this part we discuss genetic phenomena in microorganisms. Because viruses are the simplest genetic elements, we begin our discussion with them. The structure of virus particles and the nature of their nucleic acids are first considered, and we will see that the virus particle is essentially a packaging device for ensuring the transfer of virus nucleic acid from one cell to another. The infection process and the biochemical mechanisms of virus replication are next examined. We then deal with the diversity of viruses and how they are classified. The last portion of Chapter 10 discusses representative viruses that cause disease in man, higher animals, insects, and plants. Chapter 11 opens with a description of the genetic code and goes on to sketch how the information stored in the sequence of purine and pyrimidine bases in DNA is translated through the mediation of RNA into the sequence of amino acids in specific proteins. Then follows an account of mutation, the process by which changes in the genetic code arise. Chapter 12 considers the genetic hybridization processes that occur in procaryotic microorganisms. The discussion centers on the three kinds of mechanisms by which genetic recombination occurs: transformation (involving free DNA), transduction (virus-mediated genetic recombination), and conjugation (cell-to-cell contact). Chapter 13 describes genetic recombination in eucaryotic microorganisms; particular emphasis is placed here on contrasting manners of genetic recombination in eucaryotes and procaryotes. The presence in eucaryotes of genes in discrete chromosomes contained in membrane-bound nuclei is shown to require processes of genetic recombination that in some ways are more complicated than those in procaryotes but in other ways are simpler. An extensive discussion of heredity in mitochondria and chloroplasts follows, in which microorganisms are proved to be ideal experimental systems for studying inheritance of such subcellular organelles. Chapter 13 concludes with a consideration of how these subcellular organelles might have evolved.

A virus is a genetic element containing either DNA or RNA, which is able to alternate between two distinct states, intracellular and extracellular. In the extracellular or infectious state viruses are submicroscopic particles containing nucleic acid surrounded by protein and occasionally containing other components. These virus particles or *virions* are metabolically inert and do not carry out respiratory or biosynthetic functions. The role of the virion is to carry the viral nucleic acid from the cell in which the virion has been produced to another cell where the viral nucleic acid can be introduced and the intracellular state initiated. In this latter phase replication occurs, during which more nucleic acid and other components of the virus are produced. Cells that viruses can infect and in which they can replicate are called *hosts*. The host performs most of the metabolic functions necessary for virus replication.

Viruses may be considered in two ways: as agents of disease and as agents of heredity. As agents of disease, viruses can enter cells and cause harmful changes in these cells, leading to disrupted function or death. As agents of heredity, viruses can enter cells and cause permanent, hereditable changes that usually are not harmful and often may be beneficial. Not all viruses are dual agents since some act only as agents of disease and others only as agents of heredity. In many cases, which role the virus plays depends on the host cell and on the environmental conditions.

Viruses can vary widely in size, shape, chemical composition, range of organisms attacked, kinds of cell damage induced, and range of genetic capabilities. They are known to infect animals, plants, bacteria, blue-green algae, and fungi. Practically all virions are of submicroscopic size, which means that the individual particle can rarely be seen with the light microscope. This made very difficult the discovery and early study of viruses, and their properties had to be studied by indirect procedures. Virions are fairly easy to visualize with the electron microscope, however, so that our knowledge of virus structure is now greatly advanced. In fact, it was in the field of virology that the electron microscope made its first great contributions to biology.

Recognition of the existence of viruses   Viruses were first recognized as entities of disease through the observation of the harmful effects they had on plants and animals. In 1898 and 1899, studies on two diseases, foot-and-mouth disease of domestic animals and the mosaic disease of tobacco, showed clearly that these diseases were caused by submicroscopic entities, which were able to replicate in living hosts but which could not be cultured in the manner of bacteria. The work on tobacco mosaic disease was the most definitive. This disease is manifested first by a bleaching of the chlorophyll of the leaf, occurring in spots over the leaf blade, followed later by the death of part or all of the tissue of the

ten

Viruses

255

spots. The disease caused serious trouble to tobacco farmers in Holland, as the damaged leaves could not be used as cigar wrappers. In 1887 Adolf Mayer had shown that the disease was contagious by expressing sap from diseased plants and injecting it into healthy plants, which became diseased after 2 to 3 wk. The Dutch microbiologist Martinus Beijerinck first attempted to demonstrate the presence of bacteria within the sap of diseased plants, but neither microscopic nor cultural studies revealed any bacteria. He then discovered that the juice pressed from diseased plants retained its virulence even after passing through a porcelain filter fine enough to hold back all bacteria. The bacteria-free filtrate maintained its infectivity for many months.

Beijerinck showed by passing the agent successively from plant to plant that, despite the absence of culturable organisms, the infectious plant sap contained an agent that duplicated itself. Since only a small amount of juice was used to infect a plant, there was considerable dilution at each stage, but the juice from the final plant was just as infective as that of the initial plant in the series. If the disease had been caused by a non-replicating toxic agent, the toxin would have been diluted out during continued plant passage.

Beijerinck developed a method for quantitating the amount of infectious agent in a sample by determining the greatest dilution capable of initiating disease symptoms. Using this quantitation method, he then was able to show that although the causal agent replicated well in the living tobacco plant, it was completely unable to replicate in plant juice alone. Beijerinck therefore concluded that the agent reproduced only when incorporated into living protoplasm and that its reproduction was brought about by the plant cell itself. That the agent was biological was indicated by its inactivation when heated to 90 or 100°C. He concluded that the infectious agent was not cellular, but instead was a soluble contagious agent.

10.1 The virus particle or virion

The structures and chemical compositions of a variety of different types of virions have now been determined. In order to study the chemistry of a virus, the particles must first be separated from all host materials. The purification methods are usually similar to those used by the protein chemist. The first virus to be completely purified was tobacco mosaic virus (TMV), which was crystallized by Wendell Stanley in 1935. A method currently used for isolation and purification of TMV is outlined in Figure 10.1. It is necessary to demonstrate that the crystalline material does

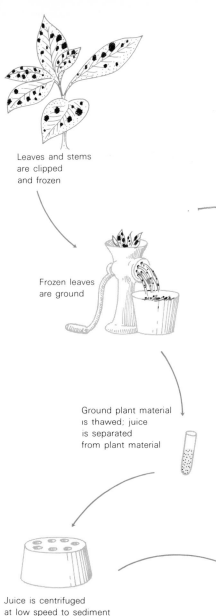

Leaves and stems
are clipped
and frozen

Frozen leaves
are ground

Ground plant material
is thawed; juice
is separated
from plant material

Juice is centrifuged
at low speed to sediment
the plant debris,
and the supernatant
is then centrifuged
at high speed to sediment the virus

Supernatant is discarded,
leaving pellet of virus

Virus

Virus is concentrated
and allowed to crystallize

Needlelike crystals are formed

indeed represent the virus by showing that, when it is dissolved in water, the solution is infectious at very high dilution, and that the symptoms are identical to those obtained upon infecting the host with crude extracts. Not all viruses can be crystallized, however. The larger viruses and those with irregular or complex structures are not crystallizable, although usually it is possible for them to be highly purified. The crystallization of TMV was historically of great importance since it showed that an agent previously thought to be living could be handled like a simple protein; indeed, this event proved to be one of the milestones in the development of molecular biology.

By now many viruses have been purified, and we know that chemically they are a heterogeneous group. Some contain RNA and others DNA, but both are never present in the same virus. We know from our earlier studies that the DNA of cells is double-stranded, containing two complementary chains that pair specifically and separate during replication (see Chapter 2). The DNA molecules of many viruses are also double-stranded, but some have single-stranded DNA. This presents for the virus certain problems in replication, which we will discuss further below. The RNA molecules of most RNA viruses are single-stranded, but a few viruses with double-stranded RNA are known. —animal virus

The naming of viruses does not follow the rules and procedures used for naming organisms. In many cases the virus is named for the disease it causes (for example, tobacco mosaic virus, polio virus), while in other cases it is named for the organ or tissue it infects (for example, adenovirus, which infects the adenoids). In the case of bacterial viruses, code numbers are usually used to distinguish several viruses infecting the same host (for example, T2, P1, and T7 of *Escherichia coli,* P22 of *Salmonella typhimurium*). From time to time there have been proposals to give viruses genus and species names, but as yet these proposals have not found extensive backing.

The structures of viruses are exceedingly diverse. With the advent of the electron microscope, it was possible to study the overall sizes and shapes

Figure 10.1 Steps in the purification of tobacco mosaic virus.

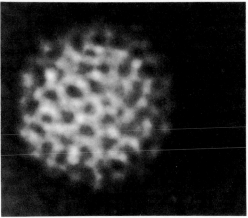
(a)

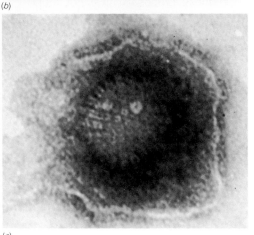
(b)

of virus particles, and in recent years new preparation methods have made it possible to study their internal structure as well. The nucleic acid is located in the center, surrounded by a protein coat called the *capsid;* the individual proteins that make up the capsid are called protein subunits, or *capsomeres*. Various morphological shapes of virus particles are illustrated in Figures 10.2 and 10.3. Many of the small viruses that resemble spheres are in reality geometrical structures called *icosahedrons*, with 20 triangular faces and 12 corners. Some virus particles are long rods, whose nucleic acid forms a central spiral surrounded by the protein capsomeres. The manner in which the capsomeres are arranged around the nucleic acid of TMV is shown in Figure 10.4. Some viruses have more complex structures, with membranous envelopes surrounding the central nucleic acid–protein core. The most complicated viruses in terms of structure are some of the bacterial viruses (Figures 10.3 and 10.6), which possess not only an icosahedral head structure containing the nucleic acid but also a tail lacking nucleic acid. Sometimes, such as in T4 virus of *E. coli*, the tail itself is a rather complicated structure. The range of sizes found in various viruses is from 30 nm, about the size of a ribosome, up to about 300 nm, the latter being resolvable with a light microscope.

10.2   Virus quantification

In any study of virus replication it is necessary to accurately count the virus particles present. With bacterial viruses, this is done with the agar layer technique, which is analogous to the colony count used for deter-

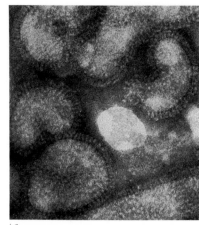
(c)   (d)

Figure 10.2   Electron micrographs of animal-virus particles of several morphological types: (a) human wart virus (magnification, 142,000×) [from W. F. Noyes: Virology 23:65 (1964)]; (b) reovirus (magnification, 670,000×) (courtesy of H. D. Mayor); (c) herpes virus (magnification, 220,000×) (courtesy of R. W. Horne); (d) influenza virus (magnification, 184,000×) [from P. W. Choppin and W. Stoeckenius: Virology 22:482 (1964)]. The viruses in (a) and (b) are naked viruses and those in (c) and (d) are enveloped viruses. In (c) note the viral core within the outer envelope.

Figure 10.3 Electron micrographs of bacterial- and plant-virus particles: (a) Escherichia coli bacterial virus (bacteriophage) T4 (magnification 30,000×) (courtesy of Thomas F. Anderson, The Institute for Cancer Research, Philadelphia); (b) TMV (magnification, 72,500×) [from J. T. Finch: J. Mol. Biol. 8:872 (1964)]. Note the complex structure of the phage in (a) and compare with Figure 10.6; compare (b) with Figure 10.4.

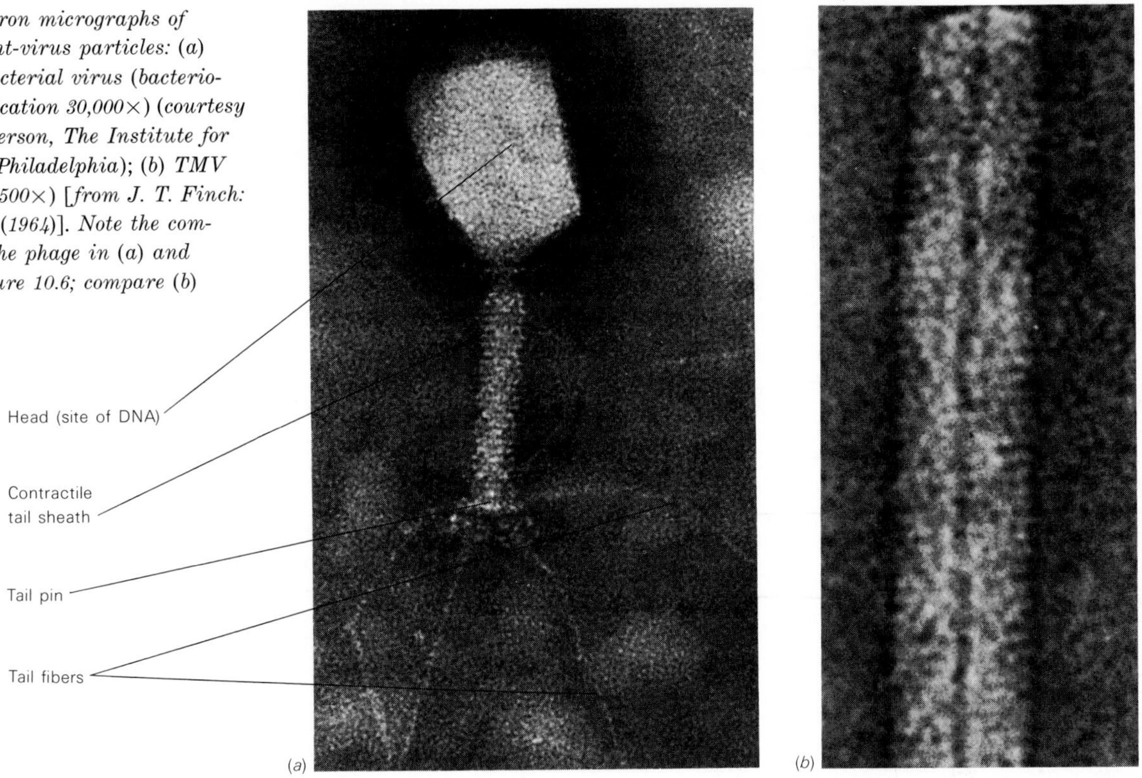

Head (site of DNA)

Contractile
tail sheath

Tail pin

Tail fibers

(a)

(b)

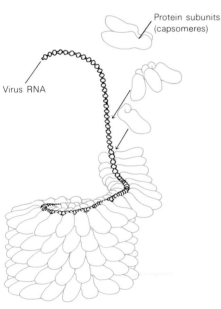

Protein subunits
(capsomeres)

Virus RNA

mining the number of viable bacteria. A dilution of the suspension containing the virus material is mixed, in a small amount of melted agar, with the sensitive host bacteria and the mixture is poured on the surface of a nutrient agar plate. The host bacteria, which have been spread uniformly throughout the top agar layer, begin to grow and after overnight incubation will form a "lawn" of confluent growth. Each virus particle that attaches to a cell and reproduces may cause cell lysis and the virus particles released can spread to adjacent cells in the agar, infect them, be reproduced, and again lead to lysis and release. This process continues for a number of generations, but since the agar prevents the new virus particles from moving too far away, a localized area of lysis develops that

Figure 10.4 Details of the structure and manner of assembly of a TMV particle. The RNA assumes a helical configuration surrounded by the protein capsomeres. The center of the particle is hollow. Compare with Figure 10.3(b). [Adapted from A. Klug and D. L. D. Caspar: Adv. Virus Res. 7:225 (1960).]

10.2   Virus quantification

259

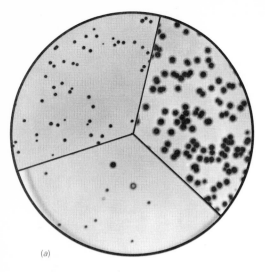

(a)

(b)

(c)

Figure 10.5 Quantification of bacterial, plant, and animal viruses: (a) plaques in lawn of E. coli induced by several strains of T4 bacteriophage; the plaque morphology is characteristic for each strain (from Dean Fraser: Viruses and Molecular Biology. New York: The Macmillan Company, 1967); (b) local lesions induced on tobacco leaves by TMV (courtesy of F. O. Holmes); (c) plaques in monolayer of monkey kidney cells induced by measles virus [from G. D. Hsiung, A. Mannini, and J. L. Melnick: Proc. Soc. Exptl. Biol. Med. 98:68 (1958)].

contains no bacteria but many virus particles, surrounded by a field of confluent bacterial growth. This local area of lysis is called a *plaque* (Figure 10.5) and represents the end result of a chain of events initiated by *one* virus particle. Thus, each plaque originated from one virus particle, just as each bacterial colony is derived from one bacterium. It is possible by means of this plaque assay technique to obtain an estimate of the total number of virus particles in a suspension. In addition, this method allows us to isolate pure strains of viruses since if each plaque has arisen from one virus particle or infected cell, it contains a clone or pure culture of this virus. Some of the particles from a single plaque can be picked and inoculated into a broth culture to make a lysate that will establish a pure line. The devising of this technique was as important for the advance of virology as was Koch's development of solid media for bacteriology.

The local lesions found on plants (Figure 10.5) and the pocks induced on some animals by viruses are analogous to the plaques of bacterial viruses. With many animal viruses it has been possible to devise animal tissue-culture systems in which virus-induced plaques can be observed (Figure 10.5), thus making it possible to count the viruses. This too has had an enormous influence on the progress of animal virology in the past 20 years (see Section 10.9).

10.3 Virus replication

The basic problems of virus replication can be put simply: The virus must somehow induce a living cell to make more of the essential components of the virus particle, these components must be assembled in the proper order, and the new virus particles must escape from the cell if they are to infect other cells. The various phases of this replication process can be summarized in six steps: (1) attachment (adsorption) of virus particle to sensitive cell; (2) penetration into cell of virus or its nucleic acid; (3) replication of the virus nucleic acid; (4) production of protein capsomeres and other essential viral constituents; (5) assembly of nucleic acid and protein capsomeres into new virus particles; (6) release of mature virus particles from the cell. A fairly complete understanding has been reached of the molecular events involved in these processes for several viruses of animals, plants, and bacteria. However, the fullest information available about these processes comes from certain bacterial viruses or bacterio-phages (-*phage* is a combining form meaning ''to eat'') that have been widely studied as model systems for virus replication. The principles developed from these studies have been applied to viruses attacking other hosts.

The structure of the bacterial virus T4, which infects *E. coli,* is shown in Figure 10.6. The particle has a *head,* within which the viral DNA is folded, and a long, fairly complex *tail,* at the end of which is a series of tail fibers. When a suspension of virus particles is added to a suspension of sensitive bacteria, the virus particles first attach to the cells by means of the tail fibers. These tail fibers then contract and the core makes contact with the cell envelope of the bacterium. Changes then occur in the surface of the cell, probably through enzymatic action, resulting in the formation of a small hole; at the same time an opening develops at the tip of the virus tail. The DNA of the virus then passes into the cell, the protein coat remaining outside. The DNA of T4 has a total length of about 50 μm, whereas the dimensions of the head of the T4 particle are 0.095 μm by 0.065 μm. This means that the DNA must be highly folded and packed very tightly within the head. Presumably this tight packing puts the

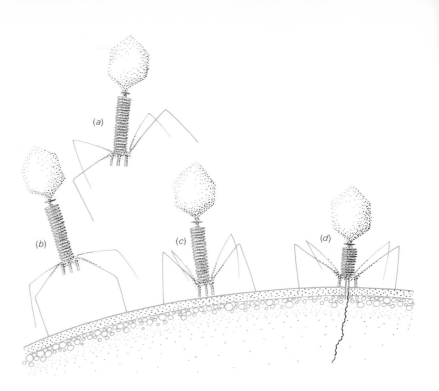

Figure 10.6 Attachment of T4 bacterio-phage particle to the cell wall of E. coli and injection of DNA: (a) unattached particle; (b) attachment to the wall by the long tail fibers; (c) contact of cell wall by the tail pin; (d) contraction of the tail sheath and injection of the DNA. Compare with Figure 10.3(a). [From L. D. Simon and T. F. Anderson: Virology 32:279 (1967).]

DNA under considerable pressure; and when the tail develops an opening, the DNA is forced rapidly out and into the bacterial cell.

Viral DNA contains genetic information for the production of a number of enzymes that the uninfected host does not make; and soon after the entry of the virus DNA, these new proteins begin to be formed (Figure 10.7). Virus enzyme formation involves the synthesis of new mRNA molecules, using the virus DNA as a template. The new mRNA molecules associate with the ribosomes of the host, and new protein production occurs by way of the host's protein-synthesizing machinery. One function of the new enzymes synthesized after infection is to form the building blocks needed for viral DNA synthesis. Thus, the synthesis of the viral DNA does not begin until these new enzymes have been formed. Once the building blocks are available, DNA synthesis proceeds and soon the cell contains a large amount of viral DNA.

During these events of virus replication, the metabolic machinery of the host continues to function, providing the energy and small molecules needed for the biosynthesis of viral enzymes and nucleic acid. The protein-synthesizing machinery of the host is taken over virtually completely by the virus, probably because it is able to bring about inhibition of host mRNA synthesis.

Virus DNA contains information for synthesis of both the new enzymes and the head and tail proteins of the virus. Soon after virus DNA is synthesized these structural proteins can be detected, and at this stage the cell contains large amounts of the protein subunits and DNA of the virion.

The protein coat of the virus particle forms spontaneously from the protein capsomeres by a *self-assembly* process, the nucleic acid serving as a kernel around which the protein capsomeres associate. In the case of T4, which has several kinds of proteins in its coat, assembly occurs stepwise, with the DNA-containing head forming first and the tail tube, tail sheath, and tail fibers being added subsequently. The release of the

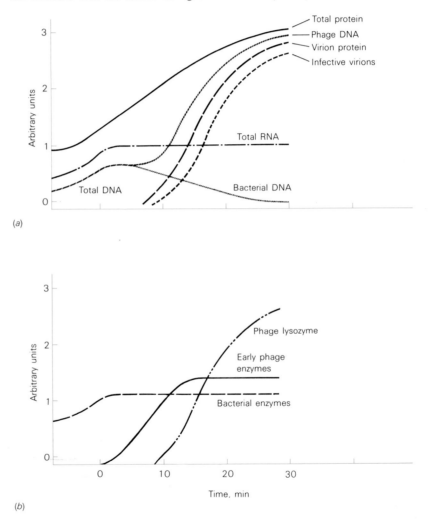

Figure 10.7 *Biochemical changes in E. coli cells after infection with bacteriophage T4. At time 0, T4 bacteriophage was added to the growing culture. (From S. E. Luria and J. E. Darnell, Jr.: General Virology, 2nd ed. New York: John Wiley & Sons, Inc., 1968.)*

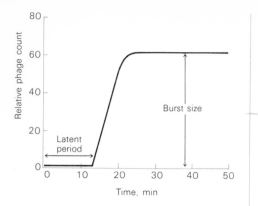

Figure 10.8 A one-step growth experiment. During the latent period, whose length is characteristic of each system, no free virus is present. The burst size represents the average number of virus particles released upon lysis of each infected cell. (From S. E. Luria and J. E. Darnell, Jr.: General Virology, 2nd ed. New York: John Wiley & Sons, Inc., 1968.)

virions occurs upon breakdown of the cell wall through the action of a lysozymelike protein whose synthesis is also directed by the virus DNA.

From the above discussion we see that the virus particle contains DNA, which behaves as hereditary material in that it contains information for the synthesis of a number of virus-specific proteins and in addition is able to control its own synthesis.

A study of the rate of virus replication can be carried out by a procedure called a *one-step growth experiment*. After adsorption of virus to host, the suspension is diluted to such an extent that virus particles released after the first round of replication cannot attach to uninfected cells; thus only one round of replication can occur. If a series of samples of this diluted suspension is plated, we find that there is a short interval, the *latent period*, during which there is no increase in titre (Figure 10.8). This is followed by a rapid increase in titre, which reflects the release of mature virus particles when the infected cells burst. A plateau is soon reached, which indicates that further synthesis, assembly, and release of virus does not occur. The number of virus particles liberated by an infected cell, called the "burst size," is about 100 virus particles per infected cell for T4 infection of E. coli.

If the infected cells are broken open at intervals and examined for virus particles, there is an interval during which it is not possible to detect any virus. This interval, called the *eclipse period,* occurs because the nucleic acid of the infecting particle has been injected into the host; no longer surrounded by its protein coat, the free virus nucleic acid is ordinarily incapable of infecting another cell. As soon as complete particles are formed, these are detected in cell extracts as infectious units. The eclipse is also seen with plant and animal viruses and is due to the separation of the nucleic acid from the protein coat after the virus has entered the cell.

Single-stranded DNA viruses

The DNA of some viruses is single-stranded instead of double-stranded. The best-studied case of a single-stranded bacterial virus is ϕX174, a small icosahedral virus that infects E. coli. Upon entry of the DNA into the cell, the first event to occur is the alteration of the single-stranded DNA to a double-stranded form, the so-called replicative form. This alteration makes use of an enzyme already present in the host cell, which directs the synthesis of an additional single strand of DNA adjacent and complementary to the initial strand. If the incoming DNA is called the *plus* strand, the complementary strand synthesized in the cell can be called the *minus* strand. This *minus* strand, as part of the double-stranded replicative form, serves as the template for the synthesis of new plus strands of viral DNA as well as mRNA used in the synthesis of new phage enzymes (Figure 10.9).

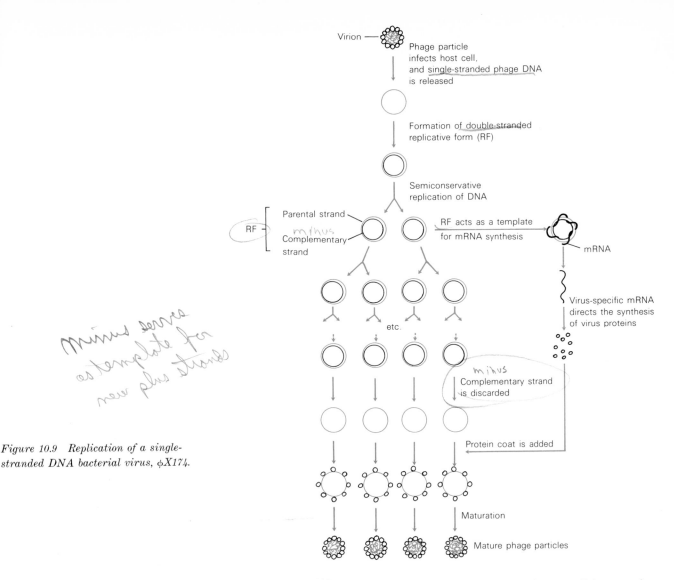

Virion — Phage particle
infects host cell,
and single-stranded phage DNA
is released

Formation of double-stranded
replicative form (RF)

Semiconservative
replication of DNA

Parental strand

minus

Complementary
strand

RF — RF acts as a template
for mRNA synthesis

mRNA

etc.

Virus-specific mRNA
directs the synthesis
of virus proteins

minus
Complementary strand
is discarded

Protein coat is added

Maturation

Mature phage particles

*Figure 10.9 Replication of a single-
stranded DNA bacterial virus, φX174.*

minus serves
as template for
new plus strands

RNA viruses   The virions of a few bacterial viruses, all known plant
viruses, and many animal viruses are composed of single-stranded RNA,
and one group of animal viruses, the reoviruses, contains double-stranded
RNA. Whether the virus RNA is single- or double-stranded, the manner
of replication of RNA viruses is different from that of DNA viruses in the
initial steps. Normal cells do not synthesize RNA by means of an RNA
template, and the viruses must provide the information for the synthesis
of unique enzymes to perform this task. The virus RNA thus must first be

able to direct the synthesis of virus-specific proteins. Single-stranded RNA viruses do this by associating with the host ribosomes and acting as mRNA for the synthesis of RNA-"replicase." The replicase then associates with the infectious strand, which can be called the "plus" strand, and enables a complementary "minus" strand to be synthesized. During this duplication process, the plus and minus strands associate as a double-stranded form of RNA, called the "replicative" form, which is analogous to the replicative form of φX174 DNA. In this replicative form the minus strand of RNA then acts as a template to dictate the proper base sequence of new plus strands, which may function as mRNA for formation of virus proteins or may be incorporated as the infectious RNA into the progeny virus particles. Viral RNA is thus polyfunctional, acting as both mRNA and as template for the synthesis of replicas of itself.

While a single-stranded RNA virus is its own messenger, we do not know how the RNA of a double-stranded virus goes about directing the synthesis of progeny virions. From reovirus-infected cells there can be isolated a double-stranded RNA identical to the parental RNA and a single-stranded RNA, probably mRNA, which is similar to one strand of the parental RNA. Various possibilities for the origin of the mRNA and manner of replication of the double-stranded form have not yet been resolved.

The uniqueness of viral RNA should be emphasized, as the ordinary RNA of the cell performs no genetic or replicating function. It is possible that viral RNA differs from cellular RNA in some subtle chemical way that permits it to perform these functions.

Virus infection in animal and plant cells Although the most extensive studies on viral replication have been done with bacterial viruses, enough work has been done with animal and plant viruses to know that the same basic processes occur, though they differ in certain important details. The whole animal-virus particle usually penetrates the cell, being carried inside by phagocytic or pinocytic action (see Section 5.12) of the cell. Once inside the cell, the viral nucleic acid becomes separated from the protein coat, and replication can then proceed. Release of mature virus particles often occurs without lysis of the infected cell, perhaps by a process of reverse pinocytosis.

In the case of those animal viruses which have membranous envelopes, the assembly process may require the active participation of the host. The nucleic acid is first surrounded by the protein capsid and then this nucleocapsid becomes surrounded by an envelope that at least in some cases derives from the host cell membrane by a process of budding and envelopment (Figure 10.10). Virus release occurs soon after, but is not necessarily accompanied by cell death. Virus release may continue for many

Figure 10.10 Origin of the envelope of an enveloped virus from the host cell membrane. (Adapted from B. D. Davis, R. Dulbecco, H. Eisen, H. S. Ginsberg, and W. B. Wood, Jr.: Microbiology. New York: Harper & Row, Publishers, Inc., 1967.)

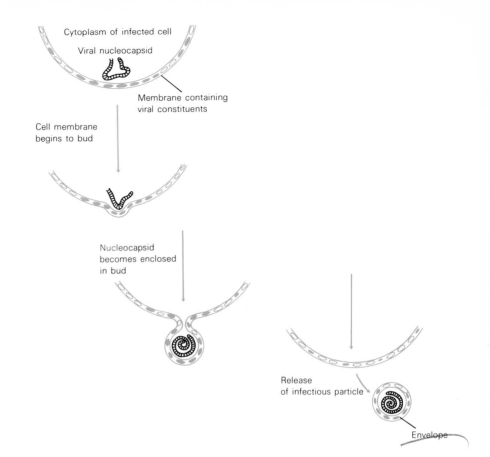

Cytoplasm of infected cell

Viral nucleocapsid

Membrane containing viral constituents

Cell membrane begins to bud

Nucleocapsid becomes enclosed in bud

Release of infectious particle

Envelope

hours or days, with a single cell liberating thousands of virus particles.

With plant viruses there seem to be no specific mechanisms to ensure penetration of virus into the cell. Plant viruses apparently enter passively through breaks or cuts in the plant surface. In experimental work with TMV, it is usual to infect leaves by rubbing them with a virus preparation suspended in an abrasive such as carborundum.

The replication of virus nucleic acid in eucaryotic plant and animal cells can occur either in the cytoplasm or in the nucleus; sites of replication in animal cells are given in Table 10.1. The site of replication can be determined experimentally by providing the infected cell with a radioactive precursor of nucleic acid and preparing autoradiographs of the infected cells to see where the radioactive nucleic acid localizes. Perhaps the most interesting case is that of the pox viruses (for example, smallpox, cowpox), which are DNA viruses that replicate in the cytoplasm (cytoplasmic DNA

Table 10.1 Site of replication of nucleic acid of animal viruses

Type of viral nucleic acid	Site of replication	
	Nucleus	Cytoplasm
DNA	Herpes Polyoma Adenovirus Most other DNA viruses	Pox viruses
RNA	None known	Poliovirus Most other RNA viruses

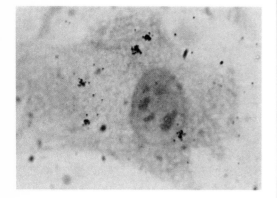

Figure 10.11 Autoradiograph of a cell infected with sheeppox virus (a DNA virus). The infected cell was incubated with tritiated thymidine. Note that the sites of labeling are in the cytoplasm rather than in the nucleus. Magnification 915×. [From A. M. Terrinha, J. D. Vigário, J. L. Nunes Petisca, J. Moura Nunes, and A. L. Bastos: J. Bacteriol. 90:1703 (1965).]

replication is not common in uninfected eucaryotes). An autoradiograph showing the cytoplasmic site of pox-virus DNA synthesis is shown in Figure 10.11. Why the nucleic acids of some viruses replicate in the nucleus and some in the cytoplasm is not known. The synthesis of viral protein in eucaryotes generally occurs in the cytoplasm. If the virus is assembled in the nucleus, viral protein then migrates from the cytoplasm to the nucleus. In the case of the enveloped viruses, assembly of viral nucleocapsid may occur in the nucleus, and the nucleocapsid then migrates to the cytoplasm and is inserted in the developing viral envelope.

Infectious nucleic acid   In recent years it has been possible to provide direct evidence that viral nucleic acid itself is infectious. The bacterial cell wall serves as a barrier to the penetration of the nucleic acid, and indeed, one of the main functions of the protein coat, tail fibers, and so on, of a bacterial virus is to provide a means of breaching the cell wall. It has been possible to infect bacterial protoplasts with purified nucleic acid of certain small bacterial viruses, but this process is very inefficient. The absence of cell walls in animal cells makes penetration by free nucleic acid easier, and nucleic acids are actively taken up by pinocytosis. Infectious nucleic acid has been isolated from a variety of small viruses containing RNA (for example, polio, western equine encephalitis) and DNA (for example, the polyoma and papilloma tumor viruses). In all cases the infectivity of the nucleic acid is less than that of the intact virus but is sufficient so that there is no doubt that the nucleic acid alone is able to direct all functions of viral replication. However, although the protein coat is not absolutely required for virus infection to occur, it plays an important role in natural virus infections. In addition to its part in the adsorption and penetration processes, the protein coat protects the nucleic acid of the virus from attack by nucleases present in the extracellular environment. Infectious nucleic acid is rapidly inactivated by nucleases, whereas whole virus particles are usually completely unaffected by them.

10.4 Specificity of virus-host interaction

Usually in nature a given virus attacks only organisms that are closely related to each other; only rarely, as in rabies virus infection, is there a wide variety of hosts for a single virus. In many cases a single virus will not even attack all strains of the same species. However, it is often possible in the laboratory to infect hosts not infected in nature. An understanding of the underlying causes of virus-host specificity may give us some clues as to how resistance to virus infection can develop.

The most common basis for virus-host specificity is the uniqueness of host adsorption sites and virus attachment sites. Often a cell has on its surface specific receptor sites to which the virus attaches. Many bacteria have in their cell walls virus-receptor sites composed of protein or lipopolysaccharide, which are specific for attachment sites such as the tail fibers of the virus. In the absence of a receptor site on the cell surface, the virus cannot adsorb, and hence cannot infect. If a host cell alters its receptor site, either through mutation or through physiological means, it may become resistant to virus infection. However, such a mutant cell is not completely immune to virus attack, as the virus can also mutate to a form that can then attack the resistant mutant! Receptor sites have also been identified for some animal viruses: The receptor site for influenza virus is a glycoprotein found on red cells and on cells of the mucous membrane, and the receptor site of poliovirus is a lipoprotein.

Not all cases of resistance result from lack of virus adsorption, however. Sometimes adsorption does occur and some later event in the infection cycle, such as nucleic acid penetration or replication, fails to take place. Animal cells frequently take up virus in a nonspecific manner by phagocytosis or pinocytosis, but this does not necessarily lead to virus replication. The host metabolic machinery may not provide all of the building blocks needed for viral nucleic acid replication, or the host cell may possess specific repressor molecules that prevent the nucleic acid from replicating.

10.5 Damage to the host

Viruses vary enormously in the degree of damage that they inflict on the host. With some viruses no effect of virus infection is detectable; these are called latent viruses, and often their existence is not even suspected. Other viruses are highly virulent and kill cells very soon after infection, perhaps even before virus replication has begun.

With some virulent viruses infection leads to an immediate cessation of host protein and nucleic acid syntheses. Such is the case for bacteriophage T4, infection with which inhibits the synthesis of host mRNA; since T4 mRNA synthesis begins soon after nucleic acid penetration, infection with T4 causes a switch from host to virus mRNA synthesis. In the case of polio virus, infection leads to a disaggregation of the polyribosomes of the host and thus to an inhibition of host protein synthesis. This effect favors viral multiplication by liberating host ribosomes for use by the virus RNA in producing virus-specific proteins. Some animal viruses containing DNA inhibit host DNA synthesis but often have little effect on host RNA and protein syntheses until late in the replication cycle.

Release of bacterial virus cannot occur without dissolution of the cell wall and cell lysis; similarly, release of animal viruses may require irreversible changes in cell structure similar to lysis. Some animal viruses induce the synthesis of specific toxins, which cause cell death. In many cases, virus replication leads to the formation of massive numbers of virus particles within the cell, and such accumulations alone conceivably could disrupt cell structure and function.

One effect of virus infection that is being increasingly well documented is the conversion of normal host cells into tumor or cancer cells. Some viruses in certain hosts act as moderate viruses that do not kill cells but induce alterations in them that lead to the cancerous state. Cancer cells are characterized by uncontrolled growth within the animal body, which apparently arises from changes in cell surface properties. The manner in which a normal cell is converted into a tumor cell through virus infection is not as yet understood.

10.6 Temperate bacterial viruses— lysogeny

Most of the viruses described above are called *virulent* viruses, since they usually kill the cells they infect. They are really agents of disease. However, many viruses are not virulent but may have much more subtle effects on cells. Such viruses are called *temperate;* they are all DNA viruses and so far have been found only in the bacteria. Temperate viruses are usually agents of heredity rather than agents of disease because their genetic material, usually duplicated along with the host genetic material at the time of cell division, is passed from one generation of bacteria to the next. Many bacteria that appear normal produce spontaneously small amounts of virus capable of infecting closely related bacterial strains. These bacteria, which have the potential of spontaneously producing temperate viruses, are called *lysogenic* (lysis-producing) bacteria.

The ability of a bacterium to reproduce virus material without lysis of the cell may seem unusual since heretofore we have associated virus replication with eventual destruction of the cell. As it turns out, if the host simply makes a copy of the virus DNA, lysis does not occur; but if complete virion particles are produced, then the host lyses. In a lysogenic bacterial culture a small fraction of the cells, 0.1 to 0.0001 percent, produce virus and lyse, while the majority of the cells do not produce virus and do not lyse. Although only rarely do cells of a lysogenic strain actually produce virus, every cell has the potentiality. Lysogeny can thus be considered a genetic trait of a bacterial strain, although this trait ordinarily is not expressed; if it is expressed, however, it is lethal.

The temperate virus does not exist in its mature, infectious state inside the cell, but rather in a latent form, often called the *provirus* or *prophage* state, which consists of the virus nucleic acid combined with the DNA of the host cell. In considering virulent viruses we learned that the DNA of the virulent virus contains information for the syntheses of a number of enzymes and other proteins essential to virus reproduction. The prophage of the temperate virus carries similar information, but in the lysogenic cell this information remains dormant and is usually not expressed. When it is expressed, virus reproduction occurs, the cell lyses, and virus particles are released. In the normal lysogenic cell, the expression of the lysogenic character is repressed through the action of a specific repressor (see Section 7.6).

A lysogenic culture can be treated so that most or all of the cells produce virus and lyse. Such treatment, called *induction*, usually involves the use of various agents, such as ultraviolet light, nitrogen mustards, and X rays. Since these agents are known to affect DNA (see Chapter 11), this is further evidence that lysogeny is genetic in nature and suggests that the rare spontaneous lysis of a cell may also be due to an alteration in the virus DNA.

Although a lysogenic bacterium may be susceptible to infection by viruses derived from other lysogenic bacteria, it cannot be infected by virus particles of the type for which it is lysogenic. In some cases, this immunity arises from inability of the lysogenic cell to adsorb particles of the virus for which it is lysogenic, whereas in other cases immunity is conferred by an intracellular repression mechanism.

What happens when a temperate virus infects a nonlysogenic organism? The virus may inject its DNA and initiate a reproductive cycle exactly as described for virulent viruses, with the infected cell lysing and releasing more virus particles. Alternatively, the virus may inject its DNA, but the cell does not lyse, and *lysogenization* occurs instead: The viral DNA becomes incorporated into the bacterial genetic material and the host bacterium is converted into a lysogenic bacterium. In lysogenization the in-

fected cell thus becomes genetically changed. Sensitive cells can undergo
either lysis or lysogenization; which of these occurs often is determined
by such nonspecific factors as nutrients, temperature, number of virus
particles per cell, and so on. The various complexities of the lysogenic life
cycle are summarized in Figure 10.12. We see that the temperate virus
can have a dual existence. Under some conditions, it is an independent
entity able to control its own replication, but under others, as when its

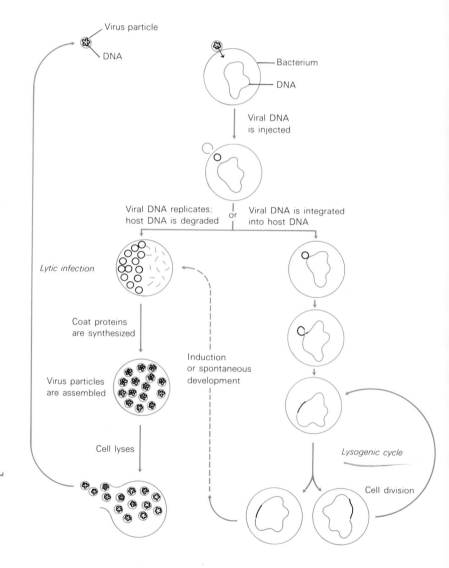

*Figure 10.12   Consequences of infection
by a temperate bacteriophage. The alternatives upon infection are integration of the
virus DNA into the host DNA (lysogenization) or replication and release of mature
virus (lysis). The lysogenic cell can also
be induced to produce mature virus and
lyse.*

DNA is integrated into the host genetic material, replication is then under the control of the host.

Variants or mutants of some temperate viruses exist that are unable to lysogenize, and hence infection of sensitive cells with these mutants leads only to the production of new virus particles and lysis. Such variants are in no way distinguishable from the virulent viruses discussed earlier. Another class of mutants is called *defective*. These viruses have the normal morphology of a bacterial virus but are unable to cause lytic infection of any host. Presumably these defective forms have lost some essential viral function or functions. The only way defective viruses can be maintained is through the lysogenic state, in which manner the viral DNA can be replicated in the host indefinitely. The only way such defective viruses can be detected is by direct examination of culture supernatants with the electron microscope for the presence of viruslike particles. It is of course possible that these particles are not really defective, but only that the proper host for them has not been found.

Lysogeny has been found in all genera of bacteria in which it has been sought. Lysogeny is as yet unknown in eucaryotic microorganisms and in higher plants, although a phenomenon similar to lysogeny may occur in certain tumor virus—animal cell interactions (see Section 10.10). The discovery of lysogeny has had considerable influence on ideas concerning the nature of viruses because it has shown that viruses are not always harmful and can have close genetic relationships with their hosts.

10.7 Interference with viral activity

Chemical inhibition   Since viruses depend on their host cells for many functions of virus replication, it is difficult to inhibit virus multiplication without at the same time affecting the host cell itself. Because of this, the spectacular successes in the discovery of antibacterial and antifungal agents (see Chapter 9) have not been followed by similar success in the search for specific antiviral agents. A few antiviral compounds are successful in controlling virus infections in laboratory situations (Table 10.2) and certain of these have been used in restricted clinical cases; but no substance has yet been found with more than limited practical use.

Interferon   Interferons are antiviral substances produced by many animal cells in response to virus infection. They are low-molecular-weight proteins that prevent viral multiplication. They were first discovered in the course of studies on virus interference, a phenomenon whereby infection with one virus interferes with subsequent infection with another virus,

Table 10.2 *Stages of virus replication at which chemical inhibition of virus action is known to occur*

Stage of replication	Chemical	Virus
Free virus	Kethoxal	Influenza virus
Adsorption	None known	
Injection of nucleic acid	Streptomycin Adamantanamine	Certain bacteriophages Influenza virus
Synthesis of virus-specific proteins	Benzimidazole, guanidine	Polio
Nucleic acid replication	5-Fluorodeoxyuridine (FUDR) 5-Iododeoxyuridine (IUDR)	Herpes virus Herpes virus
Maturation (or late protein synthesis)	Isatin-thiosemicarbazone	Smallpox virus
Release	None known	

hence the name "interferon." It has been found that interferons are formed not only in response to live virus but also to virus inactivated by radiation or to viral nucleic acid. Interferon is produced in larger amounts by cells infected with viruses of low virulence, whereas little is produced against highly virulent viruses. Apparently the virulent viruses inhibit cell protein synthesis before any interferon can be produced. Recent evidence suggests that interferon is induced by a variety of double-stranded RNA molecules, either natural or synthetic. Since double-stranded RNA does not exist in uninfected cells but exists as the replicative form in RNA-virus-infected cells, it has been suggested that double-stranded RNA serves as a signal of virus infection in the animal cell and brings into action the interferon-producing system.

Interferons are not virus-specific but host-specific; that is, an interferon produced by one type of animal (for example, chicken) in response to influenza virus will also inhibit multiplication of other viruses in the same species but will have little or no effect on the multiplication of influenza virus in other animal species. However, the uptake of interferon by cells is not species-specific; for example, chicken cells will take up mouse interferon as well as will mouse cells. Interferon has little or no effect on uninfected cells; thus it seems to inhibit specifically viral synthesis. It acts either by preventing RNA synthesis directed by virus or by obstructing the combination of virus-specific RNA with ribosomes, thus inhibiting synthesis of virus-specific proteins. It is not known precisely how interferon acts.

Because interferon is nontoxic, it would seem to be the ideal antiviral agent. However, interferon is difficult to purify, and hence large quantities are not available. Furthermore, interferon is preventive in its action rather than therapeutic; it has no effect on the course of the virus multiplication in cells already infected. Also, it is effective only during relatively short periods and thus would have to be administered continuously. Recent work has been concerned with a search for nonviral substances that might induce interferon production and thus confer immunity to virus infection. One substance with this property is a double-stranded RNA produced by a fungus. This agent, called *statolon,* is now being studied intensively as a possible protective agent.

10.8 Origin and evolution of viruses

Up to this point we have been content to discuss the phenomenology of viruses without touching on the question of whether they are living or nonliving. Let us say immediately that any answer to this question depends on how we define life itself. If we approach the question as a geneticist would, we might emphasize that viruses are undoubtedly genetic elements, capable of reproducing themselves and of controlling the synthesis of virus-specific proteins. Since self-replication is a property of living cells, the geneticist would probably consider the virus a living organism. The physiologist, concerned with the flow of materials and energy, might place most emphasis on the fact that virus replication does not occur without the agency of a cell, which provides the energy metabolism, the protein-synthesizing machinery, and the building blocks for the virus polymers; hence the physiologist would probably conclude that a virus is not living. Although both points of view are valid, neither alone is completely satisfactory. Viruses alternate between two states, the infectious, free-existing virion, and the noninfectious, replicating state of the intracellular genetic element. Neither state alone expresses the totality of the concept of a virus; only when we consider the two can we construct an acceptable definition of a virus.

Two theories for the origin of viruses have been advanced: Viruses may have arisen independently or may have been derived from cells. The second mechanism is relatively easy to imagine in DNA viruses when we consider the temperate viruses. Perhaps a series of genetic changes took place in a region of host DNA that converted this region into virus DNA by making possible the synthesis of virus proteins, the replication of virus nucleic acid, and the assembly and release of mature virus. It is much more difficult to imagine how viruses might have arisen independently of cells; it seems more likely that the viruses that exist today, which have such intimate relations to cells, arose from cells by genetic modification. The origin of RNA viruses is more difficult to explain since RNA never exhibits self-replication in normal cells.

Viruses exhibit genetic mechanisms similar to those of procaryotic micro-organisms (see Chapters 11 and 12). Both mutation and genetic recombination are known in viruses, and these processes undoubtedly play a role in viral evolution since they allow the development of new virus types. We have already seen how a host may become resistant to virus infection through modification of its virus-receptor site. The virus in turn can then mutate to a form capable of attaching to the modified receptor site. Through interactions of this type, both virus and host become progressively transformed. Furthermore, viruses can hybridize, so that genes from two different viruses can be brought together in the same virus. Viral genetics has played an important role in developing current concepts of the molecular basis of both mutation and genetic recombination (see Chapters 11 and 12).

10.9 Viruses and diseases of higher animals

A wide variety of viruses cause disease in higher animals, and our understanding of the nature and manner of action of animal viruses is advancing rapidly, spurred on by notable advances made in the development of new techniques for virus culture and assay.

The culturing of viruses in the chick embryo has become an important tool in animal-virus research since it permits researchers to grow many viruses simply and reproducibly away from the animal body. There are a number of tissues and cavities in the embryo in which viruses can be cultured, as is illustrated in Figure 10.13. The use of animal-cell cultures has also led to enormous progress in the understanding of virus-cell interaction, and has also made possible the development of new diagnostic methods and of new vaccines. Animal cells used to support virus growth may be either nongrowing, primary tissues isolated from various organs of an animal or serially propagated cell lines that are capable of growing indefinitely away from the animal body. Cell cultures are maintained in complex media containing salts, glucose, amino acids, vitamins, other growth factors, usually some natural materials such as serum or embryo extract, and antibiotics (to reduce bacterial contamination). In such media animal cells may grow or subsist for considerable lengths of time. Cell cultures may be grown in suspension or in thin sheets or monolayers on glass, agar, or plasma or fibrin clots. Monolayers are especially useful for observing pathological changes that may be induced by viruses since the cells of the monolayer can be observed directly under the microscope. Cell monolayers are used to obtain plaques suitable for virus

Figure 10.13   An embryonated egg and sites at which different animal viruses can replicate. [Adapted, in part, from G. J. Buddingh: in T. M. Rivers and F. L. Horsfall, Jr. (eds.), Viral and Rickettsial Infections of Man. Philadelphia: J. B. Lippincott Co., 1959.]

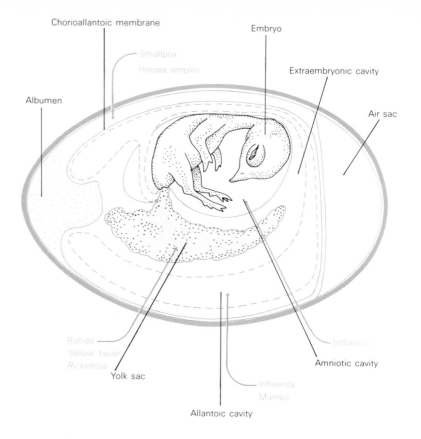

Chorioallantoic membrane

Smallpox
Herpes simplex

Albumen

Embryo

Extraembryonic cavity

Air sac

Rabies
Yellow fever
Rickettsia

Yolk sac

Influenza

Amniotic cavity

Influenza
Mumps

Allantoic cavity

quantification (see Figure 10.5). Dispersed cells may be useful for mass virus propagation and for biochemical studies of virus reproduction. Cell cultures offer the advantages over animals in being cheaper, more reproducible, and simpler; most important, they make it possible to use human cells for propagating viruses that will not grow in nonhuman hosts.

10.10   The classification of animal viruses

At one time animal viruses were classified primarily on the basis of host range, localization in specific organs and tissues, means of transmission, and disease symptoms produced. Such criteria were unsatisfactory because in many cases quite unrelated viruses came to be classified together. As an understanding of the chemistry, structure, and manner of replication of animal viruses developed, it became much easier to dis-

Figure 10-14 The classification of animal viruses.
(Adapted from J. L. Melnick: in Progress in Medical Virology, vol. 9: Basel/New York: S. Karger, 1967.)

Virus group	Nucleic acid core	Capsid symmetry	Virion: naked or enveloped	Site of capsid assembly	Site of final assembly (if enveloped)	Reaction to ether treatment	Number of capsomeres	Diameter of helix, nm	Diameter of virion, nm	Molecular weight of nucleic acid in virion, $\times 10^6$	Representative example
Picodnavirus	DNA	Cubic	Naked	Nucleus		Resistant	12 or 32		18–24	3.0–3.6	Kilham rat virus
Papovavirus	DNA	Cubic	Naked	Nucleus		Resistant	42 or 72		40–55	2–4	Polyoma tumor virus
Adenovirus	DNA	Cubic	Naked	Nucleus		Resistant	252		70–80	23	Human adenovirus
Herpesvirus	DNA	Cubic	Enveloped	Nucleus	Nuclear membrane	Sensitive	162		110	40–84	Herpes simplex virus
Poxvirus	DNA	Unknown	Complex coats	Cytoplasm		Resistant			230 × 300	160–240	Smallpox
Picornavirus	RNA	Cubic	Naked	Cytoplasm		Resistant	32		18–30	2–4	Poliovirus
Reovirus	RNA	Cubic	Naked	Cytoplasm		Resistant	92		70–75	10	Human reovirus
Arbovirus	RNA	Cubic	Enveloped	Cytoplasm		Sensitive			40–50	2	Yellow-fever virus
Myxovirus	RNA	Helical	Enveloped	Cytoplasm	Surface membrane	Sensitive		9	80–120	2–5	Influenza virus
Paramyxovirus	RNA	Helical	Enveloped	Cytoplasm	Surface membrane	Sensitive		18	100–300	8	Mumps virus
Rhabdovirus	RNA	Helical	Enveloped	Cytoplasm	Surface membrane	Sensitive		18	60 × 225	3–4	Rabies virus
Leukovirus	RNA	Unknown	Enveloped	Cytoplasm	Surface membrane	Sensitive			About 100	10	Mouse-leukemia virus
	RNA	Unknown	Enveloped	Cytoplasm	Intracytoplasmic membranes	Sensitive			About 50		Rubella and certain arboviruses

cern similarities and differences among them, and Table 10.3 presents the characteristics now used in classifying animal viruses. A number of distinct virus groups are now recognized and are outlined in Figure 10.14. The viruses within each group generally are closely related, even if they infect different hosts. In the following pages we discuss in some detail a few representative viruses, and indicate some of their characteristics and activities.

Polio   Poliovirus is the best studied and most important member of a group of viruses known as the *picornaviruses* or *enteroviruses*. Individual virus particles are very small, about 30 nm in diameter, and when highly purified they may form distinctive crystals. Electron microscopy of poliovirus particles reveals that each contains 32 protein capsomeres. The virus particles are composed of about 75 percent protein and 25 percent RNA, with complete absence of carbohydrate and lipid. A single infected cell may produce as many as 10,000 virus particles, and before release occurs, these may exist in the cell in crystalline array (Figure 10.15). Cells that produce virus usually die and lyse.

Poliovirus replicates primarily in the cells of the intestinal tract of man. It usually causes only mild symptoms that cannot be distinguished from many other trivial illnesses, and the disease may be called "summer cold" or "intestinal flu." The virus may occasionally pass into the blood or lymph system with no ill effects. In rare instances, it invades the central nervous system and causes the paralytic disease known as poliomyelitis or infantile paralysis. Thus, although infection with the virus is very common, paralytic polio disease is relatively rare.

Although the virus naturally infects only man, it can be passed to the chimpanzee, the monkey, and other primates, in which animals it can also induce paralytic disease. It has also been possible to infect mice by injecting them intracerebrally, and certain strains of virus have also been adapted to chick embryos. The greatest experimental advances in our understanding of polio came with the development of methods for growing the virus in human and monkey tissue cultures.

The virus particles are excreted in the feces in large numbers, and fecal contamination of food and water results in spreading of the virus to other individuals. In areas of poor sanitation most infants are infected early when they still possess immunity transmitted to them by their mothers. The virus remains localized in the intestinal tract but induces an active immunity which may be lifelong. Paralysis is rare, and when it does occur, it is found primarily in infants. In more civilized countries with better sanitary conditions, infection of infants is rarer. The growing infant loses its maternal immunity and may become a child or young adult without being infected. Infection that occurs later in life is more likely to lead to

Figure 10.15 Electron micrograph of a cross section through a crystal of poliovirus. Magnification, 75,000×. (Courtesy of the Virus Laboratory, University of California, Berkeley.)

the paralytic disease and hence infantile paralysis is more common in civilized countries and in older children and young adults.

The development of cell-culture methods made it possible to grow the virus in large numbers, opening the way for the production of vaccines. The benefits of vaccination are shown by the dramatic drop in the number of cases of polio after the Salk vaccine was introduced into the United States [see Figure 15.24(b)].

Influenza Frequently in the winter people suffer from a short-lived fever associated with soreness and redness of the respiratory passages and dry cough. Such a condition is commonly called "flu" or *influenza* although this name should not be applied to all respiratory infections of this type.

The influenza virus (a *myxovirus*) is round or oval shaped and 80 to 120 nm in diameter (Figure 10.2). It is an enveloped virus and fairly complex chemically; because of this it has been difficult to purify, and chemical analyses are somewhat uncertain. One test showed that the virus contained less than 1 percent RNA, about 7 percent carbohydrate, 18 percent lipid, and 75 percent protein. The RNA is contained in an inner helical core of ribonucleoprotein surrounded by a mucoprotein, and these two are held together by the lipoprotein envelope. The ribonucleoprotein is the fundamental replicating unit of the virus, while the mucoprotein apparently is involved in attachment of the virus to a cell.

When influenza virus is inoculated into the allantoic cavity of the chick embryo, rapid absorption of 50 to 70 percent of the virus occurs. There is no evidence that the nucleic acid becomes separated from the rest of the particle, although ^{32}P labelling experiments show that the nucleic acid quickly becomes associated with the cell nucleus. Within 5 to 10 hr after infection new virus is released into the fluid. There is no evidence that release of virus is accompanied by cell lysis. It has not been possible to see mature virus particles within the cell. Mature virus is seen only on the cell membrane, and it is thought that the nucleoprotein components of the virus are transferred to the cell surface where the outer units of the virus particle, the lipid and mucoprotein, are added. These latter materials may be synthesized on the cell surface, or they may be modified from substances preformed in the cell membrane (see Figure 10.10).

Human influenza virus exists in nature only in man. It is transmitted from person to person through the air, primarily in droplets expelled during coughing and sneezing. The virus infects the mucous membranes of the upper respiratory tract and occasionally invades the lungs. Localized symptoms occur at the site of infection. Systemic symptoms include an abrupt fever of 101 to 104°F for 3 to 7 da, chills, fatigue, headache, and general aching. Recovery is usually spontaneous and rapid. Most of the serious consequences of influenza infection do not occur because of the viral infection but because bacterial invaders may be able to set up severe infections in persons whose resistance has been lowered. Especially in infants and elderly people, influenza is often followed by bacterial pneumonia; death, if it occurs, is usually due to the bacterial infection.

A most interesting aspect of influenza is its occurrence in epidemics, often of worldwide proportions. Early epidemics occurred before knowledge was sufficiently advanced to make careful analysis possible, but the 1957 epidemic of the so-called Asiatic flu provided an opportunity for a careful study of how a worldwide epidemic develops. The epidemic began with the development of a mutant virus strain of marked virulence and differing from all previous strains in antigenicity. Since immunity to this strain was not present in the population, the virus was able to advance

Figure 10.16   Routes of spread of Asiatic influenza virus from its origin in China. Within six months the virus had spread throughout the world and had caused countrywide epidemics in many areas. This epidemic in 1957 and 1958 was followed by a similar one in 1968 and 1969 caused by Hong Kong influenza virus. [From Langmuir: Amer. Rev. Resp. Dis. 83, pt. 2:2 (1961).]

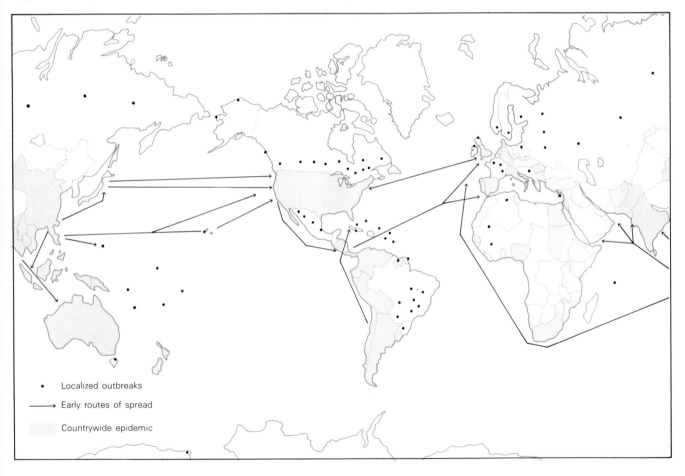

• Localized outbreaks

⟶ Early routes of spread

▓ Countrywide epidemic

10.10   The classification of animal viruses

rapidly throughout the world (Figure 10.16). It first appeared in the interior of China in late February, 1957, and by early April had been brought to Hong Kong by refugees. It spread from Hong Kong along air and naval routes and was apparently transferred to San Diego, California, by naval ships. An outbreak occurred in Newport, Rhode Island, on a naval vessel in May. Other outbreaks occurred in various parts of the United States. Peak incidence occurred in the last two weeks of October, during which time 22 million new bed cases developed. From this period on there was a progressive decline.

Smallpox, cowpox, and vaccinia   The disease smallpox has been known for centuries. In the eighteenth century Jenner recognized that the *cowpox* disease of cows was similar to the human smallpox, and he made cowpox virus the basis of vaccination procedures to be described later. Through the years cowpox virus has been grown in various animals for the production of vaccines, and the cultured virus, called *vaccinia,* now differs from naturally occurring cowpox virus. Most laboratory studies have been on vaccinia virus, so that the picture to be described below deals mostly with this virus; however, the same phenomena elicited by vaccinia virus probably also occur with smallpox and cowpox viruses.

In addition to infecting man, vaccinia virus will infect cows, sheep, rabbits, monkeys, rats, mice, guinea pigs, hamsters, and chick embryos as well as tissue cultures of cells of these animals. Thus it is much less specific in its action than the polio and influenza viruses. The virulence of the virus can be changed markedly by culturing it in different species, organs, or tissues.

In contrast to polio and influenza viruses, vaccinia virus forms inclusion bodies, consisting of aggregates of virus particles, within the cytoplasm of infected cells. These inclusion bodies can be seen with the light microscope by the use of certain staining procedures or with the phase microscope. The purified virus particles are more or less rectangular and in the electron microscope measure about 200 by 250 nm. Vaccinia is thus much larger than polio or influenza viruses and is in fact just barely visible with the light microscope. Vaccinia, the first virus to be extensively purified, is chemically complex, containing about 85 percent protein, 5 percent DNA, 5 percent lipid and 5 percent carbohydrate. The DNA is double-stranded. The vitamins biotin and riboflavin are present as integral parts of the purified virus, but the role of these constituents is unknown; they may be contaminants taken up from the host cells. Although it is a DNA virus, multiplication occurs only in the cytoplasm and may proceed at four or five localities at the same time. Virus-specific mRNA synthesis also occurs at these foci, and new virus-specific enzymes are formed. Since DNA and mRNA synthesis in eucaryotes normally occurs only in the nucleus, the synthesis

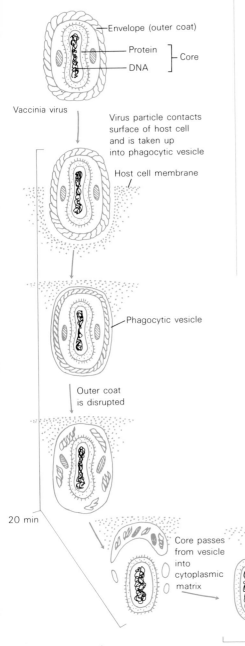

Vaccinia virus

Virus particle contacts surface of host cell and is taken up into phagocytic vesicle

Host cell membrane

Phagocytic vesicle

Outer coat is disrupted

20 min

Core passes from vesicle into cytoplasmic matrix

DNA is released

Viral DNA is replicated in cytoplasm of host cell

Structural proteins are synthesized

1 hr

3 hr

Assembly

Immature virus particles

4–6 hr

Membrane (envelope)

Maturation

6–7 hr

of vaccinia DNA and RNA in the cytoplasm is an unusual occurrence. The morphological events of virus synthesis are depicted in Figure 10.17.

Vaccinia virus, as we have stated, is a laboratory strain related to the smallpox and cowpox viruses that exist in nature. Smallpox disease has been known throughout recorded history and has occurred in epidemics many times in Western countries. The smallpox virus is found naturally only in man and monkeys, even though it can be transferred artificially to other animals or to the chick embryo. It is present in large amounts in pustules on the skin of infected people and spreads through the air, entering a new host through the upper respiratory tract. The virus first grows in the mucous membranes of the upper respiratory tract, and then multiplies in the lymphoid tissues that drain the respiratory tract. Just before the development of symptoms, the virus may be found in the bloodstream, and from there it passes to the skin, mucous membranes, and organs such as the heart, liver, spleen, and kidney. During the initial phase, fever, headache, and backache occur, followed by prostration. A rash appears on the skin, and soon after this the temperature falls to normal and the patient feels better. Vesicles teeming with virus then form crusts on the skin, which develop into the characteristic pockmarks that

Figure 10.17 Stages in the replication of vaccinia virus. The numbers indicate the time after infection of the various stages. Although they are DNA viruses, vaccinia and other pox viruses replicate in the cytoplasm. [Adapted from S. Dales: J. Cell Biol. 18:51 (1963).]

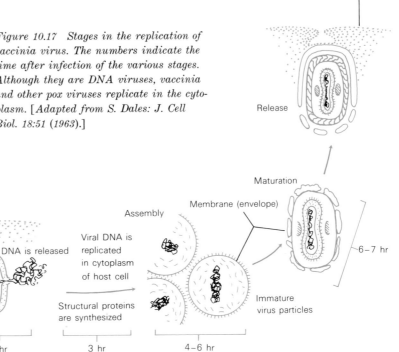

Host cell membrane

Release

10.10   The classification of animal viruses

permanently disfigure the skin. In severe cases death occurs, usually due to hemorrhage and generalized toxemia.

Recovery from smallpox infection confers complete immunity. It has been known for several centuries that material from smallpox pustules could be used to induce immunity by inoculating susceptible individuals by incision or puncture of the skin, which initiates a local lesion with mild fever and quick recovery. Even though the skin infection is only local, it causes systemic immunity to the disease. The practice of smallpox inoculation (called *variolation*, from "variola," a synonym for smallpox) was introduced into England from Turkey by a British diplomat's wife, and for a while was used extensively. However, although it was generally successful, occasionally severe and fatal cases resulted.

In 1798 Jenner reported his observations on cowpox. He had noted the clinical similarity of the pox disease of cows to smallpox of man. He also observed that persons working around cows, such as milkmaids, were often immune to smallpox without having had the disease. He reasoned that artificial inoculation with cowpox material might confer immunity to smallpox, and his experiments confirmed this. He thus introduced the process of vaccination (from *vacca*, "cow"). Cowpox virus is related to smallpox virus but causes only a mild, usually inconsequential, self-limited disease in man. Vaccination generally confers good immunity to smallpox and is the most effective control measure in existence for this highly infectious disease.

Smallpox is still endemic in India, southeast Asia, Africa, and South America. The disease is essentially nonexistent in Western Europe and North America, but because of the rapid movements of people in this airplane age, it can be readily introduced into these areas. Outbreaks in England have occurred in recent years, presumably due to introduction of virus by African or Asian immigrants. Control of the disease is by vaccination of susceptible individuals and by quarantine. No one is permitted to enter the United States who cannot present evidence that he has been vaccinated within the last three years.

Measles Measles virus grows in nature only in man, although monkeys can be infected experimentally and may also become infected naturally when kept in captivity. Only primates have been successfully infected, and because of this, research on measles progressed slowly; only since the introduction of tissue-culture methods has any significant progress been made.

Measles virus is a member of the *paramyxovirus* group, which also includes mumps and parainfluenza viruses, canine distemper virus of dogs, and the rinderpest virus of cattle and sheep. The particles are roughly spherical, about 150 nm in diameter, and contain single-stranded RNA.

The central core of the virus is an RNA-protein complex, with the RNA arranged in a helical fashion. Surrounding this core is a lipid-containing membrane.

The measles virus is transmitted from person to person by the respiratory route. The virus infects cells of the nasopharynx and then spreads throughout the body. Virus can be found in the blood and in the nasopharyngeal secretions. First symptoms are chills, followed by sneezing, running nose, redness of eyes, cough, and fever. Characteristic spots, called "Koplik spots," appear on the mucous membranes and around the salivary gland and are an early diagnostic symptom of the disease. Fever and cough become worse and a rash then appears, first on the forehead and behind the ears, then on face, neck, limbs, and trunk. There may be inflammation of the eyes and sensitivity to light. Pustules are not formed as in smallpox; the rash may be an allergic response to the presence of virus products in the body, rather than a result of direct viral multiplication in the skin cells. Recovery usually confers long-term immunity. Measles occurs in epidemics at about 3-yr intervals, and in many civilized countries nearly 90 percent of the people over 20 yr of age have had the disease. Vaccines have been developed, using viruses grown in cell cultures, which have drastically reduced the incidence of measles [see Figure 15.24(a)].

Adenoviruses   The adenovirus group illustrates well the importance of tissue cultures in the discovery of new viruses. A large number of acute respiratory diseases exist that cannot be attributed to bacterial infection, these were thought to be caused by viruses. When tissue-culture techniques were introduced (around 1953), workers discovered in respiratory secretions viruses that were able to cause destruction of cultures of human cells. A number of these viruses have been found to be closely related and were called "adenoviruses" since they were first detected in human adenoid tissues. Although adenoviruses are responsible for only a small percentage of human respiratory disease, they are of interest because they are so widespread.

There are over 50 different adenoviruses, which are related in their chemical, structural, and biological properties. These viruses have been found in nature primarily in man, although some types are found in other primates. Attempts to transfer the virus to a wide variety of experimental animals, including chick embryos, have been unsuccessful. The virus particles are icosahedrons, between 60 and 90 nm in size, which do not have outer envelopes. The nucleic acid is double-stranded DNA. The virus nucleic acid replicates within the nucleus, and at a late stage of infection crystals of virus particles can be seen in the nucleus (Figure 10.18). Pathological effects on the cells are due to inhibition of cellular macromolecular biosyntheses.

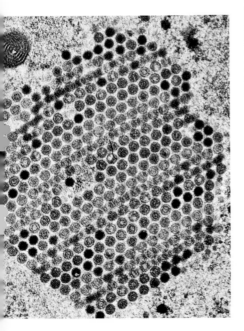

Figure 10.18 A crystal of adenovirus particles within the nucleus of a human cell, as revealed by electron microscopy. Magnification, 35,000×. (Courtesy of Councilman Morgan.)

10.10 The classification
of animal viruses

The adenoviruses are transmitted by the respiratory route and infect the tissues of the upper respiratory tract and conjunctiva. They probably remain localized in this region and are not found in the blood or other parts of the body. A number of clinical conditions due to adenoviruses have been described, the most frequent being acute respiratory disease, pharyngitis and pharyngoconjunctival fever, and conjunctivitis and keratoconjunctivitis. In most cases the symptoms are a mild, flulike illness with fever, chills, headache, malaise, loss of appetite, mild sore throat, and mild hoarseness and cough. In the latter two conditions, inflammation of the conjunctiva occurs. Certain adenoviruses can cause tumors in experimental animals. This was discovered by inoculating the animals with a wide variety of adenoviruses that had been isolated from human beings and grown in tissue culture. There is no evidence as yet that adenoviruses can cause tumors in man, however.

Rabies   Rabies virus attacks the nervous tissue of all warm-blooded animals. Rabies has been known since ancient times as a disease of dogs, but it is also common in wild animals such as rodents, foxes, skunks, and bats. The virus multiplies in the salivary gland and appears in the saliva; because of this it is readily transmitted by biting. The virus particles are about 75 nm wide and 180 nm long, with a nucleoprotein core and an outer membrane. The virus has not been purified sufficiently for biochemical studies, but cytological investigation shows that the virus particles contain RNA. Natural rabies virus is highly neurotropic; after the virus has entered the blood stream following the bite, it passes to the nervous tissue where multiplication takes place..Characteristic inclusion bodies in the cytoplasm of nerve cells, called Negri bodies, are formed during virus multiplication. These bodies contain virus protein but apparently do not contain complete virus particles. They are demonstrated by staining methods specific for protein. The nerve cells then degenerate. Infected cells of the salivary glands also show characteristic changes. The length of time before symptoms appear is highly variable, ranging from 14 to 285 da, and depends on the size, location, and depth of the wound and on the amount of virulent saliva introduced. When the virus has been passed for many generations in chick embryos, it is no longer capable of invading the central nervous system. This modified virus can then be used in the production of rabies vaccine for dogs.

Dogs may develop symptoms of two different types. In the aggressive or rabid form the dog first exhibits a change in behavior showing restlessness, a tendency to roam, and an inclination to bite, all of which make the transmissal of the virus to susceptible individuals highly probable. Later the dog becomes partially paralyzed and has difficulty in drinking. It staggers around until complete paralysis sets in, and death follows. In the dumb form of the disease, exactly the opposite symptoms develop: The animal is not vicious, has no tendency to bite or roam, and is not excitable. The most characteristic symptoms are paralysis of the lower jaw and limbs. In both forms the course of disease, once symptoms set in, is short, and death occurs in 3 to 7 da.

In human beings, the most characteristic early symptom is an abnormal sensation around the site of infection. There may be a dull, constant pain in the nervous pathways leading from the wound. The patient becomes hypersensitive, complaining of drafts, loud noises, or bright lights, and becomes increasingly nervous, anxious, and apprehensive. Muscle spasms occur, the pulse becomes rapid, and there is dilation of the pupils, crying, increased salivation, and excessive perspiration. When fluids are consumed, they cause painful, spasmodic contraction of the muscles that are involved in swallowing, and the fluids are expelled violently. After this the sight, smell, or sound of liquids induces spasms in the muscles of the throat because these suggest the act of swallowing. For this reason the name *hydrophobia* (literally, ''fear of water'') has often been applied to rabies. Death usually occurs during convulsions, although in some cases the acute excitement phase subsides, and paralysis develops, followed by stupor, apathy, coma, and death.

An animal may be suspected of having rabies from its characteristic behavior. Diagnosis of the disease in a dog thought to be infected is usually based on microscopic examination of brain tissue for Negri bodies. Intracerebral inoculation with saliva or brain tissue from the dog is performed on mice, and if the virus is present, the mice become paralyzed within 6 to 8 da.

In man, diagnosis is usually based on knowledge of exposure to the disease and on development of clinical symptoms. On autopsy Negri bodies can usually be seen. The mortality rate in man is considered close to 100 percent and no therapeutic agent exists for the disease once it has taken hold. If it is known that a person has been bitten by a rabid animal, he should immediately receive treatment to inactivate the virus while it still remains localized. Because of the relatively long incubation period, vaccination of a person who has been bitten may be successful. If the incubation period is less than 30 da, however, the development of immunity may not be rapid enough to prevent onset of the disease. Vaccine for humans is prepared from brain tissue of infected rabbits, usually by killing the virus with phenol or ultraviolet light. Since an allergic reaction to rabbit-brain tissue may occur, vaccine is rarely used unless there is good evidence of exposure to rabies.

For dogs a live virus vaccine, attenuated by long-term passage in chick embryos, is available. This vaccine does not cause sensitization and has been quite effective. It has also been used experimentally in man.

10.10   *The classification of animal viruses*

287

The ecology of rabies infection is quite complex, since the virus can maintain itself in nature only if infected animals can transmit the virus before they die. As we mentioned above, rabies virus is also found in nature in many wild animals, usually carnivores, including foxes, skunks, mink, weasels, wolves, and bats. The fact that bats carry rabies is important, since these animals may be symptomless and serve as a reservoir for the virus in nature. Insect-eating bats in the United States have attacked persons, and some of these bats have been shown to be infected. In Central and South America the vampire bat, a blood-sucking animal, frequently attacks cattle and transmits the virus to them. When infected through vampire bats, cattle develop the paralytic or dumb form of the disease rather than the rabid type. Rabies infection may be a major factor in controlling the population sizes of wolves, foxes, wild dogs, and other caninelike species: When the population of these animals builds up, an increase in incidence of rabies can bring about a drop in the population.

Tumor viruses   A tumor is composed of living cells that are modified in some way from normal body cells, such as by: (1) a decrease or total loss of some of the specialized functions; (2) an increase in a vegetative function, such as the ability to grow more vigorously; (3) the acquisition of certain new functions, such as ability to invade surrounding tissues and to continue to grow in parts of the body distant from the point of origin. Tumors are quite diverse and may vary from small warts to extensive growths that invade and bring destruction to whole organs of the body. Although tumor cells may quite likely arise in various ways, some tumors clearly have been shown to be induced by viruses, and it is these that we will consider briefly here. Both DNA and RNA viruses can cause tumors in animals.

A tumor of chickens, known as the ''Rous sarcoma,'' was the first tumor shown to be induced by a virus. Since then a number of other tumors of birds have been found to have a viral origin. Shope papilloma and Shope fibroma are virus-induced tumors of rabbits, and a mammary-gland tumor of mice has been discovered to be induced by a virus transmitted in the mother's milk. The polyoma virus (discussed below) causes a wide variety of tumors in mice and other animals. With the electron microscope it is possible to see viruses or viruslike particles within infected cells. In man it is difficult to obtain really convincing evidence of virus-induced tumors, but particles that resemble viruses can also be seen in cells from human tumors, providing considerable support to the theory that viruses may induce some tumors in man. If this is so, more effective means of prevention and treatment would be possible.

Polyoma virus, a papovavirus, has been well studied, and will illustrate

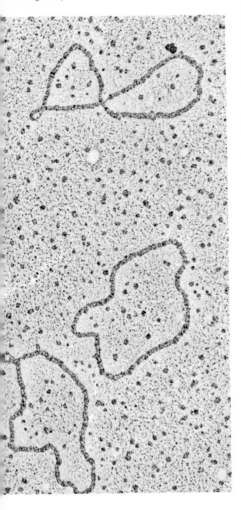

Figure 10.19 Electron micrograph of circular DNA from a tumor virus. The length of each circle is about 1.5 µm. (Courtesy of Alex van der Eb and Jerome Vinograd.)

many of the interesting aspects of the search for tumor viruses. This virus was first isolated during research on tumors in mice and has been found to cause over 20 different kinds of tumors, including those of various glands and organs, the skin, connective tissue, bone, kidney, blood vessels, liver, and nervous tissue. Because of the large variety of tumor types induced, it was called the "polyoma virus" (from *poly-,* "many," and *-oma,* "tumor"). This virus also induces tumors of varying severities and types when injected into other mammals, such as the hamster, rabbit, and rat; monkeys, however, are not affected.

The polyoma virus is icosahedral in shape and about 45 nm in diameter; it consists of only DNA and protein. The DNA has been well studied; in common with some other tumor-virus DNA's (Figure 10.19), that of polyoma is circular. When polyoma virus infects embryonic cells, it causes a virulent infection, the cells are killed and lysed, and new virus particles are produced. When the virus infects adult cells, on the other hand, it does not cause a virulent infection; the cells instead become transformed into tumor cells and no mature virus is produced. After infection, the DNA of the virus apparently enters the cell nucleus. In a virulent infection, extensive changes in the structure of the nucleus appear, and then large numbers of virus particles in crystalline array can be seen under the electron microscope; cells so infected lyse and release virus particles.

The mechanism of conversion of a normal cell into a cancer cell is uncertain but does not involve the extensive destruction just described. Virus multiplication does not occur in a cell converted to a tumor cell. The viral DNA enters the nucleus but pathogenic changes do not occur; the process resembles in many ways lysogenization of temperate bacteriophages. Certain virus-specific proteins are synthesized but virus coat proteins are not. The new tumor cell lines do not produce virus and cannot be induced to do so, but the fact that viral genes persist in these cell lines is indicated by the presence of virus-specific proteins. Isolated protein-free DNA of polyoma virus is also able to infect tissue cultures and to bring about the production of complete virus particles.

The difference between a tumor cell and a normal cell is seen in their responses to crowding. The normal cell stops growing when it comes into contact with other cells, a phenomenon called *contact inhibition*. The tumor cell, on the other hand, continues to grow even under crowded conditions. This ability may reflect the cancerous nature of the cell and suggests that some mechanism controlling the population density of normal cells is lacking in cancerous cells.

Common cold   There are a number of viruses that cause infections of the upper respiratory tract, such as influenza virus and the adenoviruses, but the symptoms of the common cold differ from those caused by these other agents. The first sign of common cold is usually a sore throat; fever and other systemic manifestations are rare, and the predominant symptom is an increased flow of nasal mucus.

Cold viruses are not highly infectious—even when nasal secretions are inoculated directly into volunteers, the percentage of takes is only 30 to 40 percent. When a volunteer is placed in an isolated room and inoculated with nasal secretions from a person with active cold infection the disease (when it appears) runs a quick course (Figure 10.20). No experimental animal except the chimpanzee has been found to be susceptible, and this animal is too expensive to keep and too difficult to work with for routine laboratory use. Attempts to culture the viruses in chick embryos have been unsuccessful, but techniques for growing the viruses in human cell cultures are now available, and many strains have been isolated in tissue culture from persons with typical colds. One group of viruses that has been implicated in causing the common cold is called the *rhinoviruses;* these are members of the picornavirus group. Although none of these cold viruses have yet been purified or even seen with the electron microscope, filtration experiments indicate a particle size of around 30 nm. Virus replication is not inhibited by the specific inhibitor of DNA synthesis, 5-fluorodeoxyuridine, and thus it is presumed that the rhinoviruses contain RNA. This conclusion is bolstered by the fact that infectious RNA has been isolated from one virus strain.

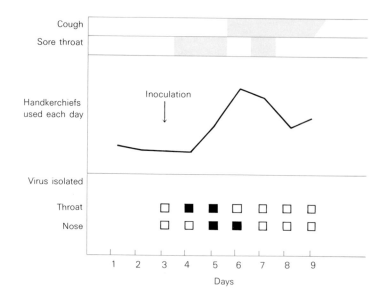

Figure 10.20   The course of a cold in a human volunteer inoculated intranasally with rhinovirus. The severity of the symptoms can be roughly quantified by the number of handkerchiefs used each day. In the lower section, the solid squares indicate positive virus culture, and the open squares, negative virus culture. [From D. A. J. Tyrrell and M. L. Bynoe: Brit. Med. J. 1:393 (1961).]

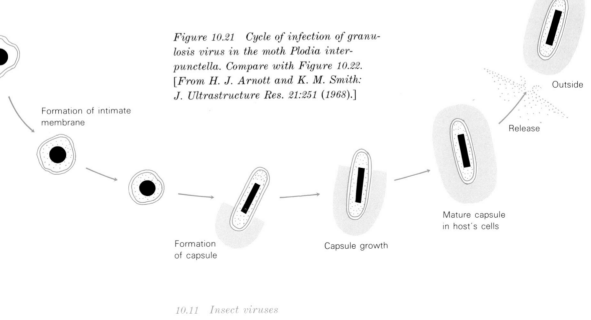

10.11 *Insect viruses*

Infection

Proliferation of virus in cytoplasm

Ordered arrays of naked virus rods associated with endoplasmic reticulum

Outer membrane formation from endoplasmic reticulum

Formation of intimate membrane

Formation of capsule

Capsule growth

Mature capsule in host's cells

Release

Outside

The first insect virus to be discovered was one causing a disease of silkworms. About 200 virus diseases are now known for insects of various taxonomic orders, and these usually infect only the larval stages. Generally virus inclusion bodies, consisting of aggregates of virus particles, are seen within either the nucleus or the cytoplasm. Much is known about the chemical composition and structure of a wide variety of insect viruses. Both rod-shaped and spherical viruses are known, and many of these have been studied under the electron microscope.

The infection cycle of one insect virus is illustrated in Figure 10.21. Rod-shaped virus particles are surrounded by two membranes, an outer developmental membrane and an inner, intimate membrane (Figure 10.22). The virus particles develop in association with the endoplasmic membrane and then become surrounded by the intimate membrane and the developmental membrane. A rod with both membranes constitutes the infectious virus that attaches to a new host cell. The developmental membrane then ruptures, releasing virus subunits from the intimate membrane, and these begin the cycle again. All insect viruses so far examined contain DNA, whether the virus develops in the nucleus or in the cytoplasm.

Figure 10.21   Cycle of infection of granulosis virus in the moth Plodia interpunctella. Compare with Figure 10.22. [From H. J. Arnott and K. M. Smith: J. Ultrastructure Res. 21:251 (1968).]

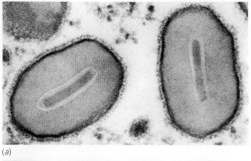
(a)

(b)

Figure 10.22   Moth granulosis virus:
(a) electron micrograph of mature virus
particle in infected insect cell (magnifica-
tion, 73,000×); (b) ordered arrays of naked
virus rods associated with endoplasmic
reticulum (magnification, 44,000×). Com-
pare with Figure 10.21. [From H. J. Arnott
and K. M. Smith: J. Ultrastructure Res.
21:251 (1968).]

Virus infection of an insect may cause rapid death or may result in a mild or latent infection. A larva may become infected when it feeds on plants that are contaminated by virus-killed individuals. Introduction of insect viruses into areas infested with certain destructive insects may be useful in their control. Virus infection in insects that are cultivated, such as the silkworm, may lead to serious economic loss.

10.12   *Plant viruses*

A large number of viruses are known that attack plants. They cause most economic damage in perennial crops or in plants raised by vegetative propagation. There is no recovery from most virus diseases, and although the plant may not be killed outright, its functions may be so seriously affected that it will not be of economic use or it may not be able to compete in nature with unaffected plants.

The most universal symptom of virus infection in plants is reduction in size or in yield of product. Many parts of the plant may be infected. Various organs may become discolored or deformed, and alteration in leaf coloring is often the first indication of virus infection. The cells may be stimulated to grow excessively, resulting in tumor-like growths. In tulips, virus infection causes a variegation of the flower color, which is thought to enhance the beauty of the blooms and hence is considered desirable. Certain varieties of tulips have been maintained for generations in the virus-infected state, with transmissal of the virus through the bulbs. This was done long before viruses were known to exist, and it was not suspected that anything infectious was involved.

Microscopic examination of virus-infected cells often reveals inclusions that are either virus crystals or crystals of material formed in response to virus infection. Cells may also show such pathological changes as small size, lack of chlorophyll, altered chloroplasts, or necrosis.

Plant viruses may be transmitted (1) by grafting healthy plants to diseased ones; (2) by inoculating healthy plants with sap or plant debris from diseased ones; (3) by insects that feed on diseased plants and then on healthy plants; (4) through seeds, tubers, or rhizomes that are taken from diseased plants and used to propagate new individuals; (5) through soil that becomes contaminated with virus by having supported growth of diseased plants.

Because relatively large amounts of virus are synthesized in some infected plants, the purification of plant viruses has advanced much more rapidly than has the purification of animal and bacterial viruses. The crys-

Table 10.4   Properties of purified plant viruses

	Shape	Size, nm	RNA, %
Tobacco mosaic	Rod-shaped	300 × 18	5
Potato X	Rod-shaped	500 × 15	5
Alfalfa mosaic	Spherical	16.5	15
Tomato bushy stunt	Spherical	30	18
Tobacco necrosis	Spherical	20	18
Southern bean mosaic	Spherical	25	21
Turnip yellow mosaic	Spherical	20	35
Tobacco ring spot	Spherical	25	40

tallization of TMV by W. Stanley stimulated work in this field, and over 20 plant viruses have now been purified and crystallized. The chemical and physical properties of some purified plant viruses are given in Table 10.4. Most of the plant viruses that have been studied extensively have been fairly small in size, and this may explain the ease in crystallizing them. A more detailed discussion of TMV structure is given below.

In general, therapeutic measures are not attempted for most virus diseases of plants, since the cost would not warrant it, although in certain plants (such as fruit trees) in which there is a long-term investment, heat therapy has been employed. Preventing the spread of virus is the most effective control method. This may be done by destroying infected plants, sterilizing infected soil, using sanitary procedures, and employing disease-free seed stocks. In the case of insect-borne viruses, elimination of the insect vector may be effective. Plant varieties that are genetically resistant to the virus in question may also be used.

Because it is relatively easy to purify (Figure 10.1) and study, TMV has served as a model for plant viruses; knowledge of its characteristics has also stimulated research in animal and bacterial viruses. The TMV particles are long and rodlike (Figure 10.3), 18 by 300 nm, with a molecular weight of 5×10^7. Each particle consists of 2,200 identical protein capsomeres surrounding a core of single-stranded RNA (Figure 10.4). Each capsomere has a molecular weight of 17,000 and is composed of 158 amino acid residues of 16 different amino acids in definite sequence. These capsomeres form a shallow spiral, making 130 turns around the central RNA core. The protein subunits can be removed from the virus by chemical means, leaving behind the RNA core, which then unwinds and forms a long, slender strand. The RNA chain has a molecular weight of 2.2×10^6 and is composed of 6,500 nucleotides. The isolated RNA is infectious, and it can be induced to reassemble with purified protein into particles that resemble the native virus. Isolated RNA can be treated with various mutagenic agents such as nitrous acid (see Chapter 11); and after it has been used to infect plants, virus mutants can be isolated. Some of these mutants show alterations in the amino acid sequence of the protein subunit.

Because of the complexities of metabolism in the plant leaf, knowledge of the biochemical processes involved in TMV replication is not so advanced as is that of bacterial viruses or even of polio virus. Apparently virus particles do not bind specifically to cells of the plant leaf. The virus infects most efficiently leaves that have been damaged; the usual procedure for experimentally infecting a plant is to rub virus into the leaves in the presence of an abrasive such as carborundum. Local lesions are obtained on infected leaves (Figure 10.5), each representing an area of dead cells where virus multiplication has occurred. The efficiency of infection is low—about a million particles must be inoculated on a leaf to obtain one lesion. In some species of susceptible plants systemic infection appears instead of only local lesions.

After separation of the RNA from the protein coat, the RNA replicates in the nucleus and then passes to the cytoplasm, where it directs the synthesis of the protein subunits. These subunits first accumulate in the cytoplasm unattached to the RNA, and then RNA and protein assemble into the complete virus particle. Under favorable conditions crystals of virus particles may occur right in the plant cell. Some of the virus formed in a cell may then spread to adjacent cells, probably through intercellular connections. If virus particles get into the circulatory system of the plant, they rapidly spread to other leaves and structures, and systemic infection results. In young developing leaves infection may not result immediately in virus multiplication, which may occur later when the leaf matures. Virus multiplication can also be obtained in excised leaves in nutrient solutions.

Tobacco mosaic disease occurs almost everywhere that susceptible plants are grown and causes considerable economic loss in tobacco-growing regions. The virus in nature is very infectious and is quite resistant to drying and other environmental changes. There are no insect vectors for TMV; the chief agent of transmissal is man. Infection of seedlings most often occurs by the hands of workers, who may transmit the virus from infected plants to healthy seedlings. Prevention is therefore best accomplished by carefully controlling sanitary conditions and by using resistant varieties.

TMV can attack over 200 species of plants in 36 botanical families, with more than one-third of the species belonging to *Solanaceae*, the family that contains potato, tobacco, and tomato. Peppers and petunias are also affected.

Supplementary readings

Casper, D. L. D., and A. Klug "Physical principles in the construction of regular viruses," *Cold Spring Harbor Symposium on Quantitative Biology.* 27:1 (1962).

Davis, B. D., R. Dulbecco, H. N. Eisen, H. S. Ginsberg, and W. B. Wood, Jr. *Microbiology.* New York: Harper & Row, Publishers, 1967.

Fenner, F. J., and F. N. Ratcliffe *Myxomatosis.* London: Cambridge University Press, 1965.

Hahon, N. (ed.) *Selected Papers on Virology.* Englewood Cliffs, N. J.: Prentice-Hall, Inc., 1964.

Haselkorn, R. "Physical and chemical properties of plant viruses," *Ann. Rev. Plant Physiol.* 17:137 (1966).

Holmes, F. O. "Trends in the development of plant virology," *Ann. Rev. Phytopathol.* 6:41 (1968).

Horsfall, F. L., Jr., and I. Tamm (eds.) *Viral and Rickettsial Infections of Man,* 4th ed. Philadelphia: J. B. Lippincott Co., 1965.

Luria, S. E., and J. E. Darnell, Jr. *General Virology.* New York: John Wiley & Sons, Inc., 1967.

Lwoff, A., and P. Tournier "The classification of viruses," *Ann. Rev. Microbiol.* 20:45 (1966).

Maramorosch, K., and H. Koprowski (eds.) *Methods in Virology,* 4 vols. New York: Academic Press, Inc., 1967–1968.

Smith, K. M. *Insect Virology.* New York: Academic Press, Inc. 1967.

Stanley, W. M., and E. G. Valens *Viruses and the Nature of Life.* New York: E. P. Dutton & Co. Inc., 1961.

Stent, G. S. *Molecular Biology of Bacterial Viruses.* San Francisco: W. H. Freeman & Co. Publishers, 1963.

Stent, G. S. (ed.) *Papers on Bacterial Viruses,* 2nd ed. Boston: Little, Brown and Company, 1965.

Weidel, W. (L. Streisinger, trans.) *Virus.* Ann Arbor, Mich.: University of Michigan Press, 1959.

Among the greatest advances in biology within the last generation has been our understanding of the chemical bases of heredity. Much of our knowledge developed from the study of inheritance in microorganisms. Although at one time scientists doubted that microbes had genetic mechanisms similar to those of higher organisms, it is now well established that at least at the chemical level the processes are similar; hence an understanding of heredity in microbes provides a useful base for studying similar processes in higher organisms.

Advantages and disadvantages of microorganisms for genetic studies Small size and rapid generation time make it possible to cultivate microbes easily and to build up large populations of genetically identical individuals. Reproduction by simple cell division means that if one begins a culture with a single cell, the population that develops consists almost exclusively of genetically identical cells. It is therefore easy to select a series of closely related strains, differing by only one or a few characteristics; such series of strains are often essential for genetic analysis. Genetic differences in related strains are reflected in the enzyme complements of these strains, and in microorganisms it is relatively easy to assess these enzymatic differences.

Procaryotes provide especially simple material for genetic analysis since there is only one DNA molecule per organism and this DNA does not have a complicated chromosomal structure, thus making it easy to isolate and purify (see Chapter 2) and to use in various kinds of genetic analyses. Certain types of genetic analyses (see the sections on transformation and transduction in Chapter 12) that greatly aid studies in molecular genetics have so far been possible only with procaryotes. Most microorganisms, both procaryotic and eucaryotic, have only one gene complement in the vegetative state (a condition called *haploid*), greatly simplifying their use in genetic analyses. Higher organisms, on the other hand, usually have two gene complements (*diploid*).

There are, however, a few disadvantages in using microbes as genetic tools. Their small size makes it difficult to study characteristics of individual cells so that these characteristics must be inferred from studies on populations. This not only greatly complicates a genetic study but also makes especially difficult the interpretation of cellular events. Another disadvantage is that sexual reproduction, an essential feature for many genetic analyses, is often lacking. Even if present, it may be difficult to control or induce, and one may have sexually and asexually reproducing cells in the same culture.

Terminology Before going farther, we shall find it desirable to define certain words used often in microbial genetics. The basic biological mate-

eleven

The genetic code

rial used by the geneticist is a *strain* or *clone,* which is a population of cells that are genetically identical. A *pure culture* is a clone, and techniques such as those already mentioned for isolating pure cultures are directly applicable to isolating clones or strains for genetic analysis. Although the term "strain" is sometimes used casually as a synonym for "organism" or "microorganism," in this chapter it refers only to a group of cells that are not only genetically identical but also genetically distinct from similar cells. *Genome* refers to the complete set of genes present in an organism. The *genotype* of an organism is its genetic complement, or the collection of genes that make up its genetic apparatus. The *phenotype,* on the other hand, comprises those characteristics of the organism observable by the investigator. Since not all the genes of an organism are expressed at any one time, (see Section 7.6) the environment plays an important role in determining which phenotype actually is expressed by a given genotype.

11.1   The genetic code

The biochemical mechanisms of protein synthesis were discussed in Chapter 7. To summarize, the first step in the process is the *transcription* of the base sequence of DNA into a complementary strand, mRNA. The information in mRNA is then *translated* into the amino acid sequence of the protein. In this chapter we will discuss the manner in which the sequence of amino acids in the protein to be formed is specified by the sequence of purine and pyrimidine bases in a segment of RNA.

It has been found that the genetic information of mRNA is *coded.* That is, four purine and pyrimidine bases (guanine, cytosine, uracil, and adenine) make up mRNA, and the order in which these bases are grouped determines the identity of each of the 20 amino acids commonly found in proteins. However, if each RNA base represented a single amino acid (singlet code), then a maximum of only four amino acids could be coded for (see Figure 11.1). If two RNA bases taken together determined a single amino acid (doublet code), then up to 16 amino acids could be coded for; whereas if three RNA bases are taken together (triplet code), a maximum of 64 amino acids could be coded for. Since there are 20 amino acids found in proteins, it seems likely that at the very least there must be a triplet code, and this deduction is now well confirmed by biochemical studies (see below). The triplet of bases in mRNA that constitutes the code for an amino acid is called a *codon,* and the triplet of bases in the tRNA that is complementary to that of the codon is called the *anticodon.*

Figure 11.1 *Number of amino acids (messages) that can be coded with a four-letter alphabet by singlet, doublet, and triplet codes.*

	Possible combinations				Possible messages
Singlet code	A	G	C	U	4
Doublet code	AA	GA	CA	UA	16
	AG	GG	CG	UG	
	AC	GC	CC	UC	
	AU	GU	CU	UU	
Triplet code	AAA	AAG	AAC	AAU	64
	AGA	AGG	AGC	AGU	
	ACA	ACG	ACC	ACU	
	AUA	AUG	AUC	AUU	
	GAA	GAG	GAC	GAU	
	GGA	GGG	GGC	GGU	
	GCA	GCG	GCC	GCU	
	GUA	GUG	GUC	GUU	
	CAA	CAG	CAC	CAU	
	CGA	CGG	CGC	CGU	
	CCA	CCG	CCC	CCU	
	CUA	CUG	CUC	CUU	
	UAA	UAG	UAC	UAU	
	UGA	UGG	UGC	UGU	
	UCA	UCG	UCC	UCU	
	UUA	UUG	UUC	UUU	

Although for many years it appeared that the cracking of the genetic code would be extremely difficult, in fact it turned out to be surprisingly easy. The method involves the use of highly simplified artificial mRNA polymers of only one, two, or three bases. When mixed with a suitable protein-synthesizing system of ribosomes, tRNA, and cofactors, these artificial mRNA's will dictate the synthesis of polypeptides containing a restricted series of amino acids.

For instance, if an artificial mRNA containing only uracil (polyuridylic acid) is used, then a polypeptide is synthesized that contains only phenylalanine (Figure 11.2). From this we infer that phenylalanine is produced when the RNA code word (codon) is a triplet of uracil bases, UUU. Since the mRNA is complementary to the DNA, this means that the code word in DNA specifying phenylalanine must have been three units complementary to UUU, or AAA (three adenines). However, it is conventional in discussing the genetic code to refer to the base sequences of mRNA rather than to those of DNA.

Not all codons are so simple to elucidate as was UUU. Some triplet codons have two different bases, others three. Fortunately, it has been possible to prepare synthetic messengers with various assemblages of bases, and in this way the RNA codons for all the amino acids common to proteins have been deduced. The procedure described will provide information concerning which bases constitute the coding triplet for each

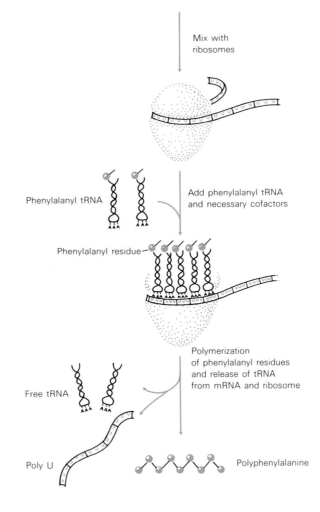

Figure 11.2 Use of artificial mRNA to deduce the genetic code. Addition of polyuridylic acid (poly U) to ribosomes leads to the synthesis of polyphenylalanine, permitting the conclusion that the triplet UUU is a code for phenylalanine.

amino acid but will not reveal the sequence of bases in the triplet. The latter information can be obtained by another technique based on the observation that, when trinucleotides are mixed with ribosomes and tRNA, the tRNA will bind to ribosomes only when the trinucleotide contains the codon corresponding to the anticodon of the tRNA. It is possible to prepare trinucleotides with any of the four bases and in any sequence, and it is thus possible to deduce not only the bases but the order in which they are present in any codon. The details of the genetic code have thus been deduced and this information is summarized in Table 11.1

11.1   The genetic code

Table 11.1 Triplet code words for amino acids

UUU	Phenylalanine	CUU	Leucine	GUU	Valine	AUU	Isoleucine
UUC	Phenylalanine	CUC	Leucine	GUC	Valine	AUC	Isoleucine
UUG	Leucine	CUG	Leucine	GUG*	Valine	AUG*	Methionine
UUA	Leucine	CUA	Leucine	GUA	Valine	AUA	Isoleucine
UCU	Serine	CCU	Proline	GCU	Alanine	ACU	Threonine
UCC	Serine	CCC	Proline	GCC	Alanine	ACC	Threonine
UCG	Serine	CCG	Proline	GCG	Alanine	ACG	Threonine
UCA	Serine	CCA	Proline	GCA	Alanine	ACA	Threonine
UGU	Cysteine	CGU	Arginine	GGU	Glycine	AGU	Serine
UGC	Cysteine	CGC	Arginine	GGC	Glycine	AGC	Serine
UGG	Tryptophan	CGG	Arginine	GGG	Glycine	AGG	Arginine
UGA	None	CGA	Arginine	GGA	Glycine	AGA	Arginine
UAU	Tyrosine	CAU	Histidine	GAU	Aspartic	AAU	Asparagine
UAC	Tyrosine	CAC	Histidine	GAC	Aspartic	AAC	Asparagine
UAG	None	CAG	Glutamine	GAG	Glutamic	AAG	Lysine
UAA	None	CAA	Glutamine	GAA	Glutamic	AAA	Lysine

* GUG and AUG, at the beginning of the mRNA, code for N-formyl methionine.

Characteristics of the code   Perhaps the most interesting feature of the genetic code is that a single amino acid is frequently coded for by several different but related triplets of bases, a phenomenon called *degeneracy*. Since there are 64 possible combinations of four bases taken three at a time, and only 20 amino acids need be coded, the other 44 combinations could of course be meaningless, and hence nonexistent. This has not been found to be the case, however. Recall that crucial to the translation of the genetic code are the tRNA molecules to which the amino acids are attached, and that an amino acid is attached to the growing polypeptide chain through the mediation of a molecule of tRNA. In each case, where there is more than one codon for a given amino acid, there is a different tRNA for each of the codons. Consider now the situation with an amino acid such as leucine, for which there are six distinct codons (Table 11.1). At one place in the mRNA, the triplet CUU may be present, at another, CUC, at a third, CUG, and so forth. Although a different tRNA will be used for each of these codons, in all cases leucine will be inserted into the growing polypeptide chain. It can be seen from Table 11.1 that almost all of the possible triplet codons specify amino acids.

Another phenomenon of the code presented in Table 11.1 is that a few triplets do not correspond to any amino acid. These triplets are called *nonsense codons*, and they may function as "punctuation" to signal the termination of the gene coding for a specific protein. If no tRNA molecules correspond to these nonsense codons, no amino acid can be inserted, and the polypeptide is terminated and released from the ribosome.

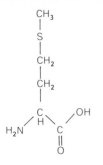

Methionine

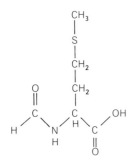

(a)   N-formyl methionine

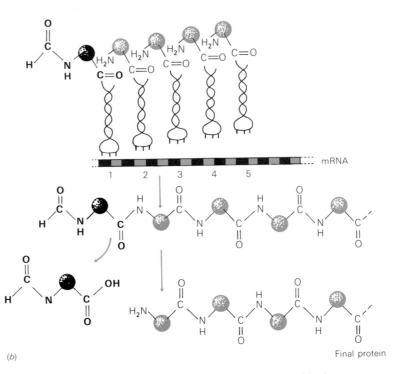

(b)

Final protein

Figure 11.3   (a) Structures of methionine and N-formyl methionine; (b) role of N-formyl methionine in the initiation of protein synthesis. The N-formyl methionine is positioned at the beginning of the peptide and is later released after the polypeptide is completed.

Nonsense codons thus may denote stopping points at the ends of genes, and their significance is that the DNA molecule (and the mRNA molecule complementary to it) is a monotonous sequence of purine and pyrimidine bases with no chemical means of signalling the ends of genes.

In addition to stopping points, *starting* points are also needed. At least one mechanism for initiation of a polypeptide chain is known, involving the amino acid N-formyl methionine (Figure 11.3). The α-amino group of this acid is blocked in such a way that it cannot be added to a growing polypeptide chain but can form only the beginning. There is a tRNA that specifically carries N-formyl methionine, and codons corresponding to this tRNA are present at the site on the mRNA where the protein is to be initiated. N-formyl methionine is juxtaposed at this site, and the next amino acid is attached by its α-amino group to the carboxyl group of N-formyl methionine. In some cases, once the polypeptide is completed, the N-formyl group is removed from the polypeptide by another enzyme so that the N-terminal amino acid of the protein is methionine. In other cases, N-formyl methionine is removed in its entirety so that the N-terminal amino acid of the protein is the second amino acid coded for by the mRNA (Figure 11.3).

The importance of having a well-defined starting point is readily visualized when it is considered that with a triplet code it is absolutely essential that translation begin at the first base of a triplet since, if translation began at the second base, the whole *reading frame* would be shifted and an entirely different protein or none at all would be formed. The problem of reading-frame shift will be considered further when we discuss mutation (Section 11.3).

Another problem in the translation of the genetic code is that error sometimes occurs. This means that a coding triplet of the mRNA may be "read" improperly and the wrong amino acid inserted. Amino acids whose codons differ by only a single base, for example, phenylalanine (UUU) and leucine (UUA), are most likely to be involved in errors because occasionally only two of the three bases are concerned with codon-anticodon recognition. In rare instances leucine may be added to the growing polypeptide instead of phenylalanine even when the codon is UUU. In the normal cell these rare errors probably occur in only a small number of all the protein molecules and hence have no detrimental effect. However, certain antibiotics, such as streptomycin, neomycin, and related substances (see Chapter 9), increase error to such an extent that many protein molecules in the cell are abnormal and the cell can no longer function properly. The anitibiotics are known to act on the ribosomes, apparently altering their structure slightly so that the mRNA codon and the tRNA anticodon do not fit properly (Figure 11.4). Other conditions that increase error are changes in cation concentration, pH, and temperature away from those optimal for cell growth.

Figure 11.4 Introduction of error into the translation process by an alteration of the ribosome and concomitant distortion of the codon-anticodon interaction.

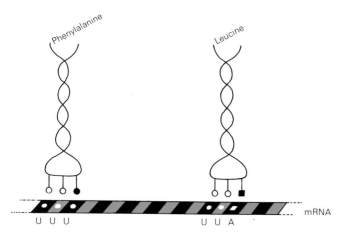

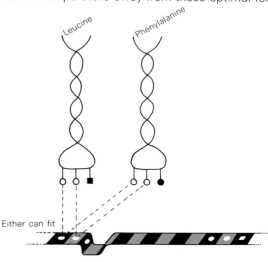

Universality of the code   The genetic code has been fully determined only in the bacterium *Escherichia coli*. By analysis of amino acid changes in proteins as a consequence of mutation (see below) and by constructing hybrid in vitro protein-synthesizing systems using mRNA from one organism and tRNA from another it is possible to deduce that the genetic code is universal—it is probably the same in man and in other organisms. It therefore seems likely that the genetic code as we know it arose early in the history of evolution (see Chapter 17) and has been maintained unchanged since then. It is of course obvious that a code, once developed, is difficult to change since changing it piecemeal would cause havoc.

One gene–one protein   One of the great theories of modern biology, which arose from studies on the biochemical genetics of the fungus *Neurospora*, was that a single gene controls the production of a single protein. However, as our knowledge of protein structure advanced to include the fact that many proteins are composed of more than one kind of polypeptide, the hypothesis has been modified to state that one gene controls the synthesis of one polypeptide chain. Each gene determines the production of a single polypeptide and the separate polypeptides then associate to form the functional protein molecule. We also know now that two proteins may have the same enzymatic function but differ markedly in amino acid sequence. So far as enzymatic function is concerned, the key thing is that the active site, where the catalytic step occurs, be maintained. Many amino acid substitutions that do not significantly alter the active site are probably possible in the protein molecule. These considerations are important for our discussion of mutation below.

11.2   Mutants and their isolation

A mutation is observed as a sudden inheritable change in the phenotype of an organism. A simple example of a mutation is the presence of rare white colonies upon agar plates seeded with inocula of the red-pigmented bacterium *Serratia marcescens* (Plate 4: Figure 11.5). If one white colony is picked and studied, it is found to be identical to *S. marcescens* in all characteristics except pigmentation. Mutations characteristically alter only one or a few traits of an organism at a time. Furthermore, if an inoculum of cells from the white-pigmented mutant is spread on agar plates, in some cases rare red colonies develop among the white ones. These are due to *reverse* or *back mutations*. The changes brought about by mutation can be distinguished from those brought about by enzyme induction or repression (see Chapter 7) by the fact that the latter represent uninheritable events resulting from changes in the environment. All of the cells in the population will show the new phenotype, whereas mutations affect only a few cells of the population. Understanding of mutations was greatly delayed since their rarity made study difficult.

Isolation of mutants We can distinguish between two kinds of mutations, selective and unselective. An example of an *unselective* mutation is that of the pigment change described above. The white-pigmented colonies have neither an advantage nor a disadvantage over the red-pigmented parent colonies when growing on agar plates (there may be a selective advantage to red-pigmented organisms in nature, however). This means that the only way we can detect such mutations is to examine large numbers of colonies and look for the "different" ones. A *selective* mutation confers upon the mutant an advantage under certain environmental conditions, so that the progeny of the mutant cell are able to outgrow and replace the parent. An example of a selective mutation is drug resistance; an antibiotic-resistant mutant can grow in the presence of antibiotic concentrations that inhibit the parent. It is relatively easy to detect and isolate selective mutants by choosing the appropriate environmental conditions.

Kinds of mutants In Chapters 2 through 9 phenotypic characteristics of various microorganisms have been discussed. Virtually any characteristic of a microorganism can be changed through mutation. Ultimately all of these changes are related to changes in enzymes or other proteins, although frequently the connection between a change in phenotype and a change in protein is difficult to determine.

Nonmotile mutants of flagellated motile bacteria frequently are isolated. Such mutants may have completely lost the ability to form flagella, or alternatively, they may form nonfunctional flagella. *Noncapsulated* mutants are often recognizable because the colonies they form on agar plates are smaller and rougher in appearance than those of the capsulated parent. Mutants that have lost the ability to form *spores* or *conidia* are widespread. In the fungi, colonies of such mutants are often recognized by the absence of pigment since it is usually the spores or conidia that make the largest contributions to colony pigmentation. A variety of other mutations in bacteria and yeasts lead to alterations in *colony morphology*. "Smooth" strains form shiny, well-defined colonies, whereas "rough" mutants form colonies that do not glisten and seem to be granular and irregular in appearance. In some species, smooth-colony morphology is directly related to the presence on the cell surface of lipopolysaccharides.

Mutants deficient in the production of enzymes of various *biosynthetic pathways* are very common. Because the defects brought about by such mutations can usually be overcome by adding specific nutrients to the medium, such mutants are usually called *nutritional mutants.* Nutritional mutants are good examples of *conditional lethal mutations,* since in the absence of the required nutrient the mutant cannot grow and will ultimately die, whereas if the nutrient is present it will grow normally. Nutritional mutants are also called *auxotrophs* (*auxo-* means "growth"), and the parents from which they are derived are called *prototrophs.* Mutants have been isolated which require many of the known amino acids, purine and pyrimidine bases, vitamins and vitamin precursors, certain sugars, organic acids, and others. Nutritional mutants can be detected by the technique of *replica plating* (Figure 11.6); using sterile velveteen cloth an imprint of the colonies is made on an agar plate lacking the desired nutrient. The colonies of the parent type will grow normally, whereas those of the mutant will not. Hence inability of a colony to grow on the replica plate will be a signal that it is a mutant. The colony on the master plate corresponding to the vacant spot on the replica plate can then be picked, purified, and characterized.

It is sometimes possible to devise ingenious means of isolating mutants that ordinarily are unselective. One procedure often used with bacteria is the *penicillin-selection method* (Figure 11.7), which makes possible the isolation of mutants requiring certain amino acids or other growth factors Ordinarily, such mutants would be at a disadvantage in competition with parent cells, as they cannot grow in the absence of the required growth factor. However, penicillin causes the death only of growing cells (Chapter 9), so that if penicillin is added to a population growing in a medium lacking the growth factor, the parent cells will be killed, whereas non-

Figure 11.6 Replica-plating method for the detection of nutritional mutants.

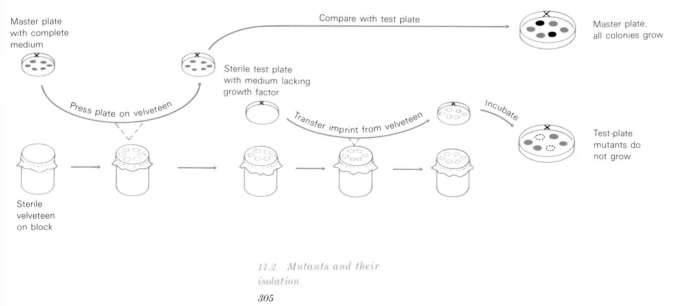

Master plate with complete medium

Press plate on velveteen

Sterile velveteen on block

Sterile test plate with medium lacking growth factor

Transfer imprint from velveteen

Incubate

Compare with test plate

Master plate, all colonies grow

Test-plate mutants do not grow

growing mutant cells will be unaffected. A method that is conceptually similar to the penicillin method is the *filtration technique* used to isolate mutants of filamentous organisms. A suspension of spores or conidia, composed of a mixture of mutant and wild-type forms, is inoculated into a medium on which only the mutant cannot grow. Wild-type spores germinate and grow, forming long filaments, whereas mutant spores remain dormant. The whole mixture is then filtered through a cheesecloth mesh, which retains the filamentous forms but allows the mutant spores to pass through. Auxotrophic mutants can also be detected on plates in which only a small amount of the growth factor is present in the medium. The mutants form small colonies, whereas the wild types form normal colonies.

Mutants unable to produce enzymes of catabolic or degradative pathways are also common. For instance, the utilization of lactose as a carbon and energy source is dependent on the presence of the enzymes *β-galactosidase* and the galactoside *permease* (see Chapter 5); mutants lacking these enzymes are known as lactose-negative (*lac⁻*) and can be recognized either by their inability to grow on lactose as sole carbon source or by their inability to ferment lactose with the production of acid when

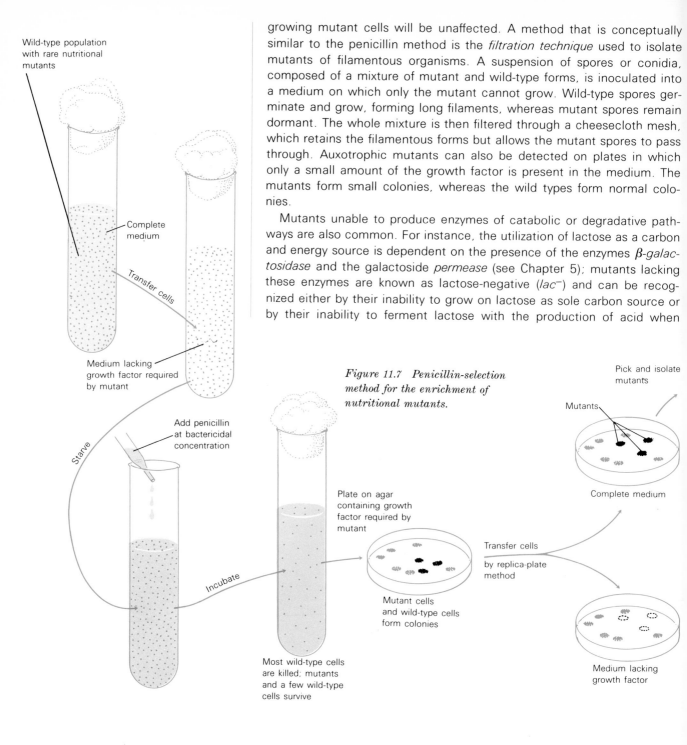

Figure 11.7 Penicillin-selection method for the enrichment of nutritional mutants.

Wild-type population with rare nutritional mutants

Complete medium

Transfer cells

Medium lacking growth factor required by mutant

Starve

Add penicillin at bactericidal concentration

Incubate

Most wild-type cells are killed; mutants and a few wild-type cells survive

Plate on agar containing growth factor required by mutant

Mutant cells and wild-type cells form colonies

Transfer cells by replica-plate method

Mutants

Pick and isolate mutants

Complete medium

Medium lacking growth factor

growing on some other energy source. Examples of other carbohydrates for which mutants are known are galactose, sucrose, glucose, maltose, and starch.

Drug-resistant mutants are some of the easiest mutants to isolate, since one need merely plate a large population of cells on agar containing a growth-inhibitory concentration of the drug; only resistant mutants present in the original population will be able to initiate colonies. Resistant mutants are known to most of the common antibiotics, such as streptomycin, penicillin, chloramphenicol, and cycloserine, as well as to the sulfa drugs and cytochrome inhibitors such as sodium azide. Some aspects of the development of drug resistance were discussed in Chapter 9. A recent phenomenon has been the isolation of strains of bacteria with multiple drug resistance. For instance, strains have been isolated that had acquired resistance to streptomycin, chloramphenicol, tetracycline, and sulfanilamide in a single step. The genetics of multiple drug resistance is of considerable interest and is discussed in Chapter 12.

Virus-resistant mutants are also easy to isolate, since one need merely expose a population of cells to a high concentration of virulent virus; only the resistant mutants survive. Virus-resistant mutants differ from the parent either in inability to adsorb virus, due to an alteration in the virus-receptor site (see Chapter 10) or because the virus DNA upon entry into the cell is either prevented from replicating or is destroyed by a host enzyme.

In addition there exists a large class of mutants that are called *temperature sensitive*. These cannot grow at some temperature at which the parent grows but grow normally at a lower temperature. This is usually because a protein of the mutant has had an amino acid substitution in its polypeptide chain that reduced its heat stability, and the protein is therefore inactivated at the temperatures at which the normal protein would be stable. Such mutants, like nutritional mutants, are also conditional lethals, and can arise for virtually any enzyme in the cell. *Cold-sensitive mutants* are also known; they cannot grow at the low temperature at which the wild type can grow. Temperature-sensitive mutants are very useful in genetic analysis since they permit the study of enzymes whose defects cannot be corrected by supplementation of the culture medium.

Mutation rate   This is expressed as the number of mutants formed per generation per unit of population. Thus a mutation rate of 1×10^{-8} would mean that when 10^8 cells divided to form 2×10^8 cells, on the average one mutant cell would be formed. This would be the mutation rate in a single gene; but of course, in the same population, mutations in more than one gene would be occurring so that in any reasonably large population there will always be a certain proportion of mutants of various kinds. Mutations are usually independent events, a mutation in a given gene neither increasing nor decreasing the probability of mutations in other genes. This means that it would be extremely unlikely that two mutational events would occur in the same cell, since the probability of a double mutant would be the product of the probabilities of each separate mutation (for example, if 10^{-8} is the mutation rate for one gene and 10^{-9} is the rate for another, the probability of a double mutant would be 10^{-17}, a very low figure indeed!). Because mutations are extremely rare events, it is difficult to measure mutation rates, especially for unselective mutants. Most of our knowledge of mutation rates, and of the factors that affect them, comes from studies on selective mutations, such as drug and virus resistance, or through measurements of the rates of back mutation from auxotrophy to prototrophy. However, even two mutations that occur in the same gene will be unlikely to occur at precisely the same DNA base, so that the mutations at individual sites in the same gene will have even lower rates of occurrence.

Mutants of independent origin   For many types of genetic analyses, it is desirable to have a series of mutants that are all within the same gene but of independent origin. But how can one be sure that two mutants are of independent origin? If we take a population of 10^9 cells and isolate two separate mutant clones from this population, we would have no way of knowing whether these mutants were independent (for example, had arisen from separate parent cells) or whether one mutant might have arisen earlier and subsequently given rise to progeny. Since these progeny would have been derived from the original mutant they would be mutant cells but would not have arisen independently. To ensure that two mutants are of independent origin it is necessary to take the original parent population, dilute it to a low population density (for example 10^3 cells per ml), and inoculate a series of culture tubes with this dilute inoculum. Because of the rarity of the mutants, it would be extremely unlikely that any of the culture tubes would have received mutant cells. When these tubes are now incubated, mutants may arise in the various tubes as independent events. If from any one tube only a single mutant is isolated, a series of mutants isolated from separate tubes would certainly have arisen independently. The importance of isolating mutants of independent origin for genetic fine-structure analysis will be discussed in Chapter 12.

Nomenclature of mutants   A mutant selection program such as is used in genetic analysis (see Chapter 12) soon results in a large collection of mutant strains that must be given names to avoid confusion. The primary name of a mutant is based on which phenotypic characteristic has been altered. Thus, mutants unable to use lactose are called *lac,* those requiring tryptophan are called *trp,* and streptomycin-resistant mutants are designated *str.* Each isolate is then given a sequential number; for instance, *trp37* would be the 37th tryptophan-requiring mutant isolated. Enzymatic and genetic analyses of a series of mutants may indicate that several different enzymes are involved, and a letter designation is then added to each mutant to indicate the gene that affects each enzyme (for example, *trpA37, trpB9,* and so on).

11.3 *The molecular basis of mutation*

Although many mutations arise spontaneously, it is possible to increase the rate of mutation many times by treatment of cells with various *mutagenic* agents. An analysis of how mutagenic agents work provides many insights into the molecular mechanisms of mutation. Further, a study of the alterations in amino acid sequence in proteins as a result of mutation provides excellent confirmation for the details of the genetic code.

Point mutations   All of the existing evidence points to mutations as resulting from alterations in the purine-pyrimidine base sequences of DNA. Several kinds of changes in DNA can be visualized. *Point mutations* spring from changes or deletions of single bases; an adenine base may be replaced by guanine or thymine by cytosine, for instance (Figure 11.8). The consequences of a point mutation will depend greatly on where in the gene the mutation occurs. It can produce no change at all in the protein, an inconsequential amino acid substitution, a serious amino acid substitution, or no protein may be formed at all. To understand these possibilities, recall first the triplet nature of the genetic code, and the fact that it is degenerate (see Section 11.2). The genetic code listed is that of mRNA; the base sequence in the gene would be complementary to this, with thymine in DNA being functionally equivalent to uracil in RNA. For convenience, we will consider only the changes in the mRNA that result from mutation in DNA since it is these that cause alterations in the protein: one should keep in mind, however, that the actual base change took place in DNA. Let us now consider the codon that specifies the amino acid glycine at a specific site in the protein tryptophan synthetase (Figure 11.9). Because of the degeneracy of the code, three of the possible base changes have no effect on the amino acid; such base changes would of course be undetectable since they would not lead to an alteration in the protein. These are called *silent mutations.* Note that these

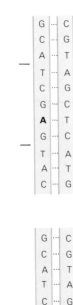

(a)

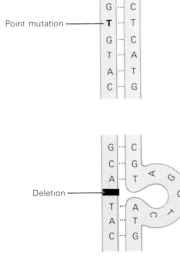

(b)

(c)

Figure 11.8   DNA base changes from wild type (a), involving point mutation (b) and deletion (c).

Point mutation

Deletion

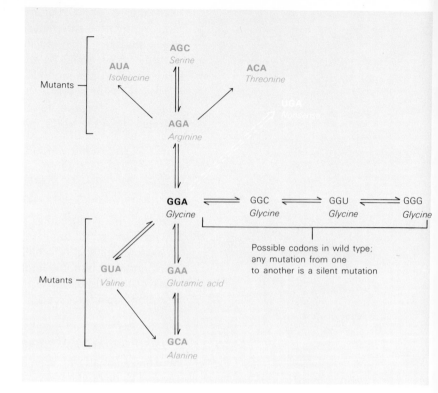

Figure 11.9 Base changes affecting the occurrence, at a single site, of amino acids in the tryptophan synthetase of E. coli. For simplicity, the changes are shown in mRNA rather than in the DNA to which it is complementary. For a means for detecting amino acid substitutions in this enzyme, see Figure 4.10.

changes occur in the third base of the codon. As seen in Table 11.1, degeneracy is found most frequently in the third base so that for many codons one-third of the mutations are not harmful. Changes in the first or second base of the triplet much more often lead to significant changes in the protein. In the example of Figure 11.9, the change from GGA to GAA causes glutamic acid (whose side chain is acidic) to be substituted for glycine, whereas a change from GGA to CGA substitutes arginine (whose side chain is basic). Since the glycine originally present at the site is a neutral amino acid, substitutions of amino acids with negative or positive charges could be expected to have marked effects on the folding of the protein and hence alter its activity. In fact, such mutations produce proteins essentially devoid of enzymatic activity. It is possible by immunological techniques (see Chapter 15) to detect these altered proteins within the cell, even though they are inactive or have only traces of normal activity. Because such altered proteins often behave immunologically like normal proteins, they are called *cross-reacting material,* or CRM (pronounced ''krim''). Another change shown in Figure 11.9, from GGA to UGA, forms a nonsense codon. As noted earlier, at the end of the gene nonsense codons probably signal the termination of polypeptide chains. With a nonsense codon in the middle of the gene, premature termination of the polypeptide chain would be effected, and hence only a fragment of the active enzyme would result. With appropriate techniques it might be possible to detect polypeptide fragments within the cell, but no CRM would be produced. All the above mutational changes are potentially reversible, and by plating a mutant population on agar plates devoid of tryptophan, we can isolate revertants. It can be shown that in these, the original amino acid, glycine, often has been restored in the protein.

Not all mutations that effect amino acid substitution necessarily lead to nonfunctional proteins. The outcome depends greatly on where in the polypeptide chain the substitution has occurred, and on how it affects the folding and the catalytic activity of the protein. If a mutation causes a substitution that does not affect enzyme activity, the altered protein might be detectable by electrophoresis studies (see Chapter 4) if the amino acid change involved a change in the electric charge of the protein (for example, substitution of glutamic acid for arginine). If the particular enzyme is examined electrophoretically in a large number of related strains, isolated independently from nature, it may be shown that a whole spectrum of enzymes exist, each differing slightly from the others in mobility, and hence in charge. These different enzymes would probably have arisen from naturally occurring mutations that had not altered the function of the enzyme.

Deletions   These are due to elimination of portions of the DNA of a gene (Figure 11.8). A deletion may be as simple as the removal of a single base, or it may involve hundreds of bases. Deletion of a single base leads to a reading-frame shift (see Section 11.1). The sequence of amino acids after the deletion will be totally different from that present in the wild-type protein (Figure 11.10), and the protein will most likely be non-

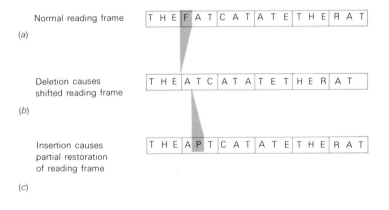

Normal reading frame

(a)

Deletion causes
shifted reading frame

(b)

Insertion causes
partial restoration
of reading frame

(c)

Figure 11.10   Consequences of reading-frame shift.

functional. If another base is inserted a little farther along the DNA chain, the reading-frame may be restored to normal, except for a small region around the site of the original change.

Deletion of a large segment of the DNA results of course in the complete loss of the ability to produce the protein. Such deletions cannot be restored through further mutations (in contrast to those inducing reading-frame shift, discussed above), but only through genetic recombination (see Chapter 12). Indeed, one way in which large deletions are distinguished from point mutations is that the latter are usually revertible through further mutations, whereas the former usually are not.

11.4 *Mechanisms of mutagenesis*

Spontaneous mutations   These probably result from a variety of causes, the word *spontaneous* merely meaning that an investigator did not *consciously* attempt to interfere with the mutation process. The accuracy with which DNA is copied during replication is very high, some studies indicating the error to be less than once in 500,000 bases copied, but since spontaneous mutations occur of the order of once in 10^8 to 10^9 DNA base pairs, random mistakes by the DNA polymerase system are still conceivable. If an error does occur during replication, it would result in the insertion of the wrong base in only one of the two new strands (Figure 11.11). This means that there will be a delay after the mutation has occurred before it will be expressed because the altered DNA must be replicated and become segregated into a cell devoid of normal wild-type DNA.

Mutagenesis by base analogs   Copy errors can be greatly increased by the use of DNA base analogs, such as 5-bromouracil (5-BU), an analog of thymine (Figure 11.12). When this base is added to the medium in which the organism is growing, it is taken up and incorporated into the DNA instead of thymine. In this way it is possible to substitute 5-BU for a large proportion of the thymine. The mRNA synthesized using this altered DNA as a template is usually normal, and adenine is in the proper place in the message. Even during DNA replication, 5-BU usually functions normally, pairing with adenine as would thymine. But occasionally— about once in several thousand bases—the 5-BU undergoes a chemical rearrangement such that it pairs with guanine instead of adenine (Figure 11.12). During the next round of replication, mutation results from substitution of a G-C pair for an A-T pair on one of the replicated DNA's. Another base analog, 2-aminopurine, can substitute for adenine in DNA

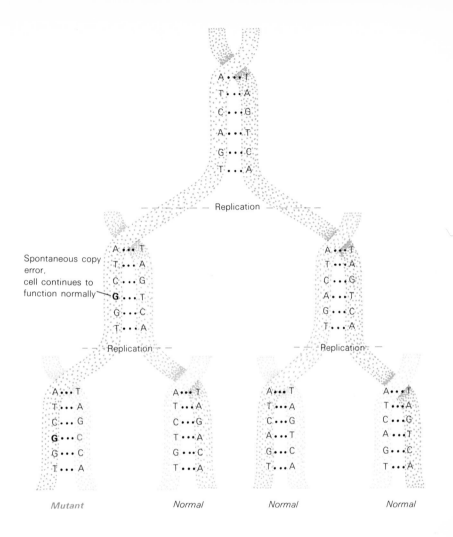

Figure 11.11 DNA replication and the consequences of an error in base pairing.

and greatly increases the probability that cytosine will be inserted in place of thymine.

Chemical mutagenesis A group of mutagenic substances reacts chemically with one or more DNA bases and alters them so that they pair improperly. Upon replication, improper pairing leads to formation of altered base pairs in one of the DNA copies. As shown in Figure 11.13, *nitrous acid* reacts with adenine and causes the formation of a G-C base pair where an A-T pair should have been. It also reacts with cytosine and causes formation of an A-T pair in place of a G-C pair. However, when nitrous acid reacts with guanine, mutants are not formed. *Hydroxylamine* reacts specifically with cytosine and causes formation of an A-T pair instead of a G-C pair. One large group of substances, the monofunctional *alkylating agents,* also react with DNA bases and alter them chemically so that during replication their pairing properties are changed. For instance, ethyl methane sulfonate ($H_5C_2OSO_2CH_3$) alkylates a nitrogen of guanine, causing it to pair with thymine instead of cytosine. These chemical agents act differently from the base analogs since the former are able to introduce direct changes even in nonreplicating DNA, whereas the base analogs must be incorporated into DNA during replication. Thus agents like nitrous acid are directly mutagenic for free DNA, whereas the base

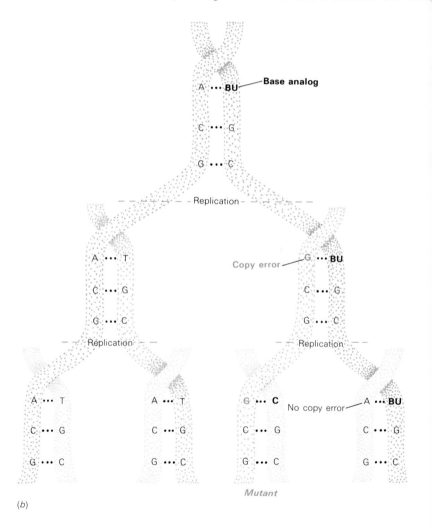

Figure 11.12 (a) Structure of 5-bromo-uracil, an analog of thymine; (b) incorporation of bromouracil into DNA in place of thymine and its mutagenic consequences. The copy error shown occurs only occasionally; usually replication of bromo-uracil DNA occurs normally. Note that the mutant does not contain the DNA double helix with bromouracil but the DNA produced after replication of the strand in which the copy error has occurred.

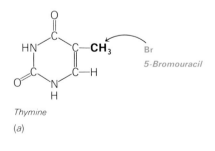

Thymine

(a)

(b)

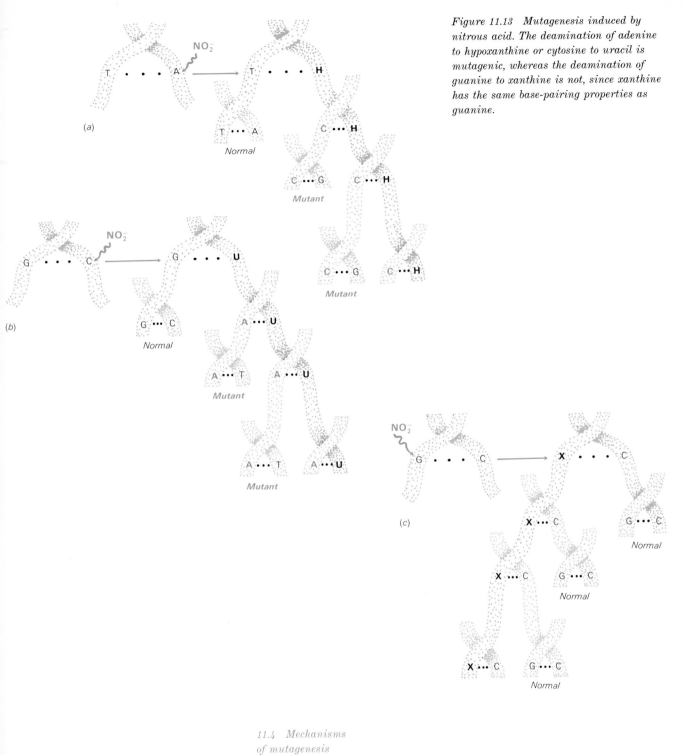

Figure 11.13 *Mutagenesis induced by nitrous acid. The deamination of adenine to hypoxanthine or cytosine to uracil is mutagenic, whereas the deamination of guanine to xanthine is not, since xanthine has the same base-pairing properties as guanine.*

analogs are not. (Mutagenesis of free DNA can be studied using transformation systems as described in Chapter 12.)

Bifunctional alkylating agents, such as the nitrogen mustards and the antibiotic mitomycin (see Chapter 9), react with both strands of the DNA, introducing cross links. These cross links prevent the unwinding of DNA in the affected region, which is essential to DNA replication. The cross-linked region can be removed as a unit through the action of a DNAse, and the cross-linking agents may thus cause deletions rather than base alterations. One of the most powerful mutagens is the alkylating agent *nitrosoguanidine,* which produces so many mutants that isolation can often be done without any selection procedure, making it possible to obtain mutants otherwise difficult to select.

Mutagenesis by reading-frame shift   The *acridines* are a group of dyes that specifically cause reading-frame shift mutations (Figure 11.14). These dyes are planar molecules capable of being inserted between two DNA base pairs (Figure 11.14) and of pushing them apart. During replication, an extra base can then be inserted where the acridine is located, lengthening the DNA by one base and shifting the reading frame. Alternatively, the presence of the acridine may inhibit the incorporation of a base that would otherwise have been inserted; the DNA thereby becomes shortened by one base, which also leads to a reading-frame shift. Because acridines cause reading-frame shifts by inducing either an addition or a

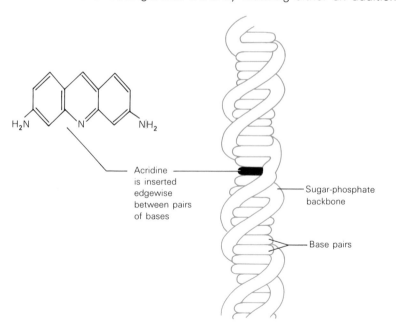

Acridine is inserted edgewise between pairs of bases

Sugar-phosphate backbone

Base pairs

Figure 11.14   Structure of an acridine and manner in which it is introduced into DNA. Intercalation by acridine causes a slight lengthening of the DNA, leading to a reading-frame shift. Compare with Figure 11.10. [Adapted from Figure 1, L. S. Lerman: J. Cell. Physiol. 64, suppl. 1:2 (1964).]

subtraction of a base, mutants induced by them can also be caused to revert by them.

Radiation   *Ultraviolet radiation* (UV) is a very effective mutagenic agent; radiations of wavelengths around 260 nm are strongly absorbed by DNA bases (see Chapter 8), leading to their chemical alteration. Among the various effects of UV on bases the best known is the dimerization of two adjacent thymine units on the same strand of DNA (Figure 11.15). Abnormal replication of the segment containing the thymine dimer causes mutations. One of the more interesting discoveries is that most organisms have at least two enzymatic means of repairing UV damage due to thymine dimer formation. One mechanism, *photoreactivation,* involves an enzyme that requires visible light for activity. When DNA containing thymine dimers is incubated with this enzyme in the dark nothing happens, but when the enzyme–DNA complex is irradiated with blue light the thymine dimer is dissociated and the original structure of the DNA is restored. Photoreactivation was discovered some years ago when it was noticed that the viability of cells inactivated by UV could be restored if they were incubated for a time in visible light. The other enzymatic mechanism of repairing UV damage in DNA, called *dark reactivation,* does not require light. A hydrolytic enzyme excises the region containing the dimer (Figure 11.15), and another enzyme catalyzes the insertion of the appropriate complementary nucleotides to fill the vacancy. These repair enzymes, whose effects are seen most dramatically in UV-treated cells, may also be active in nonirradiated cells, correcting certain analogous errors that occasionally arise spontaneously. Ultraviolet radiation can probably also cause changes in DNA bases by mechanisms not involving thymine dimers, but the mode of action is not known.

Ionizing radiations (such as X rays) and atomic radiation (see Chapter 8) are also mutagenic. These agents act in a variety of ways; they may alter bases chemically, induce breaks in the backbone of the DNA chain, or cause deletions. Cosmic rays, which are naturally occurring ionizing radiations, act similarly.

An interesting phenomenon is the so-called radioactive phosphorus *suicide.* This occurs if cells are grown in a medium containing high amounts of radioactive phosphorus, which becomes incorporated into nucleic acid. As the radioactive phosphorus atoms decay, the ^{32}P is converted to ^{32}S and the backbone of the DNA molecule is broken, causing cell death. This is called suicide because the cells die not from external effects but from changes that take place within the cell itself. Since dividing cells synthesize DNA, they incorporate radioactive phosphorus more readily than nondividing cells, and hence are more likely to exhibit suicide. This phenomenon has been useful medically in attempts to selectively kill

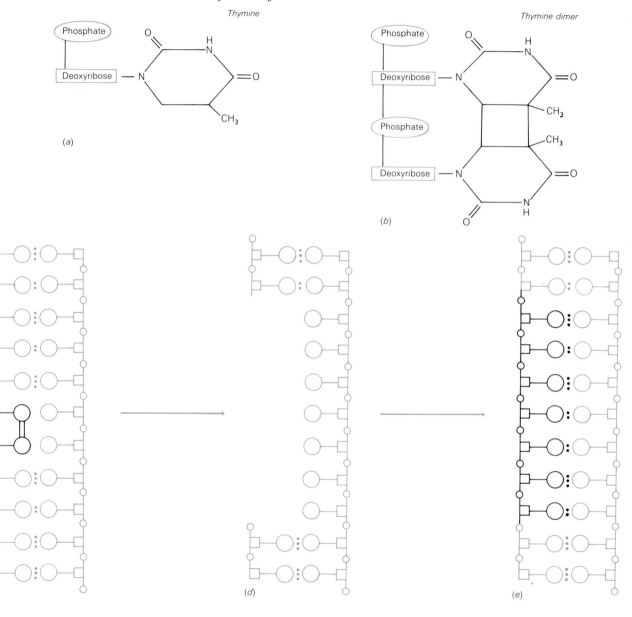

Figure 11.15 *Production of thymine dimer (b) as a result of action of UV radiation and manner in which the damage induced can be repaired: (c) Dimer cannot hydrogen-bond with complementary bases; (d) dimer and adjacent nucleotides are excised; (e) nucleotides are replaced to hydrogen-bond with complementary strand.*

Thymine

Thymine dimer

cancer cells, which divide more rapidly than normal cells. Decay of ^{32}P may also lead to mutations. A similar phenomenon occurs when highly radioactive tritiated thymidine is incorporated into DNA; the radioactive decay of the tritium leads to DNA damage and mutagenesis or cell death.

Other mutagenic effects   Some of the mutagenic agents cause *depurination,* which is the removal of a purine base without otherwise affecting the DNA chain. Upon replication of the DNA, since one strand lacks a base to provide complementarity, any of the four bases may be inserted into the newly synthesized chain.

Most mutagenic agents are lethal in higher doses. Lethality could result from gene mutations that are not correctable by changes in the environment, from multiple gene mutations, or from more drastic effects on the DNA. In some cases, mutagenic agents can also act on other cellular components, such as RNA, protein, or the cell membrane and cause cell death.

Most of the agents discussed above have rather dramatic effects. A number of agents are rather mild mutagens, causing only a small proportion of mutations. Examples are heat, Mn^{2+}, low pH, and caffeine. Some of these agents may be components of the natural environment of cells and hence may have evolutionary significance even though they are not highly mutagenic.

Some spontaneous mutations may occur when the cell produces a natural chemical mutagen, such as caffeine or certain antibiotics. Some strains of bacteria (*mutator strains*) are known to mutate more extensively than do normal strains, possibly because they produce some natural mutagenic agent in above-tolerable amounts.

An intriguing aspect of mutagen research is that many agents known to be mutagens are also *carcinogenic,* that is, they are capable of inducing cancers. This is true of UV and ionizing radiations, nitrogen mustards, and other alkylating agents. At appropriate doses most mutagens can induce lysogenic bacteria to produce phage and lyse (see Chapter 10). Although this probably affects DNA, the exact mechanism is unknown.

Mutagen specificity   Since a mutagen is incapable of being selective for certain genes, we know no way to construct a mutagen that will lead to a specific mutation. Within any given gene, however, there seem to be regions that are more sensitive to mutagens than others. These *hot spots* may exist because their reactivity is modified (increased) through the influence of surrounding bases. When the location within the gene is determined for various mutants by genetic mapping (see Chapter 12) it is usually found that mutations are quite localized.

Techniques for utilizing mutagens   Because most mutagens are also lethal agents, their experimental use must be so programmed that there will be a significant proportion of survivors after treatment. It is desirable first to determine the rate at which the cells are killed. The death rate is usually exponential. A common practice is then to choose a dose of agent and a time of treatment at which approximately 90 percent of the cells are killed. This will insure that the mutagen was indeed active but will also leave behind a sufficiently large population so that a fair number of mutants will be present. The surviving population should be allowed to undergo several cell divisions in an unselective medium to allow the mutant DNA to replicate and segregate into cells separate from the nonmutant material. It will also allow time for the mutant phenotype to be expressed. The population can then be subjected to a mutation enrichment procedure, such as penicillin treatment or filtration (see Section 11.2) and subsequently subjected to a mutant isolation process. Since it is often desirable that each mutant isolated be of independent origin, after mutagenesis the treated population should be diluted before mutants are segregated.

11.5   Back mutations or reversions

Many but not all mutations are revertible. A ''revertant'' is operationally defined as a cell in which the wild-type phenotype that had been lost in the mutant has been restored. Revertants can be of four kinds: (1) the base sequence altered in the mutant is restored to the original; (2) a mutation somewhere else in the same gene can restore enzyme function, such as in a reading-frame shift mutation; (3) a mutation in another gene may restore the wild-type phenotype (*suppressor mutation*); and (4) the production of another enzyme may occur that can replace the mutant one by introducing a metabolic pathway different from that used by the mutant enzyme. In this last type no production of the original enzyme occurs although it does in the other three types.

Suppressor mutations are new mutations that suppress the original one indirectly. One way in which a suppressor mutation can arise is through an alteration in a transfer RNA or amino acid activating enzyme so that in some of the cases the wrong (but right!) amino acid is attached to the tRNA that recognizes the mutant codon. In this way, the proper amino acid can be added to the polypeptide chain, at least some of the time. Alternatively, a mutation may occur in the anticodon of the tRNA so that it can recognize the mutant codon. It is usually characteristic of suppressor mutations of this type that normal function is not completely restored; only some of the polypeptides formed on a single mRNA are normal. Therefore, revertants arising due to suppressor mutations usually do not grow as rapidly as do the wild-type. They can also be distinguished from true revertants by genetic mapping (see Chapter 12) since the locus of the suppressor gene is in a different region of the genome from that of the original mutant gene.

11.6 Practical uses of mutagenesis

The induction and isolation of mutants is of some practical use in industries that employ microorganisms for the manufacture of desired products (see Chapter 8). Probably the most dramatic practical success of mutagenesis has been in the antibiotic industry. The original penicillin-producing isolates from nature gave only small yields of this antibiotic (about 1 to 10 μg/ml of culture fluid). Cultures of the producing organism, *Penicillium chrysogenum,* were treated with various known mutagens, and the isolates obtained were then screened for penicillin yields. The high-yielding isolates were then treated with mutagens, and the process was repeated. Eventually it was possible, by selecting the appropriate strain and by improving the culture conditions, to obtain yields of penicillin at least 1,000 times greater than those produced by the original isolate. However, it often happens that a mutant giving high yields on one medium will produce very poorly in another; therefore improvement in culture conditions goes hand in hand with mutant selection.

11.7 Mutation in diploids

Most of what we have discussed above concerns mutation in haploid organisms, primarily procaryotes. The vegetative cells of some eucaryotes are diploid, with two complements of each genome. In such organisms a mutation in a gene of one set will usually have no effect on the functioning of the equivalent gene (called an *allele*) in the other set. For this reason mutation in diploids is often *masked.* A gene that is expressed is called "dominant" to its allele, which is "recessive." If the corresponding genes on the two genomes are identical, the organism is said to be *homozygous* for this gene; if they are different the organism is said to be *heterozygous.* Only if segregation of two differing genomes occurs, as for instance during sexual reproduction (see Chapter 13), can a recessive mutant phenotype be expressed. Since many mutations are lethal, the ability of mutations to be masked in diploids makes it possible to accumulate lethal mutations that some day may be expressed. This collection of masked lethal mutant genes is called the *genetic load* of the organism. Since diploids have two copies of each gene, either or both of the genes may be transcribed into mRNA, and homozygous diploids often have slightly higher enzyme concentrations for the gene for which they are homozygous than would the heterozygote, which has only one functional gene product. Diploids are often larger and more vigorous in their metabolic and growth potentialities than are the corresponding haploids from which they were derived.

Despite the above differences, the chemical events of mutagenesis in diploid eucaryotes are the same as in procaryotes.

11.8 Heterocaryons and saltation

Heterocaryons are organisms with two or more nuclei of different genetic backgrounds present in the same cell. Many filamentous fungi are multinucleate (see Chapter 3), and among these heterocaryotic organisms are frequent; heterocaryons also occur among actinomycetes and filamentous algae. They are formed by fusion of the hyphae—but not of the nuclei—of two closely related monocaryons. The nuclei exist separately throughout vegetative reproduction and spore formation. Only when sexual reproduction occurs do the nuclei segregate to separate cells, but new hyphal fusions quickly restore the heterocaryotic state.

Often the heterocaryon has a selective advantage over the homocaryons, especially in nature, which is why many natural isolates of filamentous fungi are heterocaryotic. When such heterocaryons are grown in laboratory culture, the two kinds of nuclei often segregate into separate hyphae. If it happens that one of the homocaryons is better adapted to the laboratory environment than the heterocaryon is, it will be favored and will replace the heterocaryon. This phenomenon, which is called *saltation,* may be said to resemble mutation in some ways since it leads to the sudden appearance of a new phenotype. It is not, of course, a mutational process but is analogous to selection of one population of unicellular organisms from a mixture of two genetically diverse populations.

11.9 Mutation and microbial evolution

In most of the above discussion, we have been describing how mutation can lead to loss of function. Mutation also serves as an important mech-

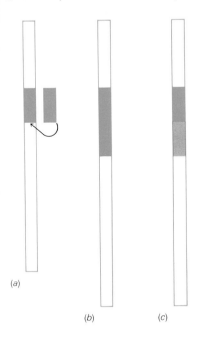

Figure 11.16   Manner in which gene evolution might occur as a result of (a) gene duplication and (b) insertion with (c) subsequent mutation of one of the two copies of the gene; in this way a former function is retained, while a new one is acquired.

(a)

(b)   (c)

anism by which an organism can acquire new characteristics by modification of the existing gene complement. One means of accomplishing this without affecting existing phenotypic characteristics is by gene duplication and subsequent alteration of only one of the two identical genes (Figure 11.16). The initial gene duplication could occur by copying the existing gene and then inserting the copy in sequence with the original. Now while one of the duplicate genes remains unchanged and continues to code for the production of the vital enzyme, the other gene can be modified by mutation in various ways without direct harm to the organism. If one of the modifications causes the production of a desirable new enzyme, the cell in which this modification occurred will have a selective advantage over the parent.

Role of environment   Environmental change is an important aspect in the evolution of some new genetic types. A mutation that might be harmful in one environment may be advantageous in another. One of the best examples of this is the increased incidence of antibiotic-resistant bacteria in man after the introduction of antibiotic therapy. Penicillin therapy was begun in the mid-1940s and at that time all strains of *Staphylococcus aureus* isolated from human beings were penicillin sensitive. Within the succeeding years, the proportion of penicillin-resistant strains of *S. aureus* isolated from patients in hospitals increased markedly, until over 90 percent of the strains being isolated were resistant. Penicillin and other antibiotics do not induce mutations, but serve merely as *selective agents* for antibiotic-resistant strains that have arisen by normal mutational processes.

The addition of an antibiotic is a violent change in the environment. Can evolutionary changes result from more subtle changes? Calculations show that even for minor environmental changes a mutant population slightly better adapted than the parent will have a selective advantage and will ultimately replace it.

In the absence of environmental change, do new strains arise? This question has been investigated using bacteria growing in a chemostat. Recall that a chemostat is a device for maintaining a growing population at a constant density and growth rate for long periods of time (see Chapter 7). Suppose we inoculate a chemostat with a pure culture at low cell density, allow this culture to grow and reach a steady state, and then sample periodically. (Assume that the original culture was optimally adapted to chemostat conditions when the experiment began.) When mutants arise that have no selective advantage over the parent, they will grow at the same rate as (or at a slower rate than) the parent and be washed out at the preestablished rate. Eventually an equilibrium will be reached for each mutant such that its frequency in the total population will be proportional to mutation rate, and the mutant remains as only a small proportion of the total population. Thus it appears that, in the absence of any environmental change selective for mutants, the population will remain genetically the same, even after long periods of time. This observation is of considerable practical significance since it means that the pure cultures we isolate and study in the laboratory will continue to breed true, thus making possible their use in genetic studies. All that is necessary is that we maintain them under defined and unchanging conditions.

Occasionally, however, a laboratory culture may be found to have undergone genetic change and to have lost some characteristic possessed by the original isolate. Such changes are probably due to the fact that the characteristic in question is one that the organism needs for survival in nature but not in the laboratory. By discarding the characteristic, the organism can probably grow faster, as the energy and nutrients previously needed to synthesize enzymes responsible for the characteristic can now be channeled into the production of essential enzymes. Since it is difficult to mimic all aspects of the natural environment in laboratory cultures, such genetic changes are inevitable. They can be obviated by maintaining cultures in the laboratory under nongrowing conditions, such as in the freeze-dried or frozen state (see Chapter 8). Fortunately for most characteristics used in genetic investigation, such uncontrolled changes in genotype do not often occur.

Why are microbes so adaptable? For a number of reasons microorganisms are rapidly adaptable to new environments, much more so than are multicellular organisms. First, the unicellular nature of microbes means that virtually every cell is a potential forerunner of a new pedigree, whereas in multicellular organisms, the forerunners of new pedigrees are only the germ cells (sperm and eggs) which are a small proportion of the total cell population. A mutation in a body (or somatic) cell of an animal or plant is not passed on to the next generation, whereas a mutation in any cell of a microbial population could be. Second, most microbes are haploid, which means that mutations can be immediately expressed, in contrast to the situation in diploids. Third, microbial cells achieve high population densities and thus there is a greater possibility that mutants will be present. Fourth, microbial growth rates are usually much faster than are the growth rates of higher organisms, making selection and development of mutant types more rapid. For these reasons microbes are admirably adapted to colonize new environments (see Chapter 17).

Adelberg, E. A. (ed.) Papers on Bacterial Genetics, 2nd ed. Boston: Little, Brown and Company, 1966.

Auerbach, C. Mutation: an Introduction to Research in Mutagenesis. Edinburgh: Oliver and Boyd, 1962.

Braun, W. Bacterial Genetics, 2nd ed. Philadelphia: W. B. Saunders Co., 1965.

Bryson, V., and H. J. Vogel (eds.) Evolving Genes and Proteins (Symposium, Rutgers Institute of Microbiology). New York: Academic Press, Inc., 1965.

Davis, B. D., R. Dulbecco, H. N. Eisen, H. S. Ginsberg, and W. B. Wood, Jr. Microbiology. New York: Harper & Row, Publishers, 1967.

The Genetic Code (Cold Spring Harbor Symposium on Quantitative Biology, vol. 31). Cold Spring Harbor, N.Y.: Cold Spring Harbor Laboratory of Quantitative Biology, 1966.

Gorini, L. and J. R. Beckwith "Suppression," Ann. Rev. Microbiol. 20:401 (1966).

Hayes, W. The Genetics of Bacteria and Their Viruses, 2nd ed. Oxford: Blackwell Scientific Publishers, 1968.

Hollaender, A. (ed.) Radiation Biology, vol. I, pts. 1 and 2, High Energy Radiation. New York: McGraw-Hill Book Company, 1954.

Hollaender, A. (ed.) Radiation Biology, vol. II, Ultraviolet and Related Radiations. New York: McGraw-Hill Book Company, 1956.

Sonneborn, T. M. "Degeneracy of the genetic code: extent, nature and genetic implications," in V. Bryson and H. J. Vogel (eds.), Evolving Genes and Proteins, p. 377. New York: Academic Press, Inc., 1965.

Watson, J. D. Molecular Biology of the Gene. New York: W. A. Benjamin, Inc., 1965.

Woese, C. R. The Genetic Code: The Molecular Basis for Genetic Expression. New York: Harper & Row, Publishers, 1967.

Genetic recombination is the process by which genetic elements contained in two separate genomes are brought together within one unit. Through this mechanism new genotypes can arise even in the absence of mutation since the elements brought together may enable the organism to carry out some new function, different components of which are under the control of the separate elements. Genetic recombination in eucaryotes is an ordered and a regular process, which is part of the sexual cycle of the organism (see Chapter 13), but in procaryotes it is largely a fragmentary or even haphazard process; under certain conditions, however, recombination may be highly efficient even in procaryotes.

Kinds of recombination In procaryotes genetic recombination involves the insertion into a recipient cell of a fragment of genetically different DNA derived from a donor cell, and the integration of this DNA fragment or its copy into the genome of the recipient cell. We shall define briefly here the three means by which the DNA fragment is carried from donor to recipient and then discuss them in detail in the rest of this chapter: (1) *transformation* is a process by which free DNA is inserted directly into a competent recipient cell; (2) *transduction* involves the transfer of bacterial DNA from one bacterium to another via a temperate or defective virus particle; (3) *conjugation* (mating) involves DNA transfer via actual cell-to-cell contact between the recipient and the donor cell. This last process most closely resembles sexual recombination in eucaryotes, but differs from it in several fundamental respects.

Genetic recombination in procaryotes usually is a rare event, occurring in only a small percentage of the population. Because of its rarity, special techniques are usually necessary to detect its occurrence. One must use as recipients strains that possess selective characteristics, as shown in Figure 12.1. Markers that are not selective can also be used, but usually not as primary genetic characteristics. Various kinds of selective and nonselective markers (such as drug resistance, nutritional requirements, and so on) were discussed in Chapter 11.

12.1 Genetic transformation

The discovery of genetic transformation in bacteria was one of the outstanding events in biology as it led to experiments proving without a doubt that DNA is the genetic material. This discovery became the keystone of molecular biology and modern genetics. Genetic transformation, as we have stated, involves the uptake into the recipient bacterial cell of a free fragment of bacterial DNA of fairly large size, and its integration into the

twelve

Genetic
recombination
in
procaryotes

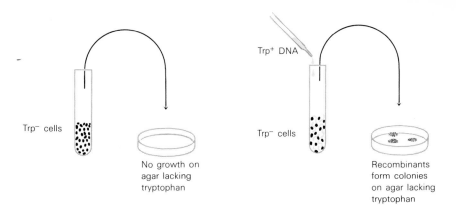

Figure 12.1 How a selective medium can be used to detect rare genetic recombinants among a large population of nonrecombinants. On the selective medium only the rare recombinants form colonies. Procedures such as this, which offer high resolution for genetic analyses, can ordinarily be used only with microorganisms.

Trp⁻ cells — No growth on agar lacking tryptophan

Trp⁺ DNA — Trp⁻ cells — Recombinants form colonies on agar lacking tryptophan

genome of the recipient. For many years it remained a puzzle as to how bacteria were actually able to take up and incorporate into their genetic material such large molecules, and this made it difficult to accept the existence of the process. Today it has been well established that transformation occurs, but the precise manner in which free DNA is incorporated into the cell is still not completely understood. Only certain genera of bacteria are transformable, including *Diplococcus pneumoniae* (the organism in which the process was originally discovered), *Hemophilus, Bacillus, Neisseria,* and *Pseudomonas.* Even within these genera, only certain strains, called ''competent,'' are transformable.

The preparation of transforming DNA Transforming DNA can be prepared by lysing cells of the donor strain with lysozyme or a detergent (Figure 12.2). Alcohol precipitates DNA, and when it is added gently as a layer on top of the preparation of lysed bacteria, the DNA accumulates at the interface between the aqueous and the alcoholic layers. A glass stirring rod is carefully inserted into the interface, and on being gently stirred, the long fibrous strands of DNA become wrapped around the stirring rod. The stirring rod containing the DNA is lifted out, the excess alcohol is drained, and the DNA is redissolved in an aqueous solution. This procedure can be repeated a number of times. Any RNA that might contaminate the preparation is removed by ribonuclease treatment, and protein can be removed by various chemical processes. The preparation ultimately obtained can be considered chemically pure as the contaminating materials represent only one-millionth part of the whole. One of the points to be taken into consideration in purifying DNA is that its destruction by the DNases present in most cells must be avoided. A very small amount of active DNase can quickly destroy the transforming activity of the DNA. However, DNase action can be averted by dissolving DNA in

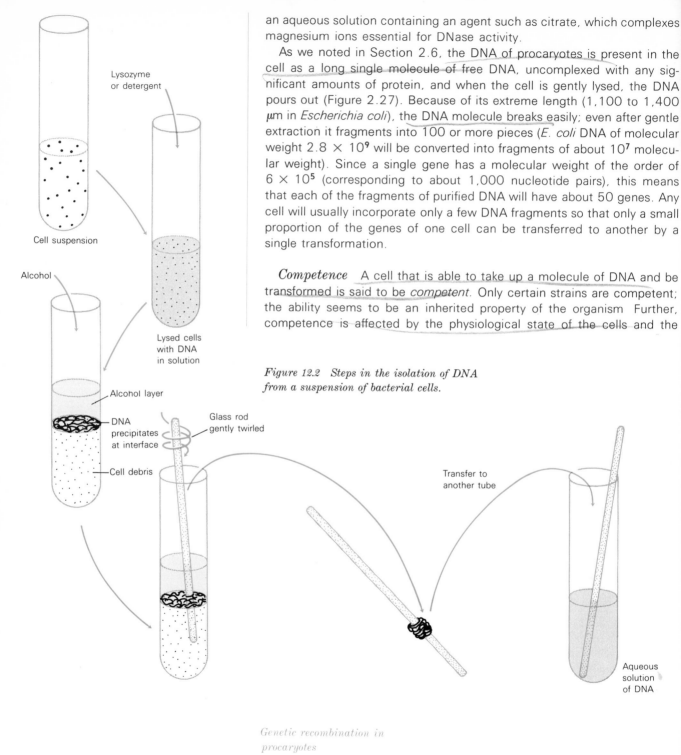

an aqueous solution containing an agent such as citrate, which complexes magnesium ions essential for DNase activity.

As we noted in Section 2.6, the DNA of procaryotes is present in the cell as a long single molecule of free DNA, uncomplexed with any significant amounts of protein, and when the cell is gently lysed, the DNA pours out (Figure 2.27). Because of its extreme length (1,100 to 1,400 μm in *Escherichia coli*), the DNA molecule breaks easily; even after gentle extraction it fragments into 100 or more pieces (*E. coli* DNA of molecular weight 2.8×10^9 will be converted into fragments of about 10^7 molecular weight). Since a single gene has a molecular weight of the order of 6×10^5 (corresponding to about 1,000 nucleotide pairs), this means that each of the fragments of purified DNA will have about 50 genes. Any cell will usually incorporate only a few DNA fragments so that only a small proportion of the genes of one cell can be transferred to another by a single transformation.

Competence   A cell that is able to take up a molecule of DNA and be transformed is said to be *competent*. Only certain strains are competent; the ability seems to be an inherited property of the organism Further, competence is affected by the physiological state of the cells and the

Lysozyme
or detergent

Cell suspension

Alcohol

Lysed cells
with DNA
in solution

Alcohol layer

DNA
precipitates
at interface

Cell debris

Glass rod
gently twirled

Transfer to
another tube

Aqueous
solution
of DNA

Figure 12.2   Steps in the isolation of DNA
from a suspension of bacterial cells.

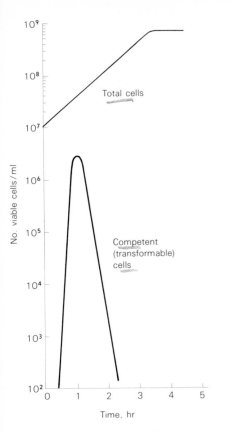

No. viable cells/ml

Total cells

Competent
(transformable)
cells

Time, hr

Figure 12.3 Appearance and disappearance of competence in Diplococcus pneumoniae (pneumococcus) during the exponential growth phase. The cells apparently undergo a physiological change and become competent when the population density reaches a specific value. [From A. Tomasz and R. D. Hotchkiss: Proc. Nat. Acad. Sci. 51:480 (1964).]

media in which they are grown, and it varies with the stages of the growth cycle. For instance, Figure 12.3 shows the proportion of competent cells present in a culture of *Diplococcus pneumoniae* at different stages of the growth cycle. There is a brief period, during late exponential phase, when the competence of the population rises dramatically and then just as rapidly falls. Good evidence exists that during this brief period of competence the surfaces of the cells change so that DNA is now able to be bound to them. This surface change can also be brought about by an enzymelike factor produced by the cells themselves, which appears in the culture medium at about the time competence appears; moreover, adding this factor to noncompetent cells of the same strain can induce them to convert into competent cells.

Uptake of DNA The DNA is bound by competent bacteria (Figure 12.4), at first in a reversible way; soon, however, the binding becomes irreversible. Competent cells bind much more DNA than do noncompetent cells—as much as 1,000 times more. However, they do not distinguish between DNA of their own species and that of other species and bind irreversibly any double-stranded DNA, although only that from genetically related species is active in transformation. Single-stranded DNA, on the other hand, is only poorly bound. Therefore, if a preparation of double-stranded DNA is heated, which causes separation of the two strands (see Figure 12.8), this DNA is both poorly bound and poorly effective in transformation. As was noted earlier, the size of the transforming fragments is much smaller than that of the whole genome, and each cell can incorporate only about 10 molecules of molecular weight 1×10^7. The DNA fragments in the mixture compete with each other for uptake, and if excess DNA that does not contain the genetic marker is added, a decrease in the number of transformants occurs. In preparations of transforming DNA, only about 1 out of 100 to 200 DNA fragments contains the marker being studied. The other fragments do not elicit transformation for this marker although they are bound and inhibit binding of the desired DNA fragment. If the frequency of transformants is plotted against the concentration of DNA, a relationship such as that shown in Figure 12.5 is obtained; at high concentrations of DNA the competition between genetically active and genetically inactive molecules results in a saturation of the system. Even under the best conditions, therefore, it is impossible to transform all of the cells in a population for a given genetic marker. The maximum frequency of transformation that has so far been obtained is about 10 percent of the population; actually the values usually obtained are between 0.1 and 1.0 percent. Another feature illustrated in Figure 12.5 is that even extremely low concentrations of fragmented DNA are able to elicit some transformants. The minimum concentration of DNA

Figure 12.4 *Electron micrograph of a portion of a DNA molecule attached to a competent cell of D. pneumoniae. [From A. Tomasz: Sci. Amer. 220:38 (1969).]*

yielding detectable transformants, which is about 0.00001 μg/ml (1 × 10⁻⁵ μg/ml), is so low that it is completely undetectable chemically.

Integration of incorporated DNA

What is the fate of the DNA after uptake? Soon after incorporation the DNA becomes converted into a single-stranded form. One of the strands is then broken down, while the other becomes integrated into the genome of the recipient. The mechanism by which incoming DNA becomes physically linked with that of the recipient is still not understood; one strand of the incoming DNA may pair with the complementary base sequence of a strand of the recipient DNA, after which the free ends of the immigrant DNA can be inserted into the recipient strand (Figure 12.6). After replication of this hybrid DNA, one parental type and one recombinant type will be formed; the latter upon segregation at cell division will cause the production of a transformed cell.

Another phenomenon exhibited in transforming systems is the transfer of *linked genetic markers,* that is, the incorporation in the same DNA fragment of genes for two or more characteristics. Since a cell has of the order of 10,000 genes and each DNA fragment has about 50 genes, the chances are not high that two genes chosen at random will exhibit linkage. However, several well-authenticated cases of genetic linkage are known. The discovery of linked markers is easiest with the use of recipients that are double mutants for two closely linked characteristics.

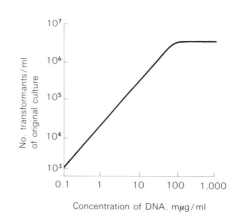

Figure 12.5 *Relationship between concentration of DNA and frequency of transformants. (From W. Hayes: The Genetics of Bacteria and Their Viruses. New York: John Wiley & Sons, Inc., 1968.)*

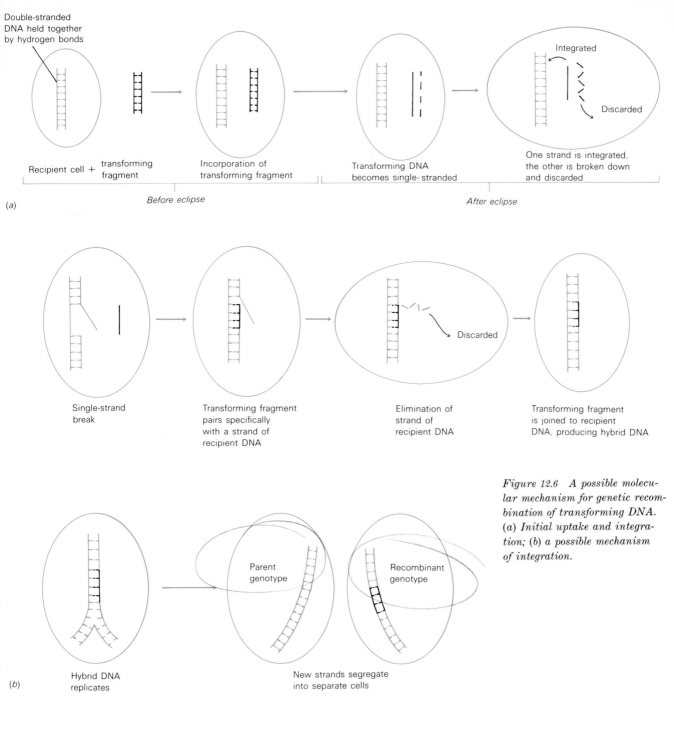

Double-stranded
DNA held together
by hydrogen bonds

Integrated

Discarded

Recipient cell + transforming
fragment

Incorporation of
transforming fragment

Transforming DNA
becomes single- stranded

One strand is integrated,
the other is broken down
and discarded

Before eclipse

After eclipse

(a)

Discarded

Single-strand
break

Transforming fragment
pairs specifically
with a strand of
recipient DNA

Elimination of
strand of
recipient DNA

Transforming fragment
is joined to recipient
DNA, producing hybrid DNA

Parent
genotype

Recombinant
genotype

*Figure 12.6 A possible molecu-
lar mechanism for genetic recombination of transforming DNA.
(a) Initial uptake and integration; (b) a possible mechanism
of integration.*

Hybrid DNA
replicates

New strands segregate
into separate cells

(b)

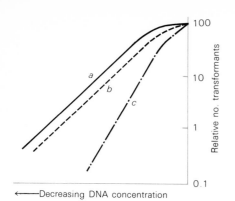

100

Relative no. transformants

10

1

0.1

a
b
c

←——Decreasing DNA concentration

Figure 12.7 Detection of genetic linkage in transformation. Curves b and c represent the number of double *transformants (for two genes) at different DNA concentrations, and curve a, which is the control, represents the number of transformants for a single gene. In curve b the two genes are linked, whereas in curve c they are not. Note that, when two genes are unlinked (curve c), the number of double transformants drops quickly as the DNA concentration is reduced since at low DNA concentrations it is unlikely that two separate DNA molecules will get into the same cell. When the two genes are linked (curve b), the drop in double transformants parallels the drop in transformants for a single gene. (From W. Hayes: The Genetics of Bacteria and Their Viruses. New York: John Wiley & Sons, Inc., 1968.)*

One need merely incubate competent cells of the double mutant with donor DNA and record cells transformed simultaneously for both characteristics. One technical difficulty arises from the fact that any cell can incorporate up to 10 DNA fragments. How can one be sure that the double transformant has arisen through a single recombinational event and not through two successive ones? This can most easily be determined by studying the relationship between number of transformants and DNA concentration, as is demonstrated in Figure 12.7. As the concentration of DNA decreases, the frequency of double transformants will fall faster than will that of singles if the two genetic markers are not linked since at low DNA concentrations there is a decreasing probability that a cell will receive two DNA fragments. If the markers are on the same DNA fragment, the frequency of double transformants will decrease at the same rate as the frequency of singles will.

Range of genetic markers transformable   Every gene of a cell is potentially transformable. The limiting factors in studying the transformation of any gene are the isolation of the appropriate recipient strain and the development of a suitable selective system for scoring transformants. The kinds of genetic markers that have been studied include drug resistance (to streptomycin, novobiocin, erythromycin, and so on), nutritional (amino acid requiring), carbohydrate utilization (mannitol, maltose), and capsule characteristics (that is, nature of the capsular polysaccharide). This latter property, which was the first marker to be used in transformation studies, will be discussed in more detail below. It is also possible to use DNA from a virus for which a competent strain is sensitive and obtain production and release of complete virus particles, a process called *transfection.*

Transformation as a tool in molecular genetics   The critical role of transformation in demonstrating that DNA is indeed the genetic material has already been mentioned. Transformation has been used in other important studies of molecular genetics. Since it is the only genetic system in which DNA can be used in the free state, it provides an excellent tool for relating the physical and chemical properties of DNA to its biological properties.

The chemical basis of mutagenesis (see Chapter 11) has been studied carefully with transforming DNA. It can be shown, for instance, that treating DNA with nitrous acid can produce mutations that may be expressed after the treated DNA has been incorporated into the recipient. At higher doses most mutagens cause complete inactivation, and it is possible to use this phenomenon to obtain some idea of the size and structure of genes. For instance, heat inactivates genetic material by

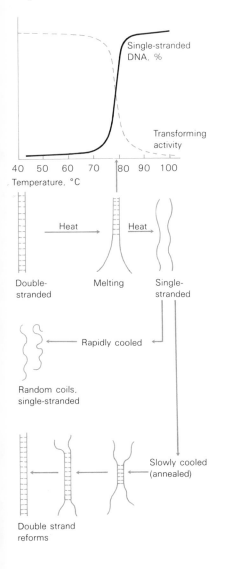

Figure 12.8 Conversion of double-stranded to single-stranded DNA by heat and the reformation of the double-stranded form upon slow cooling. Heat disrupts the hydrogen bonds that hold the two strands together. Single-stranded DNA will not function in transformation; hence the transforming ability of the DNA is lost at the same temperature at which the DNA melts.

causing depurination, the removal of purine bases (Section 11.4). By carefully heating genetically marked DNA for different lengths of time, it can be shown that certain genes are inactivated more rapidly than are others. Larger genes are inactivated more rapidly than are smaller genes and genes with higher purine contents are more heat sensitive than are those with less purine.

Perhaps one of the more dramatic uses of transformation was in studying the relationship between single-stranded and double-stranded DNA. As we already noted, single-stranded DNA does not bring about a significant amount of transformation. If a preparation of transforming DNA is heated so that its temperature rises slowly, samples that are taken at intervals and assayed for transforming ability show that within a certain narrow temperature range the transforming activity of the DNA is suddenly lost (Figure 12.8). If at the same time the physical properties of the DNA are studied, it is found to become single-stranded at the same temperature at which transforming activity is lost. (The DNA preparation converted from double to single strands by heat is said to have *melted*.) If this mixture of now single-stranded molecules is allowed to cool to a temperature just below the melting point, some of the transforming activity reappears and, concomitantly, the DNA becomes double-stranded as the complementary molecules realign and form hydrogen bonds. If the DNA is cooled rapidly, the DNA molecules lose their mobility, double strands do not reform, and ability to transform is not regained. Artificial DNA hybrids can be formed if two different kinds of DNA are mixed and heated to melting and then incubated just below the melting temperature until molecules of the two kinds have associated.

The formation of artificial hybrids occurs only if the DNA molecules are closely related. If DNA from two unrelated species is mixed, hybrids do not form and transforming activity is not regained. This experiment is a dramatic confirmation of the current theory of the complementarity of two DNA strands. It also suggests a possible means of determining whether a group of organisms are related, by noting if their DNA molecules form hybrids; the use of this method is discussed in Chapter 17. Hybrid formation between DNA and RNA also occurs, and has also been used in taxonomic studies (see Chapter 17).

The occurrence of transformation in nature So far we have been discussing bacterial transformation as a purely experimental (laboratory) phenomenon, but there is some evidence that it also occurs in nature. In fact, the phenomenon was first discovered by Fred Griffith in the process of analyzing the factors involved in the pathogenicity of *D. pneumoniae* for the mouse. It is appropriate to review these experiments here.

The pathogenicity of *D. pneumoniae* cells is directly connected with the presence on their surfaces of a capsule, which enables the organism to avoid phagocytosis by the body's white blood cells (see Chapter 15). Capsulated strains of *D. pneumoniae* are highly virulent; indeed, only a very few cells inoculated into the mouse are able to initiate an infection that results in death in several days (Figure 12.9). Noncapsulated mutants can be isolated, and these strains are nonpathogenic; massive inocula can be given without harmful effects. Thus the mouse can be said to act as a selective medium for capsulated pathogenic strains. Griffith inoculated mice with living cells of a noncapsulated strain, mixed with heat-killed cells of a pathogenic capsulated strain. To his surprise, many of the mice died, and from them could be isolated living capsulated bacteria. It was easy to show that these bacteria were neither revertants of the noncapsulated strain nor revitalized cells of the heat-killed strain. We know now that some of the heat-killed cells lysed and released DNA fragments, which

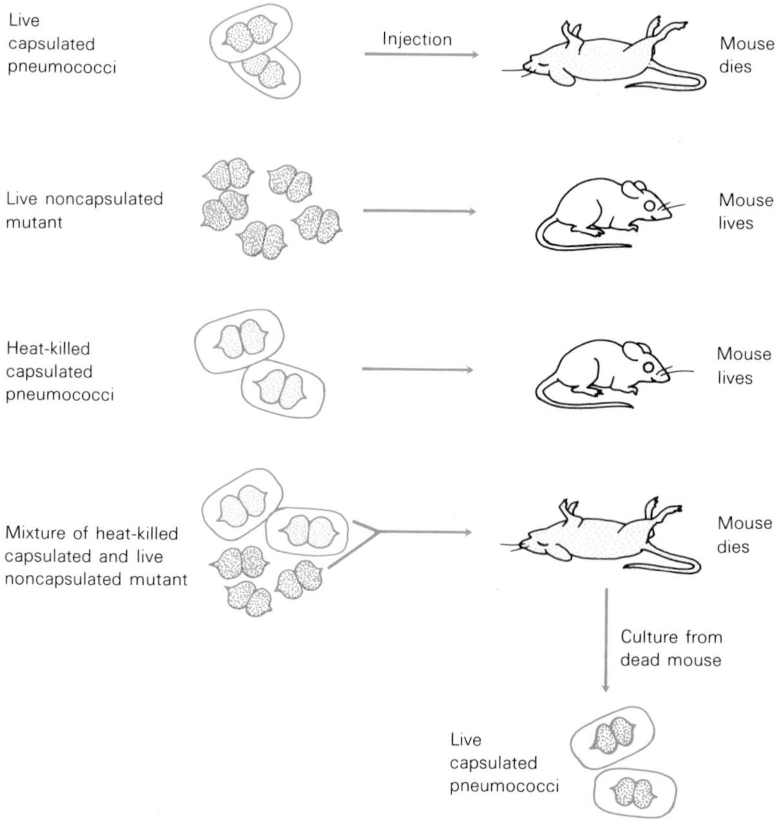

Figure 12.9 *Griffith's 1928 experiment, the first observation of bacterial transformation and a demonstration that transformation occurs in vivo.*

transformed some of the noncapsulated cells; these were then selectively favored. Transformation in the mouse has also been shown using antibiotic-resistant strains: If penicillin-resistant and streptomycin-resistant strains are inoculated in the same mouse, strains resistant to both antibiotics can be isolated.

Since transformation requires free DNA, it is not surprising to discover that it will occur most frequently in organisms that lyse spontaneously (probably through the action of autolytic enzymes). Although transformation is a rather uncertain and inefficient way of ensuring genetic recombination—since the donor cells must lose their viability and since the free DNA is unprotected from the action of DNases usually present in the animal body and other natural environments—it seems clear nonetheless that some bacteria have evolved this mechanism for inheriting the genes of their dead relatives. We will see that certain bacteria have also evolved other, more efficient means of accomplishing recombination.

12.2   Transduction

In transduction, a recombination process that has been found in a variety of bacteria, DNA is transferred from cell to cell through the agency of temperate viruses. Recall the phenomenon of lysogeny: Under certain conditions, when a temperate virus infects a cell, lysogenization takes place, and the DNA of the virus becomes integrated into the genetic apparatus of the host (Section 10.6). Lysogenization in itself represents a kind of genetic recombination, and the lysogenized cell differs genetically and phenotypically from the nonlysogenized cell. Some of the genes of the virus are expressed in the lysogenized cell, namely those which confer immunity to further infection of the cell with the same virus. However, a transducing virus can transfer not only its own genes but also some genes of the host from which it was derived.

Kinds of transduction   Genetic transfer of host genes by viruses can occur in two ways. In the first, called *specialized transduction,* a restricted group of host genes becomes integrated directly into the virus genome—usually replacing some of the virus genes—and is transferred to the recipient during lysogenization. In the second, called *generalized transduction,* host genes deriving from virtually any portion of the host genome become a part of the mature virus particle in place of, or in addition to, the virus DNA but do not become directly integrated into the virus genome. The transducing virus particle in both cases is *defective* and cannot cause lysis of the host probably because it lacks some virus genes.

Transduction has been found to occur in a variety of bacteria, including *E. coli* and species of *Salmonella, Shigella, Proteus, Pseudomonas, Staphylococcus,* and *Bacillus.* Not all temperate phages will transduce, and not all bacteria are transducible; but the phenomenon is sufficiently widespread for us to assume that it plays an important role in genetic recombination in nature.

Detailed criteria have been advanced to show that transduction is a virus-mediated process and that it is not due to transformation or conjugation. These criteria include the following: (1) Transduction does not require cell-to-cell contact between donor and recipient, proving that the process is not due to conjugation (see below). (2) The transducing ability of cell lysates is not destroyed by DNases; therefore, the process is not due to transformation. (3) The transducing agent can be purified by the same procedures used to purify virus particles. (4) Transduction occurs only with recipient cells that have receptor sites for the virus in question, and the transducing activity of a lysate can be eliminated by treating it with cells that are able to adsorb the virus. (5) Antibody that neutralizes the virus by combining with its tail protein also prevents transduction. Thus, in all cases, transducing activity and virus activity behave similarly.

Generalized transduction   In generalized transduction, virtually any genetic marker can be transferred from donor to recipient. An example of how transducing particles may be formed is given in Figure 12.10. When the population of sensitive bacteria is infected with a temperate phage, the events of the phage lytic cycle may be initiated. In a lytic infection, the host DNA breaks down into small pieces, and some of these pieces become incorporated inside the virus particles. Upon lysis of the cell, these particles are released and mingle with the normal virus particles, so that the lysate contains a mixture of normal and transducing virus particles. When this lysate is used to infect a population of recipient cells, most of the cells become infected with normal virus particles and are either lysogenized or lysed. However, a small proportion of the population receives transducing particles, whose DNA can now undergo genetic recombination with the host DNA. Since only a small proportion of the particles in the lysate are of the defective transducing type and each particle contains only a small fragment of donor DNA, the probability of a defective phage particle containing a gene for which recombination is being sought is quite low and usually only about 1 cell in 10^6 to 10^8 is transduced for a given marker. Note that transducing particles are unable either to lyse or to lysogenize sensitive cells. The molecular mechanism of genetic recombination between donor and recipient DNA is probably analogous to that occurring during transformation, as is illustrated in Figure 12.6. In fact, at the molecular level generalized transduction is a process very similar to transformation, differing mainly in the manner of transfer of DNA from donor to recipient. Furthermore, the sizes of the transducing and transforming fragments are nearly the same—about one-hundredth of the whole bacterial DNA. Only closely linked genes can be transduced in the same particle.

In the discussion above it was explained that transducing particles,

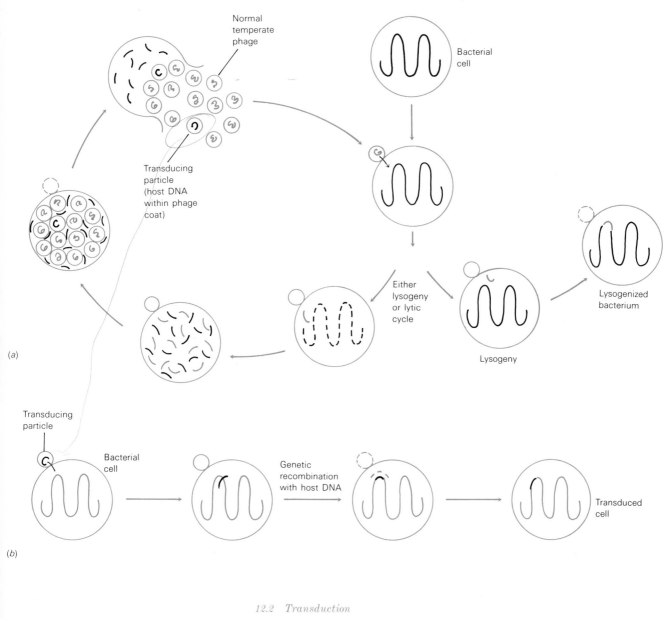

Figure 12.10 Generalized transduction: (a) one possible mechanism by which virus (phage) particles containing host DNA can be formed; (b) genetic recombination with transducing particle.

containing host DNA, are formed during lytic infection by a temperate virus. But the production of these particles does not necessarily occur immediately after infection by the temperate virus. Since a temperate virus is also capable of lysogeny, transducing particles can be formed many generations after the initial infection whenever the lysogenic cells are induced to undergo events of lytic infection. Furthermore, the transducing particles do not necessarily always contain preexisting host DNA—they may contain hybrid DNA, part derived from the host and part from the virus. For instance, in the case of phage P1 of *E. coli,* the evidence suggests that the phage DNA may become attached to the host (donor) DNA at a random site, and that during replication of the phage DNA the host DNA is also replicated; in this manner recombination between host and phage DNA can occur, producing a hybrid DNA molecule in which host genes are substituted for some of the phage genes. When this hybrid DNA is incorporated into a phage particle, a defective transducing particle is produced. In the case of P1 phage, the situation has some similarities to specialized transduction described below, except that virtually any gene of the cell can be transduced by P1.

Abortive transduction   Sometimes in generalized transduction the injected fragment does not undergo recombination with the recipient, but remains within the cell in a free state, a process called *abortive transduction.* Even if it cannot replicate, its genetic information can still be transcribed and expressed. This means that the cell with the fragment is genetically altered and can acquire a new phenotype but that at cell division only one of the daughter cells retains the unintegrated fragment. Abortive transduction, like other forms of transduction, can be detected with the use of selective media. If the agar permits only a genetically different cell to replicate, the abortively transduced cell will give rise to a microcolony (rather than a colony of normal size) because at every cell division only the one progeny containing the transducing fragment is capable of continued growth, whereas the other cell merely contributes its mass to the colony and cannot replicate under the selective conditions. Arithmetic rather than exponential growth results, and only a microcolony is formed. The consequences of abortive transduction of motility are illustrated in Figure 12.11. The type of inheritance exhibited by abortively transduced cells is called *unilinear inheritance.*

Specialized transduction   We have emphasized that transduction is a rare genetic event. However, in specialized transduction a very efficient transfer by phage of a restricted set of host genes sometimes occurs. The example we shall use, which was the first to be discovered and is the best understood today, involves the transduction of the galactose genes by the temperate phage *lambda* of *E. coli.*

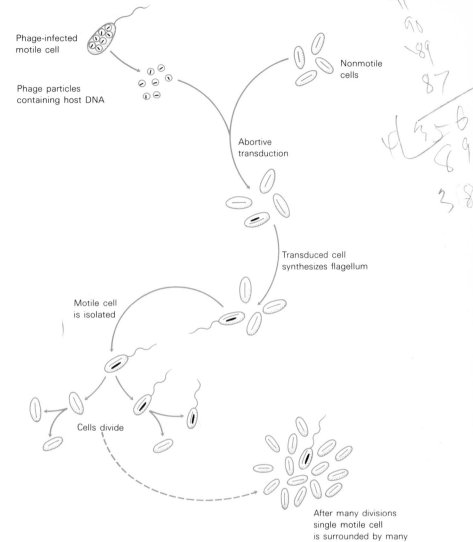

Phage-infected
motile cell

Phage particles
containing host DNA

Nonmotile
cells

Abortive
transduction

Transduced cell
synthesizes flagellum

Motile cell
is isolated

Cells divide

After many divisions
single motile cell
is surrounded by many
nonmotile cells

*Figure 12.11   Consequences of abortive trans-
duction of a gene for flagellation: The
donor DNA does not become integrated
into the recipient and consequently is not
replicated. At each cell division, only one
of the progeny contains the transduced
piece and is motile.*

As we discussed in Section 10.6, when a cell is lysogenized by a temperate phage the phage genome becomes integrated into the host DNA
at a specific site. With phage *lambda* this region is immediately adjacent
to the cluster of host genes that control the enzymes involved in galactose
utilization. The DNA of *lambda* is circular or cyclizes readily, and when
the virus DNA associates with the specific region on the host DNA with
which it is complementary, the virus DNA circle opens up and becomes

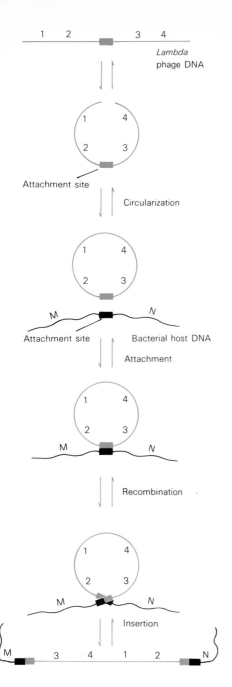

1 2   3 4

Lambda phage DNA

Attachment site

Circularization

Attachment site   Bacterial host DNA

Attachment

Recombination

Insertion

inserted into the host DNA (Figure 12.12). From then on virus DNA replication is under host control. Upon induction the virus DNA separates from the host DNA by a process that is the reverse of integration. Ordinarily (Figure 12.13) when the lysogenic cell is induced the phage DNA leaves as a unit, but under rare conditions a genetic exchange takes place between the phage genes and the host genes at induction, and the galactose cluster is inserted into the phage genome in place of some phage genes. These altered phage particles, called *lambda dg,* are defective and can never initiate the lytic process, although they can transduce the galactose genes. Thus the lysate obtained contains a few *lambda dg* particles mixed in with a large number of normal *lambda* particles; this lysate is called *LFT* (low-frequency transducing).

If a galactose-negative cell is infected with one of these *lambda dg* particles and the DNA of the defective particle becomes integrated into the host DNA, this cell now becomes galactose-positive and in addition carries in its DNA the *lambda dg* region. When a population derived from such a cell is induced, the *lambda dg* replicates in the manner of a normal temperate phage, and a lysate is obtained, all of whose particles consist of defective *lambda dg.* Since each particle can transduce the *gal* genes, this lysate is very efficient in transduction and is called *HFT* (high-frequency transduction). Note, however, that transduction can occur only for a restricted group of genes, the *gal* cluster. An HFT lysate can be produced only by induction of a population containing *lambda dg.* Although *lambda dg* is a defective phage and cannot replicate by lytic infection, it can be maintained through host replication if it is integrated into the host genome.

As we have seen, the formation of the original *lambda dg* particle involves a genetic exchange between the *lambda* DNA and the host DNA at the time of induction of a lysogenic cell. This exchange does not occur at exactly the same place in each independent induction event. Thus different cells produce *lambda dg* particles with different amounts of host and phage DNA. There is probably a maximum limit to the amount of phage DNA that can be replaced with host DNA since the attachment region must be present and sufficient phage DNA must be retained in

Figure 12.12   Events involved in the integration of the lambda temperate virus into the DNA of the host Escherichia coli. Integration always occurs at a specific site on the host DNA. Genetic recombination occurs between virus and host DNA in the attachment site, and as a result of recombination the virus DNA is linearly inserted into the bacterial DNA. Compare with Figure 12.13. [Adapted from E. R. Signer: Ann. Rev. Microbiol. 22:451 (1968). With permission of the publisher.]

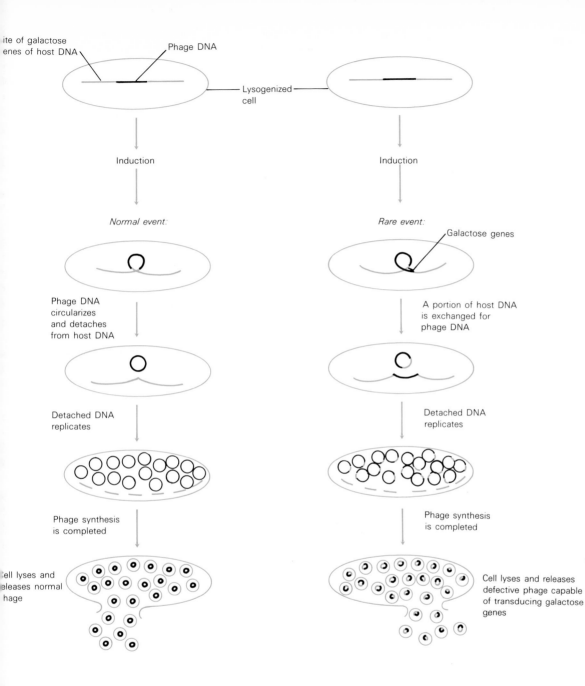

Figure 12.13 *The production of particles transducing the galactose genes in an E. coli cell lysogenic for lambda virus.*

order to provide the information for the production of the phage protein coat and for other phage proteins needed for lysis and lysogenization. One important distinction between specialized and generalized transduction is in how the transducing lysate can be formed. In specialized transduction this *must* occur by the induction of a lysogenic cell, whereas in generalized transduction it can occur either in this way or by infection of a nonlysogenic cell by the temperate phage.

Although we have discussed specialized transduction only in the *lambda-gal* system, specialized transduction is also known for phage φ80 of *E. coli*, phage P22 of *Salmonella typhimurium*, and several others.

Phage conversion   This is a phenomenon analogous in some ways to specialized transduction. When a normal temperate phage (that is, a nondefective one) lysogenizes a cell and its DNA becomes converted into the prophage state, it induces in the host cell an immunity to phage infection through the production of the phage repressor system (see Section 10.6). In certain cases other phenotypic alterations can be detected in the lysogenized cell, which seem to be unrelated to the phage immunity system. Such genetic changes, which are brought about through lysogenization by a normal temperate phage, are called *phage conversions*. Two cases have been especially well studied. One involves a change in structure of a polysaccharide on the cell surface of *Salmonella anatum* upon lysogenization with phage ε^{15}. The second involves the conversion of nontoxin-producing strains of *Corynebacterium diphtheriae* to toxin-producing, pathogenic strains, upon lysogenization with phage β. In these situations the information for the production of these new materials is apparently an integral part of the phage genome and hence is automatically and exclusively transferred upon infection by the phage and lysogenization. The conversion of a normal animal cell to a cancer cell upon infection with polyoma virus (see Chapter 10) is in many ways analogous to phage conversion.

12.3   Bacterial conjugation

Bacterial conjugation or mating is a process of genetic recombination that involves cell-to-cell contact. One cell, the donor, transmits genetic information to the other cell, the recipient. In contrast to transformation or transduction, in which only small fragments of DNA can be transferred, conjugation can involve the transfer of large portions of the genome. The discovery of bacterial conjugation caused considerable controversy, as many scientists could not believe that organisms without an organized

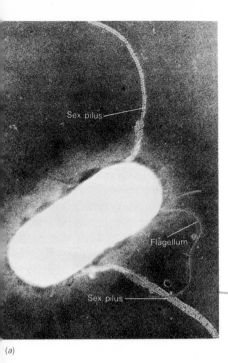

(a)

(b)

Figure 12.14   The presence of the sex pilus of E. coli Hfr is revealed by use of a bacteriophage that adsorbs specifically to the pilus: (a) whole cell with two sex pili. (Courtesy of Raymond C. Valentine.) (b) A portion of the sex pilus at higher magnification. [From R. C. Valentine and M. Strand: Science 148:511 (1965). © 1965 by the American Association for the Advancement of Science.]

nucleus could undergo a process akin to sexual reproduction. Also, initially only a single strain of *E. coli,* strain K-12, was found to show this process; other *E. coli* isolates from nature were completely unable to conjugate. Furthermore, in the initial studies the frequency of conjugation was very low, of the order of 1 recombinant per 10^6 cells, making difficult an analysis of the mechanisms of the process. That it could be studied at all was possible only by using selective genetic markers. Because of these technical difficulties, bacterial conjugation was originally misinterpreted as a sexual process analogous to that already known in eucaryotic micro-organisms (see Chapter 13), in which fusion occurs between the nuclei of the conjugating cells, resulting in formation of a diploid. We know now that bacterial conjugation differs in several fundamental aspects from that of eucaryotes.

Mechanism of conjugation   Bacterial conjugation involves a one-way transfer of genetic material from a donor strain to a recipient strain. The ability to serve as a donor or recipient is genetically determined. Recipient strains are designated as F^-, and two kinds of donor strain are known: those designated F^+, which usually donate only small portions of their genome, and strains that donate large amounts of their genome, which are called *Hfr* (from "*H*igh *f*requency of *r*ecombination"). We shall first analyze the mechanism of conjugation in *Hfr* strains and then return to the situation in F^+ strains.

The following steps in conjugation are recognized: (1) specific pairing of *Hfr* and F^- cells; (2) transfer of a DNA fragment from donor to recipient; (3) genetic recombination between the donor fragment and the recipient DNA; (4) expression of the new phenotype in the recombinant cell.

When suspensions of *Hfr* and F^- cells are mixed, specific pairing of cells of opposite type occurs. At least one specific surface component involved in pairing has been identified in *Hfr* cells. This is a specific pilus (Section 2.7), the so-called sex pilus, which is probably involved in the transfer of DNA to the F^- cell. Both F^+ and *Hfr* cells have sex pili, but F^- cells do not. Although an *Hfr* cell may have many other small pili (see Figure 2.31), it has only a few sex pili. The diameter of the hole in the sex pilus, about 2.5 nm, is just large enough for a DNA molecule to pass through lengthwise. In structure the sex pilus resembles the tails of some bacteriophages, through which DNA also passes. In addition to its role in mating, the sex pilus has another function as the receptor site for certain phages; with the electron microscope phage particles can be seen clustered about this pilus (Figure 12.14).

That *Hfr* cells transfer DNA to recipients can be shown most directly by labeling the donor DNA with a radioactive isotope and then evaluating

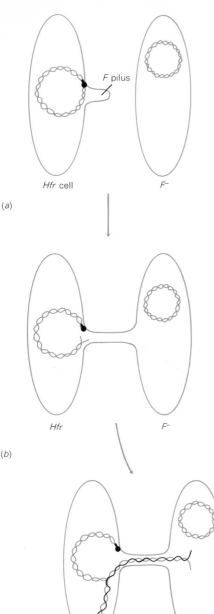

(a)

Hfr cell   *F⁻*

(b)

Hfr   *F⁻*

(c)

Hfr   *F⁻*

its transfer to the recipient by autoradiography. Experiments suggest that one of the strands of DNA transferred is newly synthesized, whereas the other strand is derived from a preexisting strand of the *Hfr*. A mechanism by which DNA transfer may occur is depicted in Figure 12.15.

The mating pairs are not strongly joined and can easily be separated by agitation in a mixer or blender. That DNA transfer is required for formation of recombinants in the *F⁻* can be shown most simply by agitating the cell mixtures at various times; the longer the time between pairing and agitation, the more genes of the *Hfr* appear in the *F⁻* recombinant. Studies of this type, called *interrupted mating experiments*, revealed another interesting aspect of bacterial conjugation: Gene transfer is an *oriented* process—a given *Hfr* strain always donates genes in a specific order, as in the experiment of Figure 12.16. We can observe that, if the suspension is agitated immediately after the cells are mixed, no recombinants result, but if mating is allowed to continue until about 8 min after the cells are mixed, recombinants containing the *thr* and *leu* genes begin to appear. After about 20 min, *gal* recombinants appear; and after 30 min, *trp*. The time of entry for each set of genes can be estimated by extrapolating each curve to the origin, and then the order in which the various genes enter the *F⁻* cell can be deduced (Figure 12.16). Figure 12.16 illustrates another feature of gene transfer from *Hfr* to *F⁻*: Genes that enter early always appear in a larger percentage of the recombinants than do genes that enter late. In addition to showing that gene transfer from donor to recipient is a sequential process, these experiments provide a method of determining the order of the genes on the bacterial DNA (mapping). The resultant series of gene loci is called a *genetic map* (see later).

Figure 12.15 Possible mechanism of DNA transfer during conjugation: (a, b) Specific pairing; circular DNA opens; (c) replication of DNA and transfer to F⁻ of one strand newly synthesized, one strand old; (d) pairs separate, and circular DNA reformed in Hfr; (e) genetic recombination in F⁻ between donor fragment and recipient DNA.

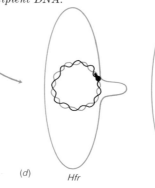

(d)
Hfr   *F⁻*

(e)   *F⁻*

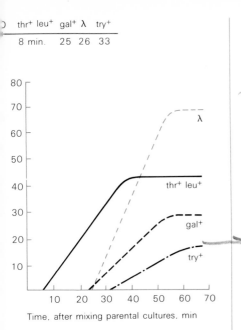

thr⁺ leu⁺ gal⁺ λ try⁺

8 min.   25 26 33

Figure 12.16 Rate of formation of recombinants containing different genes after mixing Hfr and F⁻ bacteria. The location of the genes along the Hfr chromosome is shown in the small diagram. Note that the genes closest to the origin are the first ones detected in the recombinants. The experiment is done by mixing Hfr and F⁻ cells under conditions in which essentially all Hfr cells find mates. At various intervals, aliquots of the mixture are shaken violently to separate the mating pairs and plated on selective medium on which only the recombinants can grow and form colonies. (From W. Hayes: The Genetics of Bacteria and Their Viruses. New York: John Wiley & Sons, Inc., 1968.)

Distinction between Hfr and F⁺   In the case of F⁺ strains, only small amounts of the donor genome are transferred to the recipient. However, what F⁺ strains transfer to F⁻ strains at high frequency is the F⁺ characteristic itself, and the F⁻ cells converted to F⁺ are able in turn to convert other F⁻ cells to F⁺. Thus F-plusness has the characteristic of spreading through the population in the manner of an infection. An F⁺ cell can, however, be "cured" of its F-plusness by treatment with acridine dyes or with low doses of UV radiation, the cured cells becoming F⁻.

Even though the F⁺ characteristic spreads rapidly through a population of predominantly F⁻ cells, relatively few of the original F⁻ cells become genetically changed for other characteristics. Thus the ability to be a donor is transferred independently of the other donor genes. The F⁺ factor controlling the male, or donor, characteristic thus behaves as an independent genetic particle, which is called the F factor.

What is the nature of the F factor? Several lines of evidence suggest that it is composed of DNA: (1) It is inactivated by UV radiation and other agents that affect DNA. (2) It can be eliminated from the cell by treatment with acridine dyes, which, it will be recalled (see Chapter 11), bind strongly with DNA. (3) Radioisotopic evidence shows that DNA is transferred when the F factor is transferred. (4) When the F factor is transferred to a species whose DNA differs in base composition from that of the donor, a new piece of DNA of the same base composition as the donor can be detected. (5) The F factor can under certain conditions acquire genes of the donor and be converted into what is called an F' ("F prime"), which transfers at high frequency not only itself but also the genes it has acquired. (6) The F factor causes the production of the sex pilus. (7) Only one or a few copies of the F factor are present in any cell.

The F factor therefore seems to be an independent unit of DNA, able to control its own replication and its transfer to other cells.

Under rare circumstances, Hfr strains can be isolated from F⁺ strains. Although Hfr strains transmit genes at high frequency, they usually do not convert F⁻ cells to F⁺; this suggests that Hfr strains differ from F⁺ cells through an alteration in the F factor to a state that is no longer infective. This alteration probably involves the attachment of the F factor to the donor DNA at a specific location. The origin of an Hfr from an F⁺ cell is visualized as the insertion of the F particle into the DNA, probably by a recombinational process.

A given Hfr strain always donates genes in the same order, beginning with the same position; but Hfr strains of independent origin transfer genes in different sequences. Each strain has its own characteristic pattern. When the order of gene transfer for a number of different Hfr's was determined, a surprising observation was made (see Figure 12.17). Each Hfr behaves as if it were formed by the opening of a circular DNA molecule at a random point, with either end of the opened circle serving as

12.3 Bacterial conjugation

345

Figure 12.17 Manner of formation of different Hfr strains (a), which donate genes in different orders and from different origins (b). The bacterial chromosome is visualized as a circle (c) that can open at various locations, at which F particles become attached.

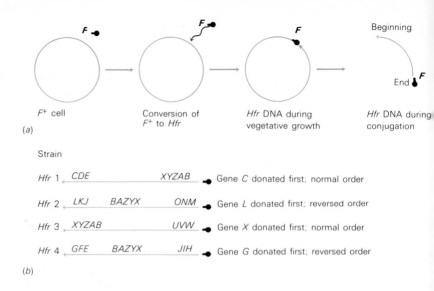

(a)

Strain

Hfr 1 ← CDE _____ XYZAB •— Gene C donated first; normal order

Hfr 2 ← LKJ   BAZYX   ONM •— Gene L donated first; reversed order

Hfr 3 ← XYZAB _____ UVW •— Gene X donated first; normal order

Hfr 4 ← GFE   BAZYX   JIH •— Gene G donated first; reversed order

(b)

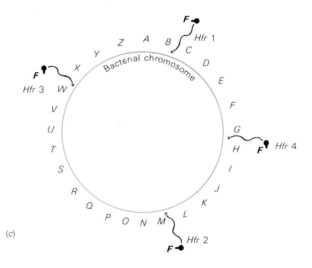

(c)

the initiation point for gene transfer. The concept of a circular DNA molecule, first developed from these genetic studies, was later confirmed by electron-microscopic and autoradiographic studies on bacterial DNA. It seems likely that the *F* particle becomes the point at which the circular DNA is attached to the cell membrane, probably near the site of the *F* pilus (Figure 12.14). During normal cell division, the DNA of the *Hfr* replicates normally; but at the time of pairing with an *F⁻* cell the circular

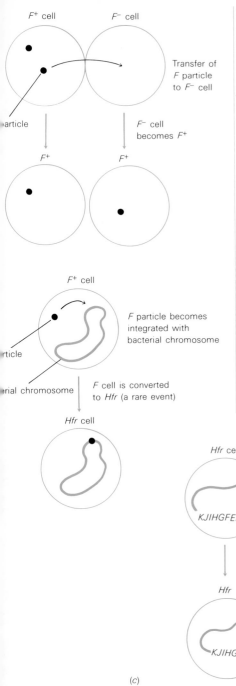

F+ cell   F- cell

Transfer of
F particle
to F- cell

F- cell
becomes F+

F+   F+

...article

F+ cell

F particle becomes
integrated with
bacterial chromosome

...ticle

...rial chromosome   F cell is converted
to Hfr (a rare event)

Hfr cell

DNA opens, and the end without the *F* particle is inserted into the *F⁻* cell, the *F* particle holding the other end of the DNA to the cell membrane. Pairs usually separate before complete gene transfer is effected, but rarely the whole DNA, including the *F* particle, is transferred. Cells that receive the whole DNA become *Hfr*, whereas those that receive only a part of the DNA lack the *F* particle and remain *F⁻* although they can, of course, become genetic recombinants for the genes that they received. The *Hfr* strain, after transfer of one copy of its DNA, still remains *Hfr*. The relationship between *F⁺* and *Hfr* is diagrammed in Figure 12.18.

Episomes and plasmids ·The *F* factor is an example of an entity sometimes called an *episome*. By definition, an episome is an independent genetic element which occurs in addition to the normal cell genome, which can be transmitted to other cells, and which can replicate either as an autonomous unit or as one integrated into the host genome. This definition fits not only the *F* factor but also temperate phages (Section 10.6).

Another group of genetic elements, called *plasmids*, never become integrated into the host chromosome but remain as independent self-replicating units. Plasmids are known that are responsible for the synthesis of certain bacteriocines and of the so-called resistance transfer factors (RTF), which confer resistance to antibiotics. Plasmids may also be transferred by cell-to-cell contact.

Thus, although we originally stated that procaryotes have only a single DNA molecule per cell, on which all genes are located, we see that this statement will have to be modified since episomes (in the unintegrated state) and plasmids are additional genetic elements.

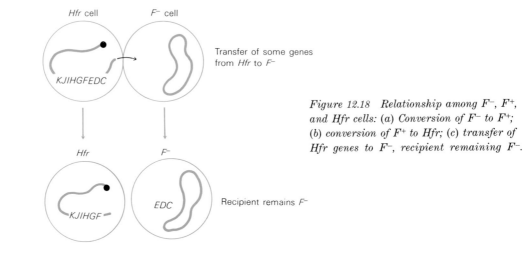

Hfr cell   F- cell

Transfer of some genes
from Hfr to F-

KJIHGFEDC

Hfr   F-

KJIHGF   EDC   Recipient remains F-

(c)

Figure 12.18   Relationship among F⁻, F⁺, and Hfr cells: (a) Conversion of F⁻ to F⁺; (b) conversion of F⁺ to Hfr; (c) transfer of Hfr genes to F⁻, recipient remaining F⁻.

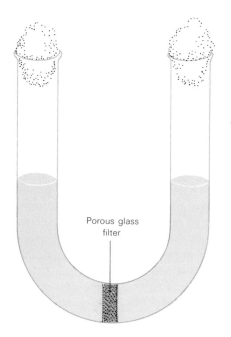

Figure 12.19 A U tube with porous glass filter, which is used to determine whether genetic recombination between two bacterial strains requires cell-to-cell contact. The glass filter permits the passage of virus particles or DNA but not of whole cells.

Porous glass
filter

The occurrence of conjugation   Mating has also been demonstrated in laboratory cultures of *Shigella, Salmonella, Pseudomonas, Serratia,* and *Vibrio.* We should anticipate that conjugation would occur in nature only if organisms can achieve the high population densities—over 10^7 cells per milliliter—necessary for cell-to-cell contact; such high population densities occur for *E. coli* in the intestinal tracts of animals. Although no direct experiments have been performed to prove that conjugation occurs outside the laboratory, many strains of *E. coli* K12 that are able to conjugate have been isolated from nature. One study showed that, of 2,000 such strains, about 50 could act as donors for a single F^- strain. About 30 percent of *E. coli* isolates from nature are F^-.

12.4   *Experimental distinction among transformation, transduction, and mating*

If recombinants are formed when cultures of two genetically distinct strains are mixed, one can determine by several simple tests whether these result from the transfer of free DNA, infection by a virus, or cell-to-cell contact: (1) If recombinants do not form when the mixture is treated with DNase, we can safely assume that transformation is involved. Some cells of the donor strain are probably lysing and releasing DNA, which is becoming incorporated into the recipient. (2) If the process is DNase-resistant, recombination could be due to either transduction or conjugation. These can be distinguished by experimentation with a U tube (Figure 12.19), which has two arms separated by a porous filter through which culture medium, DNA, and virus particles can pass but bacterial cells cannot. One of the two strains is inoculated into each arm of the U, and medium is pumped back and forth continuously. If transduction is involved, phage particles released by one strain will pass to the other arm, infect the recipients, and bring about some recombination. Since cell-to-cell contact is not possible in the U tube, no recombinants will appear if the process is due to conjugation.

12.5   *Use of genetic analysis in studies on gene structure and function*

Genetic mapping by conjugation   We have stated that the procedure of interrupted mating can be used to map the locations of the various genes. In *E. coli,* for example, different *Hfr* strains, which donate genes from different parts of the bacterial DNA (Figure 12.17), are used to locate different regions on the genetic map. By using *Hfr* strains having overlapping genes, it is possible to map the whole *E. coli* gene complement. The procedure requires, of course, the availability of *F⁻* strains that are mutant in those genes to be mapped. These mutants are isolated by the various procedures that are described in·Chapter 11, or it is possible to "construct" them by genetic recombination from other mutant strains.

In a common mapping procedure, the *F⁻* strains used are streptomycin resistant, and the *Hfr* strains are streptomycin sensitive, with the locus at the end of the DNA distant from that which is first injected. This permits the elimination, after mating, of the *Hfr* strains by plating on streptomycin agar. The *F⁻* strain must be mutant for some nutritional requirement, so that it can be eliminated by plating on a medium lacking the required nutrient. In this way only genetic recombinants will initiate colonies on the agar plates. The procedure then is to carry out an interrupted mating experiment (Figure 12.16). In addition to the selected markers, other markers present that are not selected can also be mapped by scoring each recombinant colony for the unselected ones. In this way markers that, because of their nature, cannot be selected for directly can also be mapped.

At 37°C, the time for the transmissal from *Hfr* to *F⁻* of the complete genetic complement of *E. coli* is about 90 min. "Zero time" is arbitrarily set as that at which the first gene transfer (for the *thr* gene) can be detected by strain *Hfr-H,* the first *Hfr* strain isolated. Considerably more than 200 genes have been mapped for *E. coli,* and a similar number for *Salmonella typhimurium.* An abbreviated version of the genetic map of *E. coli* is given in Figure 12.20. One minute on the genetic map represents about 50 to 100 genes, and since the whole genome represents 90 min, this means that *E. coli* may have as many as 10,000 genes. Therefore, although the number of genes mapped may seem to be large, calculations reveal that the mapped genes represent only about 2 percent of the DNA of *E. coli* under the assumption that all the DNA of *E. coli* codes for proteins. It may be of some interest that, since the total length of DNA of an *E. coli* nuclear body is about 1,100 to 1,400 μm, 1 min of DNA transfer represents about 15 μm of DNA. Another important conclusion that can be drawn from the mapping data is that *E. coli* and *S. typhimurium* (the only two bacterial species mapped to any great extent) have only a single linkage group on which are situated all the genes of the cell and that this linkage group corresponds to a single DNA molecule. This sharply contrasts with the situation in eucaryotic organisms (see Chapter 13), in which a number of linkage groups exist, each representing a single chromosome.

12.5 Use of genetic analysis in studies on gene structure and function

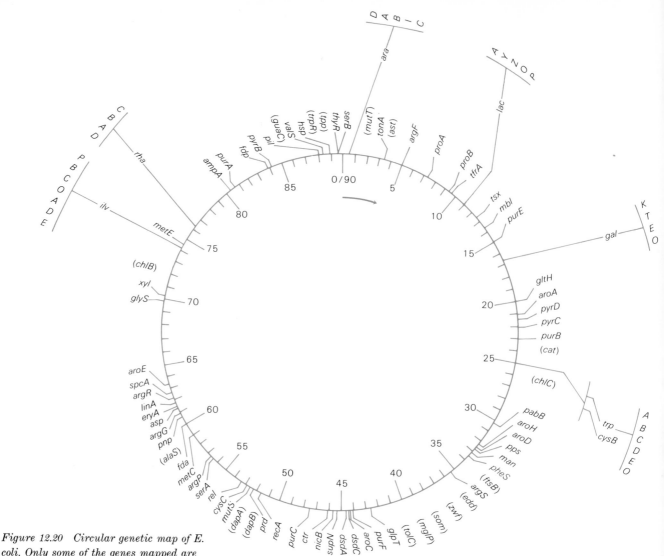

Figure 12.20 Circular genetic map of E. coli. Only some of the genes mapped are shown. Certain groups of closely related genes occur in clusters. Numbers indicate the times (minutes) of entry into the F⁻ from the Hfr of various genes, determined by interrupted-mating experiments (Figure 12.16). Parenthetical genes are approximately mapped. [Redrawn in part from A. L. Taylor and C. D. Trotter: Bacteriol. Rev. 31:332 (1967).]

Gene clusters and the operon model　It is impossible to determine the order, by interrupted mating, of genes closer than 1 min; hence the mapping of closely linked genes cannot be done by conjugation. Recall that transducing phages transfer only small pieces of DNA, each about one-hundredth of the size of the complete DNA molecule. Since the genes of a single fragment readily undergo recombination during transduction,

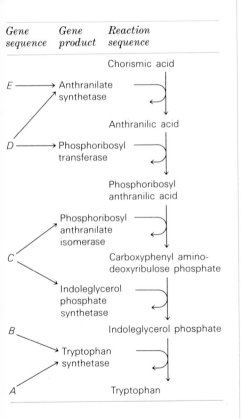

Figure 12.21   Order of genes on the tryptophan operon of E. coli (left), the enzymes coded by each gene (center), and the biochemical reaction sequence (right).

Gene sequence	Gene product	Reaction sequence
		Chorismic acid
E	Anthranilate synthetase	
		Anthranilic acid
D	Phosphoribosyl transferase	
		Phosphoribosyl anthranilic acid
	Phosphoribosyl anthranilate isomerase	
C		Carboxyphenyl amino-deoxyribulose phosphate
	Indoleglycerol phosphate synthetase	
B		Indoleglycerol phosphate
	Tryptophan synthetase	
A		Tryptophan

this method allows us to study the order of closely linked genes and to map a series of mutants all of which are within a single gene. This type of analysis, called *fine-structure genetic mapping,* has produced some important ideas about the way in which genes are organized and linked together.

The mapping of the genes that control the enzymes of a single biochemical pathway has shown that these genes are often *clustered* or closely linked. The gene clusters for several biochemical pathways on the *E. coli* chromosome are located in Figure 12.20. When the sequences of these clustered genes are studied by transductional analysis, it is frequently found not only that the genes for a given biochemical pathway are closely linked but also that their order on the chromosome parallels the order in which the enzymes are used in the biosynthetic sequence (Figure 12.21). It was found, furthermore, that all of the enzymes of a single gene cluster were affected simultaneously by induction or repression. These related observations were of considerable importance in the development of the *operon* concept, explaining not only how enzyme induction or repression can be brought about, but also how the synthesis of a series of related enzymes can be simultaneously controlled. Recall the operon model (Section 7.6): A repressor protein, produced by a regulatory gene, combines with the effector and is then able to attach to the operator region and prevent the synthesis of the mRNA. There is good evidence that, in the case of a series of clustered genes, a single mRNA is formed, which copies the information of the whole series of genes. From such a polygenic mRNA is formed a polyribosome upon which all the enzymes of the cluster are synthesized. The operon model also requires that synthesis of mRNA (gene transcription) begin at the operator region and proceed along the gene cluster since, if synthesis started at the end of the cluster opposite from the operon, there would of course be no operator control. This implies that a *polarity* in transcription should exist, the region near the operator always being transcribed before the more distant regions. Such polarity has actually been observed. Mutations in the region near the operator often affect the synthesis of proteins coded for by the more distant regions, whereas mutations in the latter regions have no effect on the function of genes in the region near the operator.

Further evidence for the correctness of the operon model is that it is possible to isolate two kinds of mutants, one defective in the repressor protein because of mutation of the regulatory gene, and the other defective in the operator locus itself. Both kinds of mutants behave as constitutives (see Chapter 7); that is, the enzymes are synthesized irrespective of the presence or absence of the effector (inducer or repressor). These two kinds of constitutive mutants can be distinguished genetically by how they behave when in *cis* or *trans* position (Figure 12.22). The regulatory gene

Figure 12.22 Genetic evidence for the operon model and the repressor gene. (a) Regulatory gene R⁺ is in cis position in relation to the operon; R⁻ gene is on a separate piece of double-stranded DNA, the F particle. (b) Regulatory gene R⁺ is on the F particle, in trans position to the operon. Regulatory gene functions in both cis and trans position. Both phenotypes are repressible by the action of the repressor on the operator. (c) Operator constitutive Oᶜ gene is in cis position to regulatory gene R⁺ and prevents repressor from functioning. Phenotype is constitutive. (d) Operator constitutive Oᶜ gene is on the separate F fragment, in trans position to the regulatory gene R⁺ and does not prevent the action of the repressor on the operon. Phenotype is repressible.

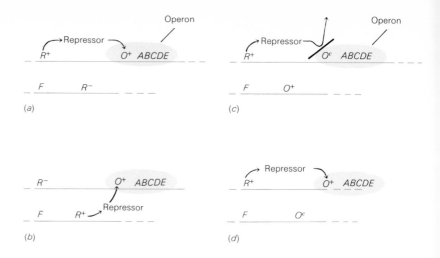

functions in both *cis* and *trans* positions because it causes the production of a cytoplasmic substance, the repressor protein; the operator gene, however, functions only in the *cis* position because its role is to serve as the initiation point for synthesis of the mRNA for the gene cluster. Mutants of the operator gene that are of this type are called "O^c" (operator closed or constitutive) mutants. Such mutants always map genetically in a small region of the DNA at one end of the gene cluster.

We therefore see that related genes might occur in clusters in order to make it possible to control the synthesis of all the enzymes of this cluster by one operator. The operon concept is of great interest to scientists studying how metabolic activities of cells are controlled. Attempts have also been made to apply the operon concept to eucaryotic organisms, but in general these have had little success because related genes of eucaryotes do not usually map in clusters, but are scattered widely on nonhomologous chromosomes (see Chapter 13).

Genetic recombination within a single gene   The high resolution of genetic analysis in bacteria makes it possible to obtain recombination even between two mutants of the same gene so long as they are not at identical sites. For instance, if two nonidentical mutants in the same gene are crossed, wild-type recombinants can be obtained because crossovers occur between the two mutant sites. Such wild-type recombinants, though possible, usually are rare because of the great improbability of crossovers between such closely linked sites. The fact that they occur at all is of considerable importance since it provides a genetic means of

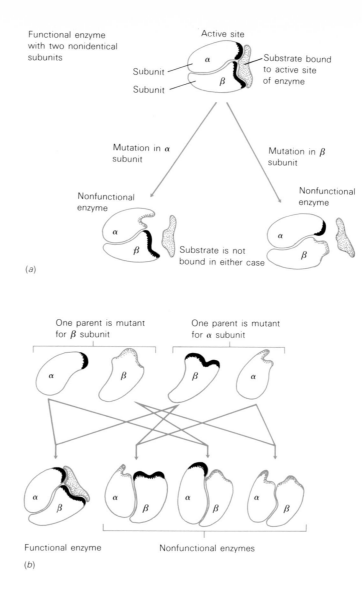

Functional enzyme with two nonidentical subunits

Subunit

Subunit

Active site

α

β

Substrate bound to active site of enzyme

Mutation in α subunit

Mutation in β subunit

Nonfunctional enzyme

α

β

Substrate is not bound in either case

Nonfunctional enzyme

α

β

(a)

One parent is mutant for β subunit

One parent is mutant for α subunit

α β β α

α β α β α β α β

Functional enzyme

Nonfunctional enzymes

(b)

*Figure 12.23   Molecular mechanism of gene complementation. (a) **Effect of mutation on function of enzyme.** (b) **Association of subunits in hybrid cell formed from cross between cells mutant for α and β subunits.***

estimating the sizes of genes. By isolating many mutants for the same gene and showing that these mutants are able to recombine with one another, it is possible to show that mutations occur at many sites within the gene. These findings have made it possible to relate genetic studies on the gene to molecular studies on the DNA molecule.

12.5   Use of genetic analysis in studies on gene structure and function

Complementation by means of mutant genes In what manner will two independent mutants of the same gene interact when they are together in the same cell? A bacterial cell is usually genetically haploid, with only one kind of gene complement, but under certain conditions it is possible to obtain partially diploid cells. Partial diploids can result from abortive transduction, in which the incoming DNA fragment is maintained in unintegrated form along with the complete recipient DNA. Cells containing unintegrated *F'* particles with their attendant host genes are also partial diploids.

Suppose we have two mutants deficient in the same enzyme. If these two mutant genes are placed in the same cell on separate particles, we might expect that this partial diploid would still behave as a mutant. However, on some occasions when this experiment is done, a wild-type phenotype appears, and the two mutations are said to *complement* each other. How are we to explain this complementation? Many enzymes are complex proteins composed of two or more polypeptide chains that can be synthesized independently of each other. One mutant may be capable of synthesizing one of these peptides, and the other mutant may be able to make the other. When two of these mutant strands coexist in the partial diploid, each can direct the synthesis of one normal polypeptide, and these can then associate in the cytoplasm to reconstitute a functional enzyme (Figure 12.23). Complementation of this type is common, presumably because of this diversity of polypeptide chains. Further, it is not essential that the two polypeptides be different for complementation to occur. Many proteins are composed of two (or more) identical subunits, but if two independent mutant genes are present in a partial diploid, each producing a defect in the polypeptide chain at a different place, the two defective polypeptide chains can sometimes associate to form a partially functional hybrid protein.

12.6 Summary

Although the discipline of genetics developed in the early 1900s, procaryotic genetics has been under study only since the late 1940s. This late start can be explained partly by the small size of procaryotic cells, making an analysis of recombinational events at the cellular level difficult, and partly by the fact that recombination in procaryotes frequently involves fragmentary transfer of genetic material by processes quite distinct from the sexual process of eucaryotes. Despite the late beginning of procaryotic genetics, it has developed rapidly and has provided most of the evidence for an interpretation of genetic phenomena at the molecular level. Probably the two most outstanding discoveries of procaryotic genetics have been the processes of transformation and transduction. The demonstration that purified DNA could transfer genetic traits from one organism to another proved conclusively that DNA is the genetic material, and the proof that viruses can function in genetic transfer has expanded our concept of the nature of a virus. Because of the high resolution of genetic recombination in procaryotes, it has been possible to carry out detailed genetic fine-structure studies, which have given us our best idea of the sizes and linkage relations of genes. Procaryotic genetics has also permitted a clear definition of the relation of gene to protein. Results of all these studies have profoundly influenced our concepts of the nature, function, and control of genes. Although it has not been so easy to carry genetic studies in eucaryotes to the molecular level as it has in procaryotes, sufficient work has been done to suggest that the underlying mechanisms are similar. As we shall see in the next chapter, however, genetic processes in eucaryotes involve some features not seen in procaryotes, and these arise at least partially from the increased structural complexity of the eucaryotic cell.

Supplementary readings

Adelberg, E. A. (ed.)   *Papers on Bacterial Genetics,* 2nd ed. Boston: Little, Brown and Company, 1966.

Braun, W.   *Bacterial Genetics,* 2nd ed. Philadelphia: W. B. Saunders Co., 1965.

Davis, B. D., R. Dulbecco, H. N. Eisen, H. S. Ginsberg, and W. B. Wood, Jr.   *Microbiology.* New York: Harper & Row, Publishers, 1967.

Dove, W. F.   "The genetics of lambdoid phages." *Ann. Rev. Genet.* 2:305 (1968).

Hayes, W.   *The Genetics of Bacteria and Their Viruses,* 2nd ed. Oxford: Blackwell Scientific Publishers, 1968.

Jacob, F., and E. L. Wollman   *Sexuality and the Genetics of Bacteria.* New York: Academic Press, Inc., 1961.

Ozeki, H., and H. Ikeda   "Transduction mechanisms," *Ann. Rev. Genet.* 2:245 (1968).

Schaeffer, P.   Transformation," in I. C. Gunsalus and R. Y. Stanier (eds.), *The Bacteria A Treatise on Structure and Function,* vol. V, p. 87. New York: Academic Press, Inc., 1964.

Spizizen, J., B. E. Reilly, and A. H. Evans   "Microbial transformation and transfection," *Ann. Rev. Microbiol.* 20:371 (1966).

Genetic recombination in eucaryotes differs in many ways from that in procaryotes. The complex nuclear organization of eucaryotes and the existence in each nucleus of a number of chromosomes lead to more regular mechanisms of gene assortment and segregation. Thus, genetic recombination in eucaryotes rarely results from fragmentary processes such as were described for procaryotes in the last chapter, but instead is more precise and predictable. Although the word "sex" can be applied to bacteria only in a very broad sense, it is used quite aptly to describe the genetic recombinational process of eucaryotic microorganisms.

13.1 Alternation of generations

It will be recalled (Section 3.6) that a chromosome is composed of a DNA molecule to which are attached variable amounts of protein and RNA. Chromosomes are variable in length throughout the cell-division cycle, from the long, thin structures of interphase to the short, thick forms of metaphase. Different eucaryotic organisms have different numbers of chromosomes, but they always have more than one. Each chromosome of the set can usually be distinguished from all others by its morphology. Eucaryotic organisms show an alternation of generations; in the *haploid* phase the number of chromosomes per cell is called *n*, and the *diploid* phase, 2*n* (Figure 13.1). Multicellular plants and animals are usually diploid, with the haploid phase present only in the germ cells (sperm and eggs), whose life span is transitory. In eucaryotic microorganisms an alternation of generations also occurs, and structures equivalent to sperm and eggs are formed, but the extent of development of the haploid and diploid phases varies. In many species the predominant vegetative phase is haploid, and the diploid stage is transitory; in others the diploid phase dominates; and in still others both haploid and diploid phases have independent vegetative existences.

Two copies of each gene are present in diploids, one on each chromosome, and if these genes differ (probably as a result of mutation), they are called *alleles*. When two alleles are present in the same cell, the expression of one may be masked by the other. The gene that is expressed is called *dominant* to the other allele, which is called *recessive*. Genetic analysis is easier in organisms whose vegetative phase is haploid because each cell contains only one copy of each gene, and gene expression is therefore unaffected by dominance or recessiveness.

Meiosis   Meiosis, or "reduction division," is the process by which the change from the diploid to the haploid state is brought about. Whereas

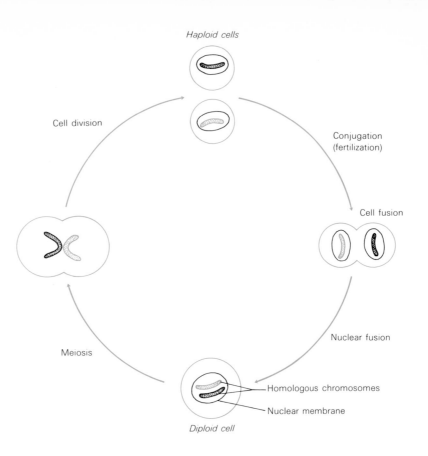

Figure 13.1 *Alternation of haploid and diploid generations in eucaryotes.*

mitosis (Section 3.6) causes duplication and doubling of the chromo-some number and yields two progeny cells with the original chromosome number, the process of meiosis reduces the number of chromosomes by one-half, from 2*n* to *n* and yields haploid cells that are precursers of germ cells. The cytological events of meiosis are well established and will be described briefly here. In the diploid cell there are two of each kind of chromosome; these pairs are known as homologues. The first step in the meiotic process is the *pairing* of these homologues at the center of the cell in the stage called *prophase* (Figure 13.2). The paired chromosomes seem to be reciprocally attracted to each other and usually show a high specificity in pairing. The chromosomes then shorten and thicken, their mutual attraction suddenly subsides, and they begin to pull apart, remain-ing attached at only a few points. At this stage each homologue appears to have replicated and the four-stranded structure that results is called a *tetrad.* It is at this stage that exchange of genes between chromosomes

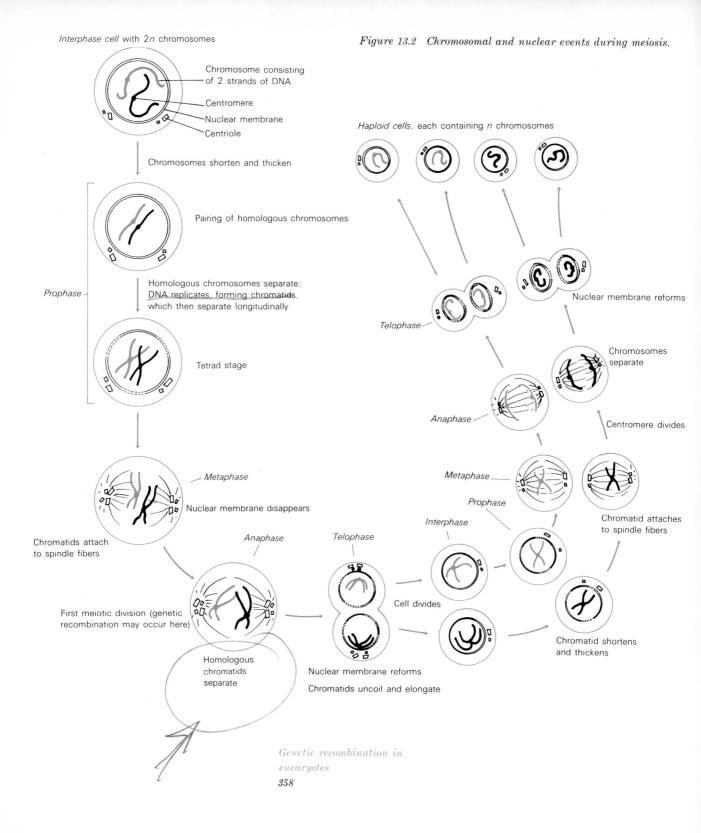

Interphase cell with 2n chromosomes

Chromosome consisting of 2 strands of DNA

Centromere

Nuclear membrane

Centriole

Chromosomes shorten and thicken

Figure 13.2 Chromosomal and nuclear events during meiosis.

Haploid cells, each containing n chromosomes

Prophase

Pairing of homologous chromosomes

Homologous chromosomes separate; DNA replicates, forming chromatids, which then separate longitudinally

Tetrad stage

Nuclear membrane reforms

Telophase

Chromosomes separate

Anaphase

Centromere divides

Metaphase

Metaphase

Nuclear membrane disappears

Prophase

Interphase

Chromatid attaches to spindle fibers

Chromatids attach to spindle fibers

Anaphase

Telophase

Cell divides

First meiotic division (genetic recombination may occur here)

Chromatid shortens and thickens

Homologous chromatids separate

Nuclear membrane reforms

Chromatids uncoil and elongate

(crossing over) probably occurs. At the next stage, *metaphase*, the nuclear membrane breaks down and a spindle apparatus forms, to which the centromeres of the divided chromosomes become attached. The centromere of each chromosome has not yet divided, however, and the duplicated chromosomes or *bivalents* migrate to the opposite poles at *anaphase*. The *first meiotic division* ends when a nuclear membrane reforms around both sets of bivalents and each of the new cells has only one-half the number of chromosomes of the original diploid. The *second meiotic division* is more like a regular mitotic division. The chromosomes, which had already divided in the first prophase, remain attached to each other by the centromeres. They line up on the metaphase plate, the centromeres divide, and half of the chromosomes of each type migrate to each pole. In the end four cells are formed, each containing a haploid number of chromosomes; these cells are the forerunners of the sperm or egg cells.

Genetic consequences of meiosis   If the diploid homologues are genetically identical, that is, if the cell is *homozygous*, the haploid cells formed by meiosis are also identical; but if the diploid is *heterozygous* for genes on one or more chromosomes, the four haploid cells will not be genetically identical. The precise genetic constitution of the haploid cells will depend on events occurring during the meiotic process itself. As shown in Figure 13.3(*a*), if the two homologous chromosomes differ by a single gene, their segregation results in the formation of haploid cells of two different genetic constitutions: For nonallelic genes, two kinds of segregation are possible—independent assortment or linkage [Figure 13.3(*b*) and (*c*)]. *Independent assortment* occurs when the dissimilar genes are located on different chromosomes; *linkage* occurs when they are on the same chromosome. In fact, all the genes present on one chromosome are linked and assort together but independently of the genes of other chromosomes, and it is generally the case that there are as many linkage groups as there are chromosomes.

Study of gene segregation at meiosis shows that *genetic recombination* also occurs between alleles on homologous chromosomes (also called ''crossing over''; see Figure 13.4), bringing together on one chromosome genes that were on separate although homologous chromosomes, so that they now segregate together. The actual frequencies of genetic recombination between two genes will be determined by how closely they are linked, since if they are closely located it is more likely that they will segregate together and recombination will be less frequent. Measuring the frequency of recombination between genes gives us a means of actually mapping the distances between genes on the chromosomes; mapping procedures are discussed in Section 13.3. Extensive genetic study of a

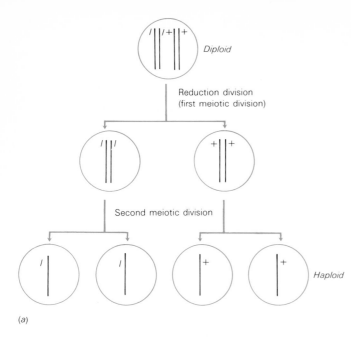

(a)

Figure 13.3 Gene segregation during meiosis: (a) segregation of a single gene (l) and its wild-type allele (+); (b) independent assortment during meiosis of unlinked genes l and m, on non-homologous chromosomes; (c) linked segregation of two genes on homologous chromosomes: (left) two mutant genes on separate chromosomes and (right) two mutant genes on the same chromosome.

wide variety of eucaryotic organisms has shown conclusively that genetic recombination occurs during meiosis at the time of chromosome pairing (Genetic recombination can also occur during mitosis, but this is rare. See Section 13.3.)

The haploid nuclei formed as a result of meiosis become incorporated into cells that are involved in the continuation of the species. These cells are called *gametes* if they participate directly in the fertilization process. In many microorganisms the haploid cells grow vegetatively for a time before becoming converted into gametes. In other microorganisms the initial haploid cells differentiate into *spores,* which often are dormant structures that play a role in the survival of the organism in nature. These spores germinate to become the forerunners of gametes or of haploid vegetative cell lines that later produce gametes. Great variations in these processes exist among different microorganisms; some of these are described later in this chapter and in Chapter 19. Despite variations, the result is the same—an alternation of generations from diploid to haploid to diploid.

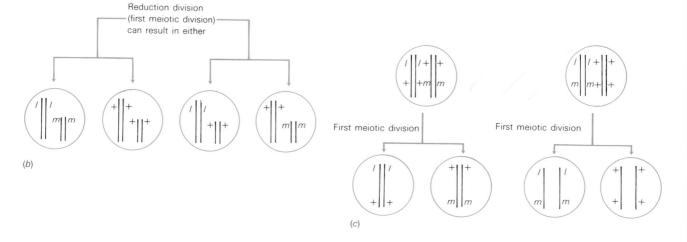

(b)

(c)

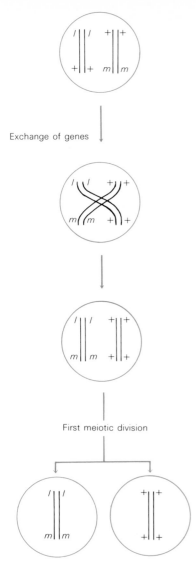

Exchange of genes

First meiotic division

Figure 13.4 Recombination between two genes that are on homologous chromosomes.

13.2 Sexuality

Eucaryotic microorganisms vary widely in the degree of sexual differentiation which they exhibit. At one extreme are the many fungi and algae that show *sexual dimorphism,* that is, the production of morphologically distinct males and females. In the most general view, a *male* is defined as an organism that produces a gamete that moves in some way towards a female gamete, the latter remaining more or less passive. Sexual dimorphism is often genetically determined; a gene or group of closely-linked genes controls maleness, and another set controls femaleness. Upon segregation at meiosis, the male genes are incorporated into one set of haploid nuclei and the female genes into another; each then produces the male or female plants. In botanical terms, organisms that show sexual dimorphism would be called *dioecious.*

A large number of microorganisms exhibit a kind of sexuality called *heterothallism.* Two genetically distinct kinds of plants (or thalli) are produced, but these are morphologically and physiologically identical and can be distinguished only by the fact that each one produces gametes that mate with the opposite type but not with its own type. Since the two plants are morphologically identical, they can hardly be called "male" and "female," and arbitrary designations such as *plus* and *minus* are usually used. In some microorganisms sexual interaction can occur between two cells derived from the same clone; this is termed *homothallism.* In these cases, either the genes controlling expression of both sexes are present in a single set of chromosomes or the gamete has two haploid nuclei, each containing the genes for one sex. Homothallism is important in making possible strict inbreeding.

Sexual differentiation, whether based on simple heterothallic mating systems or on more complex sexual differentiation (see Chapter 19), leads to *self-incompatibility* and prevents inbreeding. Sexual reproduction in self-incompatible organisms results in hybridization between genetically diverse strains and thus in the formation of new genotypes. Such outbreeding organisms are probably more adaptable to new environments than are homothallic organisms.

13.3 Genetic analysis in Neurospora and related organisms

More is known about the genetics of the fungus *Neurospora* than about any other eucaryotic microorganism. Early studies (1941 to 1950) on its biochemical genetics provided the foundation for the whole field of molecular genetics and inspired many theories that were later proved by studies on bacteria and viruses (see Chapters 10 to 12). In this section, however, we are concerned with genetic studies on *Neurospora,* which reveal some of the differences between genetic mechanisms of eucaryotes and those of procaryotes.

Life cycle of Neurospora   The life cycle of *Neurospora crassa* is given in Figure 13.5. Two mating types, designated *A* and *a,* grow vegetatively as typical multinucleate, filamentous fungi and form asexual spores, called *conidia,* which can initiate new mycelia. Either haploid mating type during vegetative growth forms a structure called a *protoperithecium,* which is the forerunner of the sexual structure. If fertilization does not occur, these protoperithecia cease developing at an early stage, but upon fertilization they develop into mature *perithecia.* Mating involves the transfer of a conidium or a piece of mycelium to the tip of a slender hypha, called a *trichogyne,* which projects from the protoperithecium. Fusion occurs between the trichogyne and the fertilizing cell, and nuclei of the latter move down the trichogyne and into the protoperithecium. Nuclei of the two mating structures then become associated in pairs and divide synchronously, and at the same time special *ascogenous hyphae* begin to grow. Pairs of unlike nuclei move into the ascogenous hyphae, and each hypha forms a hook at its tip. Septa then develop, dividing the hypha into three cells. Nuclear fusion occurs in the middle cell at the curve of the hook, and a diploid nucleus is formed (this is the only diploid stage in the life cycle of *Neurospora*). The diploid nucleus undergoes meiosis (reduction division) almost immediately, producing four haploid nuclei. At the same time, the cell undergoing meiosis elongates rapidly, forming a long tubelike cell called the *ascus.* Each of the haploid nuclei in this cell divides once mitotically, so that eight nuclei are present in the ascus; each of these is the forerunner of a dark-pigmented, heavy-walled spore, the *ascospore.* During formation of the ascus and ascospores, the protoperithecium enlarges and matures into the dark-pigmented, thick-walled perithecium. When the latter body ruptures, the many asci contained in it are extruded (Figure 13.6).

Tetrad analysis   Meiosis in *Neurospora* usually occurs in such a way that the spindle is aligned with the length of the ascus. The four strands (tetrads) of metaphase I of meiosis (see Section 13.1) are not distributed at random, but the two chromatids of each chromosome are held together at anaphase I and separate only at anaphase II. Thus two nuclei in one-half of the ascus contain chromosomes derived from one of the original homologues, and the other two nuclei contain chromosomes derived from the other homologue. After the mitotic division, the eight spores are

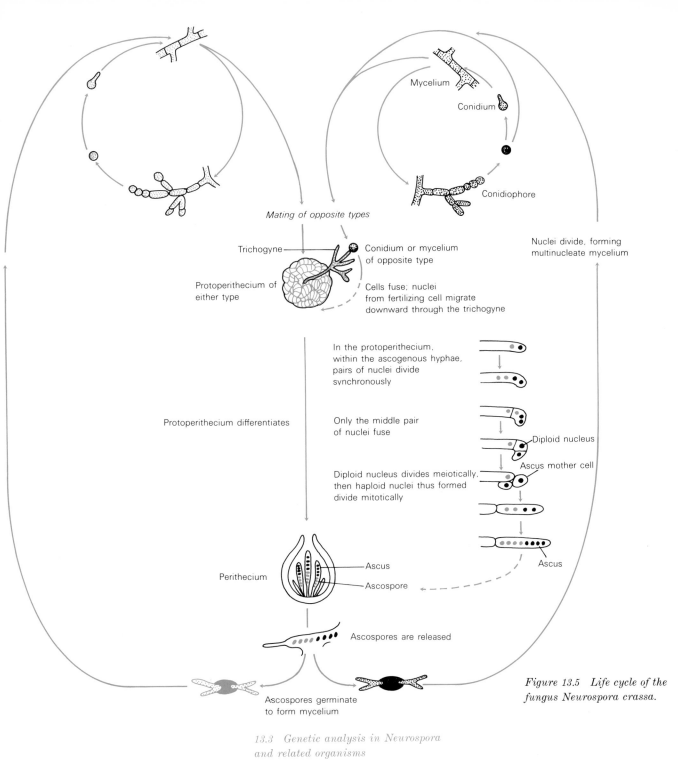

Mycelium

Conidium

Conidiophore

Mating of opposite types

Trichogyne

Conidium or mycelium
of opposite type

Protoperithecium of
either type

Cells fuse; nuclei
from fertilizing cell migrate
downward through the trichogyne

Nuclei divide, forming
multinucleate mycelium

In the protoperithecium,
within the ascogenous hyphae,
pairs of nuclei divide
synchronously

Protoperithecium differentiates

Only the middle pair
of nuclei fuse

Diploid nucleus

Ascus mother cell

Diploid nucleus divides meiotically,
then haploid nuclei thus formed
divide mitotically

Ascus

Perithecium

Ascus

Ascospore

Ascospores are released

Figure 13.5 Life cycle of the
fungus Neurospora crassa.

Ascospores germinate
to form mycelium

13.3 Genetic analysis in Neurospora
and related organisms

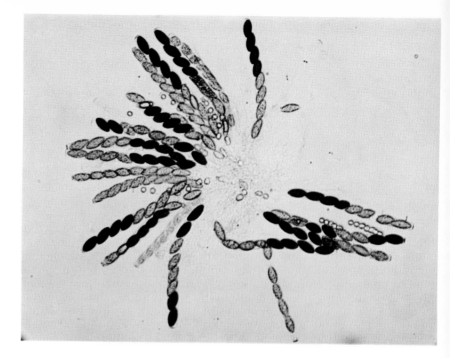

Figure 13.6 Photomicrograph of N. crassa asci from a squashed perithecium. Each ascus contains eight ascospores. Ordinarily all ascospores are dark, but these derive from a cross between wild type and a mutant that produces unpigmented ascospores. Note that the segregation of mutant and wild-type alleles is readily seen by inspection of the order of ascospores in the ascus. [From D. R. Stadler: Genetics 41:528 (1956).]

therefore arranged in groups of four. If the two original homologous chromosomes differed by a single gene, this would then be reflected in the genetic constitution of the eight ascospores in the manner shown in Figure 13.7(a); the four ascospores at one end would be of one genetic constitution, and the four at the other end would be of the other constitution. If recombination occurred at some point between the gene and the centromere, in the manner illustrated in Figure 13.7(b), a different order of the ascospores would result. The ascospores are seen to occur in an ordered sequence, which reflects the events that occurred during the meiotic divisions. The farther a gene is from the centromere, the more likely it is that recombination will take place. This likelihood is reflected in the frequency with which the genetically different spores occur in certain orders. That is, the frequency with which a wild type and its mutant gene occurs in the order 2:2:2:2, 2:4:2, or 4:4 or some other order can be used to map the gene in relation to the centromere. By carefully removing the spores in correct order from the ascus and establishing a separate culture from each spore, one can determine the genetic constitution of each spore. This process of genetically analyzing each spore and determining the frequency with which each genotype occurs in a certain order is called *tetrad analysis*. Although such a procedure can be carried out in many microorganisms, it is not ordinarily possible in multicellular organisms because all four products of mieosis are not usually available. Genetic mapping can also be done if the spores do not remain ordered after meiosis. Mapping unordered tetrads is somewhat more complicated and will not be discussed here.

When more than one gene is segregating, procedures similar to tetrad analysis can be used to determine if two genes are linked, that is, if they are on the same chromosome. For instance, in the cross $a^+ \times b^+$, if a and b are on the same chromosome they should segregate together, whereas if they are on separate chromosomes they should segregate independently. These results are shown diagrammatically in Figure 13.8. Once linkage is established, the distance between two genes can be estimated from the frequency of genetic recombination.

Neurospora crassa has a haploid chromosome number of seven, as was determined by cytological studies at meiosis (Figure 13.9) and also has seven linkage groups as determined by genetic analysis (Figure 13.10). Those chromosomes that appear to be shortest cytologically also seem to have the smallest number of genes, as might be expected.

One important result of genetic mapping in *Neurospora* deserves comment. Recall that one line of evidence in favor of the operon concept (see Section 12.5) was that in bacteria the genes controlling the enzymes

Figure 13.7   Arrangement of ascospores in the ascus as a result of meiosis: (a) without genetic recombination; (b) with genetic recombination at a point between the gene and the centromere.

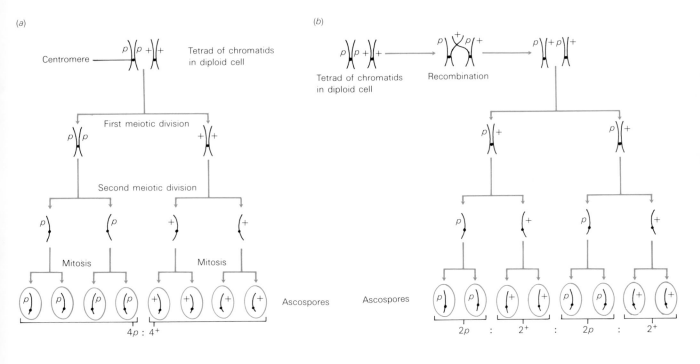

(a)

(b)

13.3   *Genetic analysis in Neurospora and related organisms*

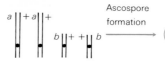

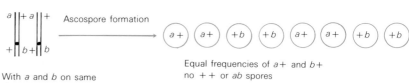

Ascospore formation

Equal frequencies of $a+$, $++$, $+b$, and ab

Markers are not linked

With a and b on separate nonhomologous chromosomes

(a)

Ascospore formation

Equal frequencies of $a+$ and $b+$ no $++$ or ab spores

Markers are linked

With a and b on same chromosome

(b)

for a specific biochemical pathway are often quite closely linked. Yet in *Neurospora,* even though similar sets of genes control the synthesis of the enzymes in these biochemical pathways, there is no evidence for close linkage of the genes; indeed, different genes are found scattered on different chromosomes. In other eucaryotic microorganisms, such as the yeast *Saccharomyces cerevisiae* and the fungus *Aspergillus nidulans,* as far as the analysis has been carried, the same situation holds true. Thus genes that are closely linked in bacteria may be widely separated in eucaryotes.

Complementation tests The phenomenon of gene complementation was discussed in Section 12.5, and we saw how gene complementation can provide insight into the manner in which protein subunits produced by different genes might interact. In bacteria, complementation tests are not easy to apply because it is difficult to arrange conditions so that both genes are present in separate DNA molecules in the same cytoplasm. Complementation tests in fungi such as *Neurospora* are easy to perform through the phenomenon of heterocaryosis. Fusions between genetically distinct hyphae even of the same mating type occur readily, leading to the formation of heterocaryons. If such heterocaryons are formed between two genetically marked strains, the nuclei mingle but do not fuse (except rarely, see below). The gene products of the nuclei can then interact in the cytoplasm. Through such investigations it has been possible to study the association of subunits of two mutant proteins, neither of which is enzymatically functional by itself.

Mitotic recombination Although genetic recombination in eucaryotes takes place normally only at meiosis, under certain special conditions recombination in some fungi can take place during mitosis. In hetero-

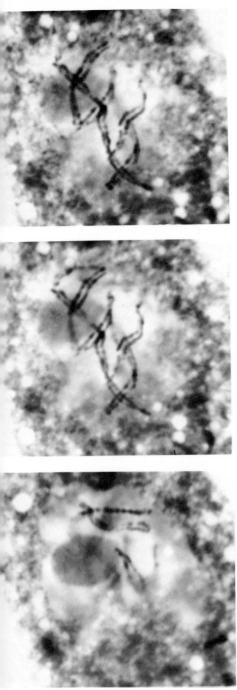

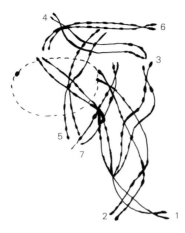

Figure 13.9 Photomicrographs at several focal planes through an orcein-stained nucleus of N. crassa undergoing meiosis and an interpretive drawing of the arrangement of the chromosomes. Note that the seven chromosomes have already paired. The large body in the lower photograph is the nucleolus. Magnification, 3,000×. Compare with Figure 13.10. (Courtesy of Edward G. Barry.)

caryons haploid nuclei occasionally fuse, and with special selection techniques one can isolate strains containing such diploid nuclei. Sometimes haploidization takes place in these strains, with some of the nuclei in the hypha reverting to the haploid state. If crossing over occurs before haploidization, the nuclei will be recombinant for the genes involved in the crossover event. This sequence of events (heterocaryon ⟶ nuclear fusion ⟶ mitotic crossing over ⟶ haploidization) has been termed the *parasexual cycle*. Mitotic recombination facilitates genetic analysis and chromosome mapping for some microorganisms that lack sexuality. Although it is probably of little importance in ensuring genetic recombination in sexually reproducing organisms, it may be of considerable importance to a number of fungi (called *imperfect* fungi) that do not reproduce sexually. The fungus *Penicillium chrysogenum,* which is used commercially in the production of penicillin, has a parasexual cycle that has been exploited to breed strains yielding higher amounts of penicillin. Evidence for mitotic recombination has not been found for *Neurospora.*

13.4   Cytoplasmic inheritance

Although the evidence is overwhelming that most characteristics of eucaryotic organisms are determined by genes located on the chromosomes, there is strong support for the theory that at least some characteristics are determined by genetic elements that replicate outside the nucleus and independently of its control. The best examples of this are the inheritance of mitochondrial and chloroplast characteristics; evidence for

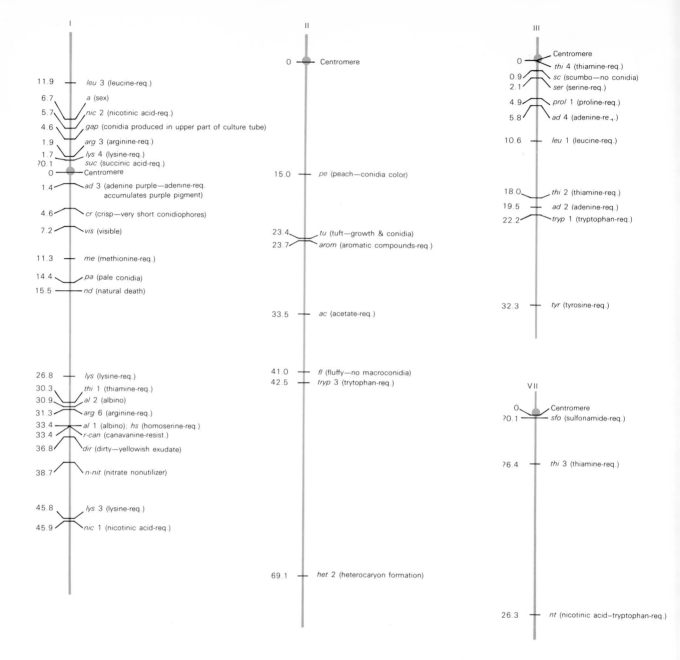

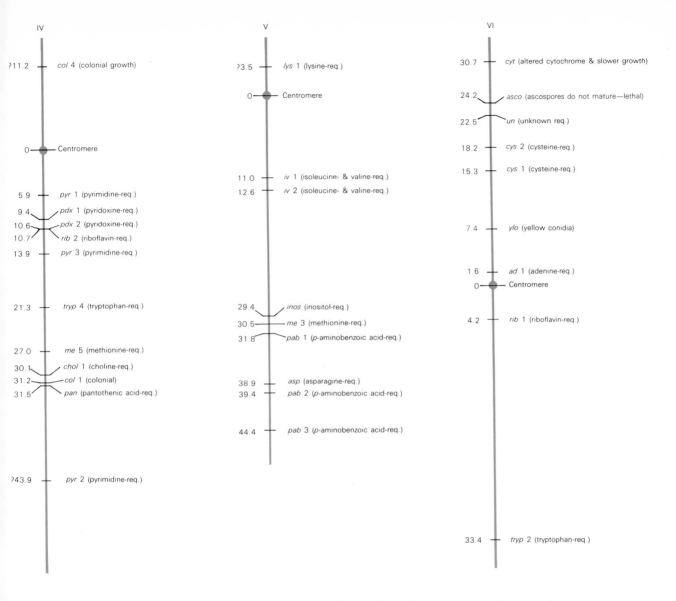

Figure 13.10 Maps of the positions of genes on the seven chromosomes of N. crassa. Many more genes than those listed have been mapped. Note that genes controlling enzymes of the same biochemical pathway are often not even on the same chromosome. This map should be contrasted with that of Escherichia coli in Figure 12.20. [From J. A. Serra: Modern Genetics, vol. I. London: Academic Press (London) Ltd., 1965.]

autonomous inheritance in other organelles has also been advanced. Although cytoplasmic inheritance has been known for many years in higher plants, it has been analyzed best (and most easily) in eucaryotic micro-organisms.

Mitochondrial inheritance in yeast   Recall that a mitochondrion is a subcellular organelle that contains the biochemical machinery for the generation of ATP through oxidative phosphorylation. Mitochondria are relatively complex membrane-bound organelles, several of which are usually present in the cell. There is some microscopic evidence that they replicate by binary fission.

Some of the clearest evidence of independent inheritance of at least some of the characteristics of mitochondria comes from studies on baker's yeast. The life cycle of a typical yeast is shown in Figure 13.11. After meiosis, the four haploid nuclei become the forerunners of spores, called *ascospores,* surrounded by a saclike structure, the *ascus.* Each of the four ascospores from a single ascus can be removed and allowed to germinate, establishing a haploid clone. Most yeasts are heterothallic and do not show sexual dimorphism, so that as long as each clone is kept separate only vegetative growth takes place. When two clones of opposite mating type are mixed, however, the cells fuse and form diploids.

In yeast cultures about 1 out of 500 cells gives rise to a colony that is smaller than normal. These smaller colonies, called *petites* (from the French word for "small"), are unable to utilize oxygen as a terminal electron acceptor and behave as anaerobes even under aerobic conditions. They presumably form smaller colonies because they synthesize a limited amount of ATP, which they can produce only by the glycolytic pathway rather than by the more efficient pathway of oxidative phosphorylation (see Chapter 4). These *petite* yeast strains also are unable to grow on organic acids such as lactate, and they are deficient in a number of important respiratory enzymes, as well as in key cytochromes (Figure 13.12). *Petite* strains contain mitochondria, as revealed by electron microscopy, but these appear malformed and structurally abnormal.

Petite yeast strains arise spontaneously at low frequency, and their frequency can be increased to close to 100 percent by growing the cells in medium containing an acridine dye. Genetic analysis reveals several kinds of *petites* which are distinguished by their behavior during crosses. One class of *petites*, termed "segregational" or "nuclear" [Figure 13.13(*a*)], behave as if they had arisen through mutation of a nuclear gene. When crossed with wild type, the *petite* and wild-type characters segregate 2 : 2 in the yeast ascus. The other type of *petite,* termed "cytoplasmic" [Figure 13.13(*b*)] shows unusual behavior during mating. All four spores formed as a result of meiosis give rise to normal, non*petite* cultures.

In the cytoplasmic *petites*, therefore, the *petite* character does not segregate at meiosis and this phenotype disappears. We may interpret these results as follows: In the cytoplasmic *petite* there has been no nuclear change, but some alteration has occurred in the inheritance mechanism of

Figure 13.11   Life cycle of a typical yeast, Saccharomyces cerevisiae.

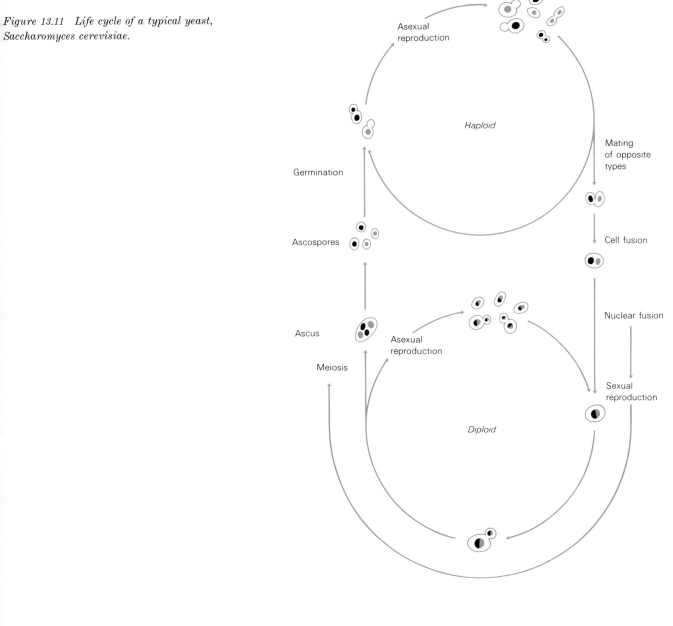

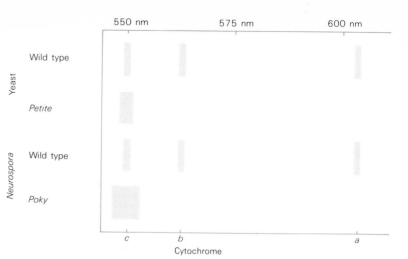

Figure 13.12 Cytochrome absorption bands in a yeast (S. cerevisiae) and in N. crassa. Contrasted are wild type and respiratory-deficient mutants. In Saccharomyces such mutants are called petite; in Neurospora they are called poky. (From R. P. Wagner and H. Mitchell: Genetics and Metabolism, 2nd ed. New York: John Wiley & Sons, Inc., 1964.)

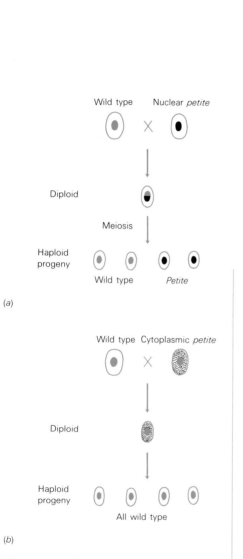

(a)

(b)

Figure 13.13 Outcome of crosses between petite and wild-type yeast: (a) nuclear petite; (b) cytoplasmic petite.

the mitochondria. Upon crossing with wild type, the cytoplasms mingle, and mitochondria from the wild type, which replicate normally, populate the cytoplasms of all spores formed at meiosis. Presumably because they are malfunctioning, the mitochondria of the *petite* are not able to compete with the wild-type mitochondria and are replaced. In the nuclear *petites* the situation is different. Here one should conclude that a nuclear gene controlling the synthesis of an essential mitochondrial component is altered. In the absence of this essential component functional mitochondria cannot be formed, even though their own hereditary machinery is in order. This suggests that some of the characteristics of the mitochondria are controlled by nuclear genes and others by self-replicating mitochondrial genes. Which biochemical defects are altered in the cytoplasmic mutants have not as yet been identified.

Cytoplasmic inheritance in Neurospora A mitochondrial mutation in *Neurospora* is called *poky*. This is a spontaneous mutation characterized by a reduced growth rate compared to that of the wild type, altered cytochromes *a* and *b,* and morphologically abnormal mitochondria. In contrast to yeast, *Neurospora* cannot grow anaerobically; hence some mitochondrial activity must remain if the mutation is not to be lethal. When crossed with wild type, *poky* shows a phenomenon usually associated with a cytoplasmic characteristic—maternal inheritance. Recall that

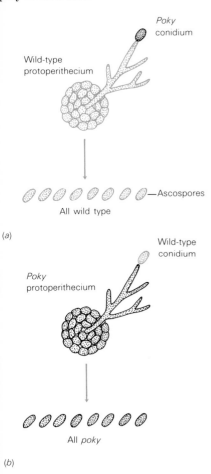

Figure 13.14 Cytoplasmic inheritance of poky in N. crassa.

Poky conidium

Wild-type protoperithecium

—Ascospores

All wild type

(a)

Wild-type conidium

Poky protoperithecium

All *poky*

(b)

in *Neurospora* a cross can be made by fertilizing the protoperithecium of one strain with conidia of the other. If the *poky* strain contributes the protoperithecium, then the characteristic is transferred to all ascospores resulting from the cross, whereas if *poky* contributes the conidia, the characteristic is lost (Figure 13.14). Presumably this is because no mitochondria are present in the conidia, and all the mitochondria of the ascospores are contributed by the strain providing the protoperithecia.

Poky differs in another way from the cytoplasmic *petite* character in yeast discussed earlier. If a heterocaryon is made by fusing *poky* and wild-type hyphae, the heterocaryon is initially normal but becomes *poky* after a number of generations. It appears as if *poky* types have some competitive advantage over wild-type mitochondria and eventually replace them. A similar situation is known for certain *petite* strains of yeast (called ''suppresive'' strains). It is possible by microinjection to insert cytoplasm from *poky* strains into hyphal tips of wild-type strains; eventually *poky* takes over. In a sense, *poky* behaves in the manner of a pathogen. These results show quite clearly the presence of a self-replicating particle in the cytoplasm of *poky* strains. They do not show, however, that this is the mitochondrion itself. It is conceivable that *poky* is a viruslike element that is able to induce pathological alterations in mitochondria.

Do mitochondria exhibit self-replication?   This question has taken on renewed significance in recent years as our understanding of molecular biology and genetics has advanced. Several lines of evidence suggest that mitochondria replicate by themselves: (1) Mitochondria contain DNA which can be seen in thin sections by electron microscopy, and this DNA differs in base composition from that of nuclear DNA. (2) Mitochondria, which synthesize proteins, have 70-S ribosomes, whereas the ribosomes of the cytoplasm and nucleus of eucaryotes are 80 S. That mitochondrial ribosomes differ in size from the cytoplasmic ribosomes of eucaryotes suggests that their protein-synthesizing machinery is under independent control. (3) Such antibiotics as erythromycin and chloramphenicol, which inhibit the replication of bacteria, also inhibit the replication of mitochondria in yeast. Since both bacteria and mitochondria contain 70-S ribosomes—and we know that the target of these antibiotics in bacteria is the ribosome—we can infer that the mitochondrial ribosomes must be functional in order for mitochondria to replicate. We have already noted that treatment of yeast with acridine dyes, which combine with DNA (see Section 11.4), results in the production of *petites* at a much higher frequency than that at which it causes nuclear mutations. (4) The replication of mitochondria has been shown directly in *Neurospora* by use of radioactive choline, which labels specifically the phospholipid of the mitochondria. If mitochondria arose de novo, one would anticipate that mitochondria formed after the radioactive choline was added would be fully labeled, whereas those present before the radioactive substance was added would be unlabeled. If mitochondria arose from preexisting mitochondria, however, one would expect to find the development of a population of mitochondria all of which were only partly labeled. Since the latter result was obtained, we conclude that mitochondria are self-replicating. (5) Finally, mitochondrial division has actually been observed in many animal cells by making time-lapse movies with the phase microscope.

Thus we may conclude that mitochondria divide, but although there is good evidence that they have at least a part of their own genetic apparatus, this does not mean that they are completely autonomous. At least one mitochondrial enzyme in yeast, cytochrome *c,* is coded for by a nuclear gene; it is synthesized in the cytoplasm and then is transferred to the mitochondrion. Microorganisms such as yeast and *Neurospora* are extremely suitable organisms for studying questions of cytoplasmic inheritance. Once the basic picture has been worked out in microorganisms, it should be possible to test whether it applies to higher animals and plants.

Genetics of chloroplasts There is good evidence that at least some characteristics of the chloroplasts in higher plants are inherited cyto-

Figure 13.15 Inhibition of chloroplast replication in Euglena gracilis and the outcome after continued cell division. Any cell that retains one chloroplast can be the forerunner of a green-cell line, but cells that have lost all chloroplasts are permanently bleached. The normal cell has about 10 chloroplasts.

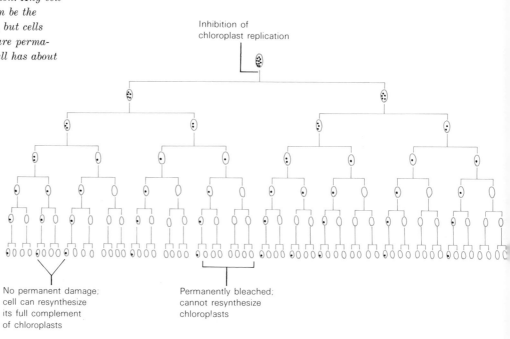

Inhibition of chloroplast replication

No permanent damage; cell can resynthesize its full complement of chloroplasts

Permanently bleached; cannot resynthesize chloroplasts

plasmically. In recent years studies on microorganisms have provided further insight into this type of inheritance. Two organisms have proved most useful, the flagellate *Euglena* and the chlorophycean alga *Chlamydomonas*. Both of these algae are easily cultured and both have the ability to grow either photosynthetically in the light or heterotrophically in the dark; the latter property enables us to study mutants that lack the photosynthetic apparatus.

Recall that chloroplasts are membrane-bound, chlorophyll-containing structures that contain a complex internal lamellar system to which chlorophyll is attached and in which the photosynthetic reactions take place (see Chapters 3 and 6). The size, shape, and number of chloroplasts in different organisms vary. *Euglena* has on the average 10 small chloroplasts scattered throughout the cytoplasm, each about the size of a bacterium; *Chlamydomonas*, on the other hand, has one large cup-shaped structure that nearly fills the cell. Many kinds of treatments can lead to permanent loss of chloroplasts in *Euglena*. In some species, continued growth in the dark leads to a bleached form which, upon return to the light, does not regain its chloroplasts. In other species, to treat chloroplast-containing cells with streptomycin, UV radiation, heat, or antihistamine drugs causes permanent bleaching because of loss of chloroplasts. These agents apparently inhibit chloroplast replication more readily than they do cell replication; continued cell division effects a dilution of the chloroplasts and leads to their eventual loss (Figure 13.15). (The reader might ponder the fact that a permanently bleached *Euglena* resembles not an alga but a protozoan. This consideration has certain implications for our discussions in Chapter 17 of both the taxonomy and the evolution of algae and protozoa.) Once a cell has lost all chloroplasts it remains permanently bleached; if only one chloroplast remains, however, return of the cell to normal environmental conditions leads to the restoration of the normal number. The implication is that the remaining chloroplast was able to divide and repopulate the cell with a complete chloroplast complement. Unfortunately, sexual reproduction is not known in *Euglena;* hence it is not possible to investigate this problem by the genetic means used with mitochondria.

The alga *Chlamydomonas* does have a sexual cycle and shows types of cytoplasmic inheritance which probably involve chloroplasts. The life cycle of *Chlamydomonas reinhardi* is shown in Figure 13.16. The organism shows two mating types and when suspensions of these two types are mixed, the cells form clumps through association of their flagellar tips. After clumping, cells pair in the region of their flagella and fuse to form a broad cytoplasmic bridge. Although the two mating types appear morphologically and physiologically identical, one of the mating types (mt^+) actually becomes dominant over the other (mt^-); for instance, the motility of the conjugating pair is taken over by the flagella of mt^+, and the cytoplasmic inheritance described below shows dominance of mt^+ over mt^-. After nuclear fusion, the fused pair becomes converted into a single thick-walled resting spore, the zygote. Under appropriate conditions the spore germinates, undergoes meiosis, and releases four haploid vegetative cells, each of which is the forerunner of a new cell line.

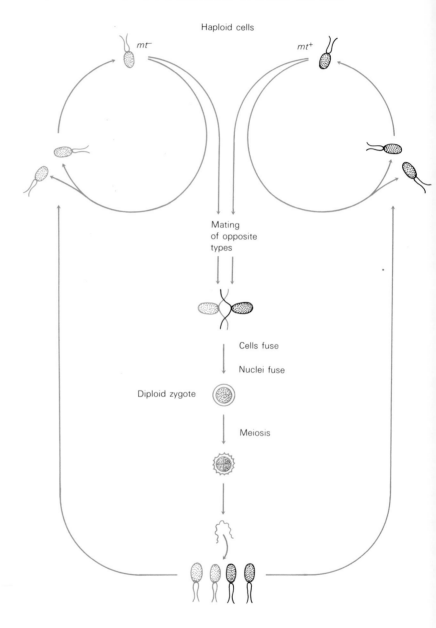

Figure 13.16   Life cycle of the green alga Chlamydomonas reinhardi.

In *Chlamydomonas* it is not possible to eliminate the chloroplast by experimental means, but it is possible to induce mutations in chloroplast genes by treatment with streptomycin. Some of these mutants will have lost the ability to synthesize chlorophyll and are yellow, whereas others will be streptomycin dependent and will retain the normal color. When one of these cytoplasmic mutants is crossed with a wild type, the resulting segregation pattern depends on whether mt^+ or mt^- was the parent containing the mutant phenotype. If the mt^- organism contains the mutant chloroplast, it does not appear in the next generation, whereas if mt^+ contains the mutant chloroplast, it shows up in the next generation, and the wild-type phenotype disappears. We can thus say that the chloroplast mutation shows inheritance through the mt^+ parent and that the chloroplast of the mt^+ parent is the forerunner of the chloroplasts of the next generation; hence the chloroplast controls at least some elements of its own heredity. Not all chloroplast characteristics are cytoplasmic, however. Nuclear mutations, showing typical meiotic segregation patterns, are also known to affect chloroplast traits. Chloroplast inheritance in *Chlamydomonas* is therefore similar to mitochondrial inheritance in yeast.

Biochemical studies on the chloroplasts of *Euglena* and *Chlamydomonas* have shown that in both organisms there is a specific chloroplast DNA that is of different base composition from nuclear DNA. This chloroplast nucleic acid can also be seen by electron microscopy. That it is synthesized in the chloroplast is suggested by the fact that radioactive thymidine becomes incorporated into the DNA of chloroplasts independently of incorporation into nuclear DNA. The algal chloroplasts also have ribosomes, and their size is 70 S like that of procaryotes rather than 80 S like that of cytoplasmic eucaryotic ribosomes. This fact is consistent with the observation that streptomycin, which specifically inhibits protein synthesis on 70-S ribosomes (see Chapter 9), also specifically inhibits chloroplast replication and function. The bleaching of *Euglena* by streptomycin takes place at concentrations of the antibiotic that do not affect growth and division of the *Euglena* cell. Thus it seems as if the chloroplast, like the mitochondrion, has the machinery to carry out its own replication; perhaps it requires the participation of the nucleus only for certain of its functions.

Infective symbiosis There are other extrachromosomal inherited particles that resemble in some ways mitochondria and chloroplasts, which have the additional property of being infective; that is, they are transmissable even in the absence of normal sexual interaction. The best-analyzed cases are for the ciliated protozoan *Paramecium*. The life cycle of *Paramecium* will be discussed in Chapter 19; it suffices to say here that the protozoan conjugates in pairs and that both nuclear and cytoplasmic exchange can take place.

Many *Paramecium* strains have associated with them certain bacteria-like organisms, designated by such Greek letters as *lambda, kappa,* or *mu.* The one we know most about, *kappa,* is a rod-shaped structure about 0.4 μm long and contains DNA, RNA, protein, and probably other normal cellular macromolecules. *Kappa* is difficult or impossible to cultivate in culture media, but it replicates readily within the appropriate strains of *Paramecium aurelia.* The ability of *kappa* to replicate within the protozoan is dependent upon the presence of a dominant gene, *K,* which probably controls the synthesis of an essential nutrient needed by *kappa. Paramecium aurelia* strains can be freed of *kappa* by being grown at high temperatures or by being treated with mutagenic agents or antibiotics such as streptomycin and chloramphenicol. (This should by now be a familiar story; these same agents are capable of affecting replication of mitochondria and chloroplasts.) Strains of *P. aurelia* cured of *kappa* can be reinfected either by addition of a suspension of *kappa* particles or by cytoplasmic exchange with a *kappa*-containing cell during conjugation. How is the presence of *kappa* recognized? It causes the production of a toxic substance, called "paramecin," which kills *P. aurelia* cells that do not contain *kappa;* for this reason *kappa*-containing cells are called *killers.* The presence of *kappa* particles in a cell thus brings about two important changes: The cells produce paramecin, and they become insensitive to paramecin action. In addition to these two dramatic effects, *kappa* probably also has more subtle effects on the nutrition of the protozoan. There is some evidence that bacterialike particles similar to *kappa* produce vitamins essential for *Paramecium.* Also, the presence of the killer trait may enable the *Paramecium* cell to eliminate other related strains of *Paramecium* with which it must compete in nature. Therefore, although *kappa* particles are self-replicating, they depend on the host for their survival. Although resembling mitochondria and chloroplasts in this respect, *kappa* particles are infectious, that is, they are transmissable from cell to cell even in the absence of conjugation.

Another kind of cytoplasmic agent, found in *Paramecium bursaria,* is a unicellular alga of the genus *Chlorella.* This alga can be eliminated by culturing *P. bursaria* in the dark; alga-free *P. bursaria* can then be reinfected with *Chlorella* and the previous state reestablished. *P. bursaria* provides a protected habitat for the growth of the alga, which through photosynthesis provides nutrients for the paramecium, an excellent example of a symbiotic relationship (see Chapter 14).

General comment on nuclear-cytoplasmic relationships From the above discussion, the reader should appreciate the fact that the more complex cellular structure of eucaryotes has led to the development of new types of hereditary mechanisms only imperfectly represented in procaryotes. We can recognize three classes of extrachromosomal hereditary units: (1) Cellular organelles such as mitochondria and chloroplasts, which are virtually essential for the life of cells, and which are completely unable to replicate outside of cells. Other examples of such cell organelles, not discussed above, are centrioles, basal bodies (kinetosomes) of flagella and cilia, the Golgi complex, and perhaps the cell membrane itself. (2) Hereditary symbionts such as *kappa* and *Chlorella,* which are found in only a restricted range of organisms, are not essential for the life of cells (although they may be significant factors in the survival in nature of the organisms containing them), and which can pass from cell to cell either by conjugation or by infection. (3) Infectious agents, such as intracellular obligate parasites, a wide variety of which cause only harm to the cells containing them. These include many important animal and human pathogens, such as the rickettsiae and the psittacosis group. These organisms, discussed in Chapter 18, may be derived from hereditary symbionts by mutation to forms that are generally harmful rather than helpful.

From time to time, attempts have been made to draw analogies between the hereditary particles of eucaryotes and the *F* particles, prophages, and other episomes of procaryotes. It should be clear, however, that these elements have little in common. The hereditary particles of procaryotes are composed mainly or exclusively of DNA, and there is no evidence that they are enclosed in membranes. The hereditary particles of eucaryotes are all complex membrane-bound structures, containing not only DNA but also RNA, ribosomes, lipids, and other biochemicals.

13.5 Comparison of genetic mechanisms in procaryotes and eucaryotes

We might summarize this chapter and Chapter 12 by comparing and contrasting the genetic mechanisms of procaryotes and eucaryotes (see Table 13.1). We have seen how these dissimilarities in genetic mechanisms arise as a result of profound differences in cellular organization between these two groups.

1 *Chromosome organization* The chromosomal material of procaryotes is composed of a single naked DNA molecule, and all of the genes are linked together. The DNA is not organized in a membrane-bound nucleus. In eucaryotes there is always more than one chromosome (the smallest number observed in a microorganism is two, and chromosome numbers greater than 100 are known), the DNA is not naked but is complexed with protein, and the chromosomes are contained in a membrane-bound nucleus. There are probably good molecular reasons

Table 13.1 *Genetic differences between procaryotes and eucaryotes*

	Procaryotes	Eucaryotes
Number of chromosomes or linkage groups	One	More than one
Chromosome composition	DNA	DNA + proteins
Nucleus	Not membrane bound	Membrane bound
Ploidy	Haploid	Haploid or diploid
Genetic exchange	Fragmentary: Transduction Transformation Conjugation	Nuclear fusion: Sexual cycle Parasexual cycle
Cytoplasmic DNA	Plasmids and episomes (not membrane bound)	Mitochondria, chloroplasts, centrioles, kinetosomes (basal bodies), Golgi bodies, microbial symbionts (all membrane bound)
Gametes	Organism itself	Organism itself or specialized meiotic product
Operons	Common	Rare (?)

why eucaryotes cannot carry all their DNA on a single chromosome. The DNA content of eucaryotes is much greater than that of procaryotes (see Table 13.2). As the length of a DNA molecule increases, replication becomes more difficult. Recall (Chapter 2) that in order to replicate, the DNA double helix must unwind. Complete unwinding of a long DNA molecule would take so long that it would be impossible for a cell to divide in any reasonable length of time. By partitioning the DNA in separate chromosomes, each capable of independent replication, this problem is overcome.

2 *Sexual reproduction and genetic recombination* In procaryotes sexual reproduction is a rare process, and even in *Hfr* strains it is fragmentary. Recombination mechanisms in procaryotes involve interaction of naked DNA molecules and probably breakage of the recipient DNA and the insertion into it of the donor fragment. In eucaryotes, sexual reproduction is usually a regular process, involving an alternation from haploid to diploid to haploid generations. Gene assortment and genetic recombination usually occur during a carefully controlled meiotic process, leading to predictable consequences. Mapping of genetic markers on linkage groups shows that the linkage groups are identical to the cytologically observable chromosomes.

Genetic recombination in eucaryotes

Table 13.2   DNA content of microorganisms

	DNA, g/haploid nucleus (logarithmic scale)	Chromosomes, haploid no.
Eucaryotes:		
Alga (*Euglena viridis*)	1.5×10^{-12}	42
Slime mold (*Physarum polycephalum*)	1.0×10^{-12} — 1×10^{-12}	50
Protozoan (*Paramecium aurelia*)	4.4×10^{-13}	38–50 (various
Alga (*Chlamydomonas reinhardi*)	3.4×10^{-13}	16 strains)
Protozoan (*Tetrahymena pyriformis*)	1.4×10^{-13}	5
	1×10^{-13}	
Fungus (*Neurospora crassa*)	4.6×10^{-14}	7
Fungus (*Aspergillus nidulans*)	4.4×10^{-14}	8
Yeast (*Saccharomyces cerevisiae*)	2.2×10^{-14}	>10
	1×10^{-14}	
Procaryotes:		
Bacterium (*Escherichia coli*)	4.3×10^{-15}	1
Bacterium (*Bacillus subtilis*)	3.2×10^{-15}	1
	1×10^{-15}	
Bacterium (*Mycoplasma gallisepticum*)	3.2×10^{-16}	1
	1×10^{-16}	

3 *Extrachromosomal inheritance* Eucaryotes have a variety of sub-cellular membrane-bound organelles, and at least some of the character-istics of these organelles are determined by genetic mechanisms residing in the structures themselves. There seems to be a progression from intra-cellular parasites to hereditary symbionts to essential organelles. The facts that structures such as mitochondria and chloroplasts contain DNA of base composition different from nuclear DNA, have 70-S ribosomes as in procaryotes rather than 80-S ribosomes as in the cytoplasm of eucaryotes, and show inhibition of replication of mitochondria and chloro-plasts by antibiotics that specifically inhibit procaryotes all suggest that these organelles may have arisen from bacterialike hereditary symbionts through mutation. Studies on the genetics of eucaryotic microorganisms have provided some of the best evidence for extrachromosomal inheri-tance.

Supplementary readings

Burdette, W. J. (ed.) *Methodology in Basic Genetics.* San Francisco: Holden-Day, 1963.

Ephrussi, B. *Nucleo-Cytoplasmic Relations in Micro-Organisms, Their Bearing on Cell Heredity and Differentiation.* London: Oxford University Press, 1952.

Esser, K., and R. Kuenen (E. Steiner, trans.) *Genetics of Fungi.* New York: Springer-Verlag, 1967.

Fincham, J. R. S., and P. R. Day *Fungal Genetics,* 2nd ed. Philadelphia: F. A. Davis Co., 1965.

Jinks, J. L. *Extrachromosomal Inheritance.* Englewood Cliffs, N.J.: Prentice-Hall, Inc., 1964.

Mortimer, R. K., and D. C. Hawthorne "Yeast genetics," *Ann. Rev. Microbiol.* 20:151 (1966).

Nanney, D. L. "Ciliate genetics: patterns and programs of gene action," *Ann. Rev. Genet.* 2:121 (1968).

Peters, J. A. (ed.) *Classic Papers in Genetics.* Englewood Cliffs, N.J.: Prentice-Hall, Inc., 1959.

Raper, J. R. *Genetics of Sexuality in Higher Fungi.* New York: The Ronald Press Company, 1966.

Roodyn, D. B., and D. Wilkie *The Biogenesis of Mitochondria.* London: Methuen & Co. Ltd., 1968.

Sager, R. "On non-chromosomal heredity in microorganisms," in *Function and Structure in Micro-organisms* (Fifteenth Symposium, Society for General Microbiology), p. 324. London: Cambridge University Press, 1965.

Strickberger, M. W. *Genetics.* New York: The Macmillan Company, 1968.

part four

Microbial
Ecology

Microbial ecology deals with the natural distribution and environmental relationships of microorganisms and their relationships to each other and to higher organisms. Chapter 14 introduces certain ecological concepts and considers the beneficial relationships of microorganisms to each other and to higher organisms. The concept of symbiosis, a mutual interaction of two organisms beneficial to both, is introduced, and the lichen symbiosis between algae and fungi is presented as an example. The normal microbial flora of the mammal is then considered in some detail, and this is followed by a discussion of germ-free animals and their significance. The role of microorganisms in the digestive processes of cows, sheep, and other ruminants is studied next, and the chapter concludes with a discussion of the relationships of microorganisms to plants, with special emphasis on interactions between microorganisms and plant roots. Chapter 15 brings into consideration some of the harmful effects of microorganisms on higher forms, with special emphasis on mammals and man. The mechanisms that permit microorganisms to invade the animal body and to cause damage are discussed, followed by an analysis of how animals are able both to resist and to develop immunity to microbial attack. Some of the means by which microbial diseases are transmitted, controlled, and prevented are then investigated. Chapter 16 examines microorganisms in aquatic and terrestrial habitats; their roles in transformations of the elements C, N, S, and O are discussed, as well as their action in soil formation and in the cycling of plant nutrients in the soil. The microbiology of water purification and sewage treatment is then considered, the chapter closing with petroleum and geological microbiology.

Microbial species rarely exist alone in nature. When two or more kinds of organisms are present in a limited space, possibilities for interactions exist, which can be beneficial or harmful to one or more of the organisms. When both organisms are benefitted by the association, the relationship is called *mutualistic* or *symbiotic.* When one organism is benefitted and the other is harmed, the relationship is *parasitic.* Occasionally, one of the organisms is benefitted and the other is unaffected; such a relationship is termed *commensalistic.* Finally, if two populations living together have no effect on each other, the relationship is called *neutralistic.* Frequently two organisms will both be dependent on some limiting nutrient or other environmental factor, and they are said to *compete.* In some cases one of the competing organisms actually attacks the other, by producing either an antibiotic or another inhibiting agent, and is said to be *antagonistic.* Some organisms may consume others as sources of nutrients; this is called *predation.* In this chapter we shall consider first the lichen symbiosis, an interaction that occurs between two microbial species, and then we shall discuss interactions between microorganisms and higher forms.

14.1 Lichens

The word *lichen* refers to any regular association of an alga and a fungus in which a functional relationship between the two partners exists. This relationship usually leads to the formation of a plantlike structure of definite shape and morphology called a *thallus,* which can usually be seen with the naked eye. Lichens are widespread in nature, and are often found growing on bare rocks, on tree trunks, on house roofs, and on the surfaces of bare soils (Plate 5: Figure 14.1). They are in many cases the predominant plants of extreme environments, such as deserts and polar regions, where other kinds of plants are not able to grow.

There is a tremendous variety of lichen types found in nature, differing in thallus structure, pigment, and habitat. Some lichen thalli are flat spreading types (*crustose*), some are raised types with definite shapes (*fruticose*), and still others are leaflike (*foliose*). The lichen thallus usually consists of a tight association of many fungus hyphae, within the matrix of which the algal cells are embedded (Figure 14.2). The shape of the lichen thallus is determined primarily by the fungal partner, of which a variety of types associate in lichens. The diversity of algal types in lichens is smaller: Many different kinds of lichen have green algae of the genus *Trentepohlia* or *Trebouxia*. Some lichens have blue-green algae, and these usually fix nitrogen. Although both the alga and the fungus can be cultivated separately in the laboratory, in nature they almost always are found

fourteen

Microbial interactions and symbiotic relationships

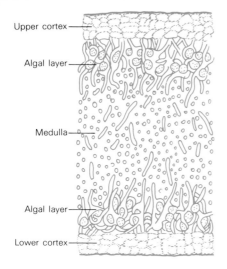

Upper cortex

Algal layer

Medulla

Algal layer

Lower cortex

(a)

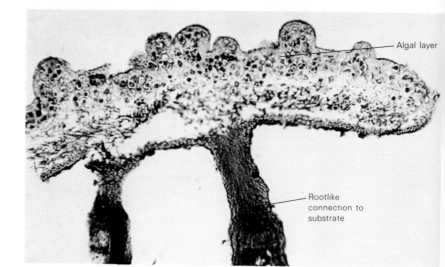

Algal layer

Rootlike connection to substrate

(b)

Figure 14.2   (a) Diagram of a cross section of a fruticose lichen thallus. The algae are localized in layers below the upper and lower surfaces. The rest of the lichen tissue is composed of fungus cells and filaments. (b) Photomicrograph of a vertical section of the fruticose lichen Parmelia conspersa. In this species, only a single algal layer is present below the upper surface and the lichen thallus is attached to its substratum by a rootlike appendage. Magnification, 100×. [Reprinted by permission of publisher from Vernon Ahmadjian: The Lichen Symbiosis. Waltham, Mass.: Blaisdell Publishing Company, A Division of Ginn and Company, 1967.]

living together. The photosynthetic alga produces organic materials from carbon dioxide, water, and other substances in the air, water, and rock; some of these organic compounds are then used as nutrients by the fungus. In many cases the fungus produces special absorptive hyphae, *haustoria,* which penetrate into the algal cells. Even in the absence of haustoria, the intimate contact between algal cells and fungal hyphae promotes nutrient exchange. The fungus thus clearly benefits from the association, but what does the alga obtain? The fungus provides a firm substratum within which the alga can grow protected from the danger of erosion by rain or wind. In addition, the fungus is probably the agent by which the inorganic nutrients needed by the alga are absorbed from the surface on which the lichen is growing.

Lichens are remarkably resistant to desiccation, heat, and cold; indeed, lichens often live in habitats exposed to direct sunlight, where they are subject to intense variations in environmental conditions. Lichens are, however, extremely sensitive to air pollutants and quickly disappear from urbanized areas. In Europe it has been possible to demonstrate a regular decrease in the extent of development of the lichen flora from the country through the suburbs to the inner city (Figure 14.3). It is felt that this great sensitivity is due to the fact that lichens absorb and concentrate elements from rainwater and possess no means of excreting them, so that lethal concentrations of toxic air pollutants can be gradually reached. The growth rates of lichens are often exceedingly slow, as is illustrated in Figure 14.4. We do not think of lichen growth in terms of hours or days but of years. Indeed, some lichens grow so slowly that they have been used by geologists to estimate the age of glacial deposits.

Many lichens produce characteristic pigments of striking color (Plate 5: Figure 14.1) and unique chemical structure; many of these pigments have antibacterial properties and conceivably function in nature as a defense against bacterial attack. For a long time it was thought that the production of these pigments was a unique expression of the lichen association, but recently it has been shown that the fungus alone is sometimes able to produce the lichen pigment in culture.

If a lichen is removed from its natural habitat and placed under favorable laboratory conditions (for example, with organic substrates or high humidity), the symbiosis breaks down; the fungus may proliferate and destroy the alga, or the alga may overgrow the fungus. It is possible to cultivate separately in the laboratory algae and fungi, but it is difficult to resynthesize the lichen from these isolated components. Resynthesis never occurs in laboratory media in which the isolated components are able to

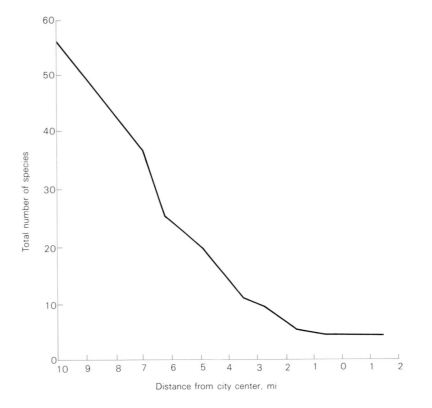

Figure 14.3 The decrease in richness of the lichen flora as the city center is approached. Data for Newcastle-upon-Tyne, an industrial city in northern England. Lichens can be used as indicators of long-term trends in air pollution. (From O. L. Gilbert: in Ecology and the Industrial Society. New York: John Wiley & Sons, Inc., 1965.)

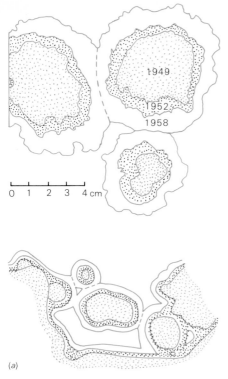

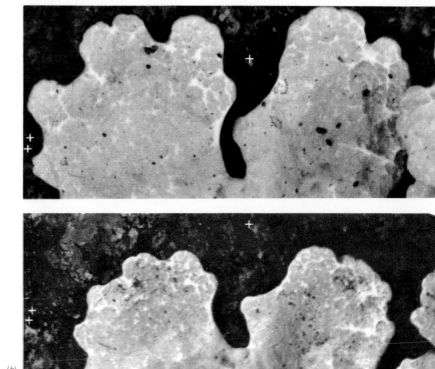

Figure 14.4 (a) Tracings of the outlines of several lichen colonies in different years. Note that the growth rate in this case is much slower than that in (b). [From M. E. Hale, Jr.: Lichen Handbook: A Guide to the Lichens of Eastern North America. Washington, D.C.: Smithsonian Institution, 1961.] (b) Rate of growth of the lichen P. caperata on a rock at Plummers Island, Maryland. The lower photograph was taken on May 22, 1968, and the upper photograph on June 22, 1968, during a period favorable to rapid lichen growth. The marks indicate identical positions on the rock. Magnifications of both photographs are the same, and their widths represent about 9 mm. The white patches are alga-free areas of fungal tissue undergoing rapid growth. (Courtesy of Mason E. Hale.)

grow, but takes place only where starvation conditions are in effect, so that the fungus and alga are "forced" to reunite. We can therefore look upon the lichen thallus in nature as an ecological response of two organisms to extremely unfavorable environments.

14.2 Microbial interactions with higher organisms

In the rest of this chapter we will consider how microorganisms may live on and possibly benefit higher organisms. Animal and plant bodies provide favorable environments for the growth of many microorganisms. They are rich in organic nutrients and growth factors required by heterotrophs, they provide relatively constant conditions of pH and osmotic pressure, and warm-blooded animals have highly constant temperatures. However, an animal or plant body should not be considered as one uniform microbial environment throughout. Each region or organ differs chemically and physically from other regions and thus provides a selective environment where certain kinds of microbes are favored over others. In the higher animal, for instance, the skin, respiratory tract, gastrointestinal tract, and so on, each provide a wide variety of microenvironments in which different microorganisms can grow selectively. However, higher animals and plants possess a variety of defense mechanisms that act in concert to prevent or inhibit microbial invasion and growth. The microorganisms that ultimately colonize successfully are thus those that have developed ways of circumventing these defense mechanisms. Interestingly, some animals and plants have developed mechanisms to encourage the growth of beneficial microorganisms.

Actually, it is often difficult to determine whether a relationship between a microorganism and a higher organism is beneficial, harmful, or neutral. This is because the outcome of an interaction may be influenced by external factors, so that under certain conditions the relationship may be beneficial, whereas under other conditions it may be neutral or harmful.

14.3 The normal microbial flora of animals

Our discussion here will emphasize warm-blooded animals, especially mammals, as it is for this group that we have the most information. Microorganisms are almost always found in those regions of the body exposed to the outside world, such as the skin, oral cavity, respiratory tract, intestinal tract, and genitourinary tract. They are not found normally in the organs and blood and lymph systems of the body; if microbes are found in any of these latter areas in significant quantities, it is usually indicative of a disease state (see Chapter 15).

Normal flora of the skin Figure 14.5 indicates diagrammatically the anatomy of the skin and suggests some regions in which bacteria may live. The skin surface itself, on top of the stratum corneum, is not a favorable place for microbial growth, as it is subject to periodic drying. Only in certain areas of the body, such as the scalp, face, ear, underarm regions, urinary and anal regions, and palms and interdigital spaces of the toes, are moisture conditions on the surface sufficiently high to support resident microbial populations; in these regions characteristic surface microbial floras do exist. Most skin microorganisms are associated directly or indirectly with the sweat glands, of which there are several kinds. The *eccrine* glands are not associated with hair follicles and are rather un-

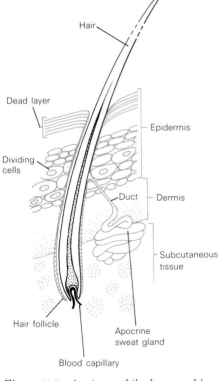

Labels: Hair, Dead layer, Epidermis, Dividing cells, Duct, Dermis, Subcutaneous tissue, Hair follicle, Apocrine sweat gland, Blood capillary

Figure 14.5 Anatomy of the human skin. Microbes are associated primarily with the sweat ducts and the hair follicles.

evenly distributed over the body, with denser concentrations on the palms, finger pads, and soles of the feet. They are the main glands responsible for the perspiration associated with body cooling. Eccrine glands seem to be relatively devoid of microorganisms, perhaps because of the extensive flow of fluid, since when the flow of an eccrine gland is blocked, bacterial invasion and multiplication do occur. The *apocrine* glands are more restricted in their distribution, being confined mainly to the underarm and genital regions, the nipples, and the umbilicus. They are inactive in childhood and become fully functional only at puberty. Bacterial populations on the surface of the skin in these warm, humid places are relatively high, in contrast to the situation on the smooth surface skin. Underarm odor develops as a result of bacterial activity on the secretions of the apocrines; aseptically collected apocrine secretion is odorless but develops odor upon inoculation with bacteria isolated from the skin. Each hair follicle is associated with a *sebaceous gland* which secretes a lubricant fluid. Hair follicles provide an attractive habitat for microorganisms; a variety of aerobic and anaerobic bacteria, yeasts, and filamentous fungi inhabit these regions, mostly within the area just below the surface of the skin. The secretions of the skin glands are rich in microbial nutrients. Urea, amino acids, salts, lactic acid, and lipids are present in considerable amounts, and vitamins, glucose, and other organic compounds occur in trace amounts. The pH of the secretions in man is almost always acidic, with the usual range being between pH 4 and 6.

The microorganisms of the normal flora of the skin can be characterized as *residents* or *transients*. Residents are organisms that are able to multiply, not merely survive, on the skin. The skin as an external organ is continually being inoculated with transients, virtually all of which are unable to multiply and usually die. The normal flora of the skin consists primarily of bacteria restricted to a few groups. These include several species of *Staphylococcus* and a variety of both aerobic and anaerobic corynebacteria. Of the latter, *Corynebacterium acnes* is ordinarily a harmless resident but can incite or contribute to the condition known as acne. Gram-negative bacteria are almost always minor constituents of the normal flora, even though such intestinal organisms as *Escherichia coli* are being continually inoculated onto the surface of the skin by fecal contamination. It is thought that the lack of success of Gram-negative bacteria is due to their inability to compete with Gram-positive organisms that are better adapted to the skin; if the latter are eliminated by antibiotic treatment, the Gram-negative bacteria can flourish. The death of many microorganisms inoculated onto the surface of the skin is thought to result primarily from two factors: the low moisture content of the stratum corneum and its organic-acid content. Those organisms which survive and grow are able to resist these adverse conditions.

The skin of the mammalian fetus *in utero* is normally sterile, but microbial colonization begins at the moment of birth. The initial source of microbes is the mother and persons involved in the birth process, but inoculation may also occur indirectly via air, inanimate objects, and so on. The pH of the skin of the newborn is initially greater than 7, because it is coated with an alkaline secretion from the mother, but frequent bathing removes this layer and the pH drops within a week to adult levels. The newborn is probably more susceptible to colonization by disease-causing organisms such as *Staphylococcus aureus* and the yeast *Candida albicans*, at least partly because it lacks a resident microbial flora that could compete with these invaders. By the end of the first week or two of age, the human infant has acquired most of the organisms found on corresponding areas of the adult.

Skin disinfection is an important procedure in medical practice. Washing with soap removes loosely attached bacteria by mechanical means but has little effect on the resident populations embedded in hair follicles. Indeed, vigorous washing with soap may actually cause an increase in the bacterial population on the surface of the skin because organisms from the hair follicles are resurrected to the surface. Soaps and lotions containing the antimicrobial agent hexachlorophene have been widely used to reduce the microbial load of the skin, but these are effective only if used repeatedly. Alcohol and iodine are relatively effective but also require repeated use and in addition tend to cause skin irritation. Some of the quaternary ammonium compounds (see Table 9.3) have been widely used, but they are readily inactivated by soap and have only limited action against Gram-negative bacteria and fungi. Newer formulations, employing quaternaries in foam wetting agents, have proved to be more effective and are probably the agents of choice.

Normal flora of the oral cavity   The oral cavity, despite its apparent simplicity, is one of the more complex and heterogeneous microbial habitats in the body. This cavity includes the teeth and tongue and the central space that they fill. As a first approximation, the teeth can be viewed only as the surface upon which saliva and materials derived from the food adsorb, rather than as a direct source of microbial nutrients. Although saliva is the most pervasive source of microbial nutrients in the oral cavity, in point of fact it is not an especially good microbial culture medium. Saliva contains about 0.5 percent dissolved solids, about half of which is inorganic (mostly chloride, bicarbonate, phosphate, sodium, calcium, potassium, and trace elements); the predominant organic constituents of saliva are proteins, such as the salivary enzymes, the mucoproteins, and some serum proteins. Small amounts of carbohydrates, urea, ammonia, amino acids, and vitamins are also present. A number of antibacterial substances have been identified in saliva, of which the most important are the enzymes lysozyme and lactoperoxidase. (Lactoperoxidase, an enzyme present in both milk and saliva, kills bacteria by an unknown reaction, involving thiocyanate [CNS⁻] and H_2O_2, two other components of these fluids.) The pH of saliva is controlled primarily by a bicarbonate buffering system and varies between 5.7 and 7.0, with a mean near 6.7. The composition of saliva varies from individual to individual, and even within the same individual variations are seen, which are influenced by physiological and emotional factors.

In the infant the lack of teeth is probably of considerable importance in determining the nature of the microbial flora. Bacteria found in the mouth during the first year of life are predominantly facultative organisms such as streptococci and lactobacilli, but a variety of other bacteria, including some aerobes, can occur in small numbers. When the teeth appear, there is a pronounced shift in the balance of the microflora from aerobes towards anaerobes, and a variety of bacteria specifically adapted for growth on surfaces and in crevices of the teeth develop. A film forms on the surface of the teeth, the so-called dental plaque, and this complex coating adheres so tenaciously to the tooth that it cannot be dislodged by ordinary brushing. Its thickness varies from about 60 μm in protected areas of the tooth to 1 to 3 μm in self-cleaning areas. Dental plaque consists mainly of Gram-positive filamentous bacteria closely packed and extending out at right angles from the surface of the tooth (Figure 14.6), embedded in an amorphous matrix. These filamentous organisms are usually classified as *Leptotrichia buccalis* (or *Fusobacterium fusiforme*). They are obligate anaerobes on initial isolation but after subculturing become microaerophilic; they ferment carbohydrates to lactic acid. Associated with these predominant filamentous organisms are smaller numbers of streptococci, diphtheroids, Gram-negative cocci, and others. The anaerobic nature of the flora may seem surprising, considering that the mouth has good accessibility to oxygen. It is likely that anaerobiosis develops through the action of facultative bacteria growing upon organic materials on the tooth since the dense matrix of the plaque decreases greatly the diffusion of oxygen onto the tooth surface. The microbial populations of the dental plaque are thus seen to exist in a microenvironment partly of their own making, and are probably able to maintain themselves in the face of wide variations in the macroenvironment of the oral cavity.

The role of the oral flora in tooth decay (dental caries) has now been well established through studies on germ-free animals (see below), although the precise mechanisms of the process are still under study. The smooth surfaces of the teeth that are exposed to frequent cleaning by the tongue, cheek, saliva, or toothbrush or to the abrasive action of food mastication are relatively resistant to dental caries. The tooth surfaces in

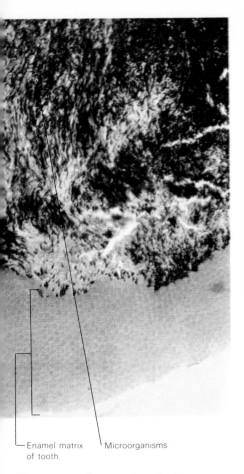

├─ Enamel matrix
│ of tooth
└─ Microorganisms

Figure 14.6   Cross section of a human
tooth showing the extensive development of
unicellular and filamentous Gram-positive
bacteria. Magnification, 575×. [From
H. E. Frisbie, J. Nuckolls, and
J. B. deC. M. Saunders: J. Amer. Coll.
Dent. 11:243 (1944).]

crevices, where food particles can be retained, are the sites where tooth decay predominates. The shape of the teeth is an important factor in the degree to which such crevices develop; dogs are highly resistant to tooth decay because the shape of their teeth does not favor the retention of food. Diets high in sugars are especially cariogenic because lactic acid bacteria ferment the sugars to lactic acid, which causes decalcification of the hard dental tissue of the tooth. Once the breakdown of the hard tissue has begun, proteolysis of the matrix of the tooth enamel occurs, through the action of proteolytic enzymes released by bacteria. Microorganisms penetrate further into the decomposing matrix, but the later stages of the process may be exceedingly slow and are often highly complex. Susceptibility to tooth decay varies greatly among individuals and is affected by inherent traits in the individual as well as by diet and other extraneous factors. The structure of the calcified tissue plays an important role in tooth decay. The incorporation of fluoride into the calcium phosphate crystal matrix makes the latter more resistant to decalcification by acid, hence the use of fluorides in drinking water or dentifrices to aid in controlling tooth decay. Although tooth decay is an infectious disease (see Chapter 15), it seems to be one of the normal burdens of civilized man, and therefore we tend to place it in a different category from other infectious diseases. However, microorganisms in the mouth can also cause infections that are more typically disease states, such as peridontal disease, gingivitis, infections of the tooth pulp, and so on.

Normal flora of the gastrointestinal tract   The gastrointestinal tract is the primary site of food digestion. It varies considerably in structure from one animal to another. In carnivores, including man, the tract is relatively simple, digestion occurring primarily in the stomach and small intestine; the large intestine serves mainly as a site of absorption of water and digestive products. In herbivores the intestinal tract is much more complicated, containing various regions adapted to the digestion of grassy and leafy foodstuffs. Because microbiological processes play a key role in digestion in herbivores, we will return to this topic later in this chapter.

In man, digestion begins initially in the mouth, where the food is mixed with salivary enzymes, and then continues in the *stomach* under the influence of gastric enzymes. The acidity of the stomach fluids is high, about pH 2. The stomach can thus be viewed as a microbiological barrier against penetration of exotic bacteria into the intestinal tract. Although the bacterial count of the stomach contents is generally low, the walls of the stomach are often heavily colonized with bacteria. These are primarily acid-tolerant lactobacilli and streptococci and can be seen in large numbers in histological sections of the stomach epithelium (Figure 14.7).

Figure 14.7 Section through a Gram-stained preparation of the stomach wall of a 14-day-old mouse, showing extensive development of lactic-acid bacteria (Lactobacillus sp.) in association with the epithelial layer. To avoid introducing changes in the bacterial flora during sample preparation, the tissue removed from the stomach was immediately frozen and was sectioned using a freezing microtome. Magnification, 1,000×. (Courtesy of Dwayne C. Savage.)

These bacteria appear very early after the birth of an animal, being well established by the first week. In man, under abnormal conditions such as cancer of the stomach that produce higher pH values, a characteristic microbial flora consisting of yeasts and bacteria (genera *Sarcina* and *Lactobacillus*) may develop.

The *small intestine* is separated into two parts, the duodenum and the ileum. The former, being adjacent to the stomach, is fairly acidic and resembles the stomach in its microbial flora, although it may lack heavy populations on the epithelium. From the duodenum to the ileum the pH gradually becomes more alkaline and bacterial numbers increase. In the lower ileum bacteria are found in the intestinal cavity (the lumen), mixed with digesting material.

In the *large intestine*, bacteria are present in enormous numbers, so much so that this region can be viewed as a specialized fermentation vessel. Many bacteria live within the lumen itself, probably using as nutrients some of the products of the digestion of food. Facultative anaerobes, such as *E. coli*, are numerous; their activities lead to the consumption of any oxygen present, making the environment of the large intestine strictly anaerobic and favorable for the profuse growth of obligate anaerobes. The proportion of anaerobes to aerobes is often grossly underestimated, due to the difficulties of cultivating the anaerobes. In recent years better anaerobic techniques have been developed, and organisms previously uncultured are now being isolated. Many of these bacteria are long, thin, Gram-negative rods, with tapering ends (called *fusiform*) and can be seen microscopically in enormous numbers packing the intestinal wall (Figure 14.8). Other obligate anaerobes include species of *Clostridium* and *Bacteroides*. The organism *Streptococcus fecalis*, a facultative organism, is almost always present in significant numbers. There may also be protozoa, although these are less often found here than in the ruminant (see below).

As water is withdrawn from the digested material, it becomes gradually more concentrated and is converted into feces. Bacteria, chiefly dead ones, make up about one-third of the weight of fecal matter. Organisms living in the lumen of the large intestine are being continuously displaced downward by the flow of material, and if bacterial numbers are to be maintained, those bacteria that are lost must be replaced by new growth. Thus, the large intestine resembles in some ways a chemostat (see Figure 7.10). The time needed for passage of material through the complete gastrointestinal tract is about 24 hr in man; the generation time of bacteria in the lumen is 1 to 2 doublings per day.

When an antibiotic is given orally it may inhibit the growth of the microorganisms present; continued movement of the intestinal contents then leads to the loss of the preexisting bacteria and the virtual sterilization of

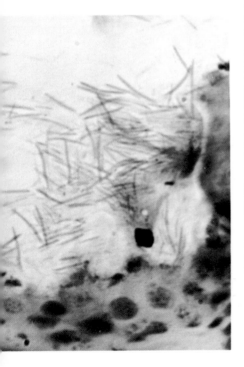

Figure 14.8 Section through the wall of the cecum (large intestine) of an adult mouse, showing the extensive development of fusiform bacteria in association with the epithelial layer. The organisms are Gram-negative. Preparation technique is the same as in Figure 14.7. Magnification, 1,100×. (Courtesy of Dwayne C. Savage.)

the intestinal tract. In the absence of the normal flora the environmental conditions of the large intestine change, and there may become established exotic microorganisms such as antibiotic-resistant *Staphylococcus, Proteus,* and the yeast *C. albicans,* which normally do not grow in the intestinal tract because they cannot compete with the usual flora. Occasionally, the establishment of these exotic organisms can lead to a harmful alteration in digestive function. The normal flora eventually becomes reestablished, but often only after considerable time.

The intestinal flora of the newborn becomes established early. In breast-fed human infants the flora is often fairly simple, consisting largely of the organism *Bifidobacterium* (formerly called *Lactobacillus bifidus*). In bottle-fed infants the flora is usually more complex. The flora is conditioned partly by the fact that the infant's main source of food is milk, which is high in the sugar lactose. The reason why the flora of breast-fed infants differs from that of bottle-fed ones is not completely understood, but it is known that human milk contains a disaccharide amino sugar that is required as a growth factor by *Bifidobacterium.* As the infant ages and his diet changes, the composition of the intestinal flora also changes, ultimately approaching that of the adult, which we described above. It should be emphasized that there are great variations in the composition of the intestinal flora among different species, and among different individuals of the same species. At present we understand only dimly why certain kinds of organisms are adapted to the intestinal environment and others are not. For instance, even though we know much about the molecular biology of *E. coli,* we have no detailed idea of why its natural habitat is the intestinal tract.

Normal flora of other body regions In the *upper respiratory tract* (throat, nasal passages, and nasopharynx) microorganisms live primarily in areas bathed with the secretions of the mucous membrane. Bacteria enter the upper respiratory tract from the air in large numbers during breathing, but most of these are trapped in the nasal passages and expelled again with the nasal secretions. The resident organisms most commonly found are staphylococci, streptococci, diptheriod bacilli, and Gram-negative cocci. These organisms are isolated by swabbing the region with a sterile swab and streaking on blood agar plates. The microenvironmental conditions under which these organisms are growing are not really understood, as it is difficult to study the respiratory tract experimentally. It is known, however, that often a person has his own characteristic flora, which may remain constant over extended periods of time. Potentially harmful bacteria, such as *S. aureus, Diplococcus pneumoniae,* and *Corynebacterium diptheriae* are often part of the normal flora of the nasopharynx of healthy individuals.

The *lower respiratory tract* (trachea, bronchi, and lungs) is essentially sterile, in spite of the large numbers of organisms potentially able to reach this region during breathing. Dust particles, which are fairly large, are filtered out in the upper respiratory tract. As the air passes into the lower respiratory tract, its rate of movement decreases markedly, and organisms settle onto the walls of the passages. These walls are lined with ciliated epithelium, and the cilia, beating upwards, push bacteria and other particulate matter toward the upper respiratory tract where they are then expelled in the saliva and nasal secretions. Since only droplet nuclei smaller than 10 μm in diameter are able to reach the lungs, the organisms most commonly causing pulmonary disease are those that can survive in tiny nuclei, such as *Mycobacterium tuberculosis* and *D. pneumoniae* (see Chapter 15).

The female *urethra* is usually sterile, while that of the male contains Gram-positive cocci and diphtheroids in the lower one-third. The *vagina* of the adult female generally is weakly acidic and contains significant amounts of the polysaccharide glycogen. A *Lactobacillus,* sometimes called "Döderlein's bacillus," which ferments glycogen and produces acid, usually is present in the vagina and may be responsible for the acidity. Other organisms, such as yeasts, streptococci, and *E. coli,* are also present. Before puberty, the female vagina does not produce glycogen, Döderlein's bacillus is absent, and the flora consists predominantly of staphylococci, streptococci, diphtheroids, and *E. coli.* After menopause, glycogen disappears and the flora then resembles that found before puberty.

14.4   Germ-free animals

As we have seen above, animals have a complex and well-developed microbial flora. Are microorganisms essential for the life of the animal? In the early days of microbiology, when it became clear that all animals were well colonized with various microbes, Pasteur advanced the idea that microbes are essential for animal life. The test of this hypothesis required the raising of germ-free animals. The conclusions from this work were that germ-free animals, given proper diet, are fully viable; indeed, colonies of germ-free animals have now been maintained for many generations. However, these germ-free individuals do differ physiologically and anatomically in many important ways from conventional animals, as we shall see below.

The culture of germ-free mammals was once a very difficult technical feat, but with the development of new equipment and techniques, it has become a rather routine matter. Establishing a germ-free animal colony is sometimes simplified by the fact that mammals frequently are sterile until birth, so that if the fetus is removed aseptically just before the time of expected birth, germ-free infants can often be obtained. These infants must then be placed in germ-free isolators (Figure 14.9), and all air, water, food or other objects entering the isolators must be sterile. Within the isolators the infants must then be hand-fed until they have developed to the stage where they can feed themselves. Once established, a germ-free colony can be maintained by continued mating between germ-free males and females. With birds, the establishment of germ-free colonies is easier, as the inside of the egg is usually sterile and the newly-hatched chick is able to feed itself immediately. To date, germ-free colonies of mice, rats, guinea pigs, rabbits, hamsters, monkeys, and chickens have been established and similarly germ-free individual lambs and pigs have been kept for considerable periods of time. The raising of large germ-free animals is obviously much more difficult than germ-free raising of small animals since large isolators are expensive and difficult to keep sterile. Germ-free mice are now available commercially from laboratory-animal supply houses; they are usually delivered directly in plastic isolators in which the desired experiments can be done directly.

How can it be ascertained that such animals are really germ-free? Microbiological culture studies using a variety of media must be performed, and body fluids and tissues of the animal should never yield microbes. However, ''germ-free'' is a negative term that implies a complete absence of microorganisms; yet all that can be determined is that with the methods used no microorganisms can be detected. This does not

Figure 14.9   Two types of germ-free isolator. The one on the right is of inexpensive plastic; and the one on the left, of more durable stainless steel. (Courtesy of Medical Audio-Visual Branch, Walter Reed Army Institute of Research, Washington, D.C.)

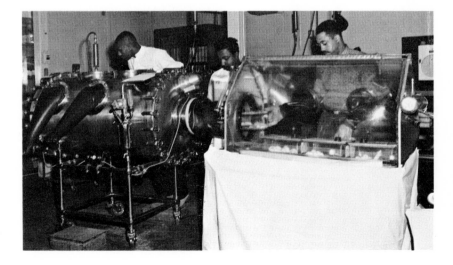

mean that, if other methods were employed, microorganisms might not be found in a presumed germ-free animal.*

Germ-free individuals differ from normal animals in several important respects. Structures involved in the defense mechanisms against bacterial invaders such as the lymphatic system, the antibody-forming system, and the reticuloendothelial system (see Chapter 15) are poorly developed in germ-free animals, and organs that have natural populations of bacteria are often reduced in size or capacity. However, the cecum of the germ-free guinea pig, rat, and rabbit is greatly enlarged (Figure 14.10). In the normal rodent the cecum is the part of the intestinal tract that has the largest bacterial population and in the germ-free state the cecum might lack the stimulus to evacuate caused by the bacteria and hence continuously fills up. Furthermore, the whole intestinal wall of the germ-free animal is thin and unresponsive to mechanical stimuli. The conclusion seems inescapable that bacteria are necessary for the normal development of the intestine.

Germ-free animals also differ in nutritional needs from conventional ones. For instance, vitamin K, which usually is not required by conventional animals in their diets, is required by germ-free animals. *Escherichia coli* synthesizes vitamin K; when this organism is established in the intestinal tract of germ-free animals, the vitamin K deficiency disappears, thus showing that *E. coli* probably is responsible for synthesizing this vitamin in the conventional animal.

Germ-free animals are much more susceptible to bacterial infections than are conventional animals. Organisms like *Bacillus subtilis* and *Sarcina lutea,* which are harmless to conventional animals, are harmful to germ-free forms. It is difficult to infect conventional animals with *Vibrio cholerae* (the causal agent of cholera) and *Shigella dysenteriae* (the cause of bacterial dysentery); yet these organisms can readily be established in germ-free animals. It is likely that the normal flora of the conventional animal has a competitive advantage in the intestinal tract, preventing the establishment of exotic organisms; with the normal flora gone, the foreign organisms can become established easily. Germ-free animals are also much more susceptible to infection by intestinal worms. On the other hand, they are resistant to infection by the intestinal amoeba that causes amoebic dysentery (*Entamoeba histolytica*). This is probably because the amoeba uses intestinal bacteria as a primary source of food and cannot grow in their absence. Also, germ-free animals do not show tooth decay (dental caries) even if they are fed a diet high in sugar, which is very conducive to tooth decay in conventional animals. However, if germ-free animals are inoculated with a pure culture of certain *Streptococcus* strains

*Frequently the term *gnotobiotic* is used instead of *germ-free.* A gnotobiotic system is one in which the composition of the microbial flora is known.

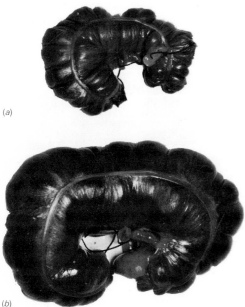

Figure 14.10 Comparison of the cecum size of a normal (a) and a germ-free (b) rodent. (Courtesy of Medical Audio-Visual Branch, Walter Reed Army Institute of Research, Washington, D.C.)

(a)

(b)

isolated from the teeth, tooth decay does occur on a high sugar diet.

Many of the characteristics of germ-free animals can be mimicked in conventional animals if they are given orally antibiotics that cause the elimination of the intestinal flora. Such animals often are more susceptible to infections than are conventional animals and show vitamin deficiencies. We can for these reasons conclude that the conventional animal does derive considerable benefit from its normal flora, even though the normal flora causes some harmful effects. The answer to our original question must therefore be equivocal: Animals *can* live without their normal microbial populations but may be better off with them. We may also draw the conclusion that, even though many microorganisms are harmful, it would not be preferable (if indeed possible) to eliminate entirely the microbial flora of the human body since this might cause more difficulties than benefits. We also see that the germ-free animal is a useful research tool for studying some of the interesting questions concerning the interrelationships of animals and microorganisms.

It has frequently been observed that normal animals grow faster when a small amount of antibiotic, such as penicillin or tetracycline, is added to their feed. Indeed, such antibiotic additions are used routinely in the raising of pigs and chickens since the more rapid growth permits the farmer to take the animal to market earlier. The means by which antibiotic feeds stimulate growth has been clarified with the use of germ-free animals. One organism often found among the normal flora is *Clostridium perfringens,* an anaerobe that produces a variety of toxins (see Table 15.1). This organism is only a minority member of the normal flora, but it is likely that the small quantities of toxins it produces cause a slowing of growth. Small amounts of antibiotics eliminate *C. perfringens* selectively without altering significantly the rest of the normal flora. That *C. perfringens* and not some other component of the normal flora causes growth retardation has been shown by isolating various microorganisms in pure culture from conventional animals and then infecting separate germ-free animals with each of these forms. Organisms such as *E. coli, Lactobacillus,* or *Streptococcus* do not cause growth delay, whereas *C. perfringens* does; in addition, the growth delay caused by the latter organism can be overcome by penicillin treatment.

14.5 Rumen symbiosis

Ruminants are animals that possess a special organ, the rumen, within which the digestion of cellulose and other plant polysaccharides occurs through the activity of special microbial populations. Some of the most important domestic animals, the cow, sheep, and goat, are ruminants. Since the human food economy depends to a great extent on these animals, rumen microbiology is of considerable economic significance.

Rumen fermentation   The bulk of the organic matter in terrestrial plants is present in insoluble polysaccharides, of which cellulose is the most important. Mammals, and indeed most animals, lack the enzymes necessary to digest cellulose, but all mammals that subsist primarily on grasses and leafy plants can break it down by making use of microorganisms as digestive agents. Unique features of the rumen as a site of cellulose digestion are its relatively large size (100 liters in a cow, 6 liters in a sheep) and its position in the alimentary tract as the organ where ingested food first goes. The high constant temperature (39°C) and the anaerobic nature of the rumen are also important factors. Food entering the rumen is mixed with the resident microbial populations and remains there on the average about 9 hr. During this time, cellulolytic bacteria and protozoa hydrolyze cellulose to the disaccharide cellobiose and the monosaccharide glucose. These sugars then undergo a microbial fermentation with the production of volatile acids, primarily acetic, propionic, and butyric, and the gases carbon dioxide and methane (Figure 14.11). The volatile acids pass through the rumen wall into the blood stream and are oxidized by the animal as its main source of energy. In addition to their digestive functions, the rumen microorganisms synthesize amino acids and vitamins that are the main source for the animal of these essential nutrients. The rumen contents after fermentation consist of many microbial cells plus partially digested plant materials, which pass through the stomach and gastrointestinal tract of the animal, where they undergo digestive processes similar to those of other animals. Microbial cells formed in the rumen and digested in the gastrointestinal tract are the main source of proteins and vitamins for the animal. Since many of the microbes of the rumen are able to grow on urea as a sole nitrogen source, it is often supplied in cattle feed in order to promote microbial protein synthesis. The bulk of this protein will end up in the animal itself. A ruminant is thus nutritionally superior to a nonruminant when subsisting on foods that are deficient in protein, such as grasses.

The rumen microorganisms   The biochemical reactions occurring in the rumen are complex and involve a wide variety of microorganisms. For many years the culture of rumen bacteria presented great difficulties. Two developments have been of most significance: (1) the perfection by R. E. Hungate and his colleagues of superior anaerobic culture techniques so that even extremely oxygen-sensitive organisms could be cultured; and (2) the discovery that many rumen bacteria require as growth factors certain

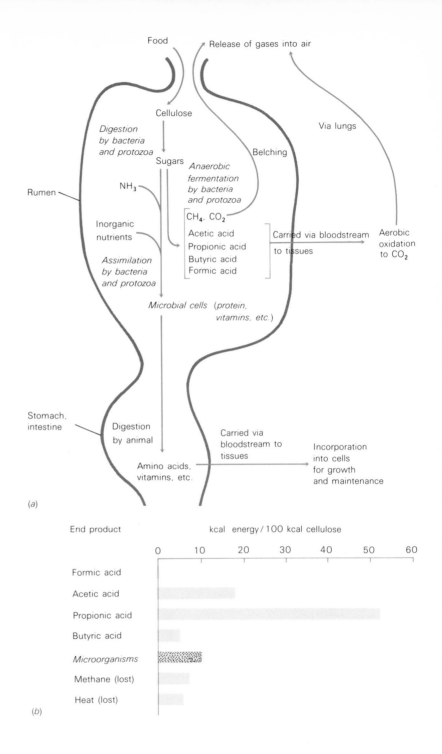

Food
Release of gases into air

Cellulose

Digestion
by bacteria
and protozoa

Via lungs

Sugars

Belching

Anaerobic
fermentation
by bacteria
and protozoa

NH_3

Rumen

Inorganic
nutrients

CH_4, CO_2

Acetic acid
Propionic acid
Butyric acid
Formic acid

Carried via bloodstream
to tissues

Aerobic
oxidation
to CO_2

Assimilation
by bacteria
and protozoa

Microbial cells (protein,
vitamins, etc.)

Stomach,
intestine

Digestion
by animal

Carried via
bloodstream to
tissues

Incorporation
into cells
for growth
and maintenance

Amino acids,
vitamins, etc.

(a)

End product
kcal energy / 100 kcal cellulose

0   10   20   30   40   50   60

Formic acid

Acetic acid

Propionic acid

Butyric acid

Microorganisms

Methane (lost)

Heat (lost)

(b)

Figure 14.11 (a) Biochemical processes in the ruminant. (b) Energy balance for the fermentation of cellulose in the rumen. The bars show the relative proportions of the energy of the cellulose that ends up in various products.

14.5 Rumen symbiosis

branched-chain acids (for example, isovaleric and isobutyric) present in rumen fluid. The bacteria of the rumen are predominantly nonsporulating Gram-negative rods and vibrios.

In addition to bacteria, the rumen has a characteristic protozoal fauna, composed almost exclusively of ciliates (Figure 14.12). Although these are not essential for rumen fermentation, the protozoa definitely contribute to the process. They are able to ferment sugars, starch, and cellulose with the production of the same volatile acids that are formed by the bacteria. Protozoal population densities are usually larger in animals on a good ration than in those fed a poor diet, and there is some reason to believe that protozoal counts could be used as indicators of the well-being of the

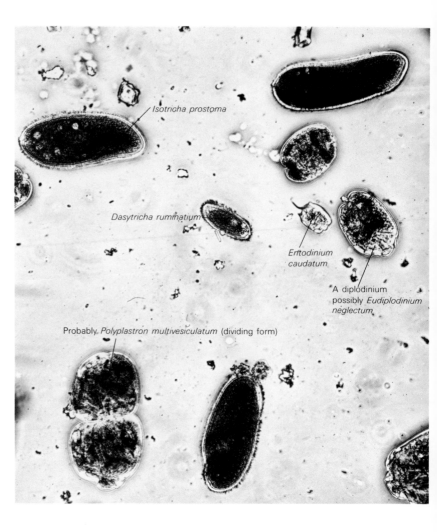

Figure 14.12 Phase photomicrograph of a variety of ciliated protozoa from the rumen of a sheep. Magnification, 290×. (Courtesy of G. S. Coleman.)

animal. It has been well established that the protozoa ingest rumen bacteria; animals without protozoa usually have higher bacterial populations, which suggests that the protozoa may control at least to some extent bacterial density.

Rumen function is often studied experimentally by cutting a hole into the side of the animal and grafting the rumen wall to the body wall. The hole—a *fistula*—is then sealed with a plug. Samples can be removed as desired, without undue disturbance to the animal. Such animals have provided much of the information on biochemical changes in the rumen in response to different feeding regimes. Much other work has been done by removing samples of rumen material and incubating them in laboratory vessels where anaerobiosis and constant temperature are maintained; this is often called the artificial-rumen technique. The flora and biochemical processes of the artificial rumen resemble those of the natural rumen initially, but they gradually diverge significantly.

Cellulose fermentation in nonruminants In nonruminant herbivores such as the horse, pig, porcupine, rat, rabbit, and guinea pig, cellulose digestion occurs in the cecum, an enlarged portion of the large intestine. The biochemical and microbiological processes are similar to those in the rumen, but because the cecum is in the central part of the large intestine below the stomach, partially digested material, microbial cells, and so on, pass out of the animal rather than entering the stomach, as occurs in ruminants. Thus the digestive process is less efficient than in ruminants. Some animals, such as rabbits and guinea pigs, partially overcome this inefficiency through the practice of *coprophagy:* They eat fecal material, thus effectively transferring material from the cecum back to the stomach. If these animals are prevented from practicing coprophagy, vitamin deficiency and other nutritional complications may arise.

14.6   Microbial symbioses with insects

Insects and microorganisms have developed a wide variety of symbiotic relationships. These symbioses are especially important to insects that obtain their food from plant sources. Plants are usually rich in carbohydrates but low in protein, and insects have in a variety of ways developed methods by which they depend on microorganisms for obtaining their nutrients.

Ambrosia beetles are wood-boring insects that live primarily on fungus mycelium, which develops in the tunnels constructed in trees by the

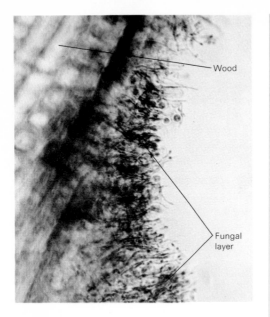

Figure 14.13 Cross section through a tunnel in an oak tree made by the pinhole borer, Platypus cylindris, showing the hyphae, conidia, and chlamydospores of the fungus Sporothrix. Magnification, 300×. [Crown copyright—reproduced by permission of the Director, Forest Products Research Laboratory. From J. M. Baker: in Symbiotic Associations (Thirteenth Symposium, Society for General Microbiology). London: Cambridge University Press, 1963.]

insect. The beetle has a special organ within which the fungus spores are stored. While the tunnel is being constructed, some of these spores are inoculated along its walls. The spores germinate and the fungus grows on nutrients seeping into the tunnel from the wood sap (Figure 14.13); the insect then ingests the fungus mycelium as its primary source of food. Each species of insect has been found to be associated with a single or a restricted number of fungus species, and there is thus a marked specificity in the insect-fungus relationship.

Certain kinds of ants have also entered into partnership with filamentous fungi. The ant, however, is actually an active cultivator of the fungus. The ants are primarily tropical leaf-cutting species and live in great colonies where chambers are constructed within which the fungi grow (Figure 14.14). Leaf fragments are carried by the ant to the nest to serve as the food for the fungus (Plate 5: Figure 14.15). The fungus mycelium ramifies extensively throughout the culture cavities; as it is grazed by the ants, it is replaced by further growth. Some termites also culture fungi in an analogous way.

It is much more common among insects for the symbiotic microorganism to actually live within the body of the insect. In some cases, the microorganism inhabits a special chamber of the intestine, analogous to the rumen, where digestive processes occur. This is true, for instance, of the protozoa that carry out cellulose digestion for the termite. In others, the microorganism is found in large numbers within the cells themselves

Figure 14.14 The fungus garden of the leaf-cutting ant Atta texana. The garden has been uncovered by excavating the nest of the ant. (Courtesy of John C. Moser, U.S. Forest Service.)

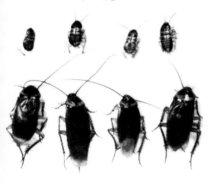

For instance, in the cockroach bacteria reside within cells of the fat body of the insect. The transfer of microbial symbionts from adult to young occurs in a variety of ways: Some insect larvae ingest the microbes in fecal material derived from the adult. In other cases, the adult smears microorganisms on the eggs as they are laid; upon hatching, the larva becomes inoculated with its symbiont. In still other cases, the microorganism is transmitted to the egg from the mother at the time the egg is formed.

What are the roles of these intracellular symbionts? They are clearly not involved in digestive functions, as they do not live in digestive organs, and their intracellular location would not make undigested food accessible to them. That they are necessary for the insect there is no doubt. If the symbiont is eliminated experimentally, by the use of chemical treatments or other means, the insect develops only very poorly (Figure 14.16). At least one function postulated for microbial symbionts is the synthesis of B vitamins. The insects that have the largest populations of intracellular symbionts are those living on vitamin-poor plant materials. In some cases symbiont loss can be compensated for by the addition to the diet of B vitamins. However, in the cockroach, B vitamins will not replace the symbiont; the microorganism must play some other role. One hypothesis is that it functions in the excretory metabolism of the insect, breaking down nitrogen compounds and converting them into products that can be excreted.

14.7 Algae and invertebrates

A wide variety of aquatic invertebrates, ranging from protozoa to higher invertebrates, harbor intracellular algal symbionts.

Paramecium bursaria is a protozoan species generally found in nature associated with an alga, a species of *Chlorella* (see also Chapter 13). In this symbiotic relationship the alga benefits by obtaining its nutrients from the protozoan, and the protozoan benefits by the photosynthetic activities of the alga. Neither organism is obligately associated with the other since each can be cultured independently, but in nature, where competition with other organisms occurs, the relationship may be of considerable benefit to both.

Multicellular invertebrates that harbor algal symbionts include sponges, coelenterates (especially corals), worms, molluscs, echinoderms, and tunicates. The algae are usually coccoid in shape and fall into three main classes: blue-green algae, *Chlorella*, and dinoflagellates (sometimes called *zooxanthellae*). *Chlorella* is a common symbiont of freshwater animals;

dinoflagellates are found primarily in marine animals (Plate 6: Figure 14.17). Initial infection of the animal occurs by ingestion of free-living algae, which are able to resist digestion and grow within the animal cells. Animals containing symbionts grow more efficiently in the light than do symbiont-free animals, and they survive for much longer periods of time in nutrient-poor environments. The algae synthesize oxygen in the light during photosynthesis, and this oxygen is at least partly consumed by the animal during respiration. The algae convert carbon dioxide into organic compounds, some of which are transferred to the animal tissue. These compounds include sugars, vitamins, and other growth factors. The algae may also contribute to the animal's well-being by aiding in the removal of waste products. Finally, in organisms such as coral the algae participate in the formation of the calcium carbonate skeleton through their ability to precipitate calcium carbonate (see Chapter 16). The algae undoubtedly benefit by association with the animal since the nutrients for algal growth pass through the animal tissues, and the animal therefore functions as an absorptive "organ" for the algae by concentrating inorganic nutrients from the surrounding water. Also, the animal by its movements or growth habit remains in shallow water, where light intensities are high, so that the algae have access to higher light intensities than might otherwise be the case.

14.8 Normal flora of plants

As microbial habitats, plant bodies are clearly vastly different from those of animals. Compared with warm-blooded animals, they vary greatly in temperature, both diurnally and throughout the year. The communication system throughout the plant is only poorly developed, so that transfer of microorganisms from one part to another is inefficient. The above-ground parts of the plant, especially the leaves and stems, are subjected to frequent drying and for this reason have developed waxy coatings that retain moisture and, incidentally, probably also keep out microorganisms. Only in tropical plants living under conditions of high rainfall and humidity do extensive microbial populations exist on leaves. The roots, on the other hand, have an environment whose moisture is less variable, and here the nutrient concentrations are higher. For this reason, the roots of plants are the main area of microbial action.

As nutrient sources, plants are high in carbohydrates but, except for the seeds, are relatively deficient in proteins. Vitamin B_{12} is not formed by plants. Many plants produce toxic chemicals that hinder or prevent microbial growth; examples are phenols produced by onions and many cereal and root plants, hydrogen cyanide produced by flax, and allyl sul fides produced by onions. Woody plants are usually resistant to microbial invasion; only if the bark is penetrated by insects or damaged by other means can microorganisms reach the interior of a tree trunk. Even then, movement through the trunk may be slow or absent.

The *rhizosphere* is the region immediately outside the root; it is a zone where microbial activity is usually high. The bacterial count is almost always higher in the rhizosphere than it is in regions of the soil devoid of roots, often many times higher. This is because roots excrete significant amounts of sugars, amino acids, vitamins, and so on, which promote such an extensive growth of bacteria and fungi that these organisms often form microcolonies on the root surface. There is some evidence that the rhizosphere microorganisms benefit the plant by promoting the absorption of nutrients, but this is probably only of minor importance. As noted above, the leaf is a less favorable place for microbial development, and a *phyllosphere* flora is extensive only on tropical plants. The phyllosphere bacteria often produce gums or slimes, which presumably help the bacterial cells to stick to leaves under the onslaught of heavy tropical rains. Both leaves and roots of freshwater and marine plants are usually well colonized with microorganisms.

14.9   Symbiotic nitrogen fixation

The legume-Rhizobium symbiosis   One of the most interesting and important symbiotic relationships is that between leguminous plants and bacteria of the genus *Rhizobium*. The legumes are a large group that contain such economically important plants as soybean, clover, alfalfa, string beans, and peas. *Rhizobium* organisms are Gram-negative motile rods. Infection of the roots of one of these legumes with the appropriate strain of *Rhizobium* leads to the formation of root nodules (Figure 14.18), which are able to convert gaseous nitrogen into combined nitrogen, a process called nitrogen fixation (see Chapter 5). Neither legume nor *Rhizobium* alone is able to fix nitrogen; yet the interaction between the two leads to development of nitrogen-fixing ability. This acquisition of a new property as a result of the interaction is one of the clearest and best examples of what is meant by a symbiotic or mutualistic relationship. Both legume and *Rhizobium* can proliferate in the absence of the other, either in nature or in the laboratory, so that the association between the two is in no way obligate. In fact, were it not known that the root nodules fixed nitrogen, it might be concluded that the *Rhizobium* was a harmful organism since the nodules resemble abnormal tumorous growths. Nitrogen fixation by the legume-*Rhizobium* symbiosis is of considerable agricul-

Figure 14.18   The nodulated root system of a soybean plant. Magnification, 3×. (From F. J. Bergersen: Trans. 9th Inter. Congr. of Soil Science, vol. II. Canberra, Australia: Commonwealth Scientific and Industrial Research Organization, 1968.)

tural importance, as it leads to very significant increases in combined nitrogen in the soil. Since nitrogen deficiency often occurs in unfertilized bare soils, legumes are at a selective advantage under such conditions and can grow well in areas where other plants cannot.

About 90 percent of all leguminous species are capable of becoming nodulated. There is a marked specificity between species of legume and strain of *Rhizobium*. A single *Rhizobium* strain may be able to infect certain species of legumes but not others. A group of *Rhizobium* strains able to infect a group of related legumes is called a *cross-inoculation group*. Even if a *Rhizobium* strain is able to infect a certain legume, it is not always able to bring about the production of nitrogen-fixing nodules. If the strain is *ineffective* the nodules formed will be small, greenish white, and incapable of fixing nitrogen; if the strain is *effective* on the other hand, the nodule will be large, reddish, and nitrogen fixing. Effectiveness is determined by genes in the bacterium that can be lost by mutation or gained by genetic transformation.

Stages in nodule formation   The stages in the infection and development of root nodules are now fairly well understood (Figure 14.19). The roots of leguminous plants secrete a variety of organic materials, which stimulate the growth of a rhizosphere microflora. This stimulation is not restricted to the rhizobia but involves a variety of rhizosphere bacteria. Around the outside of the legume root a specific membranelike layer forms, within which the rhizosphere microflora develops extensively. The purpose of the membrane is probably to prevent the root secretions from

Figure 14.19   Stages in the formation of a root nodule.

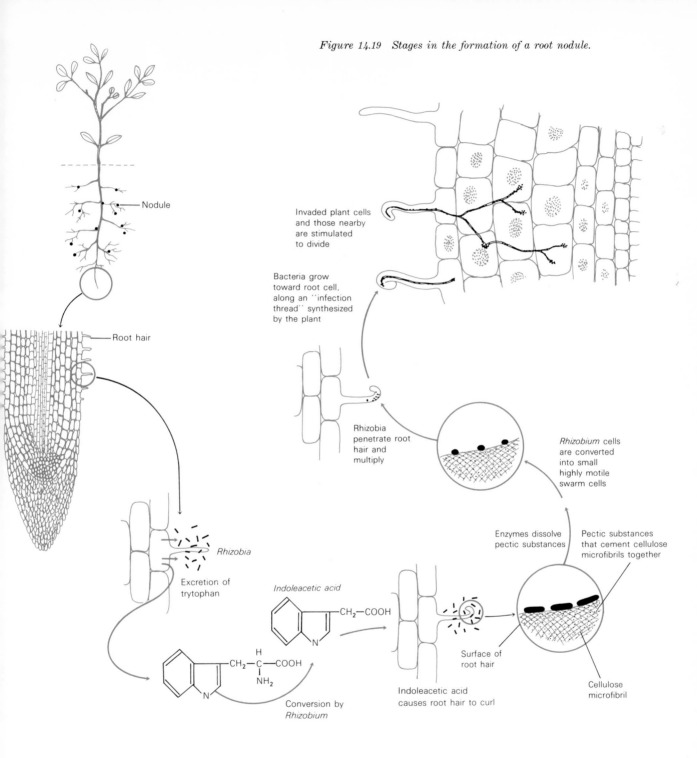

Nodule

Root hair

Rhizobia

Excretion of
trytophan

Indoleacetic acid

Conversion by
Rhizobium

Indoleacetic acid
causes root hair to curl

Surface of
root hair

Enzymes dissolve
pectic substances

Pectic substances
that cement cellulose
microfibrils together

Cellulose
microfibril

Rhizobium cells
are converted
into small
highly motile
swarm cells

Rhizobia
penetrate root
hair and
multiply

Bacteria grow
toward root cell,
along an "infection
thread" synthesized
by the plant

Invaded plant cells
and those nearby
are stimulated
to divide

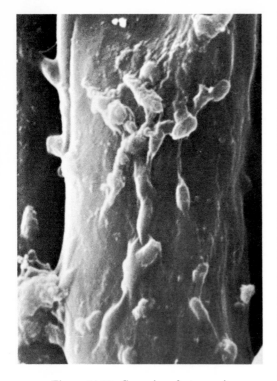

Figure 14.20 Scanning electron micro-graph of a clover root hair with Rhizobium cells on the hair surface. Magnification, 4,000×. (Courtesy of Peter J. Dart.)

leaving the region immediately adjacent to the root, and the rhizosphere flora is thus concentrated in this narrow zone. If there are rhizobia in the soil, they grow in the rhizosphere and build up to high population densities. Infection of the root occurs by way of the root hairs (Figure 14.20) which are present within the membrane-bound matrix. Recall that the cell walls of plants are composed of cellulose microfibrils arranged in a meshwork. The problem the bacteria must face is bringing about a loosening of this meshwork so that they can pass through the wall to the inside of the root hair. One of the secretions of the root is tryptophan which is converted into the plant hormone *indoleacetic acid* by the rhizobia. This hormone induces curling of some of the root hairs, a process that is a prelude to infection. At about this time enzymes may be produced that dissolve the cement holding the cellulose microfibrils together, and some of the rod-shaped rhizobia have become converted into short, almost spherical, highly flagellated "swarmer cells," which are able to slip between the opened fibrillar meshwork. After entry into the root hair, the swarmer cells proliferate and form the so-called *infection thread*, which spreads down the root hair.

A root cell adjacent to the root hair then becomes infected. If this cell is a normal diploid cell, it usually is destroyed by the infection, undergoing necrosis and degeneration; if it is a tetraploid cell, however, it can become the forerunner of a nodule. There are always in the root a small number of tetraploid cells of spontaneous origin, and if one of these cells becomes infected, it is stimulated to divide. Progressive divisions of such infected cells lead to the production of the tumorlike nodule (Figure 14.21). The bacteria multiply rapidly within the tetraploid cells and become surrounded singly or in small groups by portions of the host cell membrane (Figure 14.22). The bacteria then are transformed into swollen, misshapen, and sometimes branched forms called "bacteroids" (Figure 14.23).

The mature nitrogen-fixing nodule is red; this color results from production of a hemoglobinlike protein. Nitrogen fixation occurs only in nodules containing both bacteroids and such hemoglobin. Neither plant nor *Rhizobium* alone synthesizes hemoglobin, but formation is induced somehow through the symbiotic interaction of these two organisms. Hemoglobin probably plays a direct role in nitrogen fixation, perhaps by maintaining a reduced redox potential in the nodule environment. The biochemistry of symbiotic nitrogen fixation is not known, but probably is similar to that of nonsymbiotic systems (see Chapter 5). Nitrogen compounds produced in the nodule, which are predominantly amino acids, are transferred to the root and then to the rest of the plant, playing an important role in the nitrogen economy of the plant.

Eventually the nodule deteriorates, releasing bacteria into the soil. The

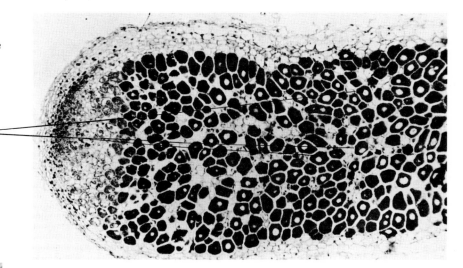

Figure 14.21 Cross section through a nodule of subterranean clover, Trifolium subterraneum. The darkly stained cells are filled with bacteria. Magnification, 75×. (Courtesy of F. J. Bergersen.)

Bacteria-filled cells

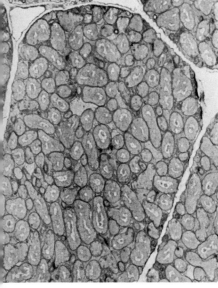

Figure 14.22 Electron micrograph of a thin section through a single bacteria-filled cell of subterranean clover. Magnification, 3,100×. (From P. J. Dart and F. V. Mercer: "Fine structure changes in the development of the nodules of Trifolium subterraneum L. and Medicago tribuloides Desr.," in Arch. f. Mikrobiol. 49:209. New York: Springer-Verlag, 1964.)

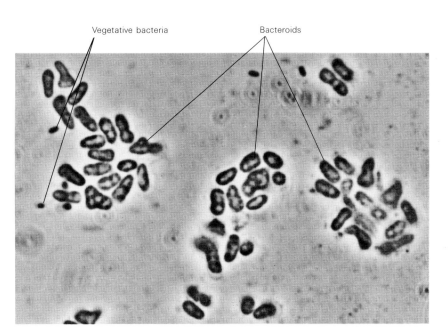

Vegetative bacteria   Bacteroids

Figure 14.23 Photomicrograph of vegetative bacteria and bacteroids from a crushed root nodule of subterranean clover, T. subterraneum, by phase-contrast microscopy. Magnification, 2,400×. (Courtesy of F. J. Bergersen.)

bacteroid forms are incapable of division, but there are always a small number of rod-shaped cells present that had lain dormant; these now proliferate, using as nutrients some of the products of the deteriorating nodule, and can initiate the infection process in other roots.

Practical aspects of the legume-Rhizobium symbiosis When a crop is planted in an area where it has not been previously raised or where it has not been grown for some time, the appropriate *Rhizobium* strain will probably be lacking in the soil. To ensure nodulation, the seed is usually inoculated with the correct strain before sowing. Whether or not seed inoculation is necessary depends on when the crop was planted last and on the nature of the soil. In some soils *Rhizobium* strains can survive for many years after the specific legume host has been planted, whereas in other soils, *Rhizobium* dies out in a few years. In nitrogen-deficient soils legumes benefit greatly from being nodulated and even in relatively rich soils some benefit is probably obtained, although the effects may not be very dramatic. The amount of nitrogen fixed annually by living organisms is about 10^8 tons for the whole world, most of which is from symbiotic sources.

Nonlegume nitrogen-fixing symbioses In addition to the legume-*Rhizobium* relationship, nitrogen-fixing symbioses occur in a variety of nonlegumes, involving microorganisms other than rhizobia. Nitrogen-fixing blue-green algae form symbioses with a variety of plants. Some tropical plants have leaf nodules, in which are found nitrogen-fixing bacteria of the genus *Klebsiella*. The alder tree has root nodules in which can be seen large numbers of spherical-shaped particles, but despite many attempts the organism has not been cultured. Alder is a characteristic pioneer tree, able to colonize bare soils in poor sites, and this is probably at least partly due to its ability to enter into a symbiotic nitrogen-fixing relationship.

14.10   Mycorrhizae

Mycorrhiza literally means "root fungus" and refers to the symbiotic association that exists between plant roots and fungi. Probably the roots of the majority of terrestrial plants are mycorrhizal. There are two general classes of mycorrhiza: *ectotrophic,* in which the fungus hyphae form an extensive sheath around the outside of the root, with only little penetration of hyphae into the root tissue itself, and *endotrophic,* in which the fungus mycelium is embedded in the root tissue.

Figure 14.24 Typical mycorrhizal root of a pine, Pinus rigida, with rhizomorphs of the fungus Thelephora terrestris. [From J. R. Schramm: Trans. Amer. Phil. Soc. 56, pt. I (1966).]

Forked
mycorrhizal
root

Fungus
rhizomorphs

Ectotrophic mycorrhizae Ectotrophic mycorrhizae are found mainly in forest trees, especially conifers, beeches, and oaks, and are most highly developed in temperate forests. In a forest, almost every root of every tree will be mycorrhizal. The root system of a mycorrhizal tree is composed of both long and short roots. The short roots, which are characteristically dichotomously branched (Figure 14.24) show the typical fungal sheath, whereas the long roots are usually uninfected. The fungi that participate in the mycorrhizal association are all Basidiomycetes (see Chapter 19) and most of them form typical mushroom fruiting bodies. The mycorrhizal fungi do not attack cellulose and leaf litter, as do many other mushrooms, but instead use simple carbohydrates for growth and usually have one or more vitamin requirements; they obtain their nutrients from the root secretions. The mycorrhizal fungi are never found in nature except in association with roots and hence can be considered obligate symbionts. On the other hand, these fungi produce plant growth substances that induce morphological alterations in the roots, causing characteristically short dichotomously branched mycorrhizal roots to be formed. Despite the close relationship between fungus and root, there is little species specificity involved. For instance, a single species of pine can form mycorrhizae with over 40 species of fungi.

The beneficial effect on the plant by the mycorrhizal fungus is best

Figure 14.25 Fungus hyphae (Boletus sp.) ramifying from the surface of a mycorrhizal root of P. virginiana. (From J. R. Schramm: Trans. Amer. Phil. Soc. 56, pt. I (1966).]

14.10 Mycorrhizae

observed in poor soils, where trees that are mycorrhizal will thrive but nonmycorrhizal ones will not. If trees are planted in prairie soils, which ordinarily lack a suitable fungal inoculum, trees that were artificially inoculated at the time of planting grow much more rapidly than uninoculated trees. However, in nutrient-rich conditions, the mycorrhizal plant does not grow any better than an uninfected control, and growth may even be slightly retarded. There is no evidence that nitrogen fixation occurs in mycorrhizal plants, but it is well established that the mycorrhizal plant is able to absorb nutrients from its environment more efficiently than does a nonmycorrhizal one. This improved nutrient absorption is probably due to the greater surface area provided by the fungal mycelium.

The relationship between the fungus hyphae on the tree roots and the mushroom fruiting bodies seen on the surface of the soil has been well established in some cases. Fungus hyphae growing out from the surface of the root (Figure 14.25) frequently form multistranded structures called *rhizomorphs* (Figure 14.24), which eventually approach the soil surface, differentiate, and form fruiting-body structures. The nutrients for the growth of the fruiting bodies come from the plant roots since rhizomorphs that have been isolated from the tree roots never form fruiting bodies.

Endotrophic mycorrhizae   In endotrophic mycorrhizae the fungus is generally confined to the root itself and does not form a sheath or spread into the surrounding soil. Endotrophic mycorrhizae occur in a much greater variety of plants than do ectotrophic mycorrhizae; but the fungi have been difficult to culture, and an understanding of the ecological relationships of these mycorrhizae is virtually lacking.

14.11   Summary

We have seen in this chapter that microorganisms can enter into a variety of beneficial relationships with higher organisms. Some of these relationships are rather casual, whereas others are highly specific. Among all groups of microorganisms—bacteria, algae, fungi, and protozoa—are species that are symbionts, and they have developed relationships with many kinds of plants and animals. Although we have a fairly clear understanding of the underlying physiological mechanisms for a few symbioses of economic significance, such as the legume-*Rhizobium* relationship or the rumen symbiosis, for many others our knowledge is meager indeed.

In symbioses between microbes and higher organisms the enormous difference in size between the two members should be emphasized. The protoplasmic mass of a microbe is almost always vanishingly small compared to the mass of the multicellular creature with which it interacts. In spite of its size, the symbiotic microbe is able to affect its partner by producing such substances as enzymes, vitamins, or other growth factors active at quite high dilutions; and as we shall see in the next chapter, the pathogenic microbe also produces potent substances. The study of the interactions of microbes with higher organisms thus provides us with some dramatic insights into the biochemical mechanisms that microbes have evolved to ensure their evolutionary success.

Supplementary readings

Ahmadjian, V. *The Lichen Symbiosis.* Waltham, Mass.: Blaisdell Publishing Co., 1967.

Annison, E. F., and D. Lewis *Metabolism in the Rumen.* New York: John Wiley & Sons, Inc., 1959.

Brock, T. D. *Principles of Microbial Ecology.* Englewood Cliffs, N.J.: Prentice-Hall, Inc., 1966.

Buchner, P. *Endosymbiosis of Animals with Plant Microorganisms,* rev. Eng. ed. New York: Interscience Publishers, 1965.

Burnett, G. W., and H. W., Scherp *Oral Microbiology and Infectious Disease,* 2nd ed. Baltimore: The Williams & Wilkins Co., 1962.

Burris, R. H. "Biological nitrogen fixation," *Ann. Rev. Plant Physiol.* 17:155 (1966).

Hale, M. E., Jr. *The Biology of Lichens.* London: Edward Arnold (Publishers) Ltd., 1967.

Henry, S. M. (ed.) *Symbiosis,* vol. 1, *Associations of Microorganisms, Plants, and Marine Organisms.* New York: Academic Press, Inc., 1966.

Henry, S. M. (ed.) *Symbiosis,* vol. 2, *Associations of Invertebrates, Birds, Ruminants, and Other Biota.* New York: Academic Press, Inc., 1967.

Hungate, R. E. *The Rumen and Its Microbes.* New York: Academic Press, Inc., 1966.

Luckey, T. D. *Germfree Life and Gnotobiology.* New York: Academic Press, Inc., 1963.

Rosebury, T. *Microorganisms Indigenous to Man.* New York: McGraw-Hill Book Company, 1962.

Rovira, A. D. "Interactions between plant roots and soil microorganisms," *Ann. Rev. Microbiol.* 19:241 (1965).

Stewart, W. D. P. *Nitrogen Fixation in Plants.* London: Athlone Press, 1966.

Symbiotic Associations (Thirteenth Symposium, Society for General Microbiology). London: Cambridge University Press, 1963.

Microbiology developed as a science out of a need to understand the manner in which microorganisms can cause harmful effects in higher organisms. It should be clear from the discussion in the preceding chapter that not all effects of microorganisms are harmful; some effects may be beneficial or even essential for the life of the animal or plant. Yet the ability to cause infectious disease is one of the most dramatic and significant properties of microorganisms, and the understanding of the physiological and biochemical bases of infectious disease has led to therapeutic and preventive measures that have had far-reaching influence on medicine and human affairs. It is the purpose of the present chapter to discuss the nature of infectious disease and its control.

A *parasite* is an organism which lives on and causes damage to another living organism, called the *host*. The relationship between host and parasite is dynamic since each modifies the activities and functions of the other. The outcome of the host-parasite relationship depends on the *pathogenicity* or *virulence* of the parasite, that is, on its ability to inflict damage on the host, but it depends equally on the resistance or susceptibility of the host. Neither the virulence of the parasite nor the resistance of the host are constant factors, however, each varying under the influence of external factors or as a result of the host-parasite relationship itself.

In all cases there is some specificity in a host-parasite relationship in that a given parasite can grow only on a restricted variety of hosts. Often this specificity is such that the parasite is restricted to only a single host species, whereas in other cases the specificity is much broader.

Infection is not synonymous with *disease* because infection does not always lead to injury of the host, even if the pathogen is potentially virulent. In a diseased state the host is harmed in some way, whereas an infection refers to any situation in which a microorganism is established and growing in a host, whether or not the host is harmed.

A disease is a process, not a thing. It is a result of the interaction between host and pathogen that leads to damage to the host. It is not the disease which is transmitted from one host to another, but the pathogen. Tuberculosis, for example, is an infectious disease, but it is the causal agent that is transmitted from man to man, not the disease. We must look upon infectious disease not as an entity but as a series of interactions that may occur upon growth of a pathogen in a host. The naming of diseases may cause certain confusions in this regard. Most infectious diseases were named before the pathogen was known, and the names often reflect the disease symptoms or organ system involved. For example, endocarditis is an infection of the endocardium of the heart, peritonitis involves infection of the peritoneum, and pneumonia refers to infection of the lungs. In each of these cases the diseased state can be

fifteen

Host-parasite relationships

initiated by more than a single organism. In other cases, however, there is a marked specificity; tuberculosis is caused only by *Mycobacterium tuberculosis,* syphilis only by *Treponema pallidum,* and brucellosis only by species of *Brucella.* In these more specific cases there is often a tendency to confuse the disease with the pathogen; yet it is important to keep in mind the distinction between these two entities. The identification of the causal agent of an infectious disease requires the use of *Koch's postulates* (already outlined in Chapter 1), which were of great significance in the development of our understanding of infectious disease.

15.1 Microbial factors in invasion

A pathogen must first gain access to host tissues and multiply before damage can be done. This requires that the organism penetrate the surface, such as the skin, mucous membranes, or intestinal epithelium, which normally act as microbial barriers. Passage through the skin into the subcutaneous layers almost always occurs through wounds; only in rare instances is there any evidence that pathogens can penetrate through the unbroken skin. Access to the interior of the body is most readily accomplished in those areas where lymph glands are near the surface, such as the nasopharyngeal region, the tonsils, and the lymphoid follicles of the intestine. Many times small breaks or lesions in the mucous membrane in one of these regions will permit an initial entry. Motility may be of some value to an invader, although many pathogens are nonmotile.

After initial entry, the organism often remains localized and multiplies, leading to the production of a small *focus of infection,* such as the boil, carbuncle, or pimple that commonly arises from *Staphylococcus* and *Streptococcus* infections of the skin. Alternatively, the organisms may pass through the lymphatic vessels and be deposited in lymph nodes. If an organism reaches the blood it will be distributed to distant parts of the body, usually concentrating in the liver or spleen. Spread of the pathogen through the blood and lymph systems can lead to a generalized infection of the body, with the organism growing in a variety of tissues. If extensive bacterial growth in tissues occurs, some of the organisms are usually shed into the bloodstream in large numbers, a condition called *bacteremia.*

However, generalized infection of this type is rare; a more common situation is for the pathogen to localize in a specific organ. One of the best-studied cases of specific organotrophy is that involving *Brucella abortus,* the organism which causes brucellosis (infectious abortion) in cows. *Brucella abortus* grows preferentially in the fetus, placenta, and birth fluids of the pregnant cow. These tissues have significant amounts

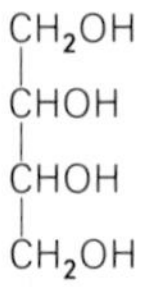

$$CH_2OH$$
$$|$$
$$CHOH$$
$$|$$
$$CHOH$$
$$|$$
$$CH_2OH$$

Erythritol

Figure 15.1

of the sugar alcohol, erythritol (Figure 15.1), a substance known to stimulate markedly the growth of *B. abortus,* and especially to enable it to grow unharmed within cells of the animal. In the cow, erythritol is absent from tissues other than those associated with the fetus. The conclusion seems inescapable that it is the presence of erythritol in the fetal tissues that is responsible for the marked specificity of the infection by *B. abortus.* Interestingly, although *B. abortus* infects man, it does not cause abortion and does not grow locally in fetal tissues, which correlates with the absence of erythritol in the human being. We have here a clear case of the biochemical basis of organ specificity; in most other cases of organ specificity the biochemical explanation is not yet known.

If it is to invade the body, a pathogen must possess mechanisms for evading the host defenses. There are a wide variety of *phagocytic cells* in the body, whose usual role is to ingest and destroy invading microorganisms (see Section 15.3). Some pathogens overcome this defense mechanism by possessing capsules or other cell surface structures that inhibit phagocytosis. The clearest case of the importance of a capsule in permitting invasion (and one that we have mentioned before) is that of *Diplococcus pneumoniae.* If only a few cells of a capsulated strain of this species are injected into a mouse, an infection is initiated that leads to death within a few days. On the other hand, noncapsulated mutants of capsulated strains are completely avirulent, and even injection of large numbers of bacteria usually causes no damage. If a noncapsulated strain is transformed genetically with DNA from a capsulated strain (see Section 12.1), both capsulation and virulence are restored at the same time. Furthermore, enzymatic removal of the capsule renders the organism noninvasive. Surface components other than capsules can also inhibit phagocytosis. Pathogenic streptococci produce on the cell surface a specific protein called "*M* protein," which apparently alters the surface properties of the cell in such a way that the phagocyte cannot act. As yet there is no biochemical explanation of how capsules and other cell surface components inhibit phagocytosis.

Some organisms are phagocytized but not killed within the phagocytes; instead they produce substances, called *leukocidins,* that destroy the phagocyte, the cells being released alive. Destroyed leukocytes make up much of the material of pus. Organisms that produce leukocidins are therefore usually *pyogenic* (pus-forming) and bring about characteristic abscesses. Streptococci and staphylococci are the most common leukocidin producers.

One group of organisms, the intracellular parasites, are readily phagocytized but are not killed, nor do they kill the phagocyte. Instead, they can remain alive for long periods of time and even reproduce within the phagocyte. Intracellular parasites are a diverse group. Some, such as

M. tuberculosis, Salmonella typhosa (cause of typhoid fever), and *Brucella melitensis* (cause of undulant fever) can live either intracellularly or extracellularly. In acute infections they multiply in the extracellular body fluids, but in chronic conditions they may live only intracellularly. When growing intracellularly the organism is protected from immune mechanisms of the host and is less susceptible to drug therapy. Some important intracellular parasites are unable to grow outside of living cells. Included in this category are the chlamydiae, the rickettsiae, and some protozoa, such as that which causes malaria; discussion of the biochemical reasons why these parasites can develop only as intracellular parasites is presented in Chapters 18 and 19.

Enzymes involved in invasion Streptococci, staphylococci, pneumococci, and certain clostridia produce *hyaluronidase,* an enzyme that promotes spreading of organisms in tissues by breaking down hyaluronic acid, a polysaccharide that functions in the body as a tissue cement. Production of this enzyme may therefore enable these organisms to spread from an initial focus. Clostridia that cause gas gangrene produce *collagenase,* which breaks down the collagen network supporting the tissues; the resulting dissolution of tissue is a factor in enabling these organisms to spread through the body. Fibrin clots are often formed by the host in a region of microbial invasion and serve to wall off the organism and prevent its spread through the body. Some organisms are able to produce fibrinolytic enzymes to dissolve these clots, thus making possible further invasion. One such fibrinolytic substance, produced by streptococci, is known as *streptokinase.* On the other hand, some organisms produce enzymes that actually promote fibrin clotting, which causes localization of the organism rather than its spread. The best-studied fibrin clotting enzyme is *coagulase,* produced by many staphylococci, which causes the fibrin material to be deposited on the cocci and may offer them protection from phagocytosis. Production of coagulase probably accounts for localization of many staphylococcal infections in boils and pimples.

15.2   *Factors in microbial pathogenicity*

The manner in which pathogens bring about damage to the host are diverse. Only rarely are symptoms due to the presence of large numbers of microorganisms per se. Although a large mass of cells can block vessels or heart valves or clog the air passages of the lungs, in most cases more specific factors are involved. Many pathogens produce toxins that are responsible for all or much of the host damage.

Exotoxins Toxins released extracellularly as the organism grows are called *exotoxins;* these are produced primarily by Gram-positive bacteria. Released from the organism that is established at a given focus of infection, these toxins may travel to distant parts of the body and hence cause damage in regions far removed from the site of microbial growth. Table 15.1 provides a summary of the properties and actions of some of the best known exotoxins.

The toxin produced by *Corynebacterium diphtheriae,* the causal agent of diphtheria, was the first exotoxin to be discovered. It is a protein of molecular weight 72,000, which differs markedly in its action on different animal species; rats and mice are relatively resistant, whereas man, rabbits, guinea pigs, and birds are susceptible. Diphtheria toxin is very potent in its action, with only 200 to 400 molecules being required to cause the death of a single cell. It binds irreversibly to the cell, and within a few hours the cell loses its ability to synthesize protein because the toxin blocks the transfer of an amino acid from a tRNA to the growing peptide

Table 15.1 *Exotoxins produced by certain bacteria pathogenic for man*

	Disease	Toxin	Action
Clostridium botulinum	Botulism	Neurotoxin	Paralysis
C. tetani	Tetanus	Neurotoxin	Paralysis
C. perfringens	Gas gangrene	α-Toxin	Hemolysis (lecithinase)
		β-Toxin	Hemolysis (lecithinase)
		γ-Toxin	Hemolysis (lecithinase)
		δ-Toxin	Hemolysis (lecithinase)
		ϵ-Toxin	?
		η-Toxin	?
		θ-Toxin	Hemolysis (cardiotoxin)
		ι-Toxin	?
		κ-Toxin	Collagenase
		λ-Toxin	Protease
Corynebacterium diphtheriae	Diphtheria	Diphtheria toxin	Inhibits protein synthesis
Staphylococcus aureus	Pyogenic infections	α-Toxin	Hemolysis
		Leucocidin	Destroys leukocytes
		β-Toxin	Hemolysis
		γ-Toxin	Kills cells
		δ-Toxin	Hemolysis, leucolysis
		Enterotoxin	Induces vomiting, diarrhea
Streptococcus pyogenes	Pyogenic infections, tonsillitis, and scarlet fever	Streptolysin O	Hemolysis
		Streptolysin S	Hemolysis
		Erythrogenic toxin	Causes scarlet fever rash

chain (see Section 7.6). Diphtheria toxin is formed only by strains of *C. diphtheriae* that are lysogenized by phage β, and its production is thought to be coded for by genetic information present in the phage genome. One interesting result of this phenomenon is that nontoxigenic and hence nonpathogenic strains of *C. diphtheriae* can be converted to pathogenic strains when infected with the phage (Section 12.2). A nongenetic factor of significance for toxin production is the concentration of iron present in the environment in which the bacteria are growing. In media containing sufficient iron for optimal growth no toxin is produced, whereas when the iron concentration is reduced to suboptimal levels, toxin production occurs. It seems that iron deprivation causes the induction of the prophage β and release of toxin.

All the symptoms of infection with *C. diphtheriae* are due to the action of diphtheria toxin, as is shown by the following lines of evidence: (1) Only toxigenic strains of *C. diphtheriae* initiate the disease; (2) upon infection of an animal with *C. diphtheriae*, the toxin can be shown to be produced in the animal; (3) all the disease symptoms of diphtheria, including death, can be induced by the injection of highly purified diphtheria toxin; (4) if injected prior to infection, specific antibody (see Section 15.4) directed against the toxin is able to neutralize disease symptoms brought about either by infection or by injection of toxin.

Tetanus toxin is one of the most potent toxins known. It is produced by *Clostridium tetani,* an obligately anaerobic bacterium that infects wounds and causes the neurologic disease tetanus. The toxin is a protein of molecular weight 67,000, which upon entry into the central nervous system becomes fixed to nerve synapses, binding specifically to the glycolipid *sphingosine*. The spastic paralysis that the toxin induces comes from its action in blocking nerve transmission. The binding of the toxin to nerve synapses is essentially irreversible, and therapy after the disease has set in is thus ineffective.

Botulinum, the toxin produced by the food-spoilage organism *Clostridium botulinum,* is an even more potent neurotoxin, and is often called the most poisonous substance known to man.

Various pathogens produce a number of proteins that are able to act on the animal cell membrane, inducing cell lysis and hence cell death. The action of these toxins is most easily detected with red blood cells, hence they are often called *hemolysins;* in probably all cases, however, they also work on cells other than erythrocytes. The production of such toxins is most readily demonstrated by streaking the organism on a blood agar plate; during growth of the colony, some of the hemolysin is released and lyses the surrounding red cells, clearing a typical zone (Plate 6: Figure 15.2). Some hemolysins have been shown to be enzymes that attack the phospholipid of the host cell membrane. Because the

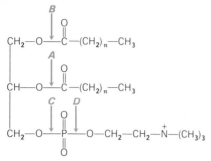

Lipases: hydrolysis at *A* or *B*

Phospholipases: hydrolysis at *C* (for example, *Clostridium perfringens* α-toxin); hydrolysis at *D* (for example, plant phospholipases)

Figure 15.3 Phospholipid and site of action of lipases and phospholipases.

phospholipid lecithin is often used as a substrate, these enzymes are called *lecithinases* or *phospholipases*. There are several sites in the phospholipid molecule that can be hydrolyzed (Figure 15.3). Hemolysins that attack the phospholipid of the cell membrane can be neutralized by adding purified phospholipid to the enzyme before cells are added. Since the cell membranes of all organisms, both procaryotes and eucaryotes, contain phospholipids, it is not surprising that hemolysins that are phospholipases sometimes destroy bacterial as well as animal cell membranes. Some hemolysins are not phospholipases, however. Streptolysin O, a hemolysin produced by streptococci, affects the sterols of the host cell membrane, and its action is neutralized by the addition of cholesterol or other sterols.

Endotoxins   Gram-negative bacteria produce as part of the outer layer of their cell walls lipopolysaccharides (Figure 15.4 and Section 2.3), which under many conditions are toxic. These are called ''endotoxins'' because they are generally cell-bound and are released in large amounts only when cells lyse. Such materials are rarely or never found in Gram-positive bacteria, which fact is probably related to the marked differences in cell-wall structure of these two groups of organisms (Section 2.3). When injected into an animal, endotoxins cause fever, diarrhea, hemorrhagic shock, and other tissue damage. Some effects of endotoxin are counteracted by steroids, hence the occasional use of steroids in the therapy of some bacterial diseases.

Allergic responses   In many cases, disease symptoms elicited in the host are the result of allergic reactions to products of the pathogen (see Section 15.5). Diseases in which many or all of the symptoms are due to such hypersensitivity or allergy include tuberculosis and two diseases initiated by streptococci: rheumatic fever and glomerular nephritis. At least some of the effects of endotoxins are allergic in nature, as they are not induced in germ-free animals. Recall (Chapter 14) that there are large numbers of Gram-negative bacteria in the normal flora of the intestinal tract; endotoxin production by these bacteria sensitizes the animal, so that later challenge with larger doses elicits an allergic reaction.

Microbial virulence   The word *virulence* refers to the relative ability of a parasite to cause disease, that is, its degree of pathogenicity. Virulence is conditioned by the *invasiveness* of the organism and by its *toxigenicity*. Both are quantitative properties and may vary over a wide range from very high to very low. An organism that is only weakly invasive may still be virulent if it is highly toxigenic. A good example of this is the organism *C. tetani*. The cells of this organism rarely leave the wound

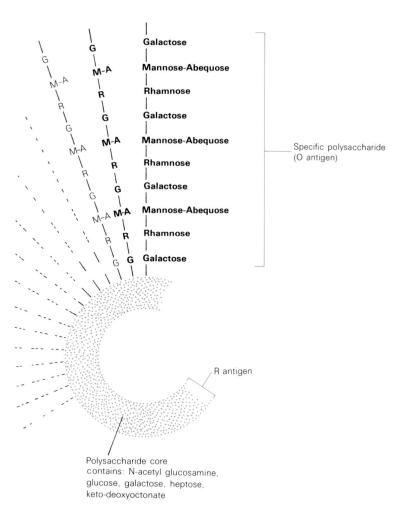

Galactose

Mannose-Abequose

Rhamnose

Galactose

Mannose-Abequose — Specific polysaccharide (O antigen)

Rhamnose

Galactose

Mannose-Abequose

Rhamnose

Galactose

R antigen

Polysaccharide core
contains: N-acetyl glucosamine,
glucose, galactose, heptose,
keto-deoxyoctonate

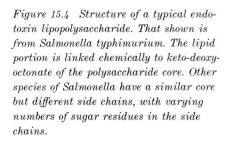

Figure 15.4 Structure of a typical endotoxin lipopolysaccharide. That shown is from Salmonella typhimurium. The lipid portion is linked chemically to keto-deoxyoctonate of the polysaccharide core. Other species of Salmonella have a similar core but different side chains, with varying numbers of sugar residues in the side chains.

where they were first deposited yet they are able to bring about death of the host because they produce the potent tetanus toxin, which can move to distant parts of the body and initiate paralysis. On the other hand, a weakly toxigenic organism may still be able to produce disease if it is highly invasive. *Diplococcus pneumoniae* is not known to produce any toxin, but is able to cause extensive damage and even death because it is highly invasive, being able to grow in the lung tissues in enormous numbers and initiate responses in the host that lead to disturbance of the functions of the lung. These two organisms exemplify the extremes of invasiveness and toxigenicity; most pathogens fall somewhere between these two extremes.

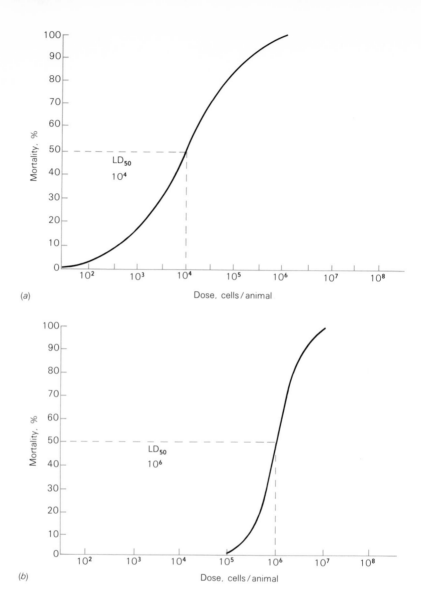

Figure 15.5   Relationship between number of bacteria injected and mortality. In general, at least 10 animals are injected with each dose. After a standard time period, the number of animals that have died is counted and the percent mortality is calculated. From the graph, the dose of bacteria that caused 50 percent mortality can be estimated. This dose, the LD_{50}, is a measure of the virulence of the bacteria. In the example given, one bacterial strain (a) is 100 times more virulent than the other (b), as its LD_{50} is 0.01 that of (b).

The virulence of a pathogen is quantified by determining the average number of cells that will cause death (or some other pathological response) in 50 percent of the animals of a group after a stated period of time. This number is called the LD_{50} (lethal dose for 50 percent) (Figure 15.5). Properties of pathogens that are involved in invasiveness and toxigenicity are genetically determined. A mutant that has lost the ability to produce a capsule, toxin, or other virulence-inducing factor would be avirulent and hence would be unable to initiate the events leading to disease. Virulence can be controlled either by the function of a single gene (for example, toxigenic versus nontoxigenic) or through the combined effects of several genes. It is often observed that the virulence of an organism can be increased by animal passage, the animal body providing a selective environment for the growth of virulent mutants present in the population.

In some cases, only a single virulent microbial cell is able to grow and establish a fatal infection, but more frequently a large number of cells is required, probably in order to overwhelm the host's defenses by releasing a toxin or some other cellular product. Once the host defense is overcome, it is then possible for the pathogen to grow rapidly and cause the pathologic changes leading to death. Agents or treatments (for example, certain drugs, ionizing radiation) that lower host resistance increase the virulence of some pathogens by making possible the successful establishment of infection from smaller inocula.

It is often observed that, when pathogens are kept in laboratory culture and not passed through animals for long periods, their virulence is decreased or even completely lost. Such organisms are said to be *attenuated*. Through successive transfers to fresh media mutants are selectively favored, and attenuation probably occurs because nonvirulent mutants grow faster. Attenuation often occurs more readily when culture conditions are not optimal for the species. If an attenuated culture is reinoculated into an animal, virulent organisms are sometimes reisolated, but in many cases loss of virulence is permanent. Attenuated strains find frequent use in the production of vaccines (see Section 15.6).

It is often found that infection with one pathogen alters the host defenses in such a way that other organisms, which ordinarily are not invasive, are now able to grow. These are called *secondary invaders*. In some cases the secondary invader may actually be responsible for many other more serious disease symptoms. For instance, infection with influenza virus ordinarily leads only to a mild self-limiting disease, yet (especially in older people) influenza infection so alters the general resistance that *D. pneumoniae* can grow and initiate a fatal bacterial pneumonia. The presence of secondary invaders frequently complicates the microbiological diagnosis of the causal agents of human diseases since the secondary invader so often predominates and the primary invader, without which the disease would not have become established, is in a great minority and is missed upon culture.

Discussions of specific diseases caused by bacteria are given in Chapter 18, by fungi and protozoa in Chapter 19, and by viruses in Chapter 10.

Animals vary widely in their susceptibility or resistance to infectious diseases. Two kinds of resistance are recognized: resistance to specific organisms or perhaps even to specific strains; and nonspecific resistance to a variety of related or unrelated pathogens. Both types of resistance may be inherent properties of the animal, or they may be acquired or induced in the animal by some specific immunization procedure. Resistance may be manifested either through a prevention of significant growth of the pathogen in the body or through a neutralization of the pathogenic or toxic properties of the pathogen without any marked effects on its growth.

Cellular factors in resistance   The invasion of a pathogen is counteracted by the action of phagocytes (literally, "cells that eat"), which ingest and destroy it (Figure 15.6). Histologists recognize a variety of phagocytic cell types involved in the destruction of microbes. One important type is the *granulocyte,* a small, actively motile cell containing many distinctly staining membrane-bound organelles called lysosomes or granules, which contain several bactericidal substances and enzymes such as lysozyme, proteases, phosphatases, nucleases, and lipases. Granulocytes are short-lived cells which are found predominantly in the bloodstream and bone marrow and which appear in large numbers during the acute phase

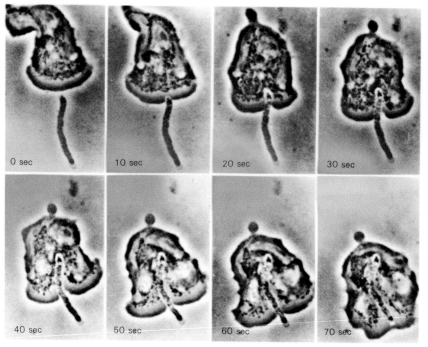

Figure 15.6   Phagocytosis: engulfment and digestion of a Bacillus megaterium cell by a human phagocyte, observed by phase-contrast microscopy. Magnification, 1,350×. [From J. G. Hirsch: J. Exptl. Med. 116:827 (1962).]

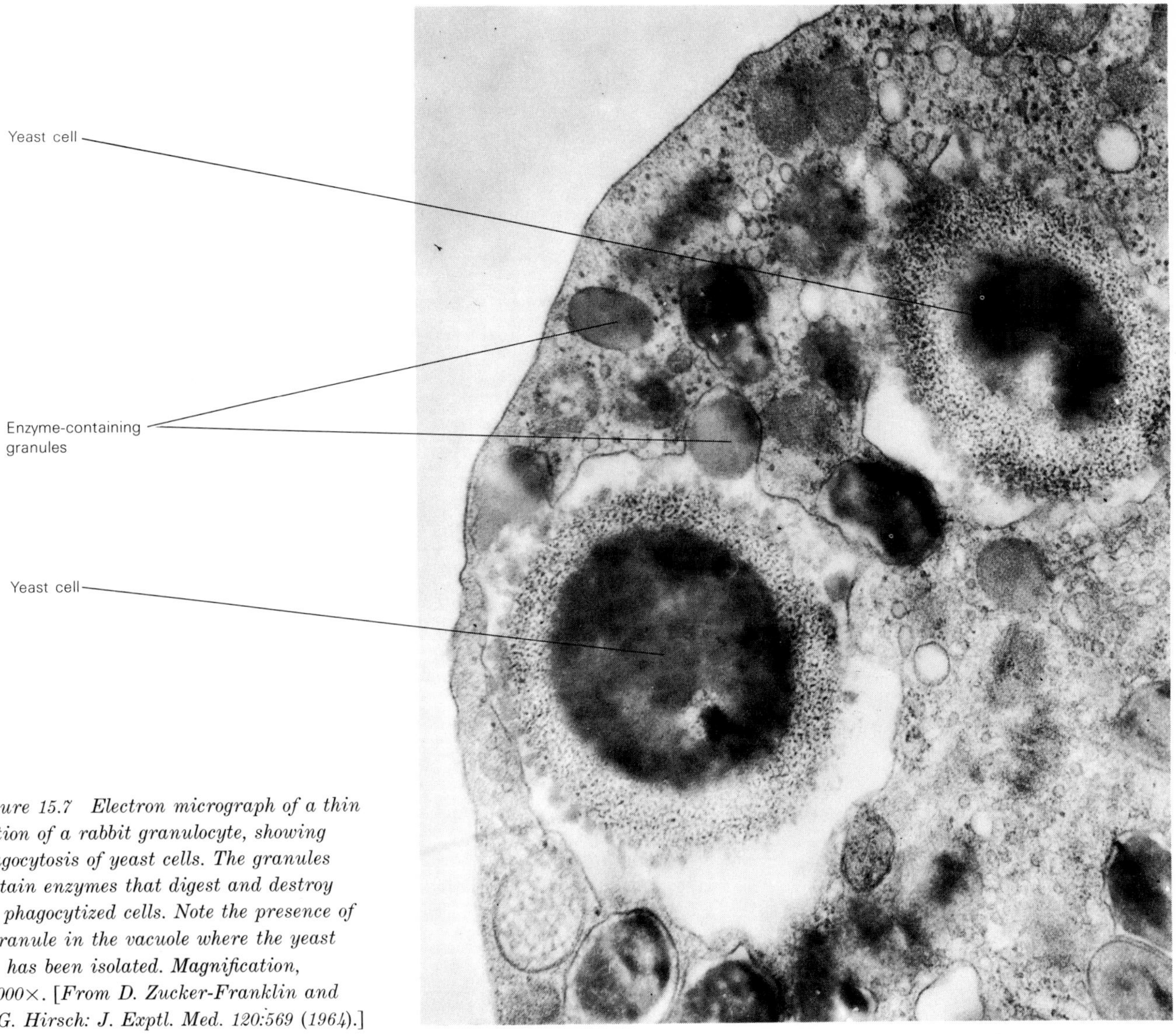

Yeast cell

Enzyme-containing
granules

Yeast cell

*Figure 15.7   Electron micrograph of a thin
section of a rabbit granulocyte, showing
phagocytosis of yeast cells. The granules
contain enzymes that digest and destroy
the phagocytized cells. Note the presence of
a granule in the vacuole where the yeast
cell has been isolated. Magnification,
33,000×. [From D. Zucker-Franklin and
J. G. Hirsch: J. Exptl. Med. 120:569 (1964).]*

of an infection. Upon ingestion a microbial cell is taken up into a region
of the phagocyte containing granules, and the latter become disrupted
and release the enzymes, which digest and destroy the invader (Figure
15.7). During the process of phagocytosis, the metabolism of the granulocyte converts from aerobic pathways to anaerobic glycolysis. Glycolysis
results in the formation of lactic acid and a consequent drop in pH; this

15.3   Mechanisms of resistance
to disease

lowered pH is partly responsible for the death of the microbe, and it also provides an environment in which the hydrolytic enzymes, all of which have acid pH optima, can act more effectively. The initial act of phagocytosis conditions a cell so that it is more efficient in subsequent phagocytic action—a cell that has recently phagocytized can take up bacteria about 10 times better than a cell that has not.

Another type of phagocytic cell is the *macrophage,* or *monocyte,* which has few lysosomes and hence does not appear granular when stained. Macrophages are of two types: *wandering cells,* which are found free in the bloostream and lymph vessels (where they are called "monocytes"), and fixed phagocytes or *histiocytes,* which are found embedded in various tissues of the body and have only a limited mobility. The system of fixed phagocytes is often called the *reticuloendothelial system* (RE system); it consists of the large macrophages of the loose connective tissue, the lymphatic tissue, spleen, and liver. The RE system can be seen experimentally by injecting intravenously into an animal an insoluble dye suspension; the dye particles are phagocytized by the various cells of the RE system, and since the dye is indigestible, it remains in place in the cells and can be seen upon autopsy. Macrophages are long-lived cells that play a role in both the acute and chronic phases of infection and they are also active in antibody formation (see Section 15.5).

Phagocytes respond chemotactically to an invading microbe and work best when they can trap a microbial cell upon a surface, such as a vessel wall, a fibrin clot, or even particulate macromolecules.

The efficiency of the phagocytic system in clearing foreign particles and microbial cells from the bloodstream and lymph system is high. For instance, 80 to 90 percent of the particles entering the liver are removed in one passage. The rate of clearance of intravenously injected bacteria in the rabbit is shown in Figure 15.8. This figure also shows the differences between virulent and avirulent strains; initially both are attacked almost equally well, but the virulent strain is eventually able to grow and overwhelm the phagocytic system.

The tissues of the animal react to infection and to a variety of other external stimuli by an *inflammatory response,* the characteristic symptoms of which are redness, swelling, heat, and pain. The initial effect of the foreign stimulus is to cause local dilation of blood vessels and an increase in capillary permeability. This results in an increased flow of blood and passage of fluid out of the circulatory system into the tissues, causing swelling (edema). Phagocytes also pass through the capillary walls to the inflamed area. Initially granulocytes appear, followed later by macrophages. Within the inflamed area, a fibrin clot is usually formed, which in many instances will localize the invading microbe. But as we noted above, those pathogens which produce fibrinolytic enzymes may be able

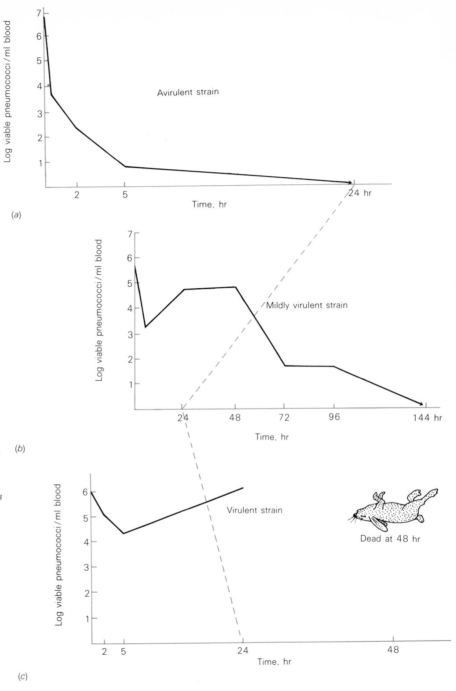

Figure 15.8 Rate of clearance of Diplococcus pneumoniae from the blood of the rabbit. At zero time, rabbits were given injections of live bacteria. At intervals, samples of blood were removed and viable counts performed. In the case of the virulent strain (c), clearance begins, but bacterial growth soon overwhelms the host defenses and the rabbit dies. With the mildly virulent strain (b), some growth occurs but the host defenses continue to act and eventually all bacteria are removed. [Data of H. D. Wright: J. Pathol. Bacteriol. 30:185 (1927).]

15.3   Mechanisms of resistance
to disease

to escape and continue to invade the body. One of the factors involved in initiating the inflammatory response is the chemical *histamine,* which is released by damaged cells and acts on the capillaries to increase their permeability. The use of antihistamine drugs in controlling inflammation is based on their ability to counteract some of the effects of histamine. However, although antihistamines may reduce the painful aspects of inflammation, they also counteract some of the beneficial effects so that their use is not an unmixed blessing.

Inflammation is one of the most important and ubiquitous aspects of host defense against invading microorganisms and is present in a small way virtually continuously. However, inflammation is also an important aspect of microbial pathogenesis since the inflammatory response elicited by an invading microorganism can result in considerable host damage. Pathogens vary in the degree to which they induce inflammation, and in some cases the host may react in a major way to only a minor microbial invasion and hence become seriously damaged.

15.4   *Specific immune mechanisms*

In addition to the properties described above that confer resistance to a variety of pathogens, higher animals possess a highly sophisticated mechanism, the *immunological response,* for developing resistance to specific microorganisms. The immunological response is a general system for the neutralization of foreign macromolecules through the formation of specific *antibodies;* since an invading microorganism contains a variety of macromolecules foreign to the host, antibodies develop against it. We shall discuss first the nature and manner of formation of antibodies and then show how they confer specific resistance to infection.

Several features of the immunological response will be listed here and then discussed in some detail below: (1) Antibodies are formed against a variety of foreign macromolecules, but ordinarily not against macromolecules of the animal's own tissues; thus, the animal is able to distinguish between its own and foreign macromolecules. (2) In virtually every case, antibody against a foreign macromolecule is not formed unless the animal is challenged with the foreign substance. (3) Many but not all foreign macromolecules elicit the immunological response; those that do are called *antigens.* (4) There is a high specificity in the interaction between antibody and antigen, although antibodies will react against closely related antigens but usually with reduced efficiency. (5) Antigen-antibody reactions are manifested in a wide variety of ways, depending on the nature of the antigen, the antibody, and the environment in which the reaction takes place. (6) Not all antigen-antibody reactions are beneficial; some, such as those involving allergy and hypersensitivity, are more or less harmful. (7) The high specificity of antigen-antibody reactions makes them useful in many research and diagnostic procedures. (8) The immunological response can be made the basis of specific immunization procedures for the prevention and control of specific diseases.

The distinction between the words "immune" and "immunological" should be clear. *Immune* refers to the ability of the animal to resist infectious disease, whereas *immunological* refers to the processes of antibody formation and antigen-antibody reaction, whether or not these lead to immunity to a particular disease. The word "immunological" was first used in the context of immunity because antibodies were first discovered through studies on the development of immunity, but as further study revealed that antibodies were involved in other situations, the meaning of the word has been broadened to cover all aspects of the antibody response.

Antigens "Antigens" are defined as substances that induce under appropriate conditions the formation in the animal of specific antibodies; antigens then react specifically with these antibodies. Antigens are distinguished from *haptens* (see below) by the fact that although the latter react specifically with antibodies they do not induce their formation. Thus, the decision as to whether or not a substance is an antigen must be operational; the material must be injected into an animal that is then shown to form antibodies. However, the antibody response is not always an automatic reaction whenever an antigen is injected. It depends very much on the species (or even strain) of the animal, its previous history, the route of injection, the frequency of injection, the concentration of antigen injected, and the presence or absence of various nonspecific antibody-stimulating agents administered with the antigen, which are known as *adjuvants*.

An enormous variety of macromolecules can act as antigens under appropriate conditions. These include virtually all proteins and lipoproteins, many polysaccharides, some nucleic acids, and certain of the teichoic acids. One important requirement is that the molecules must be of fairly large molecular weight, usually greater than 10,000. However, the antibody is directed not against the antigenic macromolecule as a whole, but only against restricted portions of the molecule that are called its *antigenic determinants*. The nature of antigenic determinants has been most clearly revealed by studies on proteins that have been chemically altered by attaching defined substituents to them. For instance, it is possible to couple an aromatic compound such as the *p*-aminobenzene arsonate derivative shown in Figure 15.9 to a normal serum protein such as

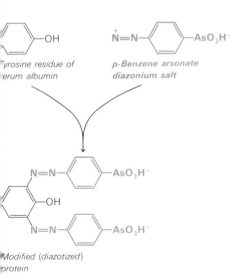

Figure 15.9 Chemical modification of a protein (serum albumin) to make an artificial antigen.

Polypeptide chain

Tyrosine residue of serum albumin

p-Benzene arsonate diazonium salt

Modified (diazotized) protein

serum albumin. When this modified protein is injected into an animal, antibodies that combine specifically with the modified protein are produced, but they do not react with the unmodified protein. (Some antibodies that react with serum albumin itself are also produced.) This shows that it is the aromatic side chain to which the antibody reacts; the side chain can therefore be called the antigenic determinant. We can demonstrate in the laboratory that if uncomplexed p-aminobenzene arsonic acid is added to the antibody before the modified protein is added, it inhibits the antigen-antibody reaction by combining with the antibody and preventing the approach of the antigen. Since the free aromatic compound itself does not induce antibody formation, it is not an antigen and is called a *hapten*.

Experiments using chemically modified proteins have been carried out extensively, and these provide some of the best information on the nature of antigenic determinants. Those recognized include sugars, amino acid side chains, organic acids and bases, hydrocarbons, and aromatic groups. Antibodies are formed most readily to determinants that project out vertically from the molecule or to terminal residues of a polymer chain. In general, the specificity of antibodies is comparable to that of enzymes (see Chapter 4) in distinguishing closely related determinants. For instance, antibodies can distinguish between the sugars glucose and galactose, which differ only in the position of the hydroxyl group on carbon 3. However, specificity is not absolute, and an antibody will react at least to some extent with determinants related to the one that induced its formation. That which induced the antibody is called the *homologous antigen* and others that react with the antibody are called *heterologous antigens*. As we shall see below, antibodies also are not uniform substances but exist in a spectrum of molecules having varying degrees of affinity for the antigenic determinant.

Antibodies   Antibodies are found predominantly in the serum fraction of the blood, although they may also be found in other body fluids, as well as in milk. Serum is the fluid portion of the blood that is left when the blood cells and the materials responsible for clotting (fibrin and cofactors) are removed. Serum containing antibody is often called *antiserum*. When serum proteins are separated by electrophoresis, four predominant fractions are seen: serum albumin and alpha, beta, and gamma globulins. Antibody activity occurs predominantly in the gamma globulin fraction (Figure 15.10), which is composed of many distinct proteins. Those gamma globulin molecules with antibody properties are called *immunoglobulins* and can be separated into five major subclasses on the basis of their physical, chemical, and immunological properties: gamma G, gamma A, gamma M, gamma D, and gamma E (Table 15.2). Even among these classes, further subdivisions are recognized, which will not

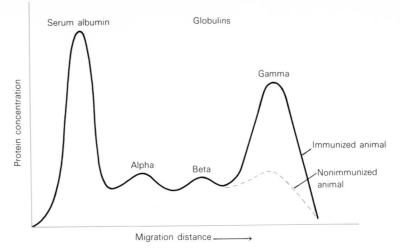

Figure 15.10 Electrophoresis of serum proteins. Antibodies are in the gamma globulin fraction; note the difference in amount of gamma globulin between the immunized and the nonimmunized animal.

concern us here. Antibody molecules specific for a given antigenic determinant are found in each of the several classes, even in a single immunized individual. Upon initial immunization, it is gamma M (the so-called *macroglobulin*), a protein with a molecular weight of about 900,000, that first appears; gamma G appears later. In most individuals about 80 percent of the immunoglobulins are gamma G proteins, and these have therefore been studied most extensively.

Gamma G immunoglobulins have a molecular weight of about 150,000 and are composed of four polypeptide chains, connected together by

Table 15.2   Properties of the immunoglobulins[a]

	Sedimentation rate, S	MW	Concentration in adult serum, g/liter	Proportion of total antibody, %	Complement-fixing ability	Distribution
Gamma A	7	160,000	0.5–1.5	10	Absent	External secretions; extracellular and intravascular fluids
Gamma G	7	160,000	5–15	80	Present	Extracellular and intravascular fluids
Gamma M	19	900,000	0.5–0.7	5–10	Present	Intravascular fluids
Gamma D	7	160,000	0.03	1–3	?	Intravascular fluids

[a]Data not yet available for gamma E.

S—S bridges (Figure 15.11). The two light chains are identical, as are the two heavy chains, and the molecule as a whole is symmetrical. When the molecule is treated with a reducing compound and the proteolytic enzyme papain under carefully controlled conditions, it breaks into several fragments (Figure 15.11). The two fragments containing the complete light chain plus the left end of the heavy chain are the portions that combine with antigen and are called "Fab" (antigen-binding fragment), whereas the fragment containing the right ends of the two heavy chains, called "Fc," does not combine with antigen, although its presence does contribute to both the specificity and the avidity of antigen-antibody conformation. Therefore, each antibody molecule of this type contains *two* antigen-combining sites. This *bivalency* is of considerable importance in the manner in which some antigen-antibody reactions occur (see below). However, not all antibodies are bivalent; some are univalent and others are multivalent. Within Fab, the antigen-combining site is localized in the heavy-chain fragment, the light-chain portion probably making only a small contribution to the antigen-antibody reaction. Within the population of gamma G molecules with a single antigen-combining site there is considerable variation in the amino acid sequence of both light and heavy chains. The immunoglobulins also contain small amounts of carbohydrate, mainly hexose and hexosamine, which are attached to portions of the heavy chain; however, the carbohydrate is not involved in the antigen-combining site.

Antigen-antibody reactions Antigen-antibody reactions are most easily studied in vitro using preparations of antigens and antisera; this is

Figure 15.11 Arrangement of the polypeptide chains of gamma G antibody molecules and sites where breakdown occurs. The Fab pieces contain the antigen-binding sites.

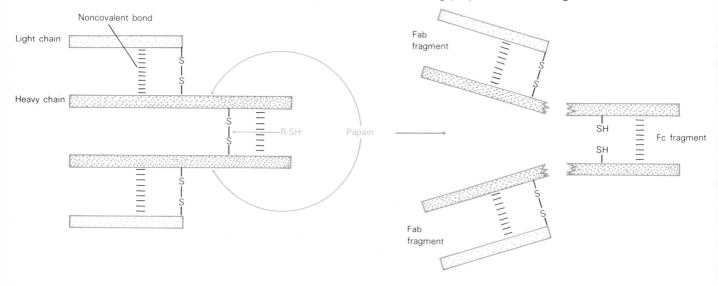

Figure 14.1 (a) Orange- and gray-colored crustose lichens growing on the surface of an Etruscan tomb at Cerveteri, Italy; (b) a foliose lichen, Letharia vulpina, on a dead lodgepole pine in Yellowstone National Park.

(a)

(b)

Figure 14.15 An ant in its fungus garden. This is Mycetosoritis hartmanii, a leaf-cutting ant. The white filaments are the fungus hyphae, which are growing on leaf fragments (yellow material) that the ant has cut and placed in its garden. (Courtesy of John C. Moser, U.S. Forest Service.)

Plate five

Figure 14.17 (far left) Sea anemone (a coelenterate). The brilliant green color results from the presence of intracellular symbiotic algae, which are dinoflagellates. The fish is a clownfish. Photograph taken on a coral reef of the Great Barrier Reef of Australia. (Photograph by Paul A. Zahl, © 1957 by National Geographic Society.)

Figure 15.2 (left) Zones of hemolysis around colonies of Staphylococcus aureus on a blood-agar plate. (Courtesy of R. K. Scherer, National Animal Disease Laboratory, Ames, Iowa.)

(a)

(b)

Figure 15.19 (a) Photomicrograph of fluorescent-antibody reaction against two bacterial species, with yellow-green indicating Clostridium septicum stained with antibody conjugated with the yellow-green fluorescing dye fluorescein isothiocyanate and red indicating C. chauvei stained with antibody conjugated with lissamine rhodamine B (for this micrograph, thanks are due to the Anaerobic Department, Wellcome Research Laboratories, Beckenham, Kent, England); (b) specific staining with fluorescent antibody to Lactobacillus helveticus (green) in thymus tissue amid cells of Streptococcus and Salmonella counterstained with pontachrome [from E. O. Hokenson and P. A. Hansen, Acta Histochemica, suppl. VII: 181 (1967)].

Plate six

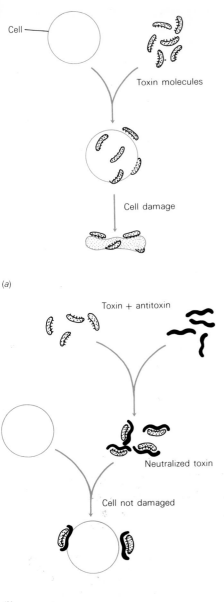

Cell

Toxin molecules

Cell damage

(a)

Toxin + antitoxin

Neutralized toxin

Cell not damaged

(b)

Figure 15.12 (a) Toxin action and (b) mechanism of neutralization of a toxin by a specific antibody (antitoxin).

called *serology*. A variety of different kinds of serological reactions can be observed, depending on the natures of the antigen and antibody and on the conditions chosen for reaction. Serology has many ramifications, only a few of which will be discussed here. We shall concentrate on those reactions that are of most importance for infectious disease.

Neutralization of microbial toxins by specific antibody can occur when toxin molecules and antibody molecules combine in such a way that the antigenic determinant on the toxin that is responsible for cell damage is blocked (Figure 15.12). Neutralization reactions of this type are known for a wide variety of exotoxins, including most of those listed in Table 15.1. Reactions of similar nature also occur between viruses and their specific antibodies. For instance, antibodies directed against the protein coats of viruses may prevent the adsorption of the viruses to their hosts. Neutralizing antibodies are also known for a variety of enzymes; these act by combining with the enzyme at or near the active site and prevent the formation of the enzyme-substrate complex. (The use of such antibodies in studies on the molecular genetics of enzymes was discussed briefly in Chapter 11.) Neutralizing antibody requires only a single antigen-combining site for action and thus can be univalent, although bivalent or multivalent antibodies can also neutralize. An antiserum containing neutralizing antibody against a toxin is sometimes referred to as an *antitoxin*.

When an antibody has two combining sites (that is, is bivalent, as for example gamma G immunoglobulin), it is possible for each site to combine with a separate antigen molecule; and if the antigen also has more than one combining site, a *precipitate* may develop, consisting of aggregates of antibody and antigen molecules (see Figure 15.13). Precipitation reactions for a wide variety of soluble polysaccharide and protein antigens are known. Because they are easily observed in vitro, precipitation reactions are very useful serological tests, especially in the quantitative measurement of antibody concentrations. Precipitation occurs maximally only when there are optimal proportions of the two reacting substances since, when either reactant occurs in excess, the formation of large antigen-antibody aggregates is not possible.

Precipitation can also be inhibited by the hapten corresponding to the antigenic determinant. In fact, hapten inhibition is an extremely useful way of studying the specificity of serological reactions since the inhibiting power of a series of related haptens can be compared quantitatively. Such studies show that although antigen-antibody reactions have a high degree of specificity, this specificity is not absolute, and antibody will almost always react with heterologous antigens. Also, antisera almost always contain mixtures of antibodies with varying specificities, and such heterogeneity must be taken into consideration in any serological study. Cross reactions can often be eliminated or minimized by allowing the antiserum

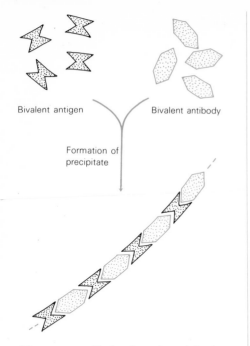

Bivalent antigen   Bivalent antibody

Formation of
precipitate

*Figure 15.13 Mechanism of precipitation
of a soluble antigen by a specific antibody.*

*Figure 15.14 Formation of precipitin
bands in agar gel due to antigen-antibody
reactions. The antiserum produced by immunization of rabbits with whole cells of
Proteus mirabilis was placed in the wells
labeled S. In wells A, B, and C are extracts of P. mirabilis cells containing soluble antigens. There are two precipitin
bands, one much stronger (a) than the
other. In well C, in (b), the extract is from
a strain lacking the antigens. [From
C. Weibull, W. D. Bickel, W. T. Haskins,
K. C. Milner, and E. Ribi: J. Bacteriol.
93:1143 (1967).]*

to react first with heterologous antigens and then removing the precipitate that forms. This will remove antibodies specific for the heterologous antigen and will leave behind in the supernatant those antibodies which do not react with heterologous antigens but react with the homologous antigen.

Precipitation reactions carried out in agar gels have in recent years proved of considerable utility in the study of the specificity of antigen-antibody reactions. Precipitation bands form in the gel in the region where antibody and antigen meet in equivalent proportions (Figure 15.14). The shapes of the precipitation bands and the distance from the wells are characteristic for the reacting substances, and it is possible to determine whether two antigens reacting against antibodies in an antiserum are identical or different by observing the bands formed when the two antigens are placed in adjacent wells near the antiserum well.

Gel diffusion is also useful in diagnostic work with pathogens as it permits a rapid characterization of the ability of cultures of suspected pathogens to produce toxins (Figure 15.15).

If the antigen is not in solution but is present on the surface of a cell or other particle, an antigen-antibody reaction can lead to a clumping of the particles, that is, to *agglutination*. When foreign cells are injected into an animal, antibodies are formed against a wide variety of their macromolecular constituents, including those in the cytoplasm as well as on the surface of the cells. Only the surface antigens are involved in agglutination, however, since only these are exposed in the intact cell. Considerable attention has been focused on the chemical nature of the surface macromolecules that might be involved in agglutination. Antibodies against both the flagella (so-called H antigens) (Figure 15.16) and the lipopolysaccharide layer (somatic or O antigens) will agglutinate cells, but

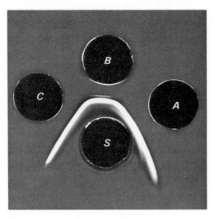

(a)

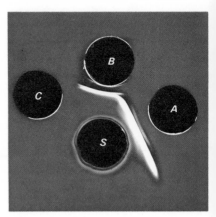

(b)

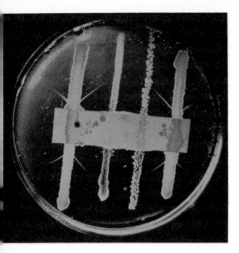

Figure 15.15   In vitro test for production of toxin by Corynebacterium diphtheriae using agar-gel precipitation. The filter-paper strip in the center was soaked with antiserum against diphtheria toxin before being applied to the agar plate. Various strains of C. diphtheriae were streaked at right angles to the filter paper, and precipitin lines are formed adjacent to toxigenic strains but not to nontoxigenic strains. [From E. O. King, M. Frobisher, Jr., and E. I. Parsons: Amer. J. Publ. Health 39:1314 (1949).]

the nature of the clump differs; that of flagellar agglutination is much looser and more flocculent (Figure 15.17).

Flagellar antibodies can induce flagellar paralysis, causing *immobilization* of the cells of a flagellated organism. Immobilization occurs at antibody concentrations considerably lower than those required for flagellar agglutination, and for this reason it is a very sensitive measure of an antigen-antibody reaction. It also is one of the few immunological reactions that can be detected with single cells.

Soluble antigens can be adsorbed to or coupled chemically to cells or other particulate structures such as latex beads or colloidal clay, and they can then be detected by agglutination reactions, the cell or particle serving only as an inert carrier. This greatly increases the ability to detect the presence of antibodies against soluble antigens since, as noted above, agglutination is much more sensitive than precipitation.

The conversion of antibody molecules into *fluorescent* substances can be effected by attaching them chemically to fluorescent organic compounds such as fluorescein isothiocyanate or rhodamine B (Figure 15.18). This does not alter significantly the specificity of the antibody but makes it possible to detect the antibody when adsorbed to cells or tissues (Plate 6: Figure 15.19), by use of the fluorescence microscope. Fluorescent antibodies have been of considerable utility in recent years in diagnostic microbiology since they permit the study of immunological reactions on single cells. The fluorescent-antibody technique is very useful in microbial ecology as one of the few methods that makes possible identification of microbial cells directly in the natural environments in which they live. Two distinct procedures, the *direct* and the *indirect* staining methods, are used. In the direct method, the antibody against the organism is itself fluorescent (Figure 15.20). In the indirect method, the presence of an antibody on the surface of the cell is detected by the use of another fluorescent antibody directed against the specific antibody itself. This is possible because immunoglobulins, as all proteins, are antigenic, and the immunoglobulins of one animal species can induce antibody formation in those of another. For this reason, fluorescent goat anti-rabbit immunoglobulin

Flagellum ⎯

Antibody molecules ⎯

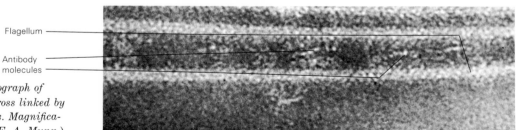

Figure 15.16   Electron micrograph of flagella of S. typhimurium cross linked by gamma M antibody molecules. Magnification, 400,000×. (Courtesy of E. A. Munn.)

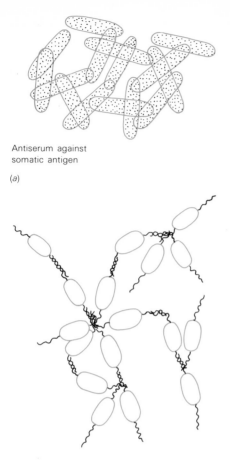

Antiserum against
somatic antigen

(a)

Antiserum against
flagellar antigen

(b)

Figure 15.17   Appearance of agglutinated bacterial cells when antibody is directed against (a) somatic (O) antigen and (b) flagellar (H) antigen.

can be used to detect the presence of rabbit immunoglobulin adsorbed to cells (Figure 15.20). One of the advantages of the indirect staining method is that it eliminates the need for making the fluorescent antibody for each organism of interest.

When antibody directed against a bacterial capsule combines with this structure, a change in the refractility of the capsule occurs, which makes it visible with the light microscope even without a stain. This phenomenon has been called *capsular swelling;* this term is misleading, however, since there is no increase in the size of the capsule, but only in its refractility.

Complement and complement fixation   In addition to the antigen-antibody reactions described above, other types of reactions exist that are of great importance in immunity. Before these reactions are discussed, however, it will be necessary to introduce another group of serum substances, called *complement.* Complement acts in concert with specific antibody to bring about several kinds of antigen-antibody reactions that would not otherwise occur. These reactions are included in Table 15.3. Although not itself an antibody, complement is found in the globulin fractions of serum. It is not a single serum protein but a complex mixture of at least eleven substances, of which some are necessary for various immunological reactions. The chemical studies on complement are very complex and will not be discussed here. We might note that the proteins of complement are not immunoglobulins but are proteins of various sizes and electrophoretic mobilities. At least one component of complement is an enzyme (esterase); perhaps others are also enzymes. Complement is very heat labile, being destroyed by heating at 55°C for 30 min; antibody, on the other hand, is heat stable. The usual procedure for studying complement-requiring antigen-antibody reactions is to destroy the complement activity of the antiserum by heating, and then add some fresh serum from an unimmunized animal as a source of complement. In this way one can set up reaction mixtures with defined amounts of complement.

Table 15.3   Types of antigen-antibody reactions

Location of antigen	Accessory factors	Reaction observed
Soluble antigen	None	Precipitation
On cell or inert particle	None	Agglutination
Flagellum	None	Immobilization or agglutination
On bacterial cell	Complement	Lysis
On bacterial cell	Complement	Killing
On erythrocyte	Complement	Hemolysis
Toxin	None	Neutralization
Virus	None	Neutralization
On bacterial cell	Phagocyte, complement	Phagocytosis (opsonization)

: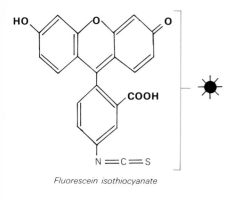

Figure 15.18 Preparation of fluorescent
antibody by coupling the fluorescent dye
fluorescein isothiocyanate to the antibody
protein.

Fluorescein isothiocyanate

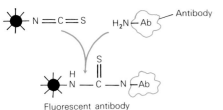

Fluorescent antibody

An important reaction in which complement participates is hemolysis, the lysing of red blood cells (Figure 15.21). If an antiserum against erythrocytes is mixed with a suspension of erythrocytes and some normal serum added as a source of complement, lysis occurs within 30 min on incubation at 37°C. Erythrocyte hemolysis is often used to test for the presence of complement in unknown sera or to measure complement fixation (see below).

Complement is necessary in the bactericidal and lytic actions of antibodies against many Gram-negative bacteria (Figure 15.22). (Interestingly, Gram-positive bacteria are not killed by specific antibody, either in the presence or absence of complement.) Probably death (including that from lysis) involves antibodies against antigens on the surface of the cell; complement perhaps brings about an actual change in the cell surface, possibly by an enzymelike reaction, after antibodies have prepared the way. No cytocidal or lytic effect is seen when cells and complement are mixed alone; whereas, if cells have adsorbed antibody first, death or lysis occurs rapidly after complement is added.

Another action of complement is in the promotion of phagocytic action against capsulated bacteria, when mixed with specific antibody. Recall that when phagocytes are mixed with capsulated bacteria, phagocytosis often does not occur because the presence of the capsule prevents the phagocyte from engulfing the bacterial cell. However, if capsulated

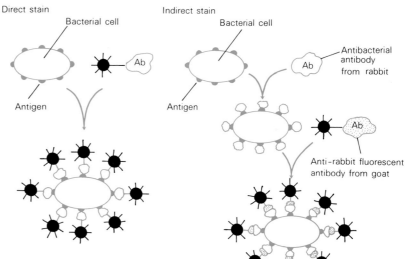

Figure 15.20 Direct and indirect methods of using fluorescent antibody to demonstrate the presence of antigen on a bacterial cell.

15.4 Specific immune
mechanisms

439

Holes in membrane

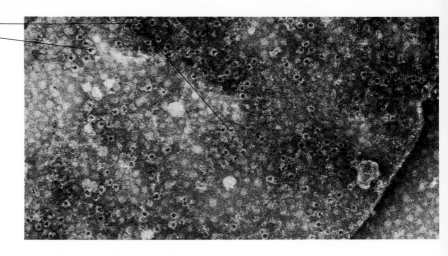

Figure 15.21 Electron micrograph of a negatively stained preparation showing holes created in a sheep erythrocyte cell membrane as a result of reaction between membrane antigens, specific antibody, and complement. It is thought that complement-induced hemolysis arises from the production of such holes. Magnification, 100,000×. (Courtesy of R. R. Dourmashkin and J. H. Humphrey.)

Holes

Figure 15.22 Electron micrograph of a negatively stained preparation of Salmonella paratyphi, showing holes created in the cell envelope as a result of reaction between cell-envelope antigens, specific antibody, and complement. Magnification, 350,000×. (Courtesy of E. A. Munn.)

bacteria are mixed with antibody specific against the capsule, in the presence of complement, phagocytosis now occurs readily. In some cases this promotion of phagocytosis by specific antibody can occur in the absence of complement, but usually complement is necessary. Promotion of phagocytosis by specific antibody (*opsonization*) is important in producing specific immunity of an animal to infectious disease. It might be noted that the involvement of complement in a variety of antigen-antibody reactions represents an economy on the part of the body since a single non-specific factor is involved in a variety of reactions initiated by a number of specific factors, the antibodies. This has probably made possible a simplification of antibody structure.

Another peculiarity of complement is that it appears to be used up in the antigen-antibody reaction. This is called *complement fixation* and is observed even in antigen-antibody reactions in which complement is not required. Although it is not completely understood, complement fixation probably involves a combination of certain components of complement with antibody after the antigen-antibody reaction has taken place. The efficiency of different kinds of antibody for binding complement vary: gamma M is about 1,000 times more efficient than is gamma G. Whatever the mechanism, the measurement of complement fixation is a sensitive way of detecting antigen-antibody reactions and is often used to detect reactions that do not produce precipitates or other visible manifestations of antigen-antibody combination. The Wasserman test for syphilis (Section 18.15) is a good example of a complement-fixation test.

Various types of antigen-antibody reactions are summarized in Table 15.3. In addition to those listed, there are a number of more specialized reactions that can be found in textbooks on immunology.

Following a single injection of an antigen, there is a lapse of time (latent period) before any antibody appears in the circulation, followed by a gradual increase in titer (which remains low at this time) and then a slow fall. This reaction to a single injection is called a *primary response* (Figure 15.23). When a second injection is made some days or weeks later, the titer rises rapidly to a maximum 10 to 100 times above the level achieved after the primary injection. This secondary injection is sometimes called a ''booster dose.'' The titer slowly drops again, but later injections can bring it back up.

Antigens vary widely in the ease with which they induce antibody formation. Insoluble antigens are almost always better than soluble ones, and the antigenicity of soluble materials can be greatly improved if they are adsorbed to alum or other particulate material. Such adsorbents are called ''adjuvants.''

Individuals of all species of mammals and birds that have so far been studied can form antibodies, but the physical and chemical properties of these antibodies may differ from one species to another. Reptiles, amphibians, and fishes are also able to produce antibody, so that the phenomenon is not restricted to warm-blooded vertebrates. Attempts have been made to immunize invertebrates such as earthworms and insects but without clear success. It appears that antibody production may be restricted to the vertebrate group. However, different species of animals differ in their ability to produce antibodies, and even different individuals of the same species may vary considerably.

The antibody response of a given animal to a given antigen is influenced

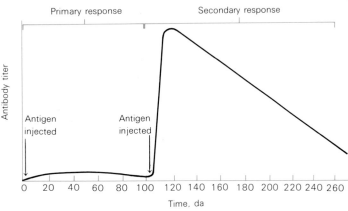

Figure 15.23 Primary and secondary antibody response.

by the nutritional state, age, hormonal balance, and many other undefined factors in the physiology of the animal. Very young animals and aged individuals are usually poor antibody producers. The antibody response can be reduced markedly by whole body irradiation with ionizing radiation. It also can be reduced in some animals by the administration of cortisone and other anti-inflammatory agents.

The mechanism of antibody formation   One of the most important questions in biology is how it is possible for higher animals to produce such a large variety of specific proteins, the antibodies, in response to invasion by foreign macromolecules. In considering the mechanism of antibody formation, we must first explore what happens to the antigen. The main sites of antigen localization in the body are the lymph nodes, the spleen, and the liver. It has been well established that antibodies are formed in both the spleen and lymph nodes; the liver seems not to be involved. If the antigen is injected intravenously, the spleen is the site of greatest antibody formation, whereas subcutaneous, intradermal, and intraperitoneal injections lead to antibody formation in lymph nodes. Fragments of lymph node or spleen from immunized animals can continue to produce antibody when placed in tissue culture or when injected into other, nonimmunized, animals.

The cells that produce immunoglobulins are a special type of lymphocyte cell called the *plasma cell.* By immunofluorescence it has been possible to show that although these cells contain and secrete antibody, they concentrate little or none of the injected antigen. Instead, the antigen is found in certain phagocytic cells, the *macrophages,* which themselves do not produce antibody but are often found surrounded by plasma cells. There is some evidence that the macrophages digest antigen and convert it into fragments, which then become transferred to the plasma cells where antibody formation takes place.

Since antibodies are protein molecules, it is assumed that they are produced by the same biochemical mechanisms by which other proteins are produced. This means that the structure of an individual antibody molecule is ultimately determined by a segment of DNA (a gene). The puzzle arises when we try to explain what role the antigen plays in this process. The hypothesis for which the most support exists is that cells capable of forming specific antibody arise as a result of random mutation from a population of undifferentiated stem cells. It is thought that a cell capable of producing a specific antibody develops further only if it comes into contact with the antigen (or antigenic determinant) with which its antibody can react; the cell is then stimulated to divide and a population of antibody-producing cells is formed. According to this model, the antigen is merely selective for a given antibody-producing cell type and is not required for the actual production of the antibody protein molecule. Thus, a population of antibody-producing cells can continue to function even in the absence of the antigen. Although in most cases an antigenic stimulus is necessary for the development of the antibody-forming cell line, some antibodies can be formed by animals which have never been in contact with the antigen.

Antigenic mimicry Some pathogenic bacteria produce cell surface polysaccharides that are identical to or closely similar in structure to polysaccharides produced by the host. For instance, some strains of *Streptococcus pyogenes* produce a hyaluronic acid capsule; this polysaccharide is also common in connective tissue. Many Gram-negative bacteria produce polysaccharides similar to the erythrocyte antigens. This phenomenon by which the antigen of a pathogen mimics a normal host component is called *antigenic mimicry*. The advantage of mimicry to the pathogen is that since the host does not normally produce antibodies against its own body components, it cannot of course produce antibodies against similar components of the pathogen. Hence the pathogen is able to escape the specific immune defense and can invade the host. Antigenic mimicry may have been of considerable significance in the evolution of microorganisms pathogenic to higher animals.

Hypersensitivity (allergy) Although immunological responses often are beneficial to the body in protecting it from infectious disease, some responses are harmful enough to cause severe symptoms or even death. These harmful immunological responses are called *hypersensitive* or *allergic reactions*. Hypersensitivity can arise in a large number of ways. In many of these the production of antibodies has been demonstrated, while in others no antibodies have so far been detected and only indirect evidence of an immunological basis exists.

The classical example of hypersensitivity is the phenomenon known as *anaphylactic shock*. A guinea pig, for example, is injected with very small amounts of a foreign protein and a period of 2 to 3 wk is allowed to elapse. If now a large amount of this same protein is injected, within a few minutes the animal shows signs of respiratory distress, due to contraction of the bronchial muscles, which prevents the exhalation of air. The animal has an acute attack of asthma, and usually dies within minutes. Although the guinea pig is unusually sensitive to anaphylactic shock, other animals and man also show similar responses. The basis of anaphylactic shock is thought to be as follows: The initial injection of foreign protein induces antibody formation, but possibly because of the small amount of antigen injected, this antibody remains fixed in the cells that produced it instead of being freed into the circulation. When the later challenge dose of antigen is received, the antigen molecules unite with this fixed antibody, and in some way this union results in a marked release of the histamine that was present within the cells. Histamine causes the contraction of the smooth muscles, thus leading to the production of the characteristic symptoms. Factors other than histamine may also be involved in anaphylactic shock, but most of the symptoms can be explained by this series of reactions.

A number of conditions occur in man which differ from anaphylactic shock but which show similar mechanisms of antigen-antibody union and histamine release. *Serum sickness* results when a relatively large amount of horse serum is injected and antibodies form against the horse serum proteins. The symptoms are usually a rash, enlargement of lymph glands, and swelling of some body tissues, all related to an increased permeability of the capillary endothelium, which is another manifestation of histamine action. *Drug allergy* and *hay fever* are types of hypersensitivity that seem to have hereditary factors. The symptoms usually result from contraction of smooth muscle and vary in severity, depending on the individual and the route by which the allergen enters. Previous sensitization to the allergen must occur, although it is usually impossible to determine when this had occurred. Allergens in pollens and foods are usually proteins. It is felt that drugs themselves are not antigens, but that they combine with some of the serum proteins, forming modified proteins that are treated as foreign substances by the animal, in a manner analogous to that described for *p*-aminobenzene arsonate (see Figure 15.9). The sensitivity of a person to a particular allergen can often be demonstrated by injecting the allergen intracutaneously; an allergic reaction will be exhibited by an immediate reddening of the skin and wheal formation.

Allergic reactions are also responsible for some of the symptoms of many microbial infections since antigenic components of many microbial cells are able to initiate the hypersensitive reaction. The initial infection may be mild or inapparent but subsequent infections cause allergic-type reactions in the infected tissues or elsewhere in the body. Usually the organisms that elicit allergic reactions cause chronic rather than acute infections. The role of hypersensitivity in producing the symptoms of tuberculosis and rheumatic fever is discussed in Sections 18.1 and 18.7.

15.6   Antibodies and immunity

Although many aspects of immunological reactions do not concern immunity to infectious disease, the major role of antibodies is in protecting the animal from the consequences of infection. The importance of antibodies in disease resistance is shown most dramatically in individuals with the inherited disorder agammaglobulinemia, in whom antibodies are not produced due to defective antibody-forming cells. Such individuals are unusually sensitive to bacterial diseases, and in the days before antibiotic therapy, few of them survived infancy. Antibodies defend against infectious diseases in a variety of ways, but it should be emphasized that the only direct function of antibodies is to *combine* with antigens; all other phenomena are secondary consequences of this primary reaction. In diseases whose symptoms spring from the action of an exotoxin, neutralization of the toxin by antibody confers resistance to the disease, even though the antibody is not directed against the pathogen itself. Once the disease symptoms are eliminated, other immune mechanisms can be brought into play to eliminate the pathogen.

Opsonization is an important function of antibody since it makes possible the phagocytosis of pathogens that would otherwise not be engulfed. Agglutination, although probably not lethal, is important in immunity as the agglutinated clumps of cells are readily filtered out in the reticuloendothelial system, thus promoting phagocytosis. Although bacteriolytic action of antibody can be readily demonstrated in the test tube, there is no evidence that it plays an important role in immunity.

The discovery of methods of inducing specific immunity to infectious diseases provided one of the first real triumphs of the scientific method in medicine, and was one of the outstanding contributions of microbiology to problems of the treatment and prevention of infectious diseases. An animal or human being may be brought into a state of immunity to a disease in either of two distinct ways. The individual may be given injections of an antigen that is known to induce formation of antibodies, which will confer a type of immunity known as *active immunity* since the individual in question produces the antibodies himself. In the second way, the individual may receive injections of an antiserum that was derived from another individual, who had previously formed antibodies against the antigen in question. This second type is called *passive immunity* since the individual receiving the antibodies played no active part in the antibody-producing process.

An important distinction between active and passive immunity is that in active immunity the immunized individual is changed fundamentally in that he is able to continue to make the antibody in question, and he will exhibit a secondary or booster response if he later receives another injection of the antigen in question. Active immunity often may remain throughout life. A passively immunized individual will never have more antibodies than he received in the initial injection, and these antibodies will gradually disappear from his body; moreover, a later inoculation with the antigen will not elicit a booster response. Active immunity is usually used as a *prophylactic* measure, to protect a person against future attack by a pathogen. Passive immunity is usually *therapeutic,* designed to cure a person who is presently suffering from the disease.

The material used in inducing active immunity, the antigen or mixture of antigens, is known as a *vaccine*. (The word derives from Jenner's vaccination process using cowpox virus; *vacca* means "cow"—see Chapter 10.) However, when inducing active immunity to toxin-caused diseases, it is clearly not desirable to inject the toxin itself; to overcome this problem, many toxins can be modified chemically so that they retain their antigenicity but are no longer toxic. Such a modified toxin is called a *toxoid*. One of the common ways of converting toxin to toxoid is by treating it with formaldehyde, which blocks some of the free amino groups of the toxin. Toxoids are usually not such efficient antigens as the original toxin but have the advantage that they can be given safely and in higher doses. When immunization against whole microorganisms is necessary, such as for endotoxin-producing organisms, several alternatives are available. The microorganism in question may first be killed by agents such as formaldehyde, phenol, or heat, and the dead cells are then injected. Endotoxin-caused diseases for which vaccines are made routinely are whooping cough and typhoid. Formaldehyde treatment is also used to inactivate viruses in preparing vaccines, such as the Salk polio vaccine.

Immunization with live cells or virus is usually more effective than with dead or inactivated material. Often it is possible to isolate a mutant strain of a pathogen which has lost its virulence but which still retains the immunizing antigens; strains of this type are called *attenuated* strains. They are used in immunization for tuberculosis, anthrax, smallpox (vaccinia), polio, and measles.

The material used in inducing passive immunity—the serum containing antibodies—is known as a *serum,* an *antiserum,* or an *antitoxin* (the last applies to a serum containing antibodies directed against a toxin). Antisera are obtained either from large immunized animals, such as the horse, or from human beings who have high antibody titers (that is, who are *hyperimmune*). The antiserum or antitoxin is standardized to contain a known amount of antibody, using some internationally agreed upon arbitrary unit of antibody titer; a sufficient number of units of antiserum must be inoculated to neutralize any antigen that might be present in the body. Sometimes the gamma globulin fraction of pooled human serum is used as a source of antibodies. This contains a wide variety of antibodies that normal people have formed through the years by artificial or natural exposure to various antigens; it is used when hyperimmune antisera are not available.

Natural passive immunity is a phenomenon that occurs when a fetus receives antibodies from the mother via the placenta and retains them for some time after birth. Later the infant must be actively immunized against various diseases since the antibodies it received from its mother gradually disappear.

The importance of immunization procedures in controlling infectious

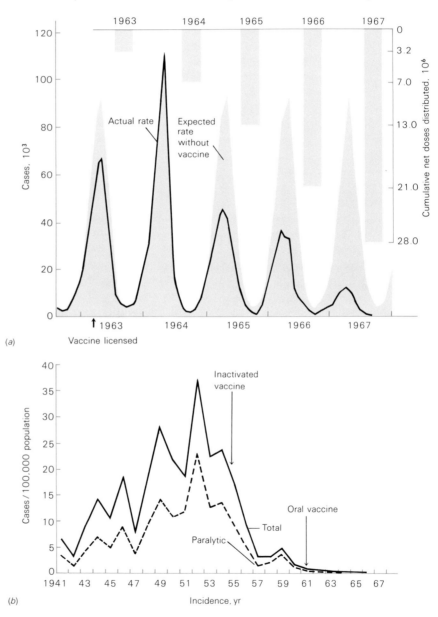

(a)

(b)

Figure 15.24 (a) *Case reports in the United States for measles over a five-year period and the consequences of the introduction of measles vaccine. The solid graph represents the expected rate of cases in the absence of vaccine, based on the mean case reports of the previous five years. [From H. B. Dull and J. J. Witte: Public Health Repts. 83:245 (1968).] (b) Annual poliomyelitis incidence rates in the United States, illustrating the consequences of the introduction of polio vaccine. [From Morbidity and Mortality Weekly Rept. 15:53 (ann. summary) (1966). Atlanta, Ga.: National Communicable Disease Center.]*

diseases is well illustrated by the data for measles and polio presented in Figure 15.24. Upon introduction of immunization procedures, the incidence of these diseases fell markedly, as was true for other diseases.

15.7   Chemotherapy

The discovery, chemical natures, and modes of action of chemotherapeutic agents were discussed in Chapter 9. Although there are presently available a wide variety of relatively nontoxic antimicrobial agents, only a restricted number actually are in use in chemotherapy. Many agents that are effective against organisms in vitro have no effect in an infected host. There are several possible explanations for this: (1) The drug might be destroyed, inactivated, bound to body proteins, or too rapidly excreted; (2) the drug might remain at the injection site or might not penetrate as far as the site of infection; (3) the parasite in vivo might be different from the parasite in vitro, perhaps showing different physiological properties or possibly growing intracellularly, where it is protected from the action of the drug. A number of considerations must enter into the decision as to whether or not to use a chemotherapeutic agent: (1) Is the organism sensitive to the agent as shown by antibiotic-sensitivity tests? (2) Are the symptoms due to the organism itself or to a toxin? If to the latter, then attack on the organism alone will not suffice to effect a cure. (3) Is it possible for the drug to reach the site of infection? Many skin infections are hard to treat because the drug will not penetrate to the infected site; again, many drugs will not enter the spinal fluid and therefore cannot be used in meningitis. (4) Is the patient allergic to the drug? (5) Will the chemotherapeutic agent cause adverse side effects or interfere with the host's defense mechanisms?

It should be emphasized that in very few infections is the drug alone responsible for a cure. The specific and nonspecific host defenses discussed above must also function to bring the treatment to a successful conclusion. This is especially true in the case of drugs that are bacteriostatic rather than bactericidal. A bacteriostatic agent will only prevent further growth of the pathogen; it remains for the host defenses to eliminate the organisms already present. With antibiotics such as penicillin, which inhibit only growing cells (see Chapter 9), an occasional problem is the presence of so-called *persisters*. These are organisms that are sensitive to the antibiotic, but exist alive in some nongrowing state in an infected region of the body. Since they are nongrowing, they will not be killed by penicillin. At a later time, after drug therapy has been stopped, they may begin to grow.

In addition to helping to cure diseases actively in progress, drugs may be given to prevent future infections in individuals who are unusually susceptible, a procedure called *chemoprophylaxis*. For example, penicillin is used to prevent streptococcal sore throats in rheumatic fever patients, since these streptococcal infections often lead to a recurrence of rheumatic-fever symptoms. Frequently, antibiotics are given prophylactically to patients who have just undergone surgery, to prevent any infections at the surgical site.

As we discussed in Chapter 11, mutations to drug resistance occur, and in the presence of the drug the mutant form has a selective advantage and may replace the parent type. Resistance can develop to virtually all chemotherapeutic agents and is known to occur in vivo as well as in vitro. The resistant strain may be just as virulent as the parent, may not be controllable by other chemotherapeutic agents and may be passed on to other individuals. Resistance to drugs can be minimized if they are used only for serious diseases and are given in sufficiently high doses so that the population level is reduced before mutants have a chance to appear. Resistance can also be minimized by combining two unrelated chemotherapeutic agents since it is likely that a mutant resistant to one will still be sensitive to the other. Some organisms, however, such as the streptococci, do not seem to develop drug resistance readily in vivo, whereas the staphlococci are notorious for developing resistance to chemotherapeutic agents.

Some microbial species normally present in the body but suppressed by the presence of the normal flora may arise once the drug-sensitive organisms have been eliminated. This phenomenon, called *superinfection,* occurs frequently in the intestinal tract when antibiotics are given orally; pathogenic yeasts proliferate, leading to diarrhea and possibly more serious symptoms (see Chapter 14). Chemotherapy of specific diseases is discussed in Chapters 18 and 19; and possibilities for chemotherapy of virus diseases, in Chapter 10.

15.8 *Epidemiology*

So far we have been considering infectious diseases only as they occur in isolated individuals. However, animals and man do not live isolated but in populations; and when we consider infectious diseases in populations, some new factors arise. The study of infectious disease in populations is part of the field of *epidemiology.*

To continue existing in nature the pathogen must be able to grow and reproduce. For this reason, an important aspect of the epidemiology of any disease is a consideration of how the pathogen maintains itself in nature. In most cases the pathogen cannot grow outside the host, and if the host dies, the pathogen will also die. A pathogen that kills the host before it is transmitted to a new host would thus become extinct. This raises the question of why pathogens occasionally kill their hosts. Actually, a well-adapted parasite lives in harmony with its host, taking only what it needs for existence, and it causes only a minimum of harm. Serious host damage most often occurs when new races of pathogens arise for which the host has not developed resistance, or when the resistance of the host changes because of nutritional or other environmental factors. Pathogens are selective forces in the evolution of the host, just as hosts are selective forces in the evolution of pathogens. When equilibrium between host and pathogen exists, both live more or less harmoniously together.

Some pathogens are primarily saprophytic (living on dead matter) and only incidentally infect and cause disease. Examples are *C. tetani* (the causal agent of tetanus) and *C. perfringens* (the causal agent of gas gangrene), whose normal habitats are the soil. Infection of man by these bacteria is not essential for their continued existence and can be considered only an accidental event; if all susceptible hosts died, these organisms would still be able to survive in nature.

Several factors influence the frequency of an infection in a population: (1) The greater the percentage of immune individuals in the population, the less frequent the infection will be. If the immunity level is very high, the infection may be completely nonexistent since the pathogen will not be able to maintain itself in the host. (2) The population density is important. In a dense population, the probability of transfer of the pathogen from one individual to another is increased. (3) A weakly virulent pathogen will usually be able to infect a population more easily than will a highly virulent pathogen because the host remains alive longer after infection and thus is able to serve as a source of inoculum for other individuals. (4) Since the pathogen must be transmitted from one individual to another if it is to spread in the population, the transmission efficiency of the pathogen is important. Different pathogens have different modes of transmission, which are usually related to the habitats of the organisms in the body. For instance, respiratory pathogens are usually airborne, whereas intestinal pathogens are spread by food or water. In the following discussion we consider some of these means of transmission.

Respiratory infections and airborne transmissal   Air usually is not a suitable medium for the growth of microorganisms; those organisms found in it derive from the soil, water, plants, animals, man, or other source. In outdoor air, soil organisms predominate. One survey of orga-

Figure 15.25 High-speed photograph of an unstifled sneeze. (Courtesy of Prof. M. W. Jennison, Syracuse University.)

nisms in the air of the city of London found that the number of particles carrying organisms able to grow on an agar medium varied from 0.38 to 14.6 particles per cubic foot. The species found were mostly Gram-positive cocci, spore-forming bacteria, actinomycetes, and an occasional yeast; very few pathogenic bacteria were found. In the same study, experiments indoors in hospital wards, industrial establishments, offices, ships, and dwellings revealed a very different picture. The total count indoors was considerably higher than that outdoors, from 6 to over 1,000 organisms per cubic foot, mostly those commonly found in the human respiratory tract.

An enormous number of droplets of moisture are expelled during sneezing (Figure 15.25), and a considerable number are expelled during coughing or even merely talking. Each infectious droplet has a size of about 10 μm and contains one or two bacteria. The speed of the droplet movement is about 100 m/sec (over 200 mi/hr) in a sneeze and about 16 to 48 m/sec during coughing or loud talking. The number of bacteria in a single sneeze varies from 10,000 to 100,000. Because of the·small size of the droplets, the moisture evaporates quickly in the air, leaving behind a nucleus of organic matter and mucus, to which the bacterial cells are attached. These very small nuclei do not settle but remain in the air until breathed in or until they attach to more rapidly settling dust particles (Figure 15.26). Experiments on the viability of various species of bacteria in droplet nuclei in the air show that most organisms found in the respiratory tract are relatively resistant to drying; those found in the intestinal tract, on the other hand, die quickly. Resistance to drying is one factor that affects the ability of a particular species to be a respiratory parasite.

The average human being breathes several million cubic feet of air in his lifetime, much of it containing microbe-laden dust, which is a potential source of inoculum for upper-respiratory infections caused by streptococci and staphylococci resistant to drying. However, because dust particles are

Figure 15.26 Possible ways of transmitting respiratory microorganisms: by coarse droplets that travel short distances, by fine droplets that travel considerably farther, and by coarse droplets settling out to be dispersed later by dust. (From C. H. Andrewes: The Natural History of Viruses. New York: W. W. Norton & Co., Inc., 1966.)

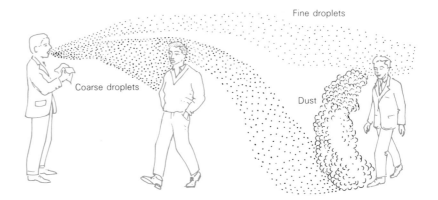

Fine droplets

Coarse droplets

Dust

15.8 Epidemiology

fairly large, they are readily filtered out in the upper respiratory tract. So in most cases the pathogen is transmitted in droplets directly from person to person. Since direct transmissal will obviously be more effective when people are crowded, this is one explanation for the increased frequency of respiratory infections in cold weather, when people more or less remain indoors. Important airborne diseases of man include diphtheria, pulmonary pneumonia, meningitis, psittacosis, and upper-respiratory infections such as the common cold and influenza. As opposed to upper respiratory infections, lung infections almost never involve organisms carried by dust; usually organisms are transmitted in minute droplet nuclei of around 5 to 10 μm in diameter which are expelled from the lungs. When air passes into the inner spaces of the lungs, it almost stops moving, and any nuclei present then have an opportunity to settle on the alveolar tissue. Pulmonary pathogens such as *D. pneumoniae* and *M. tuberculosis* exist primarily in such droplet nuclei.

A number of methods are available to reduce the microbial content of the air of a room, although these are applied systematically only in special cases. Since the microbial content of outdoor air is much lower than that of indoor air, the simplest means of reducing the content is by *ventilation*. *Sunlight* retains some bactericidal power even after it has passed through glass, so that rooms with large windows are desirable. *Ultraviolet lamps* may be installed and the air is circulated past them; UV radiation is more effective against organisms in fine particles than it is against those in coarse dust particles. Air may be circulated through fibrous *filters* or electrostatic *precipitators,* which remove a proportion of the microbial load, or it may be readily sterilized by *heating* it to 200 to 250°C in a furnace for a few seconds. This last cannot be done economically with large volumes of air, however. *Chemicals* such as hypochlorous acid, triethylene glycol, lactic acid, or resorcinol can be sprayed into the air as fine bactericidal aerosols. One of the most effective methods of reducing the organism content of the air is through *control of dust*. This may be achieved by using wet mops instead of dry mops, oiling the floors, avoiding shaking of blankets, clothing, and other objects indoors, frequently wiping surfaces with moist rags, and preventing dust accumulations under beds and in corners. *Face masks* effectively reduce the number of organisms passed into the air by a person during sneezing and coughing and may be worthwhile under special circumstances. It has been possible to apply these methods to help reduce or eliminate staphylococcal epidemics in hospitals, which at one time were a serious problem. However, it should be remembered that there may be more than one method of transmission of a pathogen, with the result that air sanitation by itself alone may not be at all sufficient for the control of the spread of the organism.

Intestinal pathogens and dispersal by water Organisms causing intestinal infections are often excreted in large numbers in the fecal material and may contaminate food or water. This contamination may occur through the agency of flies and other insects, through transfer on hands, or through accidental contamination of water supplies or foods. Although intestinal pathogens rarely if ever multiply in water, they often remain viable in it for long periods of time. One of the most significant factors in the reduction or virtual elimination of waterborne pathogens has been the development of effective methods of purifying water supplies.

Water used for domestic or industrial purposes falls into two categories: *surface waters* and *ground waters*. Surface waters are used to the greatest extent, especially by large urban centers. This kind often bears a load of potentially pathogenic microorganisms that are derived either directly or indirectly from human sources. Because of the virtually universal presence of *E. coli* in the human intestinal tract, and because of the ease with which it can be identified and counted in a water sample, the presence of this organism in a water sample is usually used as an index of fecal pollution of the water. The presence of *E. coli* is determined by standard methods based on the ability to ferment lactose with the production of acid and gas (Figure 15.27). Procedures of this sort, although highly standardized, are somewhat arbitrary and may not be trusted blindly. For instance, although *E. coli* is used as an indicator of fecal pollution, the organism may die rapidly in water, whereas some of the intestinal viruses (for example, poliovirus) may survive much longer. Thus a water may be free of *E. coli* but still be highly dangerous because it contains harmful organisms not tested for.

When water percolates through the soil, its microbial population is gradually depleted; therefore the microbial count of water from deep wells is considerably lower than that of surface water. This *ground water* available from springs or tapped wells is used primarily in small towns and in rural areas. Although it is usually devoid of pathogenic organisms, ground water cannot be assumed to be uniformly safe since in some areas geological conditions are such that the water runs directly through underground channels rather than percolating through the soil; in this case its load of potential pathogens may not be reduced.

The bacterial count of a river decreases with distance below a pollution source (Figure 16.5). This is due to a number of factors, of which the most important are (1) sedimentation of cells onto the stream bottom, (2) dilution by water entering the stream, (3) ingestion of the bacteria by protozoa and other animal predators, and (4) natural mortality. The decrease in bacterial numbers below a source of pollution such as a sewage outfall is called *self-purification* and can be an effective means of eliminating pathogens if the initial load of organisms and the organic content of

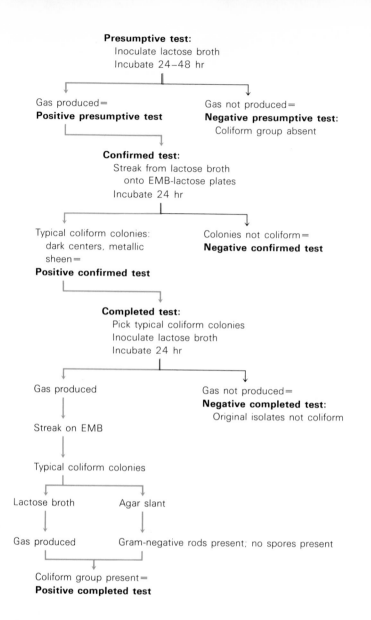

Presumptive test:
Inoculate lactose broth
Incubate 24–48 hr

Gas produced=
Positive presumptive test

Gas not produced=
Negative presumptive test:
Coliform group absent

Confirmed test:
Streak from lactose broth
onto EMB-lactose plates
Incubate 24 hr

Typical coliform colonies:
dark centers, metallic
sheen=
Positive confirmed test

Colonies not coliform=
Negative confirmed test

Completed test:
Pick typical coliform colonies
Inoculate lactose broth
Incubate 24 hr

Gas produced

Gas not produced=
Negative completed test:
Original isolates not coliform

Streak on EMB

Typical coliform colonies

Lactose broth

Agar slant

Gas produced

Gram-negative rods present; no spores present

Figure 15.27 Standard method of water analysis to detect the presence of E. coli.

Coliform group present=
Positive completed test

the water are not too high and if the water is well aerated. However, self-purification will be ineffective in highly polluted waters such as are often found in urban areas.

Microorganisms can be removed from water by mechanical means such as sedimentation and filtration. *Sedimentation* can be accelerated by introducing coagulating chemicals, such as aluminum sulfate and ferrous sulfate, which form flocculent insoluble hydroxides that trap the organisms. *Filtration* of water is usually done through sand and gravel after it has been given preliminary flocculation treatment. If the initial bacterial count is high, flocculation and filtration must be followed by treatment with the bactercidal agent chlorine. The presence of ammonia increases the efficiency of the bactericidal action of chlorine, so this gas is usually added as well. The water is usually aerated afterward to remove some of the chlorine odor and other odors that may be present. The treatment that water receives in different areas reflects local conditions, which vary widely. The water treatment carried out in a large city on the Ohio river,

Figure 15.28 Aerial photograph of the components of the water-treatment system of Louisville, Kentucky. The Ohio River, the source of water, is heavily polluted by industries and cities upstream so that the water must be intensively purified to make it suitable for drinking. (Courtesy of Billy Davis and the Courier-Journal and Louisville Times.)

which has a very polluted source of water, is illustrated in Figure 15.28. Effective sewage treatment methods (discussed in Section 16.2) are also of value in ensuring clean water, as they reduce the microbial load of raw water supplies.

Pathogens transmitted by polluted water include the casual agents of typhoid (*Salmonella typhosa*), cholera (*Vibrio cholera*), poliomyelitis, and amoebic dysentery (*Entamoeba histolytica*).

Intestinal pathogens are also transmitted by way of foods. Contamination of the food frequently occurs from infected persons, usually by way of the hands. This can be especially serious in the case of *carriers,* who continually release infective organisms though they themselves show no symptoms of disease. Another common means for infection of foods is by flies and other insects. In areas where outside toilets still exist and houses are not screened, insects moving from toilet to residence carry pathogens. Insects are the most common vectors of bacterial dysentery (caused by *Shigella dysenteriae*) and amoebic dysentery.

Direct contact   Certain pathogens die quickly outside the body and can be transmitted only by direct and intimate contact, such as sexual intercourse or kissing. Diseases transmitted by sexual intercourse are called venereal diseases; syphilis and gonorrhea are examples and are discussed in detail in Chapter 18. Nonvenereal diseases transmitted by direct contact include skin infections, such as pimples and boils. The causal organism, usually *Staphylococcus aureus,* frequently contaminates the hands of the infected individual and may be transmitted directly to another person. Organisms living in the nasal passages may also be transmitted in this way.

Transmissal on infected objects   Inanimate objects, such as pencils, toys, clothing, dishes, and so on, may become contaminated by an infected individual, with later transmissal to another individual. Organisms that resist drying are the ones most frequently transmitted in this way, such as *S. aureus* and *M. tuberculosis.*

Wound infections   In some cases a causal organism cannot penetrate the unbroken skin but can infect a wound. The organism may come from another person or it may come from some inanimate source such as soil. Examples include: pyogenic (abscess-forming) infections due to staphylococci and streptococci; gas gangrene, caused by *C. perfringens;* and tetanus, caused by *C. tetani.* Wounds that become contaminated with soil and dirt are most susceptible to clostridial infection because spores of these organisms are often found in the soil. In deep wounds and those with much tissue destruction clostridia are most likely to develop since these anaerobes cannot initiate growth in living tissues, where an adequate blood supply maintains aerobic conditions. Anaerobiosis often develops in wounds as a result of the respiratory activities of facultative organisms inoculated into the wound from the soil along with the clostridia, thus making growth of the obligate anaerobe possible. For this reason dirty wounds are more likely to become infected with clostridia than are clean wounds.

Transmissal from animal to man   Certain organisms pass from infected animals to man, causing diseases called *zoonoses*. Usually the organism maintains itself in nature in the animal species, and man is only incidentally infected. Transmissal to man may be via animal bite, as for rabies virus (see Chapter 10), via flea bite, as for *Pasteurella pestis,* the causal agent of plague, by contact with infected animal parts, as for the causal agents of tularemia and anthrax, or through infected milk, for the causal agents of brucellosis and bovine tuberculosis. The control of a zoonosis is particularly difficult because it requires the discovery and the elimination or immunization of infected animals. In wild-animal populations elimination is virtually impossible. Rabies is widespread in foxes, skunks, squirrels, and other animals, and the virus is transmitted from them to dogs, and from dogs to man. The legal requirement for vaccination of dogs helps considerably to reduce the incidence of rabies in dog and man, but does it does not eliminate rabies because of the large reservoir of the virus in wild animals.

In another group of diseases, man and animal form alternate hosts in the life cycle of the pathogen. This is true in malaria, in which the protozoan must carry out part of its life cycle in the mosquito, and part in man (see Section 19.12). Such diseases are not called zoonoses, however.

Epidemics   A disease is said to be epidemic when it occurs in an unusual number of individuals in a community at the same time. In contrast, an *endemic* disease is one which is constantly present in a population. In an endemic situation, the pathogen may not be highly virulent, or the majority of individuals may be immune, so that the frequency of disease incidence is low. However, as long as the endemic situation lasts, there will remain a few individuals who may serve as sources of infection for any newly arrived susceptible individuals.

In the development of epidemics several factors may be involved. The population, for example, may be highly susceptible to a certain organism that is not indigenous to the area. If this organism is then introduced into the population, an explosive epidemic development may occur (Figure 15.29). A perfect example of the development and spread of an epidemic is shown by myxomatosis disease in European rabbits. Myxomatosis is

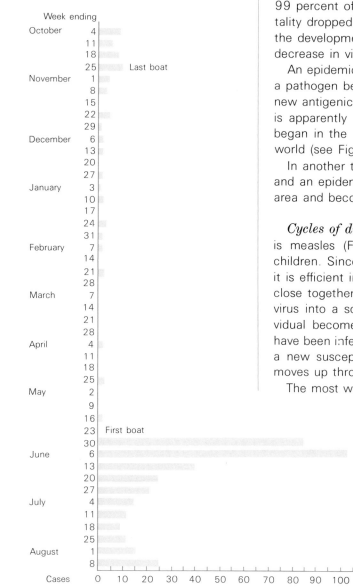

endemic in the Western hemisphere, and the Western rabbit is quite resistant; but the virus was absent from Europe, and the European rabbit was highly susceptible. Myxomatosis virus was introduced into a small population of rabbits on an isolated estate in France, and Figure 15.30 shows how rapidly the disease spread across France. The virus was later introduced into Australia to aid in controlling rabbits there; initially over 99 percent of the infected rabbits died, but within a year the case mortality dropped to 90 percent and later fell even lower, partly because of the development of increased resistance on the part of the rabbit and a decrease in virulence of the virus.

An epidemic may also develop when a population had been immune to a pathogen because of previous contact and the pathogen mutates to a new antigenic type, against which the old antibodies are ineffective. This is apparently what caused the 1957 Asiatic influenza epidemic, which began in the interior of China and within a year had spread around the world (see Figure 10.16).

In another type of situation, the disease is endemic in the population, and an epidemic arises when a susceptible population migrates into the area and becomes infected.

Cycles of disease   Certain diseases occur in cycles. A good example is measles (Figure 15.24), which occurs most commonly in schoolchildren. Since the measles virus is transmitted by the respiratory route, it is efficient in causing infection where large numbers of individuals are close together, such as occurs in a school. The introduction of measles virus into a school results in an explosive epidemic; virtually every individual becomes infected and develops immunity. After all the children have been infected and become immune, there is a quiescent period until a new susceptible population builds up, when the immune population moves up through the grades.

The most widely used methods for evaluating the extent of infection of

Figure 15.29   Incidence of colds throughout the year in an isolated community on Spitsbergen, in the Arctic Ocean. Note the sharp rise in number of cases soon after the arrival of the first boat of the summer. (From C. H. Andrewes: The Natural History of Viruses. New York: W. W. Norton & Co., Inc., 1966.)

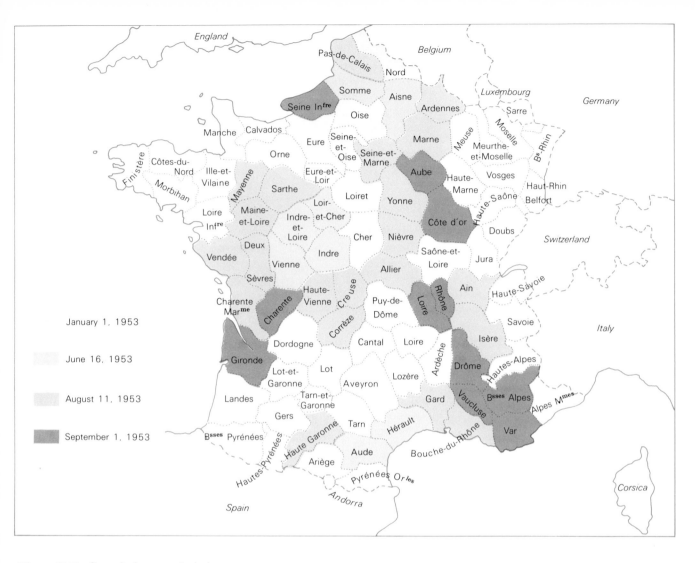

Figure 15.30 Spread of myxomatosis in wild rabbits in France in 1953, the year of the major continental extension of the disease in Europe. The virus was introduced into a single estate in Northern France and within eight months had spread throughout most of the country. (From F. Fenner and F. N. Ratcliffe: Myxomatosis. Cambridge: Cambridge University Press, 1965.)

a population are immunological ones. The presence in the serum of antibodies against a specific pathogen or toxin is presumptive evidence that this individual is now or has been infected or immunized with the pathogen of interest. For instance, the *tuberculin* test (see Section 18.7) has been used in the study of tuberculosis, the Schick test in diphtheria, and the Dick test in scarlet fever.

The control of the spread of a parasite through a population can occasionally be effected by quarantine methods. However, ideas of the value of quarantine change with time. At one time individuals were quarantined for mumps, measles, chicken pox, and so on, but now we know that the infectious agents of these diseases are so widespread in the population that a quarantine of those individuals with frank symptoms is of little value. Quarantines are still used for smallpox, however, which is a highly contagious disease. In many cases knowledge is now available that indicates the proper way to control a particular disease in a population; yet social, political, or economic considerations make impossible the application of this knowledge. Sometimes we must decide whether the effort to eliminate a disease is worth the time and expense.

15.9 Summary

The general steps in the infectious disease process can be briefly outlined: (1) transfer of the pathogen to a new host; (2) growth and establishment of the pathogen at the initial site of infection; (3) invasion of the host and spread of the pathogen through the body; (4) establishment by the pathogen of foci of infection in target organs; (5) production of disease symptoms, sometimes even death of the host; and (6) transfer of the pathogen to other hosts and continual spread of the pathogen through the population of sensitive hosts. The precise events that occur vary widely from pathogen to pathogen, and control must take into account the genetic, physiological, and biochemical properties of the pathogen.

In addition, the outcome of the infection will depend significantly on the properties of the host, and especially on its ability to resist or neutralize the pathogen. Some host properties of significance include the following: (1) the presence of intact surface barriers such as the skin, mucous membranes, and epithelial cells, which prevent the initial establishment of the pathogen; (2) the availability of appropriate nutrients and environmental conditions (for example, temperature, pH, redox potential, and so on), which the pathogen needs for growth; (3) the presence in the host of defense mechanisms of various kinds, as for example phagocytic cells that can ingest and destroy the cells of the pathogen; (4) development by the host, in response to the infection, of inflammation and other active defenses of a nonspecific nature; (5) development by the host of specific immune mechanisms through antibody formation; (6) the inherent sensitivity of the host to toxins and other harmful products of the pathogen. All these factors can play significant roles in determining whether a pathogen can initiate any or all of its characteristic disease symptoms. Different host species and different individuals within the same species may develop markedly diverse kinds of host-parasite relationships.

Supplementary readings

American Public Health Association Committee on Communicable Disease Control *Control of Communicable Diseases in Man, An Official Report,* 10th ed. New York: American Public Health Association, 1965.

Antibodies Cold Spring Harbor Symposium on Quantitative Biology, vol. 32 (1967).

Carpenter, P. L. *Immunology and Serology,* 2nd ed. Philadelphia: W. B. Saunders Co., 1965.

Cushing, J. E., and D. H. Campbell *Principles of Immunology.* New York: McGraw-Hill Book Company, 1957.

Davis, B. D., R. Dulbecco, H. N. Eisen, H. S. Ginsberg, and W. B. Wood, Jr. *Microbiology.* New York: Harper and Row, Publishers, 1967.

Dubos, R. J. *Louis Pasteur, Free Lance of Science.* Boston: Little, Brown and Company, 1950.

Dubos, R. J. *Mirage of Health: Utopias, Progress, and Biological Change.* Garden City, N.Y.: Doubleday & Company, Inc., 1959.

Gregory, P. H. *The Microbiology of the Atmosphere.* New York: Interscience Publishers, 1961.

Kabat, E. A., and M. M. Mayer *Kabat and Mayer's Experimental Immunochemistry,* rev. ed. Springfield, Ill.: A. H. Thomas Co., 1961.

Landsteiner, K. *The Specificity of Serological Reactions,* rev. ed. Cambridge, Mass.: Harvard University Press, 1945. (Reprinted. New York: Dover Publications, Inc., 1962.)

Landy, M., and W. Braun (eds.) *Bacterial Endotoxins* (Symposium, Rutgers Institute of Microbiology). New Brunswick, N.J.: Rutgers Institute of Microbiology, The State University, 1964.

Microbial Behavior, "in Vivo" and "in Vitro" (Fourteenth Symposium, Society for General Microbiology). London: Cambridge University Press, 1964.

Raffel, S. *Immunity,* 2nd ed. New York: Appleton-Century-Crofts, 1961.

Rosenberg, L. T. "Complement," *Ann. Rev. Microbiol.,* 19:285 (1965).

Roueché, B. *Eleven Blue Men, and Other Narratives of Medical Detection.* Boston: Little, Brown and Company, 1953.

Schnitzer, R. J., and F. Hawking (eds.) *Experimental Chemotherapy,* 3 vols. New York: Academic Press, Inc., 1964.

Stableforth, A. W., and I. A. Galloway (eds.) *Infectious Diseases of Animals: Diseases Due to Bacteria,* 2 vols. London: Butterworth & Co. (Publishers), Ltd., 1959.

van Heyningen, W. E. *Bacterial Toxins.* Oxford: Blackwell Scientific Publishers, 1950.

Wilson, G. S., and A. A. Miles *Topley and Wilson's Principles of Bacteriology and Immunology,* 5th ed., 2 vols. Baltimore: The Williams & Wilkins Co., 1964.

sixteen

Geochemical activities of micro-organisms

In this chapter we consider some of the activities of microorganisms in the world at large, especially in soil and in aquatic habitats. Microorganisms play far more important roles in nature than their small sizes would suggest. Some of these roles are the following: (1) Photosynthetic microorganisms are responsible for the synthesis of large amounts of organic matter used as food by higher organisms. (2) Heterotrophic microorganisms grow on soluble organic compounds and serve as food for higher organisms. (3) Heterotrophic microorganisms play important roles in the decomposition of organic material and its subsequent conversion to inorganic form, a process called mineralization. (4) Microbes are important agents in the weathering of rocks and in the formation of soil. (5) They are essential agents in the transformation of elements such as nitrogen, sulfur, and iron from one form to another, thus participating importantly in the geochemical cycles. (6) Microorganisms may be of considerable importance in the processes leading to the formation of such geologically and economically important materials as coal, oil, and sulfur, and they also participate in the destruction of some of these substances.

The ecosystem   To see properly the place of microorganisms in nature, we must first consider the concept of an *ecosystem* as it has been defined by ecologists. An ecosystem (or biocoenosis) is usually defined as a total community of organisms, together with the physical and chemical environment in which they live. Each organism interacts with its physical and chemical environment and interacts with the other organisms in the system so that the ecosystem can be viewed as a kind of superorganism that has the ability to respond to and modify the environment. A good example of an ecosystem is a lake. The sides of the lake define the boundaries of the ecosystem. Within these boundaries the organisms live and carry out their activities, greatly modifying the characteristics of the lake as well as each other.

Energy enters the ecosystem mainly in the form of sunlight and is used by photosynthetic organisms in the synthesis of organic matter. Some of the energy contained in this organic matter is dissipated by the photosynthesizers themselves during respiration, and the rest is available to herbivores, which are animals that consume the photosynthesizers. Of the energy entering the herbivores, one portion is dissipated by them during respiration and the rest is used in synthesizing the organic matter of the herbivore bodies. Herbivores are themselves consumed by carnivorous animals and these carnivores are eaten by other carnivores, and so on. At each step in this chain of events, a portion of the energy is dissipated as heat. Any plants or animals that die, whether from natural causes, injury, or disease, are attacked by microorganisms and small animals, collectively called "decomposers." These decomposers also utilize energy released by plants or animals in the form of excretory products. These reactions constitute a *food chain* or food web. Because there is a loss of energy at each stage of the food chain, ultimately practically all of the biologically useful energy that had been used to convert materials to organic matter by the photosynthesizers is dissipated; usually only very small amounts are stored. Because of this, energy is said to flow through the ecosystem. The quantitative relations are expressed by an energy flow diagram, such as the one shown in Figure 16.1.

Although the energy fixed by photosynthesizers is ultimately dissipated as heat, the chemical elements that serve as nutrients usually are not lost from the ecosystem. For instance, carbon from CO_2 fixed by plants in photosynthesis is released during respiration by various organisms of the food chain and becomes available for further utilization by the plants. Nitrogen, sulfur, phosphorus, iron, and other elements taken up by plants are also released through the activity of the decomposers, and hence are made available for reassimilation by other plants. For this reason we speak of energy *flowing* through the system, but of chemical elements *cycling* within the system. In some parts of the cycle the element is oxidized, whereas in other parts it is reduced; for many elements a *biogeochemical cycle* can thus be defined, in which the element undergoes changes in

Figure 16.1 Energy flow diagram for an ecosystem. The sizes of the boxes and channels are proportional to the relative amounts of energy at each stage. Heterotrophic microorganisms play key roles at each stage in the decomposition of dead organisms and of their excretory products. All energy that is not stored or exported from the system is ultimately degraded as heat and is lost. (Adapted from E. P. Odum: Fundamentals of Ecology, 2nd ed. Philadelphia: W. B. Saunders Co., 1959.)

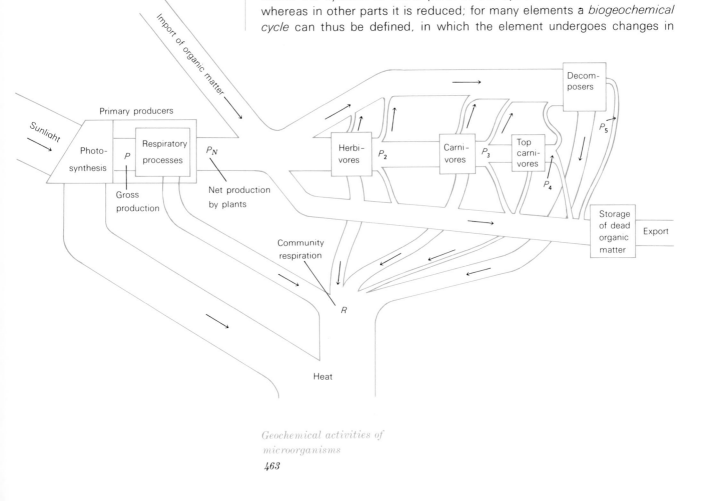

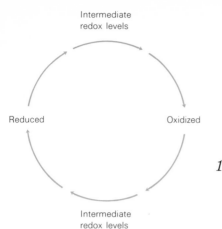

Intermediate
redox levels

Reduced   Oxidized

Intermediate
redox levels

Figure 16.2   Abbreviated biogeochemical cycle for an element, which undergoes changes in oxidation state during the functioning of the ecosystem.

oxidation state as it is acted upon by one organism after another (Figure 16.2).

In the following discussion we will attempt to examine the roles of microorganisms in the energy flow and in the biogeochemical cycles of aquatic and terrestrial ecosystems. We shall see that even though microbes are small in size, their versatile metabolic activities and rapid growth rates enable them to participate importantly in these processes.

16.1   Aquatic habitats

A wide variety of aquatic environments exist, which often differ markedly in chemical and physical properties while retaining many similarities. Water evaporates from the surface of the earth, accumulates in the atmosphere in clouds, and falls back to the surface of the earth as precipitation in the form of rain, snow, hail, and so on. Water falling on land either runs off directly into rivers and lakes or percolates through the soil until it reaches a level, the ''water table,'' from which it slowly moves, following the contours of the land, ultimately to reach the surface as springs or seepages. Most of the water eventually finds its way to the sea. Through the interactions of this hydrologic cycle, a variety of aquatic habitats are created, in which living organisms develop. A summary of some of the more important aquatic habitats is presented in Table 16.1.

Since aquatic environments differ considerably in chemical and physical properties, it is not surprising that their species compositions also differ. The predominant photosynthetic organisms in most aquatic environments are microorganisms; in aerobic areas blue-green algae and eucaryotic algae prevail and in anaerobic areas photosynthetic bacteria are preponderant. Algae floating or suspended freely in the water are called *phytoplankton;* those attached to the bottom or sides are called *benthic algae.* Because these photosynthetic organisms utilize energy from light in the initial production of organic matter, they are called the *primary producers.* In the last analysis, the biological activity of an aquatic ecosystem is dependent on the rate of primary production by the photosynthetic organisms. The activities of these organisms are in turn affected by the physical environment (for example, temperature, pH, and light) and by the kinds and concentrations of nutrients available. Open oceans are very low in production, whereas inshore ocean areas are high, with some lakes and springs being highest of all. The open ocean is infertile because the inorganic nutrients needed for algal growth are present only in low concentrations. The more fertile inshore ocean areas, on the other hand, receive extensive nutrient enrichment from rivers. There are, however,

Table 16.1 Kinds of aquatic environments

Habitat	Description	Special characteristics
Marine environments:		
Abyss	Deep ocean	No light; high pressure; low temperature
Pelagic	Surface, open ocean	Low nutrient concentrations
Coastal	Surface, inshore	Variable salinity; adequate nutrients
Intertidal	Between low and high tide	Variable salinity; adequate nutrients; variable temperature
Estuary	River meeting ocean	Highly variable salinity; excellent nutrients
Salt marsh	Shallow bay	Extensive development of emergent halophilic vascular plants; excellent nutrients
Freshwater environments:		
Lake; pond	Depression holding water from river, rain or spring	As a group, vary markedly in chemical composition; low, medium, or high in nutrients
Marsh	Shallow lake	Extensive development of emergent vascular plants; high in nutrients
Bog	Moss-filled depression	Acidic; standing water in reduced amount; low in nutrients
River	Flowing water; fed mostly by surface runoff	Variable chemical composition; variable temperature
Spring	Flowing water fed by underground source	Generally temperature, flow rate, and chemical composition are constant; as a group, vary markedly in these characteristics
Hypersaline environments:		
Salt lake	Lake basin with no outlet; occurs in arid climate	High salt concentration; restricted biological development
Saltern	Site of solar salt production	Saturated salts; very restricted biological development

some open ocean areas that are rather fertile; these are areas where winds or currents cause an extensive upwelling of deep ocean water, bringing to the surface nutrients from the bottom of the sea. It is because of such upwellings that areas off the coasts of California and Peru are so productive. The amount of economically important crop such as fish or shellfish is determined ultimately by the rate of primary production and this explains why lakes and inshore ocean areas are the richest sources of fish and shellfish.

The energy found as organic matter in primary producers reaches later stages of the food chain in several ways. Some of the organic matter is excreted in soluble form and serves as nutrient for the growth of heterotrophic microorganisms, primarily bacteria. The organic matter retained within the cells of the primary producers is a major source of food for small animals, primarily crustaceans, which are collectively called *zooplankton*. Some of the smallest zooplankton also consume bacteria as a major source of food. Zooplankton are themselves consumed by larger invertebrates, which in turn are devoured by fish. Thus a simplified food chain for an open water aquatic zone can be represented as:

Primary producers (primarily algae) $\longrightarrow$ zooplankton $\longrightarrow$
$\longrightarrow$ bacteria $\longrightarrow$

larger invertebrates $\longrightarrow$ small fish $\longrightarrow$ large fish

Since each step in the food chain involves a loss of energy, if one is interested in an end product such as fish, it is obviously important to have a short food chain in order to utilize the maximum amount of energy in synthesizing the desired product.

In inland waters, especially rivers, surrounded as they are by land areas containing large plants, much of the organic matter is derived not from primary producers but from dead leaves, humic substances, and other forms of organic detritus originating on the surrounding land. These materials are acted upon primarily by bacteria and fungi and are converted at least partially into microbial protein. In such waters the food chain may begin not with primary producers but with these heterotrophic microorganisms.

Carbon cycle   The biogeochemical cycle of carbon is shown in Figure 16.3. Such cycles occur in lakes, rivers, oceans, and other aquatic places. One important feature of many aquatic habitats is that they possess an aerobic zone in the surface layers and an anaerobic zone in the deeper waters and in the mud. As we have learned earlier (see Chapter 4), the fate of organic compounds can be quite different when they are broken down aerobically than when they are degraded anaerobically. Organic materials falling into the anaerobic zone are fermented primarily by bacteria with the production of organic acids, CH_4, H_2, and CO_2. The accumulation of acids and other inhibitory products leads to an inhibition of bacterial action, and the decomposition of organic materials ceases; the organic material then slowly accumulates and becomes buried in the mud. *Coal* is thought to have originated when enormous accumulations of organic matter in the anaerobic zones of shallow marshes and lakes of the past were subsequently modified by purely geological processes. In many coals the remains of plants are still easily visible in fossil form.

Figure 16.8 Extensive development of insoluble ferric hydroxide in a drainage ditch at the edge of a swamp. The drainage water is neutral or only slightly acidic; the bacteria associated with the iron precipitate would be Leptothrix and Gallionella.

Figure 16.9 Acid mine drainage, showing confluence of a normal river and a creek draining a coal-mining area. Bacterial oxidation of iron sulfides in the coal-mine region leads to the production of sulfuric acid and thus to the solubilization of reduced iron (Fe^{2+}). Acid-tolerant iron-oxidizing bacteria (Ferrobacillus ferrooxidans) then convert the soluble Fe^{2+} to Fe^{3+}, which precipitates as insoluble ferric hydroxide. Two detrimental effects occur in streams as a result of acid mine drainage: (1) the water becomes highly acid, making it unsuitable for drinking or fish life; and (2) the unsightly precipitate clogs the rivers and prevents the growth of aquatic plants. Photograph taken in a mining region of eastern Kentucky. (Courtesy of Bill Strode and The Courier-Journal and Louisville Times.)

Figure 16.12 Growth of algae on the surface of a sandstone rock. Algal growth on rock surfaces is very common where light and sufficient moisture are present. Organic matter contributed by the algae leads to the growth of heterotrophic bacteria, and the combined action of these algae and bacteria leads to the initial decomposition and weathering of the rock, which are the first steps in soil formation.

Plate seven

(a)

(b)

Figure 17.6 The new volcanic island, Surtsey, off the south coast of Iceland. The surface is covered with a variably thick layer of volcanic ash, and plant life had not yet become established at the time these pictures were taken: (a) the central part of the island; (b) fumaroles emitting steam show that the interior is still warm although the surface has cooled to normal temperatures; (c) frame used to hold petri dishes containing agar culture media, which were being used to obtain some idea of the extent of dispersal of microorganisms to the island.

(c)

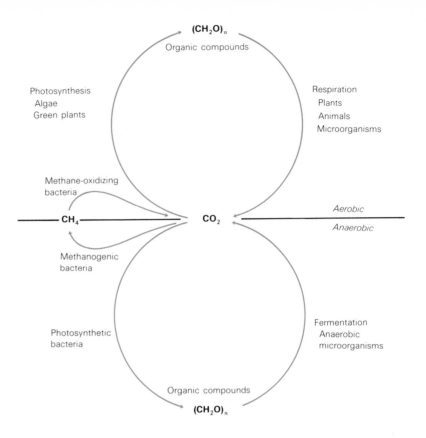

(CH₂O)ₙ
Organic compounds

Photosynthesis
Algae
Green plants

Respiration
Plants
Animals
Microorganisms

Methane-oxidizing
bacteria

CH₄

CO₂

Aerobic

Anaerobic

Methanogenic
bacteria

Photosynthetic
bacteria

Fermentation
Anaerobic
microorganisms

Organic compounds
(CH₂O)ₙ

Figure 16.3   Carbon cycle.

Petroleum is thought to have arisen in a similar way from organic deposits in marine environments. Organic matter deposited as coal or oil represents energy lost from primary producers, which is no longer available to the ecosystem. Much of this organic matter, however, is converted back to CO_2 through the combustion of these fuels by man.

In the aerobic zone of the water, decomposition usually proceeds to the complete conversion to CO_2 of organic materials such as cellulose, starch, and chitin by a wide variety of bacteria and fungi; different types of organic compounds are decomposed by different species.

Calcium carbonate deposition   The precipitation of calcium carbonate in aquatic environments is of considerable geological significance, as it leads to the ultimate formation of limestone. The extent of the process is shown by the fact that over 50 percent of the sedimentary rocks of the earth are limestone. Calcium carbonate is being precipitated today in large amounts in the warm tropical oceans, as well as in certain freshwater lakes, and the roles that living organisms play in this process have been well established. Many aquatic animals have calcium carbonate shells, for example, corals, molluscs, protozoa, and so on, and fossil shells of these creatures can be readily seen in many limestone rocks. To understand the role of microorganisms in $CaCO_3$ precipitation, one must realize that many natural waters are supersaturated with calcium ions that are held in solution as calcium bicarbonate or calcium sulfate. By altering the pH or by changing the bicarbonate ion concentration, microorganisms can cause calcium carbonate to precipitate. The equilibrium between calcium bicarbonate and calcium carbonate is influenced by the CO_2 content of the water:

$$Ca(HCO_3)_2 \rightleftharpoons CaCO_3\downarrow + H_2O + CO_2$$

and algae can cause the precipitation of $CaCO_3$ by removing CO_2 during photosynthesis, thus driving the equilibrium to the right. By their action on $CaSO_4$, sulfate-reducing bacteria are responsible for precipitation of $CaCO_3$ in anaerobic environments:

$$CaSO_4 + 4H_2 + CO_2 \longrightarrow CaCO_3\downarrow + 3H_2O + H_2S$$

Oxygen cycle A simplified oxygen cycle is shown in Figure 16.4. Green-plant photosynthesis results in a net increase in oxygen in the atmosphere. There is good reason to believe that initially the atmosphere of the earth was low in oxygen; only after the evolution of green plants did oxygen appear in significant quantities (see Chapter 17). The presence of oxygen gas in the atmosphere permitted the development of aerobic forms of life, setting the stage for the appearance of multicellular animals. Although oxygen is now one of the most plentiful gases in the atmosphere, it has limited solubility in water, and in a large water mass its exchange with the atmosphere is slow. The photosynthetic production of oxygen occurs only in the surface layers of a lake or ocean, where light is available. Organic matter that is not consumed in these surface layers sinks to the depths and is decomposed by facultative aerobic microorganisms, using oxygen dissolved in the water. Once the oxygen is consumed, the deep layers become anaerobic; here strictly aerobic organisms such as higher plants and animals cannot grow, and the bottom layers have a species composition restricted to anaerobic bacteria and a few kinds of tolerant animals. In addition, there is a conversion from a respiratory to a fermentative metabolism, with important consequences for the carbon cycle, which we described above.

Whether or not a body of water becomes depleted of oxygen depends on several factors. If organic matter is sparse, such as in unproductive lakes or in the open ocean, there may be insufficient energy available for heterotrophs to consume all the oxygen. Also important is how rapidly the

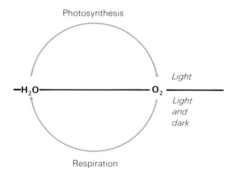

Figure 16.4 Oxygen cycle.

water from the depths exchanges with surface water. Where strong currents or turbulence occurs, the water mass may be well mixed, and consequently oxygen may be transferred to the deeper layers. In many bodies of water in temperate climates, however, the water mass becomes stratified during the summer, with the warmer and less dense surface layers separated from the colder and denser bottom layers. After stratification sets in, usually in early summer, the bottom layers become anaerobic. In the late fall and early winter, the surface waters become colder and heavier than the bottom layers, and the water "turns over," leading to a reaeration of the bottom. Most lakes in temperate climates thus show an annual cycle in which the bottom layers of water pass from aerobic to anaerobic and back to aerobic.

The oxygen relations in a river are of particular interest, especially in regions where the river receives much organic matter in the form of sewage and industrial pollution. Even though the river may be well mixed because of rapid water flow and turbulence, the large amounts of added organic matter can lead to a marked oxygen deficit. This is illustrated in Figure 16.5. As the water moves away from the sewage outfall, organic matter is gradually consumed by these organisms, and the oxygen content returns to normal.

Sanitary engineers describe the oxygen-consuming property of a body of water by a term called the "biochemical oxygen demand" (B.O.D.). The B.O.D. is determined by taking a sample of water, aerating it well, placing it in a sealed bottle, incubating for a standard period of time (usually 5 da at 20°C), and determining the residual oxygen in the water at the end of incubation. Although it is a crude method, a B.O.D. determination gives some measure of the amount of organic material in the water that could be oxidized by microorganisms. As a river recovers from contamination with a pollutant, the drop in B.O.D. is accompanied by a corresponding increase in dissolved oxygen (Figure 16.5).

We thus see that the oxygen and carbon cycles in a water body are greatly intertwined, and that heterotrophic microorganisms, mainly bac-

Figure 16.5   Chemical and biological changes in a river at various locations downstream from the source of a sewage outfall. Increased organic matter below the outfall leads to an increase in heterotrophic bacteria, a decrease in oxygen content, and an increase in NH_4^+. Farther downstream, NH_4^+ is oxidized to NO_3^- by nitrifying bacteria, and organic matter becomes oxidized by heterotrophs. Inorganic nutrients released from the decomposing organic matter make possible the growth of algae. Protozoa that feed on the bacteria are responsible for some of the decrease in bacterial numbers downstream.

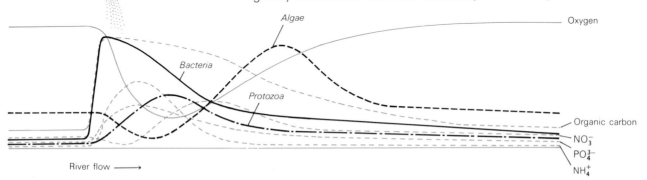

teria, play important roles in determining the biological nature and productivity of the body of water.

Nitrogen cycle Nitrogen gas (N_2) is the most abundant in the atmosphere, making up 76 percent of the weight, as compared to 23 percent for O_2. Much of the nitrogen on earth is in the atmosphere, and from this source most of the nitrogen of living organisms is ultimately derived. As we have noted in Chapter 5, N_2 is a rather stable compound. Some is fixed in the atmosphere nonbiologically, through lightning discharges and photochemical reactions, and this combined nitrogen is brought to earth in rainwater. A process equally important for producing combined nitrogen from N_2 is biological nitrogen fixation. The ability to assimilate or fix N_2 is restricted to only a few groups of organisms, primarily procaryotes. In aquatic environments, symbiotic nitrogen-fixing systems (see Section 14.9) are not common; here the bulk of the nitrogen is probably fixed by free-living blue-green algae.

An abbreviated nitrogen cycle is shown in Figure 16.6, and for some aspects of the nitrogen cycle in a polluted stream, see Figure 16.5. Nitrogen assimilated into organic matter, at the oxidation level of R—NH_2, is converted to ammonia by deamination reactions. Ammonia can be assimilated by many organisms as a sole source of nitrogen, and hence rarely is found in significant amounts in unpolluted water; but in polluted situations, large amounts of ammonia accumulate (Figure 16.5). In aerobic environments some of this ammonia is oxidized by nitrifying bacteria

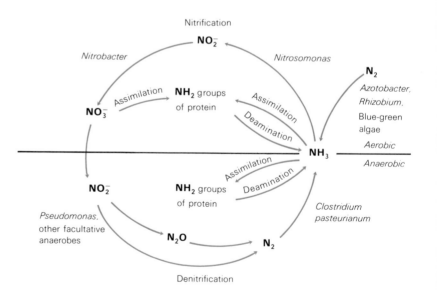

Figure 16.6 Nitrogen cycle.

and used as their primary energy source; it is ultimately converted to nitrate, a process called *nitrification*. This is of considerable importance in the recovery zone of a polluted stream, where the nitrate serves as nutrient for growth of algae (Figure 16.5). In anaerobic environments, nitrification cannot occur and if ammonia is not assimilated by microorganisms it accumulates. Also in anaerobic environments nitrate undergoes transformation to gases of nitrogen, primarily N_2 and N_2O; this process is called *denitrification*. As denitrification is of more importance in soils, it will be discussed further in Section 16.3.

Sulfur cycle   The sulfur cycle (Figure 16.7) is of considerable importance in aquatic environments. Although sulfur can exist in a number of oxidation states, only three of any significance occur in nature, sulfhydryl (R—SH) and H_2S (oxidation states $-II$), S (oxidation state 0), and SO_4^{2-} (oxidation state $+VI$). The other oxidation states are of biochemical significance in the intermediary metabolism of sulfur (see Chapters 5 and 6), but are not of direct ecological significance. Sulfate is one of the more common anions in water, and it is present in especially high amounts in sea water. Most microorganisms and plants can use sulfate as their sole source of sulfur, converting it into organic sulfhydryl compounds at the oxidation state of $-II$ (R—SH). Many microorganisms can produce H_2S, which is of considerable importance because this substance is toxic to most aerobes, and because it reacts with and precipitates many metal ions. However, H_2S is unstable aerobically and becomes oxidized to sulfur and sulfates either spontaneously or through biochemical processes. It is therefore only in anaerobic environments that H_2S accumulates. Hydrogen

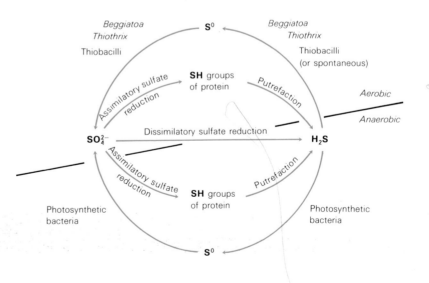

Figure 16.7   Sulfur cycle.

sulfide is produced microbiologically in two different ways: by the decomposition of organic sulfur-containing compounds ($R—SH \longrightarrow R + H_2S$), and by sulfate reduction ($SO_4^{2-} \longrightarrow H_2S$). Production of H_2S from organic compounds is of most importance in freshwater environments, where the concentration of the sulfate ion is usually low; sulfate reduction, on the other hand, is probably the most important means of H_2S formation in the sea. The organisms responsible for sulfate reduction are strict anaerobes, usually members of the genus *Desulfovibrio,* and these use sulfate as an electron acceptor instead of O_2, as was described in Section 4.5. The electron donors used in the process are organic compounds or molecular hydrogen and hence these organisms are usually found where organic concentrations are relatively high, such as in inshore marine areas or closed ocean basins like the Black Sea and the Norwegian and British Columbian fjords.

One of the consequences of H_2S production is the formation of metallic sulfides such as FeS and CuS. Iron is fairly common in all sediments, and H_2S production almost always is accompanied by FeS formation. Iron sulfide is responsible for most of the black color of anaerobic muds. Hydrogen sulfide oxidizes spontaneously in air to elemental sulfur and H_2O so that sulfur-oxidizing bacteria that carry out the same process live primarily in the region where H_2S rising from anaerobic areas meets O_2 descending from aerobic areas. If light is available, H_2S can also be oxidized anaerobically by photosynthetic bacteria (see Chapter 6), which are most often found in layers at depths where light still penetrates and where H_2S from the bottom is available (see Section 18.12).

The role of sulfur-oxidizing bacteria in the formation of economically important sulfur deposits seems well established. This conclusion is based partly on a study of the fractionation of stable (that is, nonradioactive) isotopes of sulfur. Sulfur has two major stable isotopes, [32]S, the normal isotope, which comprises 95 percent of the total in sulfate of sea water, and [34]S, which comprises 4 percent. When sulfate-reducing bacteria convert SO_4^{2-} to H_2S, they show a slight preference for the lighter [32]S isotope, and hence the H_2S formed has less [34]S than did the SO_4^{2-}. When H_2S is oxidized to S, a further preference for the light isotope is shown. During nonbiological (that is, chemical) oxidation and reduction, such a preference for the light isotope is not shown. Sulfur deposits along the Texas and Louisiana Gulf Coast have about 4 percent less [34]S than the sulfate with which they are associated, indicating that the deposits are microbial in origin.

When sulfur-oxidizing organisms convert H_2S into sulfuric acid, a marked drop in pH results, often to as low as pH 1 to 2. In coal-producing regions FeS is often present in the overburden of strip mines, and its oxidation, usually by microorganisms, leads to the formation of sulfuric acid, which leaches into the surrounding streams. These acid mine waters are toxic to fish and other higher forms of life, and are hence undesirable. At present, adequate means for the prevention of acid production in these waters are not available; perhaps research on the microbiological aspects of the sulfur cycle will suggest clues to adequate control.

Iron and manganese cycles Iron exists in two readily convertible oxidation states, ferrous (+II) and ferric (+III). The form in which iron will be found in nature is greatly influenced by pH and redox potential. At neutral pH, ferrous iron oxidizes spontaneously in air to ferric iron, which forms a highly insoluble ferric hydroxide. The reduction of iron from ferric to ferrous occurs through the action of microorganisms, which lower the redox potential of the habitat by consuming O_2. At neutral pH, therefore, ferrous iron is present in significant amounts only in anaerobic environments, such as in bogs and swamps, the bottoms of lakes, or deep wells. When water from such anaerobic environments moves to aerobic locations, the ferrous iron oxidizes spontaneously and ferric hydroxide is precipitated, leading to the formation of a brown deposit. Such deposits frequently are a serious problem in the pipes used to carry water from deep wells to domestic establishments. Iron deposits are very common at the edges of bogs (Plate 7: Figure 16.8). Many of the great iron-ore beds of the world are bog-iron deposits.

At acid pH values, ferrous iron does not oxidize spontaneously to the ferric state. It is oxidized biologically, however, by certain autotrophic sulfur and iron bacteria (*Thiobacillus ferrooxidans, Ferrobacillus ferrooxidans*) as their primary energy-generating process in the fixation of CO_2 (see Chapter 6). In the process of doing this, they precipitate large amounts of ferric hydroxide, which coats the outside of their cells. Because very little energy is generated in the oxidation of ferrous to ferric iron (see Chapter 6), these bacteria must oxidize large amounts of iron in order to grow, and consequently even a small number of bacteria can be responsible for precipitating a large amount of iron. These iron-oxidizing bacteria, which are strict acidophiles, are very common in acid mine drainages (Plate 7: Figure 16.9) and in acid springs and are probably responsible for most of the iron precipitated at acid pH values. The role of bacteria in iron transformations at neutral pH is probably nonspecific, involving mainly consumption of oxygen to render the habitat anaerobic, so that the non-biological reduction of ferric to ferrous iron can occur.

Manganese also exists in several positive oxidation states; the divalent [Mn(II)] and the tetravalent [Mn(IV)] are the most common. Divalent Mn is water soluble and predominates aerobically at pH values of less than 5.5 or at higher pH values anaerobically. At higher pH values in air, Mn(II) is oxidized spontaneously to the water-insoluble Mn(IV), which forms the insoluble MnO_2. Although there is some evidence for the specific oxidation and reduction of Mn by microorganisms, these transformations probably also occur spontaneously as a result of alterations by microorganisms of the redox conditions of the environment. Water from anaerobic portions of lakes or deep wells usually contains significant amounts of soluble Mn(II), which, when brought to the surface, may oxidize to the insoluble Mn(IV). The clogging of pipelines as a result of Mn precipitation is a problem in many parts of the world.

16.2 Sewage and waste-water treatment

Waste-water treatment is a process in which microorganisms play a crucial role, and it illustrates well some of the principles of aquatic microbiology discussed above. Waste waters are materials derived from domestic sewage or industrial processes, which for reasons of public health and for recreational, economic, and aesthetic considerations, cannot be disposed of merely by discarding them untreated into convenient lakes or streams. Rather, the undesirable materials in the water must first be either removed or rendered harmless. Removal of inorganic materials such as clay, silt, and other debris is done by mechanical and chemical methods, and microbes participate only casually or not at all. If the material to be removed is organic in nature, however, treatment usually involves the activities of microorganisms, which oxidize and convert the organic matter to CO_2. Waste-water treatment usually also results in the destruction of pathogenic microorganisms, thus preventing these organisms from getting into rivers or other supply sources. Water treatment can be carried out by a variety of processes, which may be separated broadly into two classes, anaerobic and aerobic.

Anaerobic treatment processes   Anaerobic sewage treatment involves a complex series of digestive and fermentative reactions in which the organic materials are converted into methane gas (CH_4) which can be removed and burned as a source of energy. The overall fermentation balance of an anaerobic treatment process can be represented as

$$C_6H_{12}O_6 + XH_2O \longrightarrow YCO_2 + 2CH_4$$
Hexose

Since both end products, CO_2 and CH_4, are volatile, the liquid effluent is greatly decreased in organic substances. The efficiency of a treatment process is expressed in terms of the percent decrease of the initial B.O.D. (see Section 16.1); the efficiency of a well-operated plant can be 90 percent or greater, depending on the nature of the organic waste.

Anaerobic decomposition is usually employed for the treatment of materials that have much insoluble organic matter, such as fiber and cellulose, or for concentrated industrial wastes. The process occurs in four stages: (1) the initial digestion of the macromolecular materials by extracellular polysaccharidases, proteases, and lipases to soluble materials; (2) the conversion of the soluble materials to organic acids and alcohols by acid-producing fermentative organisms; (3) the fermentation of the organic acids and alcohols to CO_2 and H_2; and (4) the conversion of H_2 and CO_2 to CH_4 by methanogenic bacteria.

An anaerobic decomposition process is operated semicontinuously in large enclosed tanks, into which the untreated material is introduced and from which the treated material is removed at intervals. The detention time in the tank is of the order of two weeks to a month. The solid residue consisting of indigestible material and bacterial cells is allowed to settle and is removed periodically and dried for subsequent burning or burial.

Aerobic treatment processes   There are several kinds of aerobic decomposition processes used in sewage treatment. A *trickling filter* is basically a bed of crushed rocks, about 6 feet thick, on top of which the liquid containing organic matter is sprayed (Figure 16.10). The liquid slowly trickles through the bed, the organic matter adsorbs to the rocks, and microbial growth takes place. As the waste passes downward, air is passed upward. The microbial film that develops is at first aerobic, but as it increases in thickness, the diffusion of oxygen to the inner portion of the film is impeded, and this portion becomes anaerobic. The bulk of the organic matter is removed in the aerobic zone. It is initially assimilated by slime-forming bacteria, filamentous bacteria, and filamentous fungi. These organisms are then eaten by protozoa, which are very active in the film, and the protozoa are in turn eaten by multicellular animals, so that there is a miniature food chain in the film. This food chain is of great importance in the efficient removal of organic matter because at each step in the food chain a portion of the organic matter is converted to CO_2 by respiratory processes and eventually the organic matter is completely eliminated.

Oxidation of organic matter in the film leads to deamination of organic nitrogen compounds and the release of NH_3, which is then converted to nitrate by the autotrophic nitrifying bacteria. Similarly, H_2S is produced by decomposition of sulfur-containing organic compounds and is converted by autotrophic sulfur-oxidizers to SO_4^{2-}. Inorganic phosphate is also formed in the film by the hydrolysis of nucleic acids and other phosphorus-containing compounds. These oxidized inorganic substances, NO_3^-, SO_4^{2-} and PO_4^{3-}, which are discarded in the effluent, are excellent

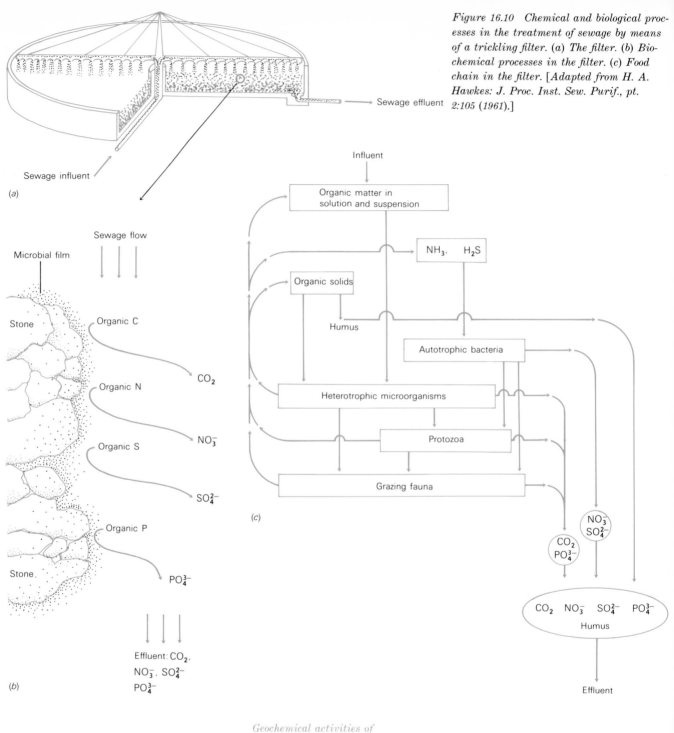

Figure 16.10 Chemical and biological processes in the treatment of sewage by means of a trickling filter. (a) The filter. (b) Biochemical processes in the filter. (c) Food chain in the filter. [Adapted from H. A. Hawkes: J. Proc. Inst. Sew. Purif., pt. 2:105 (1961).]

Sewage effluent

Sewage influent

(a)

Sewage flow

Microbial film

Stone

Organic C

CO₂

Organic N

NO₃⁻

Organic S

SO₄²⁻

Organic P

Stone

PO₄³⁻

Effluent: CO₂, NO₃⁻, SO₄²⁻ PO₄³⁻

(b)

Influent

Organic matter in solution and suspension

NH₃, H₂S

Organic solids

Humus

Autotrophic bacteria

Heterotrophic microorganisms

Protozoa

Grazing fauna

NO₃⁻ SO₄²⁻

CO₂ PO₄³⁻

CO₂ NO₃⁻ SO₄²⁻ PO₄³⁻

Humus

Effluent

(c)

nutrients for the growth of algae; indeed, rich algal blooms are common in rivers receiving effluents of sewage treatment plants. Such blooms often choke the rivers and themselves cause a significant increase in the organic content of the river. Thus they are considered detrimental. Nutrient enrichment, called *eutrophication,* has become a serious economic problem in recent years as it has resulted in deterioration of the quality of many natural waters. Considerable research is currently under way to develop ways of removing these inorganic nutrients from treated effluents.

Another aerobic treatment system is the *activated sludge* process. Here, the waste water to be treated is mixed and aerated in a large tank. Slime-forming bacteria (primarily a species called *Zoogloea ramigera*) grow and form flocs (so-called zoogloeas), and these flocs form the substratum to which protozoa and other animals attach. Occasionally filamentous bacteria and fungi also are present. The basic process of oxidation is similar to that in a trickling filter. The effluent containing the flocs is pumped into a holding tank or clarifier, where the flocs settle. Some of the floc material is then returned to the aerator to serve as inoculum, while the rest is dried and discarded. The effluent of an activated sludge plant, like that from a trickling filter, is also rich in inorganic nutrients favorable to the growth of algae. Both types of operations result in a 75- to 90-percent decrease of B.O.D.

Sewage lagoons or stabilization basins are large, open, holding reservoirs into which waste water is introduced. Oxygen is supplied either by diffusion from the atmosphere or by artificial aeration. Since light is available, heavy growth of algae and small aquatic plants occurs on the surface of the water. Depending on the depth and shape of the basin, the process may be aerobic throughout, or aerobic only at the surface and anaerobic in the depths. Bacteria grow, consume O_2, and oxidize the organic matter to CO_2; the algae photosynthesize, fix CO_2, and produce O_2. Bacteria and animals attack the algal cells and reoxidize the algal constituents to CO_2. As long as the process is in good balance, conditions remain aerobic and the B.O.D. is reduced 80 to 95 percent. Organic nitrogen is converted to nitrogen gases by denitrification in the deeper portions of the lagoon, and much of the inorganic nitrogen that would otherwise cause eutrophication can thus be eliminated. Lagoons are most useful as sewage systems that treat only small amounts of waste water since they require a considerable area of land.

Problems of sewage treatment Sugar-mill waste and some other industrial wastes are rich in carbohydrates and deficient in nitrogen and other microbial nutrients. Since waste treatment requires first that microorganisms grow, these wastes cannot be adequately treated unless the inorganic nutrients lacking are added.

Some industrial wastes contain acids, phenolic compounds, heavy metals, or other materials toxic to microorganisms. Often such wastes cannot be dealt with adequately by biological methods; frequently they pose an expensive problem. However, microbial populations show remarkable ability to adapt to new compounds. It is often observed that when a new substance is first introduced into a treatment system, it is decomposed slowly, but after a period of adaptation, decomposition is rapid (Figure 16.11). Such adaptation is probably due to the selection of rare strains able to oxidize the compound.

The organic compounds in most industrial wastes are not fermentable and hence can be treated only by aerobic processes. However, not all organic compounds are readily oxidizable by microorganisms. Some detergents are not oxidized and are discharged unchanged. Frequently, such detergents have caused extensive foaming of water, even after it has passed many miles downstream. To eliminate this problem, legislation has gone into effect requiring the use of only so-called *biodegradable* detergents, which can be handled adequately by the microorganisms in the treatment plant.

A large number of designs for a treatment plant are possible, depending on the nature of the waste water, the size of plant needed, its location,

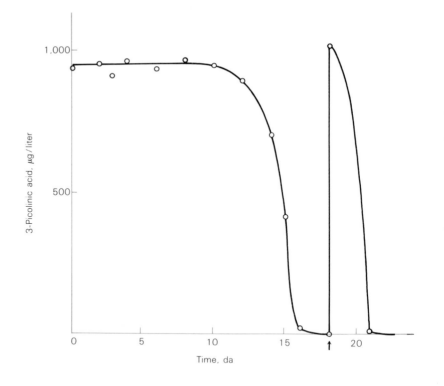

Figure 16.11 Adaptation of aquatic bacteria to the oxidation of a new compound. The graph shows the rate of disappearance of picolinic acid after being added to river water. After the initial introduction there is a long lag before the compound is decomposed. The rate of disappearance after the second addition (arrow) is quite rapid. [Redrawn from M. B. Ettinger, R. J. Lishka, and R. C. Kroner: Ind. Eng. Chem. 46:791 (1954). Courtesy of American Chemical Society.]

and other factors. A typical plant might employ an anaerobic digester for primary treatment, and its effluent might then be passed to a trickling filter for secondary treatment. A few plants now use a tertiary treatment in which nitrates and phosphates are removed before the final effluent is discharged into a river, a lake, or the ocean.

In rural areas the most common waste-treatment system is a septic tank, which is a rectangular tank buried in the ground. Raw sewage enters at one end and slowly passes through the tank. Anaerobic conditions develop, and sludge settles and is digested in the bottom, whereas soluble organic matter undergoes some decomposition in the liquid. The process, however, is quite inefficient, and only about 50 percent of the B.O.D. is removed. In addition, the tank slowly fills up with sediment, and must be cleaned every 2 to 5 yr. The effluent, still rich in organic matter, is usually dispersed in a tile field, from which the liquid enters the soil. If the soil is permeable, the liquid disperses readily and the remainder of the decomposition takes place in the soil, under the agency of normal soil organisms. In regions of clay soils, however, the liquid does not disperse well and usually channels out of the tile field in a slow trickle, which drains down into neighboring land and creates an odor and a health hazard. Septic-tank systems rarely function satisfactorily and should be discouraged since they may discharge intestinal disease-causing organisms. This is especially serious if the water supply of a residence derives from a well since septic-tank drainage may seep into the well.

16.3 Terrestrial environments

In the consideration of terrestrial environments, our attention is inevitably turned to the *soil* since it is here that many of the key processes occur that influence the functioning of the ecosystem. The process of soil development involves complex interactions between the parent material (rock, sand, glacial drift, and so on), topography, climate, and living organisms. Soils can be divided in two broad groups—*mineral soils* and *organic soils*—depending upon whether they derive initially from the weathering of rock and other inorganic material or from sedimentation in bogs and marshes. Mineral soils are predominant in most areas, and our discussion will concentrate on these.

The weathering of rock is a result of combined physical, chemical, and biological processes. A careful examination of almost any exposed rock will reveal the presence of algae, lichens, or mosses (Plate 7: Figure 16.12). These organisms are able to remain dormant on the dry rock and then grow when moisture is present. They are photosynthetic and produce

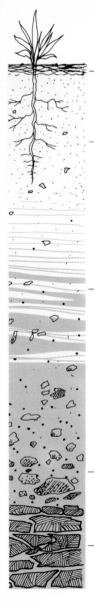

Layer of undecomposed plant materials

Soil surface (high in organic matter, dark in color, is tilled for agriculture; plants and micro-organisms grow here)

Subsoil (minerals, humus, etc., leached from soil surface accumulate here; little organic matter)

Soil base (develops directly from underlying bedrock)

Bedrock

Figure 16.13   Profile of a mature soil.

organic matter, which supports the growth of heterotrophic bacteria and fungi. The numbers of heterotrophs increase directly with the degree of plant cover of the rocks. Carbon dioxide produced during respiration by heterotrophs becomes converted into carbonic acid:

$$CO_2 + H_2O \longrightarrow H_2CO_3$$

which is an important agent in the dissolution of rocks, especially those composed of limestone. Many of the heterotrophs excrete organic acids, which also promote the dissolution of rock into smaller particles. Freezing and thawing and other physical processes lead to the development of cracks in the rock. In these crevices a raw soil forms, in which pioneering higher plants can develop. The plant roots penetrate further into crevices and increase the fragmentation of the rock, and their excretions promote the development of a rhizosphere flora (Section 14.8). When the plants die, their remains are added to the soil and serve as nutrients for an even more extensive microbial development. Minerals are further rendered soluble, and as water percolates it carries some of these chemical substances deeper. As weathering proceeds, the soil increases in depth, thus permitting the development of larger plants and trees. Soil animals become established and play an important role in keeping the upper layers of the soil mixed and aerated. Eventually the movement of materials downward results in the formation of layers, and a typical soil profile becomes outlined (Figure 16.13). The rate of development of a typical soil profile depends on climatic and other factors, but it is usually very slow, taking hundreds of years.

The nature of the soil that forms in a given area is greatly dependent on climatic conditions, which affect not only microbial action but also the kinds of plants that develop. *Podzol soils* are those that form under coniferous vegetation such as pine, spruce, and fir. The leaf litter of conifers promotes production of acids, which cause strong leaching of minerals from the *A* horizon. Bacteria, being sensitive to acid pH, do not thrive, and the microbial flora is therefore predominantly fungal. Soil animals are scarce in podzol soils, which are also relatively infertile for agricultural crops; they are usually best left to the production of trees. *Brown soils* are those developing under deciduous trees. The decomposition of the leaf litter leads to a more neutral pH; leaching is less severe, soil bacteria and animals are active, and the soil is relatively fertile. *Grassland soils* are higher in organic content than are brown soils because the grasses have extensive root systems that add large amounts of organic matter to the soil. Whereas in brown soils most of the organic matter added to the soil comes from the surface leaf litter, in grassland soils the organic matter is present to a greater depth. Decomposition of the organic matter leads to production of an extensive black humus layer, which greatly contributes

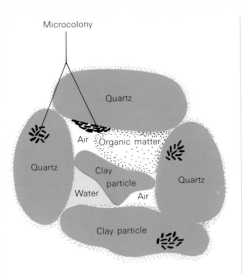

Figure 16.14 *A soil aggregate composed of mineral and organic components, showing the localization of soil microorganisms. Very few microorganisms are found free in the soil solution; most of them occur as microcolonies attached to the soil particles.*

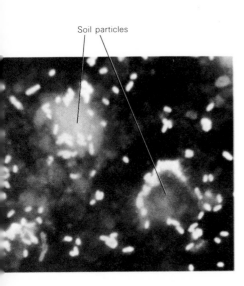

to the friability and texture of the soil. Consequently, grassland soils are excellent agricultural soils. *Tropical soils,* which develop under rain forests, are quite different from soils developing under forests in temperate climates. The warm temperatures and high rainfall of these tropical regions leads to rapid decomposition of leaf litter. However, even though an extensive forest develops, the soil contains relatively little organic matter or humus. When a tropical forest is cleared for agricultural purposes, decomposition of the remaining litter occurs rapidly, and after several years of farming the soil has been depleted of its essential nutrients. Hence, many tropical soils are not good agricultural soils, and cleared areas must be abandoned after a few years. This has led to the development of what is called *shifting agriculture,* which constitutes one reason for the lack of development of stable agricultural communities in many parts of the tropical world.

Soil as a microbial habitat   Water is a highly variable component of the soil, its presence depending on rainfall, drainage, and plant cover. It is held in the soil in two ways, by adsorption onto surfaces or as free water existing in thin sheets or films between soil particles. Different soils vary greatly in their capacity to absorb and retain water. The pore space is usually 30 to 50 percent of the total volume of a soil, and in a well-drained soil the volumes of the respective components are: soil particles, 50 percent; air, 10 percent; water, 40 percent.

Even when a soil lacks a free-water phase, each soil crumb is a surface upon which water may adsorb, and often the major part of the moisture in a soil is the portion that is adsorbed. The most extensive microbial growth takes place on the surfaces of soil particles (Figure 16.14), and because it is difficult to observe microorganisms on these opaque surfaces by ordinary microscopy, several special techniques have been developed for observing microorganisms in soil. In the slide-immersion technique a thin ditch is dug in the soil, a glass microscope slide is inserted with just its edge sticking out, and the soil is carefully packed around it. After a week or two, the slide is removed, stained, and examined under the microscope; now the shapes, sizes, and positions of the microorganisms that have grown on the slide can be readily seen. To examine soil particles directly under the microscope special reflected-light fluorescence microscopes are often used, the organisms in the soil being stained with a dye

Figure 16.15 *Bacterial microcolonies (Rhizobium japonicum) on the surface of soil particles, visualized by the fluorescent antibody technique. Note some areas with bacteria localized on soil particles. Magnification, 1,000×. [From B. B. Bohlool and E. L. Schmidt: Science 162:1012 (29 November 1968). © 1968 by the American Association for the Advancement of Science.]*

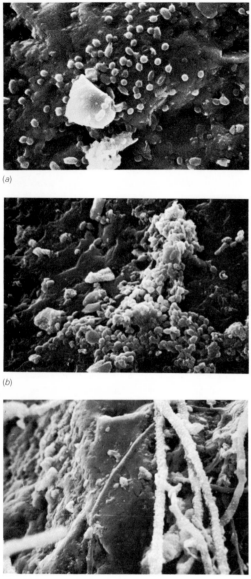

Figure 16.16 Visualization of microorganisms on the surface of soil particles by use of the scanning electron microscope: (a) bacteria (magnification, 1,600×); (b) actinomycete spores (1,390×); (c) fungus hyphae (810×). (Courtesy of T. R. G. Gray.)

that fluoresces. In effect each microbial cell is its own light source, and its shape and position on the surface of the particle can easily be seen. To observe a specific microorganism in a soil particle, fluorescent antibody staining (see Chapter 15) can be used (Figure 16.15). Microorganisms can be excellently visualized on such opaque surfaces as soil by means of the *scanning electron microscope* (Figure 16.16), which provides much higher resolution and depth of field than does the light microscope.

The water present in soils has a variety of materials dissolved in it; the whole mixture is called the soil solution. In well-drained soils air penetrates readily, and oxygen is never deficient even in the deepest portions. However, water and air compete for the space between the soil particles, and the former can drive out the latter. Thus, in a waterlogged soil the only oxygen present is that dissolved in the water, and this is soon consumed by microorganisms. Such soils quickly become anaerobic, showing profound changes in their biological properties.

Soil usually contains a variety of complex minerals, which in addition to providing the basic structure of the soil play important chemical roles. Especially significant are the clay minerals, which are complex aluminosilicates that are ionically negative and hence are capable of adsorbing cations such as sodium, potassium, magnesium, calcium, and iron. These biologically important cations are present in soil primarily associated with clay minerals and are released only by ion-exchanging reactions. A microorganism, for instance, by producing hydrogen ions, can promote the release of metal ions adsorbed to the clay mineral (Figure 16.17).

The pH values of soils vary considerably. Podzol soils may have pH values as low as 3, whereas in the alkali soils of some desert areas, where salts accumulate, pH values as high as 11 have been recorded. Most agricultural soils are slightly acidic.

Although the basic principles of the biogeochemical cycles are the same in soil as in aquatic environments, there are many significant differences in details, which arise primarily from the considerable differences in the microbial environments in these two kinds of habitat.

Carbon cycle Although in aquatic environments much insoluble organic matter falls into sediments, where anaerobic conditions promote its preservation, in terrestrial environments these materials remain near the surface, where aerobic conditions prevail and decomposition can go to completion. The initial plant litter contains some water-soluble materials such as sugars and amino acids, which are leached out and decompose readily, but the bulk of the material consists of insoluble materials such as cellulose, hemicellulose, and lignin. Litter in its initial form is not a highly favorable medium for microbial growth since it is deficient in vitamins, amino acids, and other growth factors, and its physical structure

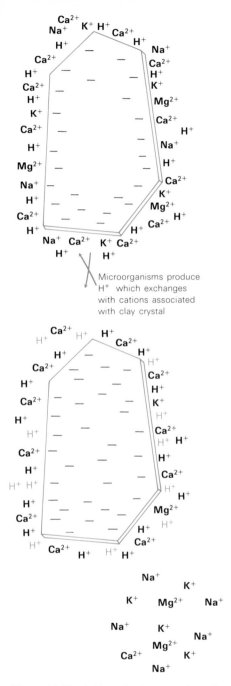

Microorganisms produce H⁺ which exchanges with cations associated with clay crystal

Figure 16.17 Action of microorganisms in mobilizing minerals from a soil particle.

makes it resistant to attack. During the initial phase of decomposition, a few organisms grow and alter the chemical composition and structure of the litter and make it more compact and better able to retain moisture. Subsequent decomposition then occurs at a more rapid rate. Animals such as earthworms, which chew the litter and increase its surface area, also promote markedly the decomposition processes. In fertile soils, the amount of organic matter is usually fairly high, and microbial numbers consequently are also high. The microbiological activity of the soil is much greater in the organically rich surface layers than it is in deeper layers.

The *food chain* in the soil is extremely complex, due to the great variety of organisms present. As would be expected, algae are not present in significant numbers in the bulk of the soil, although they are often an important part of the surface crust. The initial consumers of cellulose, hemicellulose, and lignin of the litter are most commonly fungi. Cellulose and hemicellulose are decomposed first, the more resistant lignin fraction requiring months or years to be destroyed. The fungi are consumed by bacteria that produce enzymes capable of digesting the fungal cell walls and protoplasm. Nematodes and protozoa then develop and devour the bacteria, and these in turn are preyed on by mites and other larger animals. A simplified food chain might thus be: leaf litter ⟶ fungi ⟶ bacteria ⟶ animals. Eventually the bulk of the organic matter is converted to CO_2.

The term *humus* is used to denote the organic fraction of the soil that is relatively resistant to decomposition, to distinguish it from the organic matter of the litter itself. Humus is not an homogeneous substance but is a complex mixture of materials. It is derived partly from the protoplasmic constituents of soil organisms which themselves have resisted decomposition, and partly from resistant plant material. If the plant cover is removed from a soil, as is done in agricultural practice, the humus slowly disappears. The rate of humus disappearance is greater in warmer than in cooler climates and is affected by rainfall and soil texture. Humus promotes the development of an agriculturally desirable soil texture.

Compost, an organic complex used for fertilizing and improving the texture of soils, is prepared by piling leaves and other plant materials in a heap and allowing them to undergo decomposition. The construction of a compost pile has two consequences: (1) Moisture is better retained within the plant materials, thus facilitating the growth of microorganisms; (2) the increased insulation obtained within the pile enables much of the heat produced by microbial action to be retained, and the temperature of the compost pile rises to a range best suited for thermophilic organisms. The temperature rises rapidly within the first day or two of composting, reaches a peak at about 70°C, and then slowly returns to normal as the easily oxidizable substances of the plant material are consumed.

During the process there are complex changes in the microbial populations. The organisms present on the plant material are of a variety of types and include both mesophiles and thermophiles. During the initial stages of decomposition the mesophiles are eliminated and the thermophilic fungi, actinomycetes, and bacteria flourish. Many of these thermophiles are cellulose decomposers able to break down the plant residues, which can be used beneficially as an organic supplement to soil.

The use of *herbicides* and *pesticides* has increased dramatically in the past generation. These compounds are of a wide variety of chemical types, such as phenoxyalkyl carboxylic acids, substituted ureas, nitrophenols, chlorinated organic acids, phenylcarbamates, and others. Some of these substances are suitable as carbon and energy sources for certain soil microorganisms, whereas others are not. If a substance can be attacked by microorganisms, it will eventually disappear from the soil. Such degradation in the soil is usually desirable, since toxic accumulations of the compound are avoided. However, even closely related compounds may differ remarkably in their degradability, as is shown for 2,4-dichlorophenoxyacetic acid (2,4-D) and 2,4,5-trichlorophenoxyacetic acid (2,4,5-T) in Figure 16.18. The relative persistence rates of a number of herbicides is shown in Table 16.2. However, these figures are only approximate since a variety of environmental factors, such as temperature, pH, aeration, and organic matter content of the soil, influence decomposition. Some of the chlorinated insecticides are so indestructible that they have persisted for over 10 yr.

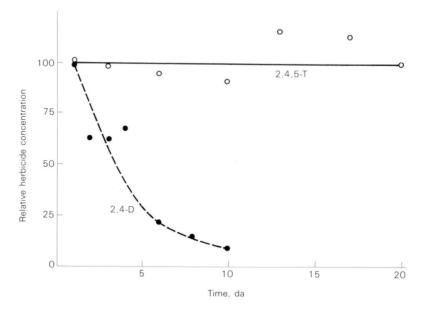

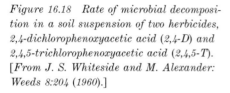

Figure 16.18   Rate of microbial decomposition in a soil suspension of two herbicides, 2,4-dichlorophenoxyacetic acid (2,4-D) and 2,4,5-trichlorophenoxyacetic acid (2,4,5-T). [From J. S. Whiteside and M. Alexander: Weeds 8:204 (1960).]

Table 16.2   Decomposition and period of persistence of several herbicides[a]

Name of compound	Abbreviation	Persistence in soil	Organisms active in decomposition
3-(*p*-chlorophenyl)-1,1dimethyl-urea	Monuron	4–12 mo	*Pseudomonas*
2,4-Dichlorophenoxyacetic acid	2,4-D	2–8 wk	*Achromobacter, Corynebacterium, Flavobacterium*
2,4,5-Trichlorophenoxyacetic acid	2,4,5-T	5–11 mo	
2-Methyl-4-chlorophenoxyacetic acid	MCPA	3–12 wk	*Achromobacter, Mycoplana*
2,2-Dichloropropionic acid	Dalapon	2–4 wk	*Agrobacterium, Pseudomonas*
Dinitro-*o*-sec-butylphenol	DNBP	2–6 mo	*Corynebacterium, Pseudomonas*
Isopropyl N-phenylcarbamate	IPC	2–4 wk	
Isopropyl N-(3-chlorophenyl) carbamate	CIPC	2–8 wk	
Trichloroacetic acid	TCA	2–9 wk	*Pseudomonas*
2,3,6-Trichlorobenzoic acid	2,3,6-TBA	>2 yr	

[a] From M. Alexander, *Introduction to Soil Microbiology* (New York: John Wiley & Sons, Inc., 1961).

The uptake of CO_2 by the plants of the entire world in one year is about $\frac{1}{35}$ of the total CO_2 of the atmosphere. This means that, were this CO_2 not replaced, eventually no CO_2 would be left in the atmosphere and plant growth would cease. The CO_2 is, of course, replaced, mainly by microbial decomposition processes in the soil; without microorganisms, the biological world as we know it would soon cease to function.

Nitrogen cycle   The amount of nitrogen present in rocks is low, and because most nitrogen compounds are soluble, volatile, or readily decomposable, there is little tendency for nitrogen to accumulate in vast deposits, as do carbon and sulfur. The nitrate deposits found on the guano-rich islands off the coasts of Chile and Peru exist only because rainfall in these areas is minimal, and the nitrate formed by bacterial nitrification of the rich fecal deposits of water birds in this region is not washed away.

Nitrogen enters soil as ammonia or nitrate from rain water, as biologically fixed nitrogen, or as a component of agricultural fertilizers. The nitrogen content of plant residues is low, a C:N ratio of 40:1 or 50:1 being usual. As the litter decomposes, the nitrogen content slowly increases, and the C : N ratio approaches the value of 10 : 1, the C : N ratio in microbial protoplasm. If an agricultural soil is allowed to return to natural grassland, the nitrogen content of the soil slowly increases, and it may double in about 50 years. Some of the increase in nitrogen that occurs in litter probably results from nitrogen fixation by free-living bacteria of the genera *Azotobacter* and *Clostridium,* and the rest may derive from microbial assimilation of nitrate and ammonia from rain water. Legumes do not occur to a significant extent in most natural habitats in temperate climates and symbiotic nitrogen fixation is consequently of relatively minor importance; in the tropics, however, many leguminous trees grow naturally and symbiotic fixation is of more consequence. Agricultural practice involving crop rotation with nitrogen-fixing legumes such as clover and alfalfa is designed at least in part to increase the nitrogen content of the soil, but in recent years farmers in areas of high agricultural productivity have been relying less and less on crop rotation and more on the use of chemically synthesized nitrogen fertilizers.

Organic nitrogen is converted to ammonia by the process of deamination. Ammonia can be assimilated directly by microbes or plants, or it can be converted to nitrate by nitrifying bacteria. *Nitrification* is an aerobic process and occurs readily in well-drained soils at neutral pH; it is inhibited, however, by anaerobic conditions or in highly acidic soils. If materials high in protein, such as manure or sewage, are added to soils the rate of nitrification is increased. Although nitrate is readily assimilated by plants, it is very water soluble and is rapidly leached from soils receiving high rainfall. Consequently, nitrification is not necessarily beneficial in agricultural practice. Ammonia, on the other hand, is cationic and consequently is strongly adsorbed to negatively charged clay minerals. Anhydrous ammonia is now used extensively in the United States as a nitrogen fertilizer, being injected directly into the soil in gaseous form. Urea (H_2N—CO—NH_2), another form of nitrogen applied to agricultural soils, is readily hydrolyzed to ammonia and CO_2 by the enzyme urease, which is formed by many microorganisms.

If a soil becomes waterlogged anaerobic conditions develop, and *denitrification* occurs. The ability of a variety of bacteria to use nitrate instead of O_2 as an electron acceptor was discussed in Section 4.5. Denitrification usually requires the presence of organic materials as energy sources, and hence should be more significant in soils rich in organic matter. Even well-drained (and hence aerobic) soils may experience some denitrification if large amounts of organic residues are applied since local regions of partial anaerobiosis will result. Denitrification causes production of two gases, nitrogen (N_2) and nitrous oxide (N_2O). Although nitrite (NO_2^-) is an intermediate in the process, it rarely accumulates in the soil; at neutral or alkaline pH values nitrite is quickly converted biologically to N_2O and

N_2, and at acid pH values it spontaneously decomposes to nitric oxide (NO). Since these products of denitrification are gaseous, they escape from the soil to the air and lead to a reduction in the nitrogen content of the soil. Clearly, denitrification is undesirable agriculturally, and farming practices attempt to combat it. The practice of tilling the soil probably inhibits denitrification by breaking up large soil clumps where pockets of anaerobiosis could develop.

Sulfur cycle and microbial corrosion   Sulfur occurs in significant quantities as sulfate in many rocks and as the gas SO_2 in the air; sulfate is also present in rain water. Plants and most microorganisms can assimilate sulfate directly and reduce it to the sulfhydryl form, as in sulfhydryl amino acids. The organic sulfur in plant proteins is returned to the soil in the litter, where it is decomposed by a large variety of microorganisms that release H_2S. Under aerobic conditions H_2S is readily oxidized by chemosynthetic sulfur bacteria, mainly *Thiobacillus*, with the formation of sulfate. Except in waterlogged soils, it is unlikely that any significant amounts of H_2S will accumulate. *Thiobacillus* also oxidizes elemental sulfur to sulfuric acid (H_2SO_4), concomitantly lowering the pH. Elemental sulfur is sometimes added to alkaline soils to effect a lowering of the pH. Many minerals are insoluble and hence unavailable to plants at alkaline pH values; they become utilizable when the pH is decreased.

Under anaerobic conditions sulfate-reducing bacteria reduce SO_4^{2-} to H_2S, and the latter is often present in fairly high amounts in flooded soils. Since H_2S is toxic to higher plants, sulfate reduction is clearly an undesirable process agriculturally.

The sulfate-reducing bacteria also play a considerable role in the corrosion of iron and steel pipes buried in the soil and in the sea. In waterlogged conditions, these bacteria oxidize metallic (elemental) iron (Fe^0) to Fe^{2+} by the series of reactions shown in Figure 16.19. There are two ways

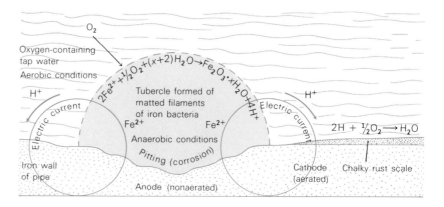

Figure 16.19   Microbial processes in the corrosion of a buried metal pipe. (Adapted from E. Olsen and W. Szybalski: Acta Chem. Scand. 3:1094 (1949).]

in which these bacteria participate: by consuming hydrogen generated spontaneously, and by producing H_2S when coupling oxidation of organic matter with sulfate reduction. The release of electrons at the iron surface is called *anodic release* because it is analogous to the process that occurs at the anode of an electrolytic battery. The acceptance of electrons by hydrogen ions and the production of H_2 is called a *cathodic reaction* since the cathode is the terminal in an electrolytic cell that accepts electrons. The oxidation of elemental Fe can be promoted by bacteria either by consuming H_2 generated at the cathode (cathodic corrosion) or by producing H_2S during sulfate reduction, which reacts with Fe^{2+} at the anode (anodic corrosion). Both these processes probably occur, but which is most important is still under debate. Microbial corrosion is of special importance because it occurs under anaerobic conditions, where spontaneous corrosion (rusting) does not take place. This problem is so serious that an unprotected pipe under anaerobic conditions can be destroyed in a few years. Soil conditions in the Netherlands are especially conducive to anaerobic corrosion; water pipes buried in soil reclaimed from the sea (whose water is high in SO_4^{2-}) had to be replaced in 2 or 3 yr even though the pipes were coated with asphalt. With iron pipes anaerobic microbiological corrosion causes such solubilization of the iron that the resulting soft matrix can be cut readily with a knife. With steel, decomposition is less severe and only localized pitting occurs.

16.4 Petroleum microbiology

The possible role of microorganisms in the formation of petroleum had been mentioned earlier in this chapter. Microbiology has made a number of contributions to our knowledge of petroleum and its uses. Among the most interesting has been the use of microorganisms in *petroleum prospecting*. Associated with liquid and solid petroleum is a gaseous fraction consisting of methane, ethane, and propane. In petroleum-producing regions these gases may seep to the surface and provide nutrients for the growth of specific hydrocarbon-utilizing bacteria. Where one finds bacteria capable of oxidizing these gases, there is a strong suggestion that a petroleum deposit is nearby. Looking for methane-utilizing organisms in searching for petroleum is not practical as methane is produced biologically in many systems that are not related to petroleum. Ethane, however, is not produced biologically in significant amounts and is almost always associated only with petroleum, so that detection of ethane-utilizing organisms can be used to discover petroleum reserves. The correlation between the presence of ethane oxidizers and the geographical

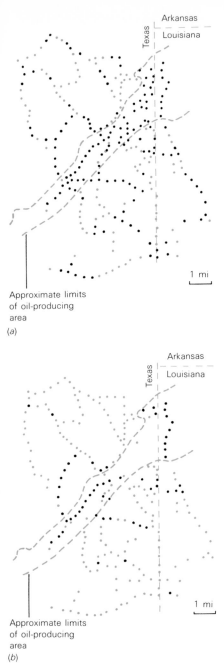

Approximate limits of oil-producing area

(a)

Approximate limits of oil-producing area

(b)

Figure 16.20   Use of hydrocarbon-oxidizing bacteria in prospecting for petroleum. Soil samples were tested for the presence of either methane-oxidizing (a) or ethane-oxidizing (b) bacteria. Positive samples are shown in black, negative samples in color. Note the good correlation between the presence of ethane oxidizers and the limits of the oil-producing area. Ethane is found in nature almost solely in petroleum deposits, whereas methane is produced as a result of many other natural processes. (From J. B. Davis: Petroleum Microbiology. Amsterdam: Elsevier Publ. Co., 1967.)

limits of an oil-producing area is shown in Figure 16.20; the same figure shows the lack of correlation between the presence of methane oxidizers and the presence of petroleum. Microbiological petroleum prospecting has as yet not found wide use since geological methods of locating petroleum deposits have thus far been adequate.

Microbial decomposition of petroleum and petroleum products is of considerable importance. Since petroleum is a rich source of organic matter and the hydrocarbons within it are readily attacked aerobically by a variety of microorganisms (see Chapter 5), it is not surprising that, when petroleum is brought into contact with air and moisture, the mixture is subjected to microbial attack. It is virtually impossible to keep moisture from bulk storage tanks; it accumulates as a layer of water beneath the petroleum. At the petroleum-water interface bacteria develop in large numbers and fungi, yeasts, and actinomycetes may also grow. Microbial growth has become an especially serious problem in the kerosene-based fuels used in jet airplanes. When such microbe-containing fuels are burned, fuel strainers become rapidly clogged, leading to power loss. In addition, microbial growth in the fuel tanks of aircraft can lead to corrosion of the tanks. Several control methods are used: (1) The fuel is filtered through membrane filters, which remove microorganisms. (2) Inhibitors of microbial growth are added. (3) Corrosion can be minimized by coating the fuel tanks with more resistant substances, such as polyurethane. (4) Aircraft fuel tanks are washed out at regular intervals with a 2 percent solution of potassium dichromate, a microbicidal compound. (5) Some hydrocarbon fractions of the fuel are more readily attacked than are others; and if these fractions can be removed from the fuel, its storage life may be lengthened.

16.5   Geochemical cycles on a global basis

We have seen in this chapter the wide variety of ways in which microorganisms participate in aquatic and terrestrial ecosystems. In the world as a whole there are complex interactions between terrestrial and aquatic

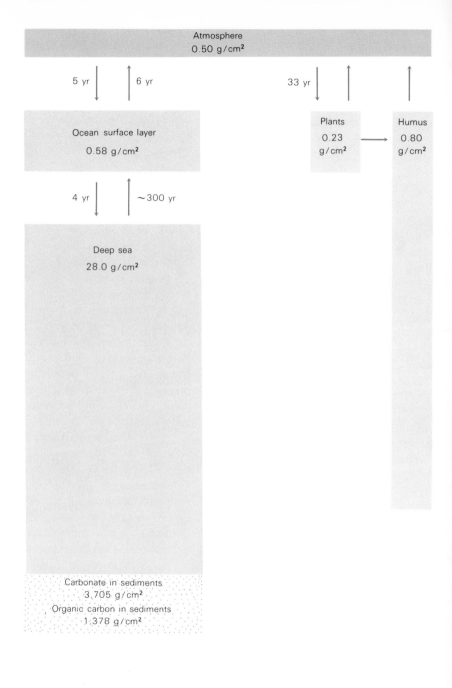

Figure 16.21 The global carbon cycle: The sizes of the carbon deposits in the various areas are reflected by the sizes of the boxes. The rate of transfer of carbon from one component to the other is expressed as the time in years necessary for complete exchange. (From C. E. Junge: Air Chemistry and Radioactivity. New York: Academic Press, Inc., 1963.)

ecosystems, which can be summarized by the description of global geo-chemical cycles. Such a cycle for carbon is shown in Figure 16.21. The rates at which different parts of this cycle take place vary markedly, but the system as a whole is integrated and has probably been in balance for millions of years—at least it has been since the time that present atmos-pheric conditions became established (see Chapter 17). In recent years, however, a possible threat to a balanced carbon cycle has been created by the activities of man in unearthing vast deposits of fossil fuels (coal, oil, and natural gas) and combusting them' to CO_2. It is still not certain whether this has led to a significant increase in the CO_2 concentration of the atmosphere on a global basis. If so, it might be anticipated that such increases would lead to increased activity of photosynthetic organisms and to the return of the carbon back to plant residues, ultimately produc-ing new fuel deposits. In the meantime there may be a transient rise in CO_2 concentration, which could have important consequences for our climate since CO_2 prevents dissipation of infrared radiation to outer space. A warming trend on earth may possibly occur. It is therefore important to understand the rates of the various transformations of the carbon cycle on a global basis and to assess the relative roles of microorganisms and of man in returning CO_2 to the atmosphere.

The conversion of an element from organic form to inorganic form, as in the conversion of organic carbon to CO_2, of organic nitrogen to NH_3 and NO_3^-, and of organic sulfur to H_2S and SO_4^{2-}, is called *mineralization;* it is the most important contribution of microorganisms to the geo-chemical cycles. Microorganisms also play key roles in the formation of such geologically important products as coal, oil, and sulfur, and partici-pate along with higher organisms in various aspects of the cycles. Thus, despite their small sizes, the impact of microorganisms on the world is great.

Supplementary readings

Alexander, M. Introduction to Soil Microbiology. New York: John Wiley & Sons, Inc., 1961.

Allen, O. N. Experiments on Soil Bacteriology. 3rd ed. Minneapolis: Burgess Publishing Co., 1957.

Beerstecher, E., Jr. Petroleum Microbiology: An Introduction to Microbiological Petroleum Engineering. New York: Elsevier Publishing Co., 1954.

Brisbane, P. G., and J. N. Ladd "The role of microorganisms in petroleum exploration," Ann. Rev. Microbiol. 19:351 (1965).

Brock, T. D. Principles of Microbial Ecology. Englewood Cliffs, N.J.: Prentice-Hall Inc., 1966.

Burges, A. Micro-organisms in the Soil. London: Hutchinson & Co. (Publishers), Ltd., 1958.

Davis, J. B. Petroleum Microbiology. New York: Elsevier Publishing Co., 1967.

Fogg, G. E. Algal Cultures and Phytoplankton Ecology. Madison Wis.: University of Wisconsin Press, 1965.

Gainey, P. L., and T. H. Lord Microbiology of Water and Sewage. Englewood Cliffs, N.J.: Prentice-Hall, Inc., 1952.

Gaudy, A. F., Jr., and E. T. Gaudy "Microbiology of waste waters," *Ann. Rev. Microbiol.* 20:319 (1966).

Gray, T. R. G., and D. Parkinson (eds.) The Ecology of Soil Bacteria (International Symposium). Liverpool: Liverpool University Press, 1968.

Hawkes, H. A. The Ecology of Waste Water Treatment. Oxford: Pergamon Press, 1963.

Hynes, H. B. N. The Biology of Polluted Waters. Liverpool: Liverpool University Press, 1963.

Ivanov, M. V. Microbiological Processes in the Formation of Sulfur Deposits (trans. fr. Russ.) Jerusalem: Israel Program for Scientific Translations, 1968. (Available from Clearinghouse for Federal Scientific and Technical Information, Springfield, Va. 22151.)

Kuznetsov, S. I., M. V. Ivanov, and N. N. Lyalikova Introduction to Geological Microbiology [C. H. Oppenheimer (ed.), Eng. ed.]. New York: McGraw-Hill Book Company, 1963.

McKinney, R. E. Microbiology for Sanitary Engineers. New York: McGraw-Hill Book Company 1962.

McLaren, A. D., and G. H. Peterson (eds.) Soil Biochemistry. New York: M. Dekker, 1967.

Mason, B. Principles of Geochemistry, 3rd ed. New York: John Wiley & Sons, Inc., 1966.

Odum, E. P. Fundamentals of Ecology, 2nd ed. Philadelphia: W. B. Saunders Co., 1959.

Parkinson, D., and J. S. Waid (eds.) The Ecology of Soil Fungi (International Symposium). Liverpool: Liverpool University Press, 1960.

Pramer, D., and E. L. Schmidt Experimental Soil Microbiology. Minneapolis: Burgess Publishing Co., 1965.

Quastel, J. H. "Soil metabolism," *Ann. Rev. Plant Physiol.* 16:217 (1965).

Strickland, J. D. H. "Phytoplankton and marine primary production," *Ann. Rev. Microbiol.* 19:127 (1965).

Waksman, S. A. Soil Microbiology. New York: John Wiley & Sons, Inc., 1952.

ZoBell, C. E. Marine Microbiology: a Monograph on Hydrobacteriology. Waltham, Mass.: Chronica Botanica Co., 1946.

In this part we emphasize the diversity that exists among microorganisms. Although, as we have seen in earlier chapters, microorganisms are constructed on a uniform biochemical ground plan, considerable variation exists in the details of microbial structure and function. We begin Chapter 17 with a discussion of the origin and evolution of life. This is placed within a framework of the probable geological evolution of the earth and of its atmosphere and hydrosphere. Then we look at present-day microorganisms, with emphasis on comparative biochemistry, after which we consider microbial taxonomy and the various approaches by which diverse microorganisms can be classified into groups of related organisms. Current ideas of molecular and numerical or computer taxonomy are brought into the discussion to show how these approaches have the potential for eventually simplifying microbial taxonomy. Chapters 18 and 19 then consider in some detail the various microbial groups. In these two chapters, the structures, life cycles, physiological processes, and ecological roles of selected groups of microorganisms are discussed, and the importance of these microorganisms for man is emphasized. Chapters 18 and 19 build heavily on material presented in earlier chapters and should demonstrate how the fundamental concepts of microbiology that we have discussed throughout this book can be applied to microorganisms of specific interest or practical significance.

It should be abundantly clear from the previous chapters that microorganisms show diverse morphological, physiological, and ecological characteristics. This variety is controlled ultimately by genetic constitution interacting with environment. Are diverse microorganisms related, and if so, how? It is clear that all existing organisms are obviously related chemically and metabolically. They all contain DNA, RNA, and protein, use ATP and NAD in their energy metabolism, and contain the same array of building blocks for their macromolecules. They all have lipid-containing plasma membranes, and are similar structurally in many other ways. Thus the diversity that obviously exists is laid upon a basic ground plan. Does this mean that the vast array of organisms we know today has arisen from a single primordial ancestor, the original cell, or is it merely that life processes are possible only if prerequisite chemical constituents—those we find in living organisms—are already present? If the creation of life requires that all appropriate substances be present, then life might have originated in several or more independent instances, and the organisms we know today may not have descended from a common ancestor. Although we obviously have no answer to this problem, most scientists feel that life arose only once and that the diversity we see today is the result of the processes of mutation and genetic recombination, with subsequent adaptation occurring to the array of environments we see on earth.

17.1 The origin and evolution of life

The manner in which life might have arisen has been an object of debate and experimentation for several hundred years. After Pasteur showed that spontaneous generation was not a regular, everyday occurrence (see Chapter 1), it became necessary to consider critically how and why life originated. Any discussion of the subject requires first that we consider both the origin and antiquity of the earth itself and its early environmental conditions (Table 17.1). Through the use of radioactive-isotope dating techniques, it is now fairly well established that the earth is about 4.5×10^9 yr old. Initially the earth may have been hot and molten, but as it cooled, rocks formed and the oceans and atmosphere separated. The oldest known rocks are about 3.3×10^9 yr old, but rocks probably formed earlier. Although the temperature of the early earth is not known, it was probably fairly high; molten volcanic lava has temperatures over 1000°C, and it begins to solidify at temperatures below this. Even today some parts of the earth are very warm, though the average temperature of the earth's surface is about 12°C.

It is now generally held that the ocean and atmosphere arose initially

Table 17.1   Time scale of evolution

	Age, $\times 10^9$ yr
Universe	12
Galaxy	7.4
Solar system	5
Earth	4.5
Oldest fossil bacteria	3

Table 17.2   Composition of Hawaiian volcanic gases[a]

Gas	Wt. %
H_2O	45.1
CO_2	20.7
CO	0.69
H_2	0.06
N_2	7.93
A	0.30
SO_2	16.7
S_2	2.5
SO_3	5.5
Cl_2	0.54
H_2S	Trace
HCl	Trace
NH_4Cl	Trace
HF	Trace
CH_4	Trace
H_3BO_4	Trace

[a] From B. Mason: *Principles of Geochemistry*, 2nd ed. (New York: John Wiley & Sons, Inc., 1958).

by outgassing of water vapor and gases from volcanoes (Table 17.2), hot springs, and other thermal environments after the surface of the earth had cooled. The lighter gases, such as hydrogen and helium, escaped the earth's gravitational pull, whereas the heavier gases, such as CO_2, N_2, argon, and water vapor, were retained. Water vapor condensed and fell as rain, ultimately forming the oceans, whereas the other gases remained in the atmosphere. The primordial atmosphere was quite different from the present atmosphere in at least one important respect: It had little or no O_2. This atmosphere was probably reducing (that is, anaerobic) and contained the key biological elements of C, N, S, and O in the following forms: CH_4, NH_3, H_2S and H_2O. Geological evidence for such a reducing atmosphere exists; iron-containing rocks older than about 1 to 2×10^9 yr contain predominantly ferrous iron, whereas rocks younger than 1×10^9 yr contain ferric iron. As we have seen in Chapter 16, in the presence of O_2 at neutral pH ferrous iron would oxidize spontaneously to the ferric state. By about 0.8×10^9 years ago, the atmosphere had become fully oxidizing, and we will consider below the crucial importance of oxygen-producing photosynthetic organisms in converting the atmosphere from a reducing to an oxidizing condition.

Chemical evolution   There is good fossil evidence that microbial life existed at least 3×10^9 yrs ago; hence life must have arisen in a reducing environment. It is now well established that the synthesis of biologically important molecules can occur if reducing atmospheres such as were present on the primitive earth are subjected to intense energy sources. Of the energy sources available on the primitive earth, the most important was UV radiation from the sun (Table 17.3), but lightning discharges, radioactivity, and thermal energy from volcanic activity were also available. If gaseous mixtures of methane, ammonia, and water are irradiated with UV light or subjected to electric discharge, a wide variety of biochemically important compounds can be made: sugars, amino acids, purines, pyrimidines, and fatty acids. If phosphate is present, nucleotides (including ATP) can be formed; whereas if H_2S is present, sulfur-containing amino acids can be made. Since the conditions used in these experiments are the kinds thought to have occurred on the primitive earth, it thus seems likely that the synthesis of these biochemically important compounds could have happened nonbiologically on the primitive earth. It has also been shown that under prebiological conditions some of these biochemical building blocks could have been induced to polymerize, leading to the formation of polypeptides, polynucleotides, and the like. We can therefore imagine that on the primitive earth a rich mixture of organic compounds accumulated. In the absence of living organisms, there is no reason why these compounds should not have persisted for countless

Table 17.3   Energy sources for chemical evolution[a]

	Amount of energy available, cal/cm²-yr
Ultraviolet radiation from sun ($<2{,}500$ nm)	570
Electric discharges (lightning)	4
Radioactivity	0.8
Volcanoes (heat)	0.13

[a] Data from Cyril Ponnamperuma.

years, and thus with time there should have been an extensive accumulation of organic materials. From such an organic soup to the beginning of life is a big step. (Not the least of the problems encountered was that this step could not have been taken on an earth that was being subjected to intense UV radiation.) Therefore, the next step in evolution must have been the development of an atmospheric shield for all but the long-wavelength ultraviolet. This shield, which we still have with us today (see Chapter 8), is composed of ozone (O_3), which arises through chemical reactions brought about by short-wavelength ultraviolet radiation in the upper atmosphere. Once this ozone shield had developed, life could have evolved (Figure 17.1).

Primitive organisms   There are two basic features that primitive organisms must have had: (1) metabolism, that is, the ability to accumulate, convert, and transform nutrients and energy, and (2) an hereditary mechanism, that is, the ability to replicate and produce offspring. Both of

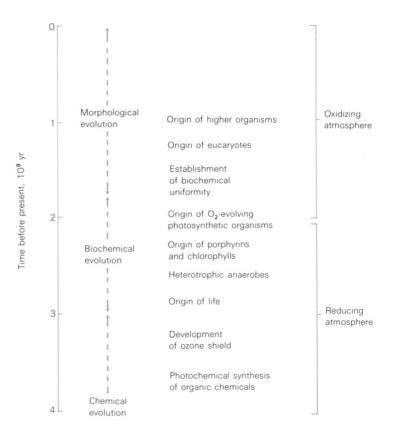

Figure 17.1   Possible course of biological evolution. The positions of the various stages are only approximate.

these features require the development of a cellular structure. Such structures probably arose through the spontaneous coming together of lipid and protein molecules to form membranous structures, within which were trapped polynucleotides, polypeptides, and other substances. This step may have occurred countless times to no effect; but just once the proper set of constituents could have become associated, and a primitive organism arose. This original organism would have found itself surrounded by a rich supply of organic materials usable as nutrients for energy metabolism and growth. From here on, evolution was relatively simple and perhaps inevitable, resulting in our present biological diversity, including man.

What was the nature of the first primitive organism? It must have been anaerobic and heterotrophic. The closest living relative is probably a non-sporulating, anaerobic organism without a cell wall, most likely resembling a mycoplasma (Section 18.16). Such an organism was probably simple, in the sense of possessing very few enzymes. No cytochrome or photosynthetic system would have been present, no cell wall, no flagella, no ability to sporulate, no citric acid cycle. It would have had a variety of growth-factor requirements. As time went on, however, growth and reproduction of this organism would have depleted the organic nutrients in its environment. Mutation and selection could then have resulted in the appearance of new organisms with greater biosynthetic capacities, which hence were better adapted to the changing environment. To make possible survival in osmotically varying environments, a cell wall probably appeared next. A key step was the development of the first cytochromelike porphyrin since this enabled the organism to carry on oxidative phosphorylation, and therefore it opened up the possibility of utilizing a variety of energy sources not available from substrate-level phosphorylation. In an environment in which preformed organic compounds were being depleted, this development no doubt provided a distinct selective advantage. Since conditions were still anaerobic, an electron acceptor other than oxygen was needed. This most likely was sulfate, a substance synthesized chemically by reaction of H_2S and ozone. Thus this second stage might be represented by an organism such as *Desulfovibrio,* which can use SO_4^{2-} as a terminal electron acceptor.

Once iron-containing porphyrins had evolved, the next step might conceivably have been the substitution of magnesium for iron, producing chlorophyll. With photosynthesis then possible, a great explosion of life could have occurred because of the availability of the enormous amount of energy from the visible light of the sun. The first photosynthetic organism was no doubt anaerobic, probably using light only for ATP synthesis and using reduced compounds from its environment—such as H_2S—as source of reducing power. Such an organism may have resembled *Chlorobium, Chromatium,* or one of the other photosynthetic bacteria.

The next step probably involved the development of the second light reaction of green-plant photosynthesis, making it possible to use the plentiful supply of H_2O as an electron donor. Since both ATP and reduced pyridine nucleotides (NADH, NADPH) could now be made photosynthetically, light energy could be used more efficiently. Clearly such an organism had considerable competitive advantage over other photosynthetic organisms. The first of this type was probably similar to one of the present day blue-green algae, some of which can grow completely anaerobically, although they produce O_2 photosynthetically. The evolution of an oxygen-producing photosynthetic apparatus had enormous consequences for the environment of the earth since, as O_2 accumulated, the atmosphere changed from a reducing to an oxidizing type. With O_2 available as an electron acceptor, aerobic organisms evolved; these were able to obtain much more energy from the oxidation of organic compounds than could anaerobes (see Chapter 4). More energy was made available, and higher population densities could develop, increasing the chances for the appearance of new types of organisms. There is good evidence from the fossil record that, at about the time that the earth's atmosphere became oxidizing, there was an enormous burst in rate of evolution, leading to the appearance of eucaryotic microorganisms and from them to higher animals and plants.

Origin of eucaryotes   As we have emphasized frequently in this book, there are vast differences between procaryotic and eucaryotic organisms, which override their fundamental similarities at the biochemical level. The reader should refer back to Table 3.1 for a summary of these differences. Because of the increased structural complexity of eucaryotes, it seems reasonable to suppose that they arose later than, and presumably from, procaryotes. How can we account for this development? Recall the discussion in Chapter 13 on the possible independent hereditary capabilities of the eucaryotic cellular organelles, especially mitochondria and chloroplasts. Many people believe that these organelles arose initially as a result of invasion of a large procaryotic cell by a smaller procaryotic cell. A photosynthetic cell (that is, a blue-green alga) may have been the forerunner of a chloroplast, whereas an aerobic but not a photosynthetic bacterium may have been the forerunner of a mitochondrion. This association of two kinds of cells might well have been a kind of symbiosis (see Chapter 14), of mutual advantage to both host and infecting cell, and the two associating units could then have continued to develop together. Through successive mutational changes, followed by selection, the associates could have become obligatorily dependent on each other. The host

cell may have taken over the nuclear functions and produced the eucaryotic nucleus, whereas the other cell may have taken over the energy-
yielding functions and became the organelle. Many further changes would
of course have had to take place to lead to the eucaryotic cells that we
know today. None of these present-day eucaryotes need be very similar
in structure to this primitive eucaryote since extensive genetic variation
and adaptation could have occurred. The important point is that by some
such process a division of labor was developed within the cell, which was
beneficial and made further evolution possible.

However, in order to evolve to the eucaryotic stage, a cell necessarily
had to sacrifice certain procaryotic features such as genetic plasticity,
structural simplicity, and the ability to adapt rapidly to new environments.
The greater complexity of the eucaryotic cell means that it faces difficulties in the ability to grow in extreme environments, such as thermal areas,
where the procaryote is preeminent (see Chapter 8). For these reasons
the evolution of the eucaryote did not toll the death bell for the procaryote. Both types of cells continued to evolve, serving as the forerunners of the various species we know today (Figure 17.2).

A point that deserves emphasis here is that none of the organisms
living today is primitive. They are all modern organisms, well adapted
to, and successful in, their ecological niches. Certain of these organisms
may indeed be very similar to primitive organisms and may represent
stems of the evolutionary tree that have not changed for millions of years;
in this respect they are related to primitive organisms, but they are not
themselves primitive. Furthermore, we must distinguish between primitive
organisms and simple organisms, that is, those whose cell structures and
biochemical potentialities are uncomplicated. The latter may represent
organisms that evolved late and became simple through reduction and
loss of properties possessed by their ancestors. The best example of this
is probably the psittacosis group, which contains obligate intracellular
parasites with very limited biochemical potentialities (Section 18.18).
These organisms grow only within the cells of vertebrates, and hence
could not have evolved until the appropriate host existed. Once established within the vertebrate cell, certain functions would no longer be
necessary and conceivably may be lost through mutation.

Evolutionary processes   We might summarize here certain features
of evolutionary processes, keeping in mind that evolution always results
not from genetic processes alone but from interaction between these and
selective environmental factors. (1) *Radiation into new environments:* A
new environment presents a challenge to existing organisms since here
is a virgin field for colonization, where competition is lacking. Chance
probably is very important in determining what organisms arrive first.

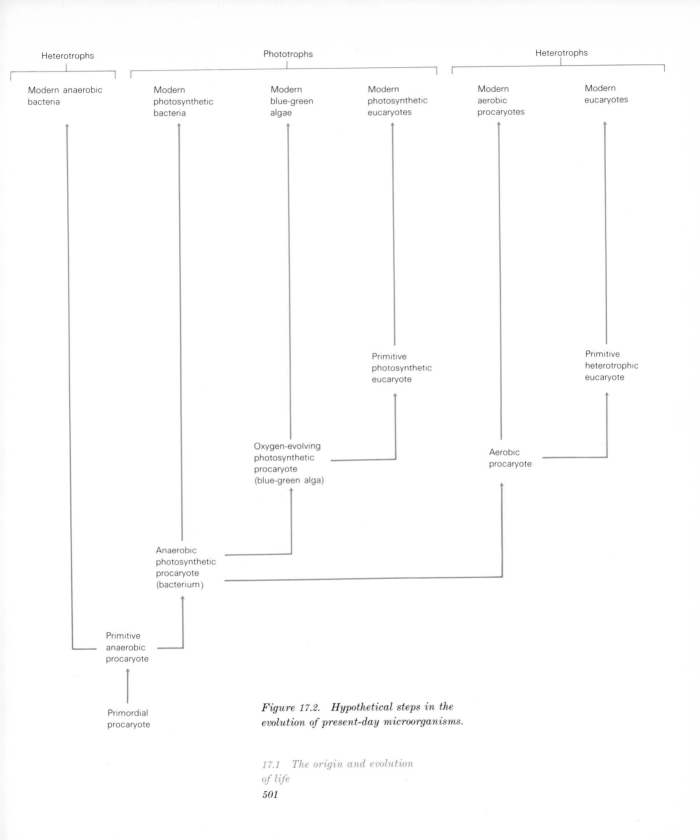

Figure 17.2. *Hypothetical steps in the evolution of present-day microorganisms.*

Initially even a poorly adapted organism may be able to colonize a new environment since it meets with no competition. As time goes on, however, evolutionary processes lead to the development of better adapted organisms; eventually all the organisms present in this environment become highly adapted to it and (probably) poorly adapted to other environments, including the one from which their ancestors came. (2) *Divergence:* If two environments that differ significantly are colonized by the same organism, adaptation to each of these environments by offspring of the forerunners will lead to divergence of characteristics. The organisms will be clearly related to each other in many ways, but will differ in ways that are ecologically relevant to their success in each environment. (3) *Convergence:* If two similar environments arise in geographically distinct areas, the colonizing species in each environment may be quite different. As evolution proceeds, however, these distinct organisms may evolve similar characteristics and converge towards a common type, insofar as the characteristics of adaptive value in the new environment are concerned. Thus the organisms may be functionally and ecologically related, but may be genetically unrelated. (4) *Parallelism:* Each stem of a converging pair may be a forerunner of an evolutionary line in which the same kinds of adaptive changes are occurring. The two lines are said to be parallel. (5) *Sexual reproduction and evolution:* In microorganisms sexual reproduction is rarely obligatory for the reproduction of the genotype although it usually is for multicellular organisms. Since microorganisms can often reproduce indefinitely in the vegetative state, evolution may proceed primarily by accumulation of selectively beneficial mutations, whereas in multicellular organisms only mutations that occur in germ cells can be passed on to succeeding generations.

If sexual reproduction in a microorganism is possible, this of course provides a means of acquiring new functions by bringing together through genetic recombination elements of two diverse genomes; for this reason sexual reproduction can lead to more rapid acquisition of new functions than can mutation.

17.2   Fossil microorganisms

If we are to put a time scale on evolution we must have some way of dating the times at which different kinds of organisms appeared. Since relatively precise methods for dating ancient rocks by means of radioisotope decay exist, if we could find evidence of the remains of organisms in dated rocks we would have an approach to determining the rate of evolution. This means we need a fossil record of the organisms in which we are interested. Fossils have been highly useful in studying the evolution of multicellular organisms, and we now have a fairly accurate record of the times at which different groups of plants and animals first appeared on earth. Until recently, scarcely any evidence of fossil microorganisms had been found in the older rocks, but new techniques have now been developed for observing fossil microorganisms; by the use of these techniques certain ancient rocks rich in microbial fossils have been discovered. Thus we can now look forward to a rapid advance in our knowledge of microbial evolution.

The fossilization process   First we must remember that we are discussing the preservation of microbial remains for over a billion years. Since most organic constituents of an organism are unstable, they slowly disappear. Even relatively stable materials such as the microbial cell wall will ultimately disintegrate, and once cellular structure is lost, there is no way of ascertaining the former presence of microbes. Two ways in which cellular structure can be preserved are known: If the organism secretes an inorganic wall or test, as is done by some bacteria (for example, iron bacteria), algae (for example, diatoms), and protozoa (for example, foraminifera), such structures under appropriate conditions can become incorporated into rock and hence may be preserved; the other way is by the deposition upon or within the cell of a stable mineral, which preserves the cellular structure even after all organic constituents have disappeared. For instance, in aqueous environments high in silica, cellular remains can be fossilized by a replacement, molecule by molecule, of their original organic material by silica. Often, these silicified remains pick up colored minerals that serve as a stain, so that the original structure may sometimes be visible even in microscopic detail. One may be familiar with petrified wood, which is formed in a similar way.

Techniques for study of microbial fossils.   In the study of microbial evolution, we are most interested in the oldest rocks, such as those that are found in the so-called Precambrian shields in large areas of Eastern Canada and Northern United States, South and Central Africa, Australia, Scandinavia, and Brazil. These rocks are older than one billion years, and the oldest rocks with biogenic deposits are around three billion years old. Granites and other rocks formed by the cooling of volcanic lava cannot of course have organisms preserved in them, but in sedimentary rocks, which are formed in aqueous environments, organisms drift into the bottom sediments, and become covered, preserved, and ultimately fossilized. The kind of rock most productive of microbial fossils is called *chert;* it is a siliceous rock with high content of organic matter.

Microscopy cannot of course be done with whole rocks. Two methods are used to prepare a sample for use under the microscope: thin sectioning and the carbon-replica technique. To make a *thin section* suitable for high-power microscopy, the rock is first cut with a diamond saw. The slab formed is then glued to a microscope slide and is carefully ground down until it is so thin that light will pass through. The slide containing the thin section is then examined under oil immersion for the presence of fossil microorganisms. Examples of the kinds of structures seen are shown in Figure 17.3. *Electron microscopy* permits the examination of the material at higher magnification and hence the more detailed study of cellular structure. For electron microscopy, the preparation of replicas is necessary. A thin section of rock is carefully treated with hydrofluoric acid, which etches away some of the rock, leaving the fossils slightly raised above the rock surface. A carbon replica of this relief is then made, and the replica examined in the electron microscope.

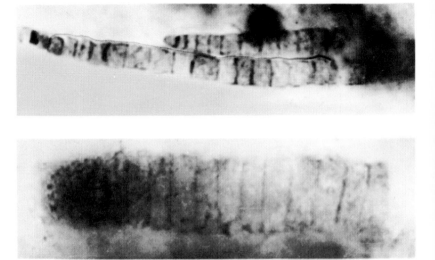

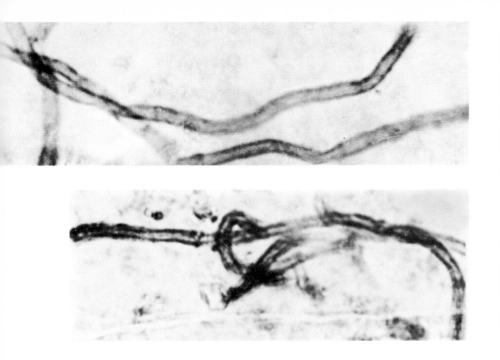

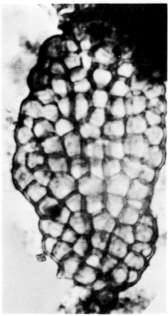

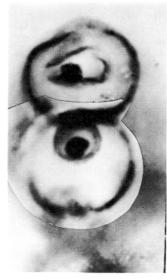

*Figure 17.3   The six photographs at left
(magnification, 2,250×) and above (1,030×)
show fossil procaryotic microorganisms,
found in the Bitter Springs formation, a
rock formation in central Australia about
1 billion (10⁹) years old. These forms bear
a striking resemblance to modern blue-
green algae or their colorless counterparts
(see Chapter 18). The two photographs at
right (2,250×) show fossils possibly of a
eucaryotic alga. The cellular structure is
remarkably similar to that of certain
modern chlorophyta, such as Chlorella sp.
These are from the same rock formation as
are the procaryotic organisms. [From
J. W. Schopf: J. Paleontol. 42, pt. I:651
(1968).]*

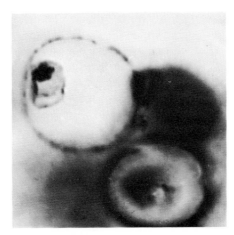

Microbial fossils and their interpretation Many of the structures seen in these ancient rocks are filamentous forms, and it is easy to convince oneself that these filamentous structures can be derived from nothing but microorganisms since the resemblance to some present-day organisms is striking. Yet we must always be aware that structures seen in an ancient rock could be artifacts, derived from a completely nonbiological source. It is also obvious that we can tell little about the physiology and biochemistry of the organisms from which the fossils (if such they are) were derived. It would be tempting to conclude that some of the organisms in Figure 17.3 are photosynthetic blue-green algae; yet there are many nonphotosynthetic forms, such as *Leucothrix* and *Beggiatoa* (Section 18.13), that are structurally similar to blue-green algae. Thus although the discovery of these ancient fossils is a notable advance, their interpretation must be made cautiously. Of major significance is the fact that these fossils enable us to set an approximate date for the first appearance of life as we know it. Currently, there are fairly convincing microbial fossils from rocks about 3×10^9 yr old. Since the earth itself is 4.5×10^9 yr old, this means that life probably evolved on earth within the first billion years.

In less ancient rocks fossil evidence of larger microorganisms is abundant, especially for certain groups that have hard shells or tests. Indeed, micropaleontology using larger microorganisms is a well-developed field of study, of considerable interest to the geologist. Algae that secrete or induce the formation of calcium carbonate are prominent as fossils in some limestones, and the resistant silica shells of diatoms are readily preserved. Fossils of filamentous fungi are found frequently in later geological periods, as parasites or epiphytes of fossil plant leaves and stems (Figure 17.4), although they are sparse or absent in older rocks that were deposited before land plants evolved. Among the protozoa, the foraminifera with their highly complex calcium carbonate shells (see Chapter 19) and the silicoflagellates are present in fossil form. The soft-bodied protozoa, such as the amoebas and ciliates, are preserved poorly and are not seen in fossil form.

A brief summary of the geological ages at which different microbial groups first appear in the fossil record is shown in Figure 17.5, with indications also of the times when higher organisms are thought to have first appeared.

Organic chemistry of ancient rocks Another approach to the study of evolution is to examine ancient rocks for the presence of organic constituents that might be of biological origin. Most biochemical molecules are not sufficiently stable to have remained for these long times, but hydrocarbons and certain amino acids are. By careful analysis with

Figure 17.4   Fossil fungi that grew on well-preserved leaves of tropical plants. From a clay bed in western Tennessee. The leaves are of the Eocene age, about 50 million years old. The fungi are so well preserved that they can be easily classified with their modern counterparts in the class Ascomycetes, order Microthyriales: (a) germinating two-celled ascospore (magnification, 600×); (b) typical branching septate hyphae (1,500×); (c) and (d) typical fruiting bodies (perithecia) (400× and 600×). [From D. L. Dilcher: Paleontographica, Abt. B, 116:1 (1965).]

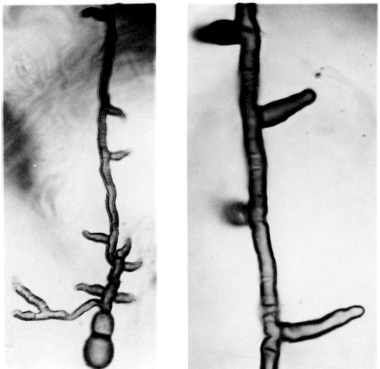

(a)   (b)

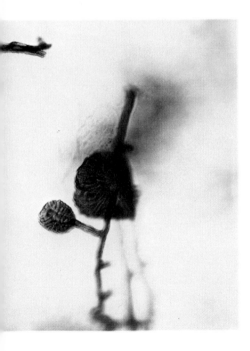

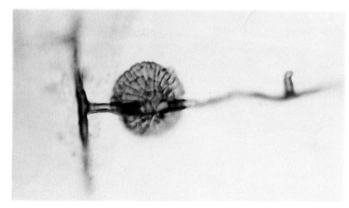

(d)

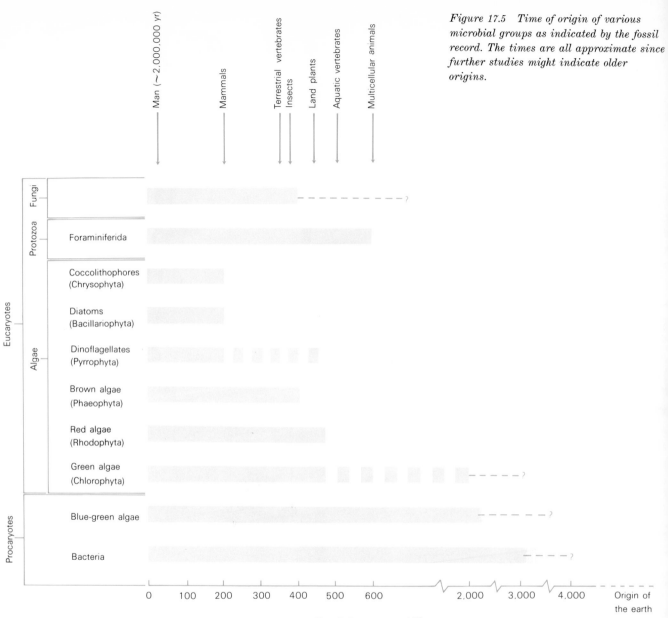

Figure 17.5 Time of origin of various microbial groups as indicated by the fossil record. The times are all approximate since further studies might indicate older origins.

highly refined techniques, it is possible to detect small amounts of many amino acids in some ancient rocks. Hydrocarbons of the kinds associated with living organisms are also present. Since these materials can also be formed by nonbiological means (see Section 17.1), their presence in ancient rocks does not prove the existence of life at that time; however, the organic chemical data, taken together with the fossil data, do present a pattern consistent with the presence of life at those times.

17.3 Colonization of new habitats

Microorganisms are often pioneer colonizers of new habitats because of their ease of dispersal and their wide adaptability to environmental conditions. Occasionally, due to geological processes, a new habitat arises that is essentially devoid of microorganisms, and often it is possible to study its colonization from the very beginning. Krakatoa, a volcanic island in the Indonesian archipelago, became famous in 1883 when it erupted so violently (blowing off the whole top of the island) that volcanic ash was scattered worldwide. Although the eruption completely eradicated all visible life on Krakatoa, 3 yr later, when the first scientific visit was made, the surface of the island showed a heavy growth of blue-green algae, which had formed mats on the surface of the ash and rocks. (One would expect that bacteria were associated with the algae, but these were not looked for.) Within a few more years, higher plants were seen, and after a decade or two the flora had returned essentially to normal. It is not surprising that blue-green algae were the first observed colonizers since these organisms are readily dispersed in the air, can grow rapidly autotrophically, and often fix N_2 as well as CO_2. The rapid colonization of such a new environment clearly reveals to us the rapidity with which it might have been possible for the earth to have been covered with living forms once life began.

More recently, another volcanic eruption has provided an opportunity to study colonization. This occurred off the south coast of Iceland beginning in 1963 and continuing through 1967, which led to the formation of the new island Surtsey. The surface of Surtsey, initially devoid of obvious life (Plate 8: Figure 17.6), began to be colonized in a year. Marine algae began to grow on the shore, and unicellular algae carried to the island by gulls and other sea birds grew in the shallow lagoon on the island, probably benefitting from the fertilization of this lagoon by bird droppings. Higher plants are now becoming established, and within another few years the organic nutrients contributed by them should promote microbial growth and soil formation.

In recent years interest has dramatically increased in the possible existence of extraterrestrial life, which quite probably was stimulated by the developing programs for the exploration of outer space. The discovery of native life on another planet would be of enormous significance since it would indicate that life is not a phenomenon unique to earth and would suggest that life is a universal phenomenon, whose development might be inevitable whenever the proper environmental conditions arose. For this reason, a considerable amount of effort in the space programs has gone into developing techniques for detecting life on our neighboring planets. First it must be decided whether we wish to detect life in any possible manifestation, or whether we will be interested only in detecting life in forms resembling earthlike organisms. Conceivably life on another planet is based not on carbon but on silicon or some other element. Even if carbon-based, DNA may not be the genetic material, or ATP may not be the energy carrier. If we are required to detect life that differs in such fundamental ways from life on earth, our problems will be immense since we have no way of devising diverse life-detection systems intelligently. The initial studies will be done on Mars since this is the planet most readily available to us on which environmental conditions reasonably favorable for earthlike life exist. Because the payload on Martian flights will be severely restricted, many procedures of value must be eliminated, and it has been decided that attempts will be made to detect only microbial life that has an earthlike metabolism. Several procedures for remote life detection by automatic instruments can be envisaged. Chemical analyses of Martian soil for the presence of DNA, proteins, and lipids could be done. Samples of this soil could be inoculated into culture media and incubated, with growth being detected by a turbidity increase, pH drop, gas production, or other means. The ability to oxidize ^{14}C-labeled carbon substrates to $^{14}CO_2$ could be measured by inoculating Martian soil into media containing the radioactive carbon compound and measuring the amount of radioactive CO_2 released. To make more certain that a substrate utilizable by a Martian organism is available, a wide variety of carbon compounds could be used, such as sugars, organic acids, and amino acids.

How can results of these studies be interpreted? A negative answer does not mean that life does not exist on Mars; it may merely mean that we have not used the proper life-detection system. Unfortunately, a positive answer is also equivocal since it may arise from earthborn contamination, malfunction of equipment, or unexplained causes. Successful repetition of initially positive observations would of course be encouraging. Ultimately, however, it will be necessary to return samples of Martian soil to the earth, where more precise study can be possible.

Several positive results have already come out of life-detection studies: (1) They have forced us to examine more carefully what we mean by life and how it can be recognized. (2) They have provided a stimulus for the development of miniaturized instruments and those that are very precise and reliable. (3) The automatic methods may find considerable use here on earth, as for instance in continuously monitoring the microbial contamination of a polluted river. (4) They have stimulated us to study and think more critically about life on earth and how it may have evolved.

17.5 Comparative biochemistry

Although all microorganisms are built on the same ground plan involving DNA, RNA, and protein, differences in metabolic pathways and biochemical constituents are common. These variations have arisen in the course of evolution either as part of the adaptation of organisms to new environments, or as a result of the development of evolutionary lines of descent. It is thus of considerable interest to examine the biochemical characteristics of the organisms that exist today and to study the ways in which they differ. Comparative biochemistry is still an infant field, but already it has revealed many interesting relationships, selected examples of which we cite here.

Enzyme differences   One of the most interesting areas of comparative biochemistry is a study of the molecular properties of enzymes that carry out the same catalytic function in different organisms. A comparison of the properties of several key enzymes of carbohydrate metabolism from yeast and from a mammal is shown in Table 17.4. We see that even though the enzymes from these two sources are similar in catalytic function, differences in molecular weight exist. This may be explained in terms of mutational changes in the genes controlling these enzymes as the organisms diverged in evolution from one common ancestor, though they still retained the catalytic function of the enzyme molecule. One enzyme in Table 17.4, *aldolase,* shows differences not only in molecular weight but in catalytic activity and cofactor requirements. Studies of the aldolases from a wide variety of organisms have shown that aldolases can be divided into two groups: type I and type II. Type I aldolase, the mammalian type, is characterized by a molecular weight of 140,000, a lack of requirement for metal-ion cofactors, and a broad pH optimum. Type II aldolase, the microbial type, has a molecular weight about one-half that of Type I, a requirement for metal ions, and a narrow pH optimum.

Table 17.5 shows the aldolase type found in a wide variety of organisms. Type I aldolase is found in all animals and plants, including algae, whereas type II aldolase is found in bacteria and fungi. How do we explain the evolution of these two types? One possibility is that the organisms producing each type represent two branches of an evolutionary tree that diverged quite early, after which each enzyme showed conservation of properties during further evolution. Since both bacteria and fungi have type II aldolase, this sort of reasoning would suggest that the fungi represent a group that arose from the bacteria early in life. Two organisms listed in Table 17.5, *Euglena gracilis* and *Chlamydomonas reinhardi*, have both aldolase types. These two organisms are flagellated photosynthetic microorganisms that have affinities with both the algae and the protozoa. Possibly these two unusual organisms produce type I aldolase under control of a nuclear gene and type II aldolase under control of a gene in a self-replicating organelle, such as the mitochondrion or chloroplast (see Chapter 13). This interpretation is consistent with the idea that mito-

Table 17.4   Comparative properties of various enzymes[a]

	Molecular weight	Cofactor	Optimum pH	Specific inhibitor
Glyceraldehyde-3-phosphate dehydrogenase:				
Yeast	122,000	NAD	8.4	p-Chloro-mercuri-benzoate
Mammal	120,000	NAD	8.8	p-Chloro-mercuri-benzoate
Alcohol dehydrogenase:				
Yeast	150,000	Zn^{2+}, NAD	7.9	Metal-binding agent
Mammal	73,000	Zn^{2+}, NAD	8.0	Metal-binding agent
Phosphoglyceric acid mutase:				
Yeast	122,000	2,3-Phospho-glyceric acid	5.9	Zn^{2+}
Mammal	64,000	2,3-Phospho-glyceric acid	5.9	Zn^{2+}
Enolase:				
Yeast	66,000	Mg^{2+}	7.7	F^-
Mammal	77,000	Mg^{2+}	6.7	F^-
Aldolase:				
Yeast (type II)	70,000	Zn^{2+}, K^+	7.0–7.5	Metal-binding agent
Mammal (type I)	140,000	None	6.5–8.5	None

[a] Adapted from William J. Rutter and William E. Groves: "Coherence and variation in macromolecular structures in phylogeny," in Charles A. Leone (ed.), *Taxonomic Biochemistry and Serology* (New York: The Ronald Press Company, copyright © 1964).

Table 17.5 *Distribution of type I and type II aldolases in organisms of different taxonomic groups*

Type I	Type II	Mixture of type I and type II
Eucaryotic algae:	Bacteria:	*Euglena gracilis*
Prototheca	*Clostridium perfringens*	*Chlamydomonas reinhardi*
Polytoma	*Escherichia coli*	Red algae
Scenedesmus	*Bacillus megaterium*	
Gonyaulax	*Veillonella gazogenes*	
(dinoflagellate)	*Erwinia carotovora*	
Protozoa:	Fungi:	
Tetrahymena	Yeast	
Animals:	*Aspergillus niger*	
Rabbit	Blue-green algae	
Frog		
Crayfish		
Clam		
Earthworm		
Cow		
Higher plants:		
Pea		
Ginkgo		

chondria and chloroplasts may have arisen from symbiotic procaryotes, which would, of course, have had type II aldolase. Comparative studies like these have been done in less detail for a variety of enzymes, and in general the results indicate that enzymes with the same function in two distantly related organisms are more likely to show divergences in structure than are enzymes from closely related organisms.

Lipid biosynthesis and O_2-requiring biosynthesis   As we discussed in Chapter 5, the synthesis of fatty acids in all organisms proceeds by the condensation and subsequent reduction of two-carbon units from acetyl-CoA. However, the synthesis of unsaturated fatty acids, which have one or more carbon-carbon double bonds, proceeds by quite different pathways in various microbial groups. In one mechanism, molecular oxygen (O_2) reacts with the saturated fatty acid and NADPH (Figure 17.7). Since molecular O_2 is required, this reaction can of course proceed only in aerobic organisms. In anaerobic bacteria and facultative aerobes the synthesis of unsaturated fatty acids proceeds by insertion of a double bond *before* the fatty-acid chain is complete, by a dehydration reaction of an hydroxyl group (Figure 17.7). One consequence of this difference is that anaerobic bacteria do not form unsaturated fatty acids with more than one double bond, whereas aerobic organisms, by introducing double bonds in the preformed fatty acid, can make polyunsaturated fatty acids.

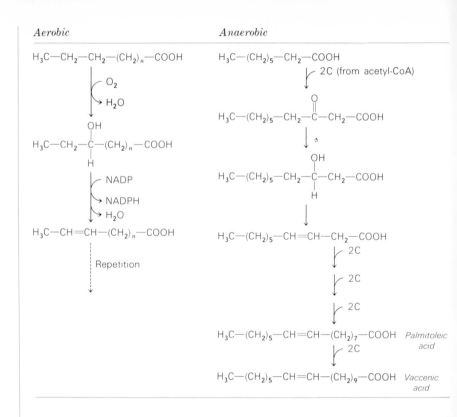

Figure 17.7 The introduction of double bonds into fatty acids by aerobic and anaerobic pathways, leading to the formation of unsaturated fatty acids. Under aerobic conditions double bonds are introduced into the preformed fatty acid, and multiple double bonds can be introduced. Anaerobically the double bond is introduced before the fatty-acid chain is complete and only one double bond can be formed.

However, not all aerobes use the pathway that involves O_2 (Table 17.6).

Two other biosynthetic sequences involving O_2 are also of evolutionary interest. (1) The synthesis of sterols always requires O_2 as a reactant. (2) The synthesis of the amino acid tyrosine in animals is O_2-requiring and involves formation of a hydroxyl by addition of oxygen to the preformed benzene ring of phenylalanine; in plants and microorganisms, however, the hydroxyl group of tyrosine is added long before the benzene ring is even closed and occurs anaerobically (Table 17.6). The anaerobic mechanisms of these biosyntheses could be considered biochemical fossils, that is, relics of the time when O_2 was absent from earth and life was completely anaerobic. Many of the modern-day aerobic bacteria are facultative organisms, and are probably faced with the necessity of growing occasionally under anaerobic conditions; therefore it may be of adaptive value for them to retain the anaerobic pathway. Higher organisms are almost exclusively aerobic and hence capitalize on an aerobic mechanism.

Lysine biosynthesis We have already noted in Chapter 5 that there are two diverse biochemical pathways for lysine biosynthesis, both starting with α-ketoglutarate but using entirely different enzymes for the interme-

Table 17.6 *Aerobic and anaerobic mechanisms for tyrosine, sterol, and unsaturated-fatty-acid synthesis in different groups of organisms*

| | Conditions for synthesis | | |
	Unsaturated fatty acids	Sterols	Tyrosine
Higher animals	O_2-requiring	O_2-requiring	O_2-requiring
Protozoa	O_2-requiring	O_2-requiring	O_2-requiring (?)
Higher plants	O_2-requiring	O_2-requiring	Anaerobic
Eucaryotic algae	O_2-requiring	O_2-requiring	Anaerobic
Fungi	O_2-requiring	O_2-requiring	Anaerobic
Bacteria:			
Actinomycetes	O_2-requiring	Sterols not synthesized	Anaerobic
Pseudomonads	Anaerobic	Sterols not synthesized	Anaerobic
Enteric bacteria	Anaerobic	Sterols not synthesized	Anaerobic
Obligate anaerobes			
(clostridia)	Anaerobic	Sterols not synthesized	Anaerobic
Blue-green algae	Both O_2-requiring and anaerobic	Sterols not synthesized	Anaerobic

diate reactions. It seems likely that these differences represent early evolutionary divergences. Table 17.7 summarizes the groups in which the two different pathways occur. The aminoadipic pathway is present in some fungi and in *Euglena;* certain other fungi and all other microorganisms possess the diaminopimelic (DAP) pathway. It is of interest that those fungi possessing the DAP pathway are the only members of the lower fungi that have cellulose cell walls and produce biflagellate motile cells; these observations have been used by some to suggest that these fungi had a separate origin from the rest of the fungi. A group that has had two or more different evolutionary origins is called *polyphyletic,* in contrast to a *monophyletic* group, which has had only a single origin.

Table 17.7 *Comparison of lysine biosynthetic pathways in different groups of organisms*

Aminoadipic pathway	*Diaminopimelic pathway*
Fungi:	All bacteria
All Basidiomycetes	Blue-green algae
All Ascomycetes	Most eucaryotic algae
Lower fungi:	Higher plants
Chytridiomycetes	Lower fungi:
Hyphochytridiomycetes (Blastocladiales)	Oomycetes
Zygomycetes	Hyphochytridiomycetes (Hyphochytridiales)
Euglenoid algae	

Let us now turn from the subject of the origin of life and some aspects of its evolutionary development to the problems involved in recognizing various microorganisms and assembling them into groups so that they can be more conveniently discussed and studied. One of the goals of *taxonomy* is to discover order in the apparent chaos of biological diversity. Just as the chemist examines the properties of the chemical elements and is able to see relationships that make possible the construction of the periodic table, so the taxonomist attempts to construct a convenient classification scheme for organisms. No implication of evolutionary relationships among the organisms of a single taxonomic group need be raised, although frequently such relationships do appear likely. One of the most useful facts to emerge from taxonomic studies is the discovery that two or more characteristics are often coupled. For instance, after we compare the property of motility with that of flagellation, we see that in a large number of cases these two properties occur together. This of course suggests the occurrence of a functional relationship between these two characteristics, which we can then go on to explore in more detail. Another goal of taxonomy is to discover a property or set of properties that makes it possible for us to recognize a given organism when we see it again. This is especially important in applied microbiology, where we wish to be able to recognize the same organism quickly and simply.

It is important to distinguish between taxonomy and *nomenclature,* which is the naming of organisms. The latter is in itself a trivial matter that need be done only after the taxonomic work is completed. The name is merely a convenient handle, which expresses in a kind of shorthand the collective properties of the organism. Often students feel that if they know the name of an organism, then they know a lot about it. Far from it! Do we know anything significant about a person we have just met after merely learning his name? The important thing is to be aware of the collective set of properties of the organism, and the name tags along afterward.

Before we begin classifying organisms, we must decide what precisely are the basic units with which we shall be dealing. The taxonomic unit of microbiology is the clone or strain, which is a population of genetically identical cells derived from a single cell (see Chapter 11). It is very easy to isolate from nature a large number of clones, which reflect an enormous diversity of microbial types. In a taxonomic study one cannot deal with the totality of microorganisms; one must first decide whether one's interest lies in bacteria, algae, protozoa, or fungi, and then concentrate on the preferred group. But even these groups are too large, and preliminary decisions formed on arbitrary grounds are usually made to select a smaller group of apparently related organisms for further study. Ultimately, though, it is necessary to compare the organisms selected with those of other groups, to be sure that the initial selection of organisms for study did not introduce some bias.

Approaches to microbial taxonomy There are many ways in which microbial classification can be approached, depending on the group of organisms to be studied and the facilities available. The first might be called the *classical approach* since it is the way microbial taxonomy has been carried out for over 100 yr. A variety of characteristics of different organisms are measured, and these traits are then used in the separation of groups. A group of strains that have most or all characteristics in common will be classified as a single species, and related species are classified in the same genus. In general, structural and morphological characteristics are more useful as primary characteristics in taxonomy than are physiological and biochemical characteristics. There are several reasons for this. First, the morphology of an organism is the result of the expression of a large number of genes that control enzymes, which in turn determine the synthesis of the various structural components. Second, morphology has a certain degree of independence from environmental influences, and related organisms usually have much the same shapes and structures in a variety of environments in which they grow. Third, morphology is usually a genetically stable characteristic, and does not undergo wide changes as a result of single gene mutations. Fourth, the morphology of an organism is fairly easy to determine, especially in the larger microorganisms, which are easily seen under the microscope. Structural distinctions between eucaryotes are relatively easily made, but in procaryotes fine structural differences can be assessed only in the larger bacteria and blue-green algae. For smaller procaryotes we are usually restricted to making general statements about cell shape, stainability, and flagellation, although the electron microscope is now making possible the assessment of structural differences among the smaller procaryotes.

Nonmorphological characteristics of taxonomic value that are widely used can be mentioned here: cell wall chemistry, cell inclusions and storage products, capsule chemistry, pigments, nutritional requirements, ability to use various carbon, nitrogen, sulfur, and energy sources, fermentation products, gaseous needs, temperature and pH requirements and tolerances, antibiotic sensitivity, pathogenicity, symbiotic relationships, immunological characteristics, and habitat. Some of these characteristics may not be applicable or useful with a particular group, and one must judge how extensive a compilation of data is needed to effect a good classification. The classical approach is casual and nonsystematic, but it has proved very useful for some microbial groups, especially those that are morphologically complex—such as the eucaryotes and the larger procaryotes.

A second approach has been called *numerical taxonomy*. It resembles the classical approach in that a large number of characteristics are measured for the different organisms, but instead of placing more weight on some characteristics than on others and inspecting the data casually, an electronic computer is used to compare all the characteristics of each organism with those of others. The computer is admirably suited to this task and is able to analyze the data for a large number of organisms and to recognize similarities and differences. Numerical taxonomy is further discussed later in this chapter.

A third and perhaps the most fundamental is the *genetic* or *molecular approach,* which aims to ascertain the degree of genetic relatedness of different organisms. The ultimate goal could be to determine the DNA base sequences of the complete genome of the organisms since the better the sequences of two organisms match, the closer is their genetic relatedness. Although this last goal is not yet attainable, several interesting approaches to it can be carried out; these will be discussed in more detail below.

17.7   Molecular and genetic taxonomy

Molecular and genetic taxonomy make use of studies designed to show directly or indirectly that the DNA base sequences of two organisms are similar or identical. One should review briefly the ideas on the genetic code, the nature of mutation, and genetic recombination (Chapters 11, 12 and 13) before beginning this section.

If two organisms are able to mate and show genetic recombination, they are probably closely related, since recombination requires a close homology between the DNA molecules of the participating genomes. Unfortunately, sexual or parasexual reproduction occurs infrequently or not at all in many microbial groups, especially the procaryotes, so that detailed genetic studies often are not possible. However, several simplifying alternatives are available.

DNA base compositions   Although the sequence of bases in the DNA of an organism is as yet impossible to determine, the proportions of the various DNA bases in the total DNA of an organism can be assayed. In double-stranded DNA the proportion of adenine always equals the proportion of thymine and that of guanine equals that of cytosine, but considerable variation in the frequency of adenine-thymine and guanine-cytosine base pairs occurs among various organisms, and these variations in base ratios are of value in taxonomic studies. By convention, the base composition of a DNA preparation is expressed as the mole percentage of guanine and cytosine (GC) of the total. Thus if the GC content is 40 percent, adenine and thymine (AT) would be 60 percent, since GC + AT = 100 percent. The mole percent GC of the purified DNA of an organism can be assessed directly by chemical analysis or indirectly by measuring its density or the temperature at which the DNA double strands separate (Section 12.1). The DNA base compositions of different microorganisms vary widely, with values ranging from 22 to 74 percent. Because of the degeneracy of the genetic code (Section 11.1), the GC content can be altered slightly without affecting the amino acid sequences of proteins. But extensive differences in the GC content will be reflected in different amino acid sequences. It has been calculated that, if two organisms differ in GC content by more than 10 percent, there will be few base sequences in common; hence two organisms having such different GC contents would not be considered to be closely related in the sense that they are direct descendents of a common ancestor.

Base compositions of DNA have been determined for a wide variety of bacteria, and several correlations can be observed: (1) Organisms with similar phenotypes often possess similar DNA base ratios. (2) If two organisms thought to be closely related are found to have widely different base ratios, closer examination usually indicates that these organisms were not so closely related as had been supposed. (3) However, two organisms can have identical base ratios and yet be quite unrelated since a variety of base sequences is possible with DNA of the same base composition.

The study of DNA base compositions has been especially useful in studies on bacterial taxonomy and has permitted detection of significant differences between organisms that had previously been thought to be closely related. For instance, the genus *Bacillus* is composed of aerobic Gram-positive spore formers, which for many years had been considered a homogeneous group. Examination of DNA revealed, however, that different members of this genus diverged widely in base composition (Figure 17.8), and studies on genetic relatedness by transformation (see below) also showed that many species thought to be related are not. Thus a complete reclassification of the genus *Bacillus* may be necessary. On the other hand, the genera *Escherichia, Salmonella,* and *Shigella,* which have long been known to be related physiologically and genetically, have DNA base compositions that are nearly the same (Figure 17.8), so that the studies of DNA base composition confirm the close relationships of these organisms as determined by other means.

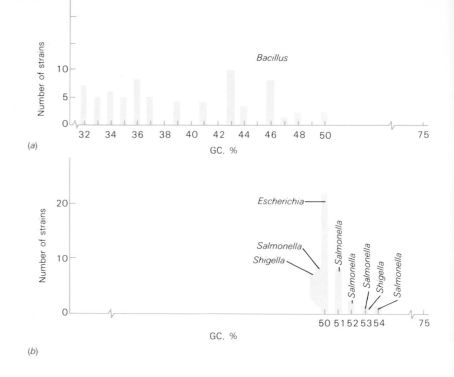

(a)

(b)

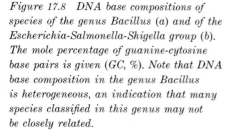

Figure 17.8 DNA base compositions of species of the genus Bacillus (a) and of the Escherichia-Salmonella-Shigella group (b). The mole percentage of guanine-cytosine base pairs is given (GC, %). Note that DNA base composition in the genus Bacillus is heterogeneous, an indication that many species classified in this genus may not be closely related.

Nucleic acid hybridization and homologies Nucleic acid hybridization provides a method for detecting similarities in base sequence between different organisms. Recall (Section 12.1) that, if a double-stranded DNA preparation is heated, the two strands separate into single-stranded molecules that are complementary to each other. If the mixture is cooled slowly, the two complementary strands reassociate and the original double-stranded complex is reformed. Since reassociation requires the base sequences to be complementary, the presence of identical or nearly identical base sequences in two different organisms can be detected by measuring the degree to which their nucleic acids are able to interact specifically (Figure 17.9). Since the base sequence of RNA is complementary to that of one of the strands of the DNA, DNA-RNA hybridization can also be used to reveal similarities or differences among nucleic acids of organisms. This provides a powerful tool for studying the genetic relatedness of the organisms.

The degree of nucleic acid hybridization is less when the organisms are not so closely related even if the overall base compositions are the same.

Thus hybridization between DNA molecules of two *Escherichia coli* strains is high, approaching 100 percent, whereas hybridization between *E. coli* and *Salmonella* DNA is less, about 70 percent.

The most convenient way to study DNA-DNA or DNA-RNA hybridization is to use radioactive nucleic acid from one strain, and nonradioactive nucleic acid from the other. The radioactive nucleic acid is first stirred rapidly to fragment it since these smaller pieces are able to form hybrids more readily. After denaturation by heating, the preparations are mixed

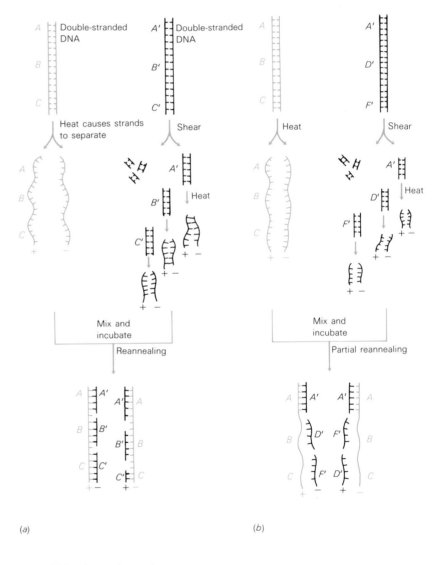

Figure 17.9 Hybridization procedure used to detect similarities in base sequence between nucleic acids of two different strains (compare with Figure 12.8). One of the nucleic acids is made radioactive and its degree of hybridization with the other nucleic acid is determined by measuring the amount of radioactivity bound. In (a), the two nucleic acids are closely related, whereas in (b), only a portion of the nucleic acids are homologous. The hybridization technique works more efficiently if one of the hybridizing nucleic acids is in fairly small pieces, hence one of the nucleic acids is subjected to high-speed shearing treatment in a blender. RNA can also be used instead of the second DNA since RNA is complementary to one strand of the DNA.

(a)

(b)

17.7 *Molecular and genetic taxonomy*

Table 17.8   DNA homology between two members of the Enterobacteriaceae and some other organisms[a]

| Source of DNA | % Carbon 14-labeled DNA bound relative to the homologous DNA | |
	Escherichia coli B	*Aerobacter aerogenes*
Escherichia coli B	100	49
Escherichia coli K12(λ)	101	
Salmonella typhimurium	71	60
Shigella dysenteriae	71	45
Aerobacter aerogenes 211	51	100
Aerobacter aerogenes 13048	45	105
Klebsiella pneumoniae	25	54
Proteus vulgaris	14	2
Serratia marcescens	7	11
Pseudomonas aeruginosa	1	2

[a] Data from B. J. McCarthy and E. T. Bolton: *Proc. Natl. Acad. Sci.* 50:156 (1963).

and allowed to renature, and the percentage of radioactive nucleic acid bound is measured. An example of the kind of data obtained is shown in Table 17.8.

Clearly, nucleic acid hybridization has great potential in elucidating the genetic relatedness between organisms since it can be performed with organisms in which sexual interaction or genetic recombination is not possible. The data so far available demonstrate clearly that this technique is capable of providing much new information on microbial taxonomy. For instance, it has been found that bacteria as a group are quite diversified and that many organisms differ from each other by over 95 percent of their genomes. Mammals, on the other hand, are much more closely related as a group since even the most diverse mammals have about 20 percent of their nucleic acid sequences in common.

Although unrelated organisms show only little or no similarities for most genes, it has been found that the genes controlling the synthesis of ribosomal RNA are often the same in unrelated organisms. Thus it appears that, even though most genes of unrelated organisms have diverged, the genes controlling the ribosomal RNA have been conserved during evolution.

Protein (enzyme) homologies   Nucleic acid homologies give us a broad picture of genetic relatedness of organisms. However, even distantly related organisms may still have some genes in common, and this can be ascertained by examining the relatedness of proteins that have the same function. There are many methods available for comparing proteins, of which the best is the determination of the amino acid sequence of the purified protein. Since this sequence is directly related to the base sequence of the gene controlling its synthesis, if we find similar amino acid sequences in these functionally similar proteins, we can infer that the organisms have similar base sequences for these genes. The results so far available show that similar or identical amino acid sequences occur for the same protein when the organisms are closely related and that differences are greater when the organisms are more distantly related. However, the amino acid sequences around the active site of the enzyme are often similar or identical even in enzymes from distantly related organisms.

If it is not feasible to determine amino acid sequences, several indirect approaches are possible. One method is to study the electrophoretic mobility of the same enzyme from different organisms. Recall that electrophoresis can be used to distinguish enzymes on the basis of their net electrical charge (see Section 4.1). Since charge is determined by the nature of the amino acid side chains on the protein, two proteins that differ in amino acid composition will usually show different electrophoretic mobilities. It is also possible to distinguish two enzymes with the same catalytic properties by measuring differences in their heat or acid stability. Another approach derives from the fact that enzymes with the same catalytic function may have different pH optima, substrate specificities, or inhibitors, and these properties can be used to obtain clues concerning whether the same enzyme from two or more organisms is the same or a different protein. For instance, studies of this type have shown that the malic dehydrogenases of the enteric bacteria (*Escherichia, Salmonella, Aerobacter,* and *Serratia*) are very similar but that they differ from those of the bacteria of other families. Enzymes with similar functions can also be distinguished immunologically (Section 15.4); this method has been used to show that, in the family of the enteric bacteria, the alkaline phosphatase of *Escherichia* differs slightly from that of *Aerobacter* but markedly from that of *Serratia* and *Proteus.*

Studies concerning protein homologies have shown that considerably greater diversity occurs among the bacteria than among higher organisms. For instance, lobsters and chickens appear to have some enzymes of related characteristics, even though they are of diverse families, whereas bacteria of different families have little or no common enzyme classes.

Genetic recombination as a taxonomic tool   In organisms that possess mechanisms for genetic recombination, relatedness may be shown through studies on the efficiency of genetic exchange. However, since even closely related organisms may not mate with each other for various reasons, genetic analysis makes possible the recognition of similarities between organisms but not of differences since, if recombination does not occur, it is not usually possible to ascertain whether this is for trivial reasons or because the organisms are unrelated. In eucaryotes the limiting factor for recombination is often the inability of the chromosomes to pair at meiosis. In procaryotes recombination is virtually always a fragmentary process, and hence it is difficult to study the relatedness of whole genomes.

In procaryotes DNA-mediated transformation has proved useful for taxonomic studies. The DNA can be extracted from the strain to be used as a potential donor and can then be used in attempts to transform a competent recipient. This approach has been used extensively in studies on the genus *Bacillus,* using *B. subtilis* as a recipient. The DNA of *Bacillus* species that have overall base compositions different from that of *B. subtilis* were found not to transform it, whereas some species that have base compositions like *B. subtilis* transform it at very high frequencies. Others transform it at lower frequencies. DNA from *Bacillus stearothermophilus* does not transform *B. subtilis* at all, even though it has the same base composition. This indicates that a similarity in base composition is necessary but is not sufficient to guarantee interspecies transformation. Presumably, the lack of recombination results from the absence of homologous DNA base sequences.

In the genus *Neisseria,* interspecies transformation is possible between all species except *N. catarrhalis,* which can be transformed by its own DNA but acts neither as a donor nor as a recipient with other *Neisseria* species. On the basis of these results, *N. catarrhalis* was examined more closely, and taxonomists concluded that it should probably not be classified as *Neisseria* at all. Its DNA base composition is 40 percent GC, while that of the other *Neisseria* species is 57 percent GC, and *N. catarrhalis* differs from the others in many physiological characteristics. Indeed, the only thing *N. catarrhalis* has in common with the other *Neisseria* species is that it is an aerobic Gram-negative coccus. Recently, what was called *N. catarrhalis* has been reclassified into the genus *Acinetobacter.* Thus genetic analysis in this case has proved of considerable aid in taxonomic studies.

Summary of molecular and genetic taxonomy We can now suggest briefly how molecular and genetic methods might be applied to the classification of microorganisms. These suggestions are of most use for procaryotes but may also be applicable to some eucaryotic groups. (1) Examine the morphological characteristics of the collection of organisms and place together those organisms which seem to be similar. (2) Determine the DNA base compositions of these morphologically similar organisms. Those which are similar in both characteristics are likely to be closely related. (3) Examine the physiological and biochemical properties of those organisms grouped together in suggestions 1 and 2. These organisms should be fairly similar in many ways. Subgroups can be formed on the basis of any major physiological and biochemical differences noted. (4) Carry out nucleic acid hybridization studies of a group of physiologically and biochemically similar strains. (5) Examine protein homologies of key enzymes for these strains.

All of these studies should permit the development of a unified picture of a microbial group, and any new strains discovered should be easily categorized, perhaps without the necessity for a complete analysis.

17.8   Numerical taxonomy

Once broad taxonomic groupings are defined by morphological and molecular criteria, it is necessary to consider how the finer divisions are constructed. These are usually based on physiological and biochemical characteristics, environmental responses and tolerances, and ecological characteristics. Usually a large number of separate tests are run, often more than 50, and each strain is listed as giving positive or negative results. With a large number of tests and a large number of strains, the mental problem of collating and analyzing the data is considerable. It is in precisely such studies that the electronic computer is valuable since it is able to compare each strain quickly with all other strains and to detect similarities and differences.

The characteristics to be studied in the assemblage of strains should be ones that are easily scored on a large number of organisms. Again, scoring should permit the results to be recorded as plus or minus, rather than as quantitative values. The characteristics are usually not weighted; each characteristic is considered as important as any other. Once the tests have been performed on a group of organisms, the results are compiled in tabular form (see Table 17.9).

The manner in which similarities are determined can be described by a simple example. Suppose we compare two strains from Table 17.9 and we find that for some attributes both strains are positive, for others both are negative, and for others one strain is positive and the other is negative. These possibilities are illustrated in Table 17.10. A *similarity coefficient* for these two strains is then calculated, as shown in the table. The higher the coefficient, the more similar are the two organisms. However, characteristics for which both strains are negative are not counted in calculating the similarity coefficient because the lack of a characteristic in two organisms does not necessarily indicate relationship. (For example, both *Pseudomonas* and *Escherichia* are non-spore-formers, but this does not make them related.)

Table 17.9   Part of a coded data table[a]

Character states	Strains			
	A	B	C	D
1	+[b]	+	−	0
2	+	+	+	+
3	+	+	+	−
4	−	+	0	0
5	+	+	+	+
6	+	+	−	+
7	+	+	−	0
8	0	−	+	+
9	+	+	+	+
10	+	+	+	−
11	+	0	−	0
12	+	+	+	−

[a] From P. H. A. Sneath: G. C. Ainsworth and P. H. A. Sneath (eds.), *Microbial Classification* (Twelfth Symposium, Society for General Microbiology) (London: Cambridge University Press, 1962).
[b] +, character present; −, character absent; 0, unscorable.

Table 17.10 *Comparison of two strains for similarity and difference, and procedure for calculation of similarity coefficient S*

Second strain	First strain	
	+	−
+	a^a (32)	b (4)
−	c (6)	d (8)

^{a} *a, b, c,* and *d* represent the number of characters or attributes in which each strain is similar to, or different from, the other strain. The numbers in parentheses are for an arbitrary example, of which the similarity coefficient is calculated:

$$S = \frac{a}{a + b + c} \qquad S = \frac{32}{32 + 4 + 6}$$

$$= \frac{32}{42} = 76 \text{ percent similarity}$$

The characteristics of only two organisms can be compared easily by inspection, but when a large number of characteristics must be compared, an electronic computer is essential. The data for each strain are coded, and similarity coefficients then calculated by the computer. The details of computer analysis will not be discussed here. Suffice to say that the calculation of similarity coefficients is a simple task for the computer programmer to set up.

The computer constructs a similarity matrix that indicates the similarity of each strain with the other strains. The matrix can then be rearranged so that similar organisms are placed together, and on the basis of this modified matrix it is then possible to arrange the strains in a hierarchy, as is shown in Figure 17.10, where strains are separated into groups (*phenons*), according to their similarity. Strains with a high percentage of similarity (for example, 90 percent or greater) are usually classified as a single taxonomic unit (for instance, species).

Numerical taxonomy has the advantage over other taxonomic procedures of yielding results that are less biased, and because it is based on an analysis of many characteristics, it provides categories that are more stable than those obtained by methods based on a few characteristics. Thus numerical taxonomies are unlikely to need radical revisions. Do the categories resulting from numerical taxonomy correlate with those

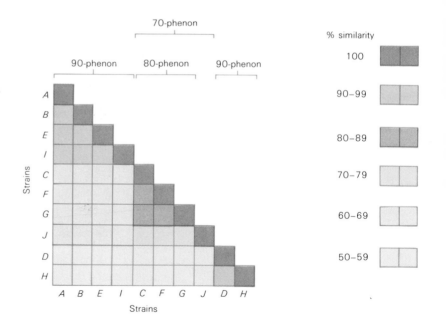

Figure 17.10 *A similarity matrix in which 10 isolates are compared. After computer analysis to determine degree of relatedness of the isolates, the matrix is constructed in such a way as to place closely related strains together. A cluster of strains with a considerable degree of relatedness is called a phenon. On the basis of relatedness at the 80- or 90-percent level, three phenons are recognized, consisting of strains A, B, E, and I, strains C, F, and G, and strains D and H. If desired, these groups of closely related strains could be named as separate species. [From P. H. A. Sneath: in G. C. Ainsworth and P. H. A. Sneath (eds.), Microbial Classification (Twelfth Symposium, Society for General Microbiology). London: Cambridge University Press, 1962.]*

Microbial evolution and taxonomy

of traditional taxonomy? In some groups correlation is very good, as in the enteric bacteria and the actinomycetes. In others correlation is poor, whereas in most correlation is moderate. Lack of correlation could mean either that the traditional taxonomy or the numerical taxonomy is faulty or that the group selected was not sufficiently homogeneous.

17.9   Current microbial taxonomies

We have discussed some of the approaches to microbial taxonomy, emphasizing the more recent concepts. We now discuss briefly some practical aspects of the classification of microorganisms. Microbial taxonomy is especially complicated because different microbial groups fall within the province of specialists from different disciplines. The algae and fungi traditionally have been the province of botanically oriented scientists, whereas the protozoa are in the domain of the zoologist. The bacteria are the responsibility of the bacteriologists, but curiously the blue-green algae, although procaryotic, have been classified by the botanists. Each discipline has its own rules for naming organisms, which have been collected into international codes, one for bacteria, one for plants, and one for animals. These international codes differ in various ways, although they have much in common.

The primary goal of any classification procedure is to collect related organisms into taxonomic groups called *species*. It is not easy to define a species. In microbiology "species" most commonly refers to a collection of similar strains or clones that differ significantly from other groups of clones. Unfortunately, this definition is so vague that it leaves much room for subjective interpretation. Furthermore, if this definition is followed, species distinctions will continually change as our knowledge of organisms broadens. Initially a group of organisms may appear quite similar, but as we get to know them better we begin to see differences, which we may eventually conclude are significant enough to warrant subdividing the group. On the other hand, as we isolate more and more organisms we may find some that are intermediate between two previously well-defined species. Ultimately, we may see not several clear-cut species but only a continuum of strains, grading from one extreme to another.

Groups of species are collected in *genera*, and genera are grouped into *families*. Related families are placed in *orders*, orders are placed in *classes*, and classes in *phyla*. This simple statement of taxonomic rank covers a multitude of complications, and lifetimes have been spent pondering the minutiae necessary for arriving at satisfying conclusions. Unfortunately, even the construction of these higher taxa is not absolute, and changes of opinion or the entry onto the scene of a new generation of scientists may lead to an upheaval. For the purposes of this book, we need not worry about the details of these matters. It will be sufficient that we know there are groups of genera that are obviously related which can be validly collected into higher taxa.

We emphasized at the beginning of this discussion that taxonomy comprises much more than mere naming of organisms. At this point we must note that nomenclature and taxonomy merge when we turn to the task of naming the species and genera that we have recognized. Each species traditionally is given a binomial name, which consists of both a genus name and a species name. For instance, the name *Bacillus subtilis* refers to a particular species that was recognized through various taxonomic criteria described above. This name tells us something about the organism since *Bacillus* refers to a large group of bacteria included in a single genus. The name *Bacillus subtilis* thus implies that this species is related to all the other species grouped in the genus *Bacillus*. Therefore genus and species names carry certain taxonomic implications that force us to consider not only nomenclature but taxonomy. The biggest problem raised is that names are difficult to change, whereas our taxonomic decisions may change as our knowledge of microbial groups expands. We may come to the conclusion that the members of the genus *Bacillus* are all really too heterogeneous to be collected into one genus, and that this genus should be subdivided. As a taxonomic operation, subdivision would not be especially difficult, but as a nomenclatural operation it is often traumatic as it may require us to discard a name we have become familiar with and introduce a new one. Throughout the past 100 yr or so of microbial taxonomy, many such revisions have occurred, resulting in a residuum in the scientific literature of outmoded names. To avoid at least part of the difficulty, whenever a reclassification requires moving a species from one genus to another, the species name is retained. Thus, it is generally the specific epithet (for example, *subtilis*) that is the defining word of the species name, rather than both genus and species names taken together.

In the following two chapters we shall consider in some detail the biology of microorganisms of various taxonomic groups.

Supplementary readings

Barghoorn, E. S. and S. A. Tyler. "Microorganisms from the Gunflint chert," *Science* 147:563 (1965).

Blum, H. F. *Time's Arrow and Evolution,* 2nd ed. Princeton, N.J.: Princeton University Press, 1955.

Microbial evolution and taxonomy

Keosian, J. The Origin of Life, 2nd ed. New York: Reinhold Publishing Corp., 1968.

Marmur, J., S. Falkow, and M. Mandel "New approaches to bacterial taxonomy," *Ann. Rev. Microbiol.* 17:329 (1963).

McAlester, A. L. The History of Life. Englewood Cliffs, N.J.: Prentice-Hall, Inc., 1968.

Microbial Classification (Twelfth Symposium, Society for General Microbiology). London: Cambridge University Press, 1962.

Oparin, A. I. The Origin of Life on the Earth, 3rd ed. New York: Academic Press, Inc., 1957.

Sokal, R. R., and P. H. A. Sneath Principles of Numerical Taxonomy. San Francisco: W. H. Freeman Co., Publishers, 1963.

In the preceding chapters we have learned something about the structure, function, and evolution of microorganisms. We have seen that they can be divided into two broad groups, procaryotes and eucaryotes, and that the members within each group have many properties in common. We discovered in the last chapter, however, that diversity exists within each group; species that are adapted to different ecological niches have evolved. In the present chapter we will describe in some detail the diversity among the procaryotes. Thus rather than discussing the features they have in common we consider the ways in which they differ. Our discussion builds on the information presented in the previous chapters, and one may find it useful from time to time to review biochemical, ecological, and genetic concepts presented earlier. Ideally, this chapter should not be read as a unit but rather a little at a time and not necessarily in sequence. Perhaps the best approach would be to study selected sections of this chapter along with earlier chapters, to illuminate concepts presented there.

Procaryotic microorganisms are a fascinating group, and their members play important roles in nature and in human affairs. Whole lifetimes have been spent studying single genera, yet even today there are great gaps in our knowledge of many procaryotes. The classification of procaryotes, especially the bacteria, is a difficult matter. It is fairly easy to recognize species and genera, but difficulty arises when an attempt is made to collect related genera into families and orders. A few genera lend themselves well to classification into higher taxa, but most do not; only after much more information is available, especially using molecular techniques, will a satisfying classification scheme for the bacteria become available. In what follows we have tried to group related genera together, but we have not used family and order names except in a few places where the situation is clear. It should not be disturbing that we have kept our classification vague: it is better to admit ignorance than to try to hide it behind a misleading scheme.

eighteen

Represen- tative procaryotic groups

18.1 Lactic acid bacteria

A wide variety of microorganisms produce lactic acid as a product of anaerobic metabolism of sugars, but when we speak of the *lactic acid bacteria,* we are referring to a specific group of closely related bacteria. These are characterized as Gram-positive, usually nonmotile, nonsporulating bacteria that produce lactic acid as a major or sole product of fermentative metabolism. Members of this group lack porphyrins, a cytochrome system, and oxidative phosphorylation, and hence obtain energy only by substrate-level phosphorylation. All of the lactic acid bacteria grow anaerobically. Unlike many anaerobes, however, most of these are not sensitive to O_2 and can grow in its presence as well as in its absence; thus they are facultative anaerobes. Some strains are able to take up O_2 through the mediation of flavoprotein oxidase systems, producing H_2O_2. No ATP is formed in this process, but the oxidase system can be used for the reoxidation of NADH. Most lactic acid bacteria can obtain energy only from sugars and related compounds, and hence are usually restricted to habitats in which sugars are present. They usually have only limited biosynthetic ability, and their complex nutritional requirements include need for amino acids, vitamins, purines, and pyrimidines. Many strains require more amino acids for growth than does man. Because of these nutritional requirements, lactic acid bacteria are commonly employed in microbiological assays for small concentrations of growth factors.

Homo- and heterofermentation   One important difference between subgroups of the lactic acid bacteria lies in the nature of the products formed during the fermentation of sugars. One group, called *homofermentative*, produces virtually a single fermentation product, lactic acid, whereas the other group, called *heterofermentative*, produces other products in addition to lactic acid (see Table 18.1). Abbreviated pathways for the fermentation of glucose by a homo- and a heterofermentative organism are shown in Figure 18.1; the differences observed in the fermentation products are determined by the presence or absence of the enzyme *aldolase*, one of the key enzymes in glycolysis (see Chapter 4). The heterofermenters, lacking aldolase, cannot break down hexose diphosphate to triose phosphate. Instead, they oxidize glucose-6-phosphate to 6-phosphogluconate and then decarboxylate this to pentose phosphate, which is broken down to triose phosphate and acetyl phosphate. Triose phosphate is converted ultimately to lactic acid with the production of one

Table 18.1   *Representative fermentation balances of homo- and heterofermentative lactic acid bacteria*[a]

Product	*Streptococcus liquefaciens* (homofermentative)	*Lactobacillus lycopersici* (heterofermentative)
Lactic acid	173[b]	80
Acetic acid	23	17
Formic acid	6.6	—
CO_2	6.6	85
Ethanol	2.0	70
Glycerol	14.1	40

[a] Adapted from W. A. Wood: in I. C. Gunsalus and R. Y. Stanier (eds.), *The Bacteria: A Treatise on Structure and Function*, vol. II (New York: Academic Press, 1961).
[b] Millimoles formed per 100 mmoles of glucose fermented.

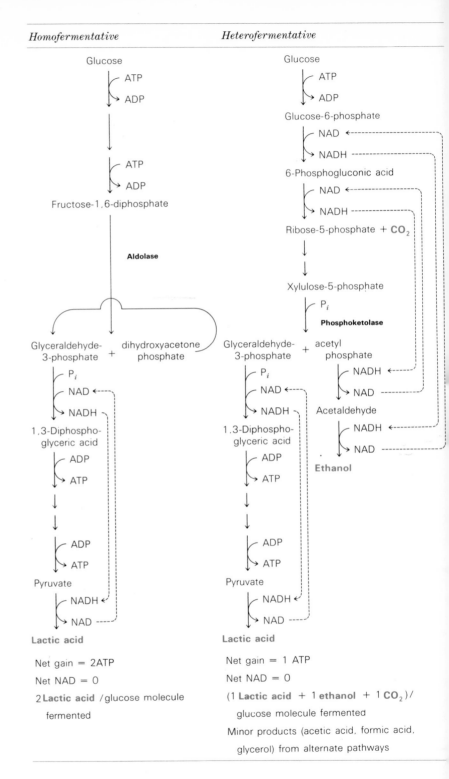

Figure 18.1 The fermentation of glucose in homofermentative and heterofermentative lactic acid bacteria.

Table 18.2 *Differentiation of the principal genera of lactic acid bacteria*

	Cell form and arrangement	Fermentation	DNA base composition, % GC
Streptococcus	Cocci in chains	Homofermentative	38–40
Leuconostoc	Cocci in chains	Heterofermentative	39–41
Pediococcus	Cocci in tetrads	Homofermentative	38
Lactobacillus	Rods, usually in chains	Homofermentative	32–53
	Rods, usually in chains	Heterofermentative	32–53

mole of ATP, while the acetyl phosphate accepts electrons from the NADH generated during the production of pentose phosphate and is thereby converted to ethanol without yielding ATP. Because of this, heterofermenters produce only one mole of ATP from glucose instead of two moles as are produced by homofermenters. This difference in ATP yield from glucose is reflected in the fact that homofermenters produce twice as much cell mass as heterofermenters from the same amount of glucose. Because the heterofermenters decarboxylate 6-phosphogluconate, they produce CO_2 as a fermentation product, whereas the homofermenters produce little or no CO_2; therefore one simple way of detecting a heterofermenter is to observe the production of CO_2 in laboratory cultures.

The various genera of the lactic acid bacteria have been defined on the basis of cell morphology and on the type of fermentative metabolism, as is shown in Table 18.2. Recent studies on DNA base compositions suggest that certain of the genera defined on traditional grounds are correctly identified, whereas other genera may need to be reassigned. As can be seen from Table 18.2, members of the genera *Streptococcus* and *Leuconostoc* have fairly similar DNA compositions; in addition, there is very little variation from strain to strain. The genus *Lactobacillus*, on the other hand, has members with widely diverse DNA compositions and hence does not constitute a homogeneous group. Further work on the taxonomy of the genus *Lactobacillus* will undoubtedly show that this group should be separated into a number of genera.

Streptococcus and other cocci   The genus *Streptococcus* contains a wide variety of members with quite distinct habitats, whose activities are of considerable practical importance to man. Some members are pathogenic to man and animals, and in preantibiotic days constituted one of the leading causes of death in man. As producers of lactic acid, certain streptococci play important roles in the production of buttermilk, silage, and other fermented products.

Streptococci are quite easy to isolate from natural materials. Use is made of the fact that since they lack a cytochrome system, their growth is not inhibited by sodium azide (NaN$_3$), which is a fairly specific inhibitor of cytochromes (Section 4.4). Thus media containing about 0.05 percent NaN$_3$, glucose, and peptone or meat extract to provide amino acids and growth factors, which is well buffered to pH 7.5, can be used to enrich for streptococci from most natural materials. Since many streptococci pathogenic to man and animals are hemolytic, use is often made of blood-agar media to which azide is added; the colonies appearing on such a medium can be immediately characterized as hemolytic or nonhemolytic. The genus *Streptococcus* is subdivided into a number of groups of related species on the basis of a series of characteristics that are enumerated in Table 18.3. Hemolysis shown on blood agar is of considerable importance in the subdivision of the genus. Colonies of those strains producing streptolysin O or S (Section 15.2) are surrounded by a large zone of complete hemolysis, a condition called "β hemolysis." On the other hand, many streptococci that do not produce hemolysins cause the formation of a greenish or brownish zone around their colonies, which is due not to hemolysis but to the discoloration of the red cells, a condition called "α hemolysis" (this is not a true hemolysis, however). The streptococci are also divided into immunological groups based on the presence

Table 18.3 Differential characteristics of streptococci

	Pyogenes group (β-hemolytic streptococci)— for example, S. hemolyticus	Viridans group (α-hemolytic streptococci)— for example, S. viridans	Fecal group (enterococci)— for example, S. faecalis, S. faecium	Lactic group— for example, S. lactis
Antigenic groups	A, B, C, F, G	Not grouped	D	N
Type of action on blood agar	β^a	α^b	α, β	γ^c
Good growth at:				
10°C	—	—	+	+
45°C	—	+	+	—
Optimum temperature, °C	37	37	35–37	25
Survive 60°C for 30 min	—	—	+	+
Growth in broth with:				
6.5% NaCl	—	—	+	—
pH 9.6	—	—	+	—
0.1% methylene blue	—	—	+	+
40% bile	—	—	+	+

^{a} β action, lysis of red blood cells.
^{b} α action, discoloration of red blood cells but not lysis.
^{c} γ action, no detectable effect on red blood cells.

Representative procaryotic groups

Table 18.4 Extracellular products of Group A streptococci

	Distribution	Comment
Erythrogenic toxin	Produced by most strains	Causes scarlet fever rash
Streptolysin S	Produced by most strains	Hemolysin (affects phospholipid of membrane)
Streptolysin O	Produced by most strains	Hemolysin (affects cholesterol of membrane)
NAD nucleotidase	Production more common among some antigenic types than among others	Decomposes NAD
Streptokinase	Produced by most strains	Causes dissolution of blood clots
Deoxyribonuclease	Produced by all strains	Hydrolyzes DNA
Ribonuclease	Produced by most strains	Hydrolyzes RNA
Hyaluronidase	Produced by most strains	Hydrolyzes hyaluronic acid in connective tissue
Streptococcal proteinase	Produced by most strains	Nonspecific protease
Amylase	Variable, but produced by many strains	Hydrolyzes starch

of specific carbohydrate antigens. These antigenic groups are designated by letters; A through O are currently recognized. Those β-hemolytic streptococci found in human beings usually contain the group A antigen, which is a cell-wall polymer containing N-acetylglucosamine and rhamnose. The fecal streptococci contain the group D antigen, a glycerol teichoic acid containing glucose side chains. Group B streptococci are usually found in association with animals and are a common agent of mastitis in cows. Streptococci found in milk, the so-called lactic streptococci, are of group N. The *pyogenic* streptococci, which are β hemolytic, are most frequently associated with disease of man. *Pyogenic* means "pus-forming" and refers to the characteristic symptoms induced by these organisms when infecting the skin or peripheral areas of the body. These symptoms result from the production of a variety of bacterial enzymes (Table 18.4) that cause destruction of phagocytic and other cells. These destroyed cells accumulate at the site of infection, leading to the formation of pus. Not all strains of pyogenic streptococci produce all the products listed in Table 18.4, but every strain produces a good proportion of them. One product, the erythrogenic toxin, is responsible for the characteristic rash of scarlet fever. Group A or pyogenic streptococci have been classified into a number of antigenic *types*, based on the immunological nature of a cell-surface protein called "M protein." As was noted in Section 15.2, most pathogenic streptococci produce M protein, which is associated with a resistance of the organism to phagocytosis. There are over 40 M protein types, which are given numerical designations, that is, type 1, type 12, type 14, and so on. The study of M protein types is of use in following the spread of a specific strain of streptococcus through a human population during an epidemic. Some types are often associated with a particular disease condition; for instance, type 12 is usually the causal agent of glomerular nephritis, a disease of the kidney.

A wide variety of disease syndromes are associated with streptococcal infection. These include sore throat (in many manifestations), mastitis, peritonitis, pneumonia, puerperal sepsis (a disease of the uterus following childbirth), erysipelas (a generalized infection of the body, sometimes called blood poisoning), glomerular nephritis, and rheumatic fever. In most streptococcal infections, the causal organism is transmitted by droplets, and infection is initiated in the upper respiratory tract. If unchecked the organism may spread to other parts of the body where more generalized infections can commence. Individuals vary greatly in their susceptibility to streptococcal infection. About 10 percent of the population are symptomless carriers of group A streptococci in their upper respiratory tracts and may serve as reservoirs for the infection of more susceptible individuals. Virtually all strains of group A steptococci are highly sensitive to penicillin and other antibiotics active against Gram-positive bacteria, and resistant strains rarely develop, so that most acute streptococcal infections are readily treated. In the preantibiotic days, however, group A streptococcus diseases were among the most frequent causes of death in man. The one streptococcal disease that is still of major concern is rheumatic fever. The pathogenesis of this disease is still obscure, but the symptoms seem to be allergic in nature, and about 3 percent of the population is susceptible. Although treatment of the symptoms of rheumatic fever is not possible, control of the disease can be effected by continuous administration of penicillin to susceptible individuals, which hinders the initiation of minor respiratory infections and thus prevents the onset of the allergic reactions.

The *viridans streptococci* are α-hemolytic strains commonly found in the mouth. They all produce a polysaccharide capsule, which makes it possible for these bacteria to adhere to the surface of the tooth, and they may be involved in tooth decay (Section 14.3).

The group D streptococci are often called the *fecal streptococci* since they are generally found in the intestinal tract of man and animals, although they are not restricted to this habitat. As can be seen in Table 18.3, the group D streptococci are considerably more tolerant to environmental extremes than are the group A streptococci. In addition, they are more resistant to antibiotics and other antibacterial agents. Because of their frequent occurence in fecal matter and their ease of isolation on azide broth or agar, the group D streptococci are often used as indicators of fecal pollution in water supplies, in the same manner as the coliforms are (Section 15.8). The group D streptococci are more resistant to environmental influences than are the coliforms, and they survive longer in water; hence they cannot be used as indicators of recent fecal pollution but only of fecal pollution at some unspecified time in the past. Certain strains of group D streptococci are used in the buttermilk fermentation.

The group N streptococci, commonly called the *lactic streptococci,* play important roles in the dairy industry since they are the organisms usually used in starter cultures (Section 8.5) for the production of buttermilk, cheese, and other fermented dairy products. In nature these organisms live primarily in association with plant materials, from which they become inoculated into milk. Through years of laboratory selection, the strains used commercially have been selected from these naturally occurring strains.

The *pneumococci (Diplococcus pneumoniae)* are a group we have discussed previously in sections on genetic transformation (Chapter 12) and on the mechanisms of invasion and pathogenicity (Chapter 15). The pneumococci are α hemolytic and are closely related to the streptococci in morphology, physiology, and genetics; indeed, reciprocal transformation between pneumococci and streptococci is possible. The most clear-cut difference between the two groups is that pneumococci tend to autolyze readily, either spontaneously in the late logarithmic phase or after the addition of bile salts. (The pneumococci are hence said to be ''bile soluble,'' although this is not a particularly accurate term.)

Placed in the genus *Leuconostoc* are organisms closely related to streptococci but which are heterofermentative. It seems possible that *Leuconostoc* arose from *Streptococcus* through mutational loss of aldolase, a key enzyme in glycolysis. *Leuconostoc* strains are often used in starter cultures together with streptococci. Although *Leuconostoc* produces less acid than does *Streptococcus,* it produces more of the desired flavoring ingredients, such as biacetyl, ethanol, and acetic acid, of fermented dairy products. Some strains of *Leuconostoc* produce large amounts of dextran polysaccharides (α-1,6-glucan) when cultured on sucrose, due to the action of the enzyme sucrose transglucosylase:

$$n\text{(Glucose-fructose)} \longrightarrow \text{(glucose)}_n + n\text{(fructose)}$$

    Sucrose   *Dextran*

This is one of the few examples of a biological polymerization in which the sugar in the polymer is not first activated by a nucleotide (Section 5.1); the energy of the bond linking the two sugars as a disaccharide is conserved in the polymer. Dextrans produced by *Leuconostoc* have found some medical use as plasma extenders in blood transfusions. Some strains of *Leuconostoc* produce fructose polymers called "levans."

Lactobacillus   In contrast to the streptococci, the genus *Lactobacillus* contains a more heterogeneous assemblage of organisms, which are similar only in that they are nonmotile, nonsporulating, Gram-positive rods lacking a respiratory metabolism. Most species are homofermentative, but some are heterofermentative. The genus has been divided into three major subgroups: thermobacterium, streptobacterium, and betabacterium (Table 18.5). Although these names are not recognized as valid genera, they serve as a useful means of subdividing this heterogeneous group. Lactobacilli are often found in association with dairy products, and some strains are used in the preparation of fermented products. For instance, *L. bulgaricus* is usually used in the preparation of yogurt, and other species are involved in the production of sauerkraut, silage, and pickles. The lactobacilli are usually more resistant to acidic conditions than are the other lactic acid bacteria, being able to grow well at pH values around 5. Because of this, they can be selectively isolated from natural materials by use of carbohydrate-containing media of acid pH, such as tomato juice-peptone agar. The acid resistance of the lactobacilli enables them to continue growing during natural lactic fermentation when the pH value has dropped too low for the other lactic acid bacteria to grow, and the lactobacilli are therefore responsible for the final stages of lactic acid fermentations. The lactobacilli are rarely or never pathogenic.

One organism, sometimes classified as *Lactobacillus bifidus*, is a common member of the intestinal flora of breast-fed human infants. In recent classification schemes, this organism has been removed from the genus *Lactobacillus* and placed in a separate genus, *Bifidobacterium*. It is often found in the intestine in essentially pure culture during the first week or so of life. One of the characteristics of this species is the forked appearance of the cells (hence the name *bifidus*). This form has a specific require-

Table 18.5   Characteristics used in subdividing the lactobacilli and typical species in each subgroup

	Homofermentative	
Heterofermentative, betabacterium group	*Thermobacterium group (growth at 45°C; no growth at 15°C)*	*Streptobacterium group (poor or no growth at 45°C; growth at 15°C)*
L. fermenti	L. helveticus	L. casei
L. buchneri	L. bulgaricus	L. plantarum
L. brevis	L. lactis	
	L. acidophilus	
	L. delbrueckii	

ment for an oligosaccharide or polysaccharide containing N-acetyl-D-glucosamine, which probably is a precursor of the cell-wall peptidoglycan. The bifid morphology apparently arises from a derangement of cell-wall synthesis when the growth factor occurs in reduced amounts. Human milk contains the growth factor, but cow's milk does not—hence the development of the organism only in breast-fed infants.

18.2   Propionic acid bacteria

The propionic acid bacteria (genus *Propionibacterium*) were first discovered as inhabitants of Swiss (Emmentaler) cheese, where their fermentative production of CO_2 produces the characteristic holes; the presence of propionic acid is at least partly responsible for the unique flavor of this product. Although this acid is produced by some other bacteria, its production by the propionic acid bacteria is a distinguishing characteristic of the genus. The bacteria in this group are Gram-positive, pleomorphic non-sporulating rods, which are nonmotile and anaerobic. They ferment lactic acid, carbohydrates, and polyhydroxy alcohols with the production of propionic acid, succinic acid, acetic acid, and CO_2. Their nutritional requirements are complex and they usually grow rather slowly.

The enzymatic reactions leading from glucose to propionic acid are now understood (Figure 18.2). The initial catabolism of glucose to pyruvate follows the Embden-Meyerhof pathway as in the lactic acid bacteria, but the NADH formed is reoxidized as one part of a cycle in which propionic acid is formed. Pyruvate accepts a carboxyl group from methylmalonyl-CoA by a transcarboxylase reaction, leading to the formation of oxalacetate and propionyl-CoA. The latter substance reacts with succinate in a step catalyzed by a CoA transferase, producing succinyl-CoA and propionate. The succinyl-CoA is then isomerized to methylmalonyl-CoA, and the cycle is complete. Reoxidation of NADH occurs in the steps between oxalacetate and succinate, and the oxidation-reduction balance is restored.

Most propionic acid bacteria also ferment lactate with the production of propionate, acetate, and CO_2. The anaerobic fermentation of lactic acid to propionate is of interest because lactic acid itself is an end product of fermentation for many bacteria. The propionic acid bacteria are thus able to obtain energy anaerobically from a substance that other bacteria are producing. One aspect of this propionic acid fermentation of considerable biochemical interest is that, although CO_2 is produced, some CO_2 is also fixed by the carboxylation of pyruvate to oxalacetate, the latter being a required compound in the propionic acid cycle (Figure 18.2). Such a

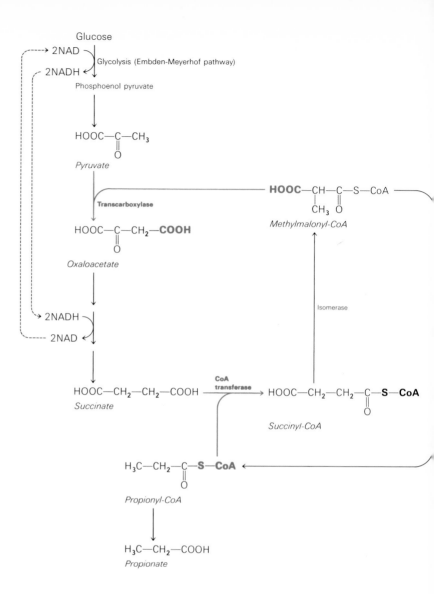

Figure 18.2   The formation of propionic acid by Propionibacterium.

CO_2-fixation reaction is of historical interest in that it was the first proved case of heterotrophic CO_2 fixation, a process we now know occurs in most heterotrophs (Section 5.1). Until heterotrophic CO_2 fixation was discovered in propionic acid bacteria, it had been thought that CO_2 fixation occurred only in autotrophs.

It is the fermentation of lactate to propionate that is important in Swiss-cheese manufacture. The starter culture consists of a mixture of homofermentative streptococci and lactobacilli, plus propionic acid bacteria. The initial fermentation of lactose to lactic acid during formation of the curd is carried out by the homofermentative organisms. After the curd has been drained, the propionic acid bacteria develop rapidly and usually reach numbers of 10^8 per gram by the time the cheese is 2 mo old. Swiss cheese ''eyes'' are formed by the accumulation of CO_2; the gas diffuses through the curd and gathers at weak points. The number and size of the eyes depends on the physical characteristics of the curd and the rate at which gas forms. It is more desirable to have a smaller number of large eyes, a condition that results when the curd possesses sufficient elasticity so that it stretches but does not break. Many defects can develop, primarily in formation of the eyes, and careful microbiological control of the process is important.

Although the propionic acid bacteria resemble lactobacilli morphologically and physiologically, they are a distinct group, as was shown recently by molecular taxonomic studies. The base composition of the DNA of propionic acid bacteria is 60 to 66 percent GC, considerably higher than that of any of the lactic acid bacteria. All strains have been isolated from dairy products, primarily Swiss cheese, but undoubtedly they occur in other habitats.

18.3 Aerobic cocci

Gram-positive cocci The Gram-positive cocci are not really a natural taxonomic assemblage since they include bacteria with widely differing physiological and genetic characteristics that are arbitrarily grouped together because they all have the same morphological characteristics. The most important genus is *Staphylococcus*, which contains common parasites of man and animals, including some that occasionally cause serious infections. Staphylococci are nonmotile, nonsporulating, Gram-positive cocci, which carry out cell division in three planes and hence can form irregular clumps. They are facultative organisms: When grown aerobically, they have a complete cytochrome system; but when grown anaerobically, cytochromes are not synthesized, and the organisms grow by fermenting sugars. Despite the absence of spores, staphylococci are relatively resistant to drying (as are most other Gram-positive cocci) and hence can be readily dispersed in dust particles through the air. They are able to grow on media containing a high concentration (7.5 percent) of sodium chloride, and this property is exploited in their selective isolation from natural materials since very few other bacteria from nonsaline environments show such tolerance. In man, two distinct forms are recognized, *S. albus* (or

S. epidermidis), a nonpigmented, nonpathogenic form that is usually found on the skin or mucous membranes, and *S. aureus,* a yellow-pigmented form that is most commonly associated with pathological conditions, including boils, pimples, pneumonia, osteomyelitis, carditis, meningitis, and arthritis. Those strains of *S. aureus* most frequently causing human disease produce a number of extracellular enzymes or toxins. At least four different hemolysins have been recognized, a single strain often producing more than one, and the production of these is responsible for the β hemolysis seen around colonies on blood agar plates. Another substance produced is *coagulase,* an enzymelike factor that causes fibrin to coagulate and form a clot (Section 15.2). The production of coagulase is generally associated with pathogenicity. It seems likely that blood clotting induced by coagulase results in the accumulation around the cocci of fibrin and thus renders them resistant to phagocytosis. In addition, the formation of such fibrin clots results in the walling off of the bacteria, making them less likely to come into contact with host defense agents. Strains are tested for coagulase production by mixing 0.5 ml of a dense suspension of bacteria with 5 ml of human citrated plasma and incubating at 37°C. A coagulase-positive strain will usually cause clotting after 3 hrs or after overnight incubation. Most *S. aureus* strains also produce *leukocidin,* a complex factor that causes the destruction of leukocytes; if an *S. aureus* cell is phagocytized, release of leukocidin can lead to the destruction of the leukocyte, and the *S. aureus* cell is enabled to escape unharmed. Production of leukocidin in skin lesions such as boils and pimples results in much cell destruction and is one of the factors responsible for pus formation. Other extracellular factors produced by some strains include proteolytic enzymes, hyaluronidase, fibrinolysin, lipase, ribonuclease, and deoxyribonuclease.

The most common habitat of *S. aureus* is the upper respiratory tract, especially the nasopharyngeal passages, and many people are carriers throughout most of their life; however, the strain being carried probably varies from time to time. Most infants become infected with *Staphylococcus* during the first week of life and usually acquire the strain associated with the mother or with another close human contact. In most cases these strains do not cause pathological conditions, and serious staphylococcal infections occur only when the resistance of the host is low because of hormonal changes, debilitating illness, or treatment with steroids or other anti-inflammatory drugs. Hospital epidemics have occurred in recent years, which have usually involved antibiotic-resistant strains. As we have emphasized elsewhere (see Chapter 9), extensive use of antibiotics has often resulted in the selection of resistant strains, and for some reason *S. aureus* has developed these strains more readily than have most other bacteria. One reason for this may be the high incidence of this bacterium so that antibiotic treatment for other infections has at the same time provided conditions in the nasopharynx selective for antibiotic resistance. Penicillin resistance in *S. aureus* is associated with the production of the enzyme penicillinase. Hospital epidemics with antibiotic-resistant staphylococci often result when patients whose resistance to infection is lowered (owing to other diseases, surgical procedures, or drug therapy) become infected from hospital personnel who have become normal carriers of antibiotic-resistant staphylococci. Control of such hospital epidemics requires careful attention to the usual rules of aseptic technique and the elimination of personnel known to be carriers of drug-resistant strains. In recent years, bacteriophage typing of *Staphylococcus* cultures has been widely used. Various bacteriophages have been isolated that show restricted host specificities, and a series of bacteriophages of different specificities can be used to type new isolates. Bacteriophage typing of *Staphylococcus* has been of considerable use in the study of hospital infections with this organism, and it has permitted the demonstration that frequently the patient acquires the initial infection from a nurse, physician, or other hospital employee who is a healthy carrier.

Some strains of *S. aureus* cause an acute gastroenteritis under special conditions. These strains usually grow only to a limited extent in the intestine and are prevented from extensive proliferation due to their inability to compete with other intestinal organisms. When these other organisms are eliminated by oral antibiotic therapy, antibiotic-resistant *S. aureus* can multiply, causing enteritis, which is occasionally serious or even fatal.

The most common type of *food poisoning* is that due to enterotoxin-producing strains of *S. aureus*. The enterotoxin acts upon the intestinal epithelium, causing severe diarrhea and vomiting, but its action is usually not fatal. Enterotoxin-producing strains of *S. aureus* live in the upper respiratory tract of many individuals, causing them no apparent harm. Staphylococcal food poisoning usually involves foods such as chicken salad, ham salad, cream desserts, and so on, all of which are excellent habitats for the growth of staphylococci. In addition, they are often kept nonrefrigerated after preparation. Enterotoxin-producing strains of staphylococci are usually inoculated into the food by a cook or food handler who is a healthy carrier of the organism. If the food is then kept for a number of hours in a warm place, the staphylococcus grows and produces enterotoxin. When the food is eaten, food poisoning results. Usually such foods show no obvious signs of spoilage. Since the organism does not grow in the intestine, the disease is self-limiting and recovery is usually complete in 24 to 48 hr. Food poisoning of this type is most common in the summer months, when warm temperatures prevail and the organism can grow rapidly; often it occurs in outbreaks, when a number of persons attending a banquet, party, or family reunion all eat the same infected food.

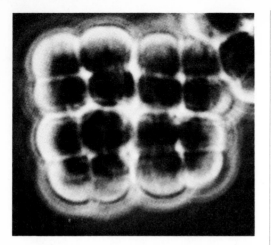

Figure 18.3 Sarcina maxima, a large Gram-positive coccus that forms regular packets of cells. Magnification, 3,800×. [From S. C. Holt and E. Canale-Parola: J. Bacteriol. 93:399 (1967).]

The Gram-positive cocci of the genera *Micrococcus, Sarcina,* and *Gaffkya* are all common soil organisms that occur frequently as airborne contaminants of various products. They are usually pigmented; and as we noted in Section 8.7, such pigmentation is related to the ability of these organisms to be dispersed through the air. Many of these cocci form packets of cells during growth, often of considerable regularity (Figure 18.3).

Gram-negative cocci   The Gram-negative cocci comprise a diverse group of organisms, some of which are common soil and water organisms and others of which are pathogenic to man. Although at one time most aerobic Gram-negative cocci were classified in a single genus, *Neisseria,* recent studies in molecular taxonomy have shown that there is in fact very little relation among many of these organisms. Those Gram-negative cocci which live in soil and water are nutritionally extremely versatile, grow well at low temperatures, and have few or any growth-factor requirements; they are currently classified in the genus *Acinetobacter* (another name sometimes used is *Moraxella).* The Gram-negative cocci that live in man are nutritionally highly specialized, grow well only near body temperature, and are extremely sensitive to inhibitory materials; these organisms are retained within the genus *Neisseria.* There are two species that are pathogenic for man, *Neisseria gonorrhoeae,* the causal agent of gonorrhea, and *N. meningitidis,* the most frequent causal agent of spinal meningitis. These two organisms are often given the colloquial names of gonococcus and meningococcus, respectively.

Gonorrhea is a venereal disease that apparently occurs only in human beings; despite numerous attempts, experimental animals have not been successfully infected. It is one of the most widespread human diseases, and in spite of the availability of excellent therapeutic drugs, it still is a common disease even in countries where the cost of drugs is no economic problem. As opposed to the other important venereal disease of man, syphilis, gonococcus infection rarely results in serious complications or death. The disease symptoms are quite different in the male and female: In the female the symptoms are usually a mild vaginitis that is difficult to distinguish from vaginal infections caused by other organisms, and the infection may easily go unnoticed; in the male, however, the organism causes a painful infection of the urethral canal, and the disease is often given the colloquial name of "strain" or "clap." In addition to gonorrhea, the organism also causes eye infections in the newborn and adult.

N. gonorrhoeae is an extremely fastidious organism and is quite difficult to culture on initial isolation from pathological material. Many components of ordinary culture media are inhibitory, of which fatty acids and trace metals are probably the most important. Starch, serum, or heated whole blood, when added to culture media, adsorbs these toxic materials and makes growth possible. On initial isolation most strains require an atmosphere containing 2 to 10 percent CO_2, and their temperature limits are very narrow—growth does not occur below 30°C or above 38.5°C. The organism is killed quite rapidly by drying, sunlight, and UV light, and this extreme sensitivity probably explains in part the venereal nature of the disease, the organism being transmitted only from person to person by intimate direct contact. The toxicity of the organism is probably due exclusively to an endotoxin, and no extracellular products significant in pathogenicity seem to be produced. Invasiveness is determined primarily by a polysaccharide surface antigen (K antigen), which inhibits phagocytosis; gonococcus infection is thus strictly extracellular.

The organism enters the body by way of the mucous membranes of the genitourinary tract, customarily being transmitted from a member of the opposite sex. Treatment of the infection with penicillin is highly successful, a single injection usually resulting in elimination of the organism and complete cure. Strains moderately resistant to penicillin have arisen but have not presented any special problem since even resistant strains can be controlled with higher doses and repeated injections of the antibiotic. Penicillin therapy is generally more rapidly effective in the male than in the female. Despite the ease with which gonorrhea can be cured with penicillin, the incidence of gonococcus infection remains relatively high. The reasons for this are twofold: (1) acquired immunity does not exist so that repeated reinfection is possible; (2) the symptoms in the female are such that even at the height of the disease it may go unrecognized, and a promiscuous infected female can serve as a reservoir for the infection of many males. The disease could be controlled if the sexual contacts of infected persons could be identified and treated, but it is often difficult to obtain the appropriate information and even more difficult to arrange treatment. The incidence of gonorrhea correlates closely with the promiscuity of the society, and despite notable advances in drug therapy, gonorrhea remains a social rather than a medical problem.

Neisseria meningitidis is also a parasite only of human beings, but its natural habitat is the nasopharynx rather than the genitourinary tract. The organism possesses much the same fastidious nature as *N. gonorrhoeae,* and similar culture conditions are used in its isolation. Meningococcus is also sensitive to drying, heat, and other adverse environmental conditions, but it is sufficiently resistant to provide the possibility for transmissal from person to person via the respiratory tract, although relatively close contact is required. The carrier rate of meningococcus in the upper respiratory tract is high, most carriers having infections with no noticeable symptoms. In certain instances the organism invades the bloodstream from the nasopharynx and sets up a generalized infection of the body, and in some of these cases the organism invades the central nervous system and becomes established in the meninges (the membranes surrounding the brain and spinal cord), leading to symptoms of meningitis: severe headache, muscular spasm, stiff neck, and exaggerated reflexes; if untreated, these are followed by convulsions, coma, and death. When these symptoms develop, the organism can usually be cultivated from samples of the cerebrospinal fluid. Pathogenicity results from the presence of an antiphagocytic capsule and the production of a characteristic endotoxin. The organism is extremely sensitive to sulfonamides, penicillin, and most other antibiotics, and drug-resistant strains have rarely developed. Successful treatment of meningitis requires the use of a drug that will penetrate the meningeal membrane, and the sulfonamides are drugs of choice because they possess this property. However, penicillin can also be used; although it does not normally penetrate to the spinal fluid in healthy persons, it does so when the meninges are acutely inflamed.

Because of the high incidence of inapparent infections of the nasopharynx, it is virtually impossible to control the spread of *N. meningitidis* through the human population. However, the clinical disease meningitis occurs only rarely, although it seems to erupt in cycles of 8 to 10 yr, and it is also fairly common in military camps and barracks, especially in new recruits. It is thought that overcrowding in such situations frequently results in transmission of the organism from carriers to susceptible individuals, and the reduced resistance consequent to fatigue and other factors attendant on the training process may result in onset of the systemic and nervous-system infections. The frequent presence of antibodies bactericidal to *N. meningitidis* in the serum of carriers suggests that these antibodies may play a role in preventing the spread of the organism from the nasopharynx to the bloodstream and nervous system.

18.4   Spore-forming bacteria

We have emphasized in a number of places in this book the uniqueness of the bacterial endospore, especially its resistance to heat and other deleterious agents. Two main genera of endospore-forming bacteria are recognized, *Bacillus,* the species of which are aerobic or facultatively anaerobic, and *Clostridium,* which contains the strictly anaerobic species. (Heat-resistant spores are also produced by a few Gram-positive cocci and certain thermophilic actinomycetes, but we shall not discuss these forms.) The genera *Bacillus* and *Clostridium* consist of Gram-positive or Gram-variable rods, which are usually motile, possessing peritrichous flagella. Members of *Bacillus* produce the enzyme catalase, which converts hydrogen peroxide (H_2O_2) into H_2O and O_2. Clostridia do not produce this enzyme, and it is thought that one reason they are obligately anaerobic is that they have no way of getting rid of the toxic H_2O_2 produced. Recent molecular taxonomic studies of the genus *Bacillus* (Section 17.7) have shown that it is extremely heterogeneous and can hardly be considered an assemblage of closely related organisms. The DNA base compositions of sundry *Bacillus* species vary from about 30 to 50 percent GC (see Figure 17.8), and studies on nucleic acid homologies by hybridization and genetic transformation also suggest considerable genetic heterogeneity. Although less molecular work has been done with members of the genus *Clostridium,* data from DNA base compositions suggest less genetic heterogeneity in this group; values from 26 to 36 percent GC have been reported. Because of the complex series of enzymatic steps involved in sporulation (Section 7.7), it seems reasonable to hypothesize that the ability to form endospores arose only once during evolution, and that a primitive spore former was, by evolutionary divergence, the forerunner of the variety of spore-forming bacteria known today.

Although not closely related genetically, all the spore-forming bacteria are ecologically related since they are found in nature primarily in the soil. Even those species that are pathogenic to man or animals are primarily saprophytic soil organisms, infecting hosts only incidentally. Possible exceptions to this are *Bacillus anthracis,* the causal agent of anthrax, which is further discussed below, and the *Bacillus* species pathogenic to insects.

Spore formation is advantageous for a soil microorganism because the soil is a highly variable environment. Although at some times nutrient supply is in excess, at other times it is deficient. Soil temperatures can be quite high in summer, especially at the surface. Thus a heat-resistant dormant structure should offer considerable survival value in nature. On the other hand, the ability to germinate and grow quickly, when nutrient becomes available, is also of value as it enables the organism to capitalize on a transitory food supply. Under conditions of limited nutrient, spores may germinate and form vegetative cells and then without cell division convert back into spores.

The longevity of bacterial endospores is noteworthy. Records of spores surviving in the dormant state for over 50 yr are well established, and there are reports of a few cases, which are of course difficult to confirm, of spores surviving for thousands of years. This longevity inspired the following suggestion of how life arose on earth, an hypothesis called *panspermia:* Life may first have come to earth from some distant planet as bacterial spores, which drifted through outer space for millions of years. Panspermia is not a widely accepted hypothesis today because the radiation in outer space is considered too intense for the survival even of bacterial spores.

Bacillus   Members of the genus *Bacillus* are easy to isolate from soil or air and are among the most common organisms to appear when soil samples are streaked on agar plates containing various nutrient media. Spore formers can be selectively isolated from soil, food, or other material by exposing the sample to 80°C for 10 to 30 min, a treatment that effectively destroys vegetative cells while many of the spores present remain viable. When such pasteurized samples are streaked on plates and incubated aerobically, the colonies that develop are almost exclusively of the genus *Bacillus*. Bacilli usually grow well on synthetic media containing sugars, organic acids, alcohols, and so on, as sole carbon sources and ammonium as the sole nitrogen source; some isolates have vitamin requirements. Many bacilli produce extracellular hydrolytic enzymes that break down polysaccharides, nucleic acids, and lipids, permitting the organisms to use these products as carbon and energy sources. Many bacilli produce antibiotics, of which bacitracin, polymyxin, tyrocidin, gramicidin, and circulin are examples. In most cases antibiotic production seems to be related to the sporulation process, the antibiotic being released when the culture enters the stationary phase of growth and after it is committed to sporulation (see Fig. 7.18). An outline of the subdivision of the genus *Bacillus* is given in Table 18.6.

The only human disease caused by a *Bacillus* is anthrax, from *B. an-*

Table 18.6   Subdivision of the genus Bacillus and representative species

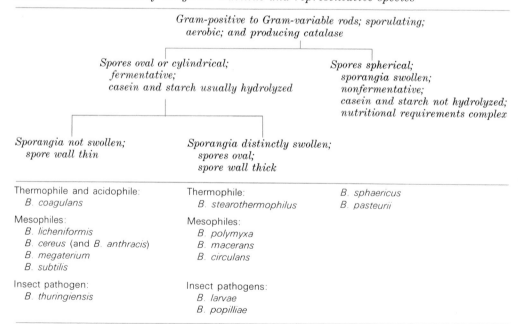

Gram-positive to Gram-variable rods; sporulating; aerobic; and producing catalase

Spores oval or cylindrical; fermentative; casein and starch usually hydrolyzed

Spores spherical; sporangia swollen; nonfermentative; casein and starch not hydrolyzed; nutritional requirements complex

Sporangia not swollen; spore wall thin

Sporangia distinctly swollen; spores oval; spore wall thick

Thermophile and acidophile: B. coagulans	Thermophile: B. stearothermophilus	B. sphaericus B. pasteurii
Mesophiles: B. licheniformis B. cereus (and B. anthracis) B. megaterium B. subtilis	Mesophiles: B. polymyxa B. macerans B. circulans	
Insect pathogen: B. thuringiensis	Insect pathogens: B. larvae B. popilliae	

Figure 18.4 Characteristic appearance of the edge of a Bacillus anthracis colony.

thracis. This organism is sometimes recognized only as a variety of *B. cereus.* Anthrax was the first disease shown conclusively to be caused by a bacterium, and its study by Koch (see Chapter 1) provided one of the foundations for the development of microbiology as a science. It is primarily a disease of farm animals and only occasionally is transmitted to human beings. *Bacillus anthracis* is a very large bacillus, 1 to 1.5 μm wide by 4 to 8 μm long, and the cells usually remain attached after division; long chains are so formed [Figure 1.6(*a*)] on agar to create colonies of characteristic morphology (Figure 18.4). Strains of *B. anthracis* form a chemically unique capsule, a polypeptide of D-glutamic acid that protects the organism from phagocytosis. Probably because of its capsule, *B. anthracis* is highly invasive, growing well throughout the body; in later stages of the disease large numbers of bacilli are found in the blood. The organism produces in the body an exotoxin that is responsible for most of the disease symptoms. This same toxin is produced by cultures if large amounts of bicarbonate are in the culture medium; the bicarbonate apparently promotes the release of the toxin from the cells into the medium. Anthrax toxin is a heat-labile protein complex that can be separated into two fractions, neither of which is very toxic alone, so that both fractions are needed for toxicity. The toxin acts by causing a physiologic shock syndrome, with profound electrolyte imbalance, edema, hemoconcentration, and acute kidney failure.

The organism infects human beings only when they come in contact with diseased animals or their hides or animal products. Many kinds of wild and domestic animals are susceptible, including cattle, sheep, horses, goats, pigs, mink, dogs, deer, birds, and even frogs and fish, although not all animals are equally susceptible. An animal usually becomes infected by ingesting bacilli or spores from an infected carcass. At death large numbers of bacilli are present in the blood and tissues, and if the carcass is opened, sporulation can occur (frequently in warm countries, where high environmental temperatures favor this process). If sporulation has occurred, the organism can remain viable in bone for long periods of time and is transmitted to uninfected animals when they eat the bones, animals usually being attracted to a carcass because of their craving for salt or their need for bone; the latter especially is observed among animals living on forage in phosphate-deficient areas. In dry soil the anthrax spores can also remain alive for very long periods of time, but in most agricultural soils the spores do not persist for more than a few years. The organism is not able to grow saprophytically in nature.

Infection in man is primarily an occupational hazard in the meat-packing and tanning industries. Invasion most often occurs by way of the skin, usually through a small scratch or abrasion, and the primary lesion develops as an inflamed pustule or blister. In most cases the disease in man

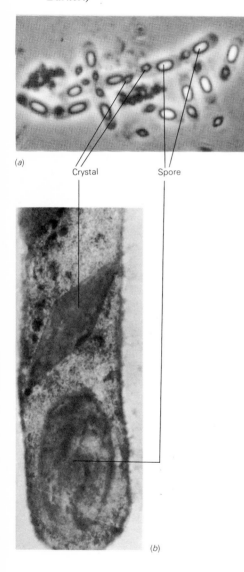

Figure 18.5 (a) Phase-contrast photomicrograph of Bacillus thuringiensis sporangia, showing the position of the spore and the toxic crystalline parasporal body. (Courtesy of John N. Aronson and Faith Moquin.) (b) Electron micrograph of a thin section of a sporangium of B. thuringiensis. (Courtesy of John N. Aronson and Pat Bowe Barker.)

(a)

Crystal   Spore

(b)

is self-limited, but the organism may spread from the initial site of infection, and initiate a fatal systemic infection. Another form of anthrax in man, called "woolsorter's disease," is a respiratory infection resulting from the inhalation of spores.

Immunity to anthrax develops through production of antibodies against the toxin. The first attenuated vaccine ever produced was the anthrax vaccine of Pasteur and his colleagues. A famous and successful trial vaccination of sheep, carried out publically by Pasteur, set the stage for the extensive vaccine research that ultimately produced live vaccines for a variety of infectious diseases. Despite Pasteur's early success, subsequent work has shown that proper attenuation is difficult, and even today no anthrax vaccine is known that is considered safe for man. Immunization of animals has also been unsatisfactory. Control of the disease in animals is primarily by eradication measures aimed to eliminate diseased animals from herds and to destroy carcasses. The infection of man by means of contaminated shaving brushes made from animal hair has been controlled by sterilizing the brushes. The human disease can be effectively treated with penicillin, the tetracyclines or erythromycin, provided the symptoms are detected before bacteremia has developed.

A number of bacilli are *insect pathogens,* and in recent years there has been considerable interest in these organisms because of their potential use in the biological control of insect infestations of plants. These insect pathogens form a crystalline protein during sporulation, called the "parasporal body," which becomes deposited within the sporangium but outside the spore. These crystal-forming bacilli cause fatal diseases of moth larvae of such forms as the silkworm, the cabbage worm, the tent caterpillar, and the gypsy moth.

The most widely studied insect pathogen is *B. thuringiensis.* The toxic crystal of *B. thuringiensis,* which is formed only when sporulation begins, is a regular diamond-shaped structure (Figure 18.5). It is composed of pure protein and its biological activity is destroyed by heat. Only 0.5 μg of crystalline material is able to cause paralysis of a silkworm larva. The toxin is active against over 100 species of moths. It is insoluble in water but dissolves in alkaline solutions. The midgut pH of susceptible moth larvae is usually quite high, 9.0 to 10.5, so that, when the toxin reaches the intestinal tract, it dissolves and becomes adsorbed to the intestinal epithelium, where it induces loss of permeability. Fluids from the highly alkaline intestine pass into the blood and cause its pH to rise from around 6.8 to greater than 8. This increase in alkalinity of the blood leads to a generalized paralysis of the insect, followed by death (Figure 18.6). Moth larvae that do not have a highly alkaline intestine are not susceptible because the toxin passes out of the intestine without dissolving.

Bacillus thuringiensis toxin is now being used as an insecticide. The

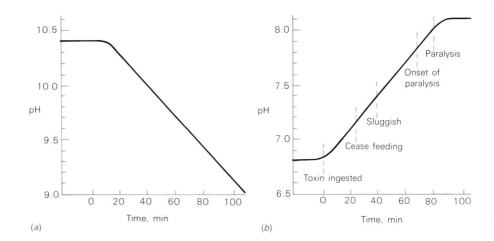

(a)

Time, min

(b)

Time, min

Figure 18.6   Changes in pH of the (a) midgut (intestine) and (b) hemolymph (blood equivalent) of a silkworm larva after ingesting crystals of B. thuringiensis toxin. The toxin dissolves in the intestine and causes damage to the intestinal wall, increasing its permeability. The highly alkaline fluids of the intestine flow into the hemolymph and raise its pH, which induces toxic symptoms in the larva, leading ultimately to death. [From A. M. Heimpel and T. A. Angus: Bacteriol. Rev. 24:266 (1960).]

organism is grown in large fermentation vessels and a concentrate containing spores and crystals is obtained and formulated as a dust or spray, for application to diseased plants. There is no danger to the health of human beings or livestock because of the high specificity of the toxin, and the only limitation is that it cannot be used in areas where silkworms are raised. More than 10 different insect pests are currently being controlled with these products. We have in the use of toxin-forming bacilli an entirely new approach to insect control, which may have a great future.

Clostridium   The clostridia are all obligate anaerobes, not only being unable to grow in the presence of air but also usually being killed by O_2 unless they exist in the spore form. In general, however, the clostridia are not so sensitive to O_2 as are the methane-forming bacteria and other strict anaerobes of the rumen, intestinal tract, and anaerobic sewage plant so that extreme precautions with anaerobic culture conditions (Section 8.6) are often not necessary with clostridia. Exclusion of air by use of N_2 or sealed tubes, addition of thioglycollate or other sulfhydryl compounds that maintain a low redox potential, and boiling culture media before inoculation to drive off O_2 are usually the only precautions necessary to obtain successful cultures of most clostridia. For many clostridia adequate anaerobiosis can be obtained by the use of jars into which H_2 or illuminating gas is introduced and burned electrically in order to consume the O_2 present.

The clostridia lack a cytochrome system and a mechanism for oxidative phosphorylation, and hence they obtain ATP only by substrate-level phosphorylation. A wide variety of anaerobic energy-yielding mechanisms are known in the clostridia; indeed, the separation of the genus into subgroups is based primarily on these properties and on the nature of the

Table 18.7 *Principal physiological subgroups among the anaerobic spore formers (genus Clostridium)[a]*

Subgroup	Nature of energy-yielding metabolism	Other characters	Typical species
Butyric acid bacteria	Fermentation of sugars, starch, and pectin to yield acetic and butyric acids, CO_2, and H_2; sometimes acetone, isopropanol, and butyl alcohol also formed	Many can fix nitrogen	*C. butyricum, C. perfringens, C. pasteurianum, C. acetobutylicum*
Cellulose fermenters	Cellulose fermented with formation of acetic and succinic acids, ethyl alcohol, CO_2, and H_2		*C. cellulosae-dissolvens*
Amino acid fermenters	Fermentation of pairs of amino acids, or of single amino acids with formation of fatty acids, NH_3, CO_2, and sometimes H_2	Many are actively proteolytic; some can also perform a butyric acid fermentation of sugars; many produce exotoxins and are pathogenic	*C. sporogenes, C. tetani, C. botulinum, C. histolyticum*
Purine fermenters	Uric acid and other purines fermented with formation of acetic acid, CO_2, NH_3, and other products	Cannot attack sugars or amino acids	*C. acidi-urici*
Clostridium kluyveri	Fermentation of mixture of acetic acid and ethyl alcohol, with formation of higher fatty acids	Cannot attack sugars, amino acids, or purines	(Only species)

[a] From R. Y. Stanier, M. Doudoroff, and E. A. Adelberg: *The Microbial World*, 2nd ed. (Englewood Cliffs, N.J.: Prentice-Hall, Inc., 1963).

energy sources used (Table 18.7). A number of clostridia ferment sugars, producing as a major end product butyric acid. Some of these also produce acetone and butanol, and at one time acetone-butanol fermentation by clostridia was of great industrial importance as it was the main commercial source of these products. Today, however, the chemical synthesis of acetone and butanol from petroleum products has mostly replaced the microbiological process. Some clostridia of the acetone-butanol type fix N_2; these strains are usually classified in the species *C. pasteurianum* and probably are the organisms responsible for most anaerobic nitrogen fixation in the soil. One group of clostridia ferments cellulose with the formation of acids and alcohols, and these are the main organisms decomposing cellulose anaerobically in soil. Sugar- and cellulose-fermenting clostridia produce ATP through the glycolytic pathway (see Chapter 4).

Some clostridia that hydrolyze pectin play a key role in *retting*, a process used in the preparation of a variety of fibers from natural plant material (such as in the production of linen from flax). Flax fibers are cemented together in the stem of the plant by pectin, a polymer of galacturonic acid. Through clostridial action, the pectin is hydrolyzed, making possible the separation of the fibers from the plant stem and their subsequent processing into a fabric. The plant material is tied in bundles that are immersed in water, the plant material becomes anaerobic, and butyric-acid producing clostridia develop. The process takes about 10 to 15 da. The bacterial inoculum is rather ill defined and derives from flora normally present on the plant or in the water. Not all retting is carried out through clostridial action; the process described above is the one used in the production of Irish linen.

Another group of clostridia obtain their energy by fermenting amino acids. Some strains do not ferment single amino acids, but when two amino acids are present in the medium, one functions as the electron donor and is oxidized, while the other acts as the electron acceptor and is reduced. For instance, *C. sporogenes* will attack a mixture of glycine and alanine. The alanine is first deaminated to pyruvate:

$$H_3C—\underset{\underset{NH_2}{|}}{CH}—COOH + H_2O + NAD \longrightarrow H_3C—CO—COOH + NH_3 + NADH$$

The pyruvate is converted to acetyl phosphate and CO_2:

$$H_3C—COCOOH + PO_4^{3-} \longrightarrow H_3C—CO—PO_3H + CO_2$$

Acetyl phosphate is then converted to acetate and ATP by the enzyme phosphotransacetylase:

$$H_3C—CO—PO_3H + ADP \longrightarrow CH_3COOH + ATP$$

so that 1 mole of ATP is formed for each mole of alanine fermented. The reduced NADH formed in the first reaction is then reoxidized by a reaction involving glycine:

$$\underset{\underset{NH_2}{|}}{CH_2}COOH + NADH \longrightarrow CH_3COOH + NH_3 + NAD$$

The end products of the two series of reactions are thus NH_3, CO_2, and acetic acid. Some of the more complicated amino acids such as glutamic acid can be fermented singly, with some of the molecules being oxidized and others reduced. It is usually found that each group of clostridia is specific in the kinds of substances it can ferment, although there are strains that can ferment both sugars and amino acids.

The main habitat of clostridia is the soil, where they live primarily in anaerobic "pockets," of which the anaerobiosis is probably created primarily by facultative organisms acting upon various organic compounds present. As was discussed in Chapter 15, several clostridia that live primarily in soil are capable of causing disease in man under specialized conditions. These are *C. botulinum,* causing botulism, *C. tetani,* causing tetanus, and *C. perfringens,* and a number of other clostridia, both sugar and amino acid fermenters, causing gas gangrene. These pathogenic clostridia seem in no way unusual metabolically, but are distinct in that they produce specific toxins or, in those causing gas gangrene, a group of toxins (Table 15.1). An unsolved ecological problem is what role these toxins play in the natural habitat of the organism. Many gas-gangrene producers also cause diseases in domestic animals, especially sheep and cattle. Tetanus also occurs in domestic animals, and botulism occurs in sheep and ducks.

Botulism is a food poisoning rather than an infection. In food poisoning the microorganism grows not in the host but in the food itself, producing a toxin, and it is this ingested toxin that causes the disease symptoms. Botulism, which is fortunately rare, is the most serious type of food poisoning known. The causal organism, *C. botulinum*, lives normally in soil or water, and its spores often contaminate raw foods. Since the organism is an anaerobe, it can grow only in anaerobic foods, such as canned materials or smoked meats. If the food is properly sterilized, no problem arises, but if viable spores of *C. botulinum* are present, they may germinate and grow in the anaerobic environment, producing toxin that may later be ingested. Usually canned food contaminated by *C. botulinum* appears obviously spoiled and would be discarded; if well cooked, the toxin may be destroyed by the heating process. Thus only rarely does botulism arise from canned foods. A more serious source of trouble in recent years has been smoked meat, especially fish. The fish acquire *C. botulinum* spores from the habitat in which they live. Smoking does not destroy the spores, and smoked fish is often kept for long periods of time unrefrigerated; moreover, it is often eaten without further cooking.

18.5 Other Gram-positive organisms

The genus *Corynebacterium* comprises a group of aerobic, nonmotile, nonsporulating Gram-positive rods. They can be differentiated from the lactobacilli (which they frequently resemble structurally) by the fact that they possess a cytochrome system and are obligate aerobes. The *Corynebacterium* cell is frequently pleomorphic and often has swollen ends so that the rod has a club-shaped appearance—hence the origin of the name (*koryne* is the Greek word for "club"). Members of the genus are widespread: Some are common saprophytes in the soil, others are causal agents of plant diseases, and still others are organisms commensal with, or pathogenic to, man and higher animals. One species, *C. acnes*, is a common inhabitant of the human skin and has been implicated in the skin condition acne.

The best known and most widely studied species is *C. diphtheriae*, the causal agent of diphtheria. We have discussed certain aspects of diph-

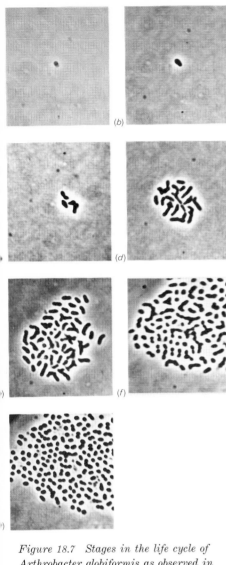

Figure 18.7 Stages in the life cycle of Arthrobacter globiformis as observed in slide culture: (a) single coccoid element; (b–e) conversion to rod and growth of microcolony consisting predominantly of rods; (f–g) conversion of rods to coccoid forms. Magnification, 1,480×. [From H. Veldkamp, G. van den Berg, and L. P. T. M. Zevenhuizen: Antonie van Leeuwenhoek 29:35 (1963).]

theria in Chapter 15, especially the relationship of diphtheria toxin to the disease. At one time diphtheria was a major cause of death in children, but today it is quite rare. Diphtheria is of historical significance because it was the first infectious disease whose symptoms were shown to be due to an exotoxin, and it was the first disease to be controlled by immunization procedures. The organism is strictly an inhabitant of the respiratory tract, being unable to invade other parts of the body. It is transmitted from person to person by the respiratory route. In nature it is found only in man, although it can be transmitted experimentally to other vertebrates. Upon establishment in the upper respiratory tract, the organism multiplies on the mucous membranes and produces the toxin that causes necrosis of adjacent cells and creates a favorable microenvironment for further replication. The inflammatory response of the throat tissues to infection results in the formation of a characteristic pseudomembrane consisting of altered tissue cells and bacteria. This pseudomembrane can result in mechanical blockage of the throat, leading to death by suffocation. In most cases, however, death results from toxemia arising from the spread of the toxin throughout the body. Thus, even though *C. diphtheriae* is weakly invasive, its powerful toxin makes it highly virulent.

Administration of formalin-treated toxin (toxoid) results in the formation of antitoxin antibodies. These antibodies completely neutralize the action of the toxin but do not prevent the establishment of the organism in the upper respiratory tract. Inapparent infections are therefore common, and carriers can be the source of infection of unimmunized individuals. Occasional small epidemics of diphtheria can arise owing to the migration of unimmunized individuals into an area where the population is predominantly immunized but carries inapparent infections. Diphtheria therapy is best effected by the use of antitoxin since this will neutralize any circulating toxin. Antibiotic therapy can also be used but never alone since the antibiotic will not affect toxin already circulating in the body.

Other Gram-positive rods occasionally associated with infectious disease in animals and man are *Listeria monocytogenes,* the causal agent of *listeriosis,* a glandular fever resulting in swollen lymph nodes, and *Erysipelothrix rhusiopathiae,* the causal agent of swine *erysipelas,* a widespread and serious disease of pigs that is occasionally transmitted to man.

An interesting Gram-positive organism widespread in soil and water is *Arthrobacter,* an organism that shows a morphological variation from coccoid to rodlike, depending upon culture medium and growth phase. When a coccoid *Arthrobacter* cell is placed on an agar medium fairly rich in organic matter, it enlarges and forms a rod that divides to form a colony of rather pleomorphic rods. These rods, upon reaching the stationary phase, revert to coccoid forms (Figure 18.7). When grown on media low in organic matter, only coccoid forms are seen in the colonies. The conversion from a rodlike to a coccoid form may be related to a change in the degree of cross-linking of the cell-wall peptidoglycan. The coccoid forms are resistant to drying and presumably tide the organism over in times of desiccation. Arthrobacters are a heterogeneous group that have considerable nutritional versatility, and strains have been isolated that decompose herbicides and other unusual organic compounds.

18.6   Actinomycetes

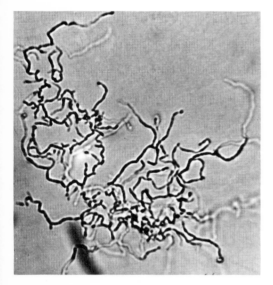

Figure 18.8 A young colony of an actinomycete, Nocardia corallina, showing typical filamentous cellular structure. Magnification, 890×. (Courtesy of Hubert and Mary P. Lechevalier.)

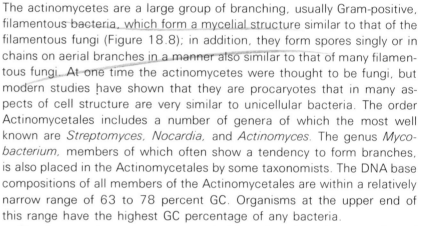

The actinomycetes are a large group of branching, usually Gram-positive, filamentous bacteria, which form a mycelial structure similar to that of the filamentous fungi (Figure 18.8); in addition, they form spores singly or in chains on aerial branches in a manner also similar to that of many filamentous fungi. At one time the actinomycetes were thought to be fungi, but modern studies have shown that they are procaryotes that in many aspects of cell structure are very similar to unicellular bacteria. The order Actinomycetales includes a number of genera of which the most well known are *Streptomyces, Nocardia,* and *Actinomyces.* The genus *Mycobacterium,* members of which often show a tendency to form branches, is also placed in the Actinomycetales by some taxonomists. The DNA base compositions of all members of the Actinomycetales are within a relatively narrow range of 63 to 78 percent GC. Organisms at the upper end of this range have the highest GC percentage of any bacteria.

Streptomyces is a genus represented in nature by a large number of species and varieties. *Streptomyces* filaments (sometimes called *hyphae* in analogy with the fungi) are usually 0.5 to 1.0 μm in diameter and of indefinite length, are multinucleate, and often lack cross walls in the vegetative phase. Growth occurs at the tips of the filaments and is often accompanied by branching so that the vegetative phase consists of a complex, tightly woven matrix, which, because of its funguslike nature, is called a *mycelium* (Section 19.6). Growth of a *Streptomyces* colony on an agar surface results in a compact convoluted colony. As the colony ages, characteristic aerial filaments (sporophores) are formed, which project above the surface of the colony and give rise to spores (Figure 18.9). *Streptomyces* spores, usually called *conidia,* are not related in any way to the endospores of *Bacillus* and *Clostridium* since the streptomycete spores are produced by the formation of cross walls in the multinucleate aerial filaments followed by conversion of the individual cells directly into spores (Figure 18.10). The surface of the conidial wall often has convoluted projections, the nature of which is characteristic of each species (Figure

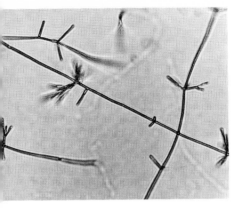

Figure 18.9 *Photomicrographs of several spore-bearing structures of streptomycetes: (a) a monoverticillate type (magnification, 400×) (courtesy of Peter Hirsch); (b) a spiral type 1,000×) (courtesy of Hubert and Mary P. Lechevalier).*

(b)

Figure 18.10 *Formation in Streptomyces virido-chromogenes of conidia from aerial hypha, as shown by electron microscopy: (a) early stage in spore-wall formation (magnification 86,000×); (b) late stage, the spines on the outside of the spores having already formed 80,000×). [From M. W. Rancourt and H. A. Lechevalier: Canadian Journal of Microbiol. 10:311 (1964). Reproduced by permission of the National Research Council of Canada.]*

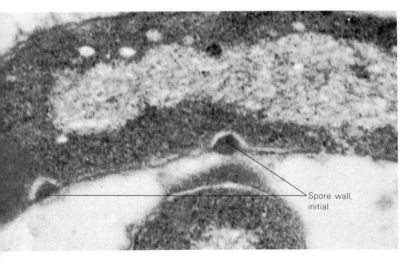

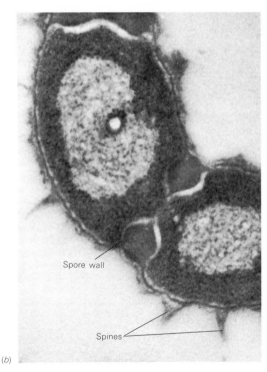

(b)

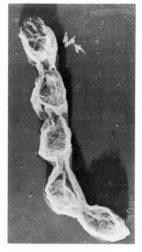

(a)

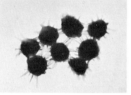
(b)

Figure 18.11 Surface configuration of streptomycete conidia: (a) Chain of spores of a streptomycete, Streptomyces hygroscopicus var. decoyicus. This is an electron micrograph of a carbon replica. Magnification, 11,700×. [From A. Dietz and J. Mathews: Appl. Microbiol. 10:258 (1962).] (b) Electron micrograph of an untreated preparation of spiny conidia of Streptomyces violaceus. Magnification, 6,000×. [From H. A. Lechevalier and A. S. Tikhonienko: Mikrobiologiya 29:43 (1960).]

18.11). Differences in shape and arrangement of aerial filaments and spore-bearing structures of various species are among the fundamental features used in separating the *Streptomyces* groups (Figure 18.12). The conidia and aerial filaments are often pigmented and contribute a characteristic color to the mature colony (Plate 9: Figure 18.13); in addition, pigments are produced by the substrate mycelium and contribute to the final color of the colony. The dusty appearance of the mature colony, its compact nature, and its color make the detection of *Streptomyces* colonies on agar plates relatively easy.

Although some streptomycetes can be found in fresh waters and a few inhabit the ocean, they are primarily soil organisms. In fact, the characteristic earthy odor of soil is due to the production of a streptomycete metabolite called *geosmin*. One species, *Streptomyces scabies*, is the plant pathogen that causes potato scab. Alkaline and neutral soils are more favorable for the development of *Streptomyces* than are acid soils. The isolation of large numbers of *Streptomyces* from soil is relatively easy: A suspension of soil in sterile water is diluted and spread on selective agar medium, and the plates are incubated at 28 to 30°C. A medium often selective for *Streptomyces* is one containing the usual inorganic salts to which starch, asparagine, or calcium malate is added as a carbon source and undigested casein as a nitrogen source. After incubation for 2 to 7 da the plates are examined for the presence of the characteristic *Streptomyces* colonies, and spores of interesting colonies can be streaked and pure cultures isolated. If one wishes to observe the structure of the aerial mycelium and the arrangement of the spores, slide cultures are desirable (see Figure 19.21); in these the organism is grown on a small block of agar with a cover slip. The sporulating structures project horizontally and thus are easily observed undisturbed under the microscope.

Nutritionally, the streptomycetes are quite versatile. Growth-factor requirements are rare and a wide variety of carbon sources, such as sugars, alcohols, organic acids, and amino acids, can be utilized. Most isolates produce extracellular enzymes that permit utilization of polysaccharides (starch, cellulose, hemicellulose), proteins, and fats, and some strains can use hydrocarbons, lignin, tannin, or rubber. A single isolate may be able to break down over 50 distinct carbon sources. Streptomycetes are strict aerobes, whose growth in liquid culture is usually markedly stimulated by forced aeration. Sporulation usually takes place not in liquid culture but only when the organism is growing on the surface of agar or another solid substrate; it can occur, however, when organisms form a pellicle on the surface of an unshaken liquid culture.

Some work on the genetics of *Streptomyces* has been carried out. The frequent occurrence of cell fusions during mycelial growth makes possible the mingling of nuclei from genetically distinct strains, resulting in the

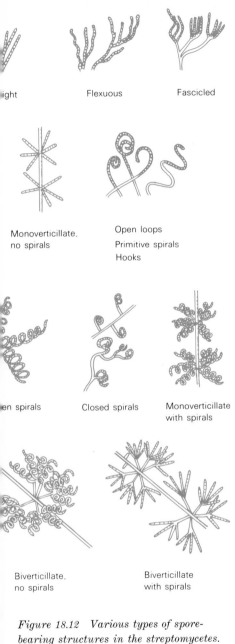

ight   Flexuous   Fascicled

Monoverticillate,
no spirals

Open loops
Primitive spirals
Hooks

en spirals   Closed spirals   Monoverticillate
with spirals

Biverticillate,
no spirals

Biverticillate
with spirals

Figure 18.12 Various types of spore-bearing structures in the streptomycetes. [From T. G. Pridham, C. W. Hesseltine, and R. G. Benedict: Appl. Microbiol. 6:52 (1958).]

formation of heterocaryons (Section 11.8). In certain species nuclear fusion and genetic interchange can occur in the heterocaryon. When a streptomycete isolate is maintained in vegetative growth in pure culture for many generations, genetic changes often take place, probably due to the segregation of nuclei into separate filaments. The formation of heterocaryons and the genetic consequences thereof are found in procaryotes only among the streptomycetes and related organisms; many natural isolates of streptomycetes are probably heterocaryotic.

Perhaps the most striking property of the streptomycetes is the extent to which they produce antibiotics. Evidence for antibiotic production is often seen on the agar plates used in the initial isolation of *Streptomyces*: Adjacent colonies of other bacteria show zones of inhibition (Figure 9.6). In some studies close to 50 percent of all *Streptomyces* isolated have proved to be antibiotic producers. Because of the great economic and medical importance of many streptomycete antibiotics, an enormous amount of work has been done on these producers. Over 500 distinct antibiotic substances have been identified as produced by streptomycetes, and a large number of these have been studied chemically (see Chapter 9). Some organisms produce more than one antibiotic, and often the several kinds produced by one organism are not even chemically related. The same antibiotic may be formed by different species found in widely scattered parts of the world. A change in nutrition of the organism may result in a change in the nature of the antibiotic produced. The organisms are usually resistant to their own antibiotics, but they may be sensitive to antibiotics produced by other streptomycetes.

More than 50 streptomycete antibiotics have found practical application in human and veterinary medicine, agriculture, and industry, including such important agents as streptomycin, chloramphenicol, novobiocin, nystatin, and the tetracyclines. (Penicillin, the first practical antibiotic agent, is produced not by a streptomycete but by a filamentous fungus.) The search for new streptomycete antibiotics continues since, as was discussed in Section 15.7, many infectious diseases are still not adequately controlled by existing antibiotics, and the development of antibiotic-resistant strains requires the continual discovery of new agents. Despite the extensive work on antibiotic-producing streptomycetes, the ecological relationships of these organisms to their natural habitats are poorly understood.

The genus *Nocardia* comprises a group of aerobic, branching, initially filamentous organisms that generally do not form conidia although the aerial mycelium may break up into rod-shaped or coccoid fragments. *Nocardia* isolates do not hydrolyze complex proteins or polysaccharides, but many utilize hydrocarbons for growth. Some species can grow autotrophically at the expense of H_2. Although many have been isolated from diseased human beings and animals, the disease syndromes these isolates may cause are not well understood.

The genus *Actinomyces* contains organisms that are morphologically similar to *Nocardia,* but these are anaerobic or microaerophilic. *Actinomyces bovis* is the causal agent of lumpy jaw of cattle and *A. israelii* is the causal agent of a similar disease in humans. A subacute or chronic infection of the face or trunk develops, and the organism grows preferentially in the connective tissue, causing an inflammatory reaction with the production of a hard, granulated swelling. The hard texture of the lesion is probably due to the precipitation of $CaCO_3$ granules by the organism. Some strains of *Actinomyces* are common residents of the oral cavity and may be responsible for some disease conditions of the gums or teeth.

The Actinoplanaceae are a very interesting family of organisms that are related to the actinomycetes but differ in spore structure and manner of spore formation. The actinoplanes are mainly aquatic organisms and often grow on submerged leaves. The organism produces a mycelium similar to *Streptomyces;* but the spores, instead of being formed in chains on aerial hyphae, are formed within spherical structures called *sporangia.* A number of spores are formed within each sporangium. They are motile,

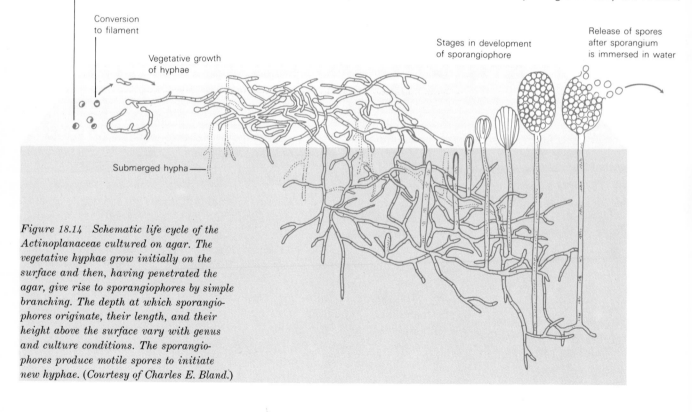

Motile spores

Conversion to filament

Vegetative growth of hyphae

Stages in development of sporangiophore

Release of spores after sporangium is immersed in water

Submerged hypha

Figure 18.14 Schematic life cycle of the Actinoplanaceae cultured on agar. The vegetative hyphae grow initially on the surface and then, having penetrated the agar, give rise to sporangiophores by simple branching. The depth at which sporangiophores originate, their length, and their height above the surface vary with genus and culture conditions. The sporangiophores produce motile spores to initiate new hyphae. (Courtesy of Charles E. Bland.)

having a tuft of polar flagella, and are released through the sporangium wall. At first the spore swims around, then it settles, loses its flagella, and produces a minute germ tube that grows and branches to form a mycelium (Figure 18.14).

18.7 Mycobacterium

The genus *Mycobacterium* contains organisms that form a rudimentary mycelium and thus are usually classified with the actinomycetes; however, mycobacteria differ from most actinomycetes in that they are characteristically *acid-fast*. This is a property by which cells stained with such dyes as basic fuchsin resist decolorization with dilute acid, whereas cells of other bacteria are readily decolorized. Acid fastness, which is found only in the mycobacteria and a few species of *Nocardia,* is conferred by the presence of large amounts of waxy substances on the cells, to which the dye molecules bind tightly. The acid-fast stain is thus a differential stain of great value for identifying mycobacteria in natural materials or in cultures. In addition to exhibiting acid fastness, mycobacteria are nonmotile, nonsporulating, aerobic rods. The genus *Mycobacterium* contains a variety of species that occur in soil, including a number of hydrogen and hydrocarbon utilizers, but the most important species are those pathogenic to man, *M. tuberculosis,* the causal agent of tuberculosis, and *M. leprae,* the causal agent of leprosy.

Tuberculosis has been one of the great scourges of mankind. Pulmonary tuberculosis, also called ''consumption'' or ''phthisis,'' has been recognized as a disease entity for hundreds of years. Robert Koch first showed in 1882 that tuberculosis was caused by a bacterium when he successfully cultured *M. tuberculosis* in the laboratory and succeeded in establishing an experimental infection in guinea pigs. The organism, which is an obligate aerobe, has simple nutritional requirements and will grow on a synthetic medium containing acetate or glycerol as a sole carbon source and ammonium as a sole nitrogen source. Growth is stimulated by lipids and fatty acids, and egg yolk is often added to culture media to achieve more luxuriant growth, a glycerol–egg yolk medium often being used in the primary isolation of the organism from infected materials. Perhaps because of the high lipid content of its cell walls the organism is able to resist such chemical agents as alkali or phenol for considerable periods of time. This property is used in the selective isolation of *M. tuberculosis* from sputum and other materials that contain other bacteria, the sputum first being treated with 1 *N* NaOH for 30 min before being neutralized and streaked on the isolation medium. Since avirulent and weakly virulent

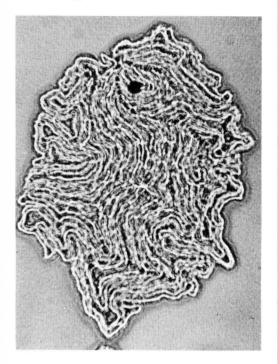

strains are widespread, any isolate must be tested to confirm virulence, and for this the guinea pig is the animal of choice because it is highly susceptible to *M. tuberculosis* infection. Pathological material or a culture suspension of virulent organisms injected subcutaneously will cause the formation of characteristic tubercle nodules at the site of injection, and the spread of the organism from the initial site may result within 4 to 5 wk in secondary tuberculosis nodules in the spleen or peritoneal cavity. Death of the animal usually occurs in 6 wk to 3 mo, and autopsy will reveal lesions in organs throughout the body. Because of the length of time required for a virulence test, many studies have been carried out in vitro to identify properties of cultures that correlate with virulence. The most consistent finding has been that virulent strains growing on agar or in liquid medium form long cordlike structures (Figure 18.15), due to side-to-side aggregation and intertwining of long chains of bacteria. Growth in cords reflects the presence on the cell surface of a characteristic lipid, the "cord factor," which is a glycolipid (Figure 18.16). *Mycobacterium tuberculosis* also produces a wide variety of other lipids, up to 20 to 40 percent of the dry weight, and these lipids are concentrated in the cell wall, altering its surface properties and making the cells very difficult to wet; this may be at least partly responsible for the slow growth rates of virulent strains. Avirulent and saprophytic strains, which have less lipid, grow considerably more rapidly. Also, the lipid coating is probably one of the main factors responsible for the resistance of tubercle bacilli to chemical agents.

The interaction of the human host and *M. tuberculosis* is an extremely complex phenomenon, being determined in part by the virulence of the strain but probably more importantly by the specific and nonspecific resistance of the host. It is convenient to distinguish between two kinds of human infection, primary and postprimary (or reinfection). Primary infection is the first infection that any individual receives and for tuberculosis

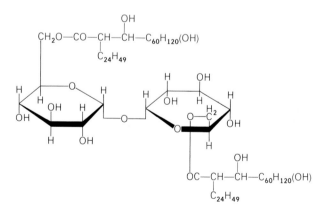

Figure 18.16   Structure of "cord factor," a mycobacterial lipid: 6,6'-dimycolyltrehalose.

Representative procaryotic groups

often results from inhalation of droplets containing viable bacteria derived from an individual with an active pulmonary infection. Another source of primary infection is dust particles that have become contaminated from saliva of tuberculous individuals. The bacteria settle in the lungs, grow, and become surrounded by macrophages that ingest the bacteria. In a few individuals with low resistance the bacteria are not effectively controlled and an acute pulmonary infection is set up, which leads to the extensive destruction of lung tissue, the spread of the bacteria to other parts of the body, and death. Susceptibility to acute infection is more common in non-Caucasians, in infants and children, or in individuals whose nutritional status is poor.

In most cases, however, acute infection does not occur; rather, an immune response develops during the initial infection, and as a result, the macrophages become ''activated'' and effectively destroy the bacteria. The infection remains localized and is usually inapparent; later it subsides. But this initial infection makes the individual hypersensitive to the bacteria or their products and consequently alters the response of the individual to subsequent infections. If *tuberculin,* a protein fraction extracted from the bacteria, is injected intradermally into the hypersensitive individual, it elicits a localized immune reaction at the site of injection, which is characterized by hardening and reddening of the site 1 to 2 da after injection. An individual exhibiting this reaction is said to be *tuberculin positive,* and many adults give positive reactions due to inapparent infections earlier in life. A positive tuberculin test does not indicate active disease but only that the individual has been exposed to the organism at some time. It is in tuberculin-positive individuals that the postprimary type of tuberculosis infection occurs. When renewed pulmonary infections occur in tuberculin-positive individuals, they are chronic types that involve destruction of lung tissue, followed by partial healing and a slow spread of the lesions within the lungs. X-ray examination may reveal spots of destroyed tissue, but only in individuals with extensive tissue destruction are viable bacteria found in the sputum or stomach washings. In many cases infections in tuberculin-positive individuals are a result of reactivation and growth of bacteria that have remained alive and dormant in the lungs for long periods of time. Malnutrition, overcrowding, stress, and hormonal unbalance often are factors predisposing to secondary infection.

Drugs such as streptomycin, isonicotinic acid hydrazide (INH), and *p*-aminosalicylic acid (PAS) are the most effective antituberculosis drugs, the latter two being almost specific in their action against mycobacteria. Treatment with these drugs either singly or in combination results in reduced infection rate, although long-term therapy is necessary to eliminate the organism from all the sites within destroyed lung tissue where it has become established. Drug therapy does not of course lead to the direct healing of damaged tissue, but by destroying the bacteria, it prevents the establishment of new sites of infection and new tissue damage.

The severity of tuberculosis disease had begun to subside in the late nineteenth century, even in the absence of chemotherapeutic measures, due to the improved nutritional and socioeconomic conditions that developed in Europe and North America. Most people became tuberculin positive at an early age, and so the disease was of the postprimary or reinfection type. The introduction of chemotherapeutic measures in the post–World War II era has changed this situation considerably. The elimination of the bacteria from infected individuals has greatly reduced the infection rate in the population as a whole, and today tuberculin-positive individuals comprise a smaller proportion of the population than they once did. At one time extensive surveys were made using X rays and the tuberculin test to detect infected individuals who could then be treated, but the incidence is now so low that the cost of such extensive surveys is hardly justified by the number of new cases detected.

A live vaccine, the BCG strain of *M. tuberculosis,* has been available for many years, but the desirability of its use has been the subject of much debate. Vaccination of tuberculin-negative individuals with BCG can convert them to tuberculin-positive status, with an increase in resistance to infection, but the vaccination confers the same hypersensitive state induced by natural infection, and so does not eliminate the chance of the postprimary type of infection. Since vaccination renders the individual tuberculin-positive, it also eliminates the possibility of using the tuberculin test at a later date to detect a new infection. Vaccination in many countries is now recommended only for tuberculin-negative individuals who have a high probability of infection, such as children of tubercular parents.

Tuberculosis in cattle at one time was a serious disease. The bovine strains of *M. tuberculosis* are also highly virulent for human beings, and in the days before public-health measures were widespread, many people became infected from contaminated dairy products. Because the organism entered the body by way of the gastrointestinal tract, the site of infection was usually not the lungs but the lymph nodes, and the organism subsequently localized in the bones and joints. Pasteurization of milk and other dairy products and the elimination of diseased cattle have virtually eliminated this type of human tuberculosis in Europe and North America. Another species, *M. avium,* causes a tuberculosislike disease of chickens and other birds, but this species is completely avirulent for man.

The disease *leprosy* (Hansen's disease) is caused by another mycobacterium, *M. leprae,* which has been difficult to culture although it grows profusely but slowly in intracellular sites in leprous lesions. Two types of infection are known: *cutaneous,* in which the organism primarily affects the skin, eventually causing extensive disfiguration, and *neural,* in which the organism infects the peripheral nerves, leading to loss of sensation. Both types of disease may be present in the same patient. Leprosy is not a highly contagious disease and the incidence in most parts of the world is very low, although in ancient times it was apparently much more common, perhaps due to crowding and poor sanitation. Therapy with the drug diaminodiphenylsulphone is effective, but treatment must be prolonged. Because of the slow and chronic nature of the disease, moreover, it is difficult to evaluate the efficacy of treatment.

18.8   *Enteric bacteria*

The enteric bacteria, or Enterobacteriaceae, comprise a relatively homogeneous group characterized as follows: Gram-negative, nonsporulating rods, nonmotile or motile by peritrichously occurring flagella, facultative anaerobes, relatively simple nutrition, fermenting sugars to a variety of end products. The DNA base compositions vary from 50 percent GC in *Escherichia* to 58 percent GC in *Serratia,* but within the same genus the DNA of various strains is nearly the same. Among the enteric bacteria are many strains pathogenic to man or animals as well as other strains of practical importance. Probably more is known about *E. coli* than about any other species of bacteria.

One of the key taxonomic characteristics separating the various genera of the enteric bacteria is the range and proportion of end products produced by the anaerobic fermentation of glucose. Two broad patterns are recognized, the *mixed-acid* fermentation and the *butylene glycol* fermentation (see Table 18.8). The importance of distinguishing between these two was discussed in Section 15.8 in relation to water analysis: *E. coli,* an indicator of fecal pollution, is a mixed-acid fermenter, whereas *Aerobacter aerogenes,* a common soil organism that also ferments lactose, is a butylene glycol fermenter. In mixed-acid fermentation, three acids are formed in significant amounts—acetic, lactic, and succinic—and ethanol, CO_2 and H_2 are also formed, but not butylene glycol. In butylene glycol fermentation, smaller amounts of acid are formed, and butylene glycol, ethanol, CO_2, and H_2 are the main products. Mixed-acid fermentation yields equal amounts of CO_2 and H_2 whereas that of butylene glycol yields more CO_2 than H_2. This is because mixed-acid fermenters produce CO_2 only from formic acid by means of the enzyme system formic hydrogenlyase:

$$HCOOH \longrightarrow H_2 + CO_2$$

Table 18.8 *A comparison of the typical end products formed in the mixed-acid and butylene glycol fermentations*[a]

	Moles product/100 moles glucose fermented	
	Mixed-acid fermentation (Escherichia)	*Butylene glycol fermentation (Aerobacter)*
CO_2	44	172
H_2	43	36
Formic acid	2	18
Acetic acid	44	0.5
Lactic acid	84	3
Succinic acid	29	0
Ethyl alcohol	42	70
2,3-Butylene glycol	0	66.5
Total acids	159	21.5
Total neutral products (alcohol + butylene glycol)	42	136.5
$CO_2 : H_2$	1 : 1	5 : 1
Acidic : neutral products	4 : 1	1 : 6

[a] In part, from R. Y. Stanier, M. Doudoroff, and E. A. Adelberg: *The Microbial World*, 2nd ed. (Englewood Cliffs, N.J.: Prentice-Hall, Inc., 1963).

and this reaction results in equal amounts of CO_2 and H_2. The butylene glycol fermenters also produce CO_2 and H_2 from formic acid but they produce additional CO_2 during the reactions that lead to butylene glycol. In addition to distinguishing mixed-acid from butylene glycol fermenters by performing a complete fermentation balance, the distinction can be made more simply by the following tests: (1) determination of the final acidity reached in the culture after complete glucose fermentation since the mixed-acid fermenters produce more acid and thus cause a lower pH than do the butylene glycol fermenters; (2) determination of the ratio of $CO_2 : H_2$; (3) determination of the formation of *acetoin,* an intermediate in the formation of 2,3-butylene glycol, whose presence can be determined fairly easily chemically. (A pink color is formed when the culture medium containing acetoin is made alkaline and the chemicals creatin and α-naphthol are added. This is the so-called Voges-Proskauer test.)

Other characteristics used in defining the various genera of enteric bacteria are shown in Table 18.9. Although these are somewhat arbitrary, in most cases they seem to define genera that have considerable homogeneity since the members of each genus are found to have many other characteristics in common, even to the extent of showing close genetic relationships. In addition to the characteristics shown, a variety of selective and differential media have been developed for isolating and characterizing specific genera of the enteric bacteria. Most of these media

Table 18.9 *Differentiation of the major subgroups of the enteric bacteria*

	DNA % GC	Lactose utilization	Indole production	Methyl red test	Acetyl-methyl carbinol production	Citrate utilization	H$_2$S production	Motility	Pigmentation	Gas from glucose	Urease
Mixed-acid fermentation:											
Escherichia	50	+[a]	+	+	–[b]	–	–	+(–)[c]	–	+	–
Shigella	50–52	–	+	+	–	–	–	–	–	–	–
Salmonella	50–52	–	–	+	–	+	+	+	–	+(–)	–
Erwinia	50–56	+	–	+	–	+	–	+	Yellow	+	–
Proteus	37–53	–	+(–)	+	–	+(–)	+(–)	+	–	+	+
Butylene glycol fermentation:											
Aerobacter (and *Klebsiella*)	51–55	+	–	–	+	+	–	+(–)	–	+	–
Serratia	54–60	–	–	–	+	+	–	+	Red	+(–)	–

[a] Characteristic present in all strains.
[b] Characteristic absent in all strains.
[c] Characteristic variable among the strains.

contain peptone or meat infusion, lactose, agents such as bile salts that inhibit the growth of bacteria of other genera, agar, and a pH indicator to measure the extent of production of acid from lactose. Once organisms are isolated, complete identification can be carried out by using media in the following way: (1) An agar medium containing iron chloride will indicate the production of H_2S by the formation of black iron sulfide; (2) motility can be indicated by making a stab in semisolid agar; (3) the hydrolysis of gelatin can be detected when gelatin is substituted for agar as a solidifying agent; (4) the production of urease can be shown with media containing urea and a pH indicator (the action of the enzyme on urea releases NH_3 with a concomitant rise in pH). In the diagnostic laboratory a wide variety of even more specialized media are used. Finally, immunological tests based on studies of both somatic (O) and flagellar (H) antigens can be used to characterize individual strains, and for some genera bacteriophage typing is also used.

Escherichia   Members of the genus *Escherichia* are almost universal inhabitants of the intestinal tract of man and warm-blooded animals, although they are by no means the dominant organisms in these habitats. As we noted in Section 14.3, *Escherichia* may play a nutritional role in the intestinal tract by synthesizing vitamins, particularly vitamin K. As a facultative anaerobe, *Escherichia* probably also helps consume oxygen, thus rendering the large intestine anaerobic. *Escherichia* strains rarely show any growth-factor requirements and are able to grow on a wide variety of carbon and energy sources such as sugars, amino acids, organic acids, and so on. Only rarely is *Escherichia* pathogenic, and then only when host resistance is low. Some strains have been implicated in diarrhea in infants, occasionally occurring in epidemic proportions in children's nurseries or obstetric wards, and *Escherichia* may also cause urinary tract infections in older persons or in those whose resistance has been lowered by surgical treatment or by exposure to ionizing radiation. Most strains are susceptible to streptomycin, the tetracyclines, and chloramphenicol.

Salmonella   Although *Salmonella* and *Escherichia* are very closely related (we compared the genetic maps of one strain of each genus in Chapter 12), in contrast to *Escherichia,* members of the genus *Salmonella* are usually pathogenic, either to man or to other warm-blooded animals. In man the most common diseases caused by salmonellae are typhoid fever and gastroenteritis. The salmonellae are characterized immunologically on the basis of three cell-surface antigens, the O, or cell-wall (somatic), antigen, the H, or flagellar, antigen, and the Vi antigen, the latter found primarily in strains of *Salmonella* causing typhoid fever. The O antigens are complex polysaccharides, the precise immunological reaction depending on the kinds of sugars and the manner of linkage. Although some of the sugars of the O antigen are common ones, which we have met with frequently in this book, four of the sugars are unusual kinds called dideoxyhexoses, which have no hydroxyl groups on carbons 3 and 6. There are dideoxyhexoses corresponding to each of the normal sugars D-galactose, L-galactose, D-mannose, and D-glucose. The structure of the O antigen of one species, *S. typhimurium,* was illustrated in Figure 15.4; the outer portion of this polysaccharide, containing the sugars rhamnose, mannose, galactose, and abequose (the latter is 3,6-dideoxy-D-galactose), determines the antigenic specificity of the polysaccharide. Over 700 distinct *Salmonella* types are known with different antigenic specificities in their O antigens. Additional antigenic subdivisions are based on the antigenic specificities of the flagellar H antigens. There is little or no correlation between the antigenic type of a *Salmonella* and the disease symptoms elicited, but antigenic typing makes possible the tracing of a single strain involved in an epidemic. The pathogenicity of salmonellae is due primarily to the action of an endotoxin, which is a cell-surface lipopolysaccharide attached to, or part of, the O antigen (see Figure 15.4).

The important disease *typhoid fever* is caused by strains of *Salmonella* that produce the Vi antigen, these strains usually being given the species designation *S. typhosa.* The Vi antigen is a simple polysaccharide composed of repeating units of N-acetylgalactosaminuronic acid, whose acidic groups make the cell surface highly negatively charged. Perhaps because of this negative charge, *S. typhosa* is able to resist digestion after phagocytosis and can live and reproduce intracellularly in macrophages and other phagocytic cells. The organism is transmitted from person to person in food or water usually contaminated from fecal sources. Initial replication of the organism occurs in the intestinal tract, and once it is established there some of the organisms enter the lymph system draining the intestine, travel to the bloodstream, and become disseminated throughout the body. The organism grows especially well in the biliary tract, the spleen, and the lymph nodes, but may reproduce in many other organs of the body. After the organism is established in the tissues, the characteristic fever develops, probably as a result of the release of endotoxin. Diarrhea is not a common symptom of typhoid fever (as it is with *Salmonella* gastroenteritis—see below). One diagnostic sign of the disease is the transitory appearance of *rose spots* on the trunk. The production of specific antibody occurs at the end of the first week of the disease, and diagnosis can be effected by measuring the agglutination titer of the serum against an authentic strain of *S. typhosa.* As we noted in Section 15.6, a rising antibody titer is usually indicative of active disease. The intestinal phase of the disease can be treated with a variety of antibiotics active against Gram-negative bacteria. However, even if the organism is eliminated from the intestinal tract by antibiotic therapy, organisms derived from the biliary tract may continually appear in the feces; this is probably the situation during the chronic carrier state. The chronic intracellular phase can be treated effectively only with chloramphenicol, an antibiotic that penetrates phagocytic cells readily. Since this antibiotic is only bacteriostatic, antibiotic therapy must be continued long enough so that the host defenses can eliminate all viable bacteria, and indeed chloramphenicol is not always able to eliminate the chronic carrier state. Typhoid fever is a disease that has been controlled most dramatically by such public health measures as pasteurization of milk, treatment of sewage, purification of water, and elimination of chronic carriers as food handlers. However, even today typhoid fever is not completely eradicated, and a breakdown in water-supply treatment could easily lead to a return of the disease. Sporadic cases are still reported around large cities, especially during the summer months. A killed-bacteria vaccine for *S. typhosa* is available and is widely used for the immunization of military personnel and travelers to parts of the world where epidemics are still common.

Gastroenteritis caused by salmonellae other than *S. typhosa* is often called "food poisoning" since infection occurs most frequently after eating contaminated food. It is not, however, a true food poisoning but an infection derived from food since the symptoms develop as a result of the multiplication of the organism in the intestinal tract. In contrast to *S. typhosa*, the enteritis-inducing strains rarely leave the intestinal tract. Between 8 and 48 hr after inoculation, onset occurs, which is usually sudden, with headache, chills, vomiting, and diarrhea, followed by a fever that lasts a few days. The disease is usually self-limiting, but antibiotic therapy may be desirable. Diagnosis is from symptoms and by culture of the organism from the feces. Many antigenic types have been implicated. The disease may be controlled by the use of public health measures, especially in monitoring food handlers and food preparation.

Two important *Salmonella* infections of chickens are known, fowl typhoid (causal agent, *S. gallinarum*) and pullorum disease (causal agent, *S. pullorum*), both of which have high fatality rates. *Salmonella* infections also occur in horses, sheep, goats, dogs, cats, cattle, and pigs, occasionally with extensive economic loss. The strains of *Salmonella* that infect these animals rarely if ever infect man, but the physiological reasons for this species specificity are not known.

Shigella   The shigellae are closely related to *Escherichia* and *Salmonella;* in fact these are so similar that they are able to undergo genetic recombination with each other and are susceptible to some of the same bacteriophages. As was noted in Table 18.9, *Shigella* differs from *Escherichia* in being unable to utilize lactose; this inability is related to a lack of either β-galactosidase or galactoside permease. Again, *Shigella* is unable to produce gas during the fermentation of sugars because it lacks the enzyme formic hydrogenylase. Shigellae are also nonflagellated and nonmotile. All three differences could easily have arisen from loss of the genes controlling the production of three proteins. In contrast to *Escherichia*, however, *Shigella* is commonly pathogenic to man, causing a rather severe gastroenteritis usually called "bacterial dysentery" (to distinguish it from amoebic dysentery, discussed in Chapter 19). The pathogenicity of some strains of *Shigella* seems related to their ability to produce a powerful endotoxin. In nature *Shigella* is found almost exclusively in man and primates, infection of other animals being rare. Communities whose sanitation measures are poor are often affected, and quite frequently military troops in combat zones, where proper sanitation is unavailable, are stricken. Since the organism apparently is incapable of invading the bloodstream as *Salmonella typhosa* often does, the disease is confined to the intestine and is usually self-limited; rarely is it fatal, spontaneous cure occurring within a few days.

Other enteric bacteria   The genus *Erwinia* is related to *Escherichia* but is characterized by a lower temperature optimum. In keeping with this property, it is never pathogenic to warm-blooded animals but is instead a plant pathogen. One species, *E. amylovora,* causes fire blight of apples and pears, and another species, *E. carotovora,* causes soft rot of vegetables by producing a pectinolytic enzyme that causes a breakdown of pectin, the intercellular cement of plant cells.

The genera *Aerobacter* and *Klebsiella* are closely related and are often classified together as a single genus. *Klebsiella* differs from *Aerobacter* primarily in producing a capsule and in being nonmotile. However, many aerobacters also produce capsules when grown at temperatures below their optima. *Aerobacter* is a common organism in soil or on plant material and is never pathogenic. One species of *Klebsiella, K. pneumoniae.* is a rare cause of pneumonia in humans, and one other species is a symbiont of the leaves of certain tropical plants. Both *Aerobacter* and *Klebsiella* strains fix N_2, a property not found among other enteric bacteria. The genus *Serratia* is also related physiologically to the *Aerobacter-Klebsiella* group, but differs in certain biochemical properties and in forming a series of red pyrrole-containing pigments called *prodigiosins,* the structure of one of which is shown in Figure 18.17. This pigment is of interest because it contains the pyrrole ring also found in the pigments involved in energy transfer: porphyrins, chlorophylls, and phycobilins. There is no evidence that prodigiosin plays any role in energy transfer, however, and its exact function is unknown.

Nonpigmented mutants of *Serratia* can be fairly easily isolated (see

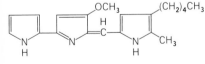

Figure 18.17   Prodigiosin, a red pigment produced by Serratia marcescens.

Section 11.2), and several of these excrete pigment precursors that can be converted to the final product by other mutants. Organisms resembling *Serratia* in physiological and biochemical traits but which are nonpigmented can occasionally be isolated from nature. *Serratia* is usually found in the soil and is rarely pathogenic to man. As noted in Table 18.9, the *Aerobacter-Klebsiella-Serratia* group has a DNA base composition a few percentages higher in GC than that of the *Escherichia-Salmonella-Shigella* group. In keeping with this difference, no genetic recombination between the two groups occurs, and nucleic acid homologies as revealed by hybridization studies are also low (see Table 17.8). However, it is possible to transfer an *F* particle from *Escherichia* to *Serratia;* the particle replicates in the latter but never becomes integrated into the chromosome.

The genus *Proteus* is characterized by its rapid motility and by its production of urease. It is a frequent cause of urinary-tract infections in man and rarely may cause enteritis. The species of *Proteus* probably do not form a homogeneous group, as is indicated by the fact that DNA base compositions vary over a fairly wide range (Table 18.9). Because of the rapid motility of *Proteus* cells, colonies growing on agar plates often exhibit a characteristic swarming phenomenon. At the edge of the growing colony the cells are more rapidly motile than are those in the center of the colony, and these move a short distance away from the colony in a mass, then undergo a reduction in motility, settle down, and divide, forming a new crop of motile cells that again swarm. As a result, the mature colony appears as a series of concentric rings with higher concentrations of cells alternating with lower concentrations.

Pasteurella pestis The causal agent of bubonic plague, *Pasteurella pestis,* has in recent years been found to be closely related genetically to *Escherichia,* although traditionally it had not been classified in the same family. *Pasteurella pestis* is sensitive to many of the bacteriophages that attack *E. coli,* and molecular taxonomic studies also suggest a close relationship between these two organisms. Until the twentieth century, bubonic plague had been one of the greatest scourges of mankind and had been responsible for epidemics that killed millions of people. In the Middle Ages, a plague epidemic caused the death of one-quarter of the population of Europe; epidemics much less extensive have occurred in various parts of the world down to the present. There is good reason to believe that the organism was considerably more virulent in the extensive epidemics of the Middle Ages than it is today, but why this should be so is not known. The disease in the Middle Ages was often associated with severe hemolysis, producing dark skin in the dying person—hence the name ''black death'' for bubonic plague. This symptom is not seen today. The disease is still endemic in parts of Asia and Africa, and in a modified form, called *sylvatic plague,* it is found in rodents in the western United States.

Plague is a disease more common in rodents than it is in man. The causal organism is transmitted from one animal to another by the bite of the flea. In areas of the world where man lives in close proximity to rodents, the flea may transmit the organism from an infected rodent to man, but rarely is the organism passed directly from one person to another. As a result of fleabite, the bacteria are inoculated into the lymph system and move to the regional lymph nodes, where they multiply and cause the formation of enlarged lymph glands (called *buboes,* hence the name "bubonic" plague). In the later stages of the disease the bacteria become disseminated throughout the body. The fatality rate of bubonic plague is high in untreated cases, often approaching 100 percent. Death results from production by the bacteria of a potent endotoxin, but other factors probably also are involved. Invasiveness is determined at least in part by a capsular protein that prevents phagocytosis. Interestingly, the optimum temperature for growth of *P. pestis* is 28°C rather than the normal human body temperature of 37°C, but at 28°C the virulence factors are not produced. The bacteria also multiply in the flea; possibly the low temperature optimum encourages multiplication in the insect. In some cases in man the bacteria become established in the lungs, leading to the condition called pneumonic plague. This type is highly contagious, the bacteria being transmitted from man to man by droplets.

Recovery from plague involves the formation of opsonizing antibodies that seem to confer lifelong immunity. The disease is treated with antibiotics such as the tetracyclines, chloramphenicol, and streptomycin, and if antibiotic therapy is begun early enough, recovery may be assured. Control of the disease in man is effected by the elimination of rats, the chief reservoir of the organism in urban areas. Elimination of sylvatic plague in wild rodents is virtually impossible, and these animals thus constitute a reservoir from which the organism could move into urban centers if effective rat control were not carried out.

18.9 Gram-negative nonmotile rods

A variety of Gram-negative, nonmotile rods are associated with infectious diseases of animals and man. The taxonomic position of many of these organisms is still uncertain, and much more work needs to be done. In this section we mention only some of the more common or medically important forms. We have already mentioned one species of *Pasteurella,* *P. pestis,* the causal agent of plague, as a relative of the enteric bacteria.

Pasteurella septica, (sometimes called *P. multocida*) a species of uncertain affinity to *P. pestis,* causes an acute systemic disease of cattle, buffalo, sheep, and pigs. The same species also causes a disease of birds, fowl cholera, which was first studied in some detail by Pasteur and served as a model for some of his experiments on the attenuation of virulent bacteria and on the induction of immunity by vaccination with such attenuated organisms.

Franciscella The causal agent of the disease *tularemia* is sometimes classified among *Pasteurella* and occasionally among *Brucella* (see below), but more recently a new genus, *Franciscella,* has been defined. Tularemia, a widespread disease of wild rodents such as squirrels and rabbits, was first discovered in 1911 in Tulare County, California, whence it receives its name. It acutely affects rodents, the organism being transmitted by blood-sucking insects. Only occasionally is the organism transmitted to man, usually through the handling of an infected carcass; the disease is most common in hunters, who acquire the infection during the dressing of a rabbit or squirrel—hence the common name "rabbit fever." *Franciscella tularensis* enters the body through the skin, and infection is concentrated in the lymph glands, producing symptoms of headache, body pain, and fever; it sometimes leads to death. The symptoms are probably due to the action of an endotoxin. Antibiotics active against Gram-negative bacteria can be used in therapy, streptomycin being the drug of choice because of its bactericidal action. The disease is best prevented by avoiding contact with the viscera of wild rodents and by using rubber gloves when cleaning the animals.

Hemophilus Organisms of the genus *Hemophilus* are characterized by a growth requirement for heme or a related porphyrin, hence the origin of the genus name. Many isolates also require NAD or NADP. A number of species of *Hemophilus* are known, all of which are found in association with animals or man. The most common of these, *H. influenzae,* received its name because it was first erroneously described as the causal agent of influenza (which in reality is a viral disease; see Chapter 10). *Hemophilus influenzae* is a capsulated organism that is invasive because the capsule prevents phagocytosis. It is an occasional cause of meningitis and also causes pneumonia and nasopharyngitis. All members of the genus *Hemophilus* are genetically transformable (Section 12.1); interspecies transformation at a reduced efficiency is also possible, thus showing the close relationship among the various species.

An organism morphologically and physiologically similar to *Hemophilus* is *Bordetella pertussis,* the causal agent of whooping cough. It is also capsulated but does not require heme or NAD as a growth factor. The organism is only weakly invasive, becoming established in the upper respiratory tract but seldom penetrating to the bloodstream. Its growth in the respiratory tract induces violent coughing, effecting transmissal of the organism from person to person by droplets. A vaccine composed of virulent bacteria that have been killed by an organic mercurial is extremely effective in controlling the disease. This vaccine is usually given to infants in a combined injection with diphtheria and tetanus toxoids.

Brucella   The genus *Brucella* consists of small, Gram-negative, non-sporulating coccobacilli that are usually pathogenic to animals or man. Several species are closely related, being differentiated primarily by certain cultural characteristics and by the host in which they are most commonly found. *Brucella abortus* has its main reservoir in cattle, *B. melitensis* in goats and sheep, and *B. suis* in pigs, but any of the species may be found in all these animals or in man. Any disease caused by a brucella is called "brucellosis." Various disease syndromes are induced, of which the most important are infectious abortion in cattle and undulant fever in humans. The organisms are aerobic and have a complex nutrition, and certain species require an atmosphere of 5 to 10 percent CO_2 for growth. Pathogenic isolates produce a surface antigen that enables them to grow intracellularly in phagocytes, and in the body an intracellular site is the natural habitat. In cattle, *B. abortus* shows a marked specificity for the reproductive tract, and as we have discussed in Section 15.1, this is due to the presence in the fetal tissues of erythritol, which greatly stimulates the growth of *B. abortus*. Brucellosis is widespread in cattle, and these animals can be infected in a variety of ways, such as by the venereal route, by ingestion, by the skin, and by inhalation. Once introduced into a herd, the organism spreads readily and may eventually infect most of the animals. Not all pregnant cows abort, but 30 to 50 percent abortion is not uncommon. Many animals recover completely, but a recovered cow may continue to harbor virulent organisms.

The organism is transmitted from cows to man most commonly in contaminated milk. Multiplication occurs in the mucous membranes of the gastrointestinal tract, and the organism then invades the lymphatics and the bloodstream, becoming selectively localized in the spleen, liver, lymph nodes, bone marrow, and kidneys. Erythritol is not present in man and the organism shows no tendency to localize in the reproductive tract. Disease symptoms are due primarily to the presence of an endotoxin, the most characteristic signs being chills, fatigue, headache, and backache. Another symptom is a fever that increases at night and drops in the daytime, from which is derived the name "undulant fever." Therapy with streptomycin and the tetracyclines is effective but must be prolonged because the intracellular localization of the organisms results in their being protected from the antibiotic. Antibiotic therapy often results in an aggravation of symptoms immediately following administration, due apparently to the induction by the antibiotic of bacterial lysis and the sudden release of large amounts of endotoxin. Agglutinating antibody develops readily in response to *Brucella* infection, and a rise in antibody titer is used in the diagnosis of the disease since the symptoms themselves are often too ill defined to permit a precise identification of the disease. Recovery does not result in complete immunity, however, and reinfection may occur. The best diagnostic procedure is to culture the organism, usually from a blood sample. The brucellae are rarely transmitted from man to man, so that control of the disease can be effected through elimination of infected cattle. After serological surveys of dairy herds to indicate the presence of *Brucella,* infected animals are segregated from the herd or slaughtered. Uninfected herds are vaccinated with a live attenuated vaccine. In the United States a federal project to eliminate brucellosis in cattle was initiated in 1934, at which time about 10 percent of all cattle were infected. By 1957 the incidence of *Brucella* infection in cattle had dropped to 1.7 percent. Pasteurization of milk is also an important procedure in preventing the spread of the organism to man.

Other Gram-negative rods   The genera *Bacteroides* and *Fusobacterium* contain obligate anaerobes that are widespread in the oral cavity and in the gastrointestinal tract of mammals. As was discussed in Section 14.3, these organisms constitute a significant part of the normal flora of the body, but they have not been associated with any particular disease conditions. Another genus, *Achromobacter,* comprises motile or nonmotile rods that do not produce pigment and usually show only limited fermentative ability, a trait that distinguishes them from the enteric bacteria. *Achromobacter* is frequently isolated from soil and water; marine forms are also common. Gram-negative, nonmotile rods of the genus *Flavobacterium* form yellow colonies due to the production of carotenoid pigments. These organisms are also frequently isolated from soil and water and seem to be a heterogeneous group. Flavobacteria are very commonly isolated from the marine environment. They are nutritionally versatile, different strains being able to utilize a variety of organic compounds. Some isolates are even able to hydrolyze lignin, a property more usually associated with fungi. Flavobacteria have sometimes been confused with the yellow-pigmented gliding bacteria of the genus *Cytophaga* (see Section 18.13), some of which lack the gliding habit under certain cultural conditions. Another group, classified as the genus *Agarbacterium,* has the property of digesting agar. Since agar is a product of marine algae, it is perhaps not surprising that agarbacteria are found almost exclusively in marine environments. Another genus, *Alcaligenes,* comprises motile rods that do not ferment sugars. The genus name is derived from the fact that during growth the medium becomes alkaline instead of acidic. *Alcaligenes faecalis* is a frequent isolate from fecal matter; because of this, it has been classified occasionally with the enteric bacteria. However, it is probably unrelated to the enterics since it has a DNA base composition of over 60 percent GC.

18.10 Azotobacter and Beijerinckia

The genus *Azotobacter* comprises large, Gram-negative, obligately aerobic rods that are capable of fixing N_2 nonsymbiotically. The first member of this genus was discovered by the Dutch microbiologist M. Beijerink early in the twentieth century, using an enrichment-culture technique with a medium devoid of a combined-nitrogen source. Although capable of growth on N_2, *Azotobacter* grows more rapidly on NH_3; indeed, adding NH_3 actually represses nitrogen fixation. Much work has been done in seeking to evaluate the role of *Azotobacter* in nitrogen fixation in nature, especially in comparison to the anaerobic organism *Clostridium pasteurianum* and the symbiotic organisms of the genus *Rhizobium*. *Azotobacter* is also of interest because it has the highest respiratory rate (measured as the rate of O_2 uptake) of any living organism. In addition to its ecological and physiological importance, *Azotobacter* is of interest because of its ability to form an unusual resting structure called a *cyst* (Figure 18.18).

Azotobacter cells are rather large for bacteria, many isolates being almost the size of yeasts, with dimensions of 4 to 6 μm (Figure 18.18). Pleomorphism is common and a variety of cell shapes and sizes have been described. Some strains are motile by peritrichously located flagella. When grown on carbohydrate-containing media, extensive capsules or slime layers are produced. The cysts are characteristic of the genus but are formed only under certain conditions; these are discussed further below. *Azotobacter* is able to grow on a wide variety of carbohydrates, alcohols, and organic acids. The metabolism of carbon compounds is strictly oxidative, and acids or other fermentation products are rarely produced. All members fix nitrogen but growth also occurs on simple forms of combined nitrogen: ammonia, urea, and nitrate. In unshaken liquid culture growth occurs as a pellicle on the surface of the liquid, but forced aeration greatly accelerates growth. The organism is rather sensitive to acid pH and usually does not grow at pH values of less than 6.0; this explains why *Azotobacter* is absent in acid soils, although it is present in most well-drained neutral or alkaline soils.

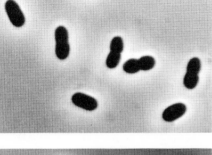

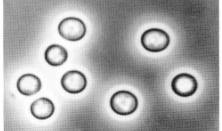

Figure 18.18 Azotobacter vinelandii: vegetative cells (a) and cysts (b), by phase-contrast microscopy. Magnification, 2,760×. (Courtesy of L. P. Lin and H. L. Sadoff.)

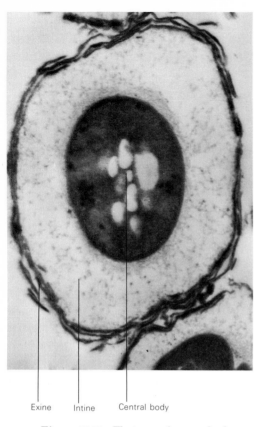

Exine   Intine   Central body

Figure 18.19   Electron micrograph of a thin section of an Azotobacter cyst. Magnification, 40,000×. (Courtesy of L. P. Lin and H. L. Sadoff.)

The structure of the *Azotobacter* cyst is illustrated in Figure 18.19. The cyst is a spherical structure that contains: (1) a dense central body resembling a vegetative cell in fine structure, (2) a less dense homogeneous layer, the *intine,* and (3) a rough, layered outer coat, the *exine.* Like bacterial endospores, cysts show negligible endogenous respiration and are resistant to desiccation, mechanical disintegration, and UV and ionizing radiation. In contrast to endospores, however, they resist heat only a few degrees greater than that which vegetative cells can withstand, and they are not completely dormant since they rapidly oxidize exogenous energy sources. Treatment of cysts with metal-binding agents such as citrate results in a solubilization of the intine and exine, presumably because the structure of these outer layers is stabilized by metals. The central body is then liberated in viable form. The central body does not possess the resistance characteristics of the cyst, thus suggesting that it is the cyst coat that confers resistance. The carbon source of the medium greatly influences the extent of cyst formation, butanol being especially favorable; although cysts will form in carbohydrate-containing media, the process is slower and fewer are formed. Glucose inhibits cyst formation when added to butanol-containing medium. Compounds related to butanol, such as β-hydroxybutyrate, also promote cyst formation. Not only must synthesis of new polymers occur, of which the exine and intine are constituted, but probably also changes take place in the structure or cross linking of the cell wall to account for the change in shape. The controls of cyst formation may be analogous to those of endospore formation (see Chapter 7) although undoubtedly the details of the process will be different. The ecological role of the cyst is probably to enable the *Azotobacter* cell to live through periods of drying and to survive radiation effects during its airborne dispersal from soil to soil. Germination of the cysts occurs when they are in a medium favorable for growth. Steps in the germination process include the enlargement of the central body, disappearance of the intine, formation of breaks in the exine, and emergence of the young vegetative cell from the remains of the disintegrating cyst structure.

Considerable research has been done on the role of *Azotobacter* in the nitrogen economy of soils. When large quantities of carbohydrates are incorporated into soil, significant increases in combined nitrogen occur, most likely due to the action of these carbohydrates in promoting the growth of *Azotobacter.* However, the gains in nitrogen seen are rather small in relation to the amount of carbohydrate added, so that carbohydrate enrichment of soil would not be an economical agricultural practice. Although inoculation of soils with *Azotobacter* has been used as a means to increase soil nitrogen, data showing the advantages of such inoculation are not convincing. Since *Azotobacter* is ubiquitous, if conditions in a soil are favorable for its growth, probably it is already there, and if conditions are not favorable for its growth, inoculation would be of little value. In forest soils, where extensive amounts of organic matter are being added in the form of leaf litter, considerable aerobic nitrogen fixation occurs, but *Azotobacter* has not yet been implicated. Further work on the ecology of this interesting genus is very desirable.

Another genus of aerobic free-living, nitrogen-fixing bacteria is *Beijerinckia,* which is considerably more common in tropical than in temperate soils. Although members of this genus were at one time considered species of *Azotobacter,* they differ from *Azotobacter* in several important ways: (1) They have smaller cells, comparable to those of most true bacteria; (2) they do not form cysts; (3) they show marked tolerance to acids; and (4) they form organic acids during carbohydrate metabolism.

18.11 *Gram-negative, polarly flagellated rods*

Bacteria with this general morphology are widespread in soil and water, and are found to be exceedingly diverse physiologically and biochemically. Table 18.10 provides a brief digest of various taxonomic groups. None of these organisms forms spores, and cells of different genera are either straight or curved.

In the following discussion we consider the more common and important members of this heterogeneous assemblage.

The aerobic pseudomonads The aerobic pseudomonads constitute a large and fairly diverse array of bacteria. Some genera have been separated on the basis of special physiological characteristics. For example, *Xanthomonas* includes plant-pathogenic, yellow-pigmented strains: *Hydrogenomonas* are those growing on H_2 as sole energy source, and *Methylomonas* are those growing on methane. The majority of the aerobic pseudomonads do not show such specialized characteristics and are subdivided on the basis of other physiological characteristics (see Table 18.11). Within each subgroup the individual species are separated on the basis of further nutritional and physiological characteristics.

Many aerobic pseudomonads, as well as a variety of other Gram-negative bacteria, metabolize glucose via the Entner-Doudoroff pathway, which differs in many respects from the Embden-Meyerhof pathway described in Chapter 4. As outlined in Figure 18.20, in the Entner-Doudoroff pathway glucose is converted to 6-phosphogluconate, which is then dehydrated to 2-keto-3-deoxy-gluconate-phosphate (KDGP). The KDGP is then split to pyruvate and glyceraldehyde-phosphate and

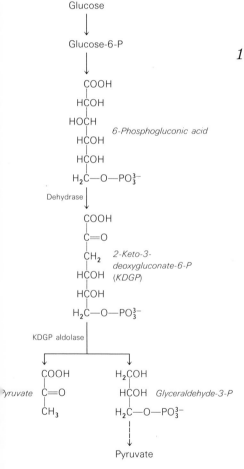

Figure 18.20 Entner-Doudoroff pathway.

Table 18.10   Groups of Gram-negative, polarly flagellated rods

	Genera, etc.	Characteristics, etc.
Aerobic pseudomonads	*Pseudomonas, Xanthomonas, Hydrogenomonas, Comamonas, Acetomonas, Alginomonas, Cellulomonas, Cellvibrio, Methylomonas*	Aerobic; not fermentative or photosynthetic; use O_2 as terminal electron acceptor; may use NO_3 anaerobically; chemoorganotrophs; some use H_2 or CH_4 as energy sources; many pigmented; can use many organic compounds; some are animal and plant pathogens; DNA, 58–69% GC
Fermentative pseudomonads	*Aeromonas, Zymomonas, Photobacterium*	Facultative aerobes, ferment sugars with formation of acids; DNA, 50–60% GC
Parasites of other bacteria	*Bdellovibrio*	Obligate or facultative parasites of other bacteria; DNA, 51–55% GC
Plant symbionts or pathogens	*Rhizobium, Agrobacterium*	Polar or peritrichous flagellation; aerobic; nonfermentative; facultative parasites or symbionts of plants; DNA, 58–65% GC
Acetic acid bacteria	*Acetobacter*	May not oxidize organic compounds completely to CO_2; produce acetic acid; DNA, 55–65% GC

the latter converted to pyruvate via the reactions of the lower part of the Embden-Meyerhof pathway. Although the overall result of both the Entner-Doudoroff pathway and the Embden-Meyerhof pathway is the conversion of one glucose molecule to two pyruvates, many of the intermediate steps are different. Two key enzymes of the Entner-Doudoroff pathway are 6-phosphogluconate dehydrase and KDGP aldolase (see Figure 18.20). A survey for the presence of these enzymes in a wide variety of bacteria has shown that they are absent from all Gram-positive bacteria (except a few *Nocardia* isolates) and are generally present in bacteria of the genera *Pseudomonas, Xanthomonas, Chromobacterium, Rhizobium,* and *Agrobacterium* as well as in some isolates of several other genera of Gram-negative bacteria. It has been suggested that bacteria possessing the Entner-Doudoroff pathway may represent a separate evolutionary line from other bacteria.

	Genera, etc.	*Characteristics, etc.*
Curved and helical rods	Vibrio, Spirillum	Facultative aerobes; a few pathogenic to man and animals; DNA, 30–65% GC
Sulfate-reducing bacteria	Desulfovibrio	Obligate anaerobes; use SO_4^{2-} as electron acceptor; DNA, 45–60% GC
Chemolithotrophs	Thiobacillus, Ferrobacillus, Nitrobacter, Nitrosomonas	Obligate or facultative lithotrophs; DNA 51–68% GC
Stalked bacteria	Caulobacter, Asticcacaulis	Produce stalks that enable attachment to solid substrates; DNA, 65% GC
Budding bacteria	Hyphomicrobium	Cell division by formation of a bud or thin hypha; DNA, 60–68% GC
Sheathed bacteria	Leptothrix, Sphaerotilus	Complex life cycle; nonmotile sheathed filaments releasing flagellated swarmer cells; DNA, 70% GC
Photosynthetic bacteria	Green sulfur bacteria, purple sulfur bacteria, nonsulfur purple bacteria	Photosynthetic; often obligate anaerobes; do not produce O_2 photosynthetically; use sulfur or organic compounds as electron donors; DNA, 45–73% GC

One of the striking properties of the aerobic pseudomonads is the wide variety of organic compounds used as carbon and energy sources. As is shown in Figure 18.21, some strains utilize over 100 different compounds, and only a few strains utilize fewer than 20 compounds. As an example of this versatility, a single strain of *P. aeruginosa* can make use of many different sugars, fatty acids, dicarboxylic acids, tricarboxylic acids, alcohols, polyalcohols, glycols, aromatic compounds, amino acids, and amines, plus miscellaneous organic compounds not fitting into any of the above categories. On the other hand, these organisms are usually unable to break down polymers into their component monomers. The aerobic pseudomonads are ecologically important organisms in soil and water and are probably responsible for the degradation of many soluble compounds derived from the breakdown of plant and animal materials.

The degradation of a single organic compound may occur by different

18.11 Gram-negative, polarly flagellated rods

Table 18.11 Characteristics of the subgroups and species of aerobic pseudomonads

	Species	Characteristics, etc.
Fluorescent subgroup	P. aeruginosa, P. fluorescens, P. putida	Produce water-soluble, yellow-green fluorescent pigments; do not form poly-β-hydroxybutyrate (PHB)
Acidovorans subgroup	P. acidovorans, P. testosteroni	Nonpigmented; form PHB
Alcaligenes subgroup	P. alcaligenes, P. pseudoalcaligenes	Nonpigmented; do not utilize sugars
Pseudomallei subgroup	P. pseudomallei	Causal agent of melioidosis in man and animals
	P. mallei	Causal agent of glanders in horses and other animals

biochemical pathways in different species of aerobic pseudomonads. For example, the steps in the breakdown of *p*-hydroxybenzoic acid by members of the fluorescent and acidovorans groups are contrasted in Figure 18.22. We can see that the mechanism by which the benzene ring is cleaved differs, leading to different end products. It can be assumed that the ability to utilize *p*-hydroxybenzoic acid evolved separately in the two groups of pseudomonads, with different enzymes being involved in the

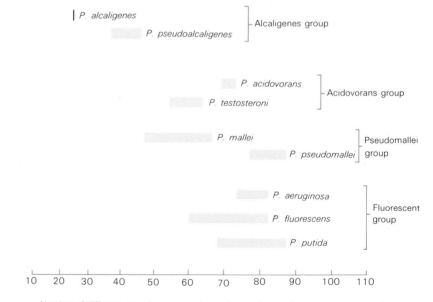

Figure 18.21 Range of organic compounds utilizable as sole carbon and energy sources by certain aerobic pseudomonads. The lengths of the lines indicate the number of compounds utilized by various strains of each species. (Adapted from R. Y. Stanier, N. J. Palleroni, and M. Doudoroff: "The Aerobic Pseudomonads," J. Gen. Microbiol. 43:159. New York: Cambridge University Press, 1966.]

Number of different organic compounds used as carbon and energy sources

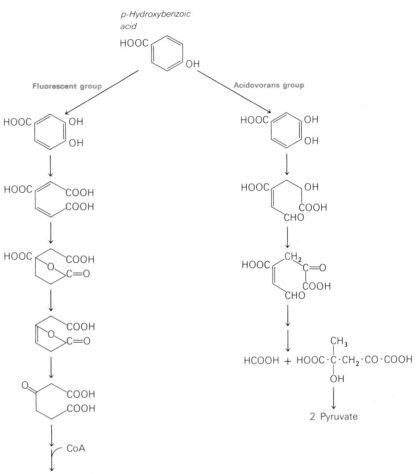

Figure 18.22 Contrasting pathways for the utilization of a single compound, p-hydroxybenzoic acid, in two different groups of aerobic pseudomonads. (Adapted from R. Y. Stanier, N. J. Palleroni, and M. Doudoroff: "The Aerobic Pseudomonads," J. Gen. Microbiol. 43:159. New York: Cambridge University Press, 1966.]

later stages of the breakdown. Further comparisons of this sort may lead to better classification of the aerobic pseudomonads using biochemical criteria, which provide a firmer basis than do the traditional taxonomical systems.

A few aerobic pseudomonads are pathogenic to man or animals. Among the fluorescent pseudomonads, the species *P. aeruginosa* is frequently associated with infections of the urinary tract in man. *Pseudomonas aeruginosa* is able to grow at temperatures of 37 to 41°C, whereas the nonpathogenic species of the fluorescent group can grow only at lower temperatures. *Pseudomonas aeruginosa* is not an obligate parasite, however, since it can be readily isolated from soil, and as a denitrifier it plays an important role in the nitrogen cycle in nature. As a

18.11 Gram-negative, polarly flagellated rods

pathogen it appears to be primarily an opportunist, initiating infections in individuals whose resistance is low. In addition to urinary infections it can also cause systemic infections, usually in individuals who have had extensive skin damage due to burns. The organism is naturally resistant to many of the widely used antibiotics, so that chemotherapy is often difficult. Polymyxin, an antibiotic that is not ordinarily used in human therapy because of its toxicity, is effective against *P. aeruginosa* and can be used with caution.

The pathogen *P. pseudomallei,* the causal agent of *meloidosis,* is also primarily a soil organism, being isolated frequently in tropical areas, but it is able under special conditions to initiate infection in man and in a variety of animals, including cows, pigs, sheep, goats, cats, dogs, rodents, and horses. The site of infection and the symptoms of meloidosis are highly variable and diagnosis can be made only when the causative organism is isolated. On the other hand, *P. mallei,* the causal organism of *glanders,* is primarily a parasite; it is not isolated from soil or other nonanimal substrate. Glanders is a disease found almost solely in horses, donkeys, and mules, although the organism can be transmitted from one of these animals to man and initiate infection. Two clinical syndromes are known in horses: glanders, in which the primary focus of the organism is the lungs, and farcy, in which the organism usually enters by way of the skin and spreads to the lymphatics. Glanders has been effectively eliminated in North America and western Europe through the destruction of infected animals and through quarantine. The elimination of horse-drawn vehicles as a major mode of transportation has undoubtedly played a part in the successful control of the disease in man.

Chromobacterium comprises motile rods that are usually deeply pigmented; the best known is *C. violaceum,* which produces a beautiful purple water-insoluble pigment, violacein (Figure 18.23). This pigment is an indole derivative formed by the oxidation of the amino acid tryptophan. Chromobacteria are found to show some relationship to the aerobic pseudomonads.

Luminescent bacteria   A number of Gram-negative, polarly flagellated rods possess the interesting property of emitting light (luminescence). Some of these bacteria have been classified in a separate genus, *Photobacterium,* but they resemble pseudomonads in respects other than luminescence; a few *Vibrio* isolates are also luminescent. Most luminescent bacteria are marine forms, usually found associated with fish. Some deep-sea fish possess a special organ in which luminescent bacteria grow. The light emitted by the bacteria may enable the fish to navigate better in the eternal darkness of their environment, in which case the bacteria-fish relationship would be symbiotic. Other luminescent marine bacteria live

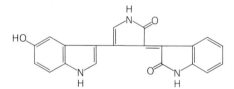

Figure 18.23   Violacein, the purple pigment of Chromobacterium violaceum.

saprophytically on dead fish. A good way of isolating luminescent bacteria is to incubate a dead marine fish for 1 or 2 da at 15 to 20°C; the luminescent bacterial colonies that usually appear on the surface of the fish can be easily seen and isolated. (To see luminescence readily, one should observe the material in a completely dark room after the eyes have become adapted to the dark.)

Although some luminescent bacteria are facultative anaerobes, they are luminescent only when O_2 is present. The amount of O_2 required is quite low, however, and in fact bacterial luminescence can be used as a relatively sensitive way of detecting small amounts of O_2 in a solution. If bacteria are incubated anaerobically for a while and air is then introduced into the culture, there results a bright flash of light, following which the light intensity decreases to a low steady-state value. It is thought that under anaerobic conditions a component required for luminescence (probably NADH; see below) accumulates in amounts higher than normal and that its rapid utilization when O_2 is introduced produces the flash.

The biochemistry of bacterial luminescence has been fairly well worked out from studies on cell-free extracts. The light-emitting cell component is probably flavin mononucleotide (FMN), a riboflavin derivative (see Figure 4.14). Other components required are an enzyme called *luciferase,* a long-chain aliphatic aldehyde (for instance, dodecanal), O_2, and an electron donor, either NADH or NADPH (Figure 18.24). The precise role of the long-chain aldehyde is not known, but it is thought to react with and modify the FMN. The modified FMN is then reduced and undergoes a reaction with O_2, leading to the formation of a peroxide in an excited state. As the FMN peroxide returns to the ground state, light is emitted and FMN is reformed. As shown in Figure 18.24, bioluminescence competes with normal electron transport for the electrons of NADH. One consequence of this competition is that, if the activity of the cytochrome system is blocked by cyanide or some other inhibitor, the intensity of luminescence is increased.

An unsolved question is what benefit, if any, an organism derives from emitting light. Since the energy used for light emission is unavailable for ATP synthesis, light emission would appear to waste energy. In the bacteria living symbiotically with fish, the light emission seems advantageous; but its advantages to free-living and saprophytic bacteria are less obvious.

Luminescence is not restricted to bacteria but is found in certain fungi and dinoflagellates (Section 19.4), as well as in fireflies and some jellyfish, worms, and other invertebrates. The biochemical mechanism of luminescence in fireflies and dinoflagellates is known to be quite different from that in bacteria—which suggests that the processes in these different organisms had independent origins and represent an example of convergent evolution.

18.11 *Gram-negative, polarly flagellated rods*

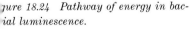
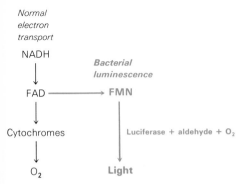

Figure 18.24 Pathway of energy in bacterial luminescence.

Methane-oxidizing bacteria   The ability to utilize methane, methanol, methylamine, and formate, all one-carbon compounds, and obtain all needed energy from these is a unique property restricted to only a few bacteria, most of which (but not all) are related to the pseudomonads; the pseudomonaslike forms are given the genus name *Methylomonas*. In all cases, methane is first oxidized to methanol, so that all methane utilizers are also methanol utilizers. Some of these organisms can also utilize multicarbon compounds, and hence are facultative one-carbon utilizers, whereas others are obligately so. The pathway by which methanol is utilized differs in the obligate and facultative one-carbon utilizers. Methanol is oxidized by the latter to formaldehyde or formate, either of which reacts with glycine to form serine, and serine then becomes the starting material for cellular syntheses. Energy is obtained by oxidative phosphorylation during the oxidative reactions of methane. In the obligate one-carbon utilizers, on the other hand, carbon from methanol is incorporated into allulose phosphate, a novel pentose sugar that is then oxidized via a pentose phosphate cycle. This process is somewhat reminiscent of the manner in which another one-carbon compound, CO_2, is assimilated by autotrophic organisms (see Chapter 6) and hence suggests an analogy between the one-carbon utilizers and the autotrophs. Because ATP can be produced only by oxidative phosphorylation, methane utilizers are usually obligate aerobes. Bacteria capable of utilizing methane are widespread in aquatic and soil environments. It should be recalled that another group of bacteria that are obligate anaerobes produce methane (see Chapter 16), especially in aquatic environments, and it is likely that microbially produced methane is the main source of this compound for the methane oxidizers.

Hydrogen-oxidizing bacteria   A number of bacteria can grow chemolithotrophically on H_2 with CO_2 as a sole carbon source, including certain mycobacteria, actinomycetes, and some micrococci, but the most widespread are Gram-negative, polarly flagellated rods that are related to the aerobic pseudomonads and are usually included in the genus *Hydrogenomonas*. No isolate of this genus so far obtained has been an obligate chemolithotroph; all strains grow well on certain organic acids, alcohols, sugars, and amino acids, and in this respect they resemble typical aerobic pseudomonads. Thus, the only characteristic which distinguishes *Hydrogenomonas* from *Pseudomonas* is the additional ability to use H_2 as a sole energy source. The enrichment procedure for H_2-oxidizing bacteria is carried out by inoculating a medium containing only mineral salts with a sample of soil or water and incubating in a gaseous atmosphere containing H_2, O_2, and CO_2. Since a mixture of H_2 and O_2 containing greater than 4 percent H_2 is potentially explosive, considerable caution must be exercised to avoid flames or sparks. After good growth is obtained in the enrichment culture, pure cultures can be isolated by streaking on agar plates containing only a mineral-salts medium and then incubating in the $H_2 + O_2 + CO_2$ atmosphere. To prove that the isolates obtained are indeed H_2-oxidizers and are not growing on traces of organic matter contaminating the medium, it should be shown that no growth occurs in the same system in the absence of H_2.

Energy is obtained by hydrogen-oxidizing bacteria from oxidative phosphorylation, H_2 serving as the electron donor and O_2 as the electron acceptor. Hydrogen is first activated by the enzyme hydrogenase, the two atoms of H_2 being split and used to reduce NAD. The NADH formed is an electron donor for the familiar electron transport system. Carbon dioxide is fixed by the series of reactions described as the Calvin cycle (see Chapter 6), and hence the chemolithotrophic hydrogen-oxidizing bacteria contain an enzymatic machinery similar to that of photosynthetic organisms.

Hydrogen-oxidizing bacteria are widespread in soil and water. Recall that enteric bacteria and clostridia produce H_2 during the anaerobic fermentation of a variety of substrates, and this is probably the main source of H_2 for the hydrogen-oxidizing bacteria.

Carbon monoxide-oxidizing bacteria Another group of pseudomonaslike bacteria are able to grow with carbon monoxide (CO) as a sole energy source, oxidizing CO to CO_2. Carbon monoxide is very toxic to most aerobes because of its ability to inhibit the activity of the cytochrome system, so that the CO-oxidizing bacteria are a rather interesting group; however, they have been little studied. Since CO is odorless and highly toxic to man, the study of CO-oxidizing bacteria in the laboratory must be carried out with considerable caution.

Bdellovibrio In 1962 a new polarly flagellated organism was discovered, which has the unique property of parasitizing other bacteria, using as nutrients the cytoplasmic constituents of their hosts. These bacterial parasites are small, highly motile cells, which stick to the surfaces of their host cells. Because of the latter property, they have been given the name *Bdellovibrio* (bdello- is a combining form meaning "leech"). A number of strains of *Bdellovibrio* have been found, and each shows some host specificity, attacking some species of bacteria but not others. In general, Gram-positive bacteria are not attacked. After attachment of a *Bdellovibrio* cell to its host, the parasite penetrates into the cytoplasm, replicates, and lyses the host cell, releasing progeny. The stages of attachment and penetration are shown in Figure 18.25. These are processes that require the active metabolism of both host and parasite

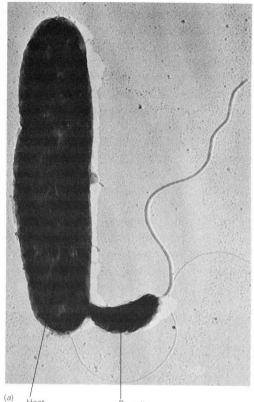

(a) Host
(*Pseudomonas*) Parasite
(*Bdellovibrio*)

Figure 18.25   Stages of attachment and penetration of a host cell by Bdellovibrio. (a) Electron micrograph of a shadowed whole-cell preparation, showing B. bacteriovorus attacking Pseudomonas. Magnification, 24,700×. [From H. Stolp and H. Petzold: Phytopath. Z. 45:373 (1962).] Electron micrographs of thin sections of a Bdello-vibrio attacking Escherichia coli: (b) initial attachment (magnification, 81,000×); (c) early penetration (59,000×); (d) late penetration (59,000×); (e) complete penetration (63,000×). The Bdellovibrio cell is enclosed in a membranous infolding of the host cell. [From J. C. Burnham, T. Hashimoto, and S. F. Conti: J. Bacteriol. 96:1366 (1968).]

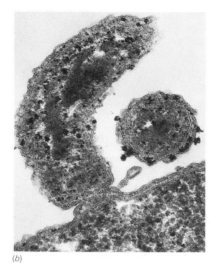

(b)

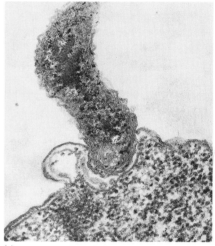

(c)

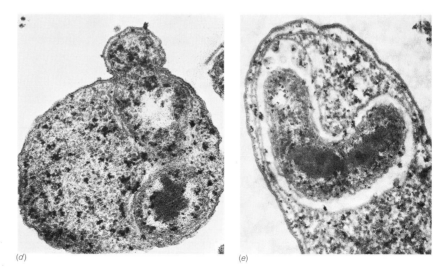

(d)   (e)

since they will not occur if the host cells are killed or if their biosynthetic capacities are inhibited by antibiotics.

Members of the genus *Bdellovibrio* are widespread in soil and water. Their detection and isolation requires methods somewhat reminiscent of those used in the study of bacterial viruses: Host bacteria are spread on the surface of an agar plate to form a lawn, and the surface is inoculated with a small amount of a soil suspension that has been filtered through a membrane filter, which retains most bacteria but allows the small *Bdellovibrio* cells to pass. Upon incubation of the agar plate, plaques analogous to those produced by bacteriophages [see Figure 10.5(c)] are formed at locations where *Bdellovibrio* cells are growing. Pure cultures of *Bdellovibrio* can then be isolated from these plaques. Initial *Bdellovibrio* isolates grow only in the presence of living bacteria and are thus obligately parasitic, but mutants can be isolated that will grow saprophytically.

Bdellovibrio cultures have been obtained from a wide variety of soils and they are thus common members of the soil population. It is surprising that they escaped the attention of soil microbiologists for so many years. As yet, the ecological role of *Bdellovibrio* is not known, but it seems likely that they play some role in regulating the population densities of their hosts.

Rhizobium and Agrobacterium   The genus *Rhizobium* contains those Gram-negative flagellated rods that are capable of entering into symbiosis with legumes and of fixing free nitrogen during this symbiosis. Studies on DNA hybridization indicate that these genera have some relationship to the pseudomonads. This genus and its symbiotic relationships are discussed in Section 14.9. The genus *Agrobacterium* comprises forms pathogenic to plants, and recent studies in molecular taxonomy suggest that they are sufficiently closely related to *Rhizobium* that eventually they may be considered part of this genus. *Agrobacterium tumefaciens* causes crown-gall, a tumorlike disease of the roots of many plant species, while *A. rhizogenes* causes hairy-root disease of apple and other plants.

Acetic acid bacteria   The acetic acid bacteria are Gram-negative rods that have some resemblance to the aerobic pseudomonads but are distinguished from them most clearly by the fact that the acetic acid bacteria do not oxidize alcohols completely to CO_2. For instance, if ethanol is oxidized, there accumulates acetic acid (for which the group is named). Another property of these bacteria is their relatively high tolerance of acidic conditions, most strains being able to grow at pH values lower than 5. This acid tolerance is probably related to their acid-forming ability. The role of the acetic acid bacteria in the production of vinegar was discussed in Section 8.5.

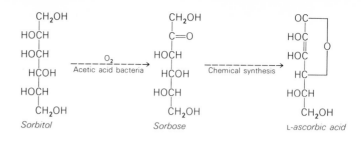

Figure 18.26   Use of acetic acid bacteria in the industrial synthesis of ascorbic acid (vitamin C).

These organisms also lack the ability to oxidize completely such organic compounds as higher alcohols and sugars. For instance, glucose is oxidized only to gluconic acid, galactose to galactonic acid, arabinose to arabonic acid, and so on. This property of underoxidation is exploited in the manufacture of ascorbic acid (vitamin C). Ascorbic acid can be formed from sorbose, but sorbose is difficult to synthesize chemically. It is, however, conveniently obtainable microbiologically by the use of certain acetic acid bacteria, which oxidize sorbitol only to sorbose (Figure 18.26). Using acetic acid bacteria makes the manufacture of ascorbic acid economically feasible.

Another interesting property of some acetic acid bacteria is their ability to synthesize cellulose. This cellulose does not differ significantly from that of plants, but instead of being a part of the cell wall the bacterial cellulose is formed as a matrix outside the wall, and the bacteria become embedded in the tangled mass of cellulose microfibrils. Acetic acid bacteria developing in an unshaken vessel form a surface pellicle of cellulose, in which the bacteria develop; since these bacteria are obligate aerobes, the ability to form such a pellicle may be a means by which the organisms are assured of remaining at the surface of the liquid, where oxygen is readily available.

Vibrio and Spirillum   In fresh water and in the sea one commonly sees rapidly moving curved cells. These often prove to be of the genera *Vibrio* or *Spirillum,* both of which consist of Gram-negative, polarly flagellated rods. Most are nonpathogenic, but some vibrios are causal agents of disease in animals and man. *Vibrio fetus,* for example, causes an infectious abortion in cattle, sheep, goats and swine.

Vibrio cholerae is the specific cause of the disease *cholera* in man; the organism does not normally infect other hosts. Cholera is one of the most common infectious human diseases and one that has had a long history. The organism is transmitted almost exclusively via water, and studies on its distribution in the nineteenth century played a major role in demonstrating the importance of water purification in urban areas. Today the disease is virtually absent from the western world, although it is still common in Asia. *Vibrio cholerae* is an aerobe that has relatively simple nutritional needs, being able to grow in a medium containing asparagine as a sole carbon and nitrogen source. It is quite insensitive to alkaline conditions, being able to grow at pH 9.0 to 9.6, at which most other intestinal bacteria are inhibited. This resistance to high pH makes possible the use of a simple selective medium for primary isolation of *V. cholerae*. Although the bacteria grow readily in the intestinal tract, they do not invade the rest of the body. The organism produces two substances that affect the intestinal mucosa: (1) the enzyme *neuraminidase,* which hydrolyzes N-acetyl neuraminic acid, a mucopolysaccharide that serves as an intercellular cement for the intestinal cells, and (2) a *toxin,* which affects the sodium-potassium transport system of the intestinal mucosa and thus causes a profound electrolyte imbalance and water loss. The combined action of these two materials results in an enormous loss of water through the large intestine, up to 10 to 12 liters per day, causing extreme dehydration followed by shock and death. However, in many cases the disease can be mild and self-limited, running its course in a week or so. Antibiotic therapy (with streptomycin or the tetracyclines) is effective only in the very early stages of the disease. More important in treatment are measures taken to correct fluid imbalance, such as intravenous injection of isotonic fluids, which usually effect a dramatic recovery. Control of cholera is primarily by sanitation measures, such as water purification and sewage treatment.

Desulfovibrio This genus contains polarly flagellated, obligately anaerobic vibrioids that use SO_4^{2-} as an electron acceptor and produce H_2S. The ecological importance of *Desulfovibrio* in the sulfur cycle and in corrosion was discussed in Chapter 16. Not all sulfate reducers belong to the genus *Desulfovibrio,* but isolates from nature are most frequently species of this genus. These organisms can use H_2 or organic compounds as energy sources, transferring the electrons through a cytochrome system to sulfate, with the production of ATP by oxidative phosphorylation. Before it can accept electrons, sulfate must be converted to adenosine-phosphosulfate (APS) (Section 5.7). Although *Desulfovibrio* can use H_2 as an energy source, it is not autotrophic as it lacks the CO_2 fixation system of autotrophs and obtains the bulk of its carbon from organic compounds present in the medium. Studies on DNA base compositions have shown that strains of *Desulfovibrio* can be separated into three categories: group 1 (61 percent GC); group 2 (52 percent GC); and group 3 (45 percent GC). What is now called *Desulfovibrio* therefore probably constitutes a heterogeneous group, which eventually will be separated into at least three genera.

18.11 Gram-negative, polarly flagellated rods

Figure 18.27 Steps in the oxidation of different sulfur compounds by thiobacilli.

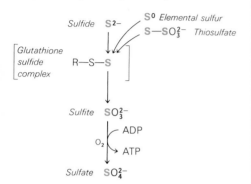

Sulfide S^{2-} S^0 *Elemental sulfur*
S—SO_3^{2-} *Thiosulfate*

$\begin{bmatrix} \text{Glutathione} \\ \text{sulfide} \\ \text{complex} \end{bmatrix}$ R—S—S

Sulfite SO_3^{2-}

O_2 — ADP
— ATP

Sulfate SO_4^{2-}

Thiobacillus The genus *Thiobacillus* contains those Gram-negative, polarly flagellated rods that are able to derive their energy from the oxidation of sulfur compounds, principally sulfides, elemental sulfur, and thiosulfate (Table 18.12). Many isolates are obligate autotrophs, unable to grow on organic compounds, but a few thiobacilli can also grow on organic compounds such as glutamic acid. When growing heterotrophically, these latter organisms are of course similar to pseudomonads. Thiobacilli are widespread in soil and water and their importance in the sulfur cycle was discussed in Chapter 16. Some of the most interesting members of the genus are those that are highly acid tolerant since they are able to grow at pH values of 1 to 2, which are lower than those at which any other procaryotes can grow. When oxidizing sulfur or other reduced sulfur compounds, the organisms produce sulfuric acid and thus lower the pH of the medium.

The biochemical steps in the oxidation of various sulfur compounds are summarized in Figure 18.27. The oxidation of sulfide and sulfur involve first the reaction of these substances with the sulfhydryl group of the tripeptide glutathione (glutamyl-cysteinyl-glycine), leading to the formation of a sulfide-glutathione complex. The sulfur is then oxidized to sulfite (SO_3^{2-}) by the enzyme sulfide oxidase. Sulfite is oxidized to sulfate by a

Table 18.12 Physiological characteristics of some species of Thiobacillus and Ferrobacillus

	Autotrophic energy sources	*Heterotrophic growth*	*Range of pH for growth*	*DNA base composition, % GC*
Obligate autotrophs: *Thiobacillus* thioparus	H_2S, metal sulfides, S^0, $S_2O_3^{2-}$	No	Neutral	62–66
T. thiooxidans	S^0, $S_2O_3^{2-}$	No	Acid	52
T. denitrificans	H_2S, S^0, $S_2O_3^{2-}$ (facultative anaerobes; use NO_3^- as anaerobic electron acceptor)	Some strains	Neutral	64
T. ferrooxidans	Metal sulfides, S^0, $S_2O_3^{2-}$, Fe^{2+}	No	Acid	57
Ferrobacillus ferrooxidans	Fe^{2+}	No	Acid	?
Facultative autotrophs: *T. novellus, T. trautweinii*	$S_2O_3^{2-}$	Yes	Neutral	66–68

cytochrome-linked sulfite oxidase, with the formation of ATP via oxidative phosphorylation. Thiosulfate ($S_2O_3^{2-}$), which can be viewed as a sulfide of sulfite (SSO_3^{2-}), is split into sulfite and sulfur. The sulfite is oxidized to sulfate with the production of ATP and the other sulfur atom is converted into elemental sulfur and sulfides. Thus, when oxidizing thiosulfate, the thiobacilli produce elemental sulfur but when oxidizing sulfides they do not. The elemental sulfur produced can itself later be oxidized when the thiosulfate supply is exhausted.

Enrichment cultures of thiobacilli are easy to prepare. To a basal salts medium with NO_3^- or NH_4^+ as a nitrogen source and bicarbonate as a carbon source is added sulfur or thiosulfate, and the medium is then inoculated with a sample of soil or mud. After aerobic incubation at room temperature for a few days, the liquid should appear turbid due to the growth of thiobacilli. If thiosulfate has been used, droplets of amorphous sulfur will also be present. Pure cultures may then be obtained by streaking from the enrichment culture onto a solidified medium of the same composition. However, pure cultures are fairly difficult to obtain since the organism grows slowly, and heterotrophic contaminants grow on small amounts of organic matter released by the thiobacilli. Another problem in obtaining pure cultures of thiobacilli that grow best at neutral pH, such as *T. thioparus*, is that sulfur oxidation leads to sulfuric acid production and a drop in pH, resulting in the death of the culture. Thus, highly buffered media and frequent transfers of the culture are necessary. On the other hand, isolation of the acidophilic *T. thiooxidans* is relatively easy since this organism is resistant to the acid it produces and most heterotrophic contaminants cannot grow at the low pH values (2 to 5) where *T. thiooxidans* thrives.

Many isolates of sulfur-oxidizing organisms can also oxidize ferrous iron. The geochemical aspects of iron oxidation were discussed in Chapter 16; recall that, at acid pH, ferrous iron is not readily oxidized spontaneously, and that the acid-tolerant thiobacilli are the main agents in nature for the oxidation of ferrous iron in acidic environments. There is still some doubt about whether the genus *Ferrobacillus* is indeed a valid group as most iron-oxidizing, acid-tolerant bacteria can also oxidize sulfur compounds. Those which oxidize both iron and sulfur compounds are currently classified as the species *T. ferrooxidans*, the species *T. thiooxidans* being restricted to those acid-tolerant thiobacilli which cannot oxidize iron.

The acid-tolerant thiobacilli are mainly responsible for producing acidic conditions in the waters that drain coal-mine areas. Oxidation of iron sulfide in the mine wastes yields sulfuric acid, with a consequent lowering of the pH, and ferrous iron becomes solubilized. Water draining these waste heaps is thus usually acidic and rich in ferrous iron, providing ideal conditions for the development of iron oxidizers, which multiply and change the iron to the ferric state. Ferric iron is insoluble and precipitates

18.11   Gram-negative, polarly flagellated rods

593

as orange ferric hydroxide, which colors the waters of many streams in coal-mine areas (Plate 7, Figure 16.9). Their low pH values make the waters unsuitable for the growth of aquatic plants and fish or for domestic uses. The reason that thiobacilli can survive and grow at pH values at which other bacteria are killed is unknown. One hypothesis is that they are impermeable to H^+ ions, but how this impermeability would be effected is also unknown.

Iron-oxidizing bacteria of the species *T. ferrooxidans* also play an important role in the leaching process used in the solubilization of *copper* from low-grade copper sulfide ores. That bacteria play a crucial role in the process has been shown in laboratory experiments; if the leaching process is carried out under aseptic conditions, the amount of copper released is negligible, whereas if the system is inoculated with a pure culture of *T. ferrooxidans,* the rate of release of copper approaches that of the natural process. The importance of this leaching process is such that many copper mining companies now use it to recover copper from their low-grade waste ores.

Nitrifying bacteria Bacteria able to grow autotrophically at the expense of reduced inorganic nitrogen compounds are called "nitrifying bacteria." Several genera are recognized on the basis of morphology and the particular steps in the oxidation sequences that they carry out. It is likely that these genera are not all closely related, but they have traditionally been classified in one group because of their unique autotrophic way of life (see Table 18.13).

Of these genera, the most studied have been *Nitrosomonas, Nitrosocystis,* and *Nitrobacter.* They are common organisms in soil and water, especially in habitats whose pH is alkaline. They do not develop well under acid conditions and hence are usually missing from acidic soils. The nitrifying bacteria are also common and active agents in sewage treatment systems; their roles in these habitats and in other parts of the nitrogen

Table 18.13 Characteristics of the nitrifying bacteria

Ammonia oxidizers	Nitrite oxidizers
Nitrosomonas: Gram-negative, polarly flagellated rods; DNA, 52% GC	*Nitrobacter:* Gram-negative, polarly flagellated rods; DNA, 65% GC
Nitrosococcus: Gram-positive, nonmotile cocci	*Nitrocystis:* Gram-negative, encapsulated rods aggregating into clumps
Nitrosospira: Gram-negative, spiral-shaped	
Nitrosocystis: Gram-negative, polarly flagellated, cysts formed	
Nitrosogloea: Gram-negative, coccoid, aggregate into encapsulated clumps	

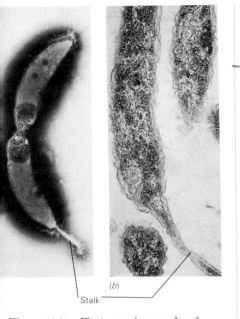

Stalk

Figure 18.28 Electron micrographs of Caulobacter cells: (a) negatively stained preparation of a cell in division (magnification, 16,600×); (b) a thin section—note that the cytoplasmic constituents are present in the stalk region (37,000×). (From G. Cohen-Bazire, R. Kunisawa, and J. S. Poindexter: J. Gen. Microbiol. 42:301. New York: Cambridge University Press, 1966.)

cycle were discussed in Chapter 16. Also, some aspects of the energetics of nitrifying bacteria were discussed in Chapter 6. Most isolates are obligate autotrophs although a few strains may be able to grow slowly at the expense of acetate and other simple organic compounds. Nitrifying bacteria are easy to enrich for by the use of media containing ammonia or nitrite as the sole energy source. Considerable difficulty is usually experienced in isolating pure cultures, however, since the nitrifying bacteria grow fairly slowly and rapidly growing heterotrophic contaminants that can grow at the expense of organic matter released by the nitrifiers are difficult to eliminate.

Stalked bacteria The stalked bacteria comprise a group of Gram-negative, polarly flagellated rods that possess a stalk, an organ by which they attach to solid substrates. Most members of this group are classified in the genus *Caulobacter*. Stalked bacteria are frequently seen in aquatic environments attached to particulate matter, plant materials, or other microorganisms; often they are found attached to microscope slides that have been immersed in lake or pond water for a few days. Electron-microscopic studies reveal that the stalk is not an excretion product but an outgrowth of the cell since the stalk contains cytoplasm surrounded by cell wall and plasma membrane (Figure 18.28). The holdfast by which the stalk attaches the cell to a solid substrate is at the tip of the stalk, and once attached the cell usually remains permanently fixed. Cell division occurs by elongation of the cell and fission, a single flagellum forming at the pole opposite the stalk. The flagellated cell so formed, called a ''swarmer,'' separates from the nonflagellated mother cell, swims around, and settles down on a new surface, forming a new stalk at the flagellate pole, and the flagellum then disappears (Figure 18.29). Stalk formation is a necessary precursory phenomenon to cell division. The cell-division cycle in *Caulobacter* is thus more complex than is simple binary fission since the stalked and swarmer cells have polar differentiation, and the cells themselves are structurally different. When many *Caulobacter* cells are present in the suspension, stalked cells often attach to each other, leading to the formation of rosettes (Figure 18.30).

Caulobacters are heterotrophic aerobes, usually showing one or several vitamin requirements, but they are able to grow on a variety of organic carbon compounds as sole sources of carbon and energy. Various amino acids serve as nitrogen sources. The enrichment culture of caulobacters makes use of the fact that they occur quite commonly in the organic film that develops at the surface of an undisturbed liquid. If pond, lake, or sea water is mixed with a dilute organic material such as 0.01 percent peptone and incubated at 20 to 25°C for 2 to 3 da, a surface film consisting of bacteria, fungi, and protozoa develops, and in this microbial film

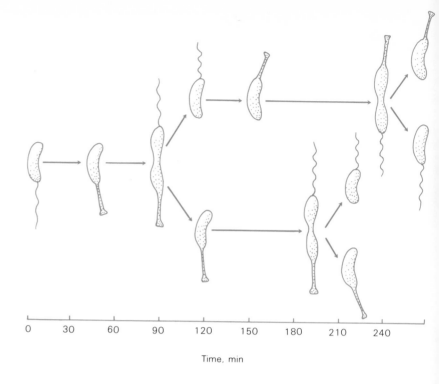

Figure 18.29 Stages in the Caulobacter cell
cycle. [From J. L. Stove and R. Y. Stanier:
Nature 196:1189 (1962).]

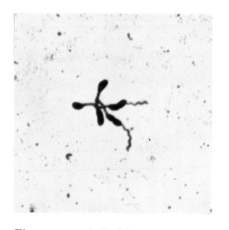

Figure 18.30 A Caulobacter rosette. The
five cells are attached by their stalks. Two
of the cells have divided and the daughter
cells have formed flagella. Magnification,
2,000×. (Courtesy of Einar Leifson.)

caulobacters are common. A sample of the surface film is then streaked
on an agar medium containing 0.05 percent peptone, and after 3 to 4
da the plates are examined under a dissecting microscope for the presence of Caulobacter microcolonies. These colonies are then picked and
streaked on fresh medium containing a higher concentration of organic
matter (for example, 0.5 percent peptone + 0.1 percent yeast extract),
and the resulting colonies are examined microscopically for stalked bacteria. Water low in organic matter is a good source of caulobacters; tap
or distilled water that has been left undisturbed usually shows good
Caulobacter development in the surface film. The ability to grow in organically dilute media is a common property of organisms that attach to
solid substrates; in the case of stalked organisms such as Caulobacter the
stalk itself may function as an absorptive organ, making it possible for the
organism to acquire larger amounts of the restricted supply of organic
nutrients than organisms lacking these appendages.

A number of other bacteria have recently been isolated that have appendages similar to the Caulobacter stalk but which have no holdfast.
Such appendages thus cannot be called "stalks," but as absorptive or-

gans they may be functionally equivalent. Many of these bacteria produce holdfasts at some point on the cell surface other than the stalk. One genus of bacteria of this latter type is *Asticcacaulis*. Another group, the budding bacteria, is sometimes confused with the stalked bacteria because of the presence of similar appendages. The budding bacteria are quite unrelated, however, since, as described below, they show a very distinct mode of cell division.

Another stalked organism that is sometimes classified with the caulobacters is *Gallionella*. This organism forms a twisted stalk containing ferric hydroxide. However, the stalk of *Gallionella* is not an integral part of the cell but is excreted from the cell surface. It contains an organic matrix to which the ferric hydroxide accumulates. *Gallionella* is a frequent organism in the waters draining bogs, iron springs, and other habitats where ferrous iron is present, usually being present in association with sheathed bacteria such as *Sphaerotilus* (see below). In very acidic waters containing iron, *Gallionella* is not present and acid-tolerant thiobacilli replace it.

Budding bacteria   Certain bacteria are unique in that they multiply by budding rather than by binary fission as do other bacteria. The two best-studied genera are *Hyphomicrobium*, which is heterotrophic, and *Rhodomicrobium*, which is photosynthetic. The process of reproduction in a budding bacterium is illustrated in Figure 18.31. The mother cell, which is often attached by its base to a solid substrate, forms a thin outgrowth that lengthens to become a hyphalike structure, and at the end of the hypha a bud forms. This bud enlarges, forms a flagellum, breaks loose from the mother cell, and swims away. Later, the daughter cell loses its flagellum and after a period of maturation forms a hypha and buds. Further buds can also form at the hyphal tip of the mother cell. Many variations on this cycle are possible. In some cases the daughter cell does not break away from the mother cell but forms a hypha from its other pole. Complex arrays of cells connected by hyphae are frequently seen (Figure 18.32). In some cases a bud begins to form directly from the mother cell without the intervening formation of a hypha, whereas in other cases a single cell forms hyphae from each end (Figure 18.32). The hypha is a direct cellular extension of the mother cell (Figure 18.33), containing cell wall, cytoplasmic membrane, ribosomes, and occasionally DNA. The resemblance of hyphae of *Hypomicrobium* to the stalks of *Caulobacter* (see above) is striking, as are resemblances in the various stages of their respective life cycles. One marked difference is that in *Caulobacter* the holdfast is formed at the end of the stalk, whereas in *Hyphomicrobium* if a holdfast is formed at all, it is found at the pole of the cell opposite the hypha. Like *Caulobacter* cells, those of *Hyphomicrobium* can aggregate and form rosettes, although because of the location of the holdfast the *Hypho-*

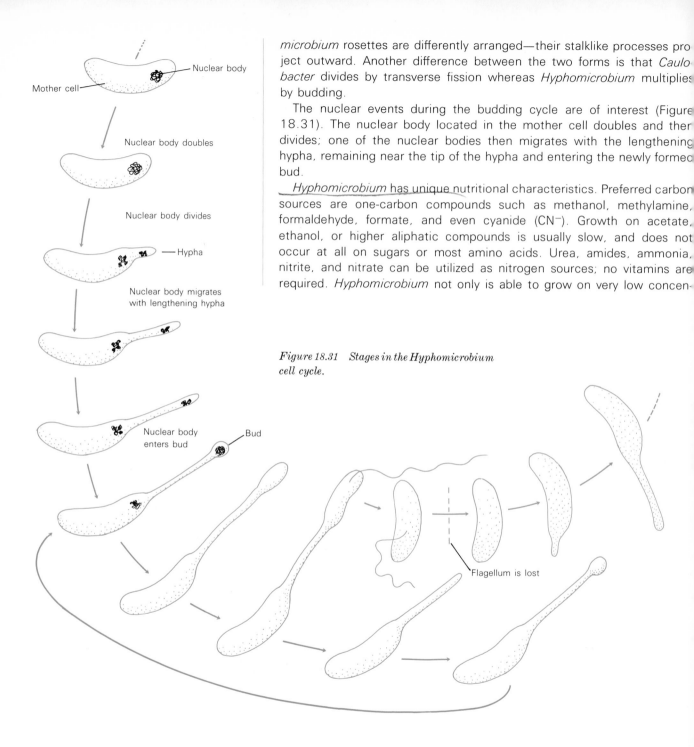

microbium rosettes are differently arranged—their stalklike processes project outward. Another difference between the two forms is that *Caulobacter* divides by transverse fission whereas *Hyphomicrobium* multiplies by budding.

The nuclear events during the budding cycle are of interest (Figure 18.31). The nuclear body located in the mother cell doubles and then divides; one of the nuclear bodies then migrates with the lengthening hypha, remaining near the tip of the hypha and entering the newly formed bud.

Hyphomicrobium has unique nutritional characteristics. Preferred carbon sources are one-carbon compounds such as methanol, methylamine, formaldehyde, formate, and even cyanide (CN^-). Growth on acetate, ethanol, or higher aliphatic compounds is usually slow, and does not occur at all on sugars or most amino acids. Urea, amides, ammonia, nitrite, and nitrate can be utilized as nitrogen sources; no vitamins are required. *Hyphomicrobium* not only is able to grow on very low concen-

Figure 18.31   Stages in the Hyphomicrobium cell cycle.

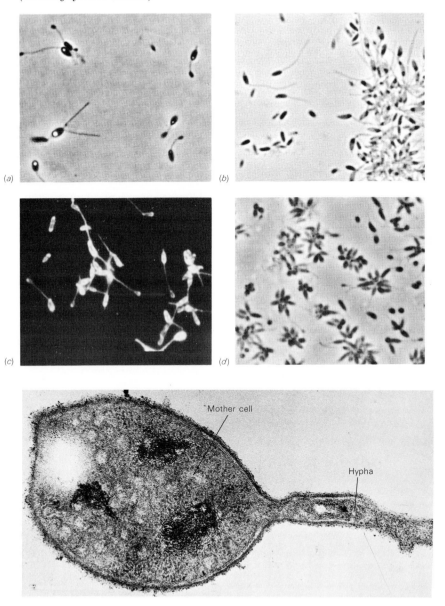

(a)

(b)

(c)

(d)

Mother cell

Hypha

Figure 18.33 Electron micrograph of a thin section of a single Hyphomicrobium cell. Magnification, 57,000×. [From S. F. Conti and P. Hirsch: J. Bacteriol. 89:503 (1965).]

18.11   *Gram-negative, polarly flagellated rods*

trations of carbon, but it will also grow in a liquid medium in the complete absence of any added carbon source, apparently obtaining its energy and carbon from volatile compounds present in the atmosphere. This has frequently led to the erroneous conclusion that *Hyphomicrobium* is a chemolithotroph, but there is no evidence that the Calvin-cycle enzymes for CO_2 fixation (see Chapter 6) are present. *Hyphomicrobium* is widespread in freshwater, marine, and terrestrial habitats, and one strain has also been isolated from the human nasopharynx, although it is not known whether this strain was part of the resident flora. Initial enrichment cultures can be prepared using a mineral-salts medium lacking organic carbon and nitrogen, to which a sample of natural material is added. After several weeks' incubation, the surface film that develops is streaked out on agar medium containing methylamine or methanol as a sole carbon source. Colonies are then checked microscopically for the characteristic *Hyphomicrobium* morphology.

Rhodomicrobium, the photosynthetic counterpart of *Hyphomicrobium,* is an obligate anaerobe that grows photosynthetically on a mineral medium containing lactate, ethanol, or other organic compound as an electron donor. It is unable to grow nonphotosynthetically on organic compounds, but otherwise it resembles the nonsulfur purple bacteria physiologically and in the nature of its chlorophyll (see below).

Sphaerotilus and Leptothrix   Members of these genera are filamentous sheathed organisms with a unique life cycle involving the formation of flagellated swarmer cells. They are common in freshwater habitats that are rich in organic matter, such as polluted streams, trickling filters, and activated sludge plants, usually being found primarily in flowing waters. In habitats where reduced iron or manganese compounds are present the sheaths may become coated with a precipitate of ferric hydroxide or manganese oxide. Not all forms cause precipitation of these metals; those that do have frequently been given the genus name *Leptothrix,* whereas the other forms are called *Sphaerotilus.*

The *Sphaerotilus* filament is composed of a chain of rod-shaped cells with rounded ends, which are enclosed in a closely fitted sheath. This thin and transparent sheath is difficult to see when it is filled with cells, but when the filament is partially empty, the sheath can easily be seen by phase-contrast microscopy (Figure 18.34) or by staining. The cells within the sheath divide by binary fission, and the new cells pushed out of the end of the sheath synthesize new sheath material. Thus sheath material is always formed at the tips of the filaments. The cells are 1 to 2 μm wide by 3 to 8 μm long and stain Gram-negatively. Individual cells are liberated from the sheaths, probably when the nutrient supply is low. These free

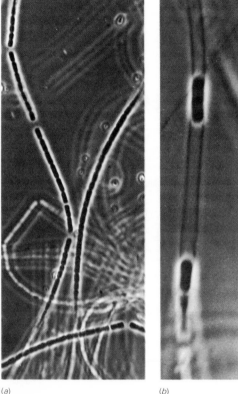

(a)   (b)

Figure 18.34   Phase-contrast photomicrographs of Sphaerotilus natans collected from a polluted stream: (a) active growth stage (magnification, 690×); (b) swarmer cells leaving the sheath (1,715×).

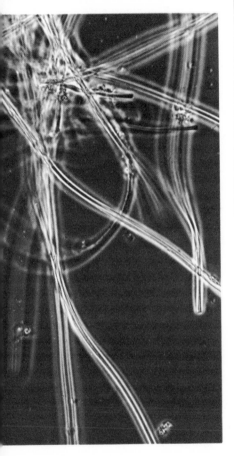

Figure 18.36 Phase-contrast photomicrograph of empty iron-encrusted sheaths of Leptothrix collected from seepage at the edge of a small swamp. Magnification, 1,875×.

cells are actively motile, the flagella being arranged lophotrichously (in a bundle at one pole). Probably the flagella are synthesized before the cells leave the sheath and, if so, may even aid in their liberation. It is thought that the swarmer cells then migrate, settle down, and begin to grow, each swarmer being the forerunner of a new filament. The sheath, which is devoid of muramic acid or other components of the peptidoglycan cell wall, is a protein-polysaccharide-lipid complex, possibly analogous to the capsules formed by many Gram-negative bacteria, but differing in that it forms a linear structure.

Sphaerotilus cultures are nutritionally versatile and are able to use a wide variety of simple organic compounds as carbon and energy sources, with inorganic nitrogen sources. Many strains require vitamin B_{12}, a substance which is frequently needed by aquatic microorganisms. Befitting its habitat in flowing waters, *Sphaerotilus* is an obligate aerobe. Its DNA base composition is 69 to 70 percent GC, which is the highest percentage known for any Gram-negative organism.

As we noted, *Sphaerotilus* is widespread in nature in aquatic environments receiving rich organic matter. Its filaments are the main component of a microbial complex that sanitary engineers call "sewage fungus." *Sphaerotilus* blooms often occur in the fall of the year in streams and brooks when leaf fall causes a temporary increase in the organic content of the water (Plate 10: Figure 18.35). In activated sludge plants *Sphaerotilus* growth is often responsible for a detrimental condition called "bulking." The tangled masses of *Sphaerotilus* filaments so increase the bulk of the sludge that it does not settle properly, thus presenting difficulties in sludge clarification. Considering the widespread distribution of *Sphaerotilus*, its importance in polluted habitats, and its interesting morphogenetic cycle, it is a pity that more extensive research has not been done on this interesting organism.

Leptothrix, as a result of its oxidation of iron or manganese compounds, causes precipitation and encrustation of insoluble salts of these metals upon the sheaths. When such deposition has occurred, the sheaths may be stable for extended periods of time; such empty structures are common in many habitats where iron oxidation is occurring (Figure 18.36). However, there is no evidence that any *Leptothrix* can grow autotrophically by oxidizing iron or manganese. At neutral pH (where these organisms live) iron oxidation will also take place spontaneously. To prove that an organism can live autotrophically at the expense of iron oxidation it must be shown that the organism will grow in the complete absence of organic compounds or other possible energy sources, and to date, no *Leptothrix* strain has been cultured under conditions of the absence of organic matter.

18.11 Gram-negative, polarly flagellated rods

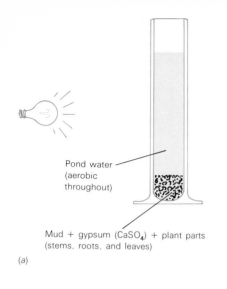

Pond water
(aerobic
throughout)

Mud + gypsum ($CaSO_4$) + plant parts
(stems, roots, and leaves)

(a)

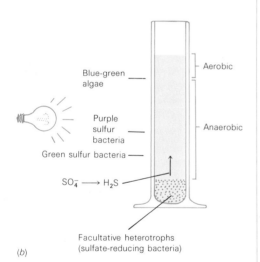

Blue-green
algae

Purple
sulfur
bacteria

Green sulfur bacteria

$SO_4^- \longrightarrow H_2S$

Aerobic

Anaerobic

Facultative heterotrophs
(sulfate-reducing bacteria)

(b)

*Figure 18.37   Winogradsky column, used
for enrichment of photosynthetic bacteria:
(a) initial state; (b) final state.*

18.12   Photosynthetic bacteria

In Chapter 6 we discussed the photosynthetic process in bacteria and compared it with that of green plants. Recall that photosynthesis in the bacteria differs in the following ways: (1) The chlorophyll pigment is of different structure and absorption spectrum; (2) only one light system is present instead of two; (3) the process is obligately anaerobic; (4) O_2 is not produced; (5) electron donors other than H_2O are used—for example, H_2S, organic compounds. Although traditionally all photosynthetic bacteria were considered to be taxonomically related, they are really quite a heterogeneous group, and the relationships among many of these bacteria may be merely ecological rather than taxonomic. At present, three major divisions of the photosynthetic bacteria are recognized, based on pigment systems and on the nature of photosynthetic electron donors; these are outlined in Table 18.14.

The photosynthetic bacteria are generally found in anaerobic zones of aquatic habitats, often where H_2S accumulates. Under certain conditions heavy blooms may occur, but often the bacteria are present in such small numbers that they can be detected only upon enrichment. By choosing appropriate conditions different kinds of photosynthetic bacteria can be isolated, as is outlined in Table 18.15. Frequently it is desirable to set up a preliminary enrichment culture in which a variety of photosynthetic bacteria can grow before one attempts to isolate a specific organism. For this purpose, the Winogradsky column (Figure 18.37) is often used: At the beginning facultative heterotrophic bacteria in the column oxidize some of the organic matter and consume O_2, making the lower parts of the column anaerobic. Sulfate-reducing bacteria then develop and produce H_2S, which diffuses toward the top of the column, creating an H_2S gradient. Green and purple photosynthetic bacteria then proliferate, frequently forming colonies on the walls of the vessel, with different species often being present at different levels in the column. Samples can then be transferred from the column to special enrichment media. In recent years improved methods of culture have made possible the isolation and growth in the laboratory of a wide variety of photosynthetic bacteria previously difficult to culture.

As we noted, the photosynthetic bacteria are commonly found in anaerobic aquatic environments such as moist and muddy soils, ditches, pools, ponds, lakes, rivers, sewage lagoons, sulfur springs, and marine habitats. The nonsulfur purple bacteria rarely occur in massive accumulations, but the green and purple sulfur bacteria are often found in such abundance that they impart a greenish or purplish cast to the water (Plate 11: Figure 18.38). In thermally stratified lakes, where the deeper water

Table 18.14   Outline of subdivisions of the photosynthetic bacteria

	Green sulfur bacteria (Chlorobacteriaceae)	Purple sulfur bacteria (Thiorhodaceae)	Nonsulfur purple bacteria (Athiorhodaceae)
Major pigment system	BChl c, abs. max. 747 nm BChl d, abs. max. 725 nm	BChl a, abs. max. 820 nm BChl b, abs. max. 1,025 nm	BChl a, abs. max. 820 nm BChl b, abs. max. 1,025 nm
Cell morphology	Motile or nonmotile rods, cocci, pleomorphic; some with gas vacuoles	Motile or nonmotile rods, cocci, pleomorphic; some with gas vacuoles	Motile rods or spirals; some multiply by budding
Photosynthetic electron donors	H_2S, thiosulfate, H_2 (organic compounds by some strains)	H_2S, thiosulfate, H_2 (organic compounds by some strains)	H_2, organic compounds (H_2S usually toxic)
Sulfur deposition	Always outside the cell	Usually inside the cell, except Ectothiorhodaceae	None
Aerobiosis	Obligate anaerobes	Obligate anaerobes	Facultative; grow in the dark aerobically, with photosynthetic growth anaerobically
Growth-factor requirements	B_{12} or none	B_{12} or none	Usually complex
DNA base composition, % GC	48–58	46–67	61–73

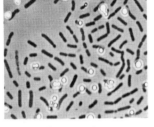

Green sulfur bacterium: *Chlorobium limicola* (magnification, 1,000×). The refractile bodies are sulfur granules deposited outside the cell.

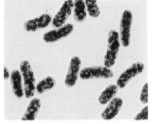

Purple sulfur bacterium: *Chromatium okenii* (magnification, 600×). Note the sulfur granules deposited inside the cells.

Nonsulfur purple bacterium: *Rhodospirillum fulvum* (magnification, 1,000×). Note the absence of sulfur granules.

(*Photographs courtesy of Norbert Pfennig.*)

does not mix with the surface layers, anaerobic conditions in the bottom often provide extremely favorable conditions for the development of the green or purple sulfur bacteria, and massive blooms occur, which are hidden from the observer who looks only at the surface of the lake. These accumulations are usually found at depths to which significant radiation still penetrates and where H_2S is high. In very deep lakes H_2S in appropriate concentrations may be found only at great depths, where sufficient radiation does not penetrate, in which case photosynthetic bacteria do not develop blooms. The student may note that the Winogradsky column (Figure 18.37) mimics in miniature the conditions prevalent in a stratified lake. The photosynthetic bacteria are able to adjust their locations in a lake to achieve optimal conditions of H_2S and light for growth. Motile species do this by being negatively tactic to O_2 (and probably positively tactic to light) so that they always remain at the most favorable position. Nonmotile species can adjust their positions by means of flotation provided them by gas vacuoles.

Morphological diversity among the photosynthetic bacteria is great (Table 18.14; Figure 18.39). Actually, most of the species and genus distinctions have been made on the basis of morphological characteristics

Table 18.15   Enrichment-culture conditions for different photosynthetic bacteria[a]

| | Conditions[b] | | | |
	Sulfide concentration	Light	Temperature	Other
Green sulfur bacteria: *Species without gas vacuoles*	$Na_2S \cdot 9H_2O$ 0.05–0.1%	350–760 nm 20–100 ft-c	20–30°C	pH 6.6–6.9
Species with gas vacuoles	0.02–0.05%	10–20 ft-c	10–20°C	
Purple sulfur bacteria: *Species without gas vacuoles*	0.05–0.1%	800–900 nm 20–100 ft-c	20–30°C	pH 7.3–7.6
Species with gas vacuoles	0.02–0.05%	10–20 ft-c	10–20°C	
Species depositing sulfur externally	0.05–0.1%	20–100 ft-c	30–40°C	Halophilic forms, 20% NaCl
Nonsulfur purple bacteria: *Species with BChl a*	No sulfide; organic electron donors, e.g., acetate, succinate, benzoate, or H_2	Visible light, 20–100 ft-c	25–30°C	Vitamins and growth factors; pH 7.0–7.5
Species with BChl b		Infrared radiation, 1,000–1,050 nm		

[a] Courtesy of Norbert Pfennig.
[b] All enrichment cultures contain basal mineral salts medium plus bicarbonate as source of CO_2 and vitamin B_{12}. Optimum pH range, 6 to 8. Incubation is anaerobic in sealed bottles or under N_2. NaCl at 1 to 3 percent must be added for enrichments from marine environments.

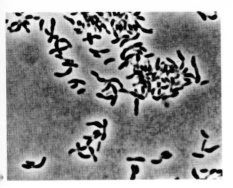

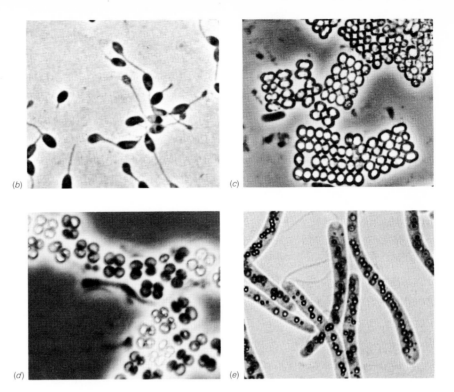

Figure 18.39 Photosynthetic bacteria:
(a) Rhodopseudomonas palustris, nonsulfur purple bacteria (magnification,
1,360×); (b) Rhodomicrobium vannielii,
budding nonsulfur purple bacteria (2,000×);
(c) Thiopedia sp., purple sulfur bacteria
(1,360×); (d) Thiocapsa sp., purple sulfur
bacteria (1,360×); (e) Thiospirillum jenense,
purple sulfur bacteria (1,200×). (Photographs (a), (c), and (d) courtesy of Peter
Hirsch, photographs (b) and (e) courtesy of
Norbert Pfennig.)

often discovered not with pure cultures but with material collected from nature. Since cell structure frequently is greatly affected by environmental conditions of growth, detailed taxonomic studies of many photosynthetic bacteria must await comparative cultural studies. In virtually all cases motile forms are polarly flagellated, and this has prompted the suggestion that the photosynthetic bacteria are related to the pseudomonads.

Considerable interest exists in the fine structure of the photosynthetic membrane systems of these bacteria. In some cases the photosynthetic membrane system is in layers parallel to the plasma membrane [Figure 18.40(a)], whereas in other organisms the photosynthetic membrane system consists of round tubes or vesicles [Figure 18.40(b)]. At high light intensity both the amount of chlorophyll and the extent of the photo-synthetic membrane system are decreased.

In the nonsulfur purple bacteria, pigment and membrane content are additionally influenced by O_2. We noted that these bacteria can grow photosynthetically under anaerobic conditions and heterotrophically in the dark when O_2 is present. In the latter case the pigment content becomes very low, with cultures appearing almost colorless and possessing comparatively few peripheral invaginations of the cytoplasmic membrane. Thus the photosynthetic bacteria seem to have some mechanism for controlling simultaneously the quantity of both pigment and membrane, but the molecular nature of this control is not yet known. It may seem paradoxical that the photosynthetic pigments, which are formed only in the light, are found in greater concentrations in cells grown in dim than in bright light. The ecological explanation for this is that in dim light it is of advantage to the cell to have a large amount of light-gathering pigment in order that more of the light reaching the cell can be captured and converted to chemical energy. When growing aerobically in the dark on organic compounds, pigment synthesis is repressed so that energy made available by oxidative reactions is not diverted into unnecessary materials.

We discussed in Chapter 16 the various nutrient cycles carried out by natural ecosystems. However, the role of photosynthetic bacteria in nutrient cycling in nature is not completely understood. We do know, however, that in lakes where extensive blooms occur much of the primary production of organic matter may be carried out by photosynthetic bacteria; CO_2 fixation by these organisms often exceeds manyfold that of the aerobic plants of the surface waters.

Figure 18.40 (a) Electron micrograph of a thin section of Ectothiorhodospira mobilis, a purple sulfur bacterium. Note the arrays of photosynthetic lamellae associated in flat sheets. Magnification, 40,000×. [From C. C. Remsen, S. W. Watson, J. B. Waterbury, and H. G. Trüper: J. Bacteriol. 95:2374 (1968).] (b) Electron micrograph of a thin section of Chromatium sp., strain D, a purple sulfur bacterium. The photosynthetic membranes are individual long tubes or vesicles. Magnification, 79,000×. (Courtesy of Jeffrey C. Burnham and S. F. Conti.)

(a)

(b)

A variety of bacteria show gliding motility. These organisms have no flagella, and we do not know how they are able to move across a solid surface. Although the gliding bacteria are nonchlorophyllous, certain members of this group are definitely related morphologically to the blue-green algae (many of which also move by gliding), but with others the relationship is tenuous. Indeed, it seems unwise at this time to consider all the gliding bacteria as related, especially since there may be different mechanisms of motility among the gliding bacteria. Many organisms that show gliding motility have been little studied, and the taxonomy of the group as a whole is in a state of flux. However, all the gliding bacteria that have been studied possess a typical Gram-negative cell wall. Table 18.16 gives a brief outline of some of the genera of gliding bacteria. We shall now discuss the biology of several of the better-studied genera.

Beggiatoa Organisms of this genus are morphologically very similar to blue-green algae, resembling especially closely the chlorophyllous pro-

Table 18.16 Characteristics of some genera of gliding bacteria

	Properties
Cellulose and chitin digesting: Cytophaga (no microcysts; DNA, 33–39% GC); Sporocytophaga (microcysts formed, DNA, 35% GC)	Unicellular, rod-shaped, chemoorganotrophic
Not digesting cellulose or chitin: Flexibacter (DNA, 30–47% GC)	Unicellular, rod-shaped, chemoorganotrophic
Saprospira (DNA, 35–48% GC)	Helical or spiral-shaped, chemoorganotrophic
Vitreoscilla (DNA, 44–45% GC); Microscilla (DNA, 32–44% GC)	Filamentous, chemoorganotrophic
Beggiatoa	Filamentous, chemoorganotrophic, producing S granules from H_2S
Leucothrix (DNA, 46–50% GC); Thiothrix	Filamentous, chemoorganotrophic or chemolithotrophic, life cycle involving gonidia and rosette formation
Simonsiella	Cells in chains or short filaments, chemoorganotrophic, occur in oral cavity of humans and animals
Fruiting myxobacteria (DNA, 67–71% GC): Myxococcus, Sorangium, Chondromyces, Archangium, Podangium	Unicellular, rod-shaped, life cycle involving aggregation, fruiting-body formation and microcyst formation

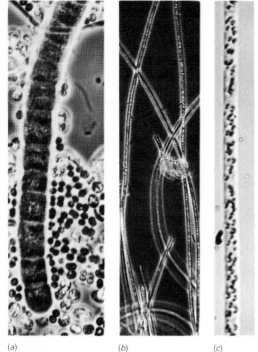

(a)   *(b)*   *(c)*

caryote *Oscillatoria*. The filaments of *Beggiatoa* are usually quite long and in addition to moving by gliding they can flex and twist so that many filaments may become intertwined to form a complex tuft. *Beggiatoa* is found in nature primarily in habitats rich in H_2S, such as sulfur springs, decaying seaweed beds, mud layers of lakes, and waters polluted with sewage, and in these habitats the filaments of *Beggiatoa* are usually filled with sulfur granules (Figure 18.41). It was with *Beggiatoa* that Winogradsky first demonstrated that a living organism could oxidize H_2S to S and then to SO_4^{2-}, leading him to formulate the concept of chemolithotrophy. However, all pure cultures of *Beggiatoa* so far isolated grow heterotrophically on organic compounds such as acetate, succinate, and glucose, and when H_2S is provided as an energy source, they still require organic substances for growth. Thus it seems that *Beggiatoa* can use H_2S as an energy source but cannot use CO_2 as a sole carbon source. Organisms that can use inorganic compounds or light as energy sources but cannot use CO_2 as sole carbon source have been called *mixotrophs*.

Leucothrix and Thiothrix   These two genera are related in cell structure and life cycle. *Thiothrix,* a chemolithotroph that oxidizes H_2S (see Figure 6.9), is probably an obligate autotroph, although this has not been proved with pure cultures. *Leucothrix* is the chemoorganotrophic counterpart of *Thiothrix,* and since its members have been amenable to cultivation, the details of its life cycle and physiology are fairly well established. *Leucothrix* is a filamentous organism that has been found in nature only in marine environments, where it grows most commonly as an epiphyte on marine algae (Figure 18.42). *Leucothrix* filaments are usually 2 to 5 μm in diameter and may reach lengths of 0.1 to 0.5 cm. The filaments have clearly visible cross walls, and cell division is not restricted to either end but occurs throughout the length of the filament. The free filaments never glide (thus distinguishing them from *Beggiatoa*) although they occasionally wave back and forth in a jerky fashion. Under environmental conditions unfavorable to rapid growth, individual cells of the filaments become round and form ovoid structures called *gonidia,* which are released individually, often from the tips of the filaments (Figure 18.43). The gonidia are able to glide in a jerky manner when they come into contact with a solid surface. They settle down on solid surfaces, synthesize a holdfast, and through growth and successive cell divisions form new filaments. Presumably, in nature the gonidia are elements of dispersal, enabling the organism to spread to other areas. If there are high concentrations of gonidia, individual cells may aggregate, probably because of reciprocal attraction; they then synthesize a holdfast that causes their ends to adhere in a rosette, and new filaments then grow out (Figure 18.44). Rosette formation is found in both *Leucothrix* and *Thiothrix* and

Figure 18.41   (a) Phase-contrast photomicrograph of a portion of a filament of a large species of Beggiatoa collected from a small pond rich in organic matter. Magnification, 1,060×. Note the resemblance to Oscillatoria (Table 18.17). The small cells around the filament are Thiocapsa sp., a photosynthetic bacterium. (Courtesy of Peter Hirsch.) (b) Beggiatoa. Magnification, 350×. This species has filaments of considerably smaller diameter than those of (a). Organisms were collected from the sediment of a small river receiving additions of treated sewage and observed with phase-contrast microscope. The filaments are packed with sulfur granules that appear as brightly shining inclusions. (c) The same species of Beggiatoa as in (b), photographed by Nomarski interference contrast to emphasize the sulfur granules. Magnification, 1,575×.

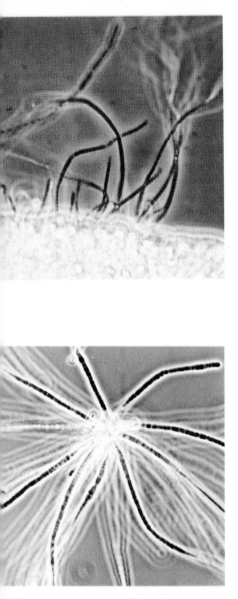

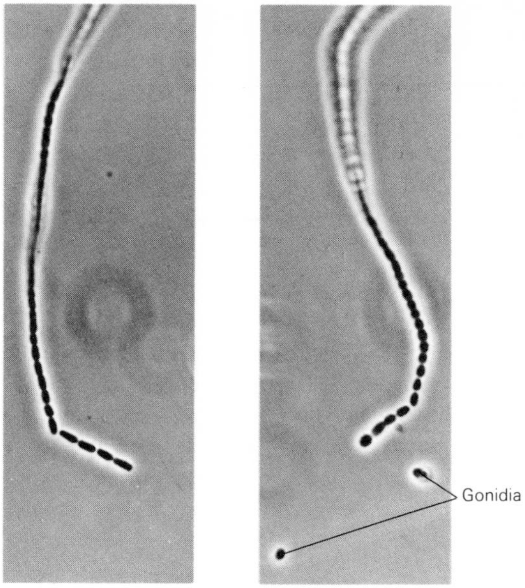

Figure 18.43   Release of gonidia from filaments of Leucothrix mucor. Magnifications, 725×.

Gonidia

Figure 18.44   Rosette of Leucothrix mucor. Magnification, 630×.

is an important means of distinguishing these organisms from many other filamentous bacteria. The life cycle of *Leucothrix* is summarized in Figure 18.45. Another interesting characteristic of *Leucothrix* is the ability of filaments to grow in such a way that knots are formed (Figure 18.46). These occur mainly when the organism is growing in rich liquid culture media, where filamentous growth is rapid. Although knot formation occurs sporadically in some filamentous algae, it is frequent enough in *Leucothrix* to be a distinguishing characteristic. Actually, knots are formed as a part of the growth process: Faster growth on one side of the filament than on the other causes the filament to form a loop, through which the tip of the filament can pass. Once the knot is formed it cannot be untied; rather, the cells in the region of the knot eventually fuse and form a bulb that later is released from the rest of the filament, thus separating the long filament into two shorter filaments. Knots are also seen occasionally in *Thiothrix* filaments in sulfur springs, where large accumulations of this organism appear.

Leucothrix is a strict aerobe that grows best under conditions of good aeration. In nature it is most usually found associated with seaweeds along open sea coasts where wave action or current is high. Most strains are psychrophilic; *Leucothrix* is rarer in warm tropical waters. Most strains of *Leucothrix* have no vitamin requirements and can grow on simple sugars or amino acids as sole carbon and energy souces. However, better growth is usually seen in media containing yeast extract or other complex organic materials.

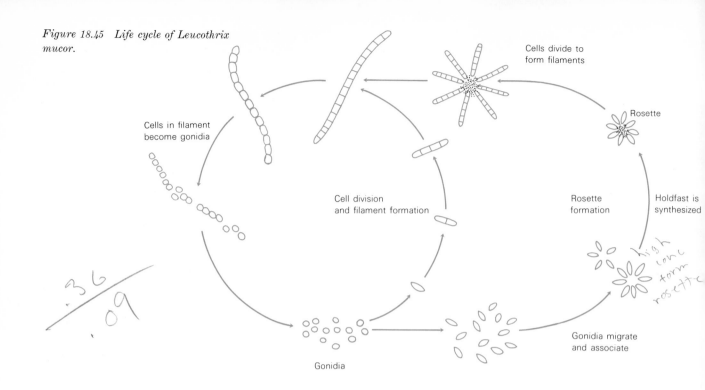

Figure 18.45 Life cycle of Leucothrix mucor.

Cells divide to form filaments

Rosette

Cells in filament become gonidia

Cell division and filament formation

Rosette formation

Holdfast is synthesized

Gonidia

Gonidia migrate and associate

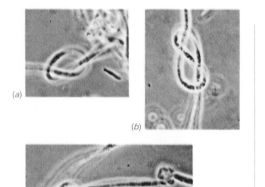

(a)

(b)

(c)

Figure 18.46 Knots formed by filaments of Leucothrix mucor. Magnifications, 670×. (From T. D. Brock: Science 144:870-872. Copyright 1964 by the American Association for the Advancement of Science.)

Leucothrix is isolated by streaking freshly collected marine algal fronds on a seawater agar medium to which 0.1 percent glutamate is added as a sole carbon, nitrogen, and energy source. If synthetic seawater is used, a small amount of phosphate (50 μg/ml) must also be added. The plates are incubated at 20 to 25°C and are examined after 1 da for the presence of *Leucothrix* colonies. Colonies can usually be recognized with a dissecting microscope by the filamentous nature of the units present. Purification is effected by restreaking on fresh agar plates.

Fruiting myxobacteria   The fruiting myxobacteria exhibit the most complex behavioral patterns and life cycles of any procaryotic organisms. The vegetative cells of the fruiting myxobacteria are simple nonflagellated, Gram-negative rods that glide across surfaces and obtain their nutrients primarily by causing the lysis of other bacteria. Under appropriate conditions a swarm of vegetative cells aggregate and construct "fruiting bodies," within which some of the cells become converted into resting structures called *myxospores*. (These structures have also been called *cysts* or *microcysts*, but the word "myxospore" is now preferred.) It is the ability to form complex fruiting bodies that distinguishes the myxobacteria from all other procaryotes. Since the vegetative cells of fruiting myxobacteria look like those of nonfruiting gliding bacteria, it is only through observation of the fruiting bodies that these organisms can be identified. (Myxobacterial fruiting bodies might be confused with those of the slime molds—see Chapter 19—but the latter are usually much larger and structurally more complex.)

The fruiting bodies of the myxobacteria vary from simple globular masses of myxospores in loose slime to complex forms with a fruiting-body wall and a stalk (Plates 12 and 13: Figure 18.47). The fruiting bodies are often strikingly colored. Occasionally they can be seen with a dissecting microscope on pieces of decaying wood or plant material. Fruiting bodies of myxobacteria often develop on dung pellets (for example, those of the rabbit) that have been incubated for a few days in a moist chamber [Figure 18.47(g)]. Although the vegetative cells are common in soils, the fruiting bodies themselves are not formed there. An effective means of isolating fruiting myxobacteria is to prepare petri plates of water agar (1.5 percent agar in distilled water with no added nutrients) on which is spread a heavy suspension of any of several bacteria that the myxobacteria can lyse and use as a source of nutrients (for example, *Sarcina lutea*). In the center of the plate a small amount of soil, decaying bark, or other natural material is placed. Myxobacteria in the inoculum lyse the *Sarcina* cells and use their liberated products as nutrients; as they grow they swarm out across the plate from the inoculum site. After several days to a week, the plates are examined under a dissecting microscope for fruiting bodies, and pure cultures are obtained by transferring to organic media cells from the fruiting body heads or from the edge of the swarm.

The life cycle of a typical fruiting myxobacterium is shown in Figure 18.48. The vegetative cells are typical Gram-negative rods and do not reveal in their fine structure any clue to either their gliding motility or to their ability to aggregate and form "fruits." The vegetative cells of many strains grow poorly or not at all when dispersed in liquid medium but grow well on the surface of solid substrates. A vegetative cell usually excretes a slime, and as it moves across a solid surface, it leaves a slime trail behind [Figure 18.49(a) and (b)]. This trail is preferentially used by other cells in the swarm so that a characteristic radiating pattern is soon created with cells migrating along slime trails. The vegetative cells obtain their nutrients primarily by lysing and digesting other bacteria. A wide variety of Gram-positive and Gram-negative bacteria as well as fungi, yeasts, and algae can be used as food sources. A few fruiting myxobacteria can also use cellulose. Many myxobacteria can be grown in the laboratory on media containing peptone or casein hydrolysate, which provides organic nutrients in the form of amino acids or small peptides;

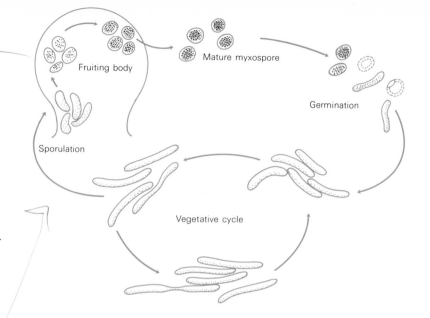

Figure 18.48 Life cycle of Myxococcus xanthus. (Courtesy of Hans Reichenbach and Martin Dworkin.)

Fruiting body

Mature myxospore

Germination

Sporulation

Vegetative cycle

Figure 18.49 (a) Photomicrograph of a swarming colony (9-mm diameter) of Myxococcus xanthus on agar; (b) single cells of M. fulvus from an actively gliding culture, showing the characteristic slime tracks on the agar. Magnification, 205×. (Courtesy of Hans Reichenbach.) (c) fruiting body of Stigmatella aurantiaca. [From H. Reichenbach: J. Gen. Microbiol. 58:1. London: Cambridge University Press, 1969.]

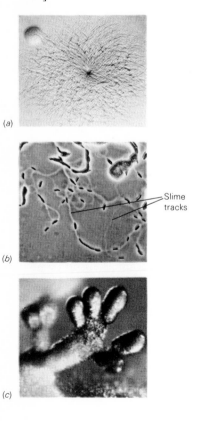

(a)

(b)

Slime tracks

(c)

carbohydrates do not ordinarily promote or stimulate growth, and vitamins are usually not required. Inability to utilize carbohydrates reflects the lack of certain enzymes of the Embden-Meyerhof pathway. The organisms are typical aerobes with a well-developed citric acid cycle and cytochrome system.

Fruiting-body formation does not occur so long as adequate nutrients for vegetative growth are present, but upon the exhaustion of amino acids, the vegetative swarms are induced to fruit. Cells aggregate, possibly through a chemotactic response, with the cells migrating toward each other and forming mounds or heaps. As the cell mounds become higher, the differentiation of the fruiting body into stalk and head begins (Figure 18.49). The stalk is composed of nonliving slime, within which a few cells may be trapped. The majority of the cells accumulate in the fruiting-body head and undergo differentiation into *myxospores*. These are refractile cells that are either spherical or rod-shaped, depending on the species. Accompanying these changes in cell structure is the accumulation on the outside of the myxospore of a large capsule (Figure 18.50). The myxospore lacks the highly differentiated appearance of the bacterial endospore or of the *Azotobacter* cyst (see Section 18.10). Compared to the vegetative cell, the myxospore is more resistant to drying, sonic vibration, UV radiation, and heat, but the degree of heat resistance is much less than that of the bacterial endospore. It seems likely that the main function of

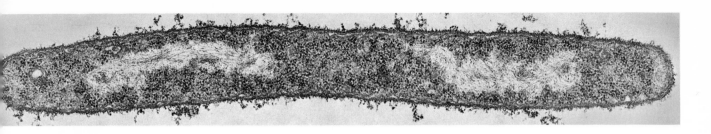

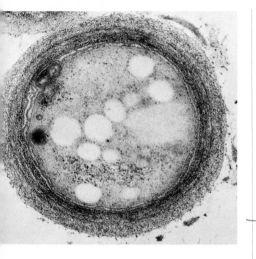

Figure 18.50 (a) Electron micrograph of a thin section of a vegetative cell of M. xanthus (magnification, 40,000×); (b) myxospore (microcyst) of M. xanthus, showing the multilayered outer wall (56,000×). (Courtesy of Herbert Voelz.)

the myxospore is to prevent desiccation during organism dispersal. The myxospore germinates by a localized rupture of the capsule, with the growth and emergence of a typical vegetative rod.

Myxobacteria are invariably colored by carotenoid pigments. Pigment formation is promoted by light, and at least one function of the pigment is photoprotection, as was described for other bacteria in Chapter 8. Since in nature the myxobacteria usually form fruiting bodies in the light, the presence of these photoprotective pigments is understandable.

The relationship of the fruiting myxobacteria to nonfruiting gliding bacteria has been under investigation for some time. The genus *Cytophaga* contains nonfruiting forms that are frequently associated with cellulose and chitin digestion in soil and water. Related to *Cytophaga* is the genus *Sporocytophaga*, which forms myxosporelike structures but not fruiting bodies. Recent studies on DNA base compositions strongly suggest, however, that neither *Cytophaga* nor *Sporocytophaga* is closely related to the fruiting myxobacteria since all the latter have base compositions in the range of 67 to 71 percent GC, whereas *Cytophaga* and *Sporocytophaga* have base compositions of 33 to 42 percent GC.

The fruiting myxobacteria exhibit the most complicated behavioral phenomena seen among procaryotes. They provide experimental material for the study of a number of interesting problems in developmental microbiology, microbial ecology, and microbial evolution, and their fruiting bodies are esthetically attractive objects.

18.14   Blue-green algae

The procaryotic nature of the blue-green algae is well established (see Chapter 2), but their photosynthetic mechanism resembles that of eucaryotes rather than that of photosynthetic bacteria (see Chapter 6). The blue-green algae, or Cyanophyta, comprise a large and heterogeneous group of organisms, which have been studied primarily by algologists. Both unicellular and filamentous forms are known. The fine structure of

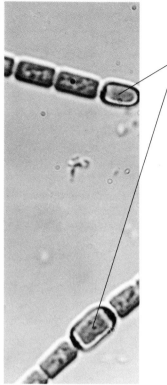

(a)

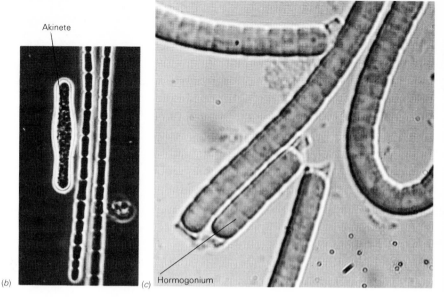

Akinete

(b)

Hormogonium

(c)

Figure 18.51 Characteristic cellular structures of blue-green algae:
(a) heterocysts of Anabaena sp. by Nomarski interference contrast
(magnification, 1,575×); (b) akinete (resting spore) of Anabaena sp.
by phase contrast (720×); (c) hormogonium of Oscillatoria sp. by
Nomarski interference contrast (1,575×). Note that the cells at
both ends of the hormogonium are rounded.

the cell wall of some of these algae is similar to that of Gram-negative bacteria. Many blue-green algae produce extensive mucilaginous envelopes, or sheaths, that bind groups of cells or filaments together. The photosynthetic lamellar membrane system is often complex and multilayered (see Figure 2.23). Blue-green algae have only one form of chlorophyll, chlorophyll a, and all of them also have characteristic biliprotein pigments (see Chapter 6), which are responsible for the blue-green color and function as accessory pigments in photosynthesis. Among the cytoplasmic structures seen in many blue-green algae are membrane-bound *gas vacuoles*, which are especially common in species that live in open waters (planktonic species). Their function is probably to provide the organism with flotation so that it may remain where light is best. (Gas vacuoles are also found in some photosynthetic bacteria—see Section 18.12.) Some blue-green algae form *heterocysts*, which are rounded, seemingly more or less empty cells, usually distributed individually along a filament or at one end [Figure 18.51(a)]. Heterocysts arise from vegetative cells and may be resting cells although they have rarely been observed to germinate. All nitrogen-fixing species of blue-green algae form heterocysts, and evidence is accumulating that these structures are sites of nitrogen fixation. The heterocysts have intercellular connections with adjacent vegetative cells, and there is mutual exchange of materials between these cells, with products of photosynthesis moving from vegetative cells to heterocysts and products of nitrogen fixation probably moving from heterocysts to vegetative cells. Many, but by no means all, blue-green algae exhibit gliding motility; flagella have never been found. Among the filamentous blue-green organisms, fragmentation of the filaments often occurs when sections of filaments become *hormogonia* [Figure 18.51(*c*)], which break away from the filaments and glide off. In some species resting spores, or *akinetes* [Figure 18.51(*b*)], are formed, which tide the organism over periods of drying or freezing. These are cells with thickened outer walls; they germinate through the breakdown of the outer wall and the outgrowth of a new vegetative filament. However, even the vegetative cells of many blue-green algae are relatively resistant to drying or low temperatures.

The nutrition of blue-green algae is simple. Most species tested are obligate phototrophs, being unable to grow in the dark on organic compounds. Vitamins are not required, and nitrate or ammonia is used as nitrogen source. Nitrogen-fixing species are also common. Some obligately autotrophic blue-green algae can assimilate simple organic compounds such as glucose and acetate, if light is present. Apparently they are unable to make ATP by oxidation of organic compounds, but if ATP is provided by means of photosynthetic phosphorylation, organic compounds can be utilized as carbon sources.

Blue-green algae are widely distributed in nature in terrestrial, freshwater, and marine habitats. In general they are more tolerant to environmental extremes than are eucaryotic algae and are often the dominant or sole photosynthetic organisms in hot springs (Section 8.1), saline lakes, and other extreme environments. Many members are found on the surfaces of rocks or soil. In desert soils subject to intense sunlight, blue-green algae often form extensive crusts over the surface, remaining dormant during most of the year and growing during the brief winter and spring rains. They are common inhabitants of the soils of greenhouses. In shallow marine bays, where relatively warm seawater temperatures exist, blue-green algal mats of considerable thickness may form. Freshwater lakes, especially those that are fairly rich in nutrients, may develop blooms of blue-green algae (Plate 14: Figure 18.52). A few algae are symbionts of liverworts, ferns, and cycads; a number are found as the algal component of lichens (Section 14.1). In the case of the water fern *Azolla* it has been shown that the blue-green endophyte (a species of *Anabaena*) fixes nitrogen that is available to the plant. Some of the algal symbionts of corals and other invertebrates also are blue-green algae (Section 14.7).

The isolation and culture of blue-green algae makes use of the same methods as for the eucaryotic algae; these methods are discussed in Section 19.3. One particular problem with the blue-green algae is the frequent difficulty in obtaining cultures free of contaminating heterotrophic bacteria. This difficulty arises from the fact that many blue-green algae form heavy gelatinous sheaths or matrixes within which bacteria grow. With actively gliding forms, purification may be effected by allowing the alga to move across the surface of a nonnutrient agar, the mechanical action of the motion sometimes cleansing the algal surface of unwanted bacteria. Antibiotics for obtaining pure cultures of eucaryotic algae usually are not successful since the blue-green algae are sensitive to the same antibiotics as are bacteria (see Chapter 9). One somewhat successful technique is the use of UV radiation, to which the blue-green algae are more resistant than are most heterotrophs, so that the latter may be selectively killed.

The taxonomy of blue-green algae has been based almost exclusively on morphological characteristics of material collected from nature. Since environmental conditions may markedly alter morphology, comparative studies of many isolates grown under reproducible conditions are greatly needed, and it may be anticipated that many aspects of blue-green algal taxonomy will be altered as cultural studies become more widespread.

Several taxonomic orders of blue-green algae have been recognized, which are described and illustrated in Table 18.17.

DNA base compositions have been determined on a variety of blue-green algae. Those of the Chroococcales vary from 35 to 71 percent GC, a range so wide as to suggest that this order contains many members with little relationship to each other. On the other hand, the values for the Nostocales vary much less, from 39 to 51 percent GC. It is interesting to note that *Leucothrix*, a colorless filamentous organism thought to be related to the blue-green algae, has a base composition of 49 percent GC, which is in the same range as that of the filamentous blue-green algae. Because of the paucity of physiological characteristics that can be used in classifying blue-green algae, techniques of molecular taxonomy will probably play a major role in their elaboration and classification.

18.15   Spirochetes

The spirochetes are bacteria with a unique morphology and mechanism of motility. They are widespread in aquatic environments and in the bodies of warm-blooded animals. Many of them cause diseases of animals and

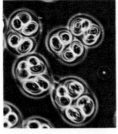

(b)

(d)

(f)

Without hormogonia:

1 Chroococcales Unicellular forms, sometimes occurring in irregular packets or colonies; cell division by binary fission, the cells showing no differentiation from apex to base. Examples: *Chroococcus* [micrograph (a), magnification, 380×], *Gloeocapsa* [(b), 480×].

2 Chamaesiphonales Unicellular, filamentous, or colonial epiphytes or lithophytes; cells exhibit marked polarity from apex to base, usually having a holdfast that permits attachment to solid substrata; cell division by internal septation or by the formation of spherical cells (gonidia) at the ends of filaments. Examples: *Chamaesiphon, Dermocarpa.*

3 Pleurocapsales Filaments showing differentiation into aerial and nonaerial elements; no heterocysts; cell division by simple cross-wall formation or by internal septation. Example: *Pleurocapsa.*

With hormogonia:

4 Nostocales Filaments showing no differentiation into aerial and nonaerial elements; unbranched or showing false branching; heterocysts and akinetes frequently present.

Oscillatoriaceae Unbranched; no akinetes or heterocysts. Examples: *Oscillatoria* [(c), 240×], *Spirulina* [(d), 360×].

Nostocaceae Unbranched, heterocysts; akinetes frequently produced. Examples: *Nostoc, Anabaena* [(e), 360×].

Rivulariaceae Unbranched or false branching; filaments tapering from base to tip; heterocysts usually at the base; akinetes in some species. Examples: *Rivularia, Calothrix.*

Scytonemataceae False branching; heterocysts often at the point of branching. Examples: *Tolypothrix* [(f), 360×], *Scytonema.*

5 Stigonematales Filaments showing aerial and nonaerial differentiation; often show true branching; heterocysts; pit connections between cells; akinetes rare. Examples: *Mastigocladus, Stigonema.*

man, of which the most important is syphilis, caused by *Treponema pallidum.* The spirochete cell is typically a slender, flexuous body in the form of a spiral, often of considerable length. Motility probably occurs by means of a unique structure, the axial filament, or axostyle, which is a fiber or bundle of fibers attached at the cell poles and wrapped around the cell in a spiral fashion (see Figure 2.36). It is thought that the tension exerted on the flexible cell by the axial fibril forces the cell into a spiral configuration and that the contractions and relaxations of this axial filament cause the cell alternately to shorten and lengthen, thus forcing the cell through the water (Figure 18.53). This pattern of motility is quite

18.15 Spirochetes

617

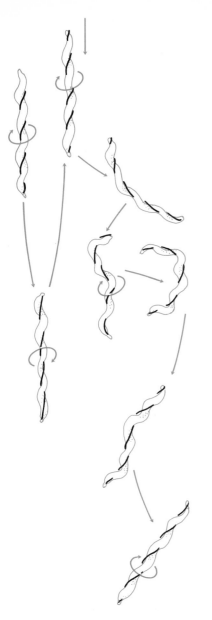

Figure 18.53   Manner of movement of a spirochete.

different from that of spiral-shaped flagellated bacteria. (*Spirillum*), which move through the water in a rapid whiplike pattern; the spirochetes move in the manner of a snake. Spiral-shaped gliding bacteria of the genus *Saprospira* have often been confused with spirochetes. However, *Saprospira* cells lack axial filaments and show gliding motility as described in section 18.13. Cell division in spirochetes is by transverse fission, the axial filament usually being the last structure to separate (Figure 18.54).

The spirochetes are classified primarily on the basis of cell size, number of spirals, structure of the axial filament, habitat, and pathogenicity. Because they are often difficult to cultivate, they have not been classified on the basis of physiological properties, and consequently the genera currently recognized may be subdivided at some future time. Table 18.18 lists the major genera and some of their characteristics.

Spirochaeta and Cristispira   One species of the genus *Spirochaeta* is *S. plicatilis,* a fairly large organism that was the first spirochete to be discovered; it was reported by C. G. Ehrenberg in the 1830s but has not yet been successfully cultured. It is a common organism in mud of both freshwater and seawater habitats and therefore is probably anaerobic. Another species, *S. stenostrepta,* has been cultured and is shown in Figure 18.55. The genus *Cristispira* contains organisms with a unique distribution, being found in nature only in the crystalline style of certain molluscs, such as clams and oysters. The crystalline style is an amorphous gel extruded from a sac and rotated against a hard surface of the digestive tract, thereby mixing with and grinding the small particles of food. Being large spirochetes, the cristispirae can readily be seen microscopically within the style as they rapidly rotate forward and backward in corkscrew fashion. The axial filament of *Cristispira* is composed of a large number of fibrils (Figure 18.56). *Cristispira* may occur in both freshwater and marine molluscs, but not all species of molluscs possess them. Unfortunately, *Cristispira* has not been cultured, so that the physiological reason for its restriction to this unique habitat is not known. There is no evidence that *Cristispira* is harmful to its host; in fact, the organisms may be more common in healthy than in diseased molluscs.

Treponema   Of the species of *Treponema, T. pallidum,* the causal agent of syphilis, is the best known. The cell of *T. pallidum* is extremely thin, and for this reason it is almost invisible with the ordinary light microscope, whether unstained or colored by conventional aniline dyes. When stained by the silver impregnation method the thickness of the cell is increased sufficiently so that it can be seen with the light microscope. Living cells are visible unstained by use of the dark-field microscope or after staining with fluorescent antibody; for many years dark-field micro-

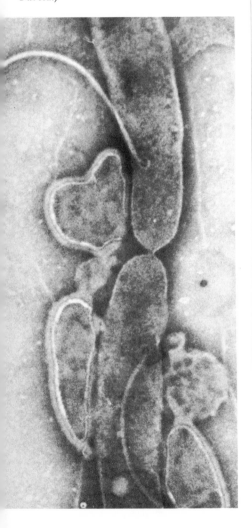

Figure 18.54 Spirochete cell (Spirochaeta aurantia) in division. This is an electron micrograph of a negatively stained preparation. Note the insertion of the axial filament. Magnification, 39,000×. (Courtesy of R. Joseph, S. C. Holt, and E. Canale-Parola.)

Table 18.18 Major genera of spirochetes and their characteristics

	Dimensions, μm	General morphology	Number of fibrils in axial filament	Habitat	Diseases
Cristispira	40–120 × 0.5–3.0	Coarse spirals; axial filament easily visible by phase-contrast microscopy	>100	Crystalline style (digestive organ of molluscs)	None
Spirochaeta	30–500 × 0.25–1.0	Large, tight spirals; axial filament not easily visible	2	Aquatic, free living	None
Treponema	3–18 × 0.25–0.5	Coarse or tight spirals, often invisible with light microscope; cannot be stained with ordinary aniline dyes	2–15	Man and other mammals; some aquatic	Syphilis, yaws
Borrelia	8–16 × 0.25–0.5	Coarse or tight spirals, often invisible with light microscope; can be stained with ordinary aniline dyes	Not known	Man and other mammals; arthropods	Relapsing fever
Leptospira	5–10 × 0.1–0.2	Tightly coiled spirals, often invisible with light microscope; cannot be stained with ordinary aniline dyes	2	Man and other mammals	Leptospirosis

scopy has been used to examine exudates from suspected syphilitic lesions. In nature *T. pallidum* is restricted to man, although artificial infections have been established in rabbits and monkeys. Although the causal agent of syphilis has never been successfully cultured, nonpathogenic organisms morphologically resembling *T. pallidum,* which were derived

Figure 18.55 Spirochaeta stenostrepta, by phase-contrast microscopy. Magnification, 2,670×. [From E. Canale-Parola, S. C. Holt, and Z. Udris: "Isolation of free-living, anaerobic spirochetes," Archiv für Mikrobiologie, 59:41. New York: Springer-Verlag, 1967.]

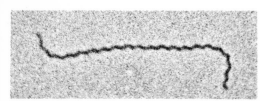

from syphilitic patients, have been cultured. The cultured forms may be nonpathogenic variants of *T. pallidum,* or they may be unrelated species. Those treponemas which have been cultured are obligate anaerobes requiring complex media, and all grow slowly, with generation times of 4 to 18 hr. Some relevant information on the pathogenic *T. pallidum* can be obtained by studying the effects of various environmental conditions on the motility of organisms taken directly from syphilitic lesions. By such studies it has been shown that *T. pallidum* is quite sensitive to increased temperature, being rapidly killed by exposure to 41.5 to 42.0°C. This had been the basis for increasing the body temperature of the patient in order to kill the spirochete, a procedure now supplanted by antibiotic therapy. The heat sensitivity of *T. pallidum* is also reflected in the fact that the organism becomes most easily established in cooler sites of the body, such as the male genital organs, although once established in other areas of the body it will multiply there. Infection of rabbits is also most extensive in cooler sites such as the testicles or skin; indeed, artificial cooling of rabbit skin results in a dramatic increase in the number of lesions. The organism is rapidly immobilized by heavy metals, which were formerly used in therapy, and by drying. As a matter of fact, the extreme sensitivity to drying of the treponema at least partially explains why transmissal between persons is only by direct contact, usually sexual intercourse.

The disease in humans exhibits variable symptoms. The organism does not pass through unbroken skin, and initial infection most probably takes place through tiny breaks in the epidermal layer. In the male, initial infection is usually on the penis, whereas in the female it is most often in the vagina, cervix, or perineal region. In about 10 percent of the cases infection is extragenital, usually in the oral region. During pregnancy, the organism can be transmitted from an infected woman to the fetus; the disease acquired in this way by an infant is then called congenital syphilis. The organism mutiplies at the initial site of entry and a characteristic primary lesion known as a *chancre* is formed within 2 wk to 2 mo. Dark-field microscopy of the exudate from syphilitic chancres often reveals the actively motile spirochetes. In most cases the chancre heals spontaneously and the organisms disappear from the site. Some, however, spread from the initial site to various parts of the body, such as the mucous membranes, the eyes, joints, bones, or central nervous system, and extensive multiplication occurs. A hypersensitive reaction to the treponema takes place, which is revealed by the development of a generalized skin rash; this rash is the key symptom of the *secondary* stage of the disease. At this stage the patient's condition may be highly infectious, but eventually the organisms disappear from secondary lesions and infectiousness ceases. The subsequent course of the disease in the absence of treatment is highly variable. About one-fourth of the patients undergo

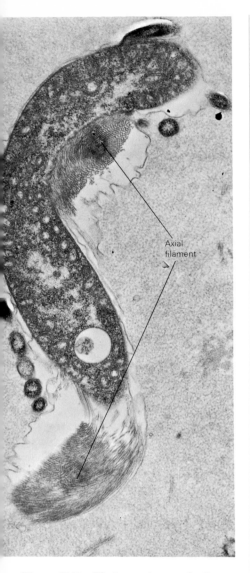

Axial
filament

Figure 18.56 Electron micrograph of a thin section of Cristispira, a very large spirochete. Note the numerous fibrils in the axial filament. Magnification, 16,900×. [From A. Ryter and J. Pillot: Annales de l'Institut Pasteur 109:552 (1966).]

a spontaneous cure and another one-fourth do not exhibit any further symptoms, although the infection may persist. In about half of the patients the disease enters the *tertiary* stage, with symptoms ranging from relatively mild infections of the skin and bone to serious infections of the cardiovascular system or central nervous system. Involvement of the latter is the most serious since it may lead to a generalized paralysis or other neurological damage. In the tertiary stage only very few organisms are present, and most of the symptoms probably result from hypersensitivity reactions to the spirochetes.

The immunological phenomena involved in syphilis are quite complex. Several kinds of antibody have been identified in sera of patients infected with *T. pallidum*. One of these, the "Wasserman antibody," reacts with cardiolipin, a diphosphatidylglycerol esterified with fatty acids and extracted from beef heart. The reaction is detected by complement fixation (Section 15.4) and can also be measured by precipitation, which is more rapidly performed but may be less specific than is complement fixation. It is not known if Wasserman antibody is formed initially in response to a *T. pallidum* antigen or if it develops against a normal tissue component that is released as a result of hypersensitivity; however, the presence of Wasserman antibody correlates highly with *T. pallidum* infection. On the other hand, Wasserman antibody has also been found in patients who have never been infected with *T. pallidum* but have had malaria, lupus erythematosis, leprosy, or one of several other diseases, thus suggesting that this antibody is not directed against a unique *T. pallidum* antigen. Despite the lack of complete specificity, serological tests involving the Wasserman antibody for the preliminary detection of syphilitic individuals are widely used because the inability to culture *T. pallidum* has made it difficult to develop serological tests directed against antigens of the organism itself. It is possible to infect rabbits with *T. pallidum,* and the organisms extracted from lesions can be used to detect specific *T. pallidum* antibodies in human sera. Two procedures have been used, an immobilization test and a fluorescent-antibody test. The *T. pallidum immobilization test* (TPI test) makes use of the fact that the spirochete is immobilized when it comes into contact with the specific antibody (Section 15.4). Although this test is simple in principle, it is technically difficult to perform and therefore is used more as a research tool than as a diagnostic instrument. The fluorescent-antibody test is an indirect method, in which the adsorption of *T. pallidum* antibody from human serum onto *T. pallidum* cells is detected by use of fluorescent antibody directed against human gamma globulin (Section 15.4). Serological tests are of great importance in diagnosing syphilis because of the variable symptoms exhibited by the disease and the lack of suitable culture methods.

Early methods of therapy involved administration of arsenic, mercury, or bismuth compounds, all of which are fairly toxic and relatively ineffective. Use of these drugs has now been completely superseded by highly effective penicillin therapy, and the early stages of the disease can usually be controlled by a series of injections over a period of 1 to 2 wk. In the secondary and tertiary stages treatment must extend for longer periods of time. Since penicillin kills only growing cells, the death rate is a function of how rapidly the organism is growing. Thus it is understandable that with such slowly growing organisms as *T. pallidum* penicillin therapy must be prolonged.

Despite the relative effectiveness of penicillin in curing syphilis, the disease is still common. This is due mainly to the social problems of locating and treating sexual contacts of infected individuals, such as we have already mentioned in relation to gonorrhea.

Other species of the genus *Treponema* are common commensal organisms in the oral cavity of man and can generally be seen in material scraped from between the teeth. Although they are sometimes morphologically indistinguishable from *T. pallidum,* they can be easily cultivated anaerobically on agar media containing serum. Two species, *T. microdentium* and *T. macrodentium,* have been recognized on the basis of their respective sizes.

Leptospira   The genus *Leptospira* contains a large number of diverse, widely distributed species. Some are aquatic saprophytes, others are commensals or harmless parasites of animals, and some are pathogens. Although many of the pathogenic forms can infect man, the natural reservoirs are primarily domestic and wild animals. Leptospiras can be readily cultured in complex media containing serum. Although they are obligate aerobes, the addition of a small amount of CO_2 to the atmosphere usually stimulates growth. Different strains are distinguished serologically by agglutination tests, and a large number of serotypes have been recognized. Rodents are the natural hosts of most leptospiras although dogs and pigs are also important carriers of certain strains. In man the most common leptospiral syndrome is called "Weil's disease"; in this disorder the organism usually localizes in the kidney. Leptospiras ordinarily enter the body through the mucous membranes or through breaks in the skin. After a transient multiplication in various parts of the body the organism localizes in the kidney and liver, causing nephritis and jaundice. The organism passes out of the body in the urine, and infection of another individual is most commonly by contact with infected urine. Therapy with penicillin, streptomycin, or the tetracyclines is possible but may require extended courses to eliminate the organism from the kidney; this is probably due to the slow growth and protected location of the leptospiras.

Vaccination of domestic animals against leptospirosis is done by using a killed virulent strain; dogs are usually immunized routinely with a combined distemper-leptospira-hepatitis vaccine. In man prevention is effected primarily by elimination of the disease from animals. The serotype that infects dogs, *L. canicola,* does not ordinarily infect man, but the strain attacking rodents, *L. icterohemorrhagiae,* does; hence elimination of rats from human habitations is of considerable aid in preventing the organism from reaching humans.

18.16 *Mycoplasma*

Organisms without cell walls that do not revert to walled organisms are currently classified in the genus *Mycoplasma.* Members of this genus are widespread in nature and have been frequently isolated from warm-blooded animals. Certain *Mycoplasma* are pathogenic, causing serious diseases in cattle and birds, and at least one species has been implicated in a human disease, primary atypical pneumonia. The first representative to be discovered and studied was the causal agent of pleuropneumonia in cattle, and for a long time all members of the group were called pleuropneumonia-like organisms (abbreviated PPLO), but currently "PPLO" has been supplanted by the generic name *Mycoplasma.* Since many of the members of this group are not closely related, it is likely that some new generic names will be created as scientists become increasingly familiar with the group through further research. At the moment, however, it is simplest to refer to all organisms in the group as members of the genus *Mycoplasma.*

In Chapter 2 we discussed protoplasts and showed how these structures can be formed when cell-wall digesting enzymes act on cells that are in an osmotically protected medium. Recall that when the osmotic stabilizer is removed, protoplasts take up water, swell, and burst. The mycoplasmae resemble protoplasts in their lack of a cell wall, but they are more resistant to osmotic lysis and are able to survive conditions under which protoplasts lyse. This ability to resist osmotic lysis is at least partially determined by the nature of the *Mycoplasma* cell membrane, which contains chemical substances that are absent in most other procaryotes. These compounds, probably the most important of which are sterols, make the membrane more rigid so that it can take over some of the stabilizing function normally carried out by the cell wall. *Mycoplasma* is unique among procaryotes in containing sterols in its cell membranes although it usually does not synthesize sterols itself but obtains them preformed from the medium.

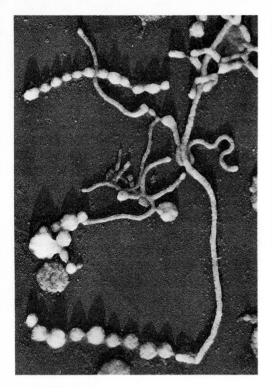

Figure 18.57 Electron micrograph of a metal-shadowed preparation of Mycoplasma mycoides. Note the coccoid and hyphalike elements. Magnification, 10,500×. (Courtesy of Alan Rodwell.)

Mycoplasma cells are usually very small, often of dimensions near the resolving power of the light microscope, and they are highly pleomorphic, which is a consequence of their lack of rigidity. A single culture may exhibit small coccoid elements, larger, swollen forms, and filamentous forms of variable lengths, often highly branched (Figure 18.57). It is from the production of filamentous, funguslike forms that the name *Mycoplasma* (*myco* means "fungus") derives. The small coccoid elements (0.1 to 0.3 μm in size) are the smallest *Mycoplasma* units capable of independent growth. These cells are so small that they easily pass through filters capable of holding back other bacteria, and it was this filterability that led some to conclude earlier in their investigation that the causal agents of some of the *Mycoplasma* diseases were viruses. Since *Mycoplasma* cells are culturable on nonliving substrates and since the cells contain both RNA and DNA, it is clear that they are not viruses, although their small size often makes it necessary to study them by methods of the virologist. The coccoid elements are the minimal reproductive units and have the smallest amount of DNA per cell of any known free-living organisms (see Table 13.2); even so, it can be calculated that there is indeed sufficient genetic information present to code for all proteins necessary for an independent existence. During multiplication the small coccoid units enlarge into spherical structures or filaments, followed by a number of rounds of DNA replication. Each nuclear body becomes the center of a region around which cell-membrane material collects, leading to the formation of many small granular regions, which, when separated by simple fission from the growing mass, become minimal reproductive units capable of initiating further growth (Figure 18.58).

The mode of growth differs in liquid and agar cultures. On agar there is a tendency for the organisms to grow so that they become embedded in the medium, and the fibrous nature of the agar gel seems to affect the division process, perhaps by promoting the separation of units from the growing mass. Colonies of *Mycoplasma* on agar exhibit a characteristic "fried-egg" appearance due to the formation of a dense central core, which penetrates downward into the agar, surrounded by a circular spreading area that is lighter in color.

Because of the small size of *Mycoplasma* cells, they scatter little light, and consequently it takes a large number of cells (close to 10^9 per milliliter) to achieve visible turbidity in liquid culture. Most other bacterial cultures will become turbid at cell densities around 10^6 to 10^7 cells per milliliter. However, growth rates of *Mycoplasma* are reasonably rapid, with generation times of 1 to 3 hr being reported for various strains. Growth of *Mycoplasma* is not inhibited by penicillin, cycloserine, or other antibiotics that inhibit cell-wall synthesis, but the organisms are as sensitive as other bacteria are to antibiotics that act on targets other than the cell

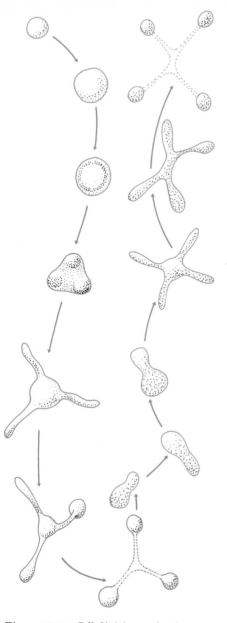

Figure 18.58 Cell-division cycle of a mycoplasma in liquid medium. [Adapted from
E. Klieneberger-Nobel: Pleuropneumonia-
like Organisms (PPLO): Mycoplasmataceae.
New York: Academic Press, Inc., 1962.]

wall. Use is made of the natural penicillin resistance of *Mycoplasma* in preparing selective media for their isolation from natural materials. The culture media used for the growth of most *Mycoplasma* have usually been quite complex. Growth is poor or absent even in complex yeast extract–peptone–beef heart infusion media unless fresh serum or ascitic fluid is added. The main constituents provided by these two adjuncts are unsaturated fatty acids and sterols. A few mycoplasmae can be cultivated on relatively simple media, however. Most organisms of this genus cannot synthesize sterols and must obtain them preformed from their environment, although a few saprophytic strains can synthesize their own sterols. Another component of serum required for growth is a protein although the protein serves not as a nutrient but as an agent in complexing and detoxifying the unsaturated fatty acids which are required for growth, which are toxic when present in large amounts, but which are slowly released from the protein–fatty acid complex. Most strains of *Mycoplasma* use carbohydrates as energy sources, and they require a range of vitamins, amino acids, purines, and pyrimidines. Except for the requirements for sterols and fatty acids, the nutritional requirements of *Mycoplasma* are no more complex than are those of the lactic acid bacteria (Section 18.1). Neither is the energy metabolism of *Mycoplasma* unique. Some species are oxidative, possessing the cytochrome system and making ATP by oxidative phosphorylation. Other species resemble the lactic acid bacteria in being strictly fermentative, producing energy by substrate-level phosphorylation and yielding lactic acid as the final product of sugar fermentation.

The lack of cell walls in the mycoplasmae has been proved by electron microscopy and by chemical analysis, the latter showing that the key wall components, muramic acid and diaminopimelic acid, are missing. However, the stability of the cell membrane is probably determined to a considerable extent by sterols since these compounds contain planar ring systems, which permit effective stacking of the molecules (a condition that is chemically stable). Strains able to grow without sterol in the membrane may use carotenoids or other lipid components to effect stability.

Once appropriate culture media and isolation techniques were devised, it was possible to show that members of the genus *Mycoplasma* are widespread in nature. Some strains can be routinely isolated from normal and pathological material of warm-blooded animals, while other strains can be isolated from plants, insects, sewage, soil, compost, and other natural materials. One of the most effective means of isolating *Mycoplasma* is to pass the material through a filter small enough to hold back most other types of bacteria, and then to streak out the filtrate on a complex agar medium containing serum and penicillin. Plates are incubated for 1 to 2 wk and examined with a dissecting microscope for colonies with the typical fried-egg appearance. For some purposes, preliminary filtration may not be desirable, and the material can be streaked directly onto plates of the selective agar medium. Another inhibitor frequently added to isolation media is thallium acetate (usually added at 250 μg/ml), a substance to which *Mycoplasma* is more resistant than are other organisms. Many isolates are saprophytes or harmless commensals, but quite a few are pathogenic. Bovine contagious pleuropneumonia, caused by *M. mycoides,* has been a serious problem to the cattle industry throughout the world, and another important disease is agalactia of sheep and goats, a disease that primarily affects the mammary glands. In man, mycoplasmae have been implicated in pneumonia, urethritis, kidney infections, and infections of the oral cavity.

One interesting question concerns the relationship between *Mycoplasma* and naturally occurring protoplastlike forms derived from other bacteria. As we discussed in Chapter 2, protoplasts can be prepared from a variety of bacteria by using either an enzyme such as lysozyme, which lyses cell walls, or an antibiotic such as penicillin, which inhibits cell wall synthesis. In the presence of an osmotic stabilizer such protoplasts can often continue to grow, especially on agar media, and they form colonies composed of spherical units that resemble *Mycoplasma*. These have been called L-forms. With the use of penicillin such L-forms can be prepared from a wide variety of both Gram-positive and Gram-negative bacteria. Two kinds of L-forms exist: reverting and stable. If penicillin is removed, the former type reverts to the normal walled bacteria, but the stable forms can never resynthesize cell walls, and they continue to grow as L-forms. Clearly, stable L-forms cannot be easily distinguished from *Mycoplasma*, and it is conceivable that some *Mycoplasma* species may have evolved from normal bacteria as a result of selection in habitats in which were present agents that act on cell walls. Most warm-blooded animals produce lysozyme and other cell-wall-degrading enzymes, and similar materials are found in soil, being produced in especially large amounts by myxobacteria. Thus, if other conditions appropriate for the growth of L-forms existed, it seems reasonable to suppose that stable L-forms might arise frequently in nature. The widespread use of penicillin in recent years might well have accentuated this evolutionary tendency. Now that techniques are available for their detection, stable L-forms are frequently being found in animals and man. However, not all stable L-forms can be equated with *Mycoplasma*. Molecular taxonomic studies (on DNA base composition and nucleic acid hybridization) have shown that many stable L-forms are related to other bacteria (for example, *Streptococcus, Proteus, Neisseria*) and are unrelated to *Mycoplasma. Mycoplasma* species have DNA base compositions with low GC content, 25 to 35 percent GC, whereas most other bacteria and their L-forms have DNA of higher GC content. The

Figure 18.13 Typical appearance of streptomycetes growing on agar slants. Varying degrees of pigmentation are shown by different species on the several culture media used; the coloration results from the production of soluble pigments that diffuse into the agar and from the production of pigmented spores: (a) Streptomyces hygroscopicus [from A. Dietz and J. Mathews: Appl. Microbiol. 16:935 (1968)]; (b) S. griseus (courtesy of A. Dietz); (c) S. purpurascens (courtesy of A. Dietz).

(a)

(b)

(c)

Plate nine

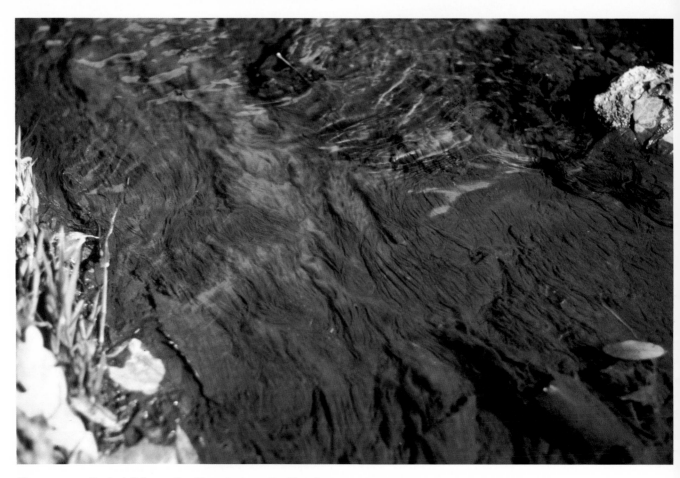

Figure 18.35   Typical Sphaerotilus bloom in a small polluted creek. The white streamers are fresh growth, and the brown streamers underneath represent older growth, primarily empty sheaths encrusted with debris and diatom frustules.

(a)

(b)

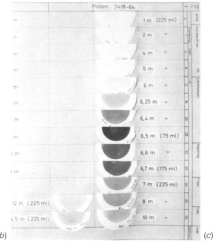

(c)

Figure 18.38   (a) Bloom of purple sulfur bacteria, Thiopedia sp., in a small pond rich in organic matter, Yankee Springs, Michigan. (Courtesy of Peter Hirsch.) (b) Water sample from a depth of 7 m in a stratified Norwegian lake. The green color is due to Pelodictyon luteolum, a green sulfur bacterium. (Courtesy of Norbert Pfennig.) (c) Series of membrane filters through which were passed water samples taken from varying depths in a stratified Norwegian lake. The green sulfur bacterium P. luteolum reaches its highest population density between 6.5 and 7 m depth. (Courtesy of John Ormerod and the Norwegian Institute for Water Research.)

Plate eleven

(a)

(b)

(e)

Figure 18.47 Series of photographs of fruiting myxobacteria,
presented to give some idea of the range of complexity of the
fruiting bodies in different genera and species: (a) Chondromyces
apiculatus, whose fruiting body is about 270 μm high; (b) Stigmatella aurantiaca, about 150 μm high; (c) Podangium erectum,
with fruiting bodies of about 50 to 60 μm long; (d) Polyangium
velatum, with ridge-shaped fruiting bodies about 100 to 130 μm
wide; (e) Myxococcus stipitatus, 170 μm high, resting on a soil
crumb; (f) Myxococcus fulvus, with fruiting bodies 75 to 150 μm
in diameter, resting on filter paper; (g) Myxococcus virescens
fruiting bodies on rabbit-dung pellet that had been incubated
in a moist chamber. (Courtesy of Hans Reichenbach and
Martin Dworkin.)

Plate twelve

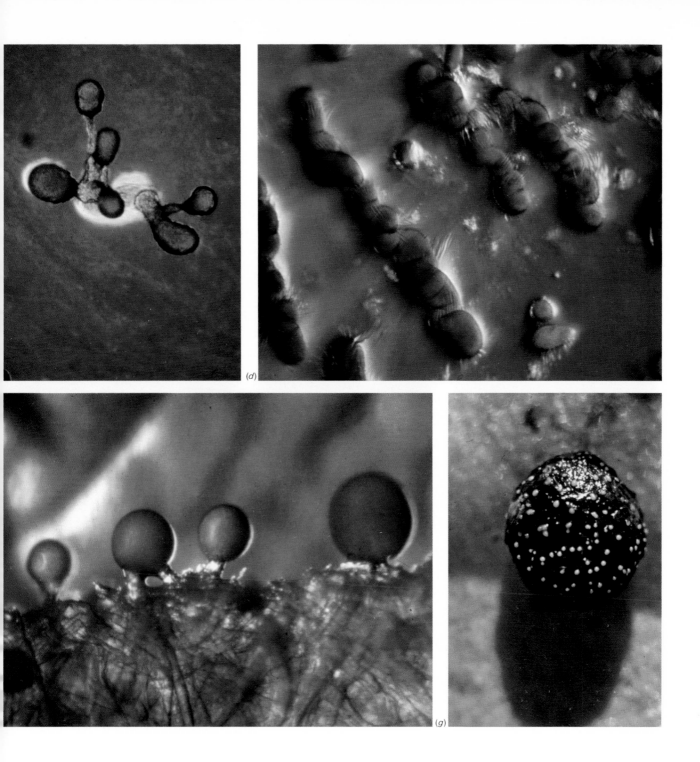

(d)

(g)

Plate thirteen

Figure 18.52 *A bloom of the blue-green alga Microcystis aeruginosa in Lake Mendota, Wisconsin. (Courtesy of A. D. Hasler.)*

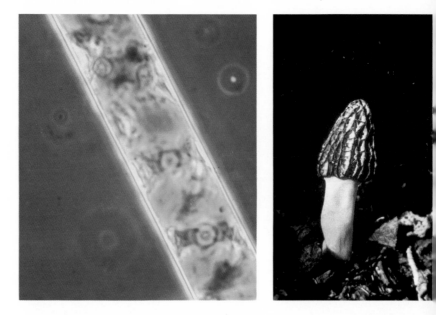

Figure 19.5 (right) *Photomicrograph of a Spirogyra filament, showing the characteristic spiral-shaped chloroplast.*

Figure 19.25 (far right) *The fruiting body of the ascomycete Morchella esculenta, the common morel. The asci are borne individually on the convoluted surface of the fruiting-body head. (Courtesy of W. Bryce Kendrick.)*

Figure 19.34 *Fungi growing on a creosoted railroad tie that had been partially buried in the ground.*

Plate fourteen

Figure 19.38 *Developmental stages of the acellular slime mold Physarum polycephalum: (a) a plasmodium that has migrated over the surface of an agar plate; (b–d) stages of fruiting-body formation; (e) mature fruiting bodies.* [*From S. Guttes, E. Guttes, and H. P. Rusch, "Some events in the life cycle of Physarum polycephalum" (film), and from J. E. Cummins and H. P. Rusch, Endeavour, 27:124 (1968).*]

(a)

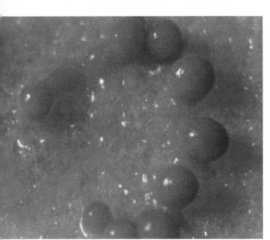

(b)

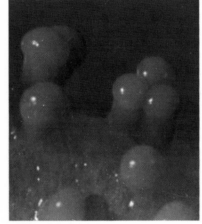

(c)

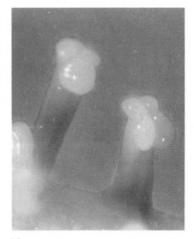

(d)

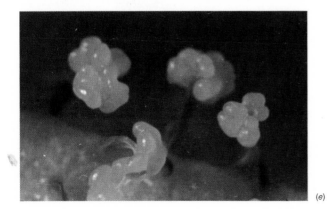

(e)

Plate fifteen

Figure 19.41 Living forami-niferal cells; their size may be compared to the eye of the needle. (Photograph by Paul A. Zahl, © 1961 by National Geographic Society.)

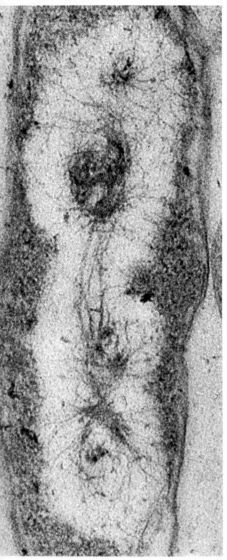

Figure 18.59 Electron micrograph of a thin section of a rickettsia, Coxiella burnetii. Note especially the well-defined cell wall and membrane. Magnification, 90,000×. (Courtesy of David Paretsky and Paul Burton.)

evolutionary relationships between bacteria, L-forms, and *Mycoplasma* are clearly of great interest and their elucidation will require much further work.

18.17 *Rickettsia*

The rickettsiae are small bacteria which have a strictly intracellular existence in vertebrates, usually in mammals, and which are also associated at some point in their natural cycle with blood-sucking arthropods such as fleas, lice, or ticks. Rickettsiae cause a variety of diseases in man and animals, of which the most important are typhus fever, Rocky Mountain spotted fever, scrub typhus (tsutsugamushi disease), and Q fever. Rickettsiae take their name from Howard Ricketts, a scientist of the University of Chicago, who first provided evidence for their existence and who died from infection with the rickettsia that causes typhus fever, *R. prowazekii*. Up to the present, rickettsiae have not been unequivocally cultured in nonliving media and hence must be considered obligate intracellular parasites, although there is nothing so unusual about their physiology as to suggest that they cannot eventually be grown in vitro. (The causal agent of trench fever has been cultured in cell-free medium, but there is some question whether this is actually a rickettsia.) Rickettsiae have been readily cultivated in laboratory animals, lice, mammalian tissue cultures, and the yolk sac of chick embryos; this latter is the host of choice for most laboratory work.

The rickettsiae are Gram-negative, coccoid or rod-shaped, with cells in the size range of 0.3 to 0.7 μm wide by 1 to 2 μm long. Electron micrographs of thin sections show profiles with a normal bacterial morphology (Figure 18.59); both cell wall and cell membrane are visible. The cell wall contains muramic acid. Both RNA and DNA are present, and the DNA is in the normal double-stranded form, with a GC content of 32.5 percent for *R. prowazekii* and 44.5 percent for *Coxiella burnetii* (the causal agent of Q fever). The rickettsiae divide by normal binary fission, with doubling times of about 8 hr. The penetration of a host cell by a rickettsial cell is an active process, requiring both host and parasite to be alive and metabolically active. Once inside the phagocytic cell, the bacteria multiply primarily in the cytoplasm and continue replicating until the host cell is loaded with parasites (Figure 18.60), at which time the host cell bursts and liberates the bacteria into the surrounding fluid.

Much attention has been directed to the metabolic activities and biochemical pathways of rickettsiae, in an attempt to explain why they are obligate intracellular parasites. Since biochemical studies must be done

Figure 18.60 Development of Rickettsia mooseri within a guinea-pig cell.

with large populations of cells, and since these populations can be obtained only by growing the parasites in animal cells, much effort has been expended on devising methods for purifying rickettsiae and for separating them from any contaminating host tissues that might confuse metabolic or biochemical studies. Purification has been made difficult by the fact that when rickettsiae are freed from host tissues they leak intracellular constituents readily and hence lose their viability. For this reason, most biochemical studies have been performed with *Coxiella burnetii* since this organism is less easily damaged during purification. The rickettsiae possess a highly distinctive energy metabolism, being able to oxidize only one amino acid, glutamate, and being unable to oxidize other amino acids, glucose, glucose-6-phosphate, or organic acids. They possess a complete cytochrome system and are able to carry out oxidative phosphorylation, using as the electron donor NADH derived from the oxidation of glutamate. They are also able to synthesize at least some of the small molecules needed for macromolecular synthesis and growth, while the rest of their nutrients they obtain from the host cell. There is some suggestion that the host also provides some key coenzymes, such as NAD and CoA. Such large coenzymes do not usually penetrate readily into bacteria that live independently, and there is evidence that the rickettsial membrane is looser and more "leaky" than those of other bacteria. For a fastidious organism capable of infecting other cells rich in nutrients and cofactors, such a permeable membrane would be advantageous. A summary of the biochemical properties of rickettsiae is given in Table 18.19.

If the membranes of the rickettsiae are indeed unusually leaky, the organisms may die quickly when out of their hosts; and this may explain why they must be transmitted from animal to animal by arthropod vectors. When the arthropod obtains a blood meal from an infected vertebrate, rickettsiae present in the blood are inoculated directly into the arthropod, where they penetrate to the epithelial cells of the gastrointestinal tract, multiply, and appear later in the feces. When the arthropod feeds upon an uninfected individual, it then transmits the rickettsiae either directly with its mouthparts or by contaminating the bite with its feces. However, the causal agent of Q fever, *C. burnetii*, can also be transmitted to the respiratory system by aerosols. *Coxiella burnetii* is the most resistant of the rickettsiae to physical damage, which probably explains its ability to survive in air.

The relationship of the rickettsiae to other bacteria is unknown although it is frequently postulated that rickettsiae evolved from other bacteria by progressive loss of function as a result of mutation, a process called "degenerative evolution." Although this would explain the obligate intracellular habitat, it would not indicate how rickettsiae are able to survive and grow within a phagocytic cell, an environment that, as we learned in

Table 18.19   Comparison of biochemical properties of rickettsiae, chlamydiae, and viruses

Property	Rickettsiae	Chlamydiae	Viruses
Structural:			
Nucleic acid	RNA and DNA	RNA and DNA	Either RNA or DNA, never both
Ribosomes	Present	Present	Absent
Cell wall	Muramic acid	Muramic acid, D-alanine	No wall
Structural integrity during multiplication	Maintained	Maintained	Lost
Biosynthetic:			
Macromolecular synthesis	Carried out	Carried out	Only with use of host machinery
ATP-generating system	Present	Absent	Absent
Sensitivity to antibacterial antibiotics	Sensitive	Sensitive	Resistant

Chapter 15, is hostile to most other bacteria. Another point is that the rickettsiae we currently know are those which cause pathological changes, but many arthropods possess other intracellular rickettsialike organisms that do not seem to be related to any disease either in the arthropod or in an alternate host. We discussed some of these briefly in Section 14.6 as possible symbionts. We also considered in brief the range of intracellular bacteria that are symbionts of protozoa (for example, *kappa* in *Paramecium*) in Section 13.4. Such intracellular symbionts may be, like rickettsiae, representatives of a much larger group of intracellular bacteria that arose from free-living forms by degenerate evolution. It does not necessarily follow, however, that all rickettsiae arose from the same free-living bacterium. The evolutionary steps may have occurred independently in quite unrelated free-living bacteria.

The mode of transmission of the major rickettsial disease agents is given in Table 18.20. Of the diseases they cause, the most important and widespread is typhus fever, also called "epidemic louse-borne typhus," caused by *R. prowazekii.* Typhus is an acute infectious disease that has a high fatality rate. It has probably afflicted mankind since ancient times. The symptoms of typhus frequently resemble those of typhoid, the disease caused by *Salmonella typhosa,* and in earlier years the two diseases were frequently confused, hence the origin of the name *typhoid,* which means "typhuslike." Typhus fever has occurred frequently in armies during military campaigns, often with disastrous consequences; indeed, this was one

Table 18.20   Major rickettsial diseases, their causal agents, and modes of transmission[a]

Disease in man	Causal agent	Mode of transmission
Epidemic typhus	R. prowazekii	. . . Man ⟶ Louse ⟶ Man ⟶ Louse . . .
Murine typhus	R. mooseri	. . . Rat ⟶ Rat flea ⟶ Rat ⟶ Rat flea ⟶ Rat . . . ↘ Man
Rocky Mountain spotted fever, boutonneuse fever, other spotted fevers	R. rickettsii	. . . Tick ⟶ Tick ⟶ Tick ⟶ Tick . . . ↘ Dog ↘ Man ↘ Man
Scrub typhus (tsutsugamushi fever)	R. tsutsugamushi	. . . Mite ⟶ Field mouse ⟶ Mite ⟶ Field mouse . . . ↘ Man
Rickettsial pox	R. akari	. . . Mite ⟶ House mouse ⟶ Mite ⟶ House mouse . . . ↘ Man
Q fever	Coxiella burnetti	. . . Tick ⟶ Small mammal ⟶ Tick ⟶ Cattle . . . Airborne ↘ Man

[a] Reprinted from J. W. Moulder: *The Biochemistry of Intracellular Parasitism* by permission of The University of Chicago Press. © 1962 by The University of Chicago.

of the reasons for the downfall of Napoleon's army in 1812. Typhus is a frequent accompaniment of famine and other human misfortune. In contrast to most of the other rickettsial diseases, typhus occurs naturally only in man. The causal agent is transmitted from person to person by two species of lice, the human body louse and the human head louse. The rickettsiae are liberated from the louse in its feces. When the louse bites, it makes a small hole in the skin and defecates at the same time. The fecal matter containing the agent gets rubbed into the puncture and is carried to the bloodstream. The organism multiplies inside endothelial cells lining the small blood vessels, and clots occur in the vessels as a result of destruction of infected cells. A characteristic symptom of typhus is a rash, appearing first in the trunk and then spreading over the whole body except for the face, palms, and soles. Fever and general torpor of the patient result, with death on the ninth or tenth day of illness. Many of the symptoms may be due to the elaboration of a toxin, but no toxin has yet been positively identified. The tetracyclines and chloramphenicol are highly effective against *R. prowazekii,* while the control of epidemics is directed primarily at the louse. Insecticides such as DDT are applied to all persons known to have been in contact with infected individuals, and in widespread epidemics all persons in the community would be so treated. Two types of vaccine are available: a killed *R. prowazekii* preparation and a live avirulent strain. Vaccination of military troops sent into areas where typhus is endemic is highly recommended. At present typhus is not a problem in western civilization, but presumably it could become so again if standards of cleanliness or economy are considerably reduced.

18.18   The psittacosis group: chlamydiae

The psittacosis group (also sometimes called bedsonia), which is comprised of the genus *Chlamydia*, probably represents a further stage in degenerative evolution from that discussed above for the rickettsiae since the chlamydiae are obligate parasites in which there has been an even greater loss in metabolic function. In fact, for many years the chlamydiae were considered to be large viruses rather than bacteria, and only since the essential nature of virus replication has been well understood (see Chapter 10) has the bacterial nature of the chlamydiae been firmly established. Many chlamydiae are smaller than some of the true viruses, such as that of smallpox, but the chlamydiae divide by binary fission, and do not replicate in the manner of viruses. Of the diseases caused by chlamydiae, psittacosis, an epidemic disease of birds that is transmitted occasionally to man, is the most important although a number of other diseases of birds and man and other mammals are known. It is not as disease entities that the chlamydiae are of most interest, however; they are intriguing because of the biological and evolutionary problems they pose. The bacterial nature of the chlamydiae was first suspected when it was discovered that, unlike viruses, they were susceptible to penicillin and other antibiotics whose action is restricted to the bacteria. When the specificity of penicillin for the bacterial cell wall was understood, the apparent viral nature of the chlamydiae was further refuted. Biochemical studies showed that the chlamydiae have typical bacterial cell walls, and they have both DNA and RNA. Electron microscopy of thin sections of infected cells shows forms that clearly are undergoing binary fission. The biosynthetic capacities of the chlamydiae are very restricted, however, even more so than are the rickettsiae. This raises the interesting question of the limits to which evolutionary loss of function can be pushed while independence of macromolecular function is still retained. Although no convincing evidence exists that all the chlamydiae are closely related, they are currently classified in a single genus; but the name ''chlamydia'' should be viewed more as a convenience than as a taxonomic entity.

The life cycle of a typical member of the genus *Chlamydia* is shown in

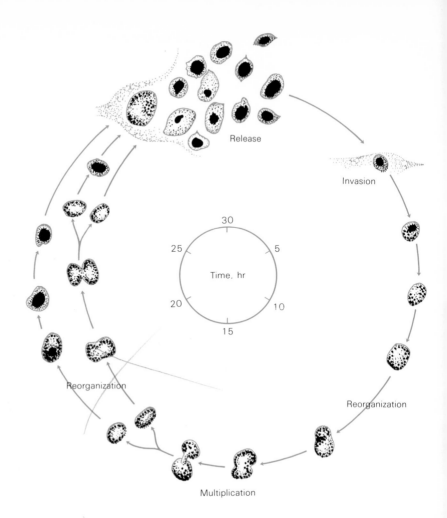

Release

Invasion

30

25   5

Time, hr

20   10

15

Reorganization

Reorganization

Multiplication

Figure 18.61   Cell-division cycle of a member of the psittacosis group. (From J. W. Moulder: The Psittacosis Group as Bacteria. New York: John Wiley & Sons, Inc., 1964.)

Figure 18.61. Two cellular types are seen in a typical life cycle: a small, dense cell, which is relatively resistant to drying and which is the agency of dispersal of the agent, and a larger, less dense cell, which divides by binary fission and which is the vegetative form of these microorganisms. Unlike the rickettsiae, the chlamydiae are not transmitted by arthropods but are primarily airborne invaders of the respiratory system—hence the significance of resistance to drying of the small cells. Recall that, when a virus infects a cell, it loses its structural integrity and liberates nucleic acid. When a chlamydia enters a cell, however, although it changes form, it remains a structural unit and enlarges and begins to undergo binary fission. A dividing form of a chlamydia is seen in Figure 18.62. After a number of divisions, the vegetative cells become converted into small,

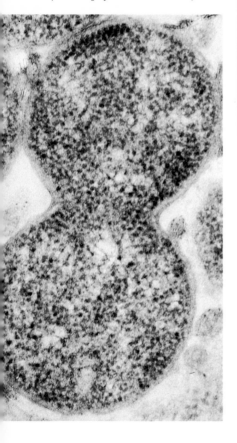

Figure 18.62   Electron micrograph of a thin section of a dividing cell of Chlamydia psittaci, a member of the psittacosis group, within a cell of a mouse tissue culture. (Courtesy of Robert R. Friis.)

dense cells that are released when the host cell disintegrates and can then infect other cells. Generation times of 2 to 3 hr have been reported, which are considerably faster than those which have been found for the rickettsiae.

As we noted, chlamydia cells have a chemical composition similar to that of other bacteria. Both RNA and DNA are present, in approximately equal proportions. The DNA content of a chlamydia corresponds to a molecular weight of approximately 4×10^8, about twice that of vaccinia virus and one-tenth that of *Escherichia coli*. The base composition is about 29 percent GC. At least some of the RNA is in the form of ribosomes, and, like those of other procaryotes, the ribosomes are 70-S particles composed of one 50-S and one 30-S unit. The cell wall of the chlamydiae contains two components ordinarily found in bacterial cell walls, muramic acid and D-alanine. As we stated above, chlamydiae are sensitive to penicillin, and when this antibiotic is added to infected tissue-culture preparations, the replicating particles become converted into spheroplasts but do not lyse so long as they remain inside the cells, presumably because they are osmotically protected by the cytoplasmic constituents of the host cell. Penicillin-resistant mutants of chlamydiae have been isolated. In keeping with the presence in the cell wall of D-alanine, chlamydiae are sensitive to inhibition by cycloserine, a D-alanine analog (see Chapter 9).

The metabolic properties of chlamydiae purified from infected cells have been studied by methods similar to those used with the rickettsiae. The biosynthetic capacities of the chlamydiae are much more limited than are those of the rickettsiae (see Table 18.19). Although macromolecular syntheses occur in the chlamydiae, no energy-generating system is present and the cells obtain their ATP from the host. They are able to oxidize glucose to pentose by means of enzymes of the pentose phosphate pathway, but they do not obtain energy from this oxidation. Chlamydiae synthesize their own folic acid, D-alanine, lysine, and probably a number of other small molecules. One interesting feature of chlamydiae is that their proteins are deficient in the amino acids arginine and histidine, which are found in most other organisms as well as in viruses. In keeping with this lack, neither amino acid is required for multiplication. From the limited biosynthetic capacities of chlamydiae it is easy to see why they are obligate parasites. Their inability to manufacture ATP means that at the least they are energy parasites on their hosts; probably they also obtain from these hosts a variety of other coenzymes, as well as low-molecular-weight building blocks. The chlamydiae have the simplest biochemical abilities known among cellular organisms.

The most important diseases caused by the chlamydiae are listed in Table 18.21. Of these *psittacosis* is the most widespread and well stu-

Table 18.21   Diseases caused by the chlamydiae

In birds (also in mammals)	In man	In other mammals
Psittacosis	Lymphogranuloma venereum	Meningopneumonitis
Ornithosis	Trachoma	Mouse pneumonitis
	Inclusion conjunctivitis	Feline pneumonitis
	Human pneumonitis	Bovine enteritis
		Bovine encephalitis
		Enzootic abortion of ewes
		Sheep pneumonitis

died, and the following discussion will be restricted to it. A large number of different species of birds can be infected, including domestic fowl, turkeys, and various species raised by bird fanciers. The disease was first discovered in psittacine birds such as parrots—hence the name *psittacosis* ("parrot fever")—and the disease was considered at one time to be restricted to them. The agent can be transferred from birds to man, and indeed birds are the primary source of human infection. Poultry and birds kept as pets or for show are probably the most important sources of infection for man, for the disease is seen most frequently in bird handlers. It is a minor public health problem; control in the United States is effected mainly by regulations on the importation of birds. The only major outbreak in the United States was in 1929 and 1930, when 170 cases of psittacosis with 33 deaths were reported; in this epidemic parrots were the principal vector. The disease both in birds and man is characterized by long periods of latency, the organism remaining alive within the body but causing no overt symptoms. In man the main foci of infection are the lungs, but other organs of the body are also affected. The organism produces a toxin that is responsible for at least some of the disease symptoms. Chlortetracycline is the drug of choice for both man and birds, although the antibiotic does not always eliminate the organism in latent infections. Immunity in human beings is incomplete after infection and active immunization is not recommended. Individuals in high-risk occupations should be checked serologically from time to time for the presence of antibodies against the psittacosis agent, and they should be subjected to antibiotic therapy if they become infected.

18.19   Summary

We have gained in this chapter some idea of the diversity of procaryotic microorganisms, which range in size from the tiny obligately parasitic rickettsiae and chlamydiae to the large, free-living blue-green algae.

However, it is not so much in morphology as in biochemical properties that the procaryotes exhibit the greatest diversity. This biochemical diversity is reflected in the outstanding ability of procaryotes to colonize different ecological niches. In any habitat where living organisms are found, procaryotes will be present. At the same time, there are habitats in which eucaryotes cannot grow but in which procaryotes not only grow but flourish.

Procaryotes are greatly restricted in the degree to which individual cells can associate to form multicellular aggregates or structures. The most complex of the multicellular structures formed by procaryotes, the fruiting bodies of the myxobacteria, are much simpler in form than many of the structures we will discuss in the next chapter. Thus procaryotes are biochemically versatile but limited in morphological diversity, from which we might conclude that the procaryotic cell structure imposes certain limitations on the evolution of morphologically complex forms. The development of the eucaryotic cell may have been necessary to permit the great degree of morphological evolution that led to the rise of multicellular plants and animals.

Supplementary readings

Barker, H. A. *Bacterial Fermentations.* New York: John Wiley & Sons, Inc., 1956.

Breed, R. S., E. G. D. Murray, and N. R. Smith (eds.) *Bergey's Manual of Determinative Bacteriology,* 7th ed. Baltimore: The Williams & Wilkins Co., 1957.

Davis, B. D., R. Dulbecco, H. N. Eisen, H. S. Ginsberg, and W. B. Wood, Jr. *Microbiology.* New York: Harper & Row, Publishers, 1967.

De Ley, J. "DNA base composition and hybridization in the taxonomy of phytopathogenic bacteria," *Ann. Rev. Phytopathol.* 6:63 (1968).

Dubos, R. J., and J. Dubos *The White Plague: Tuberculosis, Man, and Society.* Boston: Little, Brown and Company, 1952.

Dubos, R. J., and J. G. Hirsch (eds.) *Bacterial and Mycotic Infections of Man,* 4th ed. Philadelphia: J. B. Lippincott Co., 1965.

Dworkin, M. "Biology of the myxobacteria," *Ann. Rev. Microbiol.* 20:75 (1966).

Eaton, M. D. "Pleuropneumonia-like organisms and related forms," *Ann. Rev. Microbiol.* 19:379 (1965).

Frobisher, M. *Fundamentals of Microbiology,* 8th ed. Philadelphia: W. B. Saunders Co., 1968.

Gest, H., A. San Pietro, and L. P. Vernon (eds.) *Bacterial Photosynthesis* (Symposium on Bacterial Photosynthesis). Yellow Springs, Ohio: Antioch Press, 1963.

Gibbs, B. M., and F. A. Skinner (eds.) *Identification Methods for Microbiologists,* pt. A. New York: Academic Press, Inc., 1966.

Gibbs, B. M., and D. A. Shapton (eds.) Identification Methods for Microbiologists, pt. B. New York: Academic Press, Inc., 1968.

Hartman, P. A. Miniaturized Microbiological Methods. New York: Academic Press, Inc., 1968.

Heimpel, A. M. "A critical review of Bacillus thuringiensis var. thuringiensis Berliner and other crystalliferous bacteria," Ann. Rev. Entomol. 12:287 (1967).

Holm-Hansen, O. "Ecology, physiology, and biochemistry of blue-green algae," Ann. Rev. Microbiol. 22:47 (1968).

Kersters, K., and J. De Ley "The occurrence of the Entner-Doudoroff pathway in bacteria," Antonie van Leeuwenhoek 34:393 (1968).

Klieneberger-Nobel, E. Pleuropneumonia-like Organisms (PPLO): Mycoplasmataceae. New York: Academic Press, Inc., 1962.

Lang, N. J. "The fine structure of blue-green algae," Ann. Rev. Microbiol. 22:15 (1968).

Lechevalier, H. A., and M. P. Lechevalier. "Biology of actinomycetes," Ann. Rev. Microbiol. 21:71 (1967).

Moulder, J. W. Biochemistry of Intracellular Parasitism. Chicago: University of Chicago Press, 1962.

Moulder, J. W. The Psittacosis Group as Bacteria (Ciba Lectures in Microbial Biochemistry). New York: John Wiley & Sons, Inc., 1964.

Moulder, J. W. "The relation of the psittacosis group (chlamydiae) to bacteria and viruses," Ann. Rev. Microbiol. 20:107 (1966).

Peck, H. D., Jr. "Energy-coupling mechanisms in chemolithotrophic bacteria," Ann. Rev. Microbiol. 22:489 (1968).

Pfennig, N. "Photosynthetic bacteria," Ann. Rev. Microbiol. 21:285 (1967).

Pringsheim, E. G. Farblose Algen: Ein Beitrag zur Evolutionsforschung. Stuttgart: Gustav Fischer Verlag, 1963.

Schlegel, H. G. (ed.) "Anreicherungskultur und Mutantenauslese" ("Enrichment culture and mutant selection"), Zentralbl. Bakteriol., Parasitenk., Infekt. Hyg., I Abt., Suppl. 1:1 (1965).

Shinefield, H. R., and J. C. Ribble "Current aspects of infections and diseases related to Staphylococcus aureus," Ann. Rev. Med. 16:263 (1965).

Skerman, V. B. D. A Guide to the Identification of the Genera of Bacteria with Methods and Digests of Generic Characteristics, 2nd ed. Baltimore: The Williams & Wilkins Co., 1967.

Society of American Bacteriologists Manual of Microbiological Methods. New York: McGraw-Hill Book Company, 1957.

Sokolova, G. A., and G. I. Karavaiko Physiology and Geochemical Activity of Thiobacilli (trans. fr. Russ.). Jerusalem: Israel Program for Scientific Translations, 1968. (Available from Clearinghouse for Federal Scientific and Technical Information, Springfield, Va. 22151.)

Stadtman, T. C. "Methane fermentation," Ann. Rev. Microbiol. 21:121 (1967).

Representative procaryotic groups

Stanier, R. Y. "Toward a definition of the bacteria." In I. C. Gunsalus and R. Y. Stanier (eds.), *The Bacteria: A Treatise on Structure and Function,* vol. V. New York: Academic Press, Inc., 1964.

Starr, M. P., and V. B. D. Skerman "Bacterial diversity: the natural history of selected morphologically unusual bacteria." *Ann. Rev. Microbiol.* 19:407 (1965).

Thimann, K. V. *The Life of Bacteria,* 2nd ed. New York: The Macmillan Company, 1963.

van Niel, C. B. (ed.) *Selected Papers of Ernst Georg Pringsheim.* New Brunswick, N. J.: Rutgers Institute of Microbiology, The State University, 1963.

Waksman, S. A. *The Actinomycetes,* vol. I. Baltimore: The Williams & Wilkins Co., 1959.

Waksman, S. A. *The Actinomycetes,* vol. II. Baltimore: The Williams & Wilkins Co., 1961.

Waksman, S. A., and H. A. Lechevalier *The Actinomycetes,* vol. III. Baltimore: The Williams & Wilkins Co., 1962.

Waksman, S. A. *The Actinomycetes: A Summary of Current Knowledge.* New York: The Ronald Press Company, 1967.

Wilson, G. S., and A. A. Miles *Topley and Wilson's Principles of Bacteriology and Immunology,* 5th ed., 2 vols. Baltimore: The Williams & Wilkins Co., 1964.

Zdrodovskiĭ, P. F., and E. H. Golinevich (B. Haigh, trans.) *The Rickettsial Diseases.* Elmsford, N. Y.: Pergamon Press, 1960.

Zinnser, H. *Rats, Lice and History.* Boston: Little, Brown and Company, 1935.

The algae

The term "algae" refers to a large, morphologically and physiologically diverse assemblage of organisms containing chlorophyll, which carry out an oxygen-evolving type of photosynthesis. Some of these organisms are classified as procaryotes (the Cyanophyta, or blue-green algae), and these have been considered in Chapter 18. In this section we shall discuss only the eucaryotic algae. Although most eucaryotic algae are of microscopic size and hence are clearly microorganisms, a number of forms are macroscopic, some of the seaweeds growing in length to over 100 ft. It is somewhat difficult to decide how these macroscopic forms should be distinguished from such clearly nonalgal organisms as mosses and ferns. However, a characteristic of all algae is the production of spores or gametes within unicellular structures, whereas even the simplest mosses bear such reproductive cells within structures surrounded by multicellular walls.

Some algae are motile by means of flagella and seem to be related to protozoa, and thus the border separating the algae from the protozoa is rather fuzzy. This need not concern us here, however, as we will concentrate on giving some idea of the diversity of the grouping as a whole.

19.1 Characteristics used in classifying algae

Photosynthetic apparatus All algae contain chlorophyll a. Some, however, contain other chlorophylls that differ in minor structural and optical properties from chlorophyll a, and it is the presence of these additional chlorophylls that is characteristic of certain algal groups. In addition, a variety of accessory pigments are present, some of which are also characteristic of certain groups. The distribution of chlorophylls and accessory pigments in the various groups is summarized in Table 19.1. We have already discussed the roles of some of these accessory pigments in photosynthesis in Chapter 6.

All algae carry out oxygen-evolving photosynthesis, using H_2O as an electron donor. In addition, some algae can use the alternative electron donors, H_2 and H_2S, to carry out photosynthesis without yielding oxygen. Many algae are obligate phototrophs, being unable to grow in either light or dark on organic carbon compounds. A number of algae are facultative phototrophs able to assimilate and grow on simple sugars or organic acids in the dark. One of the organic compounds most widely used by algae is acetate, which can be used as a sole carbon and energy source by

nineteen

Represen-
tative
eucaryotic
groups

Table 19.1   Distribution of pigments and reserve materials among algae[a]

	Chloro-phyta	Eugleno-phyta	Xantho-phyta	Chryso-phyta	Bacillario-phyta	Phaeo-phyta	Pyrro-phyta	Crypto-phyta	Rhodo-phyta
Pigments:									
Chlorophylls:									
a	+	+	+	+	+	+	+	+	+
b	+	+	−	−	−	−	−	−	−
c	−	−	−		+	+	+	+	±
d	−	−	−	−	−	−	−	−	±
e	−	−	+	−	−	−	−	−	−
Biliproteins:									
Phycocyanin		−	−	−	−	−	−	+	+
Phycoerythrin		−	−	−	−	−	−	+	+
Reserve materials:									
Starches (α-1,4-glucans):									
True starch	+						+	+	
Floridean starch									+
β-1,3-glucans:									
Laminarin						+			
Paramylon		+	+						
Chrysolaminarin					+				
Sugars:									
Floridoside (glycerol-galactoside)									+
Sucrose	+								
Sugar alcohols:									
Mannitol						+			
Lipid			+		+	+	+		

[a] +, pigment or reserve material present; −, absent; blank, unknown.

many flagellates and chlorophytes. In addition, some algae can assimilate simple organic compounds in the light and can incorporate them into protoplasmic material, but they cannot grow on them as sole energy sources.

Reserve materials   One of the key characteristics used in the classification of algal groups is the nature of the reserve material synthesized as a result of photosynthesis. Algae of the division Chlorophyta produce a starch (α-1,4-glucan) very similar to that of higher plants. Algae of other groups, on the other hand, produce reserve materials of different composition, some of which are listed in Table 19.1.

Morphology   There is considerable variability in the vegetative structures of the algae (Figure 19.1). The simplest forms are unicellular, dividing by binary fission. Among these there is a great diversity in cell shape and size. In many cases, the two daughter cells do not separate immediately upon division but remain together, forming chains or amorphous

19.1   *Characteristics used in classifying algae*

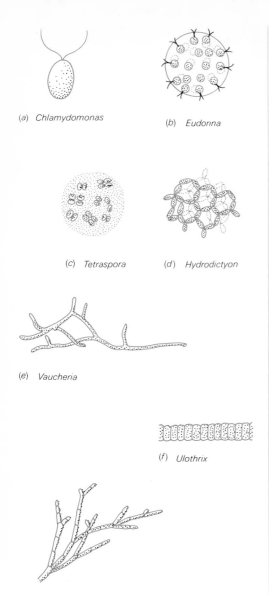

(a) *Chlamydomonas*

(b) *Eudorina*

(c) *Tetraspora*

(d) *Hydrodictyon*

(e) *Vaucheria*

(f) *Ulothrix*

(g) *Cladophora*

Figure 19.1 Various kinds of cell shapes and arrangements in eucaryotic algae: (a) unicellular, (b) colonial, (c) tetrasporal, (d) coenobium, (e) nonseptate or siphonaceous, (f) septate (unbranched filament), (g) septate (branched filament).

masses of cells. If the cells remain more or less attached to each other, the aggregate is called a *coenobium,* whereas if the cells have separated from each other but are held together by being embedded in mucilage, the arrangement is known as *tetrasporal.* Aggregates formed by many motile cells often show cooperative actions and can be considered more than simple colonies since they resemble in some ways multicellular organisms (see the discussion of *Volvox* below). Many algae are filamentous, of which several types can be recognized: (1) simple, nonseptate filaments (sometimes called *siphonaceous*); (2) simple, unbranched, septate filaments; and (3) branched filamentous forms of various degrees of complexity. Some of the siphonaceous forms grow to fairly large size and can easily be seen macroscopically. Another type, seen in many of the marine algae and in some freshwater forms, is a vegetative plant that is leafy in structure, although many differences in structure exist between the algal frond and the plant leaf. Some of the most complex seaweeds differentiate into structures that resemble leaves, stems, and roots. Many algal cells divide not by binary fission but by an internal septation process with the formation of four (or more) vegetative cells, as is shown in Figure 19.2. The vegetative cells formed in this manner are sometimes called *autospores,* although they are not true resting spores.

Motility A number of algae are motile in the vegetative phase, usually because of the presence of flagella. Cilia do not occur in the algae. Simpler flagellate forms, such as *Euglena,* have a single polar flagellum, and flagellated species of the Chlorophyta have either two or four flagella of equal length, arising from one of the poles. The dinoflagellates have two flagella of unequal length and with different points of insertion: The transverse flagellum is attached laterally and wraps partly around the transverse groove of the cell, whereas the longitudinal flagellum originates from the lateral groove but extends lengthwise from the cell (see Section 3.5). In many cases, algae are nonmotile in the vegetative state but form motile gametes.

Although they are devoid of flagella, many diatoms are motile. The mechanism of this motility is still under study, but apparently it results from some form of jet propulsion. Some desmid algae exhibit a ''creeping'' kind of gliding motility, which is thought to arise as a result of secreting mucilaginous slime. None of the eucaryotic algae demonstrates the rapid and extensive gliding motility shown by some blue-green algae (see Section 18.14).

Cell-wall structure Algae also show considerable diversity in the structure and chemistry of their cell walls. In many cases the cell wall is composed basically of cellulose (Section 3.1), but it is usually modified

Figure 19.2   Cell division cycle in
Chlorella.

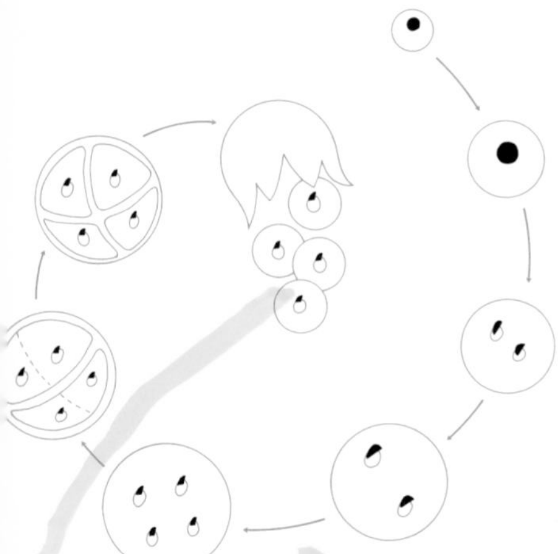

by the addition of other polysaccharides such as pectin (polygalacturonic acid), xylans, mannans, alginic acids, fucinic acid, and so on. In some algae the wall is additionally strengthened by the deposition of calcium carbonate; these are often called "calcareous" or "coralline" ("corallike") algae. Sometimes chitin is also present in the cell wall.

In the diatoms the cell wall is composed of silica, to which protein and polysaccharide are added. Even after the diatom dies and organic materials have disappeared, the external structure remains, showing that the siliceous component is indeed responsible for the rigidity of the cell. Because of the extreme resistance of these diatom frustules to decay, they remain intact for long periods of time and constitute some of the best algal fossils (see Section 17.2).

Chloroplast structure   One of the characteristics used in classifying algae into families and genera is the nature of the chloroplast, especially chloroplast shape and number. Algal chloroplasts show much greater diversity of shape than do those of higher plants: They may be discoid, ribbon-shaped, spiral, star-shaped, lobed, or branched (Figure 3.15).

19.2   Algal nomenclature and classification

Algal nomenclature follows the rules of the International Botanical Code. However, because of the great diversity of algae and the limited amount of work that has been done on many groups, many aspects of classification are still subject to considerable debate. In the algae, the largest groups are called "divisions," and are given names ending with *-phyta* (for example, Chlorophyta). Each division is separated into classes (names ending with *-phyceae*), the classes into orders (ending *-ales*), the orders into families (ending *-aceae*), the families into genera, and the genera into species. In the present discussion we shall consider only the broad outlines of algal classification and later in this section discuss certain representative forms in some detail. The algal divisions recognized by some authorities, and characteristics of these divisions, are outlined in Table 19.2.

19.3   Methods for culturing algae

Although it is possible to make many studies on algae collected from nature, cultures are essential for most detailed studies. A *unialgal* culture is one that has only one kind of alga, even though it may be contaminated with bacteria, whereas a pure (or *axenic*) culture is free of all contami-

Table 19.2 Characteristics of the major divisions of algae

Rhodophyta—the red algae   Unicellular, filamentous or leafy. Chlorophyll a. Phycobilin pigments: phycoerythrin and phycocyanin. Red color of the vegetative structures due to the presence of phycoerythrin, which masks the green color of the chlorophyll. No motile cells. Chief reserve material: floridean starch. Complex sexual cycles, with sexual organs unique to this group. Mostly marine forms, although a few live in fresh water. DNA, 49–65% GC (2 species).

Cryptophyta   Strictly unicellular, usually flagellate, with two flagella of unequal length. Chlorophylls a and c. No sexual reproduction. Reserve food: starch or starchlike compounds. Brownish in color. Phycobilin pigments. A small group with uncertain affinities.

Pyrrophyta   Strictly unicellular, flagellate organisms. Includes the class Dinophyceae (dinoflagellates). Two flagella differing in structure and in orientation. Cellulose cell walls sometimes sculptured into plates. Chlorophylls a and c. Brownish, red, or blue in color due to presence of xanthophyll accessory pigments. Reserve materials: either starch or oils. Sexual reproduction rare.

Bacillariophyta—the diatoms   A large group of organisms with a distinctive siliceous cell wall. Unicellular or colonial. Unique manner of cell division and motility. Reserve materials: β-1,3-glucan (chrysolaminarin) and oil. Chlorophylls a and c. Color typically golden brown, due to presence of carotenoid and xanthophyll accessory pigments. Vegetative cell diploid; sexual reproduction common. Source of the economically important product diatomaceous earth. Both marine and freshwater forms exist. DNA, 37—58% GC (8 species).

Phaeophyta—the brown algae   Filamentous or leafy, generally macroscopic, occasionally massive. Almost exclusively marine. Chlorophylls a and c. Brownish color due to predominance of xanthophylls over carotenoids and chlorophylls. Reserve materials: laminarin, mannitol, and fat. Complex sexual cycles. Motile reproductive cells (both asexual and sexual) of unique structure, pyriform with two lateral flagella. DNA, 59% GC (1 species).

Chrysophyta   Mainly unicellular; some colonial. Flagellate, amoeboid, or nonmotile. Reserve materials: chrysolaminarin (a β-1,3-glucan) and oil. Chlorophyll a. Color typically golden brown due to the presence of carotenoid and xanthophyll accessory pigments. Cell wall often composed of siliceous or calcareous plates. Endogenous siliceous cysts formed in some. Mainly freshwater, but some (for example, the coccolithophores) predominantly marine. A diverse group with some forms transitional with the protozoa. DNA, 58–65% GC (2 species).

Xanthophyta   Unicellular, colonial, filamentous or siphonaceous. Often classified with the Chrysophyta. Yellowish-green color. Chlorophylls a and e. Reserve materials: β-1,3-glucan (paramylon) and oil. Many with cell walls of overlapping halves, silicified. Cellulose rare or absent. Motile cells have two anterior flagella of unequal length and structure. Asexual reproduction by motile or nonmotile spores; sexual reproduction in a few genera. Mostly freshwater. DNA, 52% GC (1 species).

Euglenophyta   Strictly unicellular. Rigid cell wall lacking. Motile with one long flagellum and often an insignificant short one. Eyespot and contractile vacuole present. Color: grass-green, with chlorophylls a and b. Reserve material: β-1,3-glucan (paramylon). Many feed by ingestion (phagotrophic). Colorless phagotrophic forms occur, and the whole group is often classified with the Protozoa. Cell division longitudinal; sexual reproduction lacking. Freshwater and marine forms exist. DNA, 48–49% GC (2 species).

Chlorophyta—the green algae   A very large and diverse group. Color: grass-green, with chlorophylls a and b. Reserve material: starch. Unicellular, tetrasporal, colonial, filamentous, leafy, coenobial, siphonaceous. Motile stages are flagellate with two or four flagella of equal length, inserted anteriorly. Sexual reproduction common; asexual reproduction by motile or nonmotile cells. Freshwater and marine forms exist. DNA, 39–68% GC (20 species).

Representative eucaryotic groups

nants. For studies of the taxonomy or life cycle of an alga, unialgal cultures may be satisfactory; but for physiological and biochemical studies, pure cultures are essential. In recent years the development of culture methods for algae has been considerably advanced, and it is now possible to culture virtually any alga, even some of the seaweeds that earlier had proved difficult to grow.

Isolation methods Because of their large cell size, many algae can be isolated directly from natural materials by the use of capillary pipettes. A sample of water containing the alga of interest is placed in a dish and observed under a dissecting microscope. The capillary pipette held by a rubber suction bulb is carefully lowered into the water, and the tip is placed adjacent to an algal cell. The cell is then drawn into the capillary, and after microscopic examination of the capillary to make certain that only one cell is present, the contents of the capillary are expelled into the well of a depression slide, which contains the culture medium. The well is covered to prevent evaporation, and the slide is incubated in the light. After the alga has reproduced, the contents of the well can be transferred to a larger volume of culture medium to allow further growth.

In many cases the alga of interest may not be present in sufficient quantity in the natural material to allow its direct visualization and isolation. If a sample of natural material, such as pond water or soil, is added to a culture medium and then incubated, a variety of forms may develop that were in the minority in the original material, and those of interest can be isolated by the capillary pipette method.

In some cases one can use selective culture media to prepare enrichment cultures for species requiring particular environmental conditions. For instance, media of low pH can be used to select algae from acidic natural habitats such as bogs. Motile algae that are positively phototactic can be isolated by illuminating from only one end a long tube filled with the natural water, and removing material from the illuminated end for culture and further study. Likewise, high temperatures of incubation can be used to enrich for thermophilic algae.

Culture media and growth conditions Natural waters are usually more or less deficient in certain algal nutrients and hence cannot be used directly as culture media if large algal populations are desired. For freshwater algae a variety of culture media containing only inorganic compounds can be devised, which will permit the growth of many species. Although many algae will grow well in simple media, in many cases it is advantageous to supplement the basal salts medium with trace elements or organic growth factors. A favorite additive is *soil extract*. This is made by mixing equal amounts of ordinary garden soil (not too rich in clay or sand) with tap water and heating in a steamer for one hour or autoclaving for a few minutes. After cooling, the supernatant is decanted, filtered, and sterilized by autoclaving. Usually about 1 part of soil extract in 50 to 100 parts of culture medium is used. Soil extract is beneficial because it contains many trace elements chelated with organic humus from the soil, and it also contains vitamins that are required by some algae. Detailed studies on the nutrition of algae, however, require completely synthetic culture media, which contain known amounts of vitamins and trace elements.

Culturing marine algae had presented special problems for many years, but these problems have now been solved. Seawater is buffered only by the bicarbonate ion, and when autoclaved, CO_2 is driven off, the pH rises, and precipitates of various metal hydroxides form, leading to deficiencies in trace elements. Such problems have been overcome by using synthetic seawaters containing organic buffers, which will maintain the pH, and by using chelating agents, which will keep the important trace elements in solution. Many marine algae also require vitamins, and these must also be added. Any of the algal culture media can be made semisolid by adding agar. In most cases it is desirable to keep the agar concentration lower (1.0 to 1.2 percent agar) than for bacteria and fungi since algae usually form colonies better on less rigid agar.

An added carbon source is essential to achieve high cell yields. The most important carbon source of these autotrophs is CO_2, which is present in normal air in only trace amounts. By adding $NaHCO_3$ or by aerating cultures with air enriched with 5 percent CO_2, much heavier cell yields will usually be obtained. Addition of organic carbon sources such as acetate will often result in increased cell yield.

Light is, of course, essential. Several aspects of the illumination source must be considered: intensity, wavelength, and heat production. Very bright light is neither essential nor desirable for the development of routine cultures. Except for algae living on rocks, in shallow waters, or on the soil surface, most species are not exposed in nature to especially bright light, and indeed algae are frequently harmed by it due to photooxidative reactions (Section 8.7). The light from a north window is usually adequate for the initial isolation of cultures, or artificial light of 100 to 500 ft-c can be used. The quality of different kinds of artificial lights varies. Tungsten lamps are rich in red and infrared radiation, compared with many fluorescent lights, which are lower in red but richer in blue and near-ultraviolet radiation. Since algae differ in the absorption spectra of their pigments (Section 6.4), it is desirable to choose an artificial light that is rich in the wavelengths best utilized by the alga under investigation. When light is absorbed by the water of the culture medium, some of its energy is converted to heat. If high light intensities are used, heating of the culture to temperatures harmful to the alga may occur. The heating effect is greater with tungsten than with fluorescent lamps, since tungsten lamps produce more light of the longer wavelengths. A heat screen consisting of a vessel of water or a solution of copper sulfate (which absorbs red and infrared radiation) may be desirable. It is also sometimes desirable to immerse the culture vessels in a glass-walled, thermostatically controlled water bath to maintain a desired temperature during illumination.

Purification of algal cultures   The development of bacteria-free cultures from unialgal cultures is easy with some algae and difficult with others. If the alga has a mucilaginous sheath to which bacteria can adhere, purification is especially difficult since agar streaking methods are usually ineffective with nonmotile forms. Several methods for purification can be used. *Washing methods* are most useful for unicellular algae. They involve the washing of single cells with sterile culture medium, the cell being passed successively from one drop of culture medium to another with a capillary pipette, the process being repeated until the algal cell is free of contaminants. Motile organisms that are positively phototrophic can be induced to wash themselves by providing unilateral illumination and allowing the cells to move through a sterile medium towards the light. *Streaking methods* on agar or gelatin media can be used for those organisms which will form colonies on semisolid medium and which are hardy enough to withstand the streaking process. The principle involved here is exactly the same as that used by Koch for obtaining pure cultures of bacteria (see Chapter 1). Algae that glide across the surface of agar can often cleanse themselves by this creeping movement, through the mechanical action of the cells moving over the agar gel. Finally, to rid algal cultures of bacteria, *antibiotics* can often be used since many antibiotics are active against bacteria but are ineffective against the eucaryotic algae (see Chapter 9). After several passages through antibiotic-containing medium, the alga may then be transferred to antibiotic-free medium for maintenance and further study.

Algal culture collections   Algal cultures are usually maintained by periodic transfer to fresh media. It is the usual experience that cultures contaminated with bacteria keep better and require less frequent transfers than do pure cultures, but the reasons for this are not fully understood. Actually, some algal cultures keep well when they are free from bacteria and need be transferred only a few times a year, but other cultures die quickly and must be transferred often, perhaps as frequently as once a week. If possible, cultures should be maintained on agar slants since contamination is less frequent and is observed more easily. Upon initial subculture, the alga is allowed to grow for a time in good illumination and then is transferred to dimmer light for maintenance. Screw-capped tubes will prevent drying, and low temperatures will promote longevity of the cultures.

There are several general culture collections from which a variety of unialgal and pure cultures are available for a nominal fee. The most extensive of these collections is that of the Department of Botany of Indiana University at Bloomington, which publishes from time to time a catalog of the cultures available. Another extensive culture collection is maintained in England at Cambridge University.

19.4 *Representative algae and their life cycles*

Chlorophyta   The Chlorophyta comprises a large and diverse group of algae. We discussed the life cycle of one representative, *Chlamydomonas*, in Section 13.4. *Chlamydomonas* is the simplest form of those Chlorophyta whose vegetative cells are motile by means of flagella. In the more complex members of this group, the flagellated cells occur not singly but in colonies, and the individuality of the single cells is lost. One of the most advanced genera of these colonial algae is *Volvox;* the following discussion will deal primarily with this genus. The number of cells in a *Volvox* colony varies from around 500 in some species to over 50,000 in others. Each cell is surrounded by a gelatinous sheath and hence is physically separated from all other cells, but in some species the cells are joined to one another by cytoplasmic strands. The arrangement of the cells in a colony is quite regular (Figure 19.3). The cells are highly coordinated, and the *Volvox* colony moves through the water because of the integrated beating of the flagella of the individual cells.

The production of a new *Volvox* colony can occur in two ways, asexually or sexually (Figure 19.4). Most of the cells of a *Volvox* colony are incapable of giving rise to new colonies, but as a colony ages, a few cells in one part of the colony are converted into asexual reproductive cells called *gonidia,* each of which can be the forerunner of a new *Volvox* colony. Each gonidium is ten or more times the diameter of a vegetative cell and lies in a gelatinous sac projecting inward from the surface of the colony. The gonidium divides a number of times in a very regular sequence, forming a group of cells called an "autocolony." When the autocolony reaches a size of about 16 cells, the units are usually arranged as a hollow sphere, and divisions continue until the autocolony reaches a cell number that is characteristic for the species. The individual cells of the autocolony are at first arranged so that the pole at which the flagella will form is directed inward, but at a certain point in the develop-

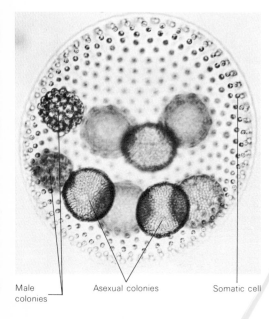

Male colonies   Asexual colonies   Somatic cell

Figure 19.3   Volvox colony with male and asexual colonies. Magnification. 125×. (Courtesy of Richard C. Starr.)

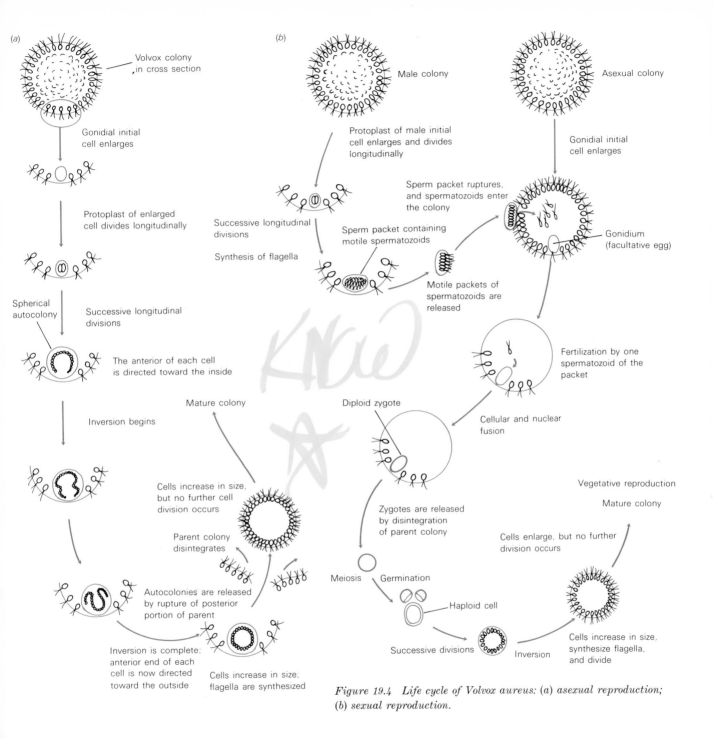

(a)

Volvox colony in cross section

Gonidial initial cell enlarges

Protoplast of enlarged cell divides longitudinally

Spherical autocolony

Successive longitudinal divisions

The anterior of each cell is directed toward the inside

Inversion begins

Mature colony

Cells increase in size, but no further cell division occurs

Parent colony disintegrates

Autocolonies are released by rupture of posterior portion of parent

Inversion is complete; anterior end of each cell is now directed toward the outside

Cells increase in size; flagella are synthesized

(b)

Male colony

Protoplast of male initial cell enlarges and divides longitudinally

Successive longitudinal divisions

Synthesis of flagella

Sperm packet containing motile spermatozoids

Motile packets of spermatozoids are released

Asexual colony

Gonidial initial cell enlarges

Sperm packet ruptures, and spermatozoids enter the colony

Gonidium (facultative egg)

Fertilization by one spermatozoid of the packet

Cellular and nuclear fusion

Diploid zygote

Zygotes are released by disintegration of parent colony

Meiosis

Germination

Haploid cell

Successive divisions

Inversion

Vegetative reproduction

Mature colony

Cells enlarge, but no further division occurs

Cells increase in size, synthesize flagella, and divide

Figure 19.4   Life cycle of Volvox aureus: (a) asexual reproduction; (b) sexual reproduction.

19.4   *Representative algae and their life cycles*

647

ment of the autocolony an inversion of the entire colony takes place, the flagellar poles now being directed outward. In some species the new colony escapes by moving through a pore on the outer side of the sac in which it has formed. An individual cell isolated from the colony can never alone initiate the formation of a new colony; thus the cells of a colony are not independent. *Volvox* cells, therefore, are more like the cells of a multicellular organism than they are like those of a microorganism.

Sexual reproduction in *Volvox* occurs by the formation of egg and sperm cells. In some species there is a sexual differentiation such that some colonies form male gametes and others form female gametes. Male gametes develop from enlarged cells resembling gonidia. Cell division ensues in a manner similar to that described above, but after inversion a colonylike mass of biflagellate gametes is liberated from the parent colony as a unit; it does not break up into individual free-swimming gametes until it penetrates a colony containing eggs. In the female colonies, only a small number of the cells develop into eggs, each egg resembling a young gonidum. Fertilization occurs through the penetration of the egg by an individual male gamete. The zygote is thick-walled and usually contains a reddish pigment; it does not germinate immediately but remains dormant until it is liberated by death or decay of the parent colony. Under appropriate environmental conditions, dormancy is broken, meiosis occurs, and a haploid biflagellate *zoospore* is produced, which becomes the forerunner of a new colony. This colony arises by a process similar to that described for asexual reproduction from gonidia.

The colonial Chlorophyta such as *Volvox* are primarily freshwater forms and are often found in large numbers in small bodies of water such as ponds, ditches, or pools. By enrichment procedures with soil-water medium they can often be isolated from the dried mud around the edges of ponds, where the zygotes accumulate. The complex differentiation of *Volvox* is of great interest, and members of this genus may provide models for understanding the nature of cell differentiation and the manner in which individual cells can enter into cooperative relationships in the manner of multicellular animals. In addition, *Volvox* colonies are highly fascinating and much pleasure can be derived from observing the behavior of these motile colonies in natural collections or in culture.

The genus *Spirogyra* is typical of a widespread group of filamentous Chlorophyta, the family Zygnemataceae. Members of this family do not produce flagellated reproductive cells, either asexual or sexual, but multiply sexually by means of nonmotile gametes. The septate filaments of *Spirogyra* are simple, unbranched, and of indefinite length. The chloroplast is a long ribbonlike structure arranged in a spiral fashion around the inner surface of the cell membrane (Plate 14: Figure 19.5; see also Figure 3.14)—hence the name of this genus. Each cell of the filament is uninucleate. Cell division may occur either at the tip or within the length of the filament; the chloroplast is severed at the point where the cross wall is formed so that filaments of *Spirogyra* often seem to have a continuous chloroplast spiralling the length of the filament. Individual filaments may fragment into shorter elements which, when dispersed, can be the forerunners of new filaments. *Spirogyra* is found primarily in quiet fresh waters, such as ponds or gently flowing streams. For most of the year, growth is primarily vegetative, sexual reproduction showing a marked periodicity and occurring usually late in the spring; probably this is associated with the rising temperature of the water. In most species, vegetative cells themselves conjugate. The cells of each filament send out conjugation tubes, which contact the conjugation tubes of the opposite filament (Figure 19.6). There is obviously some mechanism that ensures that the conjugation tubes from the adjacent filaments grow toward each other. Although this probably involves a chemotropic attraction between the two conjugation tubes, the details are not known. Once the conjugation tubes have fused, the protoplast of one cell migrates by amoeboid motion into the cell of its mate. The cell that donates the protoplast is arbitrarily called the "male," and the other cell is called the "female." Factors determining whether a cell will be male or female are not understood, and even within the same filament both male and female cells may be produced. After fusion of the nuclei, a thick wall forms around the zygote. During the ripening stage meiotic divisions occur and three of the resulting nuclei disintegrate so that the mature resting cell is haploid.

Figure 19.6 Sexual reproduction in Spirogyra.

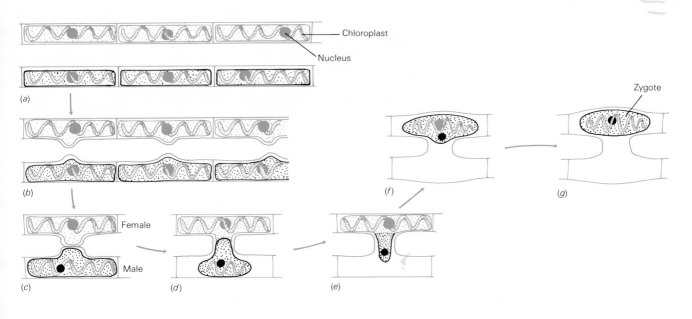

(a)

(b)

(c)   Female   Male

(d)

(e)

(f)

(g)   Zygote

Chloroplast

Nucleus

When this cell germinates, the outermost layers disintegrate, and a tubular outgrowth develops, which is the forerunner of the new vegetative filament. In some species of *Spirogyra* conjugation occurs between adjacent cells on the same filament; conjugation tubes grow out from each of the two cells and fusion takes place. There are a number of variations of the sexual cycle in the Zygnemataceae, although the basic process is that described here.

One of the most fascinating genera of the Chlorophyta is *Acetabularia*. This genus belongs to the order Siphonales, a group whose vegetative plants remain nonseptate even though they may achieve macroscopic size. Several species of *Acetabularia* are known, all of which are marine and live primarily in the warmer tropical or subtropical waters.

The vegetative plant is diploid and consists of a basal lobed rootlike structure, a cylindrical stalk several centimeters long, and an umbrella-shaped cap composed of many radially arranged compartments (Figure 19.7). The plant is somewhat rigid owing to the deposition of calcium carbonate. Although the plant is several centimeters high, it has only a single very large nucleus, situated in the base. When the plant is mature, this primary nucleus divides into a great number (10,000 to 15,000) of secondary diploid nuclei, which migrate up through the stalk into the rays of the cap. Here each nucleus becomes surrounded by cytoplasm and a membrane, and each becomes the forerunner of a *cyst*. Within the developing cyst, further nuclear divisions occur, ultimately producing a mature cyst that is a calcareous, multinucleated structure. The cap breaks down and the cysts are released into the water, where they may remain dormant for periods up to several months. Upon germination of the cyst, the diploid nuclei undergo meiosis, and each nucleus is incorporated into a single biflagellated gamete. These gametes are discharged through a pore in the cyst, conjugate, and form quadriflagellate zygotes that settle down almost immediately and attach to a substratum (rock, coral, and so on) at their flagellated ends. The zygote then loses its flagella and begins to grow, sending out rootlike projections and developing a stalk and cap. The whole life cycle (illustrated in Figure 19.7) takes 6 mo to a year.

Acetabularia has been a useful model organism for studying nuclear-cytoplasmic relationships since the single nucleus is isolated in the base of the plant throughout most of its life history. If the upper part of the plant is cut away, regeneration can occur, with the stalk growing back to its original length and forming a new cap. In nucleated plants regeneration can be made to occur repeatedly. An isolated portion of the root will be able to regenerate a normal plant as long as it retains the nucleus. Even if the nucleus is removed, regeneration can still occur and a new stalk and cap can be formed, but the extent of regeneration depends on the amount of cytoplasm remaining after amputation. It is possible to graft

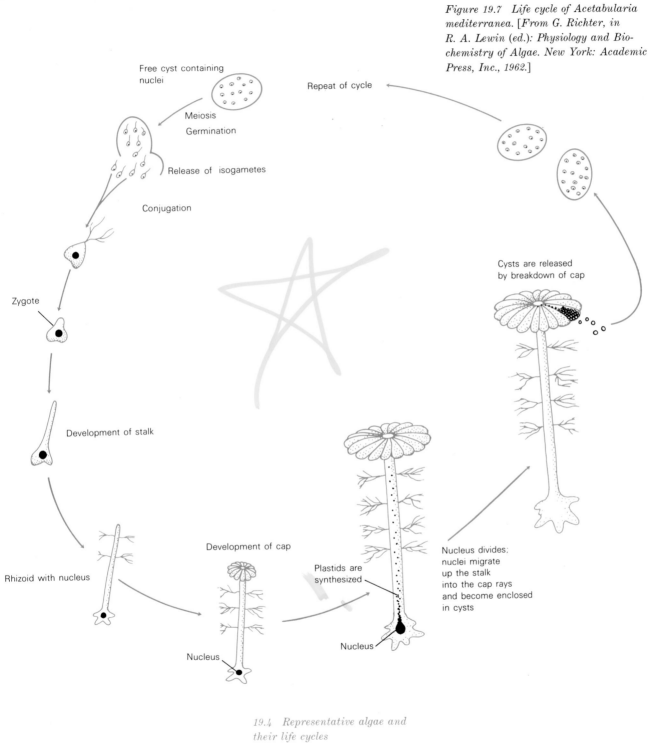

Free cyst containing
nuclei

Meiosis

Germination

Release of isogametes

Conjugation

Zygote

Development of stalk

Rhizoid with nucleus

Development of cap

Nucleus

Plastids are
synthesized

Nucleus

Nucleus divides;
nuclei migrate
up the stalk
into the cap rays
and become enclosed
in cysts

Cysts are released
by breakdown of cap

Repeat of cycle

*Figure 19.7   Life cycle of Acetabularia
mediterranea. [From G. Richter, in
R. A. Lewin (ed.): Physiology and Biochemistry of Algae. New York: Academic
Press, Inc., 1962.]*

*19.4   Representative algae and
their life cycles*

Figure 19.8 Electron micrograph of a thin section of the freshwater diatom Navicula pelliculosa. Note that the siliceous frustule occurs as two overlapping halves. Magnification, 21,500×. (Courtesy of M. L. Chiappino and B. E. Volcani.)

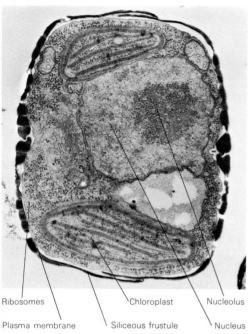

Ribosomes Chloroplast Nucleolus

Plasma membrane Siliceous frustule Nucleus

Figure 19.9 Photomicrographs of frustules of several kinds of diatoms: (a) centric diatoms; (b) pennate diatoms. (Courtesy of Johns-Manville Research and Engineering Center, Manville, N.J.)

a portion of the plant of one species of *Acetabularia* onto that of another and in this way to determine the relative contributions of nucleus and cytoplasm to the characteristics of the regenerated plant. Such experiments have shown that the nucleus retains primary control of the cap structure but that this control is maintained through stable cytoplasmic components that migrate from the nucleus to the region of the developing cap. Biochemical investigations have shown that the capacity to regenerate is closely correlated with the elaboration of RNA by the nucleus. Plants without nuclei synthesize little new RNA, and their ability to synthesize protein gradually decreases, whereas nucleated plant fragments can continue to synthesize RNA and protein and can regenerate. The role of the nucleus may be further shown by transplanting the nucleus from one plant into an enucleated fragment since this results in the continued synthesis of RNA. Although it has not been proved, it seems reasonable to conclude that the RNA synthesized by the nucleus and involved in protein synthesis and regeneration is mRNA. Such studies on *Acetabularia* provided some of the earliest evidence for the elaboration of RNA by the nucleus into the cytoplasm and its role in controlling developmental phenomena.

Bacillariophyta—diatoms   The diatoms are an interesting algal group of worldwide distribution, with freshwater, marine, soil, and aerial forms. The distinctive feature of the diatom is its siliceous frustule (cell wall) composed of two overlapping halves (Figure 19.8). The frustules have distinct markings, which are characteristic of each species (Figure 19.9; see also Figure 3.5). Because of the presence of a rigid wall, diatoms

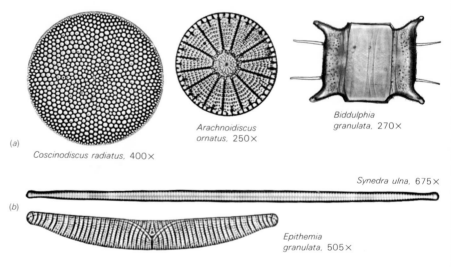

(a)

Coscinodiscus radiatus, 400×

Arachnoidiscus ornatus, 250×

Biddulphia granulata, 270×

Synedra ulna, 675×

(b)

Epithemia granulata, 505×

Figure 19.10 Cell-wall formation during cell division in a diatom. Note that, at each division, a reduction in cell size occurs.

face a special problem in cell division. The manner of cell division is illustrated in Figure 19.10, and we can see that cell size decreases with each successive division. Such progressive diminution in size is eventually compensated for by *auxospore* formation. This process can occur in a variety of ways, of which two common ones will be described.

The vegetative cells of the diatoms are diploid. One way in which an

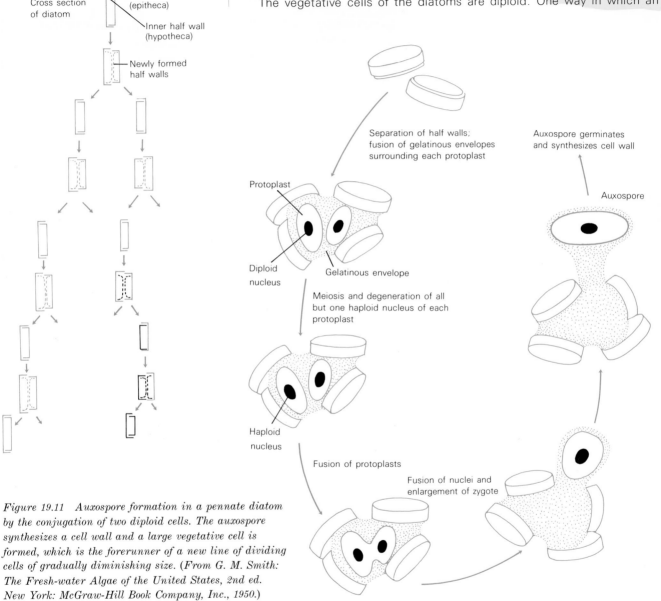

Figure 19.11 Auxospore formation in a pennate diatom by the conjugation of two diploid cells. The auxospore synthesizes a cell wall and a large vegetative cell is formed, which is the forerunner of a new line of dividing cells of gradually diminishing size. (From G. M. Smith: The Fresh-water Algae of the United States, 2nd ed. New York: McGraw-Hill Book Company, Inc., 1950.)

19.4 *Representative algae and their life cycles*

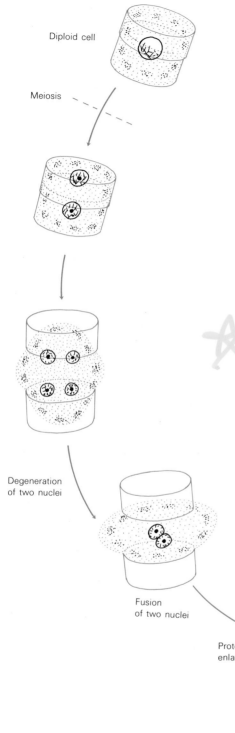

Diploid cell

Meiosis

Degeneration
of two nuclei

Fusion
of two nuclei

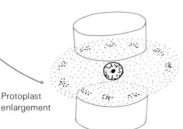

Protoplast
enlargement

Auxospore becomes
an enlarged
vegetative cell

auxospore is produced is by the conjugation of two cells. The two cells can be either sister cells or cells not derived from a common parent. The two cells are first enclosed in a common gelatinous envelope (Figure 19.11), and each cell then undergoes meiosis. Following this, three of the four resulting haploid nuclei in each cell of the conjugating pair degenerate. The protoplasts of the two cells escape from the cell walls and unite, nuclei fuse, and a diploid nucleus is again formed. The zygote then enlarges and becomes an auxospore, the size of which is larger than the maximum size of the vegetative cell. The auxospore later germinates to become the forerunner of new vegetatively reproducing diploid cells.

In other diatoms, auxospores are produced by single cells without conjugation. As is shown in Figure 19.12, the two half walls pull apart, the protoplast enlarges, and a new silicified cell wall is secreted. The auxospore germinates by dividing transversely, and the two daughter cells, which are two to three times the diameter of the original vegetative cell, are the forerunners of new cell lines. The formation of an auxospore in this manner may or may not involve meiosis followed by nuclear fusion.

Some diatoms have a sexual cycle involving small flagellated spores, called "microspores," which function as male gametes and swim to cells containing female gametes, penetrate, and fertilize them.

The diatoms are divided into two classes: those with radial symmetry, called "centric" diatoms, and those with bilateral symmetry, called "pennate" (Figure 19.13). The centric diatoms usually have many chromatophores, produce motile microspores, never show motility, and never produce auxospores by conjugation. The pennate diatoms usually have but one or two chromatophores, lack flagellated microspores, and often have motile vegetative cells, and many produce auxospores by conjugation. Diatoms are most common in colder waters. In the arctic and temperate

Figure 19.12 Auxospore formation in a centric diatom: Conjugation does not take place, but the sexual events occur in the nuclei, both of which are derived from a single cell. (From G. M. Smith: The Fresh-water Algae of the United States, 2nd ed. New York: McGraw-Hill Book Company, Inc., 1950.)

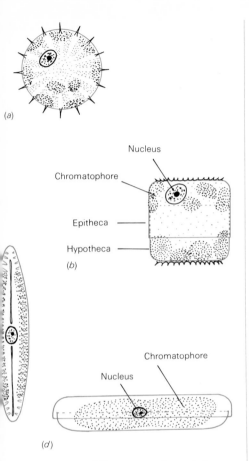

(a)

Nucleus

Chromatophore

Epitheca

Hypotheca

(b)

Chromatophore

Nucleus

(d)

Figure 19.13 Cell structure in centric (a, b) and pennate (c, d) diatoms. (a, c) valve views; (b, d) girdle views.

oceans they are the predominant primary producers of organic matter—so much so that they have often been called "grass of the sea." Consequently, the diatoms are extensively studied by marine microbiologists, but great gaps still exist in our knowledge of all aspects of their biology.

Dinophyceae—the dinoflagellates These algae are the most important members of the Pyrrophyta, and their general characteristics are described in Table 19.2. The complex cell shape of dinoflagellates was mentioned in Section 3.1, and an electron micrograph of a dinoflagellate frustule is shown in Figure 19.14. The rigid outer layer of many dinoflagellates fossilizes well, so that we have a good record of the evolutionary history of this group (see Figure 17.5). Most dinoflagellates are marine forms; just as the diatoms are characteristic of cooler seawaters, the dinoflagellates are the most important algae of warm tropical waters.

The red tide, an occasional occurrence in inshore areas, results from extensive blooms of red-pigmented dinoflagellate species. Red tides, which probably develop when the seawater becomes unusually enriched with nutrients, are of practical significance because some dinoflagellates produce fish toxins that may cause extensive and unsightly fish kills.

Many dinoflagellates are luminescent, being triggered to produce flashes of light by mechanical agitation of the water; they are the main cause of the glow of seawater, which is dramatically intensified in the wake of a ship. In most marine areas dinoflagellate densities are low, and the glow of seawater is only faint, but certain phosphorescent bays off islands in the Caribbean Sea are sites of extensive and virtually permanent blooms of luminescent dinoflagellates. The biochemistry of dinoflagellate luminescence differs from that of bacterial luminescence (discussed in Section 18.11), but we shall not give the details here. Many luminescent dinoflagellates have an endogenous rhythm for the production of light, glowing when stimulated at night but not in the daytime.

As discussed in Section 14.7, some intracellular symbionts of marine invertebrates are dinoflagellates; these symbionts are usually called *zooxanthellae*. Not all dinoflagellates are photosynthetic; some colorless forms obtain their nutrients phagotrophically. The life cycle of a common inshore dinoflagellate of temperate waters is shown in Figure 19.15.

Phaeophyta—brown algae The brown algae comprise a group of multicellular algae that is almost exclusively marine. The majority are macroscopic, and some, such as the kelps, are massive, extending in length to 100 ft or more. The life histories of many of the brown algae are more complex than are those we have discussed up until now. The life cycle of a representative genus, *Ectocarpus*, is illustrated in Figure 19.16. The vegetative plant is composed of septate branched filaments

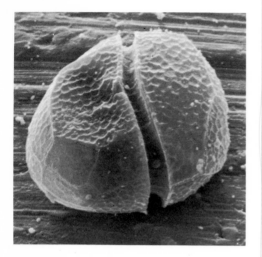

Figure 19.14 Scanning electron micrograph of the frustule of a dinoflagellate. Note the characteristic groove. Magnification, 1,300×. (© Patrick Echlin and Cambridge Scientific Instruments, Cambridge, England.)

that are attached to the substrate by rootlike appendages. Both haploid and diploid vegetative plants occur, which can be distinguished only by the reproductive structures they form. The diploid plant, or sporophyte, produces both asexual and sexual zoospores. The asexual zoospores are themselves diploid and are formed in special compartmented structures called *plurilocular sporangia*. These biflagellated motile zoospores are agents of dispersal as they can move through the water to new habitats, settle, and become the forerunners of new diploid sporophytes. The diploid plant also produces male and female haploid zoospores as a result of meiosis; these are formed in simple noncompartmented structures called *unilocular sporangia*. Although the haploid zoospores resemble gametes, they do not conjugate and hence are not true gametes. Rather, each haploid zoospore is the forerunner of a haploid vegetative plant called a *gametophyte*, which will itself produce gametes. A haploid gametophyte is ''male'' or ''female'' according to its production of male or female gametes. (A plant that produces male and female gametes in separate plants is called *dioecious*.) The male and female gametes are also biflagellate motile cells, differing in that the male gamete moves more rapidly and for a longer time than does the female. Gametes of opposite sex conjugate and form a diploid zygote, which is then the forerunner of a new diploid sporophyte. The kind of life cycle illustrated for *Ectocarpus* is one in which there is an alternation of generations with both haploid

Figure 19.15 Life cycle of a typical dinoflagellate, Gonyaulax digitalis. The motile thecate stage lives in the open water (planktonic stage) and undergoes vegetative cell divisions. In the fall of the year, as water temperatures drop, the thecate cells form cysts and fall to the bottom (benthonic stage). In the spring, when water temperatures rise, the cysts germinate, forming initially the naked gymnodinioid stage, whose cells excrete an outer wall (the theca) and convert to the summer thecate stage. [From David Wall and Barrie Dale: Micropaleontology 14:265 (1968).]

Short gymnodinioid stage

Motile thecate stage

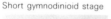
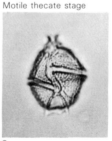
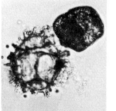
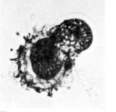
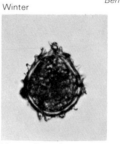

Summer

Planktonic

Benthonic

Winter

Cyst formation

Excystment

Nonmotile resting stage

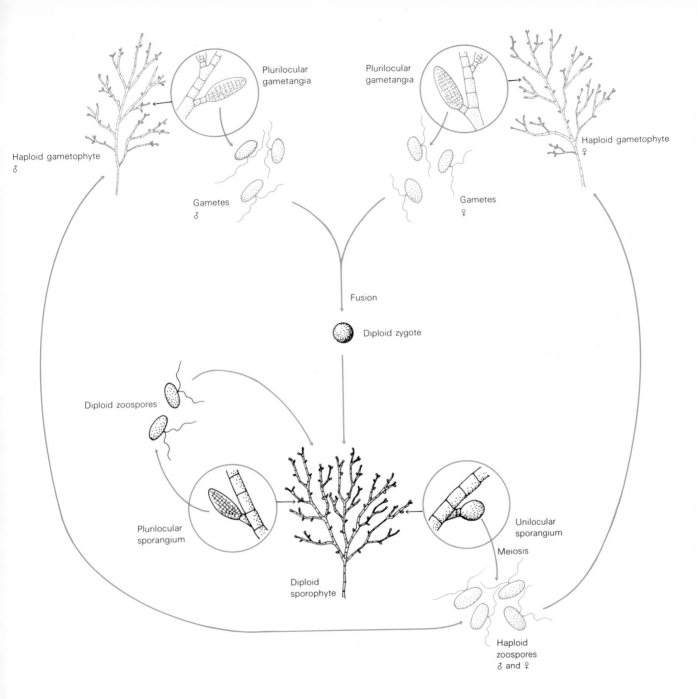

Figure 19.16 *Life cycle of a brown alga, Ectocarpus.*

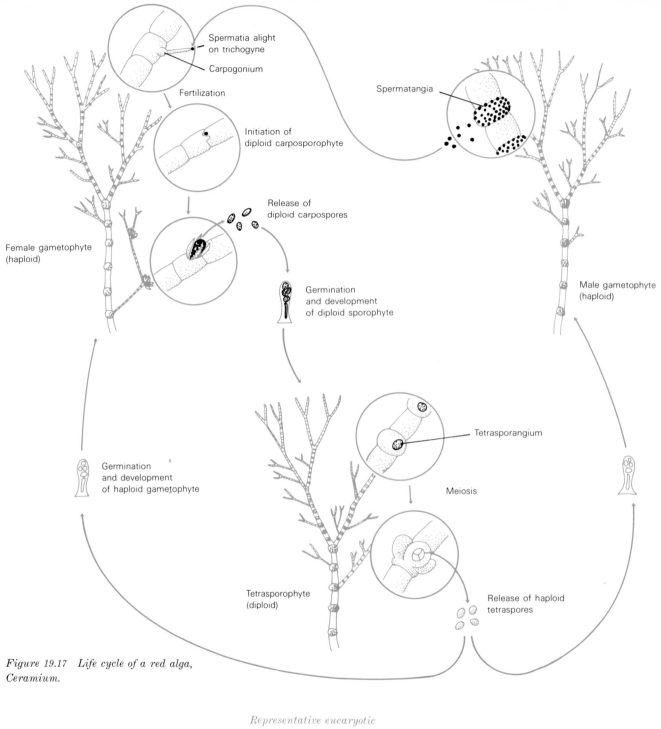

Spermatia alight
on trichogyne

Carpogonium

Fertilization

Initiation of
diploid carposporophyte

Spermatangia

Release of
diploid carpospores

Female gametophyte
(haploid)

Germination
and development
of diploid sporophyte

Male gametophyte
(haploid)

Tetrasporangium

Germination
and development
of haploid gametophyte

Meiosis

Tetrasporophyte
(diploid)

Release of haploid
tetraspores

*Figure 19.17 Life cycle of a red alga,
Ceramium.*

and diploid vegetative plants. Such a life cycle is relatively common in the more complex algae, in some of the fungi, and in mosses and ferns, but it is not found in most other microorganisms or in higher organisms.

Rhodophyta—red algae   Most of the red algae are marine organisms although there are some freshwater forms. The red algae are very common in tropical and subtropical marine waters, but they are less common in colder waters, where the brown and green algae predominate. Many of the red algae are small filamentous forms, but a number of leafy types exist, including the alga *Porphyra,* which is used as food in the Orient. The red algae are also economically important as the commercial source of the polysaccharide agar. These algae are red because of the presence of phycoerythrin; however, the freshwater forms are usually green because they have less phycoerythrin. Unlike many other algae, the Rhodophyta never produce motile reproductive cells; however, the life cycles of many red algae are even more complex than those already described. In addition to haploid gametophytes and a diploid sporophyte, there is a third stage, the diploid *carposporophyte.* Thus in the red algae there is a triphasic alternation of generations, as is illustrated in Figure 19.17. On the surface of special cells of the diploid plant, or *tetrasporophyte,* are borne small structures called ''tetrasporangia,'' in which meiosis occurs with the formation of four haploid spores. Each spore germinates and is the forerunner of a haploid gametophyte, two of the four spores producing male and the other two producing female gametophytes, which are indistinguishable except by the reproductive structures to which they give rise. The male gametophyte produces small, nonmotile cells called *spermatia,* and the female gametophyte produces a structure called a *carpogonium,* which has extending from it a thin filamentous process called a *trichogyne.* Fertilization occurs when a male spermatium fuses with the female trichogyne and the male nucleus migrates down the trichogyne and fuses with the nucleus of the carpogonium. This diploid nucleus then undergoes a number of divisions, and a series of diploid filaments develop into the carposporphyte, which is a small diploid plant growing attached to the haploid female gametophyte. The carposporophyte produces a structure called a *cystocarp,* within which diploid *carpospores* are produced. Each carpospore is the forerunner of a diploid tetrasporophyte, thus completing the life cycle.

19.5   *Summary*

The discussion above gives only a general idea of some of the complexities possible in the life cycles of algae. Manifold variations are possible, and because of the small amount of research that has been done on the marine forms, numerous details of these life cycles still remain to be discovered. Most macroscopic algae can hardly be called microorganisms; yet they are definitely related to the microscopic forms and illustrate a degree of evolutionary development possible from microbial forerunners. The importance of eucaryotic algae as primary producers of organic matter in aquatic environments is great. In the oceans they are practically the only primary producers, and they are the dominant primary producers in many lakes and streams. Because of the ease with which many algae can now be cultured, much further work on their functions and life cycles will doubtless be done.

The fungi

There is no general agreement among microbiologists regarding the limits of the group of microorganisms called *fungi*. The group is often defined as comprising those eucaryotic microorganisms that have rigid cell walls and lack chlorophyll, although by this definition certain organisms that are clearly nonchlorophyllous derivatives of algae would have to be considered fungi. In addition, many fungi have affinities with the protozoa. However, the members of two great groups of fungi, the Ascomycetes and the Basidiomycetes, cannot be related in any direct way to either the algae or the protozoa. Another group of organisms, the slime molds, is sometimes included with the fungi because they form fruiting bodies similar to those of many fungi, but the slime molds differ in so many ways from the fungi that it is easier to consider them as a separate group. Within the group classified as true fungi there are around 80,000 named species.

The habitats of the fungi are quite diverse: Some fungi are aquatic, living primarily in fresh water, and a few marine fungi are also known; most, however, have terrestrial habitats, in soil or on dead plant matter, and these types often play crucial roles in the mineralization of organic carbon (Section 16.3). A large number of fungi are parasites of terrestrial plants; indeed, fungi cause the majority of economically significant diseases of crop plants. A few fungi are parasitic on animals, including man, although members of this group are much less significant as animal pathogens than are bacteria.

All fungi are heterotrophs. Lacking chlorophyll, they of course cannot photosynthesize, and the group also seems to lack chemolithotrophic forms. When compared to the bacteria, the fungi in general have fairly simple nutritional requirements, and their metabolic and biosynthetic processes are not particularly diverse or unusual. It is in their morphological properties and in their sexual life cycles that the fungi exhibit considerable diversity; hence it is on the basis of these characteristics that the fungi are usually classified into families, orders, genera, and species. In the present section we will attempt to give some picture of this great morphological diversity, although space does not permit more than a brief account of these matters. Textbooks of mycology should be consulted for more complete discussions.

19.6　Structural characteristics of fungi

Vegetative structure　The vegetative structure of a fungus is often called a *thallus*. The most typical fungus thallus is composed of filaments, which are usually branched. Each individual filament is called a *hypha* (plural, *hyphae*), and a mass of hyphae is called a *mycelium*. In many cases the mycelium forms a structure visible to the naked eye. Mycelia come into being during growth as a result of two distinct processes, hyphal branching and fusion (Figure 19.18). If moist leaf litter on the forest floor is carefully examined, white threads can often be seen ramifying through the decomposing plant mass. These threads are composed of mycelia of various fungi. In some cases, the unit hyphae have fused so extensively that they have lost their individuality and have formed strands called *rhizomorphs*, which have a complex, differentiated structure and often exist as rootlike connections to fruiting bodies. We discussed briefly the role of rhizomorphs in the nutrient relationships of mycorrhizal fungi in Section 14.10.

Not all fungi are filamentous; some aquatic forms are unicellular, although usually these are multinucleate. The yeasts, another group of fungi we have discussed often in this book, are also unicellular, although in the case of many yeasts, both unicellular and filamentous growth is possible, depending on environmental conditions. As we described in Section 7.1, yeastlike growth occurs by a budding process, in which the daughter cell separates from the mother as soon as it has matured, whereas filamentous growth occurs by continuous extension of hyphal tips. Some fungi pathogenic for animals are dimorphic, growing either as mycelial or as yeastlike forms. When growing at temperatures around 37°C (such as are found in the animal body), the fungus is yeastlike, but when growing in laboratory media at lower temperatures the organism is myceliumlike.

In virtually every case, the vegetative cell of a fungus hypha contains more than one nucleus—often hundreds of nuclei are present. Thus, the typical fungus hypha should be properly viewed as a nucleated cytoplasmic mass contained within a system of tubes; such a system is often called *acellular*, or *coenocytic*. Usually there is extensive cytoplasmic

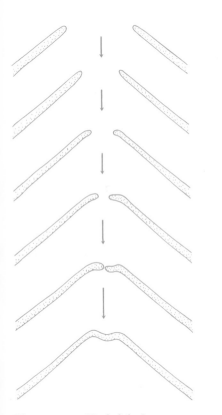

Figure 19.18　Hyphal fusion, a process frequently involved in the formation of a fungal mycelium. (From A. H. R. Buller: Researches on Fungi, vol. V. New York: Hafner Publishing Co., Inc., 1957.)

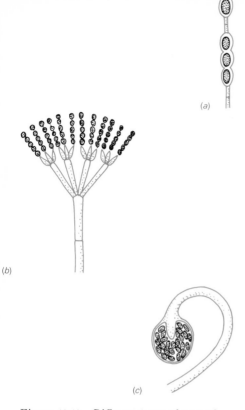

Figure 19.19   Different types of asexual spores formed by fungi: (a) chlamydospores, (b) conidiospores, (c) sporangiospores.

movement within the hypha, generally in a direction toward the hyphal tips, and the older portions of the hypha usually become vacuolated and virtually devoid of cytoplasm. Even if the hypha has cross walls (called *septa*), cytoplasmic movement is often not prevented, as there is usually a pore in the center of the septum, through which nuclei and cytoplasmic particles can move.

Spore-bearing structures   Most fungi produce spores, the function of which is either to ensure the dispersal of the species to new locations or to enable the organism to withstand extreme environmental conditions and thus to tide the species over hard times. Two general kinds of spores, asexual and sexual, are recognized. Asexual spores are usually resistant to drying or radiation but are not especially heat resistant and exhibit no dormancy; they are able to germinate when moisture becomes available, often even in the absence of nutrients. Sexual spores are usually more resistant to heat than are asexual spores, although no fungus spore shows the extreme heat resistance characteristic of the bacterial endospore. Sexual spores often exhibit dormancy, germinating only when they have been activated in some way, such as by mild heat or by certain chemicals.

Asexual spores are formed by fungi in a number of ways. In the simplest case, cells within the hypha become round and thick-walled, and then detach (Figure 19.19); such spores are called *chlamydospores*. Another kind of asexual spore is the *conidiospore,* or *conidium,* which is a spore produced by a hypha rather than being a transformed part of it. In most cases conidia are formed on distinct hyphal structures, called *conidiophores.* Often the conidia are formed not singly but in chains and clusters (Figure 19.19). The diversity of conidia-bearing structures in the fungi is tremendous, and such distinctive structures are key characteristics in the classification of fungi. In several groups of fungi asexual spores are formed inside a structure called a *sporangium* (Figure 19.19). At maturity the sporangium ruptures and the spores are liberated (Figure 19.19).

Sexual spores are formed during the process of sexual reproduction. A large variety of structures are involved in the formation of these spores. In several groups of fungi sexual reproduction occurs by the fusion of unicellular gametes, one or both of which may be motile, flagellated cells called *zoospores.* In the Zygomycetes, specialized hyphae called *gametangia* fuse, one nucleus from each fusing to form a diploid nucleus, which is incorporated into the resistant sexual spore (Figure 19.20). In one major group of fungi, the Ascomycetes, sexual reproduction involves the formation of *ascospores* within a saclike cell, the *ascus* (Figure 19.20). Usually, a group of asci are formed within a larger structure called a "fruiting body." We described in some detail the manner of formation of the ascus and ascospores of *Neurospora* in Section 13.3. In

(a)

(b)

(c)

Figure 19.20 Different types of sexual spores formed by fungi: (a) zygospore of a zygomycete, (b) ascospores in the ascus of an ascomycete, (c) basidiospores on the basidium of a basidiomycete.

another great group of fungi, the Basidiomycetes, sexual spores called *basidiospores* are produced on the outside of a specialized spore-bearing cell, the *basidium* (Figure 19.20); a number of basidia are borne on a highly organized fruiting body, which is sometimes called a *basidiocarp*. The most familiar example of a Basidiomycete fruiting body is the mushroom, but many other kinds are known. Details of some of the various types of fungus reproductive structure are described later in this chapter.

19.7 Classification of fungi

As we have noted, the predominant characteristics used in classifying fungi are structural ones. These include the following: (1) the structure of the hypha, whether septate or nonseptate, and the type of septum; (2) the manner of formation of asexual spores; (3) the manner of formation of sexual spores; (4) the structure and manner of formation of the sexual fruiting body; (5) presence or absence of motile cells, the kinds of motile cells, and the number and morphology of the flagella; (6) chemistry of the cell wall [some fungi have cellulose (β-1,4-glucan) cell walls, whereas others have chitin (a polymer of N-acetylglucosamine), and still others have β-1,3- and β-1,6-glucans; in all cases, in addition to these basal wall components, other polysaccharides are present, either as additional layers of the wall or combined chemically with the basal wall structure].

Since fungi have traditionally been studied and classified by botanists, the classification of fungi follows the rules of the International Botanical Code. Although there is no modern standard treatment of the classification of fungi such as exists for the bacteria (*Bergey's Manual of Determinative Bacteriology*), one can usually identify the family, order, and sometimes the genus to which a fungus belongs by using one of the standard mycology textbooks. It is usually necessary to refer to a monograph dealing with a particular order or genus in order to determine the species.

Since the manner of sexual reproduction is used to classify fungi, it is virtually essential to have sexual stages present in the specimen in order to identify it properly. (In some genera with highly distinctive conidiophores or other asexual structures, this requirement may not be necessary.) In many cases, however, a fungus may not form sexual stages in culture or may require special media in order to form them. Because of this, considerable difficulty is often experienced in classifying a given fungus. In a number of cases sexual stages have never been found although asexual stages are formed. Fungi in this category are often classified in an arbitrary group, the Fungi Imperfecti or imperfect fungi, and are placed in genera and species on the basis of their asexual structures. In several instances improvements in culture media have made possible the induction of sexual structures in fungi previously refractory. Once sexual structures were seen, it was often possible to assign the fungus to a previously recognized genus. For this reason, a number of common fungi are referred to by two separate generic names, one representing its imperfect (or asexual) stage and the other representing its perfect (or sexual) stage. For instance, the perfect stage of *Neurospora* was discovered only in the 1920s; before that only the imperfect stage was known and had been given the genus name *Monilia*. The latter name has now been abandoned but is seen frequently in the older literature.

19.8   Methods for the study of fungi

Although the usual methods of the microbiologist (such as have been described throughout this book) are used in studying fungi, there are in addition certain specialized mycological methods that deserve mention.

Many taxonomic studies on the larger fungi can be done using fruiting bodies collected in nature, but for any detailed study of a particular species, pure cultures are essential. Cultivation of fungi is usually not especially difficult: Nutritional requirements are generally simple, and adequate culture media can easily be prepared. (An exception to this statement must be made for the rusts, which are obligate parasites of plants.) Many fungi grow rather slowly even under optimal conditions, and investigators accustomed to studying rapidly growing bacteria may become impatient. However, the only real problems in culturing slow-growing fungi are the desiccation of the agar medium during the long incubation time and the possibility of contamination by more rapidly growing, airborne fungi. Petri dishes should be wrapped in plastic film or placed in sealed jars during incubation, thus simultaneously preventing both desiccation and the entry of contaminating spores.

Isolation methods for fungi are in general similar to those used for heterotrophic bacteria, with certain modifications. Because of the larger size of fungus spores and vegetative elements, direct isolation from natural collections is often possible, using capillary pipettes or fine steel needles. (Very fine insect pins are also very useful as they are rugged and can easily be sterilized.) A very useful technique for the isolation of aquatic fungi is the use of *baits*, which are placed in the lake, stream, or ocean water, and on which certain fungi will grow selectively. For instance, seeds of various kinds (hemp seeds are especially good) are placed in small mesh bags and are submerged in the water. After a few days or weeks the bait is retrieved and examined microscopically for the presence of fungi which, if present, can be isolated and cultured. For

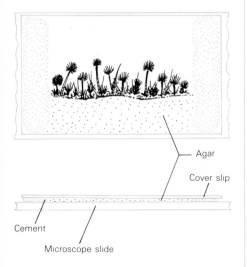

Agar

Cover slip

Cement

Microscope slide

aquatic fungi that attack wood, small wooden planks can be used as bait. A technique of some use in the study of soil fungi is the *soil-tube* technique. A small glass tube is submerged in the soil, and fungi growing up the inside of the tube can be seen with the naked eye when the tube is removed. For the isolation of fungus spores from soil, a *flotation* technique, which makes use of the fact that soil particles are considerably denser than fungus spores, can be used. Soil is suspended in a concentrated sucrose solution and centrifuged, which causes the soil particles to settle to the bottom and the fungus spores to float to the surface, from where they can be isolated and cultured. To separate fungus hyphae from fungus spores, bacteria, and other unicellular organisms, *filtration* can be done by passing a soil suspension through fine filter paper, which allows the unicellular elements to pass through but retains the hyphae. In addition, various *selective media* can also be used to isolate fungi. Since most fungi have acidic pH optima for growth, culture media of pH 3 to 4 will permit fungal growth and inhibit bacterial growth. Media containing such antibacterial agents as antibiotics (for example, streptomycin, neomycin) or dyes (for instance, rose bengal) are also useful in isolating fungi from soil. The antifungal agent cycloheximide (Actidione) is more active against common molds than it is against human pathogenic fungi and can be used to selectively isolate the latter from human tissues or secretions, which are usually contaminated with airborne spores of soil fungi.

Since taxonomic determinations of fungi are based to a great extent on the morphology of the spore-bearing structures, microscopic examination of these structures is useful. For this purpose, a *slide culture* is often prepared with a special microscope slide on which a small amount of nutrient agar is placed. The fungus is then inoculated on the agar, a cover slip is added, and the slide is incubated until the fungus has grown and formed reproductive structures (Figure 19.21). The preparation can be examined directly under the microscope without disturbing the arrangement of the reproductive structures.

19.9 Representative fungi and their life cycles

The lower fungi   At one time the lower fungi were all grouped in one class, the Phycomycetes, but recent work has shown that it is more appropriate to recognize several classes, which are enumerated in Table 19.3.

The simplest lower fungi are unicellular, producing no filaments at all. Most of the lower fungi live in aquatic habitats, although one major group, the class Zygomycetes, is strictly terrestrial. Aquatic lower fungi usually

Table 19.3 Major taxonomic divisions of the lower fungi

Class Chytridiomycetes   True mycelium lacking. Posteriorly uniflagellated zoospores. Cell wall: chitin, glucan (β-1,3; β-1,6). Lysine pathway: aminoadipic. DNA, 44–51% GC (2 species).

Class Hyphochytridiomycetes   True mycelium present. Anteriorly uniflagellated zoospores. Cell wall: cellulose, chitin. Lysine pathway: some have aminoadipic; others, diaminopimelic. DNA, 66% GC (1 species).

Class Oomycetes   True mycelium present. Motile spores biflagellate. Cell wall: cellulose, glucan (β-1,3; β-1,6). Lysine pathway: diaminopimelic.

   Order Saprolegniales   Generally aquatic, occasionally pathogenic to fish. Well-developed mycelium. Asexual spores formed on specialized structures of the mycelium. Male gamete motile, female gamete nonmotile; fertilization to form a thick-walled oospore. DNA, 27% GC (1 species).

   Order Peronosporales   Generally terrestrial, often pathogenic to plants. Well-developed mycelium. Sporangia may produce either asexual zoospores or may germinate directly to form hyphae. Both male and female gametes nonmotile.

Class Zygomycetes   Motile spores not formed. Cell wall: chitosan, chitin. Usually nonparasitic. Mycelium large and well developed; asexual spores produced in sporangia. Lysine pathway: aminoadipic. DNA, 35–49% GC (6 species).

---

produce flagellated motile cells, often called *zoospores*, and a major taxonomic division is made on the basis of whether zoospores are uniflagellate or biflagellate (see Table 19.3).

The life history of a typical water mold, *Allomyces*, is shown in Figure 19.22. Note that this fungus has two distinct vegetative thalli, a haploid thallus, which produces the male and female gametes, and a diploid thallus, which forms resting cells. On the haploid thallus, the male and female gametes are produced in special cells (*gametangia*) at the tips of hyphal branches. The gametes are uniflagellate cells, the male being distinguished from the female by its smaller size and its orange pigmentation. After union of male and female gametes, the nuclei fuse, and the resulting zygote germinates and forms a diploid thallus. This latter structure produces two kinds of sporangia, thin-walled zoosporangia, which form diploid zoospores, and thick-walled, "resistant" sporangia, which undergo meiosis and produce haploid zoospores. The diploid zoospores produced by the thin-walled sporangia germinate and form new diploid thalli, whereas the haploid zoospores from thick-walled sporangia produce haploid thalli, which are hermaphroditic and produce in turn male and female gametes. Thick-walled sporangia are called "resistant" because they do not necessarily yield zoospores immediately but may remain dormant and can resist drying. They are thus able to tide the fungus over times when the pond or stream in which it lives dries up.

The life history of a typical member of the class Zygomycetes, *Mucor mucedo,* is shown in Figure 19.23. There are two mating types, designated *plus* and *minus,* which are morphologically and physiologically

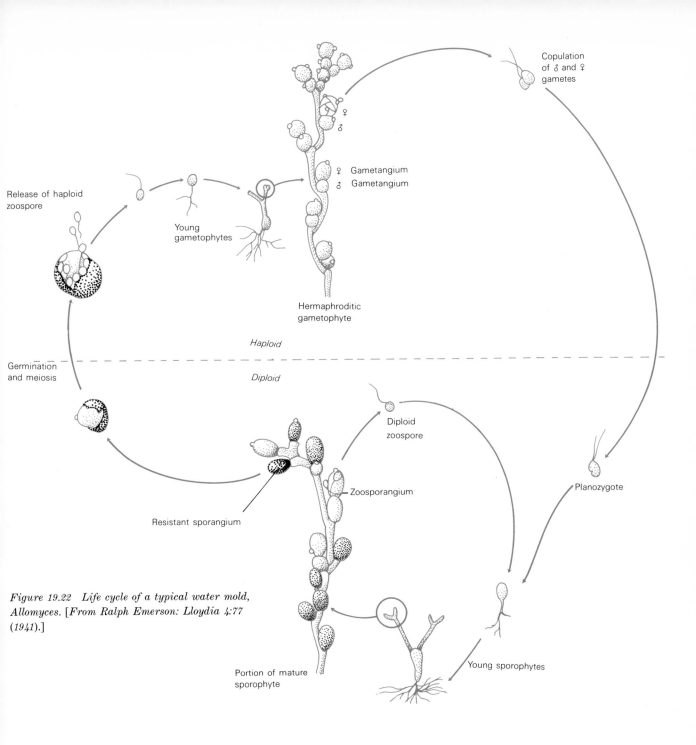

Copulation
of ♂ and ♀
gametes

Release of haploid
zoospore

Young
gametophytes

♀ Gametangium
♂ Gametangium

Hermaphroditic
gametophyte

Haploid

Germination
and meiosis

Diploid

Diploid
zoospore

Planozygote

Zoosporangium

Resistant sporangium

*Figure 19.22   Life cycle of a typical water mold,
Allomyces. [From Ralph Emerson: Lloydia 4:77
(1941).]*

Young sporophytes

Portion of mature
sporophyte

*19.9   Representative fungi and
their life cycles*

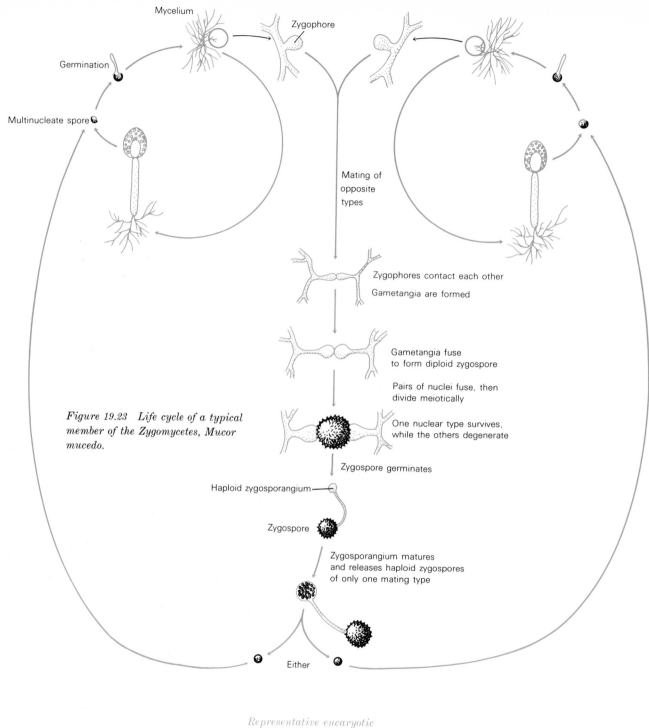

Mycelium

Zygophore

Germination

Multinucleate spore

Mating of opposite types

Zygophores contact each other

Gametangia are formed

Gametangia fuse to form diploid zygospore

Pairs of nuclei fuse, then divide meiotically

One nuclear type survives, while the others degenerate

Zygospore germinates

Haploid zygosporangium

Zygospore

Zygosporangium matures and releases haploid zygospores of only one mating type

Either

Figure 19.23 Life cycle of a typical member of the Zygomycetes, Mucor mucedo.

identical. When during growth the hyphae of each type approach those of the other, a series of sex-hormone reactions (described below) causes the mating types to interact. Special hyphae called *zygophores* form and grow toward each other. Special multinucleate structures called *gametangia* form at the tips of the zygophores; the gametangia then fuse and their haploid nuclei mingle. Nuclei from opposite mating types fuse pairwise; at the same time a dark-colored thick-walled structure, the *zygospore,* forms which now contains many diploid nuclei. Meiosis follows, reducing the nuclei back to the haploid state, after which the zygospore becomes dormant. All but one of the haploid nuclei degenerate, so that, when germination occurs, the zygosporangium, which is the forerunner of the new vegetative structure, produces haploid asexual spores of only one mating type, each of which gives rise to a haploid mycelium of this same mating type. The cycle is then complete.

The role of sex hormones in mating of *Mucor mucedo* is shown in Figure 19.24. The sex hormones are of several kinds. Each mating type produces a substance called *progamone,* which induces the opposite mating type to produce another substance called *gamone.* Thus *plus* progamone produced by mating type *plus* induces mating type *minus* to form *minus* gamone, and vice versa. *Minus* gamone then induces mating type *plus* to produce zygophores, and *plus* gamone induces mating type *minus* to elaborate zygophores. The zygophores then secrete *zygotrophic growth substances,* which act on the zygophores of the opposite mating type and cause them to grow towards each other, after which fusion and conjugation take place. Thus each mating type produces three distinct substances that are involved in the mating reaction; only the first of these, the progamone, is not induced by the opposite mating type, the others being formed in response to factors produced by the opposite mating type. In this way, the development of the structures involved in mating is controlled.

The *plus* and *minus* gamones have been isolated in crystalline form, and some aspects of their chemistry have been determined. They are highly unsaturated, aliphatic, polyhydroxy carbonyl compounds with formulas of $C_{20}H_{25}O_5$. The zygotropic substances have not been isolated, but they are known to be gaseous. Since the zygophores are aerial structures held above the surface of the medium, presumably the gaseous nature of the zygotropic substances ensures the transfer of the hormones to the zygophores of the opposite mating type. Sex hormones have also been shown to play a role in the mating reactions of a number of other fungi.

Ascomycetes The Ascomycetes are characterized by the formation of sexual spores in a saclike structure called an *ascus* (Figure 19.20). Hyphae, when present, are septate although, as we noted earlier, the septa

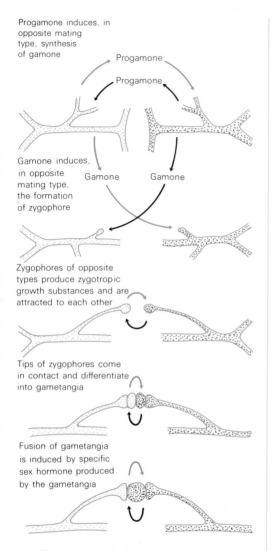

Progamone induces, in opposite mating type, synthesis of gamone

Progamone

Progamone

Gamone induces, in opposite mating type, the formation of zygophore

Gamone

Gamone

Zygophores of opposite types produce zygotropic growth substances and are attracted to each other

Tips of zygophores come in contact and differentiate into gametangia

Fusion of gametangia is induced by specific sex hormone produced by the gametangia

Figure 19.24   Role of sex hormones in the conjugation process in M. mucedo.

are rarely complete and cytoplasm can flow from one chamber to the next. Motile spores are never formed; asexual spores, if formed, develop externally on structures called *conidiophores* rather than internally in sporangia, as in the lower fungi. The Ascomycetes comprise a large and diverse group, including such typically unicellular forms as the yeasts, many of the most common molds, and even organisms such as the cup fungi, truffles, and morels, which form mushroomlike fruiting bodies. Many Ascomycetes are plant parasites, and some, such as those of the genera *Penicillium* and *Aspergillus,* are industrially important because of their economically useful products. The classification of the Ascomycetes is based primarily on the structure of the ascus and on the kinds of fruiting bodies within which the asci are borne. Table 19.4 summarizes the classification of the Ascomycetes into orders, and illustrates some of the structural differences used in this classification.

The life histories of two important members of the class Ascomycetes, the yeast *Saccharomyces cerevisiae* (subclass Protoascomycetes) and the red bread mold *Neurospora crassa* (series Pyrenomycetes), were discussed in Chapter 13. One of the most interesting organisms of the series Discomycetes is the fungus *Morchella esculenta,* the common morel, which is often confused with a basidiomycetous mushroom because it produces a macroscopic fruiting body (Plate 14: Figure 19.25). Morels are delicious, and for this reason they are much sought after by mushroom hunters. They are numerous in the spring of the year in rich forest litter or around apple orchards. Because of their brown color, the fruiting bodies often blend with the leaf litter and hence are difficult to find. Morels are especially abundant in the hardwood forests of eastern United States and often large baskets can be filled by a few hours of collecting in the right location. Although the mycelium of the morel grows among the leaf litter, it is not able to utilize cellulose or lignin and therefore probably obtains its nutrients from soluble materials leached from leaves.

The fruiting body (which is an apothecium; see Table 19.4) is composed of a mass of hyphae that differentiate into the characteristic fruiting-body structure. In the spongy material at the top of the fruiting body the asci are borne in uniform rows attached by their bases to a thin layer of tissue called the *hymenium.* The ascospores are released when the asci rupture, and presumably they are blown by the winds, alight in fresh litter, germinate, and initiate new mycelial growth.

The mycelium of *Morchella esculenta* can be easily cultured in the laboratory, and pure cultures can be made from tissue taken from the interior of a fruiting body, this tissue usually being devoid of contaminants. However, it has not been possible to induce the formation of fruiting bodies in pure culture or to produce them commercially. Thus, despite the demand for morels because of their excellent flavor, commercial raising

Table 19.4   Major taxonomic divisions of the Ascomycetes

Subclass **Protoascomycetes**   Asci arising naked, no fruiting bodies.

   Order Endomycetales   Asci arising directly from zygotes. Yeasts. DNA, 26–64% GC (51 species). [Sketch (*a*).]

Subclass **Euascomycetes**   Asci produced in a fruiting body (ascocarp).

   Series **Plectomycetes**   The ascocarp (called a *cleistothecium*) is completely closed with asci scattered within, not arising within a definite layer. DNA, 52–53% GC (2 species). [Sketch (*b*).]

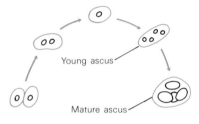

Young ascus
Mature ascus

Saccharomyces cerevisiae

(*a*)

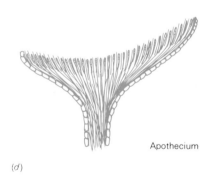

Ascus
Cleistothecium

(*b*)

   Order Taphrinales   Asci arising from binucleate cells. Plant pathogens.

Subclass **Euascomycetes**

   Series **Pyrenomycetes**   The ascocarp (called a *perithecium*) has asci in tufts or in a layer within; the ascocarp usually closed except for a pore. DNA, 53–55% GC (5 species). [Sketch (*c*).]

   Series **Discomycetes**   The ascocarp (called an *apothecium*) has asci in tufts or in a layer; the ascocarp usually widely open with asci formed on the outside. DNA, 46–50% GC (2 species). [Sketch (*d*).]

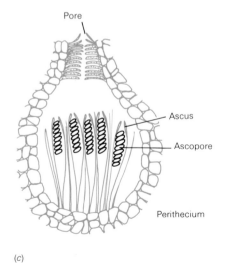

Pore
Ascus
Ascopore
Perithecium

(*c*)

Apothecium

(*d*)

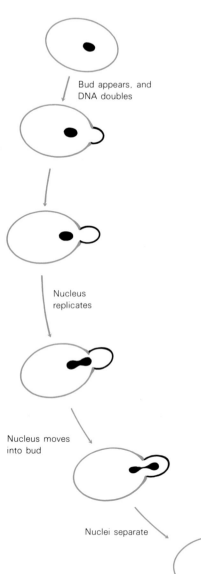

Bud appears, and
DNA doubles

Nucleus
replicates

Nucleus moves
into bud

Nuclei separate

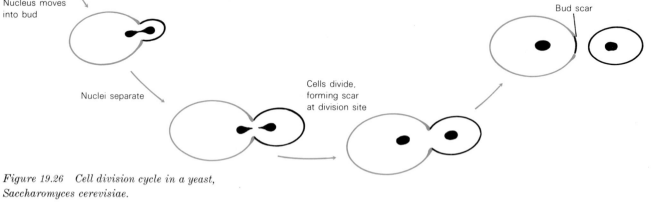

Bud scar

Cells divide,
forming scar
at division site

*Figure 19.26   Cell division cycle in a yeast,
Saccharomyces cerevisiae.*

comparable to that for true mushrooms (see below) has not been possible for morels.

Another ascomycete that produces succulent fruiting bodies is the truffle, genus *Tuber*. The fruiting bodies of truffles are produced underground and hence are very difficult to find. In Europe, however, dogs and pigs have been trained to detect the odor of truffles, making possible their discovery and commercial harvest. Although they are very expensive, truffles are occasionally served as an ingredient of gourmet meals.

Most *yeasts* are Ascomycetes and reproduce by budding. However, not all organisms that reproduce by budding are Ascomycetes since a few Basidiomycetes and Phycomycetes also grow in this manner. The budding cycle of a typical yeast is shown in Figure 19.26. Note that DNA synthesis and nuclear division are closely correlated with the budding process. This is in contrast to the situation in hyphal growth, where elongation of the multinucleate hypha is not synchronous with nuclear division. The sexual cycle of a typical yeast was described in Section 13.4. (It should be recalled that fertilization occurs through the fusion of two vegetative cells, usually of opposite mating types.) The cellular events of fusion during sexual conjugation in yeast are shown in Figure 19.27. The two cells send out protuberances along their point of contact. At the same time, fusion of the cell walls takes place, so that the cells are now permanently joined. After fusion is complete, the central portion of the cell wall separating the two cells dissolves, probably through the action of a lytic enzyme, and the cytoplasms of the two cells mingle. The two haploid nuclei move to the center, fuse, and form a diploid nucleus. The diploid cell forms as a bud at right angles to the fusion tube, and the diploid nucleus moves into this bud. The diploid cell now separates from the conjugation tube and begins to grow vegetatively; the parent cells usually disintegrate.

The process of mating in yeast is markedly different from that in bac-

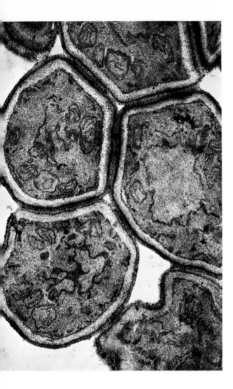

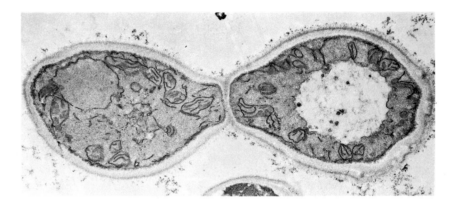

Figure 19.27 Electron micrographs of the conjugation process in a yeast, Hansenula wingei: (a) cells of opposite mating type agglutinate strongly when mixed (magnification, 15,600×); (b) two conjugating cells, which have fused at the point of contact and have sent out protuberances toward each other (12,800×); (c) late stage of conjugation (15,000×.) The nuclei of the two cells have fused and the diploid bud has formed at a right angle to the conjugation tube. This bud will eventually separate from the conjugants and become the forerunner of a diploid cell line. [From S. F. Conti and T. D. Brock: J. Bacteriol. 90:524 (1965).]

(b)

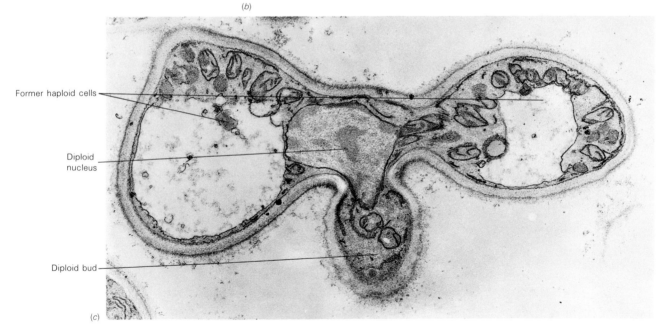

Former haploid cells

Diploid nucleus

Diploid bud

(c)

19.9 Representative fungi and their life cycles

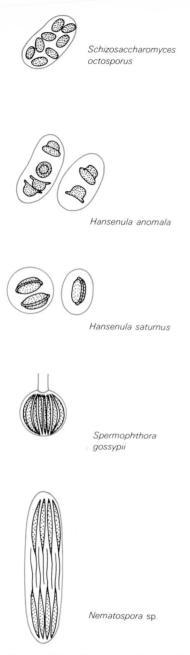

Schizosaccharomyces octosporus

Hansenula anomala

Hansenula saturnus

Spermophthora gossypii

Nematospora sp.

Figure 19.28 Shapes, numbers, and arrangements of ascospores in asci of different yeasts.

teria. In yeast both mating partners participate actively and equally in the process, become fused, and lose their individual identities, whereas in bacteria one partner is active and the other passive, and each retains its identity and continues to function afterwards.

In many cases the diploid yeast cell is able itself to grow vegetatively for many generations, so that both haploid and diploid phases can have independent existences. Under appropriate conditions, the nucleus of a diploid cell can undergo meiosis, leading to the formation of ascospores, usually four in number, which are retained within the wall of the original diploid cell (now called an *ascus*). Although most yeasts grow by budding, one genus, *Schizosaccharomyces*, grows by fission (Figure 7.2). Fission yeasts are still classified as yeasts, however, because they are unicellular fungi that produce ascospores. One of the key characteristics in the taxonomic separation of yeasts into genera is the shape, number, and arrangement of the ascospores within the ascus; representative examples are given in Figure 19.28.

Not all yeasts possess a sexual cycle. A number of species reproduce only vegetatively, including the important human pathogen *Candida albicans*. In a few cases it has been shown that these so-called asporogenous yeasts are haploid mating types of heterothallic yeasts. If the appropriate two haploid yeasts are mixed, mating and ascospore formation ensue.

Yeasts usually flourish in habitats where sugars are present, such as fruits, flowers, and the barks of trees. A number of yeast species live symbiotically with animals, especially insects, and a few species are pathogenic for animals and man. Earlier in this chapter we mentioned the transformation from yeastlike to mycelial growth of certain of these pathogenic yeasts. The most important yeasts are the baker's and brewer's yeasts, which are members of the genus *Saccharomyces*. The original habitats of these yeasts were undoubtedly fruits and fruit juices, but the commercial yeasts of today are probably quite remote from the natural strains since they have been greatly improved through the years by careful selection and breeding by industrial microbiologists. Baker's and brewer's yeasts are probably the best-known scientifically of all fungi as they have provided excellent experimental material for the study of many important microbiological and biochemical properties. Throughout this book we have had several occasions to discuss aspects of yeast biology.

Basidiomycetes The Basidiomycetes are characterized by the formation of sexual spores (basidiospores) at the tips of specialized structures called *basidia* (Figure 19.20). The basidia are usually but not always borne on fruiting bodies (basidiocarps) of macroscopic size, which are familiar to many under their common names of mushroom, toadstool, puffball, stinkhorn, and so on. The shelf fungi are Basidiomycetes that

Figure 19.29 *Manner of cell division in a dicaryotic Basidiomycete. This process ensures that each of the two unlike nuclei is present in each of the two cells. The characteristic structure formed is a clamp connection.*

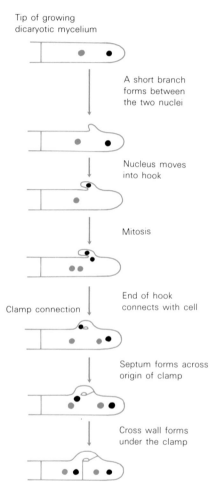

Tip of growing dicaryotic mycelium

A short branch forms between the two nuclei

Nucleus moves into hook

Mitosis

Clamp connection

End of hook connects with cell

Septum forms across origin of clamp

Cross wall forms under the clamp

produce shelflike basidiocarps, usually on the sides of trees. Some Basidiomycetes produce fruiting bodies only of microscopic size; the most important of these are the plant pathogenic rusts and smuts. The hyphae are septate and multinucleate, but at a certain stage of their life cycle a dicaryotic mycelium is produced, each cell of the dicaryon containing two nuclei, one derived from each mating type. In many Basidiomycetes, cell division in the dicaryon is preceded by the formation of a special structure, the *clamp connection,* whose structure and function are illustrated in Figure 19.29. This particular mode of cell division ensures that each new cell receives both types of nuclei. Special structures to ensure fertilization are not formed in the Basidiomycetes, mating typically occurring by simple hyphal fusions and dicaryon formation. Fusion of the two nuclei in the dicaryon occurs in the basidium just preceding basidiospore formation. Asexual spore formation occurs in some Basidiomycetes but is much less common in this group than in the lower fungi and Ascomycetes. Table 19.5 summarizes the classification of the class Basidiomycetes, and illustrates some of the key structural differences.

The most familiar examples of the class Basidiomycetes are the *mushrooms.* Many of these are mycorrhizal (Section 14.10), others live on dead organic matter in the soil, and still others live on the trunks of trees. Basidiospores borne on the fruiting body are dispersed through the air and initiate mycelial growth on favorable substrates. Although a haploid mycelium may grow extensively, it rarely initiates fruiting bodies, which are formed by a dicaryotic mycelium arising as a result of the fusion of two haploid mycelia. Many Basidiomycetes have four sexes instead of two, a condition called *tetrapolar sexuality.* About 65 percent of the Basidiomycetes are tetrapolar, the rest being simply heterothallic, that is, bipolar. In tetrapolar sexuality mating is determined by two sets of unlinked genetic factors, *A* and *B,* which assort and segregate independently at meiosis. Thus, at meiosis, four rather than two mating types are segregated. As shown in Figure 19.30, mating occurs in tetrapolar organisms only when the two mating types differ by both *A* and *B* factors. Thus, we see that each mating type recognizes self from nonself and mates only with a nonself type.

The physiology of tetrapolar sexuality is not well understood. It is known that *A* and *B* factors do not control the initial fusion reaction between hyphae of the two mating types, but do affect later events involved in nuclear association and fusion. Once hyphal fusions between compatible mating types have taken place, some nuclei of each type migrate into the hyphae of the opposite type and pair but do not fuse, producing the dicaryon. New hyphal growth now proceeds by cell division involving clamp connections (see above). The dicaryon is the forerunner of the mushroom fruiting body.

Table 19.5   Major taxonomic divisions of the Basidiomycetes

Subclass **Hemibasidiomycetes**   Basidium septate or deeply divided, or basidiospores derived from a promycelium arising from a specialized spore, the teleutospore. [Sketch (*a*).]

Order Tremellales   Fruiting bodies (basidiocarps) well developed, basidial septa longitudinal or oblique. Jelly fungi.

Order Auriculariales   Basidiocarps well developed, basidial septa transverse. Jelly fungi.

Order Uredinales   Basidiocarps absent or poorly developed. Basidiospores produced on specialized structures (*sterigmata*) on a promycelium. Rusts.

Order Ustilaginales   Basidiocarps absent or poorly developed. Basidiospores produced directly from a promycelium. Smuts.

Subclass **Holobasidiomycetes**   Basidium simple, not septate or arising from a teleutospore. [Sketch (*b*).]

Series **Hymenomycetes**   Basidia borne as a well-defined layer that becomes exposed as the basidiocarp matures. Mushrooms, shelf fungi, and tooth fungi. DNA, 44–58% GC (8 species).

Series **Gasteromycetes**   Basidiocarp remains closed with the basidia within, becoming exposed only through mechanical breakage or weathering. Puffballs, earthstars, stinkhorns, and bird's nest fungi.

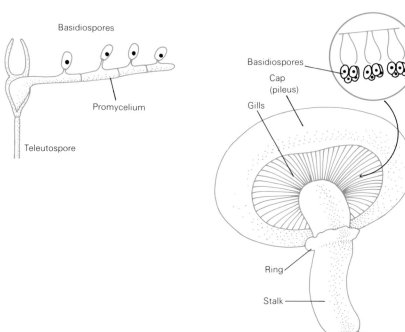

(a)

(b)

Mature mushroom

Under certain environmental conditions, dicaryotic hyphae grow together and form small buttonlike structures, which are the fruiting-body primordia. If one of these buttons is cut open the structural outlines of the mature fruiting body can be seen, but these forms are usually buried beneath the litter or in the soil and are not ordinarily in view. The buttons may remain underground for considerable periods of time until favorable environmental conditions develop, usually after heavy rains, at which time they enlarge rapidly into mature fruiting bodies (Figure 19.31). Although hyphal growth does take place, much of the expansion at this stage is due to the uptake of water. This expansion is usually so rapid that a mature fruiting body can be formed within a few hours or a few days. In most cases a number of fruiting bodies will mature at the same time in a given area of soil, producing so-called flushes of fruiting bodies; so one day a given locality can be apparently devoid of fruiting bodies, while the next day a large number may be present.

In the gill fungi, the underside of the cap of the mature mushroom is

Figure 19.30 Bipolar and tetrapolar sexuality in Basidiomycetes. Note that in the latter only one-fourth of the possible pairings are fertile.

	Bipolar	Tetrapolar
Diploid types		
Haploid types		

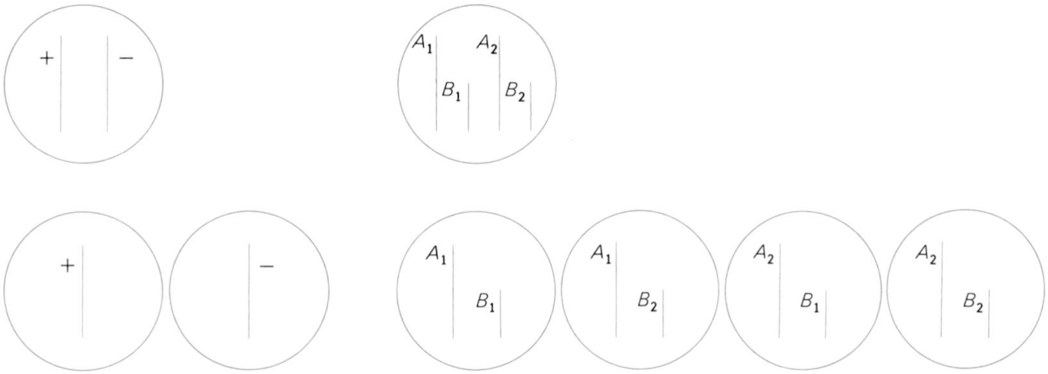

Mating combinations

	+	−
+	Sterile	Fertile
−	Fertile	Sterile

Mating only of opposite types

	A_1B_1	A_1B_2	A_2B_1	A_2B_2
A_1B_1	Sterile	Sterile	Sterile	Fertile
A_1B_2	Sterile	Sterile	Fertile	Sterile
A_2B_1	Sterile	Fertile	Sterile	Sterile
A_2B_2	Fertile	Sterile	Sterile	Sterile

Mating only when types differ at *both* A and B

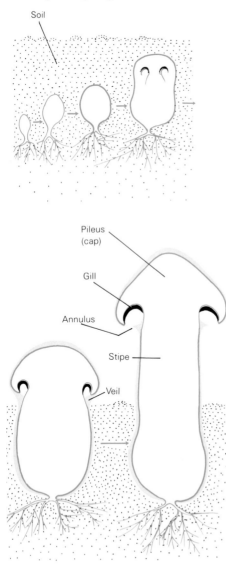

Figure 19.31 Diagrammatic stages in the development of a Basidiomycete fruiting body (mushroom). Strands are formed by branching, interweaving, parallel growth, and fusion of the substrate mycelium. From these strands an aggregate of branched hyphae arises, increases in size, and differentiates into a fruiting-body primordium. When favorable conditions occur, the primordium enlarges into the mature fruiting body.

Soil

Pileus (cap)

Gill

Annulus

Stipe

Veil

lined with lamellae or gills, along which the basidia are borne. In the pore fungi, basidia are produced in deep pores that are formed on the underside of the cap. The basidiospores of many mushrooms are pigmented and impart a characteristic color to the gills or pores. Soon after maturation, the spores are discharged from the whole undersurface of the cap; "spore prints" can be made by attaching the cap to the lid of a petri dish and allowing the spores to be discharged onto the base of the dish. Prints so formed will usually show fairly precisely the arrangement of the gills or pores of the fruiting body.

Although many mushrooms are edible and delicious, some are poisonous. Indeed, a few are so highly poisonous that the ingestion of a single mushroom may suffice to cause death. The recognition of edible and poisonous mushrooms in the field is not an easy matter since often one species of a genus will be edible and another species poisonous. For instance, most members of the genus *Amanita* are poisonous, but one, *A. caesarea,* is edible. According to legend, Julius Caesar dispatched enemies by serving them poisonous species of *Amanita,* while he would eat specimens of *A. caesarea.* Most mushroom hunters do not attempt to identify all poisonous mushrooms but rather only certain edible species. Various field manuals are published for those who wish to collect and eat wild mushrooms.

The mushroom commercially available is *Agaricus bisporus,* which is cultivated in special mushroom farms. The commercial production of mushrooms is a highly developed art and science. The organisms are grown in special beds, usually in buildings whose temperature and humidity are carefully maintained. Beds are prepared by mixing soil with a material rich in organic matter, such as horse manure, and these beds are then inoculated with mushroom "spawn." This is the dicaryotic mycelium grown in pure culture, usually on an organic-rich substrate, in large bottles. In the bed, the mycelium grows and ramifies through the substrate, and after several weeks is ready for the induction of fruiting-body formation. This is done by adding to the surface of the beds a layer of soil devoid of excess organic matter, called "casing soil," which in some way induces the formation and maturation of the fruiting bodies. Once the fruiting bodies flush (Figure 19.32), they must be collected quickly before they have matured to the point of spore formation since the flavor of the mushroom is best before spores develop. After collection, the mushrooms are packaged and kept cool until brought to market. Several flushes will take place on a single bed, and after the last flush, the bed must be cleaned out and fresh substrate added before it can be reinoculated.

Other Basidiomycetes of note are the rusts and smuts, which cause

Figure 19.32 A mushroom bed, typical of a commercial mushroom establishment. (Courtesy of Lee C. Schisler.)

many serious diseases of economically important plants. Among the most widely studied species is *Puccinia graminis,* the causal agent of wheat rust, whose life cycle is one of the most interesting and complex among the fungi. This species exhibits an alternation of hosts, carrying out part of its life cycle on wheat and the other part on the common wild barberry (*Berberis vulgaris*). During the various stages of the life cycle of *P. graminis,* five different kinds of spores are formed (see Figure 19.33). The rust is a highly specialized parasite, growing only on the plants in question, and it has been impossible as yet to cultivate it in the laboratory in any of the stages of its life cycle. Thus all the details of the life cycle have been worked out through studies on natural infections or through the experimental infection of greenhouse plants.

The incidence of the disease has been somewhat reduced in two different ways. Varieties of wheat resistant to the rust have been developed by selection and breeding and usually prove effective for a time; but eventually new strains of the fungus develop, which can infect the resistant wheat. More than 100 races of the rust are known, which are able to infect different varieties of wheat, and this necessitates a breeding program for rust resistance that is continuous and wide-ranging. Another way to reduce the occurrence of the disease is to eradicate the organism's alternate host, barberry, since this will break the life cycle. In 1918 the United States government launched an extensive program to eliminate wild barberry, which succeeded in reducing the incidence of the disease. Complete elimination by this method is not possible, however, as in warmer parts of the United States the thin-walled uredospores can overwinter, and since these spores can maintain a nonsexual cycle in wheat in the absence of the alternate host, continued infection of northern wheat can occur when uredospores blow in from the southern part of the United States. It is uneconomic to use fungicides to control an organism infecting a crop planted on such extensive acreage as is wheat.

Other rusts infect barley, oats, and rye, and some also attack wild plants such as wild blackberry. These fungi are especially interesting microorganisms because of their highly specialized habitats, their alternation of hosts, and their complex life cycles.

Another activity of many Basidiomycetes that has economic consequences is the decomposition of wood, paper, cloth, and other products derived from natural sources. Basidiomycetes that attack these products are able to utilize cellulose or lignin as carbon and energy sources. Susceptibility to attack is greatly affected by the moisture content of the products, and those which are buried or are on the surface of the soil are most subject to decay (Plate 14: Figure 19.34). A vast amount of technology is involved in developing means of treating items such as

19.9 Representative fungi and their life cycles

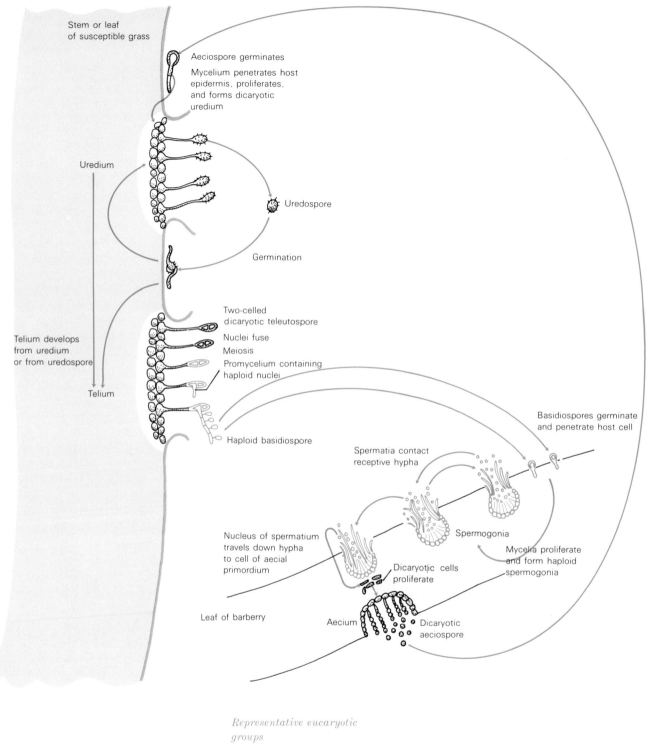

Stem or leaf
of susceptible grass

Aeciospore germinates

Mycelium penetrates host
epidermis, proliferates,
and forms dicaryotic
uredium

Uredium

Uredospore

Germination

Telium develops
from uredium
or from uredospore

Telium

Two-celled
dicaryotic teleutospore

Nuclei fuse

Meiosis

Promycelium containing
haploid nuclei

Haploid basidiospore

Basidiospores germinate
and penetrate host cell

Spermatia contact
receptive hypha

Spermogonia

Nucleus of spermatium
travels down hypha
to cell of aecial
primordium

Mycelia proliferate
and form haploid
spermogonia

Dicaryotic cells
proliferate

Leaf of barberry

Aecium

Dicaryotic
aeciospore

*Representative eucaryotic
groups*

telephone and electrical poles so that they will remain in good condition after their ends have been buried in the soil. Wood preservatives such as creosote or other organic antimicrobial agents are forced into the wood under high pressure in order to impregnate it completely. Even with the best methods, the ''life'' of a telephone pole in the ground is not indefinite. The wood and bark of coniferous trees (pines, spruces, redwoods, and so on) contain natural antimicrobial agents that prevent microbial attack and are at least in part responsible for the longevity of conifers.

The bracket fungi are a group of Basidiomycetes whose fruiting bodies are frequently seen on tree trunks. They are called ''bracket fungi'' because of the bracketlike shape of their fruiting bodies and their mode of attachment. These fungi are responsible for considerable damage to, and death of, forest trees. Once a tree is cut and converted into lumber, the chances of decay are even greater. Other Basidiomycetes cause blue mold stain of lumber, which, although not damaging to the structure of the wood, gives it an unsightly appearance. Dry rot in construction lumber, both before and after incorporation into buildings, can cause a serious decrease in strength of the wood; occasionally drastic consequences follow for the building of which it has become a component. Wood allowed to lie on the ground will rot quickly; hence lumbermen take special pains to keep their wood dry and off the ground.

Cotton and wool fabrics, being derived from natural materials, are subject to attack by a variety of fungi, mainly Basidiomycetes. Cotton products that come into contact with the soil, such as tent canvas, must be well treated with preservatives. The efficacy of these preservatives is usually evaluated by soil-burial tests. Treated and untreated samples of the product are buried for various periods of time and then are dug up and examined for deterioration. The degree of deterioration can be evaluated quantitatively by comparing the tensile strengths of fabrics treated in various ways. Synthetic fibers are quite resistant to microbial attack although they may undergo detrimental chemical oxidation reactions.

19.10   *Pathogenic fungi*

Human pathogenic fungi   In contrast to the bacteria, fungi are only rarely pathogenic for man or animals; it has been estimated that only about 50 species are pathogenic. Further, these fungi do not cause many types of serious diseases. Most pathogenic fungi are not restricted to growth in the host because their primary habitat is the soil; they are merely opportunistic invaders of persons with lowered resistance to infection. Transmission of the organism from person to person is rare so

Table 19.6 Important human mycoses

	Causal organism	Main disease foci
Systemic mycoses:		
Cryptococcosis	Cryptococcus neoformans	Lungs, meninges
Coccidioidomycosis	Coccidioides immitis	Lungs
Histoplasmosis	Histoplasma capsulatum	Lungs
Blastomycosis	Blastomyces dermatitidis	Lungs, skin
Candidiasis	Candida albicans	Oral cavity, intestinal tract
Aspergillosis	Aspergillus fumigatus	Bronchi
Superficial mycoses (dermatomycoses):		
Ringworm	Microsporum audouini	Scalp of children
Favus	Trichophyton schoenleinii	Scalp
Athlete's foot	Epidermophyton and other genera	Between toes, skin

that fungal diseases are generally not contagious. Fungal diseases of man, called *mycoses*, can be divided into two main groups (Table 19.6): (1) systemic, or deep-seated, mycoses, in which the pathogen is widely disseminated, growing in various organs and tissues, and (2) superficial mycoses, involving infections of the skin, hair, or nails. As sexual stages are unknown for most pathogenic fungi, their taxonomic position is not clear; they are, however, probably Ascomycetes.

Systemic mycoses are caused by yeasts and also by dimorphic fungi that can grow in a yeastlike manner in the body (or at 37°C in culture) or as a mycelium in culture at lower temperatures. Usually the infecting organisms reside in the soil and enter the body by being inhaled. They initiate a mild respiratory disease that has no characteristic symptoms and often resembles a cold. Subsequently the organisms become disseminated throughout the body, and a chronic infection ensues, of which the symptoms are highly variable; general debilitation is common, but fatal infections are rare. The yeast *Candida albicans* causes a variety of infectious states in the oral cavity and gastrointestinal tract. It will be recalled (see Chapters 9 and 15) that fungus infections of the latter occur frequently as a consequence of elimination of the normal intestinal bacteria by oral antibiotic therapy. Several fungi cause pulmonary infections that can mimic tuberculosis.

Certain of the deep-seated mycoses are found only in restricted geographical areas. Histoplasmosis is found in high incidence in certain parts of the midwestern United States, especially in rural areas of the Ohio and Mississippi River valleys. The causal agent, *Histoplasma capsulatum,* is found in large numbers in soils receiving droppings of chickens and other birds, and it is thought that infection of persons occurs from such contaminated soils. Although many human beings may become infected, relatively few of them develop the typical pulmonary lesions of histoplas-mosis. Coccidioidomycosis (also called "San Joaquin Valley fever") is generally restricted to desert areas of the southwestern United States. The fungus grows in desert soils, and the spores are disseminated on dry windblown particles and inhaled. In endemic areas, up to 80 percent of the inhabitants may be infected although only a small number of these develop serious symptoms. The precise factors responsible for pathogen-esis are not understood, but many of the symptoms may be allergic in nature.

Superficial mycoses are caused primarily by mycelial fungi. These or-ganisms, frequently called "dermatophytes," have an affinity for keratin-containing regions such as the skin, hair, and nails and usually have the ability to digest this protein. (For a discussion of the anatomy of the skin and the hair follicles, see Section 14.3.) Some of these fungi can grow within the superficial tissues, causing itching and reddening of the skin; some can grow on the surface of the hair; and others can grow within the hair fiber itself. The initiation of a dermatophytic infection generally depends on host factors such as hormonal balance and various other physiological influences. Age is a factor in susceptibility to two dermato-mycoses: athlete's foot, which is common in adults and rare in children, and ringworm of the scalp, which is frequent in children but rare in adults. Adults may be resistant to ringworm because of increased secretions by the sebaceous glands, especially the secretion of unsaturated fatty acids with antifungal activity; one of these antifungal substances, undecylenic acid, is now used as a component of many ointments marketed for the control of fungal infections of the skin. Moisture is an important factor in the development of many surface infections. Athlete's foot usually occurs between the toes because moisture accumulates in this region, promoting fungal growth.

There are only a few antibiotics effective for fungal infections (see Chapter 9). Griseofulvin has been used considerably for superficial my-coses. Although it is ineffective if applied directly to the skin, it is effective if given orally because the drug can accumulate in the newly synthesized keratin-containing tissues and render them resistant to fungal infection. Because such tissues gradually rejuvenate, griseofulvin therapy must be continued for prolonged periods until the newly synthesized tissues have replaced those sloughed off. Deep-seated mycoses are not affected by griseofulvin but can be treated with some of the polyene antibiotics (see Chapter 9). No immunization procedures for fungal infections are currently in use.

Plant pathogenic fungi A wide variety of fungi cause diseases of plants, and fungi are by far the most important plant pathogens. A description of one plant disease, wheat rust, was given earlier in this chapter in relation to its causal agent, *P. graminis*. Many Ascomycetes and Basidiomycetes are pathogenic for plants; among the lower fungi, many members of the order Peronosporales and a few members of the class Chytridiomycetes are also pathogens. Pathogenic fungi affect all kinds of plants, from the smallest herbs to the largest trees, although naturally most attention has been focused on diseases of crop plants. All plant parts (roots, stems, leaves, flowers, fruits, seeds) are susceptible to attack although a single fungus generally shows marked specificity for certain organs. The fungus can damage its host by producing a toxin, by blocking vessels that transport water and food, or by inducing malformed and tumorlike growth (especially of roots). In some cases, damage to the plant may be slight but may, through disfiguration or discoloration, have serious economic consequences. Mycorrhiza would be considered root pathogens were it not known that they are actually beneficial to the plant. The study of plant pathogenic fungi and the diseases they cause is such a broad subject that space is not available here to discuss them in even a cursory way.

19.11 Summary

The discussion above gives only a general idea of the basic features of the fungi. The lower fungi are primary aquatic forms, and have been only little studied. The terrestrial fungi, of which more is known, are probably the most recently evolved; many of them are adapted primarily to grow in association with terrestrial plants or their products. The mycorrhizal fungi, which are symbiotic Basidiomycetes, were discussed in Section 14.10. Many of the terrestrial fungi are of considerable economic importance as pathogens, as agents in nutrient cycling, or as instrumental agents in industrial processes. Because of the relative ease with which most fungi can be cultured in the laboratory, they are ideal organisms for a variety of scientific investigations, and much more work deserves to be done with them.

The slime molds

The slime molds are organisms that have at times been classified as both fungi and protozoa. The confusion arises because the vegetative structure of a slime mold is a protozoanlike amoeboid mass, whereas its fruiting body resembles the fruiting body of a fungus and produces spores with cell walls. Because the slime-mold fruiting body was first studied by mycologists, the slime molds have traditionally been discussed in mycology rather than protozoology texts, even though the affinities of these organisms to protozoa are just as close as they are to fungi.

The slime molds can be divided into two groups, the *cellular* slime molds, whose vegetative forms are composed of single amoebalike cells, and the *acellular* slime molds, whose vegetative forms are naked masses of protoplasm of indefinite size and shape (called *plasmodia*). Other differences between these two kinds of slime molds will be discussed below. Both types live primarily on decaying plant matter, such as leaf litter, logs, soil, and so on. Their food consists mainly of other microorganisms, especially bacteria, which they ingest in typical phagocytic fashion. One of the easiest ways of detecting and isolating slime molds is to bring into the laboratory small pieces of rotting logs and place them in moist chambers; the amoebae or plasmodia will proliferate, migrate over the surface of the wood, and eventually form fruiting bodies, which can then be observed under a dissecting microscope. The fruiting bodies of slime molds are often very ornate and colorful.

Cellular slime molds   Although a number of genera of cellular slime molds have been described, most of them have not been studied in detail. An exception is the genus *Dictyostelium,* for which the life cycle has been worked out and for which a considerable amount of information is known about the physiology and biochemistry of fruiting-body formation. The vegetative cell of *Dictyostelium discoideum* is an amoeboid cell [Figure 19.35(*a*)] that feeds on living or dead bacteria and which grows and divides indefinitely. When the food supply has been exhausted, a number of amoebae aggregate to form a pseudoplasmodium, a structure in which the cells lose some of their individuality but do not fuse [Figure 19.35(*b*) and (*c*)]. This aggregation is triggered by the production of *acrasin,* a substance recently identified as cyclic adenosine-monophosphate, which attracts other amoebae in a chemotactic fashion. Those cells which are the first to produce acrasin serve as centers for the attraction of other amoebae, and since each newcomer also produces acrasin, the centers increase rapidly in size. The swarm patterns around such aggregating masses are shown in Figure 19.35(*b*) and (*c*).

Because the pseudoplasmodium formed resembles a slimy, shell-less snail in its appearance and movement, it is called a ''slug.'' Around the outside is secreted a mucoid sheath that probably protects the periphery of the slug from drying. The pseudoplasmodial slug migrates as a unit across the substratum as a result of the collective action of the amoebae [Figure 19.35(*d*)]. Presumably there is some mechanism that coordinates

Figure 19.35   Photomicrographs of various stages in the life cycle of the cellular slime mold Dictyostelium discoideum. (a) Amoebae in preaggregation stage. Note irregular shape, lack of orientation. Magnification, 540×. (b) Aggregating amoebae. Note the regular shape and orientation. The cells are moving in streams in the direction indicated by the arrow. Magnification, 346×. (c) Low-power view of aggregating amoebae. Magnification, 12×. (d) Migrating pseudoplasmodia (slugs) moving on an agar surface and leaving trails of slime in their wake. Magnification, 14×. (e) Early stage of fruiting-body formation. Magnification, 52×. (f) A late stage of a developing fruiting body. Magnification 92×. (g) Mature fruiting bodies. Magnification, 6×. [From K. B. Raper: Proc. Am. Phil. Soc. 104:579 (1960).]

the movements of the individual amoebae so that the slug maintains its unity, but the nature of this mechanism is unknown. This subjugation of the individuality of the amoebae to the collective mass has been likened to the actions of human beings in forming societies. The pseudoplasmodium usually migrates for a period of hours, and as it is positively phototropic, it will migrate toward the light. In nature the pseudoplasmodium is presumably formed within the dim recesses of soil or decaying bark and migrates in response to light to the surface, where fruiting-body formation takes place. Thus the formation of the pseudoplasmodium and its behavioral responses are essential features to ensure that the spores formed will be borne in the air and so can be effectively dispersed.

Fruiting-body formation begins when the slug ceases to migrate and becomes vertically oriented [Figure 19.35(e) to (g)]. The fruiting body is differentiated into a stalk and a head; cells in the forward end of the slug become stalk cells and those from the posterior end become spores. The

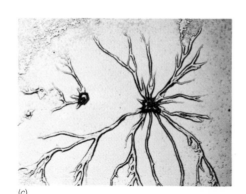

(a)

(b)

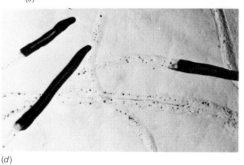

(c)

(d)

(e)

(f)

(g)

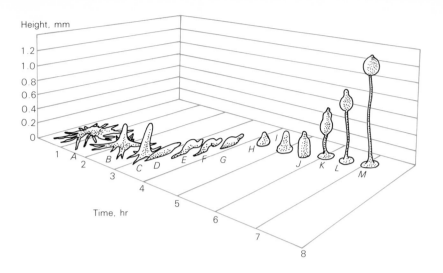

amoebae that form the stalk cells begin to secrete cellulose, which provides the rigidity of the stalk. Other amoebae, from the rear of the slug, swarm up the stalk to the tip and form the head; most of these become differentiated into spores that are usually embedded in slime. Upon maturation of the head, the spores are released and are dispersed by wind or insects. Each spore germinates and forms a single amoeba, which then initiates another round of vegetative growth. Apparently sexual reproduction does not occur in the cellular slime molds (as opposed to the acellular slime molds discussed below), and each amoeba is capable of initiating the complete series of events leading again to fruiting-body formation. The temporal and spatial relationships of fruiting-body formation are illustrated in Figure 19.36.

The developmental processes in fruiting-body formation in *D. discoideum* have been subjected to extensive biochemical analysis. The motivation behind such studies is that these processes in *D. discoideum* may be analogous to the embryological processes in higher organisms, and if so, the mechanisms in the slime mold would be much easier to analyze biochemically. Because both stalk and spore cells synthesize large amounts of cellulose and glycogen and because these polysaccharides are absent from vegetative cells, much attention has been focused on the events controlling the synthesis of these two glucose polymers. As has been mentioned, the production of acrasin and the events of fruiting-body formation are triggered by starvation: One hypothesis is that the enzymes involved in the synthesis of the polysaccharides are subject to catabolite repression (Section 7.6), and upon starvation repression is overcome. Energy for the synthesis of the new enzymes probably comes from carbohydrate and proteinaceous reserves that are degraded in the differentiation period. Uridine-diphosphoglucose (UDPG), the glucose donor for polysaccharide synthesis, accumulates soon after starvation as a result of increased activity of the enzyme UDPG synthetase. Later, the enzymes involved in the formation of the polysaccharides appear, followed by the cell-wall polymers themselves. The spore cells contain the disaccharide trehalose as an energy reserve, and the enzyme involved in its synthesis also appears at the same time as do the polysaccharide-synthesizing enzymes. These biochemical changes are outlined in Figure 19.37. There is good evidence that the synthesis of at least some of the enzymes involved in differentiation is preceded by the synthesis of mRNA specific for them, and mRNA synthesis probably begins as a result of derepression. However, not all the enzymes are synthesized at the same time: Those which are needed first appear first. This has led to the suggestion that the developmental events in fruiting-body formation are programmed in an orderly manner, perhaps by some mechanism that derepresses the relevant enzymes in sequence. Studies on the biochemistry of fruiting body-formation in *D. discoideum* have led to many interesting findings, which will probably be relevant to our understanding of developmental processes in higher organisms. The value of a microbial system such as this as a model is that it is easily controlled and manipulated and is subject to sophisticated biochemical analysis.

Dictyostelium discoideum has a DNA base composition of about 22 percent GC, one of the lowest guanine-plus-cytosine contents known for microorganisms. Other acellular slime molds also have fairly low GC values, ranging in three different species from 29 to 36 percent GC.

Acellular slime molds   In the vegetative phase the acellular slime molds exist as a mass of protoplasm of indefinite extent, which might be compared to a giant amoeba. This structure is actively motile by amoeboid motion, the plasmodium flowing over the surface of the substratum, engulfing food particles as it moves. Protoplasmic streaming usually occurs in definite strands, each surrounded by a thin plasma membrane (Plate 15: Figure 19.38), and within the strands the protoplasm moves for a while in one direction before reversing and flowing in the other direction. The net result of the cytoplasmic movement is the slow traverse of the substratum by the plasmodium due to the more extensive movement of the mass in one direction than in another during each oscillation. The acellular slime molds have provided excellent experimental material for detailed study of cytoplasmic streaming, of which we discussed some aspects in Section 3.5. The strands of protoplasm do not retain their individuality indefinitely but fuse occasionally into larger masses, which then separate again into smaller strands. Slime-mold plasmodia are often bril-

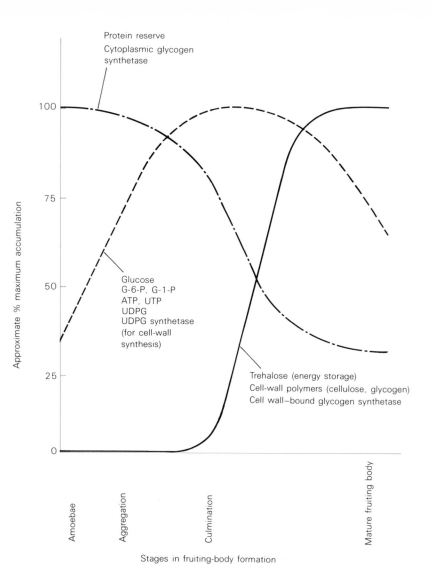

Figure 19.37 Biochemical changes during fruiting-body formation in D. discoideum. The changes in certain enzymes and carbohydrates during different stages of the process are shown. Compare with Figure 19.36. [From B. E. Wright: Science 153:830 (1966). © 1966 by the American Association for the Advancement of Science.]

liantly colored and frequently can be seen spreading across the surface of a piece of wood that has been kept moist. The fusion of two parts of a plasmodial mass to form a larger mass is species specific: Two masses of different species do not fuse. In fact, fusion of masses from the same species but from separate natural isolates does not always occur.

Two kinds of resting structures, *fruiting bodies* and *sclerotia,* are produced by the plasmodium (which is diploid) of the acellular slime molds.

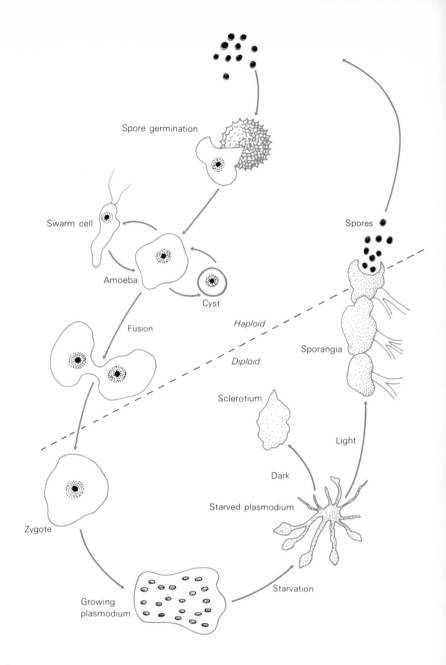

Spore germination

Swarm cell

Spores

Amoeba

Cyst

Sporangia

Haploid

Fusion

Diploid

Sclerotium

Light

Dark

Starved plasmodium

Zygote

Growing
plasmodium

Starvation

Figure 19.39   Stages in the life cycle of Physarum polycephalum. Compare with Plate 15: Figure 19.38. (Courtesy of Harold P. Rusch.)

Fruiting bodies arise as part of the sexual cycle (Figure 19.39); their production is initiated by the aggregation of the protoplasmic mass into a compact unit and culminates in the secretion by the plasmodium of a stalk, usually devoid of protoplasm, and in the formation of a complex head within which the spores develop. In some species, light is required for fruiting-body formation, and we may again presume that this requirement ensures that the fruiting body will be formed in the air above the substratum. The structures of the fruiting-body heads are often quite complex and are distinctive for different species. Soon after the formation of the head, nuclei in this portion undergo meiosis, and these haploid nuclei then become forerunners of the spores. The protoplasmic mass in the head collects around the haploid nuclei, and a thick cell wall is formed around each. The spores thus formed are discharged when the noncellular part of the fruiting body disintegrates.

These resting spores are relatively resistant to drying and other unfavorable conditions and are able to remain dormant for long periods of time. Before germination, dormancy must be broken, but the conditions necessary for this are ill defined. Once dormancy is broken, germination will occur in plain water. Each spore forms one to four biflagellate swarm cells without cell walls, which swim around for a while, feed, and may produce more swarm cells by binary fission. Eventually swarm cells conjugate in pairs, and the resulting zygote also swims around for a while before losing its flagella and becoming converted into a diploid amoeboid cell. In some species the swarm cells lose their flagella before conjugation, and it is then that the resulting amoeboid cells unite. The diploid amoeboid cell then grows and undergoes successive nuclear divisions, producing the new diploid plasmodium.

In the absence of light and under starvation conditions, some species form the other kind of resting structure, the sclerotium, an irregular, hard, dry mass of protoplasm, resistant to drying and desiccation. No sexual process is involved in this conversion, and the sclerotium, having enabled the organism to overwinter, can form a new plasmodium when conditions are favorable.

One of the interesting properties of the slime-mold plasmodium is that divisions of the nuclei within a single plasmodial mass are usually synchronized. If two plasmodial masses having synchronized nuclear divisions at different times are placed together and allowed to fuse, mitosis in the coalesced pair occurs at a time midway between those which would have occurred in each separate plasmodium. Detailed analysis of this phenomenon has shown that the synchrony is controlled by a protein that stimulates nuclear division and that this protein is released from the nuclei during the G_2 period of the cell cycle (Section 3.6) and accumulates in the cytoplasm. When a critical concentration of this protein in the cytoplasm is reached, mitosis is triggered. Further studies may lead to a better understanding of the biochemical events controlling mitosis, and the concepts developed may be of importance in understanding the cancer cell, on which normal controls of mitosis seem to be abolished. Again, we see that studies on a simple microbial system may provide important

The slime molds

insights into those phenomena which are analogous in higher organisms.

The slime molds are a fascinating group of organisms, with many aspects of their ecology, biochemistry, and genetics still undefined, and they are excellent organisms for the pursuit of many interesting microbiological studies. It is of interest that one group of bacteria, the fruiting myxobacteria (see Section 18.13), exhibits resemblances in life history to the slime molds—another example of convergent evolution.

The protozoa

The simplest definition of *protozoa* states that they are unicellular animals, but as usual, such a general definition leaves many questions open, not the least of which is: What is an animal? On the one hand, some motile, chlorophyll-containing organisms (for example, *Volvox* and *Chlamydomonas*) are classified either as algae or as protozoa, depending on whether the classifier leans to botany or zoology. On the other hand, some organisms (for instance, cellular and acellular slime molds) are classified as either fungi or protozoa. Thus the lines separating the protozoa from other groups of eucaryotic microorganisms are fuzzy. In the present discussion we shall ignore these lines and concentrate instead on typical protozoa. A typical protozoan is a unicellular organism that lacks a true cell wall and obtains its food phagotrophically. Protozoa are enormously various in size and shape. Some are as small as larger bacteria, whereas others are large enough to be seen without a lens. Protozoa are found in a variety of freshwater and marine habitats; a large number are parasitic on other animals, and some are found growing in soil or in aerial habitats, such as on the surfaces of trees.

19.12 Characteristics used in separating protozoal groups

As befits organisms that "catch" their own food, most protozoa are motile; indeed, their mechanisms of motility are key characteristics used to subdivide the phylum Protozoa into groups. Protozoa that move by amoeboid motion are called Sarcodina; those using flagella are called Flagellata or Mastigophora; and those using cilia are called Ciliophora. The Sporozoa, a fourth group, are usually nonmotile and are all parasitic on higher animals. The Sporozoa often divide by multiple fission of a single cell into a number of smaller cells called "spores"—hence the name of the group;

however, these spores are in no way identical with the spores of other microorganisms. Table 19.7 describes and illustrates the representative structural characteristics of the major protozoal groups.

Table 19.7   Major taxonomic groups of the phylum Protozoa

Subphylum Sarcomastigophora Flagellate or amoeboid motility; some forms have both means of locomotion. Single type of nucleus. Typically, no spore formation.
Superclass Mastigophora (the flagellates) Flagellate motility. Cell division by longitudinal binary fission. Sexual reproduction rare. Amoeboid forms frequent in some species. Many species parasitic. DNA, 30–56% GC (7 species). [Sketch (a).]

Subphylum Sporozoa Spores typically present. Single type of nucleus. Cilia and flagella absent except for flagellated gametes in some groups. All species parasitic. [Sketch (c).]

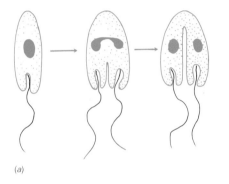

(a)

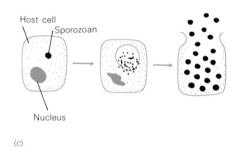

(c)

Superclass Sarcodina (the amoebae) Amoeboid motion. Flagella, when present, restricted to developmental stages. Cell body naked or with external or internal tests or skeletons. Cell division by binary fission. Most species free-living. DNA, 23–66% GC (4 species). [Sketch (b).]

Subphylum Ciliophora (the ciliates) Ciliary motility. Two types of nuclei: micronucleus and macronucleus. Cell division by transverse binary fission. Most species free-living. DNA, 22–35% GC (3 species). [Sketch (d).]

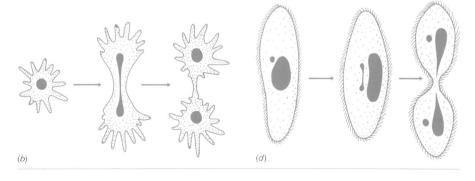

(b)

(d)

19.12 *Characteristics used in separating protozoal groups*

693

A number of the Sarcodina produce shells, or tests, which are rigid external structures either secreted by the cell or composed of foreign materials such as sand grains or diatom frustules picked up from the environment. Those sarcodines called *foraminifera* produce calcium carbonate shells, whereas the *radiolaria* have shells of either silica or strontium sulfate. A shell is not necessarily the outermost limit of the protozoal cell; in many of the foraminifera the protoplasm can extend beyond, and may even cover, the external portion of the shell.

Some protozoa are sessile, with holdfasts that permit them to attach to a substratum. The structure and function of the holdfast are distinctive and are of use in classification. Some sessile protozoa have developed a colonial existence, in which many individuals of the same species associate and form structures of characteristic shape and size. Again, many protozoa form *cysts,* which are resting structures usually resistant to drying and occasionally resistant to other noxious agents. These enable the organism to endure a period of bad times or aid in species dispersal.

Parasitic protozoa usually show marked specificity for the animals they infect, and host range is used as one characteristic in the classification of these forms. Often these parasites show complex life cycles, alternating at different stages from one host to another.

In contrast to those of the algae and fungi the life cycles of most protozoa are relatively simple. Some show no sexual reproduction at all, whereas those that reproduce sexually usually exhibit a simple life cycle, without the complexities described earlier in this chapter for algae and fungi. (The foraminifera are an exception to this statement since both haploid and diploid stages with independent existences are known.)

19.13   *Methods for studying protozoa*

Because of the large size and distinctive morphology of protozoa, many significant observations can often be made on natural collections. Many forms live in environments relatively rich in organic matter, where bacteria, their most common food, also occur. Small ponds, shallow pools, ditches, and bogs, all of which are rich in aquatic vegetation, are favorable sites for the development of protozoa. In the marine environment, seaweed beds, tide pools, and salt marshes provide similarly favorable habitats. Sewage-treatment plants, especially the activated-sludge and trickling-filter beds, are also rich in protozoa. Parasitic protozoa are, of course, found only in association with their hosts. Many protozoa live as commensals in the intestinal tracts of animals; often each species of animal has its own characteristic fauna.

For many studies cultures are essential, and the isolation and culture of free-living protozoa is often fairly easy. Infusions of plant materials, such as hay or lettuce, in pond water will give good general culture media for many freshwater species. In such infusions various bacteria usually develop and serve as food for the protozoa. If desired, a plant infusion can be sterilized and then inoculated with a pure culture of a single species of bacteria, such as *Aerobacter aerogenes,* on which a variety of protozoa seem to thrive.

The techniques used for obtaining uniprotozoal cultures are similar to those described for the algae. Much use is made of capillary pipettes for the isolation of single protozoal cells. Axenic or bacteria-free pure cultures are often more difficult to obtain for protozoa than they are for algae since many protozoa require particulate food. Some will grow well on completely soluble nutrient media, and these are of course easier to purify. Complex organic substances, such as casein digest, meat extract, or yeast extract, added to a mixture of synthetic salts usually provide the necessary nutrients. A few protozoa have been cultured in completely synthetic media containing vitamins, amino acids, and salts. Acetate can be a source of energy and of carbon for many flagellates and for a few ciliates. However, knowledge of protozoal nutrition is much less extensive than that of the nutrition of other microbial groups, and only in a few cases have adequate nutritional studies been carried out.

Most parasitic protozoa cannot be cultured outside an animal host, and laboratory animals such as the rat, mouse, or hamster can in many cases be used. Certain forms of the malaria parasite can be grown in birds; the chicken or duck is usually used. Some parasitic protozoa have also been grown in special culture media containing whole blood or serum.

The maintenance of stock cultures of protozoa usually requires periodic transfer to fresh media, although some of the smaller and less delicate forms can be preserved in the frozen state. If cyst formation can be induced in culture, the isolate may be maintained indefinitely in the encysted state. Extensive culture collections such as those that exist for bacteria, fungi, or algae are not available for the protozoa although some research workers maintain cultures of specific forms, which they will distribute to interested scientists. The Culture Index of the Society of Protozoologists (U.S.A.) lists the names of laboratories and investigators that maintain various species of protozoa.

19.14   *Representative protozoal groups*

Mastigophora—the flagellates   This is the protozoal group most closely related to the algae. Such genera as *Chlamydomonas*, *Euglena*, and *Volvox* comprise flagellated organisms that are often classified by protozoologists as plantlike Mastigophora. There are, however, a large number of flagellated forms that are devoid of chlorophyll, have an animal-type nutrition, and hence are undoubtedly protozoa. The plantlike flagellates differ from the animallike organisms in their mode of obtaining energy and in several other respects. The plantlike organisms form starch or β-1,3-glucan as their chief storage product, whereas the animallike organisms produce glycogen or oils. The former group often has a cellulose wall or cyst, whereas the latter does not. Animallike flagellates generally lack sexual stages or life cycles, whereas in the plantlike flagellates sexual reproduction is common. The conclusion seems inescapable that these two groups of flagellated organisms have little similarity except for their manner of motility, and therefore it would seem unwise to place them all in one group solely on the basis of this characteristic. However, at least one group of plantlike flagellates is clearly related to the animal forms; this is the group that contains the genus *Euglena*. Recall from Section 13.4 that it is possible by using various chemical or radiation treatments to induce a loss of chloroplasts from *Euglena*. The organism may become permanently bleached, and since it has no way of regaining its chloroplasts, it assumes an existence that is clearly animallike. Interestingly, *Euglena* and its chlorophyll-containing relatives possess the other characteristics of the animallike flagellates listed above. Furthermore, many colorless forms are known to occur naturally, perhaps having also arisen from green forms through loss of chloroplasts.

Among the flagellates that are clearly protozoa are a number that are parasitic on, or pathogenic to, animals and man. Two groups may be mentioned briefly. The order Trichomonadida contains protozoal parasites of the vertebrate alimentary and genitourinary tracts. These flagellates have three to six flagella, one of which typically is trailed behind the cell during movement. Although most of the vegetative forms divide by longitudinal binary fission, some forms divide by multiple fission after the original cell has enlarged and has become internally subdivided into a number of small cells. Most of the parasitic forms produce cysts, which are the infective stages by which the species is transferred from host to host. Cyst formation is strictly asexual, and no sexual stages are known. There are three species parasitic in man: *Trichomonas tenax* (or *T. buccalis*), found in the mouth, *T. hominis* of the colon, and *T. vaginalis* of the vagina. The last species, despite its name, is also found in the genitourinary tract of males. A surprisingly high proportion of the human population harbors one or more of these three species: *T. hominis* is found in 2 percent and *T. tenax* in 10 percent of the population; and *T. vaginalis*, in 25 percent of females and 4 percent of males. It has not been proved that these trichomonads are pathogenic to human beings, however. *Trichomonas vaginalis* has often been implicated in inflammation of the vagina in women, but whether the organism is a cause or a result of the vaginitis is not known. *Trichomonas tenax* probably is nonpathogenic, and most likely is transmitted by kissing.

The most important pathogenic Mastigophora are the *trypanosomes,* which are classified in the family Trypanosomatidae. These organisms cause a number of serious diseases of man and vertebrate animals, including the feared disease *African sleeping sickness.* The trypanosomes are highly variable in morphology, both from species to species and within a single species at various stages of the life cycle. In *Trypanosoma,* the genus infecting man, the protozoa are rather small, around 20 μm in length, and are thin, crescent-shaped organisms with a single flagellum that originates in a basal body and folds back laterally across the cell, where it is enclosed by a flap of surface membrane (Figure 19.40). Both the flagellum and the membrane participate in propelling the organism, making possible an effective movement even in blood, which is rather viscous. These organisms are usually pleomorphic in the blood, with short and broad forms, long and slender forms, and intermediate forms all seen in the same blood smear. *Trypanosoma gambiense* is the species that causes the chronic but usually fatal African sleeping sickness. In man the parasite lives and grows primarily in the bloodstream, but in the later stages of the disease invasion of the central nervous system may occur, causing an inflammation of the brain and spinal cord that is responsible for the characteristic neurological symptoms of the disease. The parasite is transmitted from man to man by the tsetse fly, *Glossina* sp., a blood-sucking fly found only in Africa. The parasite proliferates by binary fission in the intestinal tract of the fly and then invades intestinal cells, where further multiplication is accompanied by an alteration in morphology. The parasite then invades the salivary glands and mouthparts of the fly, from which sites it may be transferred to a new human host. Accompanying the morphological changes in the parasite when passing from man to insect are characteristic biochemical changes. The bloodstream form metabolizes glucose via a respiratory pathway lacking cytochromes, in which the conversion of glucose to pyruvate occurs by normal glycolysis, but the NADH is oxidized to NAD by a flavoprotein oxidase system. In the bloodstream some of the trypanosomes undergo a morphological transformation into short, stumpy types and acquire a cytochrome-linked respiratory metabolism. These cytochrome-containing forms are the ones that are infective to the insect.

The parasite is never transferred directly from one person to another but is transmitted only through the mediation of a tsetse fly (Figure 19.40). Because of this, the spread of the disease is prevented by destroying the tsetse fly. Control of the fly proved relatively difficult until the development of rather potent insecticides such as DDT; however, through their use and

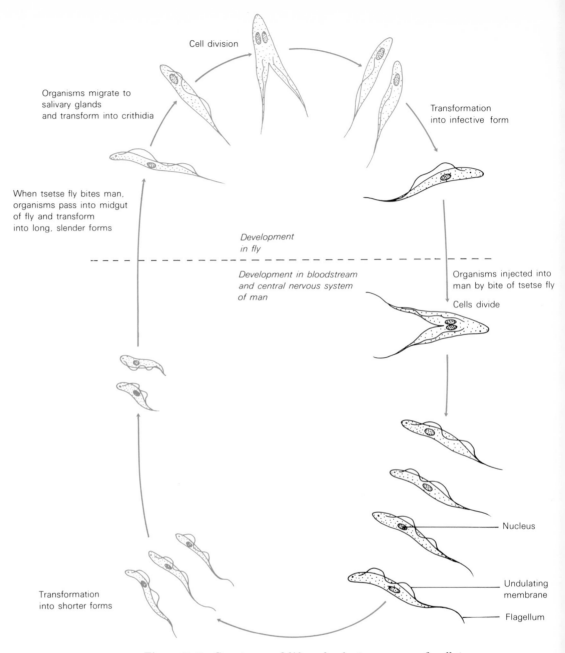

Cell division

Organisms migrate to
salivary glands
and transform into crithidia

Transformation
into infective form

When tsetse fly bites man,
organisms pass into midgut
of fly and transform
into long, slender forms

*Development
in fly*

*Development in bloodstream
and central nervous system
of man*

Organisms injected into
man by bite of tsetse fly

Cells divide

Nucleus

Undulating
membrane

Flagellum

Transformation
into shorter forms

*Figure 19.40   Structure and life cycle of a trypanosome flagellate,
Trypanosoma gambiense, the causative agent of African sleeping
sickness.*

through the clearing of underbrush, in which the flies find shelter, the tsetse fly has been eliminated from many parts of West Africa that previously had been heavily infested. Several new drugs have also been developed for treating the disease in man, and these have also reduced the incidence of the disease by reducing the number of sources from which the flies could become infected. Another trypanosome, *T. brucei*, infects domestic animals such as cattle and pigs, causing symptoms similar to African sleeping sickness and seriously restricting the ability to raise cattle in certain parts of Africa. This parasite, which is probably a genetic variant of the form that attacks man, is also transmitted by a species of tsetse fly, whose control is effected at the same time that the vector of *T. gambiense* is eliminated. Control of the tsetse fly has made possible the establishment of a cattle industry in many parts of Africa previously unsuitable for this purpose.

Sarcodina—the amoebae Although amoeboid movement is traditionally thought of in connection with cells of the superclass Sarcodina, it is not restricted to these protozoa and occurs in many specialized cells of metazoa as well as in slime molds; occasionally it is even a transitory feature of the Mastigophora. Flagellar movement also occurrs among the Sarcodina—flagella are present on some gametes—thus making a rigid definition of this group difficult. The best we can do is to include in the Sarcodina those solitary protozoa that move mainly or exclusively by amoeboid motion. The mechanism of this manner of moving was discussed in Section 3.5. Among the sarcodines are organisms that in the vegetative phase are always naked, forming a rigid wall only during encystment, whereas a large number of others secrete a shell during vegetative growth. The best known of the shelled forms are the foraminifera and the radiolaria.

Shelled sarcodines present a variety of interesting features. Tests of different species show distinctive characteristics and are often of great beauty (Plate 16: Figure 19.41). The cell is usually not firmly attached to the test, and the amoeba may extend partway out of the shell during feeding. One way in which cell division occurs in these shelled forms is a process reminiscent of budding in yeast (Section 19.26): Mitosis occurs, and one of the nuclei moves into a portion of the cytoplasm that is extruded through the aperture of the shell; a new shell is secreted around the extruded part, and the offspring separates from its parent. Thus the offspring has a whole new shell, whereas the parent retains all the old shell.

A sexual cycle occurs in many foraminifera, and in some cases a true alternation of generations occurs. The life cycle of one species is shown in Figure 19.42. The haploid cells are called *gamonts*, and the diploid

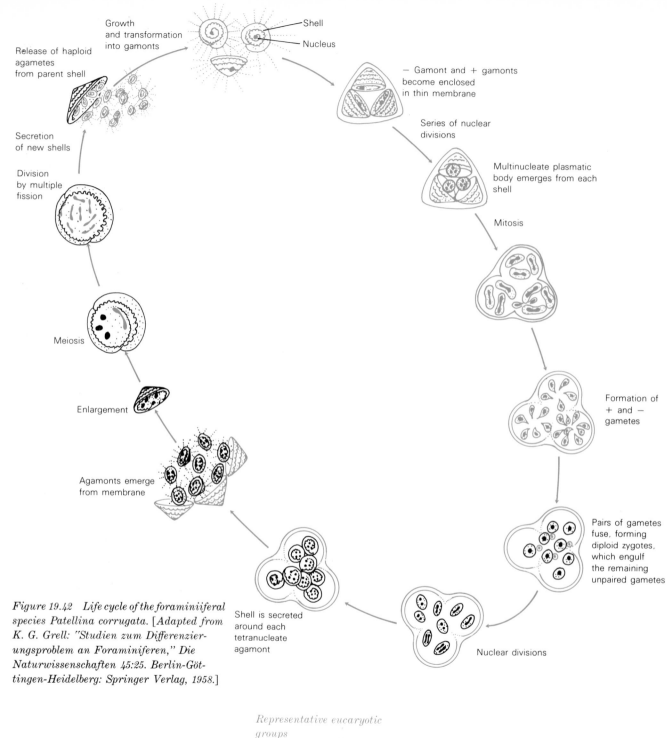

Growth
and transformation
into gamonts

Shell

Nucleus

Release of haploid
agametes
from parent shell

— Gamont and + gamonts
become enclosed
in thin membrane

Series of nuclear
divisions

Secretion
of new shells

Multinucleate plasmatic
body emerges from each
shell

Division
by multiple
fission

Mitosis

Meiosis

Formation of
+ and —
gametes

Enlargement

Pairs of gametes
fuse, forming
diploid zygotes,
which engulf
the remaining
unpaired gametes

Agamonts emerge
from membrane

Shell is secreted
around each
tetranucleate
agamont

Nuclear divisions

*Figure 19.42   Life cycle of the foraminiiferal
species Patellina corrugata. [Adapted from
K. G. Grell: "Studien zum Differenzier-
ungsproblem an Foraminiferen," Die
Naturwissenschaften 45:25. Berlin-Göt-
tingen-Heidelberg: Springer Verlag, 1958.]*

cells are called *agamonts*. Gamonts are sexually differentiated and are given arbitrary designations of + and −. In the species illustrated two + gamonts and one − gamont come together, synthesize a thin external membrane, and undergo mitosis several times. The multinucleate protoplast of each gamont emerges from its shell and remains in the space between the shell and the external membrane. The nucleus of each gamont then divides again, and a group of gametes is formed, with more + than − ones. These gametes fuse pairwise + to − (the extra + gametes do not enter the process), forming diploid nuclei and zygotes. Each zygote then undergoes two nuclear divisions to form a tetranucleate agamont. The agamonts in turn secrete shells and quit their enclosing membrane. After enlarging, an agamont undergoes meiosis and multiple fission. The sixteen agametes formed from one agamont grow and become transformed into gamonts, and the cycle is completed.

Foraminifera and radiolaria are exclusively marine organisms, the former living primarily in coastal waters and the latter in the open sea. Because of the weight of the test, the cell usually sinks to the bottom, and it is thought that the organisms feed on particulate deposits, primarily bacteria and detritus. The shells of these forms are relatively resistant to decay and hence become readily fossilized. (The "White Cliffs" of Dover, England, are composed to a great extent of foraminiferal shells.) Fossil foraminifera have proved especially useful for geologists, who utilize them to relate the ages of limestone rocks in one part of the world to those in another and to aid in the search for oil-bearing formations. Because of the excellent fossil record these organisms leave, we have a good idea of their evolutionary history (see Figure 17.5).

A wide variety of amoebae are parasites of man and other vertebrates, their usual habitat being the oral cavity or the intestinal tract. *Entamoeba histolytica* can serve as an example of these parasitic forms. This organism is found in the intestinal tract of a high percentage of individuals living in regions where sanitation is poor, and it is found occasionally even in members of higher-income groups. In many cases, infection causes no obvious symptoms, but in some individuals it produces ulceration of the intestinal tract, which results in a diarrheal condition called *amoebic dysentery*. The organism is transmitted from person to person in the cyst form. Excystment occurs in the intestinal tract, and the amoebae divide by binary fission to establish a population (Figure 19.43). Since they feed directly on intestinal bacteria, amoebae can become established only when bacteria are present, which is why it is impossible to infect germ-free animals with *E. histolytica*. Under certain conditions not yet defined, the amoebae form cysts, in which form they leave the intestinal tract in the stool. The cysts are resistant to drying and hence can survive for long periods of time outside the body. *Entamoeba histolytica* is transmitted by fecal contamination of water supplies and by house flies. The disease it causes can often be treated by the use of antibacterial antibiotics taken orally, but these probably act only indirectly on the entamoebae by reducing or eliminating the intestinal bacteria upon which they feed. Amoebic dysentery is diagnosed by microscopic examination of the stool

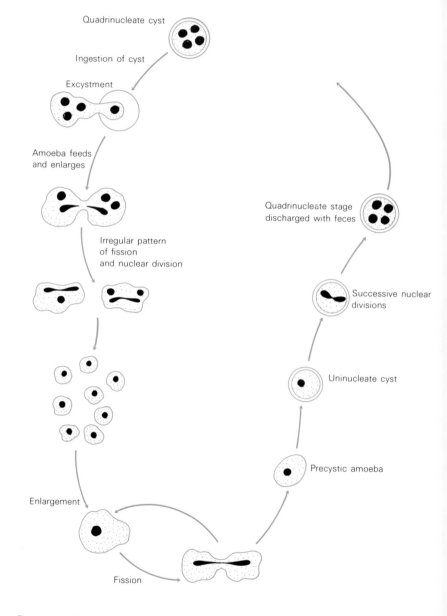

Quadrinucleate cyst

Ingestion of cyst

Excystment

Amoeba feeds and enlarges

Irregular pattern of fission and nuclear division

Quadrinucleate stage discharged with feces

Successive nuclear divisions

Uninucleate cyst

Precystic amoeba

Enlargement

Fission

Figure 19.43 Life cycle of the parasitic amoeba Entamoeba histolytica.

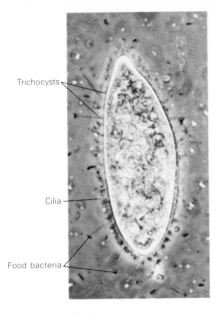

Figure 19.44 *Photomicrograph of a paramecium cell, by phase-contrast microscopy. Magnification, 300×.*

Labels on photomicrograph: Trichocysts, Cilia, Food bacteria

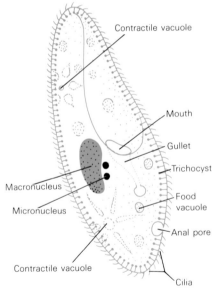

Labels on diagram: Contractile vacuole, Mouth, Gullet, Trichocyst, Food vacuole, Anal pore, Cilia, Contractile vacuole, Micronucleus, Macronucleus

Figure 19.45 *Diagram of a paramecium cell, showing the various structural features.*

for cysts, which can be separated and concentrated by centrifugation in a solution of zinc sulfate. This latter material is of high specific gravity, and the cysts float to the surface; a loopful of the surface film can be stained with iodine and examined under the microscope.

Ciliophora—the ciliates   The ciliate cell is structurally one of the most complex cells in the living world. Ciliates are defined as those protozoa which at some stage of their life cycle possess cilia. They are unique in having two kinds of nuclei: the *micronucleus*, which is concerned only with inheritance and sexual reproduction, and the *macronucleus*, which plays no role in inheritance but is responsible for the production of mRNA for various aspects of cell growth and function. Probably the best known and most widely distributed of the ciliates are those of the genus *Paramecium*, which will be used to introduce the present discussion. The overall structure of a paramecium cell is shown in Figures 19.44 and 19.45, and the nuclear and cytoplasmic events of cell division are described in Figure 19.46. Most ciliates obtain their food by ingesting particulate materials through a distinct oral region or mouth. Around this region are special cilia whose function is to pull food particles toward the mouth. The mouth itself can open and close in response to the presence of food particles. Once inside, the food particles are carried down the esophagus and into the cytoplasm, where they are enclosed in a *food vacuole,* a lysosomelike structure (Section 3.2) into which digestive enzymes are secreted. The food vacuoles move in a regular pattern through the cytoplasm and are dissipated in the region of the anal pore, where excretion of waste products occurs. Another kind of vacuole, the *contractile vacuole,* functions in eliminating water and plays an important role in osmotic regulation. The pulsation of these contractile vacuoles can often be seen within the cell, the contraction being correlated with the excretion of water. Many marine ciliates lack contractile vacuoles since they do not have the osmotic problem the freshwater forms have. Some marine forms acquire contractile vacuoles when adapted to fresh water. In addition to cilia many ciliates have *trichocysts,* which are long, thin filaments of a contractile nature, anchored beneath the surface of the outer cell layer. These enable the protozoa to attach to a surface and can aid in defense by indicating to the cell that it is being attacked by a predator. In the case of predatory ciliates, such as *Didinium,* trichocysts hold on to and paralyze the prey as a prelude to ingestion. The cilia, trichocysts, and other surface structures of the ciliates are arranged in a highly ordered pattern on the surface of the cell, as is revealed by a special silver-impregnation technique (Figure 19.47). The cilia and trichocysts are attached to basal granules called *kinetosomes,* which are self-replicating, DNA-containing organelles that are probably responsible for the synthesis of the appendages, as well

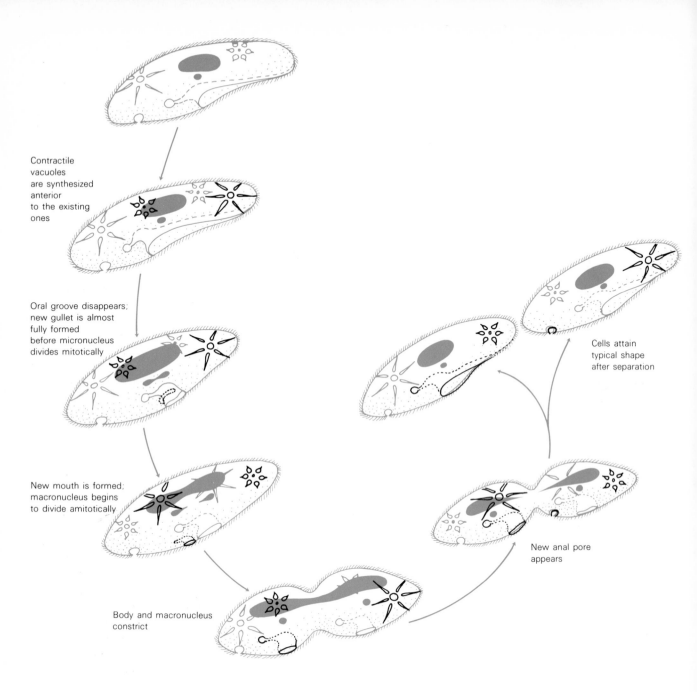

Contractile
vacuoles
are synthesized
anterior
to the existing
ones

Oral groove disappears;
new gullet is almost
fully formed
before micronucleus
divides mitotically

New mouth is formed;
macronucleus begins
to divide amitotically

Body and macronucleus
constrict

Cells attain
typical shape
after separation

New anal pore
appears

Figure 19.46   Stages in cell division in Paramecium.

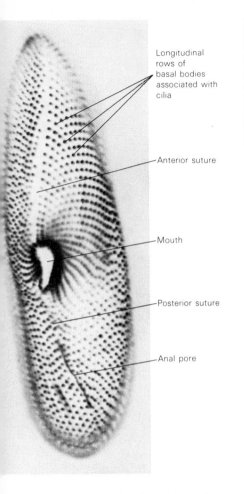

Longitudinal rows of basal bodies associated with cilia

Anterior suture

Mouth

Posterior suture

Anal pore

Figure 19.47   Organization of the peripheral region of a paramecium cell. The organism had been subjected to a special silver-impregnation technique that stains the basal bodies specifically. [From T. M. Sonneborn: in J. M. Allen (ed.), The Nature of Biological Diversity. New York: McGraw-Hill Book Company, 1963.]

as for the control of their movement. The movement of a cell by means of cilia requires a highly coordinated system to ensure that all cilia are moving in the same direction at the same time (see Figure 3.23), but the precise mechanism of ciliary coordination is not yet understood.

As we have noted, the two kinds of nuclei in the ciliates set this group aside from all other protozoa. The micronucleus is diploid, but the macronucleus is highly polyploid, containing from about 40 to over 500 times as much DNA as does the micronucleus. The macronucleus is often spherical, but in many ciliates it is a long banded or chainlike structure. The macronucleus does not divide mitotically but merely separates into two parts. The micronucleus plays no direct role in vegetative growth and cell division since strains without micronuclei can continue to divide normally; however, if the macronucleus is removed, the cell quickly dies. It is interesting to note that, if only a part of the macronucleus is left, this fragment can regenerate a whole macronucleus.

Two kinds of nuclear phenomena are known to occur in ciliates. The first, called *autogamy*, involves nuclear reorganization in the absence of sexual processes and is a means for counteracting senescence. If a culture is kept in vegetative division for many generations, aging, or senescence, occurs, apparently due to changes in the ability of the macronucleus to function; subsequently, the division rate of the cell greatly decreases. At this time the macronucleus begins to degenerate, reorganization of the micronucleus occurs, and a new macronucleus is formed (Figure 19.48). As a result of autogamy, the cell becomes homozygous for all genes because both types of new nuclei have been derived from one haploid product.

The second type of nuclear phenomenon in ciliates, sexual reproduction, involves conjugation between two cells of opposite mating type; however, the details of the process are quite different from those of the sexual cycle in any of the microorganisms discussed previously in this chapter. Each of the two conjugants is diploid and retains its identity throughout the conjugation process. The two conjugants contact each other in the region of the oral groove, and the micronucleus of each conjugant undergoes meiosis, while the macronucleus begins to degenerate (Figure 19.49). One haploid micronucleus of each conjugant then migrates across a cytoplasmic bridge that was formed in the region of contact and fuses with the stationary nucleus of its mate, producing what is called a *syncaryon* and restoring the micronuclei to the diploid state. The two conjugating cells now separate, and the syncaryons undergo a series of mitotic divisions, which result in the restoration of the typical macronucleus and micronucleus. One of the nuclei resulting from these mitotic divisions is the forerunner of the micronucleus and the other is the forerunner of the macronucleus. The net result of conjugation is the

Figure 19.48   Autogamy in Paramecium aurelia.

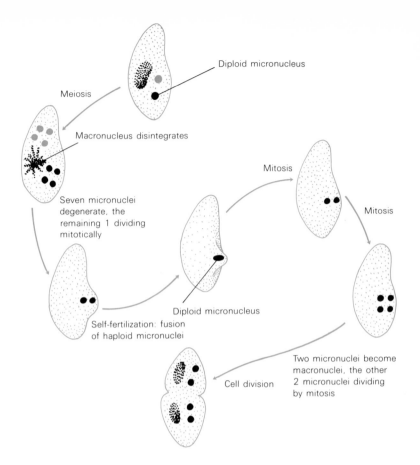

formation of exconjugants that are hybrid for the genes of the two conjugants. Since mating types exist, conjugation between two cells of the same clone will not occur. One should note that in both autogamy and conjugation similar nuclear phenomena occur.

We discussed cytoplasmic inheritance of *Paramecium* in Section 13.4.

In addition to *Paramecium,* two other ciliate genera that have been the objects of considerable study are *Tetrahymena* and *Stentor. Tetrahymena* is useful because it grows well on soluble food sources; defined culture media have been developed for it. *Tetrahymena* is being used widely in biochemical studies on growth and cell division.

Stentor is a very large ciliate. The largest species, the beautifully blue-green *Stentor coeruleus,* is 1 mm long, making it visible to the naked eye. Stentors are stalked ciliates that normally live a sessile existence, attached to convenient perches, where they feed by directing currents of

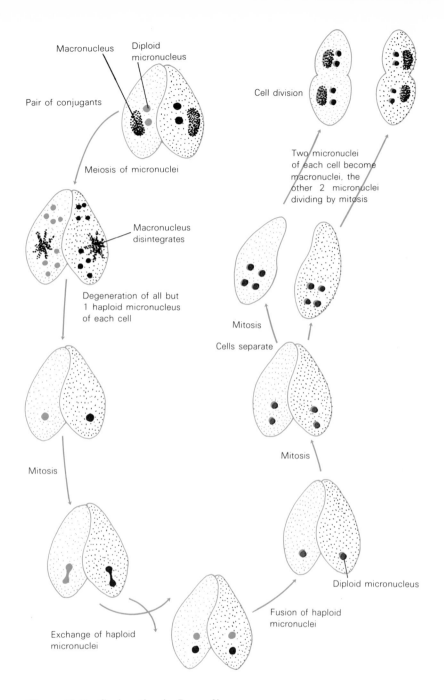

Figure 19.49   Conjugation in P. aurelia.

19.14   Representative protozoal groups

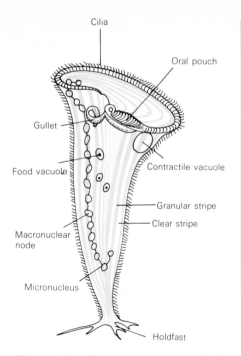

Figure 19.50 *Stentor coeruleus, a stalked ciliate. The organism is about 1 mm long. (Reprinted with permission from Vance Tartar: The Biology of Stentor. New York: Pergamon Press, 1961.)*

Figure 19.51 *Cell division in S. coeruleus. The complex morphology of the cell necessitates a complex distribution of cell parts during the division process. The new cell parts are indicated in gray. (Reprinted with permission from Vance Tartar: The Biology of Stentor. New York: Pergamon Press, 1961.)*

water into their gullets through the activity of a band of cilia lining the top of the gullet opening (Figure 19.50). Stentors have been studied to a considerable extent to determine the factors controlling the development of their complex cellular form. Cell division in *Stentor* is itself a complex process, as can be seen in Figure 19.51. Note that during division one of the newly formed cells retains the gullet of the mother and forms a new holdfast, whereas the other retains the old holdfast and forms a new gullet.

One of the properties of *Stentor* that has been subject to considerable experimental study is the ability of the cell to regenerate portions that had been experimentally removed. For instance, if the gullet is excised, the cell reforms a new one in about 8 to 10 hr (Figure 19.52). A critical stage of regeneration is the coalescence of the separate nodes of the macronucleus into a more compact structure, with the reformation of the nodulated structure at about the time regeneration is complete. Regeneration will not occur if the macronucleus is removed completely but will take place as long as one node remains. If a stentor is cut into a number of separate pieces, each will reform a whole animal as long as it contains a macronuclear node. Regeneration requires new RNA and protein synthesis, as can be shown by the fact that the process is blocked by agents which inhibit the synthesis of RNA and protein, such as actinomycin D and puromycin. Many other experimental studies with *Stentor* are possible. For instance, nuclear-cytoplasmic relationships can be analyzed by transplanting nuclei or cytoplasm from one strain to another. Because of the large size of a stentor, manipulations of this sort are relatively easy.

Although a few ciliates are parasitic to animals, this mode of existence is less extensively developed in the ciliates than it is in the other groups of protozoa. The species *Balantidium coli* is primarily a parasite of domestic animals, but occasionally it infects the intestinal tract of man, producing symptoms similar to those caused by *Entamoeba histolytica*. In ruminants there is usually a characteristic fauna of ciliates in the rumen (Figure 14.12), which are thought to play a beneficial role in the digestive and fermentative processes.

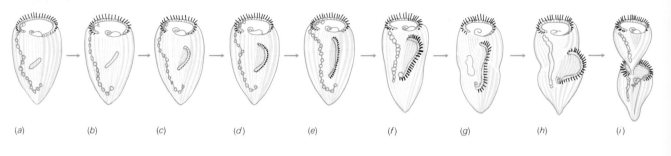

(a)   (b)   (c)   (d)   (e)   (f)   (g)   (h)   (i)

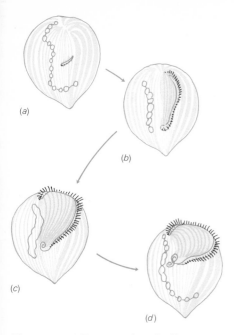

(a)

(b)

(c)

(d)

Figure 19.52   Regeneration of cell parts in Stentor. The oral region was removed experimentally with a needle. The regenerated material is shown in gray. (Reprinted with permission from Vance Tartar: The Biology of Stentor. New York: Pergamon Press, 1961.)

Sporozoa   The Sporozoa comprise a large group of protozoa, all of which are obligate parasites. They are characterized by a lack of motile adult stages and by a nutritional mode of life in which food is generally not ingested but is absorbed in soluble form through the outer wall, such as occurs in bacteria and fungi. Although the name "Sporozoa" implies the formation of spores, this group does not form true resting spores, such as do bacteria, algae, and fungi, but instead produces analogous structures called *sporozoites,* which are involved in the transmission of the species to a new host. Numerous kinds of vertebrates and invertebrates serve as hosts for Sporozoa, and in some cases an alternation of hosts occurs, with some stages of the life cycle occurring in one host and some in another. The most important members of the class Sporozoa are the coccidia, which usually are parasites of birds, and the plasmodia (malaria parasites), which infect birds and mammals, including man. Our discussion will be restricted to the plasmodia.

The malaria parasite is one of the most important human pathogens and has played an extremely significant role in the development and spread of human culture; indeed, it has even affected the human evolutionary process (see below). Four species infect man: *Plasmodium vivax, P. falciparum, P. malariae,* and *P. ovale.* These differ in the degree of severity of symptoms they cause and in certain aspects of their life cycles. We shall consider here only *P. vivax,* which is the most widespread species and the one about which we have the most information. This parasite carries out part of its life cycle in man and part in the mosquito, by which vector it is transmitted from man to man. Only mosquitoes of the genus *Anopheles* are involved; and since these inhabit primarily the warmer parts of the world, malaria occurs predominantly in the tropics and subtropics.

The life cycle of *P. vivax* is given in Figure 19.53. Man is infected by plasmodial *sporozoites,* small elongated cells produced in the mosquito, which localize in the salivary gland of the insect. When biting, the female mosquito inserts her proboscis directly into a capillary, thereby inoculating the sporozoites directly into the bloodstream. The sporozoites are carried throughout the body and are removed from the blood by the reticuloendothelial system (including lymph nodes, spleen, and liver). Replication occurs in the liver, and during this stage (termed the *exoerythrocytic* stage) the parasites are absent from the blood. The sporozoite becomes transformed to a *schizont,* which replicates by enlarging and segmenting into a number of small cells called *merozoites,* which are liberated from the liver into the bloodstream. Some of the merozoites infect the red blood cells, initiating the erythrocytic schizont stage. The cycle in the red cells proceeds as in the liver and usually occupies a definite period of time, 48 hr in the case of *P. vivax.* It is during the erythrocytic stage that

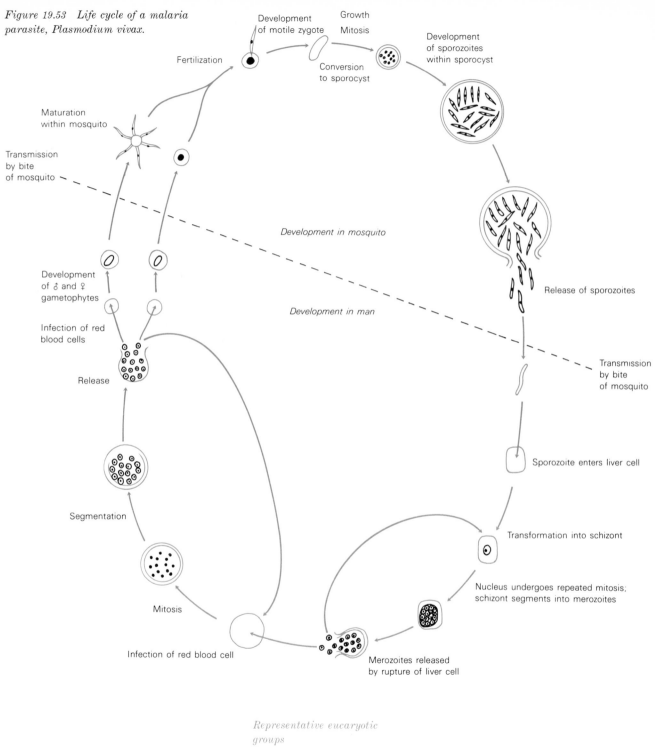

Figure 19.53 Life cycle of a malaria parasite, Plasmodium vivax.

Fertilization

Development of motile zygote

Growth
Mitosis

Conversion to sporocyst

Development of sporozoites within sporocyst

Maturation within mosquito

Transmission by bite of mosquito —

Development in mosquito

Release of sporozoites

Development of ♂ and ♀ gametophytes

Infection of red blood cells

Development in man

Transmission by bite of mosquito

Release

Sporozoite enters liver cell

Segmentation

Transformation into schizont

Mitosis

Nucleus undergoes repeated mitosis; schizont segments into merozoites

Infection of red blood cell

Merozoites released by rupture of liver cell

the characteristic symptoms of fever and chills occur, the chills occurring when a new brood of parasites is liberated from the erythrocytes. Not all the cells liberated from red cells are able to infect other erythrocytes; those which cannot, called *gametocytes,* are infective only for the mosquito. If these gametocytes happen to be ingested when another insect of the proper species of *Anopheles* bites the infected person, they mature within the mosquito into *gametes.* Two gametes fuse, and a zygote is formed, which migrates by amoeboid motility to beneath the outer wall of the insect's intestine, where it enlarges and forms a number of sporozoites. These are liberated, some of them reaching the salivary gland of the mosquito, from where they can be inoculated into another person.

This complex cycle can be broken by eradication of the *Anopheles* mosquito; destroying the vector has essentially eliminated malaria from the United States and Europe. Even so, some infected persons may continue to serve as reservoirs for the plasmodium, and a renewed spread of the disease may result if the mosquito returns to these areas. The erythrocytic phase of the disease can be cured by a variety of drugs, of which the best known are quinine, atebrine, and chloroquine. These drugs affect only the stages in the erythrocytes and therefore do not completely cure the disease. It is possible to control the symptoms of the disease by eliminating the cycles of replication in the erythrocytes. However, the exoerythrocytic stages can be eliminated by primaquine, and this drug in combination with one of the others mentioned above can effect a complete cure. *Plasmodium vivax* has been very difficult to grow in culture, and because of its specificity for man, experimental infection has also been difficult. For these reasons, most of the experimental work on malarial parasites has been done with species of *Plasmodium* that infect birds or rats, and it is with these forms that most of the studies on the development of new drugs have been carried out. Conclusive evidence for the diagnosis of malaria in man is obtained by examining blood smears for the presence of infected cells; the various stages of replication in the erythrocytic phase are easily observed. Diagnosis of the disease in the tissues (exoerythrocytic stage) is much more difficult.

As we have just noted, the plasmodia are obligate parasites and have never been successfully cultivated in artificial media. Considerable information is available, however, on biochemical aspects of the growth of the erythrocytic stage. Such studies have been possible because infected red cells, removed from the bloodstream, can be incubated in vitro under conditions in which metabolism and multiplication of the parasite can take place. The parasite has its own energy-generating system, metabolizing glucose via the Embden-Meyerhof pathway and the citric acid cycle, and synthesizes its own ATP. At least one coenzyme, CoA, is not synthesized by the parasite but is obtained preformed from the host. The parasite synthesizes folic acid but must be provided one of its building blocks, *p*-aminobenzoic acid. The parasite obtains its amino acids for protein synthesis from hemoglobin, which constitutes 90 percent of the protein of the red cell. Electron micrographs of infected cells have revealed that the parasite takes up hemoglobin by pinocytosis into food vacuoles. Upon digestion of the hemoglobin, the heme portion is removed and is accumulated, whereas the globin portion is hydrolyzed, probably to free amino acids and small peptides, which then serve as the amino acid source for protein synthesis. The mammalian red cell is devoid of DNA and has only low amounts of RNA; it is probable that the parasite synthesizes its own pyrimidines and purines, perhaps using amino acids as starting materials.

We thus see that the plasmodia possess extensive biosynthetic capabilities. Why then are they obligate intracellular parasites? One possibility is that the cell membranes of the plasmodia are especially leaky in order that large-molecular-weight cofactors, such as CoA, can be taken up, and because of this leakiness, they rapidly lose cofactors when placed in an extracellular environment. Another possibility is that, since the plasmodia use hemoglobin as a source of major building blocks for biosynthesis, they will naturally grow only in an environment, such as the red cell, in which large amounts of hemoglobin are present. Nothing is known about the biochemistry of the exoerythrocytic stages, but it is likely that their metabolism is different since they grow in tissue cells that are biochemically quite different from the red cell.

One of the most interesting discoveries about malaria concerns the mechanism by which human beings in regions of the world where it is endemic acquire resistance to plasmodium infection. Malaria is a disease primarily localized in warmer parts of the world, where its mosquito vectors are most common, and has undoubtedly been endemic in Africa for thousands of years. In Africans resistance to malaria caused by *P. falciparum* is associated with the presence in their red cells of hemoglobin S, which differs from normal hemoglobin A only in a single amino acid in each of the two identical halves of the molecule. In hemoglobin S a neutral amino acid, valine, is substituted for an acidic amino acid, glutamic acid. Red cells containing hemoglobin S have reduced affinity for oxygen, and the malaria parasite, having a highly aerobic metabolism, cannot grow as well in these red cells as it can in normal ones. An additional consequence is that individuals with hemoglobin S are less able to survive at high altitudes, where oxygen pressures are lower, but in tropical lowland Africa this disadvantage is not manifested. In West Africans resistance to malaria caused by *P. vivax* is associated with the presence of another abnormal hemoglobin, hemoglobin E. In certain Mediterranean regions where malaria is endemic, resistance to *P. falciparum* is associated with a deficiency in the red cells of the enzyme glucose-6-phosphate dehydrogenase. Presumably the abnormal cells are deficient in certain metabolites necessary for the growth of the parasite. So the malaria parasite has been a factor in the biochemical evolution of human beings. Other microbial parasites have also probably promoted evolutionary changes in their hosts; but in no case do we have such clear evidence as in the case of malaria.

19.15   Summary

The above discussion has focused on only a few representative protozoa in order to give at least some idea of the diversity of forms and of the amount and scope of study that needs to be done. Difficulties in laboratory cultivation have generally been overcome for the protozoa so that there is no reason why physiological, biochemical, and genetic studies cannot advance to the level of sophistication now common for other microbial groups. The parasitic protozoa provide some of the most interesting experimental material for studying the ecology, biochemistry, and evolution of host-parasite relationships. At the same time, increased knowledge in these areas may lead to more effective control of serious protozoal diseases. Some protozoal groups, such as the foraminifera and radiolaria, are of considerable help in evolutionary studies because their tests are well preserved in the fossil record, making it possible to trace much of their history. It is with a brief consideration of the fascinating and beautiful protozoa that we end this book, hoping that we have infected the reader with some of the delight in microorganisms that the microbiologist experiences.

Supplementary readings

Ainsworth, G. C., and A. S. Sussman (eds.)   *The Fungi: An Advanced Treatise*, vol. I, *The Fungal Cell*. New York: Academic Press, Inc., 1965.

Ainsworth, G. C., and A. S. Sussman (eds.)   *The Fungi: An Advanced Treatise*, vol. II, *The Fungal Organism*. New York: Academic Press, Inc., 1966.

Ainsworth, G. C., and A. S. Sussman (eds.)   *The Fungi: An Advanced Treatise*, vol. III, *The Fungal Population*. New York: Academic Press, Inc., 1968.

Alexopoulos, C. J.   *Introductory Mycology*, 2nd ed. New York: John Wiley & Sons, Inc., 1962.

Alexopoulos, C. J., and H. C. Bold   *Algae and Fungi*. New York: The Macmillan Company, 1967.

Bartnicki-Garcia, S. "Cell wall chemistry, morphogenesis, and taxonomy of fungi," *Ann. Rev. Microbiol.* 22:87 (1968).

Bessey, E. A. *Morphology and Taxonomy of Fungi.* Philadelphia: Blakiston, 1950. (Reprinted. New York: Hafner Publishing Co., 1961.)

Bonner, J. T. *The Cellular Slime Molds,* 2nd rev. ed. Princeton, N.J.: Princeton University Press, 1967.

Conant, N. F., D. T. Smith, R. D. Baker, J. L. Callaway, and D. S. Martin *Manual of Clinical Mycology,* 2nd ed. Philadelphia. W. B. Saunders Co., 1954.

Cook, A. H. (ed.) *The Chemistry and Biology of Yeasts.* New York: Academic Press, Inc., 1958.

Corliss, J. O. *The Ciliated Protozoa.* Elmsford, N.Y.: Pergamon Press, 1961.

Dawson, E. Y. *Marine Botany: An Introduction.* New York: Holt, Rinehart & Winston, Inc., 1966.

Dogiel, V. A. *General Protozoology,* 2nd ed. rev. by J. I. Poljanskij and E. M. Chejsin. London: Oxford University Press, 1965.

Dubos, R. J., and J. G. Hirsch (eds.) *Bacterial and Mycotic Infections of Man,* 4th ed. Philadelphia: J. B. Lippincott Co., 1965.

Emmons, C. W., C. H. Binford, and J. P. Utz *Medical Mycology.* Philadelphia: Lea & Febiger, 1963.

Florkin, M., and B. T. Scheer (eds.) *Chemical Zoology,* vol. I, *Protozoa.* New York: Academic Press, Inc., 1967.

Fritsch, F. E. *The Structure and Reproduction of the Algae,* vol. I. London: Cambridge University Press, 1935. (Reprinted. Cambridge: The Syndics of the Cambridge University Press, 1965.)

Fritsch, F. E. *The Structure and Reproduction of the Algae,* vol. II. London: Cambridge University Press, 1945. (Reprinted. Cambridge: The Syndics of the Cambridge University Press, 1965.)

Gray, W. D., and C. J. Alexopoulos *Biology of the Myxomycetes.* New York: The Ronald Press Company, 1968.

Grimstone, A. V. "Structure and function in protozoa," *Ann. Rev. Microbiol.* 20:131 (1966).

Large, E. C. *The Advance of the Fungi.* London: Jonathan Cape, Ltd. 1940. (Reprinted in paper covers. New York: Dover Publications, Inc., 1962.)

Lewin, R. A. (ed.) *Physiology and Biochemistry of Algae.* New York: Academic Press, Inc., 1962.

Manwell, R. D. *Introduction to Protozoology.* New York: St. Martin's Press, 1961.

Moulder, J. W. *Biochemistry of Intracellular Parasitism.* Chicago: University of Chicago Press, 1962.

Newton, B. A. "Biochemical peculiarities of trypanosomatid flagellates," *Ann. Rev. Microbiol.* 22:109 (1968).

O'Kelley, J. C. "Mineral nutrition of algae," *Ann. Rev. Plant Physiol.* 19:89 (1968).

Paasche, E. "Biology and physiology of coccolithophorids," *Ann. Rev. Microbiol.* 22:71 (1968).

Pitelka, D. R. *Electron-Microscopic Structure of Protozoa.* New York: The Macmillan Company, 1963.

Pringsheim, E. G. *Farblose Algen: Ein Beitrag zur Evolutionsforschung.* Stuttgart: Gustav Fischer Verlag, 1963.

Pringsheim, E. G. *Pure Cultures of Algae: Their Preparation and Maintenance.* London: Cambridge University Press, 1949.

Robertson, N. F. "The growth process in fungi," *Ann. Rev. Phytopathol* 6:115 (1968).

Smith, G. M. *The Fresh-water Algae of the United States,* 2nd ed. New York: McGraw-Hill Book Company, 1950.

Tartar, V. *The Biology of* Stentor. Elmsford, N.Y.: Pergamon Press, 1961.

Vickerman, K., and F. E. G. Cox. *The Protozoa.* Boston: Houghton Mifflin Company, 1967.

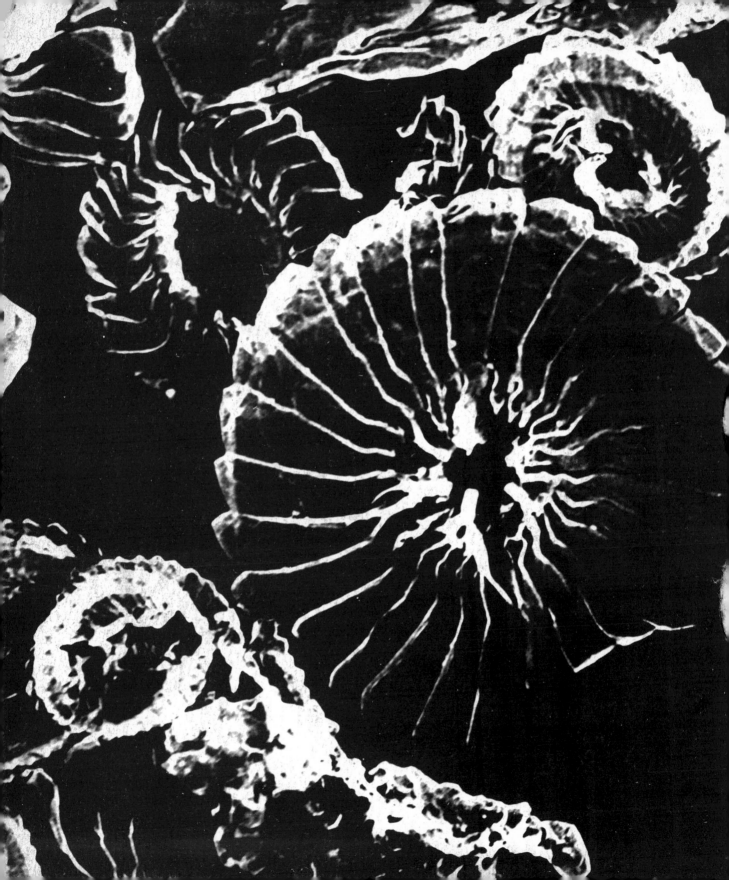